Ecological Aspects of Nitrogen Metabolism in Plants

Ecological Aspects of Nitrogen Metabolism in Plants

Editors

Joe C. Polacco
Christopher D. Todd

WILEY-BLACKWELL

A John Wiley & Sons, Ltd., Publication

This edition first published 2011 © 2011 by John Wiley & Sons, Inc.

Wiley-Blackwell is an imprint of John Wiley & Sons, formed by the merger of Wiley's global Scientific, Technical and Medical business with Blackwell Publishing.

Registered Office
John Wiley & Sons Ltd, The Atrium, Southern Gate, Chichester, West Sussex, PO19 8SQ, UK

Editorial Offices
2121 State Avenue, Ames, Iowa 50014-8300, USA
The Atrium, Southern Gate, Chichester, West Sussex, PO19 8SQ, UK
9600 Garsington Road, Oxford, OX4 2DQ, UK

For details of our global editorial offices, for customer services and for information about how to apply for permission to reuse the copyright material in this book, please see our website at www.wiley.com/wiley-blackwell.

Authorization to photocopy items for internal or personal use, or the internal or personal use of specific clients, is granted by Blackwell Publishing, provided that the base fee is paid directly to the Copyright Clearance Center, 222 Rosewood Drive, Danvers, MA 01923. For those organizations that have been granted a photocopy license by CCC, a separate system of payments has been arranged. The fee codes for users of the Transactional Reporting Service are ISBN-13: 978-0-8138-1649-4/2011.

Designations used by companies to distinguish their products are often claimed as trademarks. All brand names and product names used in this book are trade names, service marks, trademarks, or registered trademarks of their respective owners. The publisher is not associated with any product or vendor mentioned in this book. This publication is designed to provide accurate and authoritative information in regard to the subject matter covered. It is sold on the understanding that the publisher is not engaged in rendering professional services. If professional advice or other expert assistance is required, the services of a competent professional should be sought.

Library of Congress Cataloging-in-Publication Data

Ecological Aspects of Nitrogen Metabolism in Plants / Editors, Joe C. Polacco, Christopher D. Todd.
    p. cm
  Includes bibliographical references and index.
  ISBN 978-0-8138-1649-4 (hardback)
1. Plants–Effect of nitrogen on. 2. Plant ecology. 3. Plants–Metabolism. 4. Plant-microbe relationships. 5. Plant-soil relationships. 6. Nitrogen cycle. I. Polacco, Joe C. (Joe Carmine), 1944– II. Todd, Christopher D.
  QK898.N6E28 2012
  572′.5442–dc22
                                          2010052572

A catalogue record for this book is available from the British Library.

This book is published in the following electronic formats: ePDF 9780470959381; Wiley Online Library 9780470959404; ePub 9780470959398

Set in 10/11.5pt Times New Roman by SPi Publisher Services, Pondicherry, India
Printed and bound in Singapore by Fabulous Printers Pte Ltd

1 2011

Contents

Contributors vii
Preface ix

Section 1 The Nitrogen Cycle 3

1 The New Global Nitrogen Cycle 5
   Jan Willem Erisman

Section 2 Plant-Soil Microbe Interactions 17

2 Plant Associations with Mycorrhizae and *Rhizobium*—Evolutionary Origins and Divergence of Strategies in Recruiting Soil Microbes 19
   Gerben Bijl, Stéphane De Mita, and René Geurts

3 Arbuscular Mycorrhizas and N Acquisition by Plants 52
   Luisa Lanfranco, Mike Guether, and Paola Bonfante

4 Ectomycorrhiza and Nitrogen Provision to the Host Tree 69
   Michel Chalot and Claude Plassard

5 Proteins in the Rhizosphere: Another Example of Plant-Microbe Exchange 95
   Clelia De-la-Peña and Jorge M. Vivanco

6 Actinorhizal Symbioses 117
   Katharina Pawlowski

7 Two in the Far North: The Alder-*Frankia* Symbiosis, with an Alaskan Case Study 138
   Mike Anderson

8 The Path of Rhizobia: From a Free-Living Soil Bacterium to Root Nodulation 167
   Pedro F. Mateos, Raúl Rivas, Marta Robledo, Encarna Velázquez, Eustoquio Martínez-Molina, and David W. Emerich

9	Exploiting Mycorrhizae and Rhizobium Symbioses to Recover Seriously Degraded Soils *Sérgio Miana de Faria, Alexander S. Resende, Orivaldo J. Saggin Júnior, and Robert M. Boddey*	195

Section 3 Epi- and Endo-Phytic Microbes — **215**

10	Nitrogen: Give and Take from Phylloplane Microbes *Mark A. Holland*	217
11	N_2-Fixing Endophytes of Grasses and Cereals *Veronica Massena Reis, Jos Vanderleyden, and Stijn Spaepen*	231

Section 4 Arthropods — **255**

12	Effects of Insect Herbivores on the Nitrogen Economy of Plants *Leiling Tao and Mark D. Hunter*	257
13	Plant Defense Proteins That Inhibit Insect Peptidases *Carlos Peres Silva and Richard Ian Samuels*	280
14	Nutrient Acquisition and Concentration by Ant Symbionts: The Incidence and Importance of Biological Interactions to Plant Nutrition *Cynthia L. Sagers*	308

Section 5 Environmental Signalling in N Acquisition — **331**

15	The Functions of Flavonoids in Legume-Rhizobia Interactions *Oliver Yu and Yechun Wang*	333
16	Plant Hormones and Initiation of Legume Nodulation and Arbuscular Mycorrhization *Arijit Mukherjee and Jean-Michel Ané*	354
17	Nitric Oxide as a Signal Molecule in Intracellular and Extracellular Bacteria-plant Interactions *Andrés Arruebarrena Di Palma, Lorenzo Lamattina, and Cecilia M. Creus*	397

Index — 421

Contributors

Mike Anderson, Department of Biology, MacAlester College, Saint Paul, MN, USA

Jean-Michel Ané, Department of Agronomy, University of Wisconsin, Madison WI, USA

Andrés Arruebarrena Di Palma, Unidad Integrada Facultad de Ciencias Agrarias, Universidad Nacional de Mar del Plata—EEA INTA Balcarce, Balcarce, Argentina

Gerben Bijl, Wageningen University, Wageningen, The Netherlands

Robert M. Boddey, Embrapa Agrobiologia, Seropédica, Rio de Janeiro, Brazil

Paola Bonfante, Department of Plant Biology, University of Torino, IPP-CNR, Torino, Italy

Michel Chalot, Nancy-Université, Faculté des Sciences et Techniques, Vandoeuvre cedex, France

Cecilia M. Creus, Unidad Integrada Facultad de Ciencias Agrarias, Universidad Nacional de Mar del Plata—EEA INTA Balcarce, Balcarce, Argentina

Clelia De-la-Peña, Department of Horticulture and Landscape Architecture, Colorado State University, Fort Collins, CO, USA

Stéphane De Mita, Wageningen University, Wageningen, The Netherlands

David W. Emerich, Department of Biochemistry, University of Missouri, Columbia, MO, USA

Jan Willem Erisman, Energy Research Center of the Netherlands, ECN Petten, The Netherlands

René Geurts, Wageningen University, Wageningen, The Netherlands

Mike Guether, Department of Plant Biology, University of Torino, Torino, Italy

Mark A. Holland, Department of Biological Sciences, School of Science and Technology, Salisbury University, Salisbury, MD, USA

Mark D. Hunter, Department of Ecology and Evolutionary Biology, University of Michigan, Ann Arbor, MI, USA

Lorenzo Lamattina, Instituto de Investigaciones Biológicas, Universidad Nacional de Mar del Plata, Mar del Plata, Argentina

Luisa Lanfranco, Department of Plant Biology, University of Torino, Torino, Italy

Eustoquio Martínez-Molina, Departamento de Microbiología y Genética and Centro Hispano Luso de Investigaciones Agrarias, Universidad de Salamanca, Salamanca, Spain

Veronica Massena Reis, Embrapa Agrobiologia, Seropédica, Rio de Janeiro, Brazil

Pedro F. Mateos, Departamento de Microbiología y Genética and Centro Hispano Luso de Investigaciones Agrarias, Universidad de Salamanca, Salamanca, Spain

Sérgio Miana de Faria, Embrapa Agrobiologia, Seropédica, Rio de Janeiro, Brazil

Arijit Mukherjee, Department of Agronomy, University of Wisconsin, Madison, WI, USA

Claude Plassard, Nancy-Université, Faculté des Sciences et Techniques, Vandoeuvre cedex, France

Katharina Pawlowski, Department of Botany, Stockholm University, Stockholm, Sweden

Carlos Peres Silva, Departamento de Bioquímica, Centro de Ciências Biológicas, Universidade Federal de Santa Catarina, Florianópolis, Brazil

Joe C. Polacco, Department of Biochemistry, University of Missouri, Columbia, MO, USA and Visiting Professor, Centro de Biotecnologia, Universidade Federal do Rio Grande do Sul, Porto Alegre, Brazil

Alexander S. Resende, Embrapa Agrobiologia, Seropédica, Rio de Janeiro, Brazil

Raúl Rivas, Departamento de Microbiología y Genética and Centro Hispano Luso de Investigaciones Agrarias, Universidad de Salamanca, Salamanca, Spain

Marta Robledo, Departamento de Microbiología y Genética and Centro Hispano Luso de Investigaciones Agrarias, Universidad de Salamanca, Salamanca, Spain

Cynthia L. Sagers, Department of Biological Sciences, University of Arkansas, Fayetteville, AR, USA

Orivaldo J. Saggin Júnior, Embrapa Agrobiologia, Seropédica, Rio de Janeiro, Brazil

Stijn Spaepen, Center of Microbial and Plant Genetics, K. U. Leuven, Belgium

Richard Ian Samuels, Department of Entomology and Plant Pathology, Universidade Estadual do Norte Fluminense, Campos Dos Goytacazes, Rio de Janeiro, Brazil

Leiling Tao, University of Michigan, Department of Ecology & Evolutionary Biology, Kraus Natural Sciences Building, Ann Arbor, MI, USA

Christopher D. Todd, Department of Biology, University of Saskatchewan, Saskatoon, SK, Canada

Jos Vanderleyden, Center of Microbial and Plant Genetics, K. U. Leuven, Belgium

Encarna Velázquez, Departamento de Microbiología y Genética and Centro Hispano Luso de Investigaciones Agrarias, Universidad de Salamanca, Salamanca, Spain

Jorge M. Vivanco, Department of Horticulture and Landscape Architecture, Colorado State University, Fort Collins, CO, USA

Oliver Yu, Donald Danforth Plant Science Center, Saint Louis, MO, USA

Yechun Wang, Donald Danforth Plant Science Center, Saint Louis, MO, USA

Preface

Nitrogen: Too Much and Too Little of a Good Thing

Joe C. Polacco and Christopher D. Todd

Nitrogen, plants, and the environment, the three intertwined themes of this volume, are not only of interest in their own right but also can be viewed from an anthropocentric perspective: Humans and plants interact with environmental nitrogen (N) from opposite poles. We, as all terrestrial vertebrates, ingest a daily excess so that our average excretion of urinary N (16 g/d) is the equivalent of the protein N in 1000 soybeans (~250 g)—protein that the plant dedicated the major portion of its "lifetime N" to deposit. We hunt, gather, cultivate, and industrially fix our N, whereas plants are usually N-limited. N presents a challenge to plants as arguably the major limiting environmental mineral nutrient. However, anthropogenic excesses of N (e.g., in artificial fertilizer, combustion of bio- and fossil fuels, and in poorly treated waste), present other environmental challenges to plants and to the biosphere in general. The proper stewardship of plants can lead to amelioration of many environmental problems, including those brought on by too much, or too little, N. Stewardship, of course, begs an understanding of the bio/chemical transformations of N and of how plants acquire and use it.

On a dry weight basis N is the most abundant plant macronutrient derived from the soil. It is also one of the three plant macronutrients (NPK) in common compound fertilizers. In relative abundance N is significantly higher than phosphorus (P) in plant dry weight[1]. To be utilized N and P must be presented to the plant in proper chemical form, meaning that abundance of these elements in the environment is not necessarily equivalent to their availability to plants. The atmosphere is about 80% N_2, but, because of the stability of its triple bond, N_2 is usually not available as a biological N source. Further, N is relatively rare in the earth's crust so that natural and many agricultural systems are usually limited by N availability. This was especially true in the pre-industrial era before mass production of artificial fertilizers.

The biological route of N_2 to reactive N (Nr) is exclusively by reductive fixation to ammonia in some bacteria. (Nr is defined in Chapter 1 as all forms of inorganic N different from N_2, essentially N with a non-zero valence.) Since plants are anchored in the soil, which is usually relatively N-poor, and since their aerial portions do not fix N_2 (unlike their superb ability to fix the much less abundant CO_2), plants have evolved strategies to acquire, conserve, and recycle Nr. These strategies form part of the N cycle, whereby bacteria convert N_2 to NH_3 which, in turn, is sequestered in carbon skeletons as organic molecules, or which is used by soil bacteria as an electron source, oxidizing NH_3 to nitrite and nitrate, compounds for which plants have evolved efficient uptake and reduction mechanisms. Nitrate and nitrite, in turn, can be electron acceptors for denitrifying bacteria in anaerobic soils, leading to re-release of N_2 (Chapters 1 and 17), and obligate co-release of the pollutant, N_2O[2]. This very simplistic description does not do justice to the roles that plants play in the cycle, nor does it consider the massive anthropogenic inputs of Nr into the cycle and the causes and ramifications of those inputs. One goal of this book is an attempt to address aspects of plant roles in the (new) global N cycle (Chapter 1). Nr encompasses both plant nutrients and pollutants. Negative players among the latter can be carcinogens, greenhouse gases, and ozone depleters.

Within the "environmental" theme, our goal is the coverage of various operative environmental scales in plant N acquisition and assimilation. One 'micro-scale' deals with the many specialized interactions of plants with the bacterial world: associations of legumes with rhizobia (Chapters 2, 8, and 16) and of many trees with actinorhizal bacteria (Chapters 6 and 7), the latter of great importance in boreal forests. Even more primordial associations are those of plant roots with fungi: The mycorrhizae play a crucial role in N provisioning to plants, and N-providers are not limited to the ectomycorrhiza (ERM) (Chapter 4). The arbuscular mycorrhiza (ARM), well-recognized as aiding in P provision, also play a role in "virtual extension" of the plant root system, and N provision culminates from an interesting spatial and root-fungal delimitation of components of the arginine/urea cycle (Chapter 3). Indeed, plants do not excrete arginine-derived urea as do terrestrial mammals but employ it as a recyclable N intermediate.

"Never say never" is a proper adage applied to interactions in nature. Two-component symbioses are often our first view of an association, later revealed to contain other players. A case in point is the potential role(s) of bacteria in the ARM association, considered in Chapter 3. Bacterial endophytes may provide N directly to the plant, or improve the plant's ability to "mine" N. These questions on endophytes are addressed in Chapter 11, which also considers potential agronomic applications of the endophytic bacteria. And, the extensive leaf surface area provided by plants is an environment not ignored by microbes; Chapter 10 discusses ramifications of phylloplane colonization on plant N acquisition. One is that there is much nutrient and "waste" exchange on the leaf surface, and much of it influenced by phylloplane microorganisms.

Clearly, N is not only a nutrient, but also a component of simple, inorganic signaling compounds, such as nitric oxide (NO) and nitrate (NO_3^-). NO is but one of the possible forms between NH_3 (N valence –3) and nitrate (N valence +5). Such inter-conversions catalyzed by soil and endophytic bacteria are considered in Chapter 17, especially in light of regulatory effects on the plant. NO and nitrate also have profound effects on legume nodulation and root development. In general the polluting effects of excessive Nr need also to take into account the altered environmental levels of regulatory Nr molecules.

At higher trophic levels, plant interactions with insects and herbivores are worthy of increased study. Chapters 12 and 14 portray the evolutionary and environmental complexities of plant interactions with insects, from the "exotic" ant-plant associations to more general questions of N flux between plant and insect. Such a flux is highly regulated given the incredible panoply of plant proteins aimed at insect proteases, discussed in Chapter 13, a discussion also germane to "plant investment strategies" in N-containing insecticides.

On a global scale, massive anthropogenic inputs of Nr from burning of fossil and biofuels, human waste treatment, the Haber-Bosch process and concomitant intensive agriculture open up issues of equitable distribution of nitrogenous fertilizer—its overuse in Europe, North America, and China compared to under-use in much of the developing world[3]. Excessive inputs lead to Nr being not only a soil pollutant, but also a greenhouse gas and an ozone-depleter, in the form of N_2O[4]. Ice core analyses indicate rising atmospheric nitrate (derived from NO and N_2O) with the advent of the industrial age[5]. So, a new mandate, in addition to more equitable distribution of fertilizer, is to reduce fertilizer use overall (thus avoiding such negative outcomes as eutrophication, health problems, dead zones, reduced species diversity, etc.) To lower Nr in soil, water, and air requires improved cultural practices, increased biological N-fixation, and efficient means of industrial denitrification (anammox avoids N_2O evolution[2]). A more long-range strategy is to improve the N use efficiency of plants, topics broached in Chapter 1.

Badly degraded and eroded soils, of course, suffer from N-depletion, and Chapter 9 shows how recruitment of proper leguminous trees and rhizobial and ARM isolates can lead to dramatic recovery of three types of degraded soils (erosion ditches, mine sites, "detopped" mountains) in Brazil's Rio de Janeiro state. To be sure, soil is more than a semi-solid solution of nutrients; amendments of proteins, N-rich polypeptides secreted by plants and ARM, undoubtedly influence soil texture, tilth, humidity, and other factors. Roles of plant-secreted proteins in influencing the microbial rhizosphere are considered in Chapter 5. And, it has been reported that plants can consume intact microbes as nutrient sources[6].

Hundreds of millions of years of N-acquisition have led to evolved associations of plants with bacteria (rhizobial and actinorhizal) and fungi (ARM and ERM). Intriguing evolutionary questions on the origins and relatedness of these associations are discussed in Chapters 2, 6, and 16. Chapter 15 discusses roles of flavones and isoflavones in the complex chemical signaling between plants and their potential symbionts, chemical signals operative not only in the rhizosphere but in the establishment and development of N_2-fixing nodules.

The need is clear for integrating the complexities inherent in plant performance, environmental interactions, and N inter-conversions. A greater appreciation of these interactions will lead to better understanding of our place, and that of plants, in our shared environment. We thank the authors for their excellent contributions to an appreciation of the complex N-Plant-Environmental interactions. We thank Justin Jeffryes for reinforcing the need for this volume. And our interactions with Justin's Wiley-Blackwell colleagues, Shelby Allen and Susan Engelken, to name two, have been nothing short of pleasant, and productive, in bringing the volume to fruition.

Literature Cited

[1]Güsewell, S. 2004. N: P ratios in terrestrial plants: variation and functional significance (Tansley review). *New Phytologist* 164: 243–266.

[2]Canfield, D.E., Glazer, A.N., Falkowski, P.G. 2010. The evolution and future of Earth's nitrogen cycle. *Science* 330: 1519–1520.

[3]Vitousek, P.M., Naylor, R., Crews, T., David, M.B., Drinkwater, L.E., Holland, E., Johnes, P.J., Katzenberger, J., Martinelli, L.A., Matson, P.A., Nziguheba, G., Ojima, D., Palm, C.A., Robertson, G.P., Sanchez, P.A., Townsend, A.R., Zhang, F.S. 2009. Nutrient imbalances in agricultural development. (Policy Forum). *Science* 324: 1519–1520.

[4]Ravishankara, A.R., Daniel, J.S., Portmann, R.W. 2009. Nitrous oxide (N_2O): The dominant ozone-depleting substance emitted in the 21st century. *Science* 326: 123–125.

[5]Hastings, M.G., Jarvis, J.C., Steig, E.J. 2009. Anthropogenic impacts on nitrogen isotopes of ice-core nitrate. *Science* 324: 1288.

[6]Paungfoo-Lonhienne, C., Rentsch, D., Robatzek, S., Webb, R.I., Sagulenko, E., Näsholm, T., Schmidt, S., Lonhienne, T.G.A. 2010. Turning the Table: Plants Consume Microbes as a Source of nutrients. *Plos One* 5, issue 7 (July).

Ecological Aspects of Nitrogen Metabolism in Plants

Section 1
The Nitrogen Cycle

Chapter 1
The New Global Nitrogen Cycle
Jan Willem Erisman

Introduction

Almost 80% of the atmosphere is N_2. And nitrogen, in reactive forms is essential for life on earth and is crucial for sustainable development.* Nitrogen in our environment has both benefits as well as negative consequences. The primary benefit of nitrogen is the stimulation of plant growth in agriculture for food, feed, and fuel, whereas the negative aspects include almost all environmental impacts. Compared to elements such as carbon, sulfur, or phosphorus, nitrogen contributes to a variety of negative impacts, most interrelated, such as climate change, eutrophication, soil acidification, degraded human health, loss of biodiversity, etc. In 2008 we celebrated the one-hundredth anniversary of the invention of the production of ammonia by Fritz Haber in 1908. Ammonia is the basis for fertilizer production, and Carl Bosch was able to turn that ammonia production into an industrial process (Smil, 2001; Erisman et al., 2008). Bosch and Haber were awarded the Nobel Prize for their achievements. Ammonia is not only the basis for fertilizer, but also for many industrial chemicals, including explosives. So, overall, the Haber-Bosch process was beyond doubt one of the most important inventions of the twentieth century (Erisman et al., 2008). Over the past decades the production of fertilizer has become very energy and economically efficient (Kongshaug, 1998) and on a large scale has increased agricultural productivity. Without fertilizer input, the biosphere would produce 48% less food (Erisman et al., 2008). Therefore, at present, we cannot live without fertilizer. At the same time, industrialization increased the use of fossil fuels. To use the energy from fossil fuels, they are burned, resulting in a large release of oxidized nitrogen (NOx) into the atmosphere. Vehicular traffic, energy use, and industry are the principal sources of oxidized forms of Nr. The dispersion of NOx has affected human health and increased nitrogen deposition in remote areas leading to eutrophication (Erisman and Fowler, 2003).

* Reactive nitrogen, Nr, is defined here as all forms of nitrogen in the biosphere except N_2.

Ecological Aspects of Nitrogen Metabolism in Plants, First Edition. Edited by Joe C. Polacco and Christopher D. Todd.
© 2011 by John Wiley & Sons, Inc. Published 2011 by John Wiley & Sons, Inc.

This book is dedicated to plant "strategies" to acquire, assimilate, and conserve N. Much emphasis is given to plant interactions with microorganisms (*Frankia*, rhizobia, mycorrhizae). Biological N-fixation is prominently covered and is an important part of the nitrogen cycle. However, in order to establish the need for knowledge in the area of biological N fixation (and the N "conduits" that mycorrhizae provide for delivering soil N to plant), we have to address the "big picture" first, providing background on the major components of the nitrogen cycle on different scales, emphasizing the human contribution to the increased cycling of nitrogen in the biosphere and the resulting impacts on ecosystems and humans. This introductory chapter addresses these issues as a global overview of the human influence on the nitrogen cycle.

The Preindustrial Nitrogen Cycle

Nitrogen is an important element—the most abundant constituent of the atmosphere, hydrosphere, as well as the biosphere. It is also one of the essential elements for the growth of plants and animals and has a crucial role in ecology and in the environment. It is useful to look at the reactions of elements in the form of a closed cycle. Such a cycle is often termed a biogeochemical cycle because chemistry, biology, and geology all provide important inputs. Cycling of elements is often governed by kinetics and may involve the input of energy, so that chemical equilibrium states are not attained. The ultimate source of energy for driving energetically uphill reactions is the sun. The earth's surface receives an average radiation input of $100-300\,W/m^2\,\%$ day, depending on latitude. Some of this is captured by photosynthesis, and is used to produce high-energy content molecules, such as oxygen. Because of the inherently low efficiency of the photosynthetic process and the production of phytomass, energy supply from this source has low power densities and hence high land demands. Recent estimates of the global terrestrial net primary productivity (NPP) average approximately 120 gigaton (Gton) of dry biomass produced annually, and that contains some $1,800 \times 10^{18}$ joules (1,800 exajoule [EJ]) of energy (Smil, 2004). In principle, there is globally enough possible annual crop growth to produce the food needed to feed the population currently and for the coming decades. Furthermore, there is enough annual production of new biomass to cover up to four times current human annual energy use. However, in order to grow, collect, and use biomass in a sustainable way to satisfy human food and energy needs, a well-regulated and optimized process is needed.

The atmosphere contains mostly elemental di-nitrogen, along with other nitrogen containing trace gases (ammonia, nitric oxide, nitrogen dioxide, and nitrous oxide). Aquatic systems primarily contain soluble forms of nitrogen, such as nitrate and ammonia/ammonium, as well as biological nitrogen found in proteins, DNA, RNA, and other compounds that make up living systems. Since lone pairs of electrons on nitrogen are usually basic, ammonia coexists in both the protonated and de-protonated forms near neutral pH. The most important aspect of the environmental nitrogen cycle is the dynamic exchange of chemical species that occurs between the atmosphere and the surface landmasses and oceans. Figure 1.1 illustrates the nitrogen cycle.

From Figure 1.1, it is obvious that living systems are the main players in the interconversions of the reduced and oxidized forms of nitrogen. They play an important role in providing reduced nitrogen compounds for the global cycle, by denitrification processes (the conversion of nitrate to N_2 and N_2O), biosynthesis (making amino acids, DNA, and RNA), and nitrogen fixatio*n* (reduction of N_2 to NH_3 by both free-living bacteria and bacteria associated with plants, usually in root nodules, the so-called biological nitrogen fixation [BNF])

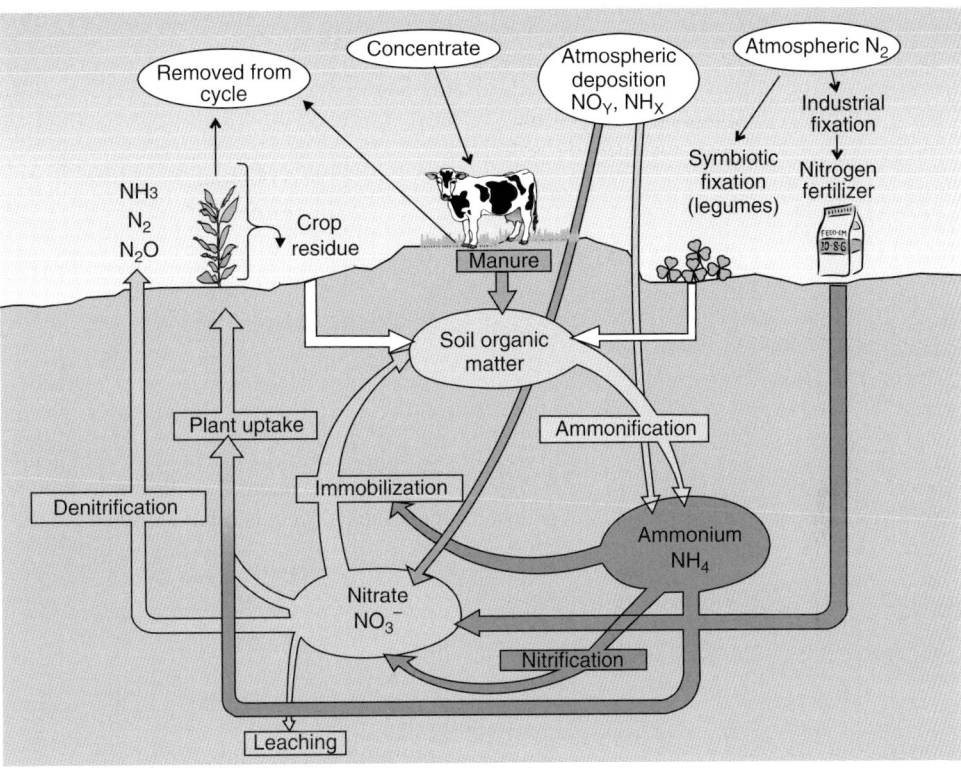

Figure 1.1. Illustration of the nitrogen cycle.

(Marschner, 1997). Nearly all organisms can use ammonia for biosynthesis (ammonia assimilation). Ammonia is also a major metabolic end product, as in the bacterial decomposition of dead organisms (ammonification). Mammals eliminate ammonia; however, the liver transforms it into the less toxic compound urea before excreting it.

Although reduced nitrogen is preferred for biosynthetic reactions, plants have also learned to capture needed nitrogen by assimilatory nitrate reduction (Marschner, 1997). This evolutionary consequence came about because nitrate is the dominant soluble form of nitrogen in aerated soils. Formation of reduced nitrogen compounds is energetically uphill. Besides plant use of nitrate, its reduction also occurs by bacterial action in oxygen-free soil and sediments. Reduced nitrogen can however be formed in a different way in the soil, where bacteria and plants work together. Symbiotic nitrogen-fixing bacteria in aerated soils provide an oxygen-free environment for their oxygen-sensitive nitrogenase within nodules induced on the roots of legumes, such as clover and alfalfa (Marschner, 1997). BNF symbioses also occur in many trees associated with gram-positive *Frankia*.

The preindustrial nitrogen cycle could be characterized as a careful preservation of the very low amount of Nr available in nature. Nr is created and distributed by natural processes such as forest fires, lightning, volcanic eruptions, and biological nitrogen fixation (Smil, 2001; Reid et. al., 2005). The natural production of Nr is estimated by many authors to be around 125 (Tg) N per year, with 5 Tg from lightning and 120 Tg from biological nitrogen fixation (Galloway et al., 2003, 2008; Schlesinger, 2009).

Human-induced Changes of the Nitrogen Cycle

In addition to the natural formation of Nr, there are also significant geophysical and anthropogenic sources. The Haber-Bosch process is the first high-pressure industrial process that fixes nitrogen to ammonia on a large scale and "helps" nature increase plant growth (Smil, 2001). The second largest bulk industrial nitrogen feedstock, nitric acid, lies at the other extreme of the redox series. Ammonia is oxidized to produce nitric acid in the Oswald process, named after its discoverer Wilhelm Oswald. It makes good chemical sense that the principal industrial nitrogen compounds lie at opposite ends of the redox series (valencies of −3 for ammonia to +5 for nitrate). With these two reagents, it is ultimately possible to synthesize any desired compound with an intermediate oxidation state for nitrogen. The major use for nitrogen compounds is as fertilizers. Industrial nitrogen fixation is estimated to be 125 Tg/yr, of which 82% is for fertilizer production, and the remaining is for industrial use and explosives (Galloway et al., 2008). Fertilizer application to cultivated plants is more than double the natural amount of BNF. Cultivation of arable land has increased BNF by about 40 million metric tons per year (Tg/yr) (Galloway et al., 2008). Fossil fuel combustion adds another 25 Tg/yr of oxidized nitrogen, released to the atmosphere (Galloway et al., 2008; Schlesinger, 2009). For 2005 the total human-induced Nr added up to 170 (Schlesinger, 2009) to 187 Tg/yr (Galloway et al., 2008). Adding the naturally produced Nr results in about 300 Tg/yr generated globally.

Figure 1.2 shows the increase in human-induced Nr production since the beginning of the last century. By 1980 human Nr production became larger than natural production and,

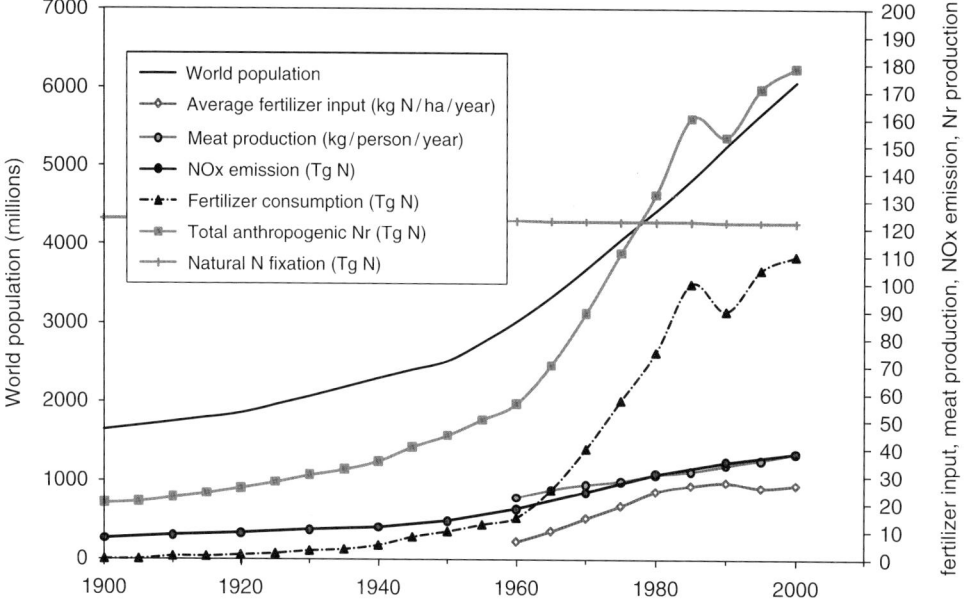

Figure 1.2. Changes in global Nr production (modified from Reid et al., 2005; Erisman et al., 2008; Van Aardenne, 2001).

until 1980, the production increase was higher than the population increase. This changed after the 1990s when the economic situation in Eastern Europe affected agricultural production in that area. Since then the increase in fertilizer use matches the population growth. One can conclude that there is room for improvement in the fertilizer use efficiency and at the same time, that as world population grows, fertilizer use will grow at least at the same rate if no measures are taken.

Too Much or Too Little of a Good Thing

There are two major problems with Nr: some regions of the world have too little and others too much. In the latter case burning of fossil fuel and inefficient incorporation of nitrogen into food has resulted in a large number of major human health and ecological problems. In the N-deficient areas, too little food, especially proteins, will cause more malnutrition eventually leading to increased mortality. The rate of change in N distribution is enormous and probably greater than that for other ecological problems. According to the Food and Agriculture Organization of the United Nations (FAO), currently more than 1 billion people, especially in the south of the world, suffer from malnutrition. In those regions of the world where there is not enough food, nitrogen availability through fertilizers might be the limiting factor, apart from availability of water and other nutrients. Obviously, fertilizer distribution depends on infrastructure and social and/or economic factors.

In Europe, the US, and more recently in China, food production is stimulated by government support (e.g., by subsidizing fertilizers). This has created a shift from good nutrient management to industrialized food production with appreciable waste of inexpensive nitrogen and its release to the environment. Global nitrogen fertilizer production occurs in a limited part of the globe, but nitrogen is transported to other regions in the form of food and feed and through the air (NOx). Human creation and use of Nr (see Figure 1.2) has created a new nitrogen cycle, especially because the efficiency of N use has decreased along with its increased, but less efficient, production. The energy efficiency of nitrogen in fossil fuel is, in essence, zero, because the NOx formed during combustion is an unwanted by-product and has no function in the production of energy. In agriculture Nr is added in the form of fertilizer to increase crop yield. However, only 5–10% of fertilizer N ends up in our food depending on the type of food (meat or vegetables), the agricultural system, and management (Smil, 2001; UNEP, 2007). Meat production is an additional inefficient step in the food N chain with further Nr losses to the environment (Steinfeld et al., 2006). Most of the nitrogen lost to the environment is through air emissions (ammonia, nitrogen oxides, nitrous oxide) and water (NO_3 run-off and/or leaching), in addition to that returned to the atmosphere as N_2 (Galloway et al., 2003). These losses cascade through the different environment compartments, exchanging forms and contributing to many different effects in time and space. This is called the "cascade effect of nitrogen" (Galloway et al., 2003).

Schlesinger (2009) estimated that the fates of natural Nr were mainly conversion to N_2 via denitrification (92 of the 125 Tg/yr), the other major flux being river flow. Figure 1.3 shows the increase in the fates of Nr from the preindustrial era to the current situation. For anthropogenic nitrogen, the biospheric increment amounts to 9 Tg, river flow increment amounts to 35 Tg, groundwater leaching to 15 Tg, denitrification to 17 Tg, and atmospheric transport to the ocean to 48 Tg per year. The fate of anthropogenic nitrogen is therefore much less as N_2 than the fate of natural nitrogen. These data are, however, highly uncertain as discussed in Galloway and

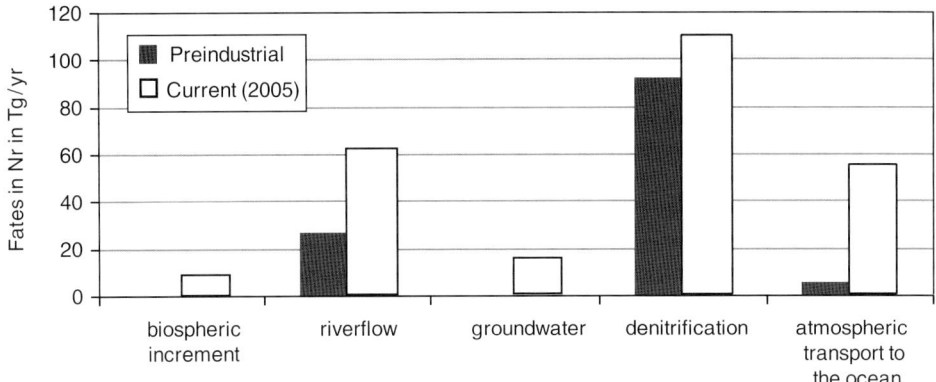

Figure 1.3. Increase in the fates of Nr currently (2005) compared to preindustrial (Tg/y).

others (2008). What is clear, though, is that the losses of Nr to the environment have increased substantially, especially in the areas of intensive fossil fuel use and agriculture.

Anthropogenic disruptions in the nitrogen cycle have led to an 1,100% increase in the flux of nonreactive atmospheric nitrogen (N_2) to Nr compounds. Once converted to a reactive state, nitrogen persists in the environment, "cascading" through various compounds (NH_3, N_2O, NOx, NO_3), resulting in impacts such as the production of ground-level ozone, acidification, eutrophication, hypoxia, stratospheric ozone depletion, and climate change (Vitousek and Melillo 1997; Mansfield et al., 1998; Langan, 1999; Cowling et al., 2002; Galloway et al., 2003, 2008).

N-limitation and Biodiversity

In the natural environment nitrogen is a limiting factor for growth, leading to a rich biodiversity of strategies for N acquisition, N "management," and N "preservation." Trees, for example, retreat their leaf nitrogen before senescence. If Nr becomes increasingly available, the fast growing species, no longer growth limited (provided other nutrients and water are sufficiently available), will overgrow slow-growing species resulting in serious loss of biodiversity, algal blooms, hypoxic zones, etc. (e.g., Goulding et al., 1998; Bobbink, 2004; Stevens et al., 2004; Dise and Stevens, 2005; Phoenix et al., 2006; Klimek et al., 2007). The effects due to high N deposition extend to fauna, a notable example of which is the disappearance of the red-backed shrike (*Lanius collurio*) in coastal dunes of the Netherlands and in other parts of Western Europe. The red-backed shrike is a good indicator of fauna diversity, since its breeding success depends highly on the availability of large insects and small vertebrates (Beusink et al., 2003). Evidence for serious adverse effects on faunal species diversity and marine systems exists, but the level of scientific understanding is rather low.

The negative impacts of Nr emissions include eutrophication of seminatural ecosystems and surface waters, soil acidification, nitrate pollution of groundwater, particulate formation leading to impacts on human health, and alterations of the earth's radiation balance, such as ozone formation leading to effects on humans and vegetation, and to climate change when N is transformed into nitrous oxide, one of the most important greenhouse gases (e.g., Galloway, 1998; Galloway and Cowling, 2002; Matson et al., 2002; Galloway et al., 2003). Although the effects of N inputs on the carbon cycle remain controversial (Houghton et al., 1998; Nadelhoffer et al., 1999; de Vries et al., 2008), there are indications of a strong relationship with stimulation of greenhouse gas emissions as well as with carbon sequestration and plant/forest growth. Increased inputs of N to aquatic ecosystems from atmospheric deposition, sewage, and agricultural runoff can cause eutrophication, including damage to fisheries in coastal ecosystems (Rabalais, 2002). The formation of N_2O during nitrification and denitrification in all systems results in tropospheric warming and stratospheric ozone depletion (Prather et al., 2001). These undesirable "cascading effects," as Galloway and others (2003) call them, of Nr moving through aquatic and terrestrial ecosystems and the atmosphere do not stop until the Nr is fixed or eventually converted back to N_2 through the process of denitrification. Therefore, one molecule of Nr can contribute to a cascade of effects, each changing natural biodiversity.

Strategies to Limit N-effects

The science of Nr in the environment has advanced beyond recognition of the seriousness of the problem. Evidence of Nr's multiple effects has been presented not only at the four international conferences on the subject but also in many papers published in refereed academic journals (see www.initrogen.org for the third [Nanjing, 2004] and fourth [Brazil, 2007] international nitrogen conferences and Erisman et al., 1998; Cowling et al., 2002). The next phase is to assess the issues in such a way that international cooperation to solve the problems becomes more likely. In 2004 The Nanjing declaration on nitrogen management was signed at the third international nitrogen conference (Erisman, 2004). The declaration affirms the need for an international protocol to reduce Nr to sustainable levels in agricultural systems, estuaries, and ecosystems. This might be achieved by improving utilization of nitrogen in food production, by limiting energy consumption, or by spurring changes in the fuels used as energy carriers. Since the Nanjing declaration, many scientific efforts have focused on management of nitrogen. In the Netherlands the introduction of regional nitrogen ceilings based on effect limits was proposed (Erisman et al., 2005). Current policies in the Netherlands that have succeeded in decreasing Nr focused on setting and meeting the targets through some freedom of choice for the stakeholders and not, as has been done mostly up till now, by prescribing what the stakeholder should do. Targets should be based on all effects (no trade-offs), on sound science, and on a clear long-term vision that takes into account the position of agriculture in relation to other sectors and human needs. We should aim to close the nutrient cycles and minimize resource use. Therefore, the scientific community should come up with simple, easy to determine indicators that can be used to implement, monitor, evaluate, and maintain policies, such as nitrogen use efficiency (NUE).

Some of the options should be agreed upon within the framework of the European Union, or even on a larger scale, because it affects countries' competitiveness. Such measures include taxes or financial grants, and setting targets for N losses. Furthermore, the closing of cycles should be done concurrently and at different scales. We are accustomed to seeking our improvements in technological options. There are potential technologies that might lead to substantial emission reduction (catalytic converters, hydrogen economy, nitrification inhibitors, fermentation of manure, etc.). Furthermore, important technology developments needed include traditional plant breeding that employs powerful new tools, such as marker-assisted selection, to identify genes associated with increased NUE, and genetic manipulation based on better understanding of the transition between nitrogen assimilation and nitrogen cycling (Good et al., 2004). These are important, but the reduction in environmental load they cause should not lead to other environmental risks or increased import of raw materials, leading to changes in cycles at supranational scales. Furthermore, other components (such as carbon) and such issues as access to freshwater should be taken into account to prevent trade-offs. If financial incentives are given, it is important to secure the period that these incentives will last, in order to guarantee a return on investment.

Galloway and others (2008) recently quantified potential Nr reductions by implementation of four interventions to limit Nr production:

- The first intervention point involves fossil fuel combustion. Using best available technologies, Nr creation during fossil fuel combustion can be decreased by about one-third from its current level. The barriers to such a decrease are primarily financial, as the scientific, engineering, and the policy instruments are all well developed. Utilizing this available technology, Nr creation by fossil fuel combustion would decrease from ~24 Tg N yr^{-1} to ~16 Tg N yr^{-1}.
- The second intervention point is to increase the NUE of crop production. Cassman and colleagues estimate that there will be a 38% increase in global cereal demand by 2025. They project that this demand can be met with a 25% decrease in N fertilizer application to cereals if the current decline in cereal harvest area is halted (−0.33% yr^{-1} over 20 years) and the crop yield response to applied N is increased by 20% (NUE increase). This 'intervention' would reduce the global Nr creation rate by ~15 Tg N yr^{-1}.
- The third intervention is improved animal feeding strategies and manure management. It is estimated that 13% of the N lost from animal waste in the EU-27 countries could be either eliminated or captured by a combination of low-protein animal feeding, barn adaptations, covered manure storage, air purification, and efficient manure applications (to crop land and grassland). Extending this to the animal populations of the world would result in a decrease of ~17 Tg N lost to the environment. In considering this intervention, it is equally important to address the fraction of meat and dairy products in human diets. In many developed countries, diets contain more animal protein than needed. Livestock are particularly effective at releasing Nr to the environment, so that eating less meat would both substantially decrease Nr losses while having potential health benefits.
- The fourth intervention point is improved sewage treatment. Each year, the global population produces ~20 Tg N in human waste of which <1% undergoes treatment that will convert it to N_2 (generally tertiary treatment). Even if only half of the 3.2 billion people who live in an urban environment had access to the required level of sewage treatment, this would result in 5 Tg N being converted to N_2.

Galloway and others (2008) estimated that these four interventions represent a potential decrease of ~50 Tg N yr^{-1} created per year, or ~25% of the total Nr created in 2005.

The Role of Understanding Plant Processes

One of the major Nr issues is the low NUE of plants. When nitrogen is limiting there is an almost linear increase in yield with increased nitrogen additions. However, this increase levels off rapidly with an optimum of nitrogen addition compared to yield increase (Yara, 2005). At this point, however, the NUE is about 30–50%. It is therefore very important to find ways to increase the NUE by plants without affecting yields (Good et al., 2004). In effect, as NUE increases, the same yield is achieved with lower N input, and therefore there is less N lost to the environment.

Mycorrhizal fungi live in symbiotic association with plants. These biotrophs have been associated with roots of most land plants, worldwide, for hundreds of millions of years. At least three chapters in this book cover the ancient plant-mycorrhizal association, one that greatly increases the ability of plants to take up nutrients (particularly phosphorous) from the soil. In return for assistance in nutrient uptake, mycorrhizae receive their carbon from the host plant. Mycorrhizae have a role in efficiently mining available ammonium (and probably nitrate that they convert to ammonium) and making it available to the plant. Mycorrhizae are therefore very important for nutrient uptake and for its uptake efficiency (Smith and Read, 1997).

This volume also covers biological nitrogen fixation by free-living, endophytic, and symbiotic (nodule-forming) bacteria (the latter well known in associations with members of the leguminosae and ulmaceae). Crop rotation systems and soil management can effectively increase the NUE. There is a need for further research on plant associations with nitrogen-fixing symbionts and how these bacteria can be effectively exploited. (There is only a very small amount of nitrogenase in the biosphere, present only in bacteria.) It may be key to increase nitrogenase (and nitrogenase expressers) to effect changes in favor of natural N fixation.

Acknowledgements

This chapter is a contribution to the International Nitrogen Initiative (www.initrogen.org), jointly sponsored by Scientific Committee On Problems of the Environment (SCOPE) and International Geosphere-Biosphere Programme (IGBP), and the European projects NitroEurope IP (European Commission), Nitrogen in Europe (NinE, European Science Foundation), and COST Action 729. I greatly appreciate the guidance and help of the editors, Joseph Polacco and Chris Todd.

References

Beusink, P., Nijssen, M., van Duinen, G.J., Esselink, H. 2003. Broed-en voedselecologie van Grauwe Klauwieren in intacte kustduinen bij Skagen, Denemarken. Stichting Bargerveen, Afdeling Dierecologie, Katholieke Universiteit Nijmegen.

Bobbink, R. 2004. Plant species richness and the exceedance of empirical nitrogen critical loads: an inventory. Report Landscape Ecology. Utrecht University/RIVM, Utrecht, pp. 1–19.

Cowling, E.B., Galloway, J.N., Furiness C.S., Erisman J.W. 2002. Optimizing Nitrogen Management in Food and Energy Production and Environmental Protection: Report from the Second International Nitrogen Conference. Ecological Society of America, Washington, DC.

de Vries W., et al. 2008. Ecologically implausible carbon response? Nature 451, E1–E3.

Dise, N.B., Stevens, C.J. 2005. Nitrogen deposition and reduction of terrestrial biodiversity: evidence from temperate grasslands. Science in China Series C, Life Sciences/Chinese Academy of Sciences 48: 720–728.

Erisman, J.W., et al. 1998. Nitrogen, the Confer-N-s, First International Nitrogen Conference 1998: Summary statement. Environmental Pollution 102: 3–12.

Erisman, J.W., Fowler, D. 2003. Oxidized and reduced nitrogen in the atmosphere. In Knowledge for Sustainable Development, An Insight into the Encyclopedia of

Life Support Systems, Volumes I, II, III, UNESCO Publishing-Eolss Publishers, Oxford, UK.

Erisman, J.W. 2004. The Nanjing declaration on management of reactive nitrogen. BioScience 54 (4): 286–287.

Erisman, J.W., Domburg, N., de Vries, W., Kros, H., de Haan, B., Sanders, K. 2005. The Dutch N-cascade in the European perspective. Science in China. Series C, Life sciences/Chinese Academy of Sciences 48: 827–842.

Erisman, J.W., Galloway, J.A., Sutton, M.S., Klimont, Z., Winiwater, W. 2008. How a century of ammonia synthesis changed the world. Nature Geoscience 1: 636–639.

Galloway, J.N. 1998. The global nitrogen cycle: changes and consequences. Environmental Pollution 102: 15–24.

Galloway, J.N., Cowling, E.B. 2002. Reactive nitrogen and the world: 200 years of change. Ambio 31: 64–71.

Galloway, J.N., Aber, J.D., Erisman, J.W., Seitzinger, S.P., Howarth, R.W., Cowling, E.B., Cosby, B.J. 2003. The nitrogen cascade. BioScience 53 (4): 341–356.

Galloway, J.N., Townsend, A.R., Erisman, J.W., Bekunda, M., Cai, Z.E., Freney, J.R., Martinelli, L.A., Seitzinger, S.P., Sutton, M.A. 2008. Transformation of the nitrogen cycle: Recent trends, questions, and potential solutions. Science 320: 889–892.

Good, A.G., Shrawat, A.K., Muench, D.G. 2004. Can less yield more? Is reducing nutrient input into the environment compatible with maintaining crop production? Trends in Plant Science 9: 597–605.

Goulding, K.W.T., Bailey, N.J., Bradbury, N.J., Hargreaves, Y.P., Howe, M., Murphy, D.V., Poulton, P.R., Wilson, T.W. 1998. Nitrogen deposition and its contribution to nitrogen cycling and associated soil processes. New Phytologist 139 (1): 49–58.

Houghton, R.A., Davidson, E.A., Woodwell, G.M. 1998. Missing sinks, feedbacks and understanding the role of terrestrial ecosystems in the global carbon balance. Global Biogeochemistry Cycles 12: 25–34.

Klimek, S., Kemmermann Gen., A.R., Hofmann, M., Isselstein, J. 2007. Plant species richness and composition in managed grasslands: the relative importance of field management and environmental factors. Biological Conservation 134 (4): 559–570.

Kongshaug, G. 1998. Energy Consumption and Greenhouse Gas Emissions in Fertilizer Production. IFA Technical Conference, Marrakech, Morocco. September/October 1998. Updated in Jenssen, T.K., Kongshaug, G. 2003. "Energy Consumption and Greenhouse Gas Emissions in Fertiliser Production."

Langan, J., editor. 1999. The Impact of Nitrogen Deposition on Natural and Seminatural Ecosystems. Kluwer Academic Publishers, Dordrecht, The Netherlands, pp. 1–251.

Mansfield, T.A., Goulding, K.W.T., Sheppard, L.J., editors. 1998. Major biological issues resulting from anthropogenic disturbance of the nitrogen cycle. New Phytologist, vol. 139. Cambridge, UK: Cambridge University Press, pp. 1–234.

Marschner, H. 1997. Mineral Nutrition of Higher Plants, 2nd edition. Academic Press Limited, London. ISBN 0-12-473542-8, pp. 889.

Matson, P., Lohse, K.A., Hall, S.J. 2002. The globalization of nitrogen: consequences for terrestrial ecosystems. Ambio 31, 113–119.

Nadelhoffer, K.J., Emmett, BA., Gundersen, P., Kjønaas, O.J., Koopmans, C.J., Schleppi, P., Tietema, A., Wright, R.F. 1999. Nitrogen deposition makes a minor contribution to carbon sequestration in temperate forests. Nature 398: 145–148.

Phoenix, G.K., et al. 2006. Atmospheric nitrogen deposition in world biodiversity hotspots: the need for a greater global perspective in assessing N deposition impacts. Global Change Biology 12: 470–476.

Prather, M., et al. 2001. Atmospheric chemistry and greenhouse gases. In: Houghton, J.T., et al. (Eds.), Climate Change 2001: The Scientific Basis, Contribution of Working Group I to the Third Assessment Report of the Intergovernmental Panel on Climate Change. Cambridge University Press, Cambridge, UK.

Rabalais, N.N. 2002. Nitrogen in Aquatic Ecosystems, Ambio 31: 102–112.

Reid, W.V., Mooney, H.A., Cropper, A., Capistrano, D., Carpenter, S.R., Chopra, K.E.A. 2005. Millennium Ecosystem Assessment. Ecosystems and Human Well-Being: Synthesis. Island Press, Washington DC.

Schlesinger, W.H. 2009. On the fate of anthropogenic nitrogen. Proceedings of the National Academy of Sciences 104:203–208.

Smil, V. 2001. Enriching the Earth: Fritz Haber, Carl Bosch and the Transformation of World Food Production. The MIT Press, Cambridge, MA, xvii and 338 pp.

Smil, V. 2004. World history and energy. In: Clevaland, C. (Ed.) Encyclopedia of Energy, Vol. 6, Elsevier, Amsterdam, pp. 549–561.

Smith, S.E., Read, D.J. 1997. Mycorrhizal Symbiosis, 2nd edition. Academic Press, San Diego and London. ISBN 0-12652840-3, 605 pp.

Steinfeld, H., Gerber, P., Wassenaar, T., Castel, V., Rosales, M., de Haan, C. 2006. Livestock's long shadow: environmental issues and options. Food and Agriculture Organization of the United Nations, Rome.

Stevens, C.J., Dise, N.B., Mountford, J.O., Gowing, D.J. 2004. Impact of nitrogen deposition on the species richness of grasslands. Science 303 (5665): 1876–1879.

UNEP. 2007. Reactive Nitrogen in the Environment—Too Much or Too Little of a Good Thing. UNEP, Nairobi, Kenia.

Van Aardenne, J.A., Dentener, F.J., Olivier, J.G.J., Klein Goldewijk, C.G.M., Lelieveld, J. 2001. A 1 × 1 degree resolution dataset of historical anthropogenic trace gas emissions for the period 1890–1990. Global Biogeochemical Cycles 15(4): 909–928.

Vitousek, P.M., Melillo, J.M. 1997. Nitrate losses from disturbed forests: Patterns and mechanisms. Forest Science 25: 605–619.

Section 2
Plant-Soil Microbe Interactions

Chapter 2
Plant Associations with Mycorrhizae and *Rhizobium*—Evolutionary Origins and Divergence of Strategies in Recruiting Soil Microbes

Gerben Bijl, Stéphane De Mita, and René Geurts

Introduction

Symbiotic relationships help plants increase access to essential nutrients. Among the oldest of such symbiotic associations is the root-based interaction with fungi of the phylum Glomeromycota. It has existed since the Devonian period (~400 million years ago) and still plays a prominent role in terrestrial ecosystems today (Remy et al., 1994; Taylor et al., 1995). This intimate symbiosis leads to the formation of an intraradical mycelium and takes shape by the formation of haustorium-like structures, named arbuscules, within root cortical cells. The interaction was named after these structures, arbuscular mycorrhizae (AM).

Nitrogen-fixing interactions between legume plants (Fabaceae, formerly known as Leguminosae) and gram-negative bacteria collectively named rhizobia, evolved much later (Sprent and James, 2007). It is also a major evolutionary breakthrough since the interaction provides most plants of this family with virtually unlimited access to nitrogen. AM and rhizobial symbioses are complementary since they provide access to different nutrients. Rhizobia specifically provide nitrogen whereas AM fungi mainly facilitate the uptake of phosphorus, though several lines of evidence suggest that the uptake of several other components is also facilitated, including NH_4^+ and NO_3^-. Supporting this complementary nutrient provisioning, most legumes are able to establish both symbioses, and do so in nature. Interestingly, these two symbioses have been described as representing two extremes in host specificity; although host-symbiont specificity is very strict in the rhizobial symbiosis, most mycorrhizal fungi can interact with hosts from very diverse phyla. In spite of this, the symbioses exhibit analogies. They both take place in the root of the plants and require intracellular accommodation of the microsymbiont, a process that is relatively rare in the plant kingdom outside of pathogenic interactions. Further, elucidation of the genetic bases of the establishment of both these symbioses revealed that they actually are homologous at the molecular (they actually share the same genes) and evolutionary (rhizobial features derived from mycorrhizal features) levels.

Ecological Aspects of Nitrogen Metabolism in Plants, First Edition. Edited by Joe C. Polacco and Christopher D. Todd.
© 2011 by John Wiley & Sons, Inc. Published 2011 by John Wiley & Sons, Inc.

In this chapter, we will examine the evolutionary origins of these two symbioses and the features of the recruitment strategies that control partner choice and specificity. Then we will review the hypotheses explaining why the rhizobial symbiosis has evolved from mycorrhizal features rather than through building a completely new genetic architecture, and what evolutionary hypotheses might explain the divergence of recruitment strategies.

Arbuscular Mycorrhization

AM Symbiosis Is Ancient

Ancestral Glomeramyceta fungi were present in Ordovician times (488 to 443 million years ago), when land flora consisted of ancestors of today's mosses, hornworts, liverworts, and even tracheophytes (Redecker et al., 2000). Fossils demonstrating that Glomeramyceta formed endomycorrhizae are more than 400 million years old. It is hypothesized this symbiosis facilitated plants' conquest of land by providing an opportunity to gain access to soil nutrients. Plant and fungal fossil records (Taylor et al., 1995), combined with phylogenetic reconstructions based on fungal DNA sequences form the basis for this hypothesis (Simon et al., 1993; James et al., 2006). The original claim was made predominantly on fossil records in the Rhynie Chert beds (near Aberdeenshire, Scotland) that provided various insights on both land plants and fungi during the early Devonian (417 million years ago). Fungal structures that resemble contemporary AM fungi were associated with early Devonian plants (Pirozynski and Malloch, 1975). Among these are fossils of the plant *Aglaophyton major* that display arbuscule-like infections (Remy et al., 1994). The infecting fungus is ascribed to the clade of Glomites, today's clade of AM fungi. Early-Devonian plants do not have roots but rather have rhizomes—horizontally oriented underground stems with a definite node and internode architecture. Mycorrhizal-like colonization occurred in the stem cortex, as is the case for today's root mycorrhization.

Despite the evolution of true roots that allowed significantly increased soil penetration, AM symbiosis persisted. Approximately 70% of today's land plant species have the capacity to establish AM symbiosis, suggesting a major selective advantage of this interaction. The major evolutionary advantage provided by endomycorrhizae is inferred from today's situation (i.e., increased accessibility of essential nutrients, phosphate in particular). In soil, a large part of the available phosphorus is bound in very stable and insoluble mineral complexes, and, as a result of biological uptake, the soil in direct vicinity of root systems is rapidly depleted of phosphorus. The mycelium of the fungal partner significantly increases the area that can be explored and exploited. In addition to phosphate, the plant can also obtain other nutrients through the mycorrhizal system, including sulfate and nitrate (Smith and Read, 1997), the latter germane to the theme of this volume.

Mycorrhizal Infection

Because mycorrhizae are difficult both to observe *in situ* and to cultivate *in vitro*, the Glomeromycota phylum is taxonomically largely uncharacterized. To date about 200 species have been identified, though additionally an overwhelming number of sequence entries from meta-sequencing of soil samples (classified as "uncultured" or "environmental") suggest that the actual number of species is much higher. Despite this moderate level of characterization, it is generally assumed that all Glomeromycota species are obligatory symbionts. While the fungus depends on carbohydrates provided by the host plant, the plant partner, in most cases, is able to survive in absence of the fungal symbiont.

AM fungal spores can germinate in the absence of potential host plants, relying on stored energetic sources in the form of triacylglyceride (Beilby and Kidby, 1980) and glycogen. The hyphal germ tube extends several centimeters in search of plant roots. If no nearby roots are found, hyphal extension ceases. Once a plant root is perceived, hyphal growth can be resumed. Depending on initial spore size and therefore the amount of available energy, such re-initiation can occur up to 10 times (Koske, 1981). Once in the vicinity of a plant root, the hypha starts branching, triggered by strigolactones present in root exudates. Generally, these molecules are excreted into the rhizosphere by the plant in extremely low quantities but increase upon phosphate starvation (Akiyama et al., 2005). Strigolactones are biologically active at subnanomolar concentrations, and their perception stimulates AM fungal mitochondrial NADH production, in turn increasing mitotic activity and hyphal growth (Besserer et al., 2006, 2008). Plant mutants, in *Arabidopsis thaliana* and pea (*Pisum sativum*), affected in strigolactone biosynthesis show reduced root colonization by AM fungi (Gomez-Roldan et al., 2008). In addition, they also display developmental effects such as increased shoot branching, shortened internodes and decreased outgrowth of lateral roots (McKay et al., 1994; Gomez-Roldan et al., 2008; Umehara et al. 2008). Since strigolactones are involved in regulating such a large number of processes, and are also produced by plant species (e.g., *Arabidopsis thaliana*) that have lost AM-symbiosis during their evolution, they are now considered to be a novel group of endogenous plant hormones. Genes involved in strigolactone synthesis are transcriptionally upregulated by auxin, a hormone important in plant development and plant architecture. Therefore, it is hypothesized that strigolactones affect auxin signaling (Hayward et al., 2009). The prominent role of strigolactones in higher plant architecture raises the heretofore unresolved question as to whether their function as an attractant for AM fungi is ancestral to a plant hormone function. Homologs of strigolactone biosynthesis genes have been identified in the bryophyte *Physcomitrella patens*, a moss without higher order architecture (Gomez Roldan et al., 2008). However, there is also no experimental evidence that bryophytes form functional, nutrient-exchanging mycorrhizal interfaces (Davey and Currah, 2006). This suggests a third, yet undiscovered function of strigolactones, one that is still present in bryophytes.

Once the fungus has reached the plant root, an appressorium is formed at the site of fungal contact with the root. Appressoria are flattened extensions of the fungal hyphae, which are positioned on the root epidermis. The appressorium, also termed the hypopodium, forms an initial interaction interface between plant and fungus. The nucleus of the underlying plant epidermal cell migrates toward the appressorium (Genre et al., 2005). Repositioning of the nucleus to the future entry point of the penetrating fungus precedes the development of the plant pre-penetration apparatus (PPA). Subsequently, the nucleus migrates to the base of the cell, creating a cytoplasmic tunnel through the epidermal vacuole. This structure consists of different cytoskeletal components (e.g., microtubules and actin microfilaments, together with endoplasmic reticulum cisternae). This cytoplasmic tunnel guides a fungal hypha that originates from the appressorium through the epidermal cell by invagination of the cell membrane and extends in the direction of the underlying cortical cell layer. The entering fungus remains surrounded by a plant-derived perifungal membrane, as well as a primary cell wall. The PPA continues to form also in the outer cortical cells, facilitating fungal passage to the inner cortex. There, within inner cortical cells, the interface for nutrient exchange between plant and fungus is created by extensive branching of the fungal hyphae into a bush-like structure, called an arbuscule after the Latin word for bush.

Two types of arbuscules can be distinguished: Paris- and Arum-type arbuscules, both named after the plant species in which the morphological type was first observed (*Paris quadrifolia* [Herb Paris] and *Arum Maculatum* [Cuckoo-Pint] [Gallaud, 1905]). Arum-type arbuscules are also known as terminal arbuscules. The penetrated hypha invades a single cell, and subsequently, by branching, an arbuscule is formed. The neighboring cell is reached by extension of the hyphae in the space between the outer- and inner-cortical layer. So, arbuscules form the terminal end of the hyphae. Alternatively, the arbuscules are not terminal, though new hyphae emerge from it entering neighboring cells. In this case, an ongoing network of hyphae and arbuscules is formed. This Paris-type of infection structure is also termed the intercalary arbuscule. Carrot (*Daucus carota* is a well-known plant species that hosts Paris-type arbuscules, whereas legumes (Fabaceae) host AM fungi that display Arum-type arbuscules. The type of arbuscule formed is shown to be dependent on both interacting partners. For a large number of plants, the type of arbuscular structure formed is described (reviewed by Smith and Smith, 1997) suggesting that the form is determined by the plant host. However, more recently it was discovered for tomato (*Solanum lycopersicum*), that inoculation with different strains of AM fungi can also lead to the formation of different types of arbuscules (Cavagnaro et al., 2001). This experiment also shows that the fungus plays a crucial role in determining the arbuscular form.

Nutrient Exchange Between Fungus and Plant

The arbuscule is the interface linking plant and fungus. This highly lobed fungal structure remains at all times surrounded by a plant-derived membrane, thereby creating a periarbuscular space between plant and fungus. Little is known of the regulatory mechanisms involved in the exchange of resources across the periarbuscular space. There is molecular and genetic evidence for the presence of plant-derived phosphate and ammonium transporters in the periarbuscular membrane (Harrison et al., 2002; Javot et al., 2007; Guether et al., 2009)). Nitrogen is transported from the extra cellular hypha to the intracellular hypha in the form of arginine. At the plant fungal interaction site, arginine is believed to be converted into ammonium. An ammonium transporter, recently identified in *L. japonicus* (Guether et al., 2009), LjAMT2;2 is specifically induced in arbuscule-filled cortical cells, and is the first AM symbiosis-specific N transporter identified. Due to its recent discovery, information related to its regulatory role is currently absent. In the case of phosphate, it is suggested that this nutrient is transported within the hyphae in the form of polyphosphate (polP) granules. In the arbuscule, polP granules are hydrolyzed into orthophosphate (Pi) and subsequently exported into the periarbuscular space (Cox et al., 1980; Ezawa et al., 2004). Plant Pi transporters are responsible for the uptake of the phosphate into the root cortex cell. Plant Pi transporters are generally divided into three groups based on phylogenetic clustering (although not all Pi transporters end up into these subclasses) (Bucher et al., 2001), two of which are known to be involved in the AM symbiosis. These include Pi transporters that are transcriptionally upregulated upon mycorrhization, though they also have a basal function in non-mycorrhized roots (subfamily III) and Pi transporters specifically expressed during mycorrhization (subfamily I). Members of both subfamilies have been characterized in a number of different legume and non-legume plant species (Rausch et al., 2001; Maeda et al., 2006; Nagy et al., 2005, 2006; Javot et al., 2007). The best studied, MtPT4 of *M. truncatula,* is specifically expressed during mycorrhization where the protein localizes in the plant-derived periarbuscular membrane. Reverse genetic experiments demonstrated that MtPT4 is essential for maintaining the fungus-plant interaction. AM fungi infecting *Mtpt4* mutant plants are not affected in appressorium formation, penetration of the cortex or arbuscule formation, though the arbuscules are not viable, possibly as a consequence of the lack of sufficient phosphate uptake by the plant. Ultimately, not only arbuscules, but also septate intracellular hyphae, degenerate and the symbiosis terminates (Javot et al., 2007). This finding underlines the importance of successful nutrient exchange between plant and fungus, and predominance of the plant in determining whether or not an interaction is sustained. Since AM fungi are obligatory symbionts, the viability of the fungus is seriously affected in a terminated interaction. The importance of successful nutrient exchange has been shown for phosphate transported from fungus to plant. Similar mechanisms likely apply for exchange of nitrogen and (in the opposite direction) the exchange of carbohydrates.

Genetic Network Controlling AM Formation

Molecular genetic analysis of AM symbiosis has been hampered due to the labor intensity of genetic screens on symbiotic phenotypes and the lack of suitable model systems. The latter is largely because *Arabidopsis thaliana*, the best-characterized plant model, does not establish AM symbiosis. Genetic dissection of the AM symbiosis in plants was opened by the development of legume models. To date, two legume species for which substantial genomic data and genetic tools have been developed are used to unravel the mycorrhizal symbiosis, namely *Medicago truncatula,* and *Lotus japonicus*. In addition to these two legumes, rice (*Oryza sativa*) and tomato (*Solanum lycopersicum*) have been employed for genetic studies.

Mutants in which the mycorrhizal interaction is disrupted have been identified in two ways. The first group was initially identified among legume mutants unable to establish rhizobial symbiosis. As mentioned above, endomycorrhizal and rhizobial symbioses share a substantial part of their signaling programs needed for infection. In addition, unbiased screens for AM-deficient plants are now being conducted; resulting mutants affected in different stages of the AM symbiosis will provide the genetic basis to unravel further the molecular network underlying AM-symbiosis. However, since such mutants have not yet been characterized at the molecular level, information presented here is therefore based exclusively on genes identified in legume mutants that are also affected in rhizobial symbiosis.

In legumes, mutations in seven genes that block AM-symbiosis have been identified, and they identify the myc⁻ mutants (Table 2.1). Homologs of these genes have been identified in various non-legume species (Zhu et al., 2006; Banba et al., 2008), and at least for some of them it has been demonstrated that they are essential for AM-symbiosis in the non-legumes (such as rice) (Banba et al., 2008). This suggests they form key components in the genetic network controlling the plant-AM interaction. The myc⁻ mutants can be placed, based on phenotype, into a small genetic network. It should be pointed out, however, that the reported phenotypes are not static; they are, at least in part, dependent on experimental setup and fungal species. For example, infection of three *M. truncatula* (*dmi1*, *dmi2*, *Mtsym16*) mutants by *Glomus mosseae* was more impaired compared to infection by *Glomus intraradices*. The latter frequently can enter the root epidermis of mutant plants nominally blocked in epidermal penetration, though with a decreased efficiency when compared to wild-type plants (Morandi et al., 2005).

Mutants with the most severe phenotype are characterized by the inability to form a PPA upon hyphal contact (Genre et al., 2005). As a consequence, hyphal penetration is blocked. Fungal appressoria formed on the root surface of these mutants appear to be more swollen, possibly as result of the inability to enter the root. At least two genes, for which mutant phenotypes have been characterized in two or more species, are essential for hyphal penetration, namely those that encode SYMBIOTIC RECEPTOR KINASE (*SYMRK*) (Ané et al., 2002; Endre et al., 2002; Stracke et al., 2002) and CALCIUM CALMODULIN KINASE (*CCAMK*)

Table 2.1. Plant mutants impaired in arbuscular mycorrhization. Indicated are mutagen (T-DNA: insertion line, EMS: ethylmethane sulfonate treatment, SCM: somaclonal mutation, FN: fast neutron bombardment, GAMMA: gamma irradiation, SM: spontaneous mutation, NEU: nitrosoethylurea treatment, NMU: nitrosomethylurea) and strains tested (G.i: *Glomus intraradices*, Gi.m: *Gigaspora margarita*, G.m: *Glomus mosseae,* G.v: *Glomus versiforme*, G. R-10: *Glomus R-10*, G.c: *Glomus clarum*). Table adapted from (Marsh and Schultze, 2001) and extended with recently published mutants.

Host	Mutant	Locus/Allele	Mutagen	Inoculum
Lotus japonicus **B-129 Gifu**				
Wegel et al., 1998	282–287	Lj sym2–1	T-DNA	G.i
Schauser et al., 1998	282–288	Lj sym2–2	T-DNA	G.i
Stracke et al., 2002	EMS61	Lj symRK–7	EMS	G.i
	cac41.5	Lj symRK–3	T-DNA	G.i
	EMS76	nup85-1/ Lj sym24	EMS	N/A
Kawaguchi et al., 2002	1-1E	nup85-2/ Lj sym73	EMS	G.i
	1-6F	nup85-3/ Lj sym85	EMS	G.i
Schauser et al., 1998	5371–22	nup133-1/ Lj sym3–1	T-DNA	G.i
Wegel et al., 1998	2557–1	nup133-2/ Lj sym3–2	T-DNA	G.i
N/A	EMS 247	nup133-3/ Lj sym3–3	EMS	G.i
Kanamori et al., 2006	cacc33.1	nup133-4/ Lj sym45	NA	NA
	10512.9	cyclops-1/ Lj sym6–1	T-DNA	G.i
	1962-124	cyclops-2/ Lj sym6-2	T-DNA	N/A
	EMS126	cyclops-3/ Lj sym30/ Lj sym6–3	EMS	G.i
	N-4	cyclops-4/ Lj sym82	EMS	N/A
	SL1347–2	cyclops-5	EMS	N/A
Schauser et al., 1998	Lj sym15–1	CCaMK-1/ Lj sym15-1	T-DNA	G.i
Demchenko et al., 2004	cac57.3	CCaMK-1/ Lj sym15-2	T-DNA	G.i
Senoo et al., 2000a	mcbep	CCaMK-3/ Ljsym72-1 CCaMK-4/ Ljsym72-2	EMS	G. R-10
Szczyglowski et al., 1998	EMS76	Lj sym24	EMS	G.i
Schauser et al., 1998	282–227	castor-1/ Lj sym4-1	T-DNA	G.i,Gi.m
Bonfante et al., 2000	EMS 1749	castor-2/ Lj sym4-2	EMS	G.i,Gi.m
Szczyglowski et al., 1998	EMS 46	castor-3/ Lj sym22-1	EMS	G.i
Senoo et al., 2000a	mcbex	castor-4/ Lj sym71-1	EMS	G. R-10
	mcbex	castor-5/ Lj sym71-2	EMS	G. R-10
Kawagushi, unpublished	N5	castor-6	EMS	N/A
	N10	castor-7	EMS	N/A
Umehara, unpublished	G00472	castor-8	SCM	N/A
	G00716	castor-9	SCM	N/A
	G00862	castor-10	SCM	N/A
Imaizumi-Anraku et al., 2005	SL3251-2	castor-12	EMS	N/A
	SL820-3	castor-13	EMS	N/A
	SL1715-2	castor-14	EMS	N/A
	SL1966-3	castor-15	EMS	N/A
	SL3160-3	castor-16	EMS	N/A
	SL6812-2	castor-17	EMS	N/A
Szczyglowski et al., 1998	EMS70	pollux-1/ Lj sym 23-1	EMS	G.i
	EMS167	pollux-2/ Lj sym 23-2	EMS	G.i
Kawagushi, unpublished	29-2A	pollux-3/ Lj sym 86-1	EMS	N/A
Imaizumi-Anraku et al., 2005	SL571-2	pollux-4	EMS	N/A
	SL3130-2	pollux-5	EMS	N/A
	SL5691-3	pollux-6	EMS	N/A
	SL1899-2	pollux-7	EMS	N/A
	SL159-3	pollux-8	EMS	N/A

Table 2.1. (cont'd)

Host	Mutant	Locus/Allele	Mutagen	Inoculum
	SL405-6	pollux-9	EMS	N/A
	SL1070-2	pollux-10	EMS	N/A
MG-20				
Umehara, unpublished	M89-27	castor-11		N/A
Lycopersicon esculentum cv 76R				
Barker et al., 1998	rmc	rmc	FN	G.i,Gi.m,G.m
Medicago sativa				
Bradbury et al., 1991	MN NN-1008	MN NN-1008	SM	G.i,G.v
	MN IN-3811	MN IN-3811	SM	G.i,G.v
Medicago trunculata cv Jemalong				
Catoira et al., 2000	C71	domi/*dmi1–1*	EMS	N/A
Penmetsa and Cook, 1997	B129	dmi1–2	EMS	N/A
	Y6	dmi1–3	EMS	G.i
Catoira et al., 2000	P1	dmi2–1	EMS	G.i
Sagan et al., 1995	TR25	dmi2–2	GAMMA	G.i,G.m
	TR26	dmi2–3	GAMMA	G.i,G.m
	TR89	*Mtsym2 (dmi2)*	GAMMA	G.i,G.m
	TRV9	Mtsym2 *(dmi2)*	GAMMA	G.i,G.m
Catoira et al., 2000	TRV25	dmi3–1	GAMMA	G.i,G.m
	TRV58	Mtsym16	GAMMA	G.i,G.m
Oryza sativa Dougjin				
Gutjahr et al., 2008	1B-08643	castor-1	T-DNA	G.i
Hwayoung				
Gutjahr et al., 2008	1C-03411	pollux-1	T-DNA	G.i
Niponbare				
Banba et al., 2008	NC6423	pollux-2	T-DNA	G.i
	ND5050	pollux-3	T-DNA	G.i
Gutjahr et al., 2008		ccamk-1		G.i
Banba et al., 2008	NE115	ccamk-2	T-DNA	G.i
Chen et al., 2007	NF8513	NF8514	T-DNA	G.i
	NG2508	NG2509	T-DNA	N.D.
Gutjahr et al., 2008	NG0782	cyclops-1	T-DNA	G.i
	NC2415	cyclops-2	T-DNA	G.i
Chen et al., 2008	NC0263	NC0263	T-DNA	G.i
	NC2713	cyclops-3	T-DNA	G.i
	NC2794	NC2794	T-DNA	G.i
Phaseolus vulgaris cv OAC Rico				
Shirtliffe and Vessey, 1996	R69	R69	EMS	G.m,G.c
Pisum sativum cv. Finale				
Gianinazzi-Pearson, 1996	RisNod24	s	EMS	G.m
cv. Frisson				
Duc and Messager, 1989	P1	a	EMS	G.i,G.m
	P2	a	EMS	G.i,G.m
	P3	a	EMS	G.i,G.m
Sagan et al., 1994	F4-1 (P53)	a/*sym30*	EMS	G.i,G.m

(continued)

Table 2.1. (cont'd)

Host	Mutant	Locus/Allele	Mutagen	Inoculum
Schneider et al., 1999	F4-141 (P55)	c/*sym19*	EMS	G.i,G.m
Duc and Messager, 1989	P4	c	EMS	G.i,G.m
Duc and Messager, 1989	P6	f	EMS	G.i,G.m
cv. Rondo				
Weeden et al., 1990	K24	c/*sym19*	EMS	N/A
cv. Sparkle				
Gianinazzi-Pearson, 1996	N/A	a/*sym30*	N/A	N/A
Balaji et al., 1994	R72	b/*sym9*	GAMMA	Gi.m
Weeden et al., 1990	NMU1	c/*sym19*	NMU	N/A
	NEU5	c/*sym19*	NEU	N/A
Albrecht et al., 1998	R19	p/*sym8*	GAMMA	Gi.m
Gianinazzi-Pearson, 1996	E140	p/*sym8*	EMS	Gi.m
Balaji et al., 1994	R25	p/*sym8*	GAMMA	Gi.m
Vicia faba				
Duc and Messager, 1989	Indian 778	sym1	SM	G.i,G.m

(Ané et al., 2002; Lévy et al., 2004; Mitra et al., 2004) (Table 2.1). *SYMRK* is a leucine-rich repeat type receptor kinase that is located at the plasma membrane, suggesting it could interact directly with fungal-derived components. However, there is no experimental evidence that this is indeed the case. The inter-species variation of the extracellular amino acid sequence of SYMRK is striking. For example, a large portion of the extracellular NEC (N-terminal extracellular region of predicted SYMRK proteins) domain is fully absent in SYMRK of monocots (Figure 2.1). Nevertheless, monocot receptors function in symbiotic signaling, when expressed in legumes (Markman et al., 2008). Unfortunately, the role of SYMRK at the molecular level is currently largely unresolved. Future studies, including identification of the ligand perceived by this receptor, should elucidate SYMRK function. The second gene, which, when mutated, is blocked in hyphal penetration, *CCaMK,* encodes a nuclear localized calcium-regulated kinase. Its nuclear localization suggested that it could interact with transcriptional regulators, and studies in heterologous systems indicate that this indeed could be the case (Smit et al., 2005; Gleason et al., 2006)).

Both *SYMRK* and *CCaMK*, are ubiquitously expressed in all cell layers of the infection zone, suggesting they could be active within the same signaling network. However the strict separation in subcellular localization of both proteins suggests that other components are required to create a functioning signaling network, since the signal must be transferred from the plasma membrane, where SYMRK is located, to within the nucleus, where CCaMK is located. CASTOR and POLLUX (Imaizumi-Anraku et al., 2005) are cation channels localized on the nuclear envelope and among the intermediate signal mediators (Table 2.1). This is supported mainly by studies in *M. truncatula* that has only one *POLLUX* ortholog, *MtDMI1* (Ané et al., 2004). Similar to *symrk* and *ccamk* mutants, in the *M. truncatula mtdmi1* mutant, the fungal hypha does not penetrate the epidermal layer. We speculate that PPA development upon fungal appressorium formation is hindered. Interestingly, in root epidermal cells, close to an AM hypha, nuclear oscillations of the calcium concentration are induced (Kosuta et al., 2008). In *M. truncatula,* these oscillations are dependent on SYMRK (MtDMI2) and POLLUX (MtDMI1), but independent of CCaMK (MtDMI3). Therefore, it is hypothesized that CCaMK activity is sensitive to (or triggered by) these calcium oscillations. CYCLOPS is a putative

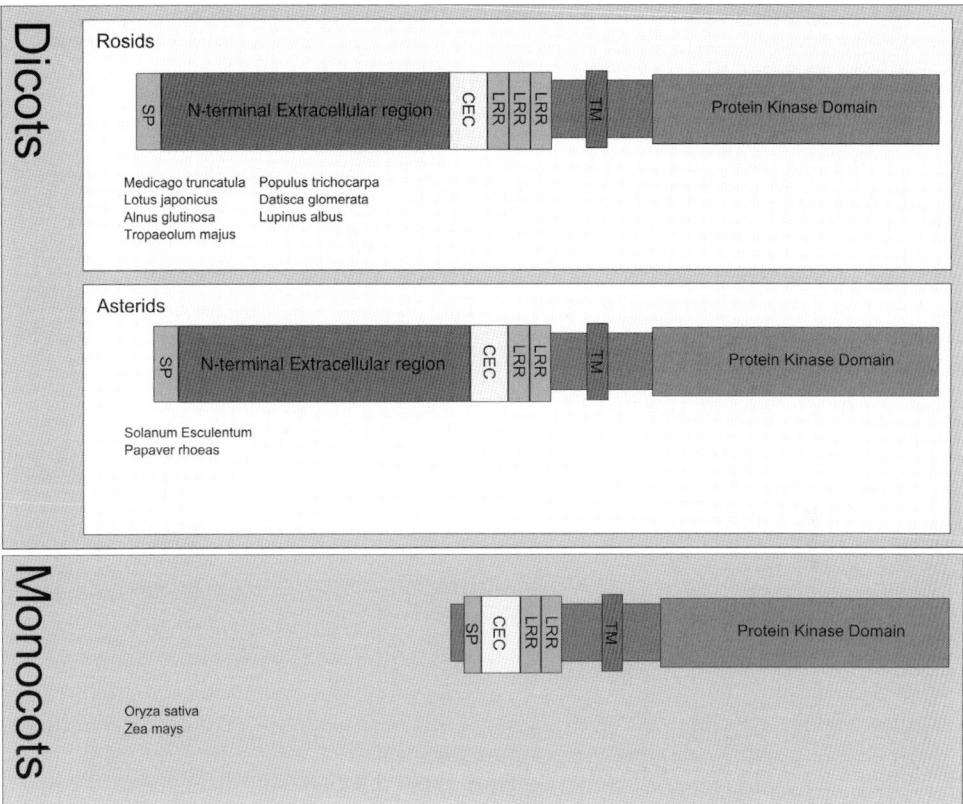

Figure 2.1. Diversification of SYMRK
Protein domain structure of the symbiotic receptor kinase SYMRK encoding a Leucin Rich Repeat (LRR) domain containing receptor kinases. Given is the protein structure for three phylogenetic clades: rosids, asterids (both part of the group of dicots), and monocots. Legumes belong to the subclass of rosids. CEC: conserved extracellular region preceding LRRs; TM: transmembrane domain; SP: signal peptide. (Figure adapted from Markmann et al., 2008).

phosphorylation target of CCaMK. Mutations in CYCLOPS affect the AM symbiosis, though less severely when compared to *ccamk* knockout mutants, indicating that it could be only one of several targets (Yano et al., 2008).

In contrast to *M. truncatula*, *L. japonicus castor* and *pollux* mutants still allow AM hyphae protrusion in the epidermal cell layer, hyphal protrusion that is subsequently blocked in entering the underlying root cortex. This difference in functioning of POLLUX in *L. japonicus* and *M. truncatula* can be the result of change of function following the loss of CASTOR in *M. truncatula*. A mutation in the single *M. truncatula* gene might result in a more severe phenotype compared to *L. japonicus* because of possible redundancy or overlap between *L. japonicus* CASTOR and POLLUX. A similar situation could apply for nuclear porin genes *nup85* (Saito et al., 2007) and *nup133* (Kanamori et al., 2006), also in which mutation results in a block in cortical infection (Marsh and Schultze, 2001). Nuclear pores are formed by a complex of 30 proteins and control nuclear-cytoplasmic trafficking of proteins, RNA, and

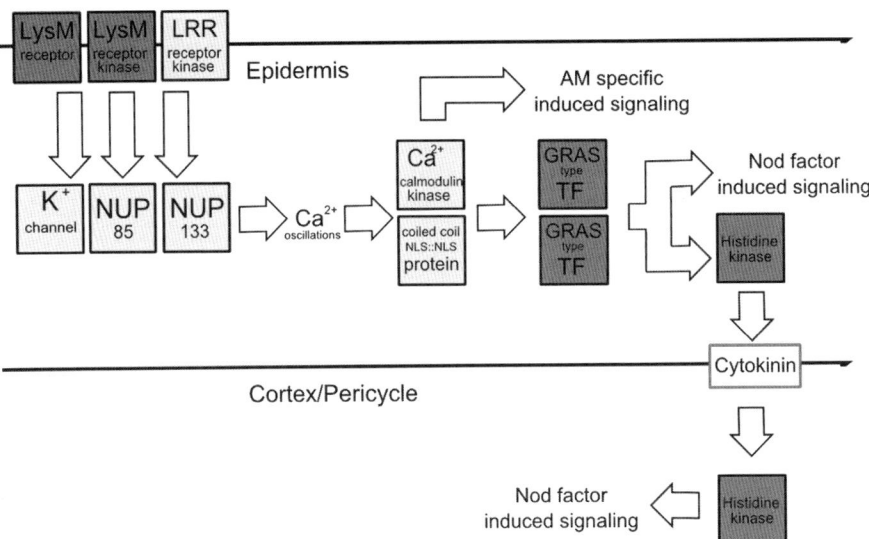

Figure 2.2. Nod factor signaling cascade. Schematic representation of Nod factor signaling cascade based on genetics and mutant phenotype.

ribonucleoprotein particles. Since nuclear pores likely have essential and numerous functions, it is unlikely that a knockout mutation of either *nup* will lead to complete loss of function. More likely, the specificity or efficiency of nuclear pore function is affected. The fact that both mutants are affected in *Rhizobium*-induced calcium signaling, as in *symrk, castor*, and *pollux* mutants, suggests their function in the genetic network upstream of CCaMK (Figure 2.2). For example, they might be responsible for the transfer of a rhizobial-signaling molecule to the nucleus.

That *SYMRK, CASTOR, POLLUX, NUP85/133, CCAMK,* and *CYCLOPS* play a prominent role in two rather distinct symbiotic interactions suggests roles in a more general pathway. The phenotype of the *ccamk* P2 mutant allele in pea led to speculation that some of these genes play a role in regulation of plant defense (Gollotte *et al*. 1993; Blilou *et al*. 1999; Ruiz-Lazano *et al*. 1999). Defense responses are triggered upon infection by a symbiotic partner, though subsequently are downregulated. However, mutants carrying the P2 allele displayed an increase in the steady state level of a number of defense related genes. Additionally, the level of salicylic acid (SA), a defense-related signaling compound, is elevated. Upon AM-fungal interaction, an increased deposition of callose and phenolic compounds occurred at points of appressorial contact, similar to the plant's response to a pathogen.

No other loci have been characterized at the molecular level in neither *L. japonicus* nor *M. truncatula*. While characterization of mutants controlling later steps of arbuscule formation would be especially valuable, all AM association mutants thus far recovered and characterized have been shown to represent new (weak) alleles of the genes described above (Senoo et al., 2000b). Novel pathway components may be identified among the remaining "uncloned" mutants, especially those that display distinct phenotypes in arbuscular development. Among current mutant candidates, some are in species other than *L. japonicus* and

M. truncatula, species that are less amenable for cloning of the affected genes. These mutants include MN-IN-3811 in alfalfa (*Medicago sativa*) (Bradbury et al., 1991), R69 in common bean *(Phaseolus vulgaris)* (Shirtliffe and Vessey, 1996), and RisNod24 in pea (*Pisum sativum*)(Gianinazzi-Pearson, 1996). In the alfalfa and common bean mutants, the fungal hyphae enter the innercortical cells, but subsequent arbuscule development is blocked. In the pea RisNod24 mutant, arbuscule formation can be initiated, but proper development is hampered leading to truncated arbuscules. A differential screen comparing RisNOD24 with wild-type pea did reveal increased transcript levels of the gene *Psam4*, a negative marker for AM symbiosis (Lapopin et al., 1999). *Pssam4* encodes a proline-rich repeat protein. Induction of this type of protein is often related to plant defense initiation. Finally, in the *mcbo L. japonicus* mutant, arbuscules are formed and have a normal appearance, but senescence seems to occur prematurely. This suggests that arbuscules in the *mcbo* mutant do not function properly and fail to provide nutrients to the host, triggering sanctions and premature senescence (Senoo et al., 2000a).

Given that there is a relatively small set of seven genes shown to be required for the early regulatory steps of AM symbiosis, namely, *SYMRK, CASTOR, POLLUX, NUP85, NUP133, CcaMK,* and CYCLOPS, it is currently hypothesized that these genes function in a linear signaling pathway as depicted in Figure 2.2, although there is little molecular evidence that this is indeed the case. These genes are also required for rhizobial symbiosis. Since the number of genetic screens dedicated to the AM-symbiosis is increasing, identification of mutants unique to this symbiosis seems but a matter of time. Cloning the identified genes will provide insights into AM-induced signaling, the molecular processes leading to arbuscule formation and nutrient exchange between both partners.

Occurrence of *Rhizobium* Symbiosis in *Fabaceae*

Symbiosis with nitrogen-fixing rhizobial bacteria is a prominent feature of the legume family. Fabaceae represent an ubiquitous and diversified plant family with a worldwide distribution ranging from tropical to temperate climates and includes annual herbs up to forest trees (Lewis et al., 2005). It consists of three subfamilies of which the largest, the Papilionoideae, contains many crop species. The vast majority of Papilionoideae and Mimosoideae legumes (Sprent, 2007) are able to use atmospheric molecular nitrogen as a substitute to soil nitrogen thanks to the symbiosis with nitrogen-fixing bacteria. In contrast, the ability to establish such symbiosis is more scattered in the other subfamily Caesalpinioideae, respectively (Figure 2.3). The rhizobia bacteria form polyphyletic group sharing the sole property of being able to form a nitrogen-fixing symbiosis with legumes. Rhizobia are found as independent clades falling in (to date) orders alphaproteobacteria and betaproteobacteria (Moulin et al. 2001; Chen et al., 2003). Rhizobial genomics unambiguously show that the capacity of nitrogen fixation, traced through the phylogeny of genes encoding the nitrogenase enzyme, and the capacity of infecting legumes hosts, mediated by genes determining signal molecules, evolved by lateral gene transfer (Wernegreen and Riley, 1999).

Rhizobia can live as free-living bacteria in soil. Although little is known about the functioning of microbial soil ecology, rhizobia are thought to be a member of the community of the rhizosphere that is the nutrient-rich interface between plant roots and the soil, where active nutrient exchange can take place. Due to horizontal gene transfers rhizobia are suspected to be able to switch between saprophytic and symbiotic lifestyles (Batista et al., 2007).

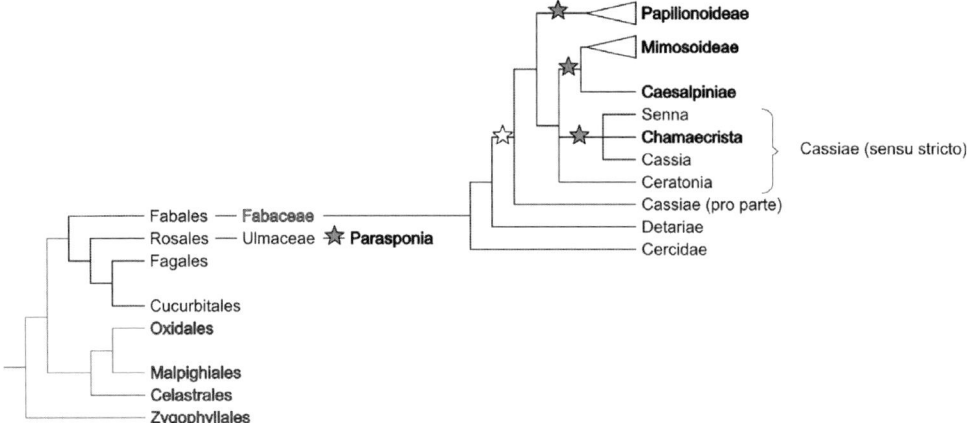

Figure 2.3. Phylogenetic tree of eurosid I clade. Consensus tree of eurosids I based on Soltis et al., 2000; Lewis et al., 2005; Tree of Life web project. Nitrogen fixation clade (phylogenetic clade containing species that can be nodulated either by *Rhizobium* or *Frankia* bacteria) indicated in black. Clades containing species that can sustain *Rhizobium* symbiosis are indicated in dark-gray. Legume family indicated in light-gray. Origin of *Rhizobium* symbiosis indicated by star; in case of a single event, white star; in case of multiple events, gray stars.

Bacterial Uptake and Infection

Symbiotic rhizobia can infect the root by two mechanisms, crack entry and root hair-based infection (Boivin-Masson et al., 2009). In case of crack entry, bacteria enter the root through cracks that arise during lateral root formation. Once inside, the bacteria can infect predisposed cortical cells. These cells contain arrays of microtubules to support bacterial penetration into the cortex layer. This will occur via so-called infection threads, which are tubular structures of bacterial cell files surrounded by a plant-derived membrane. Compared to crack entry, root hair-based infection is more sophisticated. A single bacterium redirects root hair growth in such a way that it grows around the bacterium thereby isolating it from the surrounding environment. In the created microenvironment, an infection thread is formed that subsequently grows through the epidermis. Subsequent infection in predisposed cortical cells will guide the bacteria to the newly formed nodule primordium.

In concurrence with rhizobial infection, nodule development is triggered. This requires epigenetic reprogramming of existing cortical cells, so they can enter the cell cycle and form a nodule primordium. At the moment the infection thread reaches this primordium, bacteria are released into the cells via a budding-like process. Subsequently the primordium differentiates into a mature nodule (for detailed description of nodule development see Chapter 16).

Nitrogen Fixation Within the Nodule

Rhizobia inside nodule cells are named bacteroids, whereas the units surrounded by a plant membrane are named symbiosomes. In a single symbiosome, one or more bacteroids can be present, depending on the plant species. Symbiosomes display an organelle-like nature. They divide and fill the expanding host cell. Subsequently the bacteria will start nitrogen fixation.

Rhizobia only fix nitrogen when hosted in nodules, this is in contrast to several other diazotrophic bacteria. The genes encoding the component of the nitrogenase complex, the actual protein complex converting N_2 into NH_3, are exclusively expressed in differentiated bacteroids. Legume nodules represent a favorable environment for nitrogen fixation, by meeting conditions that are not found during the "free living" lifestyle of rhizobia. A particular characteristic of nitrogenase is its sensibility to oxygen. On the other hand, rhizobia do require oxygen for respiration, especially because of the fact that nitrogen fixation requires large amounts of energy. The environment in the central zone of the nodule is isolated by a suberized endodermal layer that forms a barrier for atmospheric oxygen. The availability of oxygen underneath this barrier is regulated by the coordinated expression of leghemoglobin.

Although the ecological advantage of obtaining a virtually unlimited resource of nitrogen is undoubtable, it doesn't come without a price tag. The plant has to sustain a large number of bacterial cells (on the order of hundreds per infected cell), of which it has to prevent cheating by verifying productivity. To do so, sophisticated mechanisms for the exchange of nutrients have evolved resulting in mutual dependence. To allow efficient transfer of ammonium to the plant, bacteroids have downregulated their own ammonium assimilation. As a consequence, the plant not only has to provide sugar in the form of dicarboxylic acids, but also amino acids (Lodwig et al., 2003). This puts the plant in position to control the viability of the symbiotic relation, a mechanism that could have similarities with AM fungal control, which avoids the presence of noncooperative cheaters.

Maintenance of bacteroids is not the only sink of carbohydrates; the nitrogen fixation process itself requires 16 moles of ATP per mole of NH_3 fixed. Large polyhydroxybutyrate granules are visible in bacteroids that function as energy silos (storage). Generally such granules are found only in bacteroids that are still capable of redifferentiating in nonsymbiotic forms. Therefore, it has been speculated that these energy stocks are depleted by the end of nodule life cycle (Lodwig et al., 2005). Upon nodule senescence, a substantial fraction of bacteria that was captured in the nodule are released back into the environment, providing opportunities to populate surrounding soils.

Molecular Bases of Nitrogen Fixation in the Legume-Rhizobial Symbiosis

From the rhizobial side, the molecular determinants controlling symbiosis are better understood thanks to straightforward genetics studies. The usually modular constitution of bacterial genomes is even exacerbated in case of rhizobia. By combining the analyses of many available mutants, and more recently complete genome sequences of several rhizobia (Kaneko et al., 2000, 2002; Galibert et al., 2001; González, 2006; Young et al., 2006; Amadou et al., 2008), our understanding of the genetics and genomics of rhizobia has reached a comprehensive level.

In general the genome of a *Rhizobium* can be separated in three components: nonsymbiotic genes, common symbiotic genes, and species-specific symbiotic genes. Nonsymbiotic genes range from genes involved in cellular machinery (so-called housekeeping genes) to genes controlling ecological traits related to life outside legume nodules (e.g., for a saprophytic lifestyle). Such genes vary between different lineages of rhizobia and the divergence of these genes between different classes of rhizobia is much greater than the divergence of symbiotic genes shared between the species.

Common symbiotic genes are defined as genes that are shared by all rhizobia, and by definition encompass the core set of genes essential for symbiosis. This set of genes can be split into two subgroups: the nodulation (*nod, nol, noe*) genes and the fixation *(fix* and *nif)* genes, respectively (Perret et al., 2000). The group of nodulation genes is responsible for inducing nodule formation and infection, whereas the fixation genes are essential for fixation of nitrogen. This includes the constitution and regulation of the nitrogenase enzyme complex and cellular transport required for the actual fixation process. Additionally, it encompasses the genes operating the cellular transformation into functional symbiosomes.

Nodulation genes control essential steps for the establishment of symbiosis, such as the perception of plant-excreted flavonoids plus the synthesis and secretion of the bacterial signaling molecule, the so-called nodulation factor (Nod factor). The latter forms literally the key to enable entry into the legume host plant (see also Chapter 16).

Flavonoids are recognized by NodD proteins. *Rhizobium* NodD belongs to the class of LysR-type transcriptional regulators, a family of regulators that becomes activated upon the binding of an external signal. In case of NodD, this signal is generally a plant-secreted flavonoid or iso-flavonoid that triggers a conformational change leading to an increased binding affinity to promoters of other nodulation genes (Chen et al., 2005). Among these is the *nodABC* operon of which the encoded enzymes synthesize the core structure of Nod factors; a chitin oligomer that is substituted with a common fatty acid (generally stearic acid [18:0] or vaccenic acid [18:1]). Further structural decorations of the basic Nod factor backbone are possible, and often species-specific nodulation genes are involved in this process. For example, the terminal glucosamine residues can be modified by O-acetylation, O-carbamoylilation, or O-sulfonation. Genetic analysis of different *Rhizobium* strains demonstrated that such modifications play a crucial role in host-symbiont specificity. Nod factors are responsible for initiating most of the host responses and cause a major bottleneck during rhizobial infection, one of the early steps in the establishment of the symbiosis.

Species-specific symbiosis genes are particularly known for their role in determining host range (Van Rhijn and Vanderleyden, 1995; Perret et al., 2000). A large fraction of these genes encode enzymes that determine various structural alterations of the Nod factor. The ability of host plants to discriminate different Nod factors, and furthermore the role of Nod factors as a key to successful infection, has been demonstrated experimentally many times (Relì et al., 1994; Ardourel et al., 1995; Radutoiu et al., 2007). Nevertheless, genes encoding Nod factor-modifying enzymes are far from being the only genes controlling host range phenotype. Other, more heterogeneous collections of genes can be categorized as species-specific symbiosis genes including genes controlling the synthesis of exopolysaccharides, type three secretion systems, and other proteins commonly involved in pathogenicity (Fraysse et al., 2003; Becker et al., 2005; Deakin and Brouhgton, 2009; Kambara et al., 2009).

In spite of the conceptual division between common and specific symbiotic genes, species-specific variation can also be found within the genes belonging to the common symbiosis group. Also these differences contribute to a delineation of the host range. In particular *nodD* shows large diversity between species in copy number as well as flavonoid specificity. Binding experiments with NodD1 of *Sinorhizobium meliloti* showed that several different flavonoids bind to this protein thereby increasing promoter-binding affinity. However not all isoflavonoids that increase DNA binding ultimately trigger transcriptional activation, suggesting a competition between inducing and noninducing flavonoids that are secreted by potential host plants, and thereby narrowing down a potential host range for *S. meliloti* (Peck et al., 2006). In contrast, similar experiments with the broad host-range *Rhizobium* sp. NGR234 identified a wide range of activating isoflavonoids suggesting the absence of such competition for this species (Kobayashi et al., 2004). Another example of a common symbiotic gene that showed to determine the host range is NodC. *nodC* encodes a N-acetylglucosaminyltransferase that determines the length of the chitin backbone of the Nod factor. The precise length of chitin backbone appeared to be a determinant for host specificity on alfalfa (*Medicago sativa*)

(Roche et al., 1996). With the progress in genomics, the conceptual definitions of "common" and "specific" symbiotic genes start to be made obsolete. To date the most striking finding is the absence of the *nodABC* operon in two photosynthetic Bradyrhizobium strains, ORS278 and BTAi1. Consequently these species are unable to produce Nod factors. Nevertheless these strains are still symbiotic on a narrow host range of plants belonging to Aeschynomenes (Giraud et al., 2007). This finding illustrates the presence of a diverse pallet of genetic variation, which is the result of over 60 million years of co-evolution.

Dissecting Rhizobium *Symbiotic Signaling Networks in Plants*

The capacity to sustain a symbiotic interaction with rhizobia is one that is, with one exception, exclusively restricted to legumes. It requires the formation of a novel organ, the root nodule. Hosting prokaryotes inside plants requires tight regulation. Most likely the signaling cues for such regulatory processes have specifically evolved in legumes. To obtain insights in the molecular networks controlling this symbiotic relation, genetic studies have been conducted successfully in two legume model species, *M. truncatula* and *L. japonicus*.

Plant traits constituting the rhizobial symbiosis can be divided roughly in five categories: the infection process, nodule organogenesis, nitrogen fixation, regulation of nodule number, and nodule senescence. The initial focus was mainly on the molecular characterization of the genetic network essential for Nod factor perception and subsequent early signaling controlling infection and nodule formation. This research largely uncovered the key components in the Nod factor-signaling network. Since this signaling cascade includes a large number of genes already discussed for mycorrhizal signaling, we only describe this pathway in short: Nod factors are perceived by two structurally different LysM-domain containing receptors (for *M. truncatula*: *NFP* and *LYK3* [Limpens et al., 2003; Arrighi et al., 2006] for *L. japonicus NFR1* and *NFR5* [Radutoiu et al., 2003]) (Table 2.2), resulting in parallel activation of two symbiotic processes—one leading in mitotic activation of cortical cells and the other in bacterial uptake and progression of infection threads through the different root cell layers. Based on morphology, physiology, and marker genes, both processes can be easily discriminated. On a genetic level, however, significant crosstalk between both processes hampers untwining of the pathways involved. Both Nod factor receptors as well as a range of downstream components, including the components in common with the mycorrhizal symbiosis (Table 2.2), showed to be involved in both processes. Weak alleles and RNAi knockdown experiments of a number of these genes underline their dual role. Downstream of these "common" symbiosis genes, a number of genes specific to the legume-*Rhizobium* interaction have been identified including two GRAS-type transcription factors (NSP1, NSP2) (Kaló et al., 2005; Smit et al., 2005) and ethylene response factor (ERN1) (Middleton et al., 2007) (Table 2.2). Based on phenotype and subcellular localization, a genetic network underlying Nod factor signaling has been constructed (Figure 2.2). For more details, also see Chapter 16.

Entry of rhizobia and the role of Nod factor signaling in this process have been extensively studied. Root hair curling and subsequent infection thread formation are based on a cell autonomous response to Nod factors. The growth of young root hairs can be redirected by local application of Nod factors (Esseling and Emons, 2004). The function of the subsequent bacterial containment in the closed cavity of a root hair curl is poorly understood, but the hypothesis has been raised that this isolation could be required to resist the turgor pressure of the root hair cell during local cell wall degradation needed for infection thread formation (Geurts et al., 2005).

Table 2.2. Proteins involved in nod factor signaling. Table containing different proteins known to be essential for early Nod factor signaling. Putative pathway based on genetics and phenotype is displayed in Figure 2.2.

Protein function	M. truncatula		L. Japonicus	
	Abbrev	Full length name	Abbrev	Full length name
LysM domain containing receptor	NFP	Nod Factor Perception	NFR5	Nod Factor Receptor 5
LysM domain containing receptor kinase	LYK3	LysM receptor kinase	NFR1	Nod Factor Receptor 1
Potassium ion channel	DMI1	Doesn't make infections 1	CAS	Castor
			POL	Pollux
LRR domain containing receptor kinase	DMI2	Doesn't make infections 2	SYMRK	Symbiosis receptor kinase
Calcium calmodulin kinase	DMI3	Doesn't make infections 3	CCaMK	Calcium Calmodulin Kinase
Coiled Coil—NLS: NLS protein	IPD3	Interaction protein of DMI3	N/A	Cyclops
Nucleoporin	N/A	N/A	NUP85	Nucleoporin 85
			NUP133	Nucleoporin 133
GRAS family type transcription factor	NSP1	Nod factor signaling pathway 1	NSP1	Nod factor signaling pathway 1
	NSP2	Nod factor signaling pathway 2	NSP2	Nod factor signaling pathway 2
Ethylene response factor	ERN	Ethylene response factor functioning in nodulation	N/A	N/A

Furthermore, it allows accumulation of Nod factors to relatively high concentrations. Of all Nod factor controlled responses, successful infection through the epidermis is most stringently controlled in respect to Nod factor structure as well in respect to functional alleles of most Nod factor signaling genes.

Besides cell autonomous signaling required for root hair-based infection, perception of Nod factors also leads to relatively fast systemic signaling. Within 3 hours after perceiving Nod factors at the epidermis, the pericycle cells display transcriptional changes (Compaan et al., 2001). This marks the onset of nodule organogenesis. Activation of inner cells is most likely not a result of intracellular diffusion of Nod factors, since typical marker genes for cell autonomous Nod factor signaling, like *ENOD11*, are not activated in these underlying layers. To date little is known about the molecular nature of the signal that is relayed through several cell layers including the endodermis to activate cells in the pericycle. In *L. japonicus*, a cytokinin perception mutant, *snf2*, has been identified in which part of the signaling network involved in early Nod factor signaling is bypassed. A single C to T transition changing leucine266 into a phenylanaline results in a hypersensitive or "always active" version of the cytokinin receptor. Introduction of this mutated receptor into wild-type plants leads to the formation of nodule-like structures in absence of rhizobia (Tirichine et al., 2008). Combining this observation with the fact that Nod factor-induced nodule organogenesis can be mimicked by external application of cytokinin suggests that cytokinin could fulfill this role of intracellular messenger.

Evolution of Symbiosis in Rhizobia

As already mentioned, it rapidly became clear that the symbiotic capacity of rhizobia spreads by means of horizontal gene transfer between relatively unrelated bacterial lineages. Unambiguously the nodulation genes were transferred between species. At the moment, the most parsimonious hypothesis implies two independent events of such gene flow from Alphaproteobacteria to Betaproteobacteria. It seems however probable that nitrogen fixation is ancestral to proteobacteria, although it is not known whether this ancestral fixation was symbiotic or not (Raymond et al., 2004). The phylogeny of genes encoding the nitrogenase enzyme complex shows that the nitrogenase genes of Alphaproteobacteria and Betaproteobacteria are not related. Therefore these genes were probably not concerned by the initial event of lateral transfer (Chen et al., 2003). Consequently it is hypothesized that the set of nodulation genes allowing rhizobia to infect legume roots has been transmitted from lineage to lineage, and gained success in symbiotic sense when it met lineages already able to perform nitrogen fixation. Such mechanism is analogous to the dynamics of pathogenicity of bacteria, where transfer of plasmids mediating virulence (and ability to infect and attack a given host) are commonplace (Ziebuhr et al., 1999). The transfer of nodulation genes might be the signature of an ancient pathogenic interaction that eventually was given way to nitrogen-fixing symbiosis through the fortuitous encounter of genes allowing infection with genes controlling nitrogen fixation.

An ancestral pathogenic lifestyle driven by nodulation genes finds support in two ways: (1) The NodD driven chemotaxis involves specific host recognition that is similar as found in pathogenic interactions. (2) Though unique in structure, the backbone of rhizobial Nod factors is structurally similar to chitin tetramers or pentamers. One of the LysM-domain containing receptor kinases that perceives Nod factors in legumes, namely LjNFR1/MtLYK3, is homologous to the *CERK1* gene of Arabidopsis that mediates perception of chitin oligomers and is responsible for the induction of defense responses (Miya et al., 2007). Chitin oligomers are generated by the action of chitinases on the chitin of fungal cell wall, and therefore represent a pathogen-associated molecular pattern. At the molecular level, the transition from (undecorated) chitin perception to Nod factors seems trivial and for sure does not represent a large evolutionary challenge. However, the initial selective force that has driven rhizobia to make use of the CERK1-mediated signaling pathway as an entry route to legumes' root is all but obvious, though, endomycorrhizal fungi might represent the missing link.

Evolution of Symbiosis in Host Plants

In contrast to rhizobia, their host plants form a by far more consistent group. They are all included in the Fabaceae, with the single exception of the genus Parasponia, which is part of the Celtidaceae family. This contrast may just explained by the inability of plants to transfer genetic material between species horizontally. In the following paragraphs, we will suggest that biological nitrogen fixation has found different means to spread across plant lineages, but first we will focus on the evolution of nitrogen fixation in legumes.

Legume root nodules are novel organs that, from a morphological point of view, are only distantly related to nodules formed on actinorhizal plants to host nitrogen-fixing *Frankia* bacteria, and even more distinct to lateral roots. Nodules of all symbiotic legumes show a common ontology suggesting a single occurrence in evolution. The evolution of root nodules shortly after the emergence of the Fabaceae seems most parsimonious, since the most basal clade, containing the tribes Cercidaceae and Detarieae, does not contain any evidenced instance of nodulation (Figure 2.3) (Sprent, 2009). Based on phylogenetic studies, it is hypothesized that the most recent common ancestor to nodulation legumes lived between 59 and 58.4 million years ago (Lavin et al., 2005; Sprent, 2007). By that time legumes have split into two major branches: the subfamily Papilionoideae, of which all but a few ancestral species are nodulating, and a clade containing the Mimosoideae subfamily and some Caesalpinioideae lineages. Around 90% of Mimosoideae species can be nodulated. In contrast, nodulating Caesalpinioideae are relatively scarce, which would suggest that upon emergence of nodulation several losses would have occurred (Figure 2.3). Alternatively, it can be hypothesized that nodulation has evolved at least three times independently in the legume family (Doyle, 1998). In this latter hypothesis, the first occurrence of nodulation would have taken place at the root of the Papilionoideae, the second in a common ancestor of the Mimosoideae and part of the Caesalpinioideae (tribe Caesalpiniae and others), and the third time in the single genus of Chamaecrista, which seems to be a relatively isolated nodulating lineage beyond non-nodulated species (Figure 2.3). It should be noted that there is no good reason to assume that Chamaecrista is an old genus. As a result its specific event of evolution of nodulation should be rather recent (the order of magnitude of genus ages is 1–10 million years). Nevertheless, nodulation is a profitable but relatively expensive trait. It is also a complex trait that is most probably difficult to evolve. These two considerations favor a theory of multiple losses rather than multiple gains. The progress in understanding the molecular bases of the ability of forming nodules will probably allow to undercover the very mutations that allowed the emergence of nodulation, and thereby contribute in determining how frequently nodulation has appeared.

The genus *Parasponia*, consisting of five species, is the unique example of non-legume hosting nitrogen-fixing rhizobia. *Parasponia* is relatively distantly related to legumes. Although the orders containing Fabaceae and Celtidaceae, respectively Fabales and Rosales, are sister clades, it is still very unlikely that the ability to interact with rhizobia has been inherited from a common ancestor. An additional argument for convergent evolution is the difference in nodule ontology. *Parasponia* nodules display an ontology similar to nodules formed by actinorhizal plants to host nitrogen-fixing *Frankia* (ascomycete) bacteria. Such nodules are, although in some instances similar in shape to legume nodules, anatomically unrelated. *Parasponia* and actinorhizal nodules are basically modified lateral roots that are infected by nitrogen-fixing symbionts. In contrast with the legume-rhizobial symbiosis, the microsymbiont in the actinorhizal-*Frankia* symbiosis is a consistent (monophyletic) group whereas host plants are spread across different lineages pointing to parallel emergence of a similar symbiosis in different lineages. In evolutionary perspective, it is more parsimonious to view the *Parasponia* symbiosis as another parallel emergence of actinorhizal-type nodules, but with the difference that *Frankia* is substituted by rhizobia.

One important point to consider is that the legume-*Rhizobium* symbiosis is probably older than 50 million years while most of the instances of actinorhizal symbiosis, as well as the *Parasponia-Rhizobium* symbiosis, are likely significantly more recent and rarely spanning more than a single genus. If the hypothesis of multiple evolution of nitrogen fixation is true, it has to be assumed that legume nodules predated most, if not all, other forms of nitrogen-fixing symbiosis by at least several tens of million years.

A Confusing Pattern of Specificity in the Rhizobial Symbiosis

In symbiosis, a classical signature of host-symbiont specificity is cocladogenesis induced by coevolution and co-speciation of matching partners. Such a pattern is also frequently observed in host-pathogen interactions. In that case the supposed mechanism is that the host elaborates parasite-specific defenses and that the parasite in turn evolves specific mechanisms to evade these defenses. This process tends to promote specialization toward one host (for the pathogen). This is particularly clear for gene-for-gene systems determined by plant R genes and their corresponding antigen molecules in their pathogens. In case of *Rhizobium* symbiosis, cocladogenesis seems most obvious in the tribes Fabeae, Trifolieae, and Cicereae, and part of the Galegeae (together named galegoids) (Wernegreen and Riley, 1999). The microsymbionts of plants of this clade generally have a narrow host range. However, a pattern of cocladogenesis is far from clear for most other legume species. A wide range of different rhizobia can be isolated from nodules of a single legume species. In some cases, rhizobia from different classes (Alphaproteobacteria and Betaproteobacteria) cohabit in the nodules of a single plant (Liu et al., 2007). Although the majority of tropical legumes allow such a symbiotic promiscuity, the symbiont of a given legume species is frequently incompatible with another species in a pattern that resists cocladogenesis (Sprent, 1994). The observed patterns likely are the result of several host range shifts and specialization empowered by lateral gene transfer. Genomic characterization of the broad host range *Rhizobium* sp. NGR234 support this view. This strain has a particularly rich repertoire of nodulation genes (Freiberg et al., 1997). As a result, it produces a wide repertoire of different Nod factors, explaining, at least in part, its compatibility with a large range of host plant species, including *Parasponia*.

Determinants of Host Specificity, Comparison of Two Symbioses

Plant Determinants of Symbiotic Specificity in Legume-Rhizobium Symbiosis

The molecular determinants of symbiotic specificity are far better known from the rhizobial side than from the plant side. Bacterial Nod factors are commonly described as the main determinant of rhizobial specificity (Van Rhijn and Vanderleyden, 1995). Logically, their plant receptors should act as their counterparts. These Nod factor receptors have been identified as receptor kinases containing two or three specific LysM domains, known to bind peptidoglucan, in their extracellular region (LYK). In at least one case, identified in the Lotus genus, a single amino acid difference in such a LysM domain, was shown to control symbiont specificity (Radutoiu et al., 2007). The observed difference might have resulted from a concerted specialization event in both partners. The tight level of host specificity found in galegoid legumes suggests that similar events of co-evolution occurred in this group as well.

Beyond mechanisms that control host-partner specificity, an initially established engagement can still become an unsuccessful one. Some rhizobial strains can form nodules, but are unable (or "unwilling") to fix nitrogen efficiently. Based on cross-inoculation tests (Provorov, 1994) there is a limited relationship between successful nodule establishment and taxonomy (consistent with only a marginal pattern of cocladogenesis), but there is no dependency at all on the success of subsequent nitrogen fixation. In other words, there is as much chance that a bacterial symbiont of a species from the same host genus will fail to fix nitrogen as a symbiont of a more distantly related species. Fixing versus non-fixing most likely does not follow an on/off system, but more likely follows a continuum of nitrogen fixation efficiencies ranging from very profitable strains to complete non-fixers. It is also probable that the variation of nitrogen fixation efficiency relates to the natural selection mechanisms acting on resource allocation efficiencies of the rhizobia than on co-adaptation between the two partners.

Specificity in Endomycorrhizal Symbiosis

The endomycorrhizal symbiosis is remarkable for its extent in nature. Indeed, endomycorrhizal fungi can associate with most land plants, and, more importantly, the most recent common ancestor of their hosts dates back to the colonization of land by plants. Such an early event would have left tens of millions of years for divergence and speciation to occur. However, it is widely recognized that hosts from different high-level vegetal taxa can interact and share the same fungal symbionts. In other words, at first glance there does not seem to be any form of host-symbiont specificity in the endomycorrhizal symbiosis. However, this is somewhat overstated, given examples where preference could be demonstrated. One of these is the case of some orchid species that acquire carbon though mycorrhizal networks (Selosse et al., 2006). These plants actually act as saprophytes (if not parasites) and claim photosynthetate from surrounding trees through mycorrhizae. The interaction might still be mutualistic, for example in the case where "saprophytic" plants provide some benefit (e.g., vitamins) to other members of the networks. Alternatively, the plant might just be taking advantage of the presence of a fungus-mediated network rich in nutrients, thereby acting as a parasite. It can be argued that natural selection against such parasitic behaviour might be weak, due to the discrepancy between the cost (minimized for donor trees, relatively speaking) and the benefit (in some cases, orchids completely stop performing photosynthesis themselves). Are these orchids just cheaters in a generalist system? While this sounds plausible, the fact remains that they exhibit a marked preference with respect to the interacting fungi (Roy et al., 2009). It can be imagined that the plant is engaged in a parasitic relationship (it might parasitize the fungus, or both partners might cooperate to parasitize the host tree) that would trigger the specialization-driving process.

In general, there is growing evidence that host-symbiont preferences are present in endomycorrhizal interactions, even when they are not as marked nor occur at a very early stage as in rhizobial symbioses. Indeed, though any plant/fungus couple might be able to interact, this does not preclude that preferences act in the field, or when different fungi compete in a single root system. Studies enlightening such preference in the wild have come to support this view, suggesting that specificity might occur at a later stage, once the interaction is established (Vandenkoornhuyse et al., 2002).

Evolutionary Explanations for Specificity

Intriguingly, in rhizobial symbiosis the level of specificity is astoundingly higher when compared to arbuscular mycorrhization. It occurs at a very early stage of the symbiosis. Even before any contact is made, based on the flavonoid/NodD interaction, and Nod factor variability, each partner is able to prevent any response to a noncompatible host or symbiont. These restrictions are active on both sides. Very likely, natural selection has driven the evolution of receptors and other downstream components involved in mechanisms restricting specificity. Theoretically both interacting partners could promote less stringent variants of their specificity-enforcing proteins, starting with the rhizobial NodD protein and the legume Nod factor receptors. Reduced specificity could be favorable, since it tends to increase the number of opportunities for beneficial interactions. On the downside, a reduction in specificity likely leads to an increase in the number of species competing for interaction. Natural selection has promoted mechanisms to restrict symbiotic promiscuity in the case of the rhizobial interaction, and these mechanisms act from very first stages of the interaction and continue afterward. In contrast, in endomycorrhizal symbiosis, selection mechanisms are less strict and seem to act only once the interaction is established.

We will address three possible evolutionary explanations that could have driven early levels of specificity of both symbioses: (1) vicariance, (2) the cheater hypothesis, and (3) functional co-adaptation of partners.

Vicariance

Vicariance is a neutral (nonadaptive) process leading to functional differentiation between populations and is caused by natural geographical processes. It promotes divergence only by the action of random genetic drift, but it can be exacerbated or accelerated by specific adaptations in case the local environments differ in their characteristics. Interestingly, temperate legumes tend to present more frequent and stronger cases of increased specificity than tropical legumes (Sprent, 1994). In general, tropical ecosystems are considered stable over time, in contrast with temperate latitudes that repetitively have experienced dramatic episodes of perturbation. Furthermore, the origin of the Fabaceae is placed in tropical regions, and the family is supposed to have subsequently colonized other parts of the world (Sprent, 1994). It is likely that temperate legumes have experienced, with respect to their tropical relatives, more episodes of new land colonization, re-colonization after being excluded, and isolation due to less uniform habitats. Moreover, each event of colonization of a new biotope is usually associated, at its earliest stages, with a founder effect caused by the small size of initial founding populations leading to a strong episode of genetic drift.

It would not be surprising that, besides inducing reproductive isolation, drift-caused population differentiation leads to divergence of symbiotic signaling molecules. Assuming that natural selection favors matching of signals and receptors between legume hosts and rhizobial symbionts (in both directions), episodes of increased drift would augment the probability of fixation of a new variant of a signaling molecule or its receptor. In case of such an event, natural selection would strongly promote either the reversion of this substitution or a matching change in the interacting molecules. This would lead to concerted evolution of signal and receptor, preserving the efficiency of the interaction but losing the compatibility with other isolated populations of symbionts and hosts that have undergone an equivalent but independent process. Indeed legumes are frequently found to be unable to interact with rhizobia retrieved from allopatric locations, especially in the case of temperate lineages (Wilkinson et al., 1996).

Vicariance is in essence a passive and accidental process and, might explain why temperate legumes, that have presumably undergone more episodes of geographical isolation and founder effects, tend to exhibit more specificity than tropical groups. However, it cannot really explain the discrepancy between rhizobial and endomycorrizal symbioses, particularly in case of legumes that are able to perform both symbioses. It is possible, still, that the effect of drift could have been exacerbated in the case of the legume-rhizobial symbiosis, for instance because of the shorter generation time of bacteria, relative to fungi. More probably, the effect of isolation could have been triggered by the pre-existence of complex systems of recognition and specificity, offering a wider array of opportunities for divergence to occur than in the case of the endomycorrhizal symbiosis, which seems to lack most of those systems.

The Cheater Hypothesis

A plausible role for systems ensuring symbiotic specificity is to constitute a filter for unwanted partners, either less efficient rhizobial strains or parasites that take advantage of the entry route constituted by mechanisms allowing symbiotic rhizobial infection. The prevalence of both types of unwanted visitors is real. First, genuine rhizobia failing to perform efficient nitrogen fixation is commonplace, either because of a mutation in crucial enzyme-coding genes, a default of host-symbiont physiological compatibility, or simply a switch in strategy from cooperation toward parasitism. Second, nodules are nutrient-rich organs colonized by opportunistic bacteria (such as *Pseudomonas*) (Ibanez et al., 2009). Specificity could be the sole means for legumes to restrict root entry to their most profitable symbionts.

However, it seems unlikely that good nitrogen fixers can be distinguished from bad fixers or parasites based solely on the structure of Nod factors. There is no good reason to expect "honest signaling" from cheaters or parasites, especially since the genes controlling symbiotic properties are different than, and even in some cases not genetically linked to, genes on which the specificity filters are based. Unfortunately, compared to host-parasite interactions, little theoretical work has been devoted to the emergence and maintenance of specificity in mutualism. In a superficial "colloquial" sense, a high-speed ("runaway") process involving hosts and their "good" symbionts, if based on many loci, could possibly succeed in outrunning cheaters, or at least in limiting their prevalence and therefore their cost to the host.

If the evolutionary aim of specificity is to exclude cheating strains of rhizobia, it can explain why rhizobia produce strain-specific Nod factors. However, it does not explain at all why the rhizobia themselves are specific with respect to the legume species with which they interact. It is necessary, to resolve this issue, to reverse the latter argument and imagine the existence of cheating lineages of legumes, assimilating nitrogen but providing only a minimum of photosynthetic assimilates in return.

The problem with "the cheater hypothesis" is that, similar to the vicariance-based explanation, it doesn't introduce a factor that explains the differences between rhizobial and arbuscular mycorrhizal symbioses. There is no reason to expect more prevalence of cheaters in the case of the nitrogen-fixing mutualism than for the fungal associations. Both involve entry of a potential pathogen into host roots, mobilization of energy-resources toward the entering organism and significant investment on the part of the microsymbiont. Moreover, there is yet another weakness in this proposition, which is that posterior sanctions are obviously a more efficient way to interrupt unprofitable associations, and even apply a selective pressure on the symbiont population to favor cooperating variants (Oono et al., 2009). Mechanisms of legume sanctions against non-fixing rhizobia has been proposed and tested and seems to represent a valid explanation for the stability of both mutualisms (Kiers et al., 2003; Kiers and Van der Heijden, 2006). This leaves the question why legumes engage in "prescreening" still unanswered.

A potential answer to this question might be found by taking into account the significant difference in initial costs, in terms of nutrients and energy, between the two symbioses.

These initial costs would, in the case of *Rhizobium* symbiosis, consist of the material mobilized for nodule organogenesis and possibly an energy-rich carbon source provided to the microsymbiont during the time they are colonizing. During endomycorrhizal establishment, costs for organogenesis are arguably lower and, the fungus, being a single individual infecting only partially its hosts, is more likely to sustain itself until the symbiosis is established. In short, it is possible that both members start to pay the costs and receive the benefits at the same time in the mycorrhizal symbiosis, while legumes might have to support their rhizobial symbiont at first before having a chance of receiving a benefit. In such a scenario, any means the increases the chances that a future nodule will not be a parasitic sink of energy will be favored by natural selection. From the rhizobial side, the investment is only one or a small number of bacteria, but evolutionarily speaking, the cost of a failed interaction is total. Choosing a good host will therefore be essential. This proposition requires that "cheating legumes" exist, which has never been described in such terms, but there is no good reason to consider the idea implausible.

Functional Co-adaptation of Partners

This final hypothesis is actually not fundamentally different from the "cheater hypothesis," but introduces an essential factor, namely the physiological compatibility between partners. What has been cited before is the possibility to break symbiotic compatibility by a deleterious mutation. However, such compatibility is not necessarily an on/off process and, more specifically, there is room for host-dependent variability of symbiotic efficiency between different microsymbionts, which might explain also the difference in specificity between rhizobial and endomycorrhizal symbioses.

Nitrogen fixation is actually an energy-expensive reaction requires a relatively tight interaction between both partners (Lodwig et al., 2003). Rhizobia are differentiated and included in organelle-like symbiosomes. The complexity of the nitrogen-fixation reaction is of the same order as photosynthesis (namely, carbon fixation), and in both cases plants co-opted N and C fixation processes by engaging microsymbionts. In contrast, the function of endomycorrhizal interactions is to enhance uptake of existing nutrients, and the elaborate features that appeared in the course of evolution concern more the optimization of molecular exchanges. In contrast to nitrogen fixation, it can be conceived that mycorrhizae-enhanced nutrient uptake is less likely to fail completely and provide no output.

A successful rhizobial interaction requires matching of a number of molecular traits between host and microsymbiont, such as exchange of specific amino acids, the type of plant-provided carbohydrates and fine-tuning of dosage-dependent compounds. For example, the transporters acting in the plant-derived outer membrane should match these of the differentiated rhizobia. There can be several factors triggering co-adaptation at the physiological level between rhizobia and legumes, including vicariance. If signal molecules can co-evolve and diverge as a result of independent drift in (temporarily) differentiated populations, metabolic traits controlling nitrogen fixation and the required co-adaptations might just as well co-evolve. So, if such optimization occurs in allopatry, it might promote divergence. Also co-adaptation can serve the purpose of adaptation to specific nitrogen-fixation needs of a particular host. For example, it sounds likely that trees and herbaceous species should diverge in terms of timing of nitrogen fixation, nodule lifetime, or regulation of nitrogen fixation.

Nitrogen fixation (rate and output), and therefore also its molecular determinants, is a quantitative variable depending on the host genotype, the symbiont genotype as well as on their interaction (Heath and Tiffin, 2007). These observations indicate that mutual adaptation is rather likely and opens the possibility for a wide range of strategies: microsymbiont as well as host can evolve in a generalist or specialize with a preferred, more efficient, partner. In such a situation, partner choice based on "honest" signals (signals excreted exclusively by true symbionts) would represent an adaptive advantage. Each partner would have priority interest to interact with a genotype with which it will collaborate most efficiently, but not with others. In comparison, a cheater, which "on purpose" collaborates inefficiently with all hosts (in evolutionary terms), would have no interest in preventing an interaction with a putative host. "Honest" variation in symbiotic capabilities might coexist with cheaters failing to coadapt with any host and therefore inefficient with all hosts. The first type of variation can explain the existence of the recognition system, while the cheaters would be required to present a broad host range.

The *Rhizobium* strain NGR234 indeed "owns" a wide panoply of nodulation genes allowing it to interact with an impressive number of legume host species, but is not described as an efficient nitrogen fixer (Schump et al., 2009). The presence of such "unfair" strains might even promote a runaway process at the level of signaling molecules and recognition for "fair" couples. Fast evolution could be promoted to escape these less efficient but willingly generalist strains, a selective force susceptible to act on both "good" host and their respective "good" microsymbionts.

Summary

Arbuscular mycorrhization and *Rhizobium*-legume symbiosis are two prominent endosymbiotic plant-microbe interactions. Both symbioses facilitate plant nutrient uptake thereby expanding the area that can be colonized by the interacting species. Approximately 70% of all land plants can sustain arbuscular mycorrhization, whereas *Rhizobium*-legume symbiosis is restricted to a phylogenetically smaller group of plants, the legume family (*Fabaceae*). There are distinct differences between the two symbioses; in morphological responses, evolutionary origin and levels of host specificity, however the genetic basis of these interactions significantly overlap.

Arbuscular mycorrhization is estimated to be at least 400 million years old, a time frame that coincides with emergence of land plants. Therefore, it is suggested that this association was critical for land colonization by plants. In contrast, the *Rhizobium*-legume symbiosis is much younger. It evolved approximately 60 million years ago, dating shortly after the origin of the legume family. Based on genetic evidence, it is hypothesized that in legumes parts of the signaling machinery essential for arbuscular mycorrhization have underwent specific evolution to gain a function in *Rhizobium*-legume symbiosis. Currently more than a dozen genes are known to function in the legume-signaling cascade triggered upon *Rhizobium* perception, and seven of these genes are also essential in the establishment of arbuscular mycorrhizal symbiosis. That such genes are required for both symbioses implies shared intercellular signaling important for establishment of both symbioses. A potential role could be suppression of plant defenses, which is required for initiation of both interactions.

On comparing these two plant-microbe endosymbioses, the most striking difference is in the specificity of the interaction. For arbuscular mycorrhization, there seems to be almost no selection criteria for potential partners on either side. A single microsymbiont is known to interact with a wide range of hosts whereas a given host can interact with a broad variety of AM fungi even at a given time point. In contrast, the *Rhizobium*-legume interaction represents the other end of the specificity spectrum; most rhizobia only interact with a small, phylogenetically restricted, group of plants. The control mechanisms regulating bacterial uptake are best described as a key-lock mechanism. A sound explanation for the large difference in specificity cannot be given, but a factor that can explain (at least in part) the differences is the initial cost for setting up the symbiosis. In arbuscular mycorrhization, infection by an obligatory symbiotic fungus and subsequent arbuscule formation does not come at high costs for the host plant. Also the risk for the fungus is limited, especially since it can interact with multiple hosts. In the case of legume-Rhizobium symbioses, the plant has to invest significantly more energy before it can actually benefit from the interaction. Also the bacterium, once enclosed by the plant, has no way back. We hypothesize that this could have lead to more stringent selection of interacting partners thereby explaining the difference in specificity between these two plant-microbe interactions.

References

Albrecht, C., Geurts, R., Lapeyrie, F., Bisseling, T. 1998. Endomycorrhizae and rhizobial Nod factors both require SYM8 to induce the expression of the early nodulin genes PsENOD5 and PsENOD12A. Plant J 15: 605–614.

Ané, J.M., Lévy, J., Thoquet, P., Kulikova, O., de Billy, F., Penmetsa, V., Kim, D.J., Debellé, F., Rosenberg, C., Cook, D.R., Bisseling, T., Huguet, T., Dénarié, J. 2002. Genetic and cytogenetic mapping of DMI1, DMI2, and DMI3 genes of Medicago truncatula involved in Nod factor transduction, nodulation, and mycorrhization. Mol Plant Microbe Interact 15: 1108–1118.

Ané, J.M., Kiss, G.B., Riely, B.K., Penmetsa, R.V., Oldroyd, G.E., Ayax, C., Lévy, J., Debellé, F., Baek, J.M., Kalo, P., Rosenberg, C., Roe, B.A., Long, S.R., Dénarié, J., Cook, D.R. 2004. Medicago truncatula DMI1 required for bacterial and fungal symbioses in legumes. Science 303: 1364–1367.

Amadou, C., Pascal, G., Mangenot, S., Glew, M., Bontemps, C., Capela, D., Carrère, S., Cruveiller, S., Dossat, C., Lajus, A., Marchetti, M., Poinsot, V., Rouy, Z., Servin, B., Saad, M., Schenowitz, C., Barbe, V., Batut, J., Médigue, C., Masson-Boivin, C. 2008. Genome sequence of the beta-rhizobium Cupriavidus taiwanensis and comparative genomics of rhizobia. Genome Res 18:1472–1483.

Akiyama, K., Matsuzaki, K., Hayashi, H. 2005. Plant sesquiterpenes induce hyphal branching in arbuscular mycorrhizal fungi. Nature 435: 824–827.

Ardourel, M., Lortet, G., Maillet, F., Roche, P., Truchet, G., Promé, J.C., Rosenberg, C. 1995. In: Rhizobium meliloti, the operon associated with the nod box n5 comprises nodL, noeA and noeB, three host-range genes specifically required for the nodulation of particular Medicago species. Mol Microbiol 17: 687–699.

Arrighi, J.F., Barre, A., Ben Amor, B., Bersoult, A., Soriano, L.C., Mirabella, R., de Carvalho-Niebel, F., Journet, E.P., Ghérardi, M., Huguet, T., Geurts, R., Dénarié, J., Rougé, P., Gough, C. 2006. The Medicago truncatula lysin motif-receptor-like kinase gene family includes NFP and new nodule-expressed genes. Plant Physiol 142: 265–279.

Balaji, B., Ba, A.M., LaRue, T.A., Tepfer, D., Piché, Y. 1994. Pisum sativum mutants insensitive to nodulation are also insensitive to invasion in vitro by the mycorrhizal fungus Gigaspora margarita. Plant Sci 102: 195–203.

Banba, M., Gutjahr, C., Miyao, A., Hirochika, H., Paszkowski, U., Kouchi, H., Imaizumi-Anraku, H. 2008. Divergence of evolutionary ways among common sym genes: CASTOR and CCaMK show functional conservation between two symbiosis systems and constitute the root of a common signaling pathway. Plant Cell Physiol 49:1659–1671.

Barker, S.J., Stummer, B., Gao, L., Dispain, I., O'Connor, P.J., Smith, S.E. 1998. A mutant in Lycopersicon esculentum Mill. with highly reduced VA mycorrhizal colonization: isolation and preliminary characterisation. Plant J 15: 791–797.

Batista, J.S.S., Hungria, M., Barcellos, F.G., Ferreira, M.C., Mendes, I.C. 2007. Variability in Bradyrhizobium japonicum and B. elkanii seven years after introduction of both the exotic microsymbiont and the soybean host in a cerrados soil. Microb Ecol 53: 270–284.

Becker, A., Fraysse, N., Sharypova, L. 2005. Recent advances in studies on structure and symbiosis-related function of rhizobial K-antigens and lipopolysaccharides. Mol Plant Microbe Interact 18: 899–905.

Besserer, A., Puech-Pagès, V., Kiefer, P., Gomez-Roldan, V., Jauneau, A., Roy, S., Portais, J.C., Roux, C., Bécard, G., Séjalon-Delmas, N. 2006. Strigolactones stimulate arbuscular mycorrhizal fungi by activating mitochondria. PLoS Biol 4: e226.

Besserer, A., Bécard, G., Jauneau, A., Roux, C., Séjalon-Delmas, N. 2008. GR24, a synthetic analog of strigolactones, stimulates the mitosis and growth of the arbuscular mycorrhizal fungus Gigaspora rosea by boosting its energy metabolism. Plant Physiol 148(1): 402–413.

Beilby, J.P., Kidby, D.K. 1980. Biochemistry of ungerminated and germinated spores of the vesicular-arbuscular mycorrhizal fungus, Glomus caledonius: changes in neutral and polar lipids. J Lipid Res 21: 739–750.

Blilou, I., Ocampo, J.A., Garcia-Garrido, J.M. 1999. Resistance of pea roots to endomycorrhizal fungus or Rhizobium correlates with enhanced levels of endogenous salicylic acid. J Exp Bot 50: 1663–1668.

Boivin-Masson, C., Giraud, E., Perret, X., Batut, J. 2009. Establishing nitrogen-fixing symbiosis with legumes: how many rhizobium recipes? Trends Microbiol 17: 458–466.

Bonfante, P., Genre, A., Faccio, A., Martini, I., Schauser, L., Stougaard, J., Webb, J., Parniske, M. 2000. The Lotus japonicus LjSym4 gene is required for the successful symbiotic infection of root epidermal cells. Mol Plant–Microbe Interact 13: 1109–1120.

Bradbury, S.M., Peterson, R.L., Bowley, S.R. 1991. Interactions between three alfalfa nodulation genotypes and two Glomus species. New Phytol 119: 115–120.

Bucher, M., Raush, C., Daram, P. 2001. Molecular and biochemical mechanisms of phosphorus uptake into plants. J Plant Nutr Soil Sci 164: 209–217.

Catoira, R., Galera, C., de Billy, F., Penmetsa, R.V., Journet, E.-T., Maillet, F., Rosenberg, C., Cook, D., Gough, C., Dénarié, J. 2000. Four genes of Medicago truncatula controlling components of a Nod factor transduction pathway. Plant Cell 12: 1647–1665.

Cavagnaro, T.R., Gao, L.-L., Smith, F.A., Smith, S.E. 2001. Morphology of arbuscular mycorrhizas is influenced by fungal identity. New Phytol 151: 469–475.

Chen, W.M., Moulin, L., Bontemps, C., Vandamme, P., Béna, B., Boivin-Masson, C. 2003. Legume Symbiotic Nitrogen Fixation by β-Proteobacteria Is Widespread in Nature. J Bact 185: 7266–7272.

Chen, X.C., Feng, J., Hou, B.H., Li, F.Q., Li, Q., Hong, G.F. 2005. Modulating DNA bending affects NodD-mediated transcriptional control in Rhizobium leguminosarum. Nucl Acids Res 33: 2540–2548.

Chen, C., Gao, M., Liu, J., Zhu, H. 2007. Fungal symbiosis in rice requires an ortholog of a legume common symbiosis gene encoding a Ca2+/calmodulin-dependent protein kinase. Plant Physiol 145: 1619–1628.

Chen, C., Ané, J.M., Zhu, H. 2008. OsIPD3, an ortholog of the Medicago truncatula DMI3 interacting protein IPD3, is required for mycorrhizal symbiosis in rice. New Phytol 180: 311–315.

Cavagnaro, T.R., Gao, L.L., Smith, F.A., Smith, S.E. 2001. Morphology of arbuscular mycorrhizas is influenced by fungal identity. New Phytol 151: 469–475.

Compaan, B., Yang, W.C., Bisseling, T., Franssen, H. 2001. ENOD40 expression in the pericycle precedes cortical cell division in Rhizobium-legume interaction and the highly conserved internal region of the gene does not encode a peptide. Plant Soil 230: 1–8.

Cox, G., Moran, K.J., Sanders, F., Nockolds, C., Tinker, P.B. 1980. Translocation and transfer of nutrients in vesicular-arbuscular mycorrhizas. III. Polyphosphate granules and phosphorus translocation. New Phytol 84: 649–659.

Davey, L., Currah, R.S. 2006. Interactions between mosses (Bryophyta) and fungi. Can J Bot 84: 1509–1519.

Deakin, W.J., Broughton, W.J. 2009. Symbiotic use of pathogenic strategies: rhizobial protein secretion systems. Nat Rev Microbiol 7: 312–320.

Demchenko, K., Winzer, T., Stougaard, J., Parniske, M., Pawlowski, K. 2004. Distinct roles of Lotus japonicus SYMRK and SYM 15 in root colonization and arbuscule formation. New Phytol 163: 381–392.

Doyle, J.J. 1998. Phylogenetic perspectives on nodulation: evolving views of plants and symbiotic bacteria. Trends Plant Sci 3: 473–478.

Duc, G., Messager, A. 1989. Mutagenesis of pea (Pisum sativum L.) and the isolation of mutants for nodulation and nitrogen fixation. Plant Sci 60: 207–213.

Endre, G., Kereszt, A., Kevei, Z., Mihacea, S., Kaló, P., Kiss, G.B. 2002. A receptor kinase gene regulating symbiotic nodule development. Nature 417: 962–966.

Esseling, J.J., Emons, A.M. 2004. Dissection of Nod factor signalling in legumes: cell biology, mutants and pharmacological approaches. J Microsc 214: 104–113.

Ezawa, T., Cavagnaro, T.R., Smith, S.E., Smith, F.A., Ohtomo, R. 2004. Rapid accumulation of polyphosphate in extraradical hyphae of an arbuscular mycorrhizal fungus as revealed by histochemistry and a polyphosphate kinase/luciferase system. New Phytol 161: 387–392.

Freiberg, C., Fellay, R., Bairoch, A., Broughton, W.J., Rosenthal, A., Perret, X. 1997. Molecular basis of symbiosis between Rhizobium and legumes. Nature 387: 394–401.

Fraysse, N., Couderc, F., Poinsot, V. 2003. Surface polysaccharide involvement in establishing the rhizobium-legume symbiosis. Eur J Biochem 270: 1365–1380.

Galibert, F., Finan, T.M., Long, S.R., Puhler, A., Abola, P., Ampe, F., Barloy-Hubler, F., Barnett, M.J., Becker, A., Boistard, P., Bothe, G., Boutry, M., Bowser, L., Buhrmester, J., Cadieu, E., Capela, D., Chain, P., Cowie, A., Davis, R.W., Dreano, S., Federspiel, N.A., Fisher, R.F., Gloux, S., Godrie, T., Goffeau, A., Golding, B., Gouzy, J., Gurjal, M., Hernandez-Lucas, I., Hong, A., Huizar, L., Hyman, R.W., Jones, T., Kahn, D., Kahn, M.L., Kalman, S., Keating, D.H., Kiss, E., Komp, C., Lelaure, V., Masuy, D., Palm, C., Peck, M.C., Pohl, T.M., Portetelle, D., Purnelle, B., Ramsperger, U., Surzycki, R., Thebault, P., Vandenbol, M., Vorholter, F.J., Weidner, S., Wells, D.H., Wong, K., Yeh, K.C., Batut, J. 2001. The composite genome of the legume symbiont Sinorhizobium meliloti. Science 293: 668–672.

Gallaud, I. 1905 Etudes sur les mycorrhizes endotrophs. Rev Gen Botanique 17, 5–48, 66–85, 123–136, 223–239, 313–325, 423–433, 479–500.

Genre, A., Chabaud, M., Timmers, T., Bonfante, P., Barker, D.G. 2005. Arbuscular mycorrhizal fungi elicit a novel intracellular apparatus in Medicago truncatula root epidermal cells before infection. Plant Cell 17: 3489–3499.

Geurts, R., Fedorova, E., Bisseling, T. 2005. Nod factor signaling genes and their function in the early stages of Rhizobium infection. Curr Opin Plant Biol 8: 346–252.

Guether, M., Neuhäuser, B., Balestrini, R., Dynowski, M., Ludewig, U., Bonfante, P. 2009. A mycorrhizal-specific ammonium transporter from lotus japonicus acquires nitrogen released by Arbuscular Mycorrhizal Fungi. Plant Physiol 150: 73–83.

Gianinazzi-Pearson, V. 1996. Plant cell responses to arbuscular mycorrhizal fungi: getting to the roots of the symbiosis. Plant Cell 8: 1871–1883.

Giraud, E., Moulin, L., Vallenet, D., Barbe, V., Cytryn, E., Avarre, J.C., Jaubert, M., Simon, D., Cartieaux, F., Prin, Y., Bena, G., Hannibal, L., Fardoux, J., Kojadinovic, M., Vuillet, L., Lajus, A., Cruveiller, S., Rouy, Z., Mangenot, S., Segurens, B., Dossat, C., Franck, W.L., Chang, W.S., Saunders, E., Bruce, D., Richardson, P., Normand, P., Dreyfus, B., Pignol, D., Stacey, G., Emerich, D., Verméglio, A., Médigue, C., Sadowsky, M. 2007. Legumes symbioses: absence of Nod genes in photosynthetic bradyrhizobia. Science 316: 1307–1312.

Gleason, C., Chaudhuri, S., Yang, T., Muñoz, A., Poovaiah, B.W., Oldroyd, G.E. 2006. Nodulation independent of rhizobia induced by a calcium-activated kinase lacking autoinhibition. Nature 441: 1149–1152.

Gomez-Roldan, V., Fermas, S., Brewer, P.B., Puech-Pagès, V., Dun, E.A., Pillot, J.P., Letisse, F., Matusova, R., Danoun, S., Portais, J.C., Bouwmeester, H., Bécard, G., Beveridge, C.A., Rameau, C., Rochange, S.F. 2008. Strigolactone inhibition of shoot branching. Nature 455: 189–194.

Gollotte, A., Gianinazzi-Pearson, V., Giovannetti, M., Sbrana, S., Avio, L., Gianinazzi, S. 1993. Cellular localization and cytochemical probing of resistance reactions to arbuscular mycorrhizal fungi in a 'locus a' myc⁻ mutant of Pisum sativum L. Planta 191: 112–122.

González, V., Santamaría, R.I., Bustos, P., Hernández-González, I., Medrano-Soto, A., Moreno-Hagelsieb, G., Janga, S.C., Ramírez, M.A., Jiménez-Jacinto, V., Collado-Vides, J., Dávila, G. 2006. The partitioned Rhizobium etli genome: genetic and metabolic redundancy in seven interacting replicons. Proc Natl Acad Sci USA 103: 3834–3839.

Gutjahr, C., Banba, M., Croset, V., An, K., Miyao, A., An, G., Hirochika, H., Imaizumi-Anraku, H., Paszkowski, U. 2008. Arbuscular mycorrhiza-specific signaling in rice transcends the common symbiosis signaling pathway. Plant Cell 20: 2989–3005.

Harrison, M.J., Dewbre, G.R., Liu, J. 2002. A phosphate transporter from Medicago truncatula involved in the acquisition of phosphate released by arbuscular mycorrhizal fungi. Plant Cell 14: 2413–2429.

Hayward, A., Stirnberg, P., Beveridge, C., Leyser, O. 2009. Interactions between auxin and strigolactone in shoot branching control. Plant Physiol 151: 400–412.

Heath, K.D., Tiffin, P. 2007. Context-dependence in the coevolution of plant and rhizobium mutualists. Proc R Soc Lond B 274: 1905–1912.

Ibanez, F., Angelini, J., Taurian, T., Tonelli, M.L., Fabra, A. 2009. Endophytic occupation of peanut root nodules by opportunistic Gammaproteobacteria. Syst Appl Microbiol 32: 49–55.

Imaizumi-Anraku, H., Takeda, N., Charpentier, M., Perry, J., Miwa, H., Umehara, Y., Kouchi, H., Murakami, Y., Mulder, L., Vickers, K., Pike, J., Downie, J.A., Wang, T., Sato, S., Asamizu, E., Tabata, S., Yoshikawa, M., Murooka, Y., Wu, G.J., Kawaguchi, M., Kawasaki, S., Parniske, M., Hayashi, M. 2005. Plastid proteins crucial for symbiotic fungal and bacterial entry into plant roots. Nature 433: 527–531.

James, T.Y., Kauff, F., Schoch, C.L., Matheny, P.B., Hofstetter, V., Cox, C.J., Celio, G., Gueidan, C., Fraker, E., Miadlikowska, J., Lumbsch, H.T., Rauhut, A., Reeb, V., Arnold, A.E., Amtoft, A., Stajich, J.E., Hosaka, K., Sung, G.H., Johnson, D., O'Rourke, B., Crockett, M., Binder, M., Curtis, J.M., Slot, J.C., Wang, Z., Wilson, A.W., Schüssler, A., Longcore, J.E., O'Donnell, K., Mozley-Standridge, S., Porter, D., Letcher, P.M., Powell, M.J., Taylor, J.W., White, M.M., Griffith, G.W., Davies, D.R., Humber, R.A., Morton, J.B., Sugiyama, J., Rossman, A.Y., Rogers, J.D., Pfister, D.H., Hewitt, D., Hansen, K., Hambleton, S., Shoemaker, R.A., Kohlmeyer, J., Volkmann-Kohlmeyer, B., Spotts, R.A., Serdani, M., Crous, P.W., Hughes, K.W., Matsuura, K., Langer, E., Langer, G., Untereiner, W.A., Lücking, R., Büdel, B., Geiser, D.M., Aptroot, A., Diederich, P., Schmitt, I., Schultz, M., Yahr, R., Hibbett, D.S., Lutzoni, F., McLaughlin, D.J., Spatafora, J.W., Vilgalys, R. 2006. Reconstructing the early evolution of fungi using a six-gene phylogeny. Nature 443: 818–822.

Javot, H., Penmetsa. R.V., Terzaghi. N., Cook, D.R., Harrison, M.J. 2007. A Medicago truncatula phosphate transporter indispensable for the arbuscular mycorrhizal symbiosis. Proc Natl Acad Sci USA 104: 1720–1725.

Kaló, P., Gleason, C., Edwards, A., Marsh, J., Mitra, R.M., Hirsch, S., Jakab, J., Sims, S., Long, S.R., Rogers, J., Kiss, G.B., Downie, J.A., Oldroyd, G.E. 2005. Nodulation signaling in legumes requires NSP2, a member of the GRAS family of transcriptional regulators. Science 308: 1786–1789.

Kambara, K., Ardissone, S., Kobayashi, H., Saad, M.M., Schumpp, O., Broughton, W.J., Deakin, W.J. 2009. Rhizobia utilize pathogen-like effector proteins during symbiosis. Mol Microbiol 71: 92–106.

Kanamori, N., Madsen, L.H., Radutoiu, S., Frantescu, M., Quistgaard, E.M., Miwa, H., Downie, J.A., James, E.K., Felle, H.H., Haaning, L.L., Jensen, T.H., Sato, S., Nakamura, Y., Tabata, S., Sandal, N., Stougaard, J. 2006. A nucleoporin is required for induction of Ca2+ spiking in legume nodule development and essential for rhizobial and fungal symbiosis. Proc Natl Acad Sci USA 103: 359–364.

Kaneko, T., Nakamura, Y., Sato, S., Asamizu, E., Kato, T., Sasamoto, S., Watanabe, A., Idesawa, K., Ishikawa, A., Kawashima, K., Kimura, T., Kishida, Y., Kiyokawa, C., Kohara, M., Matsumoto, M., Matsuno, A., Mochizuki, Y., Nakayama, S., Nakazaki, N., Shimpo, S., Sugimoto, M., Takeuchi, C., Yamada, M., Tabata, S. 2000. Complete genome structure of the nitrogen-fixing symbiotic bacterium Mesorhizobium loti. DNA Res 7: 331–338.

Kaneko, T., Nakamura, Y., Sato, S., Minamisawa, K., Uchiumi, T., Sasamoto, S., Watanabe, A., Idesawa, K., Iriguchi, M., Kawashima, K., Kohara, M., Matsumoto, M., Shimpo, S., Tsuruoka, H., Wada, T., Yamada, M., Tabata, S. 2002. Complete genomic sequence of nitrogen-fixing symbiotic bacterium Bradyrhizobium japonicum USDA110. DNA Res 9: 189–197.

Kawaguchi, M., Imaizumi-Anraku, H., Koiwa, H., Niwa, S., Ikuta, A., Syono, K., Akao, S. 2002. Root, root hair, and symbiotic mutants of the model legume Lotus japonicus. Mol Plant Microbe Interact 2002 Jan;15: 17–26.

Kiers, E.T., Rousseau, R.A., West, S.A., Denison, R.F. 2003. Host sanctions and the legume-rhizobium mutualism. Nature 425: 78–81.

Kiers, E.T., van der Heijden, M.G. 2006. Mutualistic stability in the arbuscular mycorrhizal symbiosis: exploring hypotheses of evolutionary cooperation. Ecology 87: 1627–1636.

Kobayashi, H., Naciri-Graven, Y., Broughton, W.J., Perret, X. 2004. Flavonoids induce temporal shifts in gene-expression of nod-box controlled loci in Rhizobium sp. NGR234. Mol Microbiol 51: 335–347.

Koske, R. 1981. Multiple germination by spores of Gigaspora gigantea. Trans Br Mycol Soc 76: 328–3318:17 8-2-201018:17 8-2-20100.

Kosuta, S., Hazledine, S., Sun, J., Miwa, H., Morris, R.J., Downie, J.A., Oldroyd, G.E. 2008. Differential and chaotic calcium signatures in the symbiosis signaling pathway of legumes. Proc Natl Acad Sci USA 105: 9823–9828.

Lapopin, L., Gianinazzi-Pearson, V., Franken, P. 1999. Comparative differential RNA display analysis of arbuscular mycorrhiza in Pisum sativum wild type and a mutant defective in late stage development. Plant Mol Biol 41: 669–677.

Lavin, M., Herendeen, P.S., Wojciechowski, M.F. 2005. Evolutionary rates analysis of Leguminosae implicates a rapid diversification of lineages during the tertiary. System Biol 54: 574–594.

Lévy, J., Bres, C., Geurts, R., Chalhoub, B., Kulikova, O., Duc, G., Journet, E.P., Ané, J.M., Lauber, E., Bisseling, T., Dénarié, J., Rosenberg, C., Debellé, F. 2004. A putative Ca2+ and calmodulin-dependent protein kinase required for bacterial and fungal symbioses. Science 303: 1361–1364.

Lewis, G., Schrire, B., Mackinder, B., Lock, M., Eds. 2005. Legumes of the world. Royal Botanic Gardens, Kew.

Limpens, E., Franken, C., Smit, P., Willemse, J., Bisseling, T., Geurts, R. 2003. LysM domain receptor kinases regulating rhizobial Nod factor-induced infection. Science 302: 630–633.

Liu, X.Y., Wang, E.T., Li, Y., Chen, W.X. 2007. Diverse bacteria isolated from root nodules of Trifolium, Crotalaria and Mimosa grown in the subtropical regions of China. Arch Microbiol 188: 1–14.

Lodwig, E.M., Hosie, A.H., Bourdès, A., Findlay, K., Allaway, D., Karunakaran, R., Downie, J.A., Poole, P.S. 2003. Amino-acid cycling drives nitrogen fixation in the legume-Rhizobium symbiosis. Nature 422: 722–726.

Lodwig, E.M., Leonard, M., Marroqui, S., Wheeler, T.R., Findlay, K., Downie, J.A., Poole, P.S. 2005. Role of polyhydroxybutyrate and glycogen as carbon storage compounds in pea and bean bacteroids. Mol Plant Microbe Interact 18: 67–74.

Maeda, D., Ashida, K., Iguchi, K., Chechetka, S., Hijikata, A., Okusako, Y., Deguchi, Y., Izui, K., Hata, S. 2006. Knockdown of an arbuscular mycorrhiza-inducible phosphate transporter gene of Lotus japonicus suppresses mutualistic symbiosis. Plant Cell Physiol 47: 807–817.

Markmann, K., Giczey, G., Parniske, M. 2008. Functional adaptation of a plant receptor-kinase paved the way for the evolution of intracellular root symbioses with bacteria. PLoS Biol 6: e68.

Marsh, J.F., Schultze, M. 2001. Analysis of arbuscular mycorrhizas using symbiosis-defective plant mutants. New Phytol 150: 525–532.

McKay, M.J., Ross, J.J., Lawrence, N.L., Cramp, R.E., Beveridge, C.A., Reid, J.B. 1994. Control of internode length in Pisum sativum (Further evidence for the involvement of indole-3-acetic acid). Plant Physiol 106: 1521–1526.

Middleton, P.H., Jakab, J., Penmetsa, R.V., Starker, C.G., Doll, J., Kaló, P., Prabhu, R., Marsh, J.F., Mitra, R.M., Kereszt, A., Dudas, B., VandenBosch, K., Long, S.R., Cook, D.R., Kiss, G.B., Oldroyd, G.E. 2007. An ERF transcription factor in Medicago truncatula that is essential for Nod factor signal transduction. Plant Cell 19: 1221–1234.

Mitra, R.M., Gleason, C.A., Edwards, A., Hadfield, J., Downie, J.A., Oldroyd, G.E., Long, S.R. 2004. A Ca^{2+}/calmodulin-dependent protein kinase required for symbiotic nodule development: Gene identification by transcript-based cloning. Proc Natl Acad Sci USA 101: 4701–4705.

Miya, A., Albert, P., Shinya, T., Desaki, Y., Ichimura, K., Shirasu, K., Narusaka, Y., Kawakami, N., Kaku, H., Shibuya, N. 2007. CERK1, a LysM receptor kinase, is essential for chitin elicitor signaling in Arabidopsis. Proc Natl Acad Sci USA 104(49): 19613–19618.

Morandi, D., Prado, E., Sagan, M., Duc, G. 2005. Characterisation of new symbiotic Medicago truncatula (Gaertn.) mutants, and phenotypic or genotypic complementary information on previously described mutants. Mycorrhiza 15: 283–289.

Moulin, L., Munive, A., Dreyfus, B., Boivin-Masson, C. 2001. Nodulation of legumes by members of the beta-subclass of Proteobacteria. Nature 411: 948–950.

Nagy, F., Karandashov, V., Chague, W., Kalinkevich, K., Tamasloukht, M., Xu, G.H., Jakobsen, I., Levy, A.A., Amrhein, N., Bucher, M. 2005. The characterization of novel mycorrhiza-specific phosphate transporters from Lycopersicon esculentum and Solanum tuberosum uncovers functional redundancy in symbiotic phosphate transport in solanaceous species. Plant J 42: 236–250.

Nagy, R., Vasconcelos, M., Zhao, S., McElver, J., Bruce, W., Amrhein, N., Raghothama, K., Bucher, M. 2006. Differential regulation of five Pht1 phosphate transporters from maize (Zea mays L.). Plant Biol 8: 186–197.

Oono, R., Denison, R.F., Kiers, E.T. 2009. Controlling the reproductive fate of rhizobia: how universal are host sanctions? New Phytol 183: 967–979.

Peck, M.C., Fisher, R.F., Long, S.R. 2006. Diverse flavonoids stimulate NodD1 binding to nod gene promoters in Sinorhizobium meliloti. J Bacteriol 188: 5417–5427.

Penmetsa, R.V., Cook, D.R. 1997. A legume ethylene-insensitive mutant hyperinfected by its rhizobial symbiont. Science 275: 527–530.
Perret, X., Staehelin, C., Broughton, W.J. 2000. Molecular basis of symbiotic promiscuity. Microbiol Mol Biol Rev 64: 180–201.
Pirozynski, K.A., Malloch, D.W. 1975. The origin of land plants: a matter of mycotropism. Biosystems 6: 153–164.
Provorov, N.A. 1994. The interdependence between taxonomy of legumes and specificity of their interaction with rhizobia in relation to evolution of symbiosis. Symbiosis 17: 183–200.
Radutoiu, S., Madsen, L.H., Madsen, E.B., Felle, H.H., Umehara, Y., Grønlund, M., Sato, S., Nakamura, Y., Tabata, S., Sandal, N., Stougaard, J. 2003. Plant recognition of symbiotic bacteria requires two LysM receptor-like kinases. Nature 425: 585–592.
Radutoiu, S., Madsen, L.H., Madsen, E.B., Jurkiewicz, A., Fukai, E., Quistgaard, E.M., Albrektsen, A.S., James, E.K., Thirup, S., Stougaard, J. 2007. LysM domains mediate lipochitin-oligosaccharide recognition and Nfr genes extend the symbiotic host range. EMBO J 26: 3923–3935.
Rausch, C., Daram, P., Brunner, S., Jansa, J., Laloi, M., Leggewie, G., Amrhein, N., Bucher, M. 2001. A phosphate transporter expressed in arbuscule-containing cells in potato. Nature 414: 462–466.
Raymond, J., Siefert, J.L., Staples, C.R., Blankenship, R.E. 2004. The natural history of nitrogen fixation. Mol Biol Evol 21: 541–554.
Redecker, D., Kodner, R., Graham, L.E. 2000. Glomalean Fungi from the Ordovician. Science 289: 1920–1921.
Remy, W., Taylor, T.N., Hass, H., Kerp, H. 1994. Four hundred-million-year-old vesicular arbuscular mycorrhizae. Proc Natl Acad Sci USA 91: 11841–11843.
Reli, B., Perret, X., Estrada-García, M.T., Kopcinska, J., Golinowski, W., Krishnan, H.B., Pueppke, S.G., Broughton, W.J. 1994. Nod factors of Rhizobium are a key to the legume door. Mol Microbiol 13: 171–178.
Roche, P., Maillet, F., Plazanet, C., Debellé, F., Ferro, M., Truchet, G., Promé, J.C, Dénarié, J. 1996. The common nodABC genes of Rhizobium meliloti are host-range determinants. Proc Natl Acad Sci USA 93: 15305–15310.
Roy, M., Watthana, S., Stier, A., Richard, F., Vessabutr, S., Selosse, M.A. 2009. Two myco-heterotrophic orchids from Thailand tropical dipterocarpacean forests associate with a broad diversity of ectomycorrhizal fungi. BMC Biol 7: 51.
Ruiz-Lazano, J., Roussel, L., Gianinazzi, S., Gianinazzi-Pearson, V. 1999. Defense genes are differentially induced by a mycorrhizal fungus and Rhizobium sp. in wild-type and symbiosis-defective pea genotypes. Mol Plant Microbe Interact 12: 976–984.
Sagan, M., Huguet, T., Duc, G. 1994. Phenotypic characterization and classification of nodulation mutants of pea (Pisum sativum L.). Plant Sci 100: 59–70.
Sagan, M., Morandi, D., Tarenghi, E., Duc, G. 1995. Selection of nodulation and mycorrhizal mutants in the model plant Medicago truncatula (Gaertn.) after gamma-ray mutagenesis. Plant Sci 111: 63–71.
Saito, K., Yoshikawa, M., Yano, K., Miwa, H., Uchida, H., Asamizu, E., Sato, S., Tabata, S., Imaizumi-Anraku, H., Umehara, Y., Kouchi, H., Murooka, Y., Szczyglowski, K., Downie, J.A., Parniske, M., Hayashi, M., Kawaguchi, M. 2007. NUCLEOPORIN85 is required for calcium spiking, fungal and bacterial symbioses, and seed production in Lotus japonicus. Plant Cell 19: 610–624.
Schauser, L., Handberg, K., Sandal, N., Stiller, J., Thykjær, T., Pajuelo, E., Nielsen, A., Stougaard, J. 1998. Symbiotic mutants deficient in nodule establishment identified after T-DNA transformation of Lotus japonicus. Mol Gen Genet 259: 414–423.

Schneider, A., Walker, S.A., Poyser, S., Sagan, M., Ellis, T.H.N., Downie, J.A. 1999. Genetic mapping and functional analysis of a nodulation-defective mutant (sym19) of pea (Pisum sativum L.). Mol Gen Genet 262: 1–11.

Schumpp, O., Crèvecoeur, M., Broughton, W.J., Deakin, W.J. 2009. Delayed maturation of nodules reduces symbiotic effectiveness of the Lotus japonicus-Rhizobium sp. NGR234 interaction. J Exp Bot 60: 581–590.

Senoo, K., Zakaria Solaiman, M., Kawaguchi, M., Imaizumi-Anaraku, H., Akao, S., Tanaka, A., Obata, H. 2000a. Screening and characterization of mycorrhizal mutants in Lotus japonicus. In: Molecular Genetics of Model Legumes: Impact for Legume Biology and Breeding. Meeting of Euroconference Project HPCF-CT-1999–00015, John Innes Center, Norwich, UK. Abstract 28.

Senoo, K., Solaiman, M.Z., Kawaguchi, M., Imaizumi-Anaraku, H., Akao, S., Tanaka, A., Obata, H. 2000b. Isolation of two different phenotypes of mycorrhizal mutants in the model legume plant Lotus japonicus after EMS-treatment. Plant Cell Physiol 41: 726–732.

Selosse, M.A., Richard, F., He, X., Simard, S.W. 2006. Mycorrhizal networks: des liaisons dangereuses? Trends Ecol Evol 21: 621–628.

Shirtliffe, S.J., Vessey, J.K. 1996. A nodulation (Nod'/Fix$^-$) mutant of Phaseolus vulgaris L. has nodule-like structures lacking peripheral vascular bundles (Pvb$^-$) and is resistant to mycorrhizal infection (Myc$^-$). Plant Sci 118: 209–220.

Simon, L., Bousquet, J., Lévesque, R.C., Lalonde, M. 1993. Origin and diversification of endomycorrhizal fungi and coincidence with vascular land plants. Nature 363: 67–69.

Smit, P., Raedts, J., Portyanko, V., Debellé, F., Gough, C., Bisseling, T., Geurts, R. 2005. NSP1 of the GRAS protein family is essential for rhizobial Nod factor-induced transcription. Science 308: 1789–1791.

Smith, F.A., Smith, S.E. 1997. Structural diversity in (vescular)-arbuscular mycorrhizal symbiosis. New Phytol 137: 373–388.

Smith, S.E., Read, D.J. 1997. Mycorrhizal Symbiosis, Academic Press, San Diego.

Soltis, D.E., Soltis, P.S., Chase, M.W., Mort, M.E., Albach, D.C., Zanis, M., Savolainen, V., Hahn, W.H., Hoot, S.B., Fay, M.F., Axtell, M., Swensen, S.M., Prince, L.M., Kress, W.J., Nixon, K.C., Farris, J.S. 2000. Phylogeny inferred from 18S rDNA, rbcL, and atpB sequences. Bot J Linnean Soc 133: 381–461.

Sprent, J. 1994. Evolution and diversity in the legume-rhizobium symbiosis—chaos theory? Plant Soil 161: 1–10.

Sprent, J. 2007. Evolving ideas of legume evolution and diversity: a taxonomic perspective on the occurrence of nodulation. New Phytol 174: 11–25.

Sprent, J.I., James, E.K. 2007. Legume evolution: where do nodules and mycorrhizas fit in? Plant Physiol 144: 575–81.

Sprent, J. 2009. Legume nodulation. Wiley-Blackwell, Oxford.

Stracke, S., Kistner, C., Yoshida, S., Mulder, L., Sato, S., Kaneko, T., Tabata, S., Sandal, N., Stougaard, J., Szczyglowski, K., Parniske, M. 2002. A plant receptor-like kinase required for both bacterial and fungal symbiosis. Nature 417: 959–962.

Szczyglowski, K., Shaw, R.S., Wopereis, J., Copeland, S., Hamburger, D., Kasiborski, B., Dazzo, F.B., de Bruijn, F.J. 1998. Nodule organogenesis and symbiotic mutants of the model legume Lotus japonicus. Mol Plant-Microbe Interact 11: 684–697.

Taylor, T.N., Remy, W., Hass, H., Kerp, H. 1995. Fossil arbuscular mycorrhizae from the early Devonian. Mycologia 87: 560–573.

Tirichine, L., Sandal, N., Madsen, L.H., Radutoiu, S., Albrektsen, A.S., Sato, S., Asamizu, E., Tabata, S., Stougaard, J. 2008. A gain-of-function mutation in a cytokinin receptor triggers spontaneous root nodule organogenesis. Science 315: 104–107.

Tree of Life web project, www.tolweb.org.

Umehara. M., Hanada, A., Yoshida, S., Akiyama, K., Arite, T., Takeda-Kamiya, N., Magome, H., Kamiya, Y., Shirasu, K., Yoneyama, K., Kyozuka, J., Yamaguchi, S. 2008. Inhibition of shoot branching by new terpenoid plant hormones. Nature 455: 195–200.

Vandenkoornhuyse, P., Husband, R., Daniell, T.J., Watson, I.J., Duck, J.M., Fitter, A.H., Young, J.P.W. 2002. Arbuscular mycorrhizal community composition associated with two plant species in a grassland ecosystem. Mol Ecol 11: 1555–1564.

van Rhijn, P., Vanderleyden, J. 1995. The Rhizobium-plant symbiosis. Microbiol. Mol Biol Rev 59: 124–142.

Wernegreen, J.J., Riley, M.A. 1999. Comparison of the evolutionary dynamics of symbiotic and housekeeping loci: a case for the genetic coherence of rhizobial lineages. Mol Biol Evol 16: 98–113.

Weeden, N.F., Kneen, B.E., LaRue, T.A. 1990. Genetic analysis of sym genes and other nodule-related genes in Pisum sativum. In: Gresshoff, P.M., Roth, L.E., Stacey, G., Newton, W.E., Eds. Nitrogen fixation: achievements and objectives. Chapman & Hall, New York, 323–330.

Wegel, E., Schauser, L., Sandal, N., Stougaard, J., Parniske, M. 1998. Mycorrhiza mutants of Lotus japonicus define genetically independent steps during symbiotic infection. Mol Plant Microbe Interact 11: 933–936.

Wilkinson, H.H., Spoerke, J.M., Parker, M.A. 1996. Divergence in symbiotic compatibility in a legume-Bradyrhizobium mutualism. Evolution 50: 1470–1477.

Yano, K., Yoshida, S., Müller, J., Singh, S., Banba, M., Vickers, K., Markmann, K., White, C., Schuller, B., Sato, S., Asamizu, E., Tabata, S., Murooka, Y., Perry, J., Wang, T.L., Kawaguchi, M., Imaizumi-Anraku, H., Hayashi, M., Parniske, M. 2008. CYCLOPS, a mediator of symbiotic intracellular accommodation. Proc Natl Acad Sci USA 105: 20540–20545.

Young, J.P.W., Crossman, L.C., Johnston, A.W., Thomson, N.R., Ghazoui, Z.F., Hull, K.H., Wexler, M., Curson, A.R., Todd, J.D., Poole, P.S., Mauchline, T.H., East, A.K., Quail, M.A., Churcher, C., Arrowsmith, C., Cherevach, I., Chillingworth, T., Clarke, K., Cronin, A., Davis, P., Fraser, A., Hance, Z., Hauser, H., Jagels, K., Moule, S., Mungall, K., Norbertczak, H., Rabbinowitsch, E., Sanders, M., Simmonds, M., Whitehead, S., Parkhill, J. 2006. The genome of Rhizobium leguminosarum has recognizable core and accessory components. Genome Biol 7: R34.

Zhu, H., Riely, B.K., Burns, N.J., Ané, J.M. 2006. Tracing nonlegume orthologs of legume genes required for nodulation and arbuscular mycorrhizal symbioses. Genetics 172: 2491–2499.

Ziebuhr, W., Ohlsen, K., Karch, H., Korhonen, T., Hacker, J. 1999. Evolution of bacterial pathogenesis. Cell Mol Life Sci 56: 719–728.

Chapter 3
Arbuscular Mycorrhizas and N Acquisition by Plants

Luisa Lanfranco, Mike Guether, and Paola Bonfante

Arbuscular mycorrhizal (AM) symbiosis is an ancient plant-fungus association that involves most land plants ranging from liverworts to ferns, and from gymnosperms to angiosperms and whose distribution covers diverse environments (Bonfante and Genre, 2008; Parniske, 2008; Smith and Read, 2008). The success of AM fungi in time and space is mainly linked to the nutritional benefits they confer to their hosts. Investigations carried out over the last decade have provided much evidence that, on a global scale, mycorrhizal fungi may be making a significant contribution to ecosystem nutrient cycling (Girlanda et al., 2007). Physiological studies, molecular approaches that target specific genes, and, more recently, -*omics* investigations have been instrumental in unraveling the cellular and molecular bases of nutrient acquisition in AM (Balestrini and Lanfranco, 2006; Sawers et al., 2008 and reference therein). Although P_i (inorganic phosphate) acquisition has received more attention (Bucher, 2007; Javot et al., 2007), important advances in the understanding of nitrogen (N) movement in AM symbiosis have been made in recent years.

The impact of AM symbiosis on plant N uptake is not as clearly defined as that of P (phosphorus). While AM fungi can increase N uptake by plants, there are often only minor effects on the plant N content (Ames et al., 1983; Johansen et al., 1993; Tobar et al., 1994; Hawkins et al., 2000; Hodge et al., 2001). Results from field and greenhouse experiments on the effect of soil N enrichment on the degree of root colonization by AM fungi vary greatly, as increased, decreased, or no effects have been observed (Jumpponen et al., 2005 and references therein). On the other hand, clear shifts in the composition of AM fungal communities have been found in natural ecosystems following the application of organic or mineral N fertilizers to the soil (Egerton-Warburton et al., 2001; Jumpponen et al., 2005; Bradley et al., 2006; Porras-Alfaro et al., 2007; Toljander et al., 2008). Taken together, these data suggest that AM fungi are sensitive to soil N availability and might play a role in N uptake where there is competition for N.

Ecological Aspects of Nitrogen Metabolism in Plants, First Edition. Edited by Joe C. Polacco and Christopher D. Todd.
© 2011 by John Wiley & Sons, Inc. Published 2011 by John Wiley & Sons, Inc.

The aim of this chapter is to provide an overview of the current knowledge on N transfer in AM symbiosis with special emphasis on the two interfaces involved: one located between the soil and the extraradical mycelium and the other between the intraradical hyphae and the plant cortical cells (Figure 3.1). The potential of N metabolites to act as signaling molecules is also discussed.

Nitrogen Movement in AM Symbiosis

N Acquisition and Assimilation in AM Fungi

Nitrogen is found in soil in two main forms, organic and inorganic compounds. The organic compounds are simple molecules, such as urea, amino acids (AA), amines, and peptides or complex ones, such as proteins, while inorganic N compounds are mainly represented by nitrate (NO_3^-) and ammonium (NH_4^+). Plants are known to make use of all these N-compounds. Even though it has been shown that plant roots exhibit proteolytic activity and can use proteins as sole N sources (Paungfoo-Lonhienne et al., 2008), most of the N becomes available to plants via enzymatic proteolytic cleavage and the subsequent mineralization by soil-borne microbes, such as fungi and bacteria.

Mycorrhizal fungi are also known to use different N sources, depending on specific biochemical, physiological, and ecological features of the fungus involved (Girlanda et al., 2007). For example, ectomycorrhizal fungi are localized in the upper, organically enriched soil horizons and some of them have been shown to possess proteolytic capabilities and organic N uptake systems (Chapter 4); this makes them ideal candidates for organic N acquisition. In addition, the availability of the *Laccaria bicolor* complete genome has offered clear confirmation of the capacity of free-living mycelium to import organic and inorganic N sources, including nitrate, ammonium, and peptides, thanks to specific transporters (Martin et al., 2008a).

AM fungi are able to take up and assimilate ammonium (Ames et al., 1983; George et al., 1992; Johansen et al., 1992; Frey and Schuepp, 1993; Johansen et al., 1996), nitrate (Tobar et al., 1994; Bago et al., 1996; Johansen et al., 1996), and amino acids (Hawkins et al., 2000; Hodge et al., 2001; Govindarajulu et al., 2005). The translocation of different N forms in AM was first demonstrated using compartmented pot cultures where N-labeled sources were only provided to the extraradical mycelium. The data were then confirmed using monoaxenic cultures of AM fungi (*in vitro* dual AM fungus and root-organ cultures), which offer the advantage of excluding possible interference by other microorganisms (Govindarajulu et al., 2005; Jin et al., 2005; Cruz et al., 2007). The results made it possible to draw a model of the N flux in AM (Figure 3.1). The first step in N acquisition requires the activity of specific transporters located at the interface between the soil and the extraradical mycelium.

As far as inorganic N is concerned, only one gene, *GintAMT1*, which encodes a high-affinity NH_4^+ transporter, has so far been characterized (Lopez-Pedrosa et al., 2006). The heterologous expression of GintAMT1 complements the defect in a *Saccharomyces cerevisiae* mutant, and gene expression profiles suggest that the gene is involved in uptake of micromolar NH_4^+ through the extraradical mycelium from the surrounding media. NH_4^+, directly absorbed by the external hyphae, or derived from NO_3^- reduction, is rapidly assimilated through the glutamine synthetase/glutamate synthase (GS/GOGAT) cycle (Johansen et al., 1996; Breuninger et al., 2004; Govindarajulu et al., 2005), although the involvement of glutamate dehydrogenase has not been experimentally excluded.

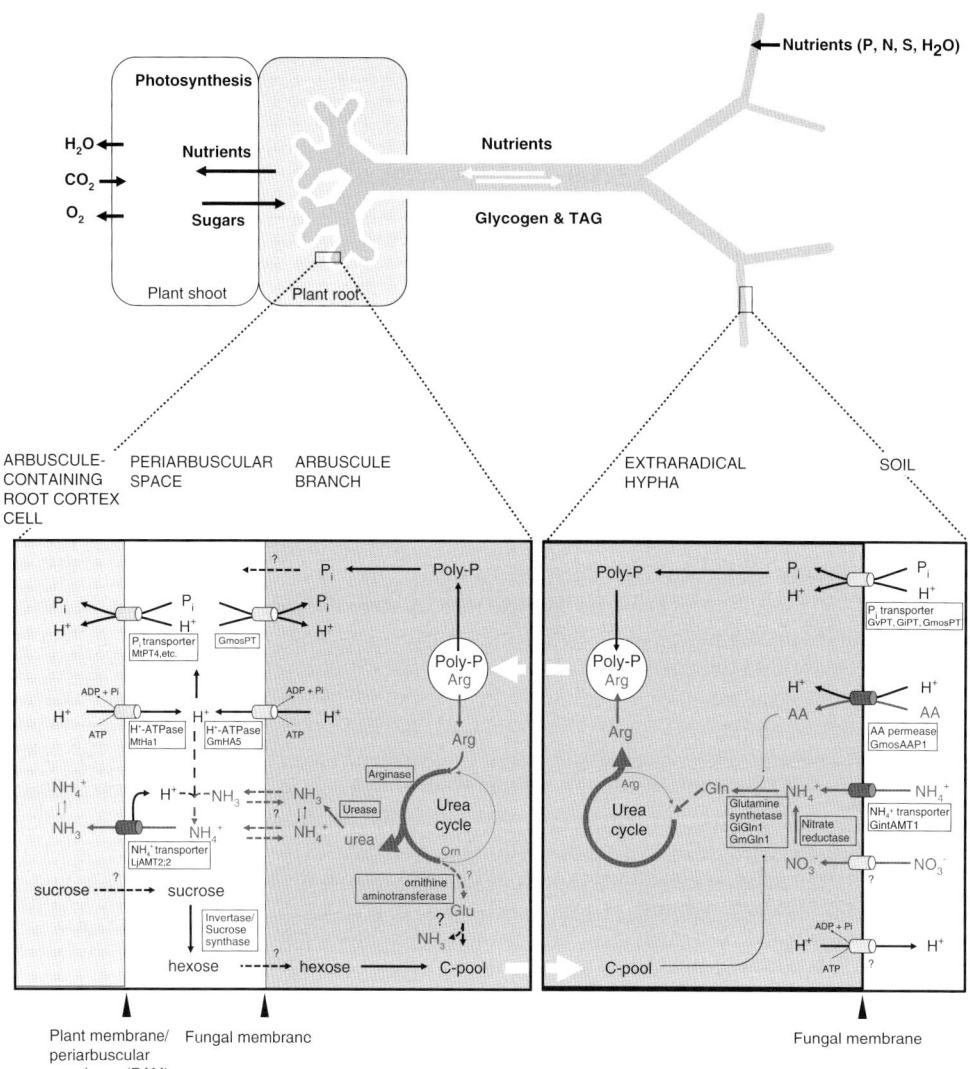

Figure 3.1. Nutrient exchange processes in the AM symbiosis with emphasis on the translocation of nitrogen (N) at the two interfaces, one located between the soil and the extraradical mycelium (ERM) and the other between the intraradical mycelium (IRM) and the plant cortical cells.

Mineral or organic forms of N and P are taken up by specialised transporters located on the fungal membrane in the ERM (*GvPT*, Harrison and van Buuren 1995; *GiPT*, Maldonado-Mendoza et al., 2001; *GmosPT*, Benedetto et al., 2005; *GintAMT1*, Lopez-Pedrosa et al., 2006; *GmosAAP1*, Cappellazzo et al., 2008). Even though not identified so far, ATP dependent proton-pumps, which generate the proton-gradient necessary for the nutrient:proton cotransport, are likely to be active in the ERM. After their import into the fungal hyphae, the three N sources—NH_4^+, NO_3^- and amino acids (AA)—are assimilated (Johansen et al., 1996; *GiGln1, GmGln1,* Breuninger et al., 2004; Govindarajulu et al., 2005). Some N enter the anabolic arm of the urea cycle generating Arg (Govindarajulu et al., 2005, Jin et al., 2005; Cruz et al., 2007). Arg is then translocated from the ERM to the IRM within vacuoles where it is likely to act as counterion of poly-phosphates (Bago et al., 2001). Within arbuscular hyphae Arg flows into the catabolic arm of the urea cycle generating

Assimilated, N, mainly as arginine (Arg), is thought to be translocated to the intraradical mycelium within tubular vacuoles along the coenocytic hyphae (Govindarajulu et al., 2005; Jin et al., 2005; Cruz et al., 2007). Ammonium, produced by the two sequential enzymatic activities of arginase and urease in intraradical structures, is the most likely form of N transferred from fungus to plant. This mechanism, which would also ensure that no molecule containing both N and C is directly transferred from the fungus to the plant, relies on differential expression of enzymes for N assimilation in the extraradical and intraradical hyphal compartments. Quantitative real-time polymerase chain reaction (PCR) assays have in fact shown that a gene of primary nitrogen assimilation (glutamine synthase) is preferentially expressed in the extraradical hyphae, whereas genes involved in arginine breakdown (urease accessory protein, ornithine aminotransferase) and NH_4^+ transfer (ammonium transporter) are more highly expressed in the intraradical mycelium (Govindarajulu et al., 2005). Transcripts for enzymes of the urea cycle as well as amino acid metabolism have recently been localized in arbuscules using a laser microdissection approach in the *Medicago truncatula-Glomus intraradices* interaction (Gomez et al., 2009). The data set also includes a putative arginase, which, being responsible for the breakdown of arginine to ornithine and urea, is considered to be a key enzyme in the release of ammonia.

In addition, biochemical analyses of enzymatic activities of glutamine synthetase, argininosuccinate synthetase, arginase, and urease carried out at the same time on extraradical mycelium and on AM roots have highlighted the synchronization of the spatially separated reactions involved in the anabolic and catabolic arms of the urea cycle. Overall, these data support the role of arginine as a key component in N translocation in the AM mycelium (Cruz et al., 2007) and point to ammonia as the N form transferred to the plant cell (Figure 3.1).

The nature of N transport across the fungal-plant interface is not known. Three mechanisms have been hypothesized for the excretion from the fungal cell to the apoplast: de-protonated ammonia (NH_3) has the potential to diffuse passively into the apoplast following a concentration gradient; alternatively, either NH_3 or NH_4^+ could be exported by the activity of specific efflux systems or by an exocytotic mechanism where intracellular ammonia-loaded vesicles could fuse to the fungal plasma membrane (Chalot et al., 2006). A further transfer of ammonia

---

Figure 3.1. (cont'd) urea and ornithine. Urea is further hydrolyzed to release ammonia (Govindarajulu et al., 2005; Cruz et al., 2007; Gomez et al., 2009). NH_3/NH_4^+ and P_i (originated from the hydrolysis of the poly-P_i) are exported to the periarbuscular space by so far unknown mechanisms and are then taken up by specific plant transporters (Balestrini et al., 2007; Javot et al., 2007). At the periarbuscular space plant and fungal H^+-ATPases are again expected to generate a proton gradient to sustain transport processes (i.e., $P_i:H^+$ cotransport (Gianinazzi-Pearson et al., 2000; *MtHa1*, Murphy et al., 1997; Krajinski et al., 2002; *GmHA5*, Requena et al., 2003). The mycorrhiza-specific ammonium transporter (AMT) (*LjAMT2;2*, Guether et al., 2009b) translocates the de-protonated form of NH_4^+: protons left in the periarbuscular space can therefore sustain the proton gradient. As hypothesized in Balestrini et al. (2007) AM fungi might control the net P_i delivery to the plant by the activation of their own P_i transporters at the periarbuscular space. A similar control mechanism by fungal NH_3/NH_4^+ transport systems might also occur.

The names of the specific genes, which have been cloned and characterized, are indicated in the picture. During the final steps of publication of this chapter a paper was published by Tian et al. (2010) who cloned 11 genes involved in the primary N assimilation and metabolism of the AM fungus *Glomus intraradices*. The time course of expression and N movement from fungal to host tissues was analyzed following nitrate supply to the extraradical mycelium of a mycorrhiza grown under N-limiting conditions. The results substantially confirm the model of N movement through the AM symbiosis described in Fig. 3.1. The authors regret that this chapter could not be updated.

from the apoplast to the plant cells would rely on NH_4^+ transporters, one of which has recently been described for the model legume *Lotus japonicus* (Guether et al., 2009b). For more information, refer to the following paragraph.

AM fungi could also be involved in the acquisition of organic N, a feature that has traditionally been associated with ectomycorrhizal fungi (Chapter 4). It has been shown that AM colonization increased the growth and total N content of plants fed with a nutrient solution containing either NO_3^-, aspartate, or serine as the sole N source (Cliquet et al., 1997; Hawkins et al., 2000) have shown that AM hyphae are able to take up glycine and glutamic acid and transport nitrogen from these sources to the plant roots. The uptake of exogenously supplied Arg has also been observed in the extraradical mycelium grown in *in vitro* cultures (Govindarajulu et al., 2005; Jin et al., 2005). Recent molecular investigations have led to the identification of *GmosAAP1*, an amino acid permease in the AM fungus *Glomus mosseae* (Cappellazzo et al., 2008). *GmosAAP1* mRNAs were detected in the extraradical fungal structures and were more abundant upon exposure to 2 mM compared to 2 µM organic nitrogen. These findings suggest that GmosAAP1 plays a role in the first steps of amino acid acquisition, allowing direct amino acid uptake from the soil and extending the molecular tools by which AM fungi exploit soil resources.

Apart from the capability of taking up amino acids, there is increasing evidence that AM fungi could increase N capture from complex organic material (Hodge et al., 2001; Leigh et al., 2009). Using compartmented microcosms, *Glomus hoi* has been shown to grow preferentially toward an organic patch rather than to a second potential host plant, suggesting a great potential in the exploitation of such soil resources. In addition, the AM fungus is also able to acquire N from this organic material, possibly in the form of inorganic N (Hodge et al., 2001). Interestingly, when two AM fungal species were compared they differed in capacity to take up N and transfer it to the host, with *G. intraradices* being more efficient than *G. hoi* (Leigh et al., 2009). This phenomenon (different efficiency in nutrient acquisition among AM fungi) has been well documented for P and has led to the concept of functional complementarity among fungal species. A fascinating hypothesis is that multiple colonizations of a root by distinct AM fungi may provide benefits not only for P but also for N acquisition (Jansa et al., 2008; Leigh et al., 2009).

N Transfer: From the AM Fungus to the Host Plant

On the basis of the knowledge gained so far, it can be stated that the inorganic N compound, NH_4^+, is the most likely N-form acquired by the plant in the AM interaction. There are two different NH_4^+ uptake systems in plants: the LATS (low affinity transport system) operative in the mM range and the HATS (high affinity transport system) active in the sub-millimolar range (Glass et al., 1997). So far only HATS transporters have been characterized at the molecular level for NH_4^+. The question whether the HATS AMTs (ammonium transporters) transport the charged (NH_4^+) or uncharged (NH_3) form is still unresolved (von Wirén and Merrick, 2004; Ludewig et al., 2007). However, there is evidence that Family 1 of plant AMTs transports net NH_4^+ and Family 2 recognizes NH_4^+ but translocates the deprotonated NH_3 (Sohlenkamp et al., 2002; Guether et al., 2009b; Neuhäuser et al., 2009). It is worth noting that NH_4^+/NH_3 translocation over membranes can also be facilitated by other transporters, such as major intrinsic proteins (MIP) (Jahn et al., 2004; Holm et al., 2005; Loque et al., 2005).

Large-scale gene expression analyses have identified mycorrhiza-induced transcripts that encode putative ammonium transporters (Frenzel et al., 2005; Güimil et al., 2005; Hohnjec et al., 2005; Gomez et al., 2009; Guether et al., 2009a). The *LjAMT2;2* gene was found to be the highest upregulated gene in a transcriptomic analysis of *Lotus japonicus* roots upon colonization with the AM fungus *Gigaspora margarita*. It has been characterized as a high affinity ammonium transporter belonging to the AMT2 subfamily. The localization of *LjAMT2;2* transcripts, via laser microdissection, has revealed their accumulation in arbusculated cells (Guether et al., 2009b). Heterologous complementation of an *S. cerevisiae* mutant, impaired in ammonium uptake, has demonstrated that the LjAMT2;2 protein is a functional ammonium transporter (Guether et al., 2009b), which depends on an acidic external pH. Other transport experiments using *Xenopus* oocytes have indicated that the LjAMT2;2 protein transports NH_3 instead of NH_4^+, suggesting that the transporter binds charged ammonium in the apoplastic interfacial compartment and releases the uncharged NH_3 into the plant cytoplasm. It has therefore been suggested that LjAMT2;2 plays a role in the acquisition of inorganic N from the interfacial apoplast (Guether et al., 2009b). In addition, the peculiar transport mechanism of LjAMT2;2 suggests a model where the protons from the NH_4^+ deprotonation process at the interface space could maintain or even reinforce the gradient for H^+-dependent transport processes, for example, for phosphate transport (Figure 3.1).

Other mycorrhiza-induced AMT genes have not been characterized extensively. Among 10 *Medicago truncatula* putative AMT genes, only one was found to be upregulated more than 5-fold in mycorrhizal roots in a microarray analysis (Benedito et al., 2010). This gene also belongs to the AMT2 family and has recently been shown to be expressed in arbusculated cells (Gomez et al., 2009). However, on the basis of a phylogenetic analysis, this gene does not seem to be the ortholog of *LjAMT2;2* (Figure 3.2). A more comprehensive characterization, in terms of spatial expression profiles, of the different AMT2 genes within a single species would clarify whether multiple AMT2 are simultaneously activated in arbusculated cells. This mechanism, which could optimize and guarantee N capture from the periarbuscular space, has already been described for phosphate transporters in tomato (Balestrini et al., 2007).

While the above findings provide additional support for the transfer of inorganic N to the plant as NH_4^+ (Govindarajulu et al., 2005; Jin et al., 2005), are there other AM-translocated inorganic N forms? Indeed, mycorrhiza-induced nitrate transporters have been identified in tomato, Medicago, and rice (Hildebrandt et al., 2002; Güimil et al., 2005; Hohnjec et al., 2005). Some members of the nitrate transporter gene family can also be downregulated during AM symbiosis (Burleigh, 2001). A similar situation of up- or downregulation of genes belonging to the same family is well known for Pi transporters (Javot et. al., 2007 and references therein). It remains unclear whether the upregulated transporters are involved in nitrate uptake at the plant-fungus interface or whether they just represent a response of the root to the improved P nutrition and to an increased demand for N for assimilatory processes.

The transfer of organic N by the AM fungi cannot be ruled out either. The upregulation of several gene sequences showing similarity to amino acid permease and peptide trasporters in mycorrhizal roots (Guether et al., 2009a) and the localization of their transcripts, mainly in arbusculated cells (Guether et al., unpublished results), is rather puzzling.

It has been proposed that different N forms can be released from the mycorrhizal fungus, depending on C availability (Hampp et al., 1999; Chalot et al., 2006); an organic N transfer would occur under high C availability, whereas inorganic N would preferentially be translocated under C depletion. The first option would be more appropriate for ectomycorrhizal fungi that have saprotrophic capabilities. AM fungi, as obligate biotrophs, instead rely on the host plant to acquire C; the release of organic N to the host could be disadvantageous because of the concomitant loss of C.

As mentioned above, MIPs (major intrinsic proteins, often misleadingly generalized as aquaporins) have been shown to transport ammonium and other nutrients such as water, boron, silicone, etc. (Maurel et al., 2008). In this context, it is again worth noting that some MIP-encoding genes are upregulated in mycorrhizal roots (Krajinski et al., 2000; Küster et al.,

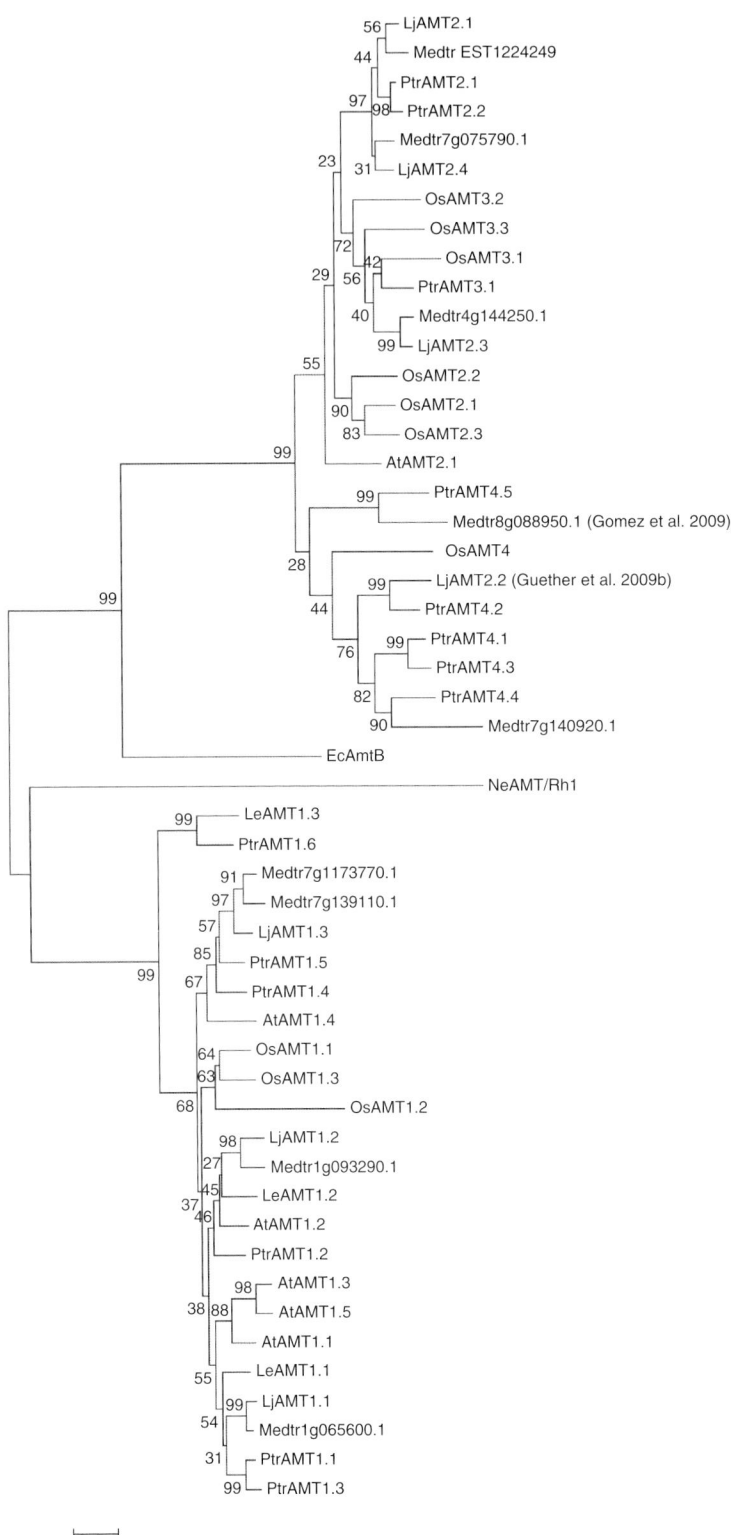

2004; Manthey et al., 2004; Liu et al., 2007; Guether et al., 2009a; Benedito et al., 2010). The activation of these MIP genes has been related to an improved tolerance to drought stress that is often observed in mycorrhizal plants (Ruiz-Lozano, 2003), but its possible role in the transfer of nutrients surely deserves further investigation (Uehlein et al., 2007).

Nitrogen as a Signaling Molecule in AM Symbiosis

Great progress has been made in understanding the process of plant and fungus N uptake and assimilation in recent years. However, much less is known about the signaling events in response to N availability. Nitrogen can in fact behave as a signaling molecule, which informs the fungus and the plant about nutrient availability in the soil (Bahn et al., 2007; Vidal and Gutierrez, 2008).

As far as the plant is concerned, lateral root elongation in response to external N is regarded as a classic indicator of N signaling (Walch-Liu et al., 2006). Recent data have pointed to a role of the specific phospholipase, phospholipase D (PLD)ε, which has been shown to promote root growth and lateral root elongation via generation of lipid messenger phosphatidic acid (Hong et al., 2009). Under severe N deprivation, PLDε plays a role in increasing the root surface area to improve N uptake and utilization, suggesting that PLDε is involved in N signaling. It would be interesting to analyze the function of this PLD in the AM symbiosis, which is also known to have an impact on the architecture of the root apparatus. Plant colonization by AM fungi is associated with the stimulation of the development of the root system, mainly through the formation of new lateral roots (Berta et al., 2002). Intriguingly, lateral root formation occurs not only at late stages of the interaction, but is also induced by diffusible compounds released by AM fungi (Oláh et al., 2005; Bucher et al., 2009).

Very little is known about the fungal side. The extraradical mycelium of AM fungi is a multifunctional compartment with a variety of structures and functions including spores, propagules for dispersal in time and space, runner hyphae, and BAS (branching absorbing structures) that explore soil and take up nutrients. In analogy with other fungi, the extraradical mycelium of AM fungi is also expected to adapt its growth pattern and metabolic capabilities to explore and exploit different soil environments (Bago, 2000; Bahn et al., 2007). Using monoaxenic

---

Figure 3.2. (cont'd)   Unrooted phylogenetic tree for the amino acid sequences of plant ammonium transporters (AMT). The dendrogram was generated by Mega 4.0 software using ClustalW for the alignment and the neighbor-joining method for the construction of the phylogeny (Tamura et al., 2007). For the bootstrap tests, 1,000 replicates were used. The branch lengths correspond to the phylogenetic distances. Abbreviations for plant species: At, *Arabidopsis thaliana*; Le, *Lycopersicon esculentum*; Lj, *Lotus japonicus*; Os, *Oryza sativa*; Ptr, *Populus trichocarpa;* Ec, *Escherichia coli*; Ne, *Nitrosomonas europaea*; Medtr, *Medicago truncatula*. The mycorrhiza-specific AMTs are marked with their corresponding publications (Gomez et al., 2009; Guether et al., 2009a or b?). The accession numbers for the plant AMTs are as follows: AtAMT1;1 (1703292); AtAMT1;2 (4324714); AtAMT1;3 (5880355); AtAMT1;4 (7450345); AtAMT1;5 (5672513); AtAMT2 (3335376); LeAMT1;1 (P58905); LeAMT1;3(Q9FVN0); LeAMT1;2 (O04161); LjAMT1;1-3 (D'Apuzzo et al., 2004); LjAMT2;1 (AAL08212), LjAMT2;2 (FJ668388); PtrAMTs (Couturier et al., 2007), OsAMTs (Suenaga et al., 2003); EcAmtB (NP_286193) and NeRh-1 (NP_840535). The sequences for the predicted *M. truncatula* AMT proteins were retrieved from the Mt3.0 Genome annotation (IMGAG) and Gene Index Project (http://compbio.dfci.harvard.edu/tgi/). One *Medicago* AMT sequence (Benedito et al., 2010) could not be included into the alignment because too short.

cultures, Bago and others (2004) have observed that *G. intraradices* is able to adapt its hyphal morphology and architecture in response to the availability of different inorganic N sources. Hyphal growth in the presence of NO_3^- was characterized by the increased formation of runner hyphae, BAS, and spores, while considerable morphological changes were noted and sporulation was suppressed in the NH_4^+-amended medium. These results clearly show the plasticity of the extraradical mycelium and the potential to switch from an assimilative to a sporulative (reproductive) metabolism, depending on the surrounding environment. So far, the molecular mechanisms governing N sensing and the subsequent morphogenetic and metabolic changes are unknown. It will be of interest to determine whether some NH_4^+ transporters in AM fungi have evolved a sensing function, as reported for the Mep2 gene in *S. cerevisiae* and *Candida albicans* (Lorenz and Heitman, 1998; Biswas and Morschhauser, 2005).

The observation that AM fungi are able to proliferate in organic patches and increase N capture from complex organic material (Hodge et al., 2001) has suggested that AM fungi also perceive and respond to organic N conditions in the environment. An untargeted approach, based on the construction of a subtractive cDNA library, has shown that *G. intraradices* extraradical structures respond to a 48-hour organic N limitation with transcriptional activation of genes involved in different functional categories and has supported the role of N as a signaling molecule in AM fungi (Cappellazzo et al., 2007).

The N compound, nitric oxide (NO), is emerging as a biological mediator that plays an important role in key physiological processes in plants systems (Besson-Bard et al., 2008, Section D3). NO is known to participate in plant-pathogen interactions (Delledonne et al., 1998; Zeidler et al., 2004; Van Baarlen et al., 2007), and evidence is emerging of NO involvement in plant symbioses, in particular, in the early (Shimoda et al., 2005; Nagata et al., 2008; Nagata et al., 2009) and late stages of the nitrogen-fixing symbiosis (NFS) that legumes establish with rhizobia (Baudouin et al., 2006; Pii et al., 2007). Despite these cellular and molecular analogies between NFS and AM (see Chapter 16), no information is currently available on the involvement of NO in AM establishment or function. Interestingly, nitrate reductase and nitrite reductase transcripts accumulate upon hyphopodium formation (Weidmann et al., 2004; Gianinazzi-Pearson et al., 2008). In addition to the crucial role of their protein products in nitrate assimilation, these enzymes have also been described as potential sources of NO (Yamasaki and Sakihama, 2000; Stöhr and Stremlau, 2006). Their activation during this early stage of the interaction suggests their involvement in NO metabolism rather than nitrogen assimilation (Gianinazzi-Pearson et al., 2008). Thanks to the use of the specific fluorescent probe 4,5-diaminofluorescein diacetate, NO accumulation has been recently detected in roots of *M. truncatula* during the first minutes following treatment with AM fungal exudates (Calcagno et al., in preparation). The analysis of mutant lines defective in the AM colonization process suggest the involvement of plant-produced NO in the signaling pathway that mediates early stages of the AM interaction.

Conclusions and Perspectives

Genomic Analysis of N Nutrition

Recent studies have unambiguously highlighted the role of the AM symbiosis in plant N nutrition. Several aspects still have to be investigated concerning both the fungal and plant sides. There is a clear need to investigate the spatial and temporal distribution of mycorrhiza-induced plant transporters and to analyze the specific contribution of their activity in plant N acquisition. This will also allow exploration of possible parallels with mycorrhiza-specific Pi transporters. Interestingly, P and N are transferred by similar mechanisms in AM roots. Polyphosphates (poly-P) and Arg are stored in vacuoles, translocated along hyphae from the extra- to the intra-radical mycelium and ionic forms (Pi and NH_4^+) are released to the plant apoplast where they are then taken up by mycorrhiza-induced plant transporters (Javot et al., 2007). The observation that an AM fungal Pi transporter is also regulated by the presence of N (Olsson et al., 2005) supports possible links between P and N. In addition, Arg has been proposed to act as a counter-ion for the co-translocation of poly-P (Bago et al., 2001).

An overarching significance of N transfer in AM symbiosis, again in analogy to what has been proposed for Pi, can also be hypothesized for N (Fitter, 2006; Helgason and Fitter, 2009). Is the carbon allocation from the plant to the fungus influenced by the capability of the fungus to deliver N?

Knowledge of the interdependence between P and N transfer and its relationship with C allocation in AM symbiosis will surely be increased by exploiting plant mutants defective in specific transport activities.

On the fungal side the *Glomus intraradices* genome project will offer a valuable chance of increasing our knowledge on the N metabolism of AM fungi, in spite of the experimental difficulties inherent in the biological peculiarities of AM fungi (Martin et al., 2008b). Based on what has emerged from the genome sequencing of two ectomycorrhizal fungi (Martin et al., 2008a; Martin et al., 2010) it can be can predicted that the catalogue of the protein-coding genes involved in N uptake and assimilation will shed light on some of the most crucial mechanisms active during AM symbiosis.

Possibile Bacterial Roles in the AM-Plant Association

It is worth considering that in natural conditions, bacteria associate with mycorrhizal fungi, colonize the surface of extraradical hyphae (Scheublin et al., 2010) and, at least in some fungal taxa, live in the cytoplasm as endobacteria. Bacteria therefore seem to represent the third component (at last count) of mycorrhizal associations and most likely play a role in mycorrhizal function (Garbaye, 1994; Bonfante and Anca, 2009). Garbaye (1994) opened this field with the widely acknowledged term "helper bacteria," which defined those bacteria that support mycorrhizas establishment. However, new knowledge and insight have been added, leading to a new scenario. There is in fact increasing evidence that bacteria-fungi interactions are more widespread than expected and that their dynamics may be crucial in ecosystems (Bonfante and Anca, 2009). Although the exact nature of the complex interspecies/interphylum interactions remains unclear, a great deal of evidence supports the current trend to view mycorrhizas as tripartite associations. A first attempt to analyze this issue was made by Toljander and other (2008) who showed changes in community composition of AM fungi and bacteria after long-term application of different organic and mineral N fertilizers. It is possible to speculate that N bioavailability to AM fungi may be affected to a great extent by mycorrhiza-associated bacteria. Taken as a whole, the presence of these bacteria may offer a new level of similarity between AMs and N fixing symbioses (Chapter 2). It can be predicted that this knowledge will have an important and positive impact on the use of AM fungi in sustainable agrosystems.

Acknowledgements

Contributions to this review were partly funded by the project Converging Technologies-BioBITs, funded by CIPE- Regione Piemonte and by PRIN-MIUR 2008 to PB, and by a University grant (60%) to LL. MG was funded by the European Project Integral and by BioBITs.

References

Ames, R.N., Reid, C.P.P., Porter, L.K., Cambardella, C. 1983. Hyphal uptake and transport of nitrogen from two ^{15}N-labelled sources by *Glomus mosseae*, a vesicular-arbuscular mycorrhizal fungus. New Phytologist 95: 381–396.

Bago, B. 2000. Putative sites for nutrient uptake in arbuscular mycorrhizal fungi. Springer, Dordrecht, PAYS-BAS.

Bago, B., Cano, C., Azcón-Aguilar, C., Samson, J., Coughlan, A.P., Piché, Y. 2004. Differential morphogenesis of the extraradical mycelium of an arbuscular mycorrhizal fungus grown monoxenically on spatially heterogeneous culture media, Vol 96. Mycological Society of America, Lawrence, KS, ETATS-UNIS.

Bago, B., Pfeffer, P., Shachar-Hill, Y. 2001. Could the urea cycle be translocating nitrogen in the arbuscular mycorrhizal symbiosis? New Phytologist 149: 4–8.

Bago, B., Vierheilig, H., Piché, Y., Azcón-Aguilar, C. 1996. Nitrate depletion and pH changes induced by the extraradical mycelium of the arbuscular mycorrhizal fungus *Glomus intraradices* grown in monoxenic culture. New Phytologist 133: 273–280.

Bahn, Y.S., Xue, C., Idnurm, A., Rutherford, J.C., Heitman, J., Cardenas, M.E. 2007. Sensing the environment: lessons from fungi. Nature Review Microbiology 5: 57–69.

Balestrini, R., Gomez-Ariza, J., Lanfranco, L., Bonfante, P. 2007. Laser microdissection reveals that transcripts for five plant and one fungal phosphate transporter genes are contemporaneously present in arbusculated cells. Molecular Plant-Microbe Interactions 20: 1055–1062.

Balestrini, R., Lanfranco, L. 2006. Fungal and plant gene expression in arbuscular mycorrhizal symbiosis. Mycorrhiza 16: 509–524.

Baudouin, E., Pieuchot, L., Engler, G., Pauly, N., Puppo, A. 2006. Nitric oxide is formed in *Medicago truncatula-Sinorhizobium meliloti* functional nodules. Molecular Plant-Microbe Interations 19: 970–975.

Benedetto, A., Magurno, F., Bonfante, P., Lanfranco, L. 2005. Expression profiles of a phosphate transporter gene (GmosPT) from the endomycorrhizal fungus *Glomus mosseae* Mycorrhiza 15: 620–627.

Benedito, V.A., Li, H., Dai, X., Wandrey, M., He, J., Kaundal, R., Torres-Jerez, I., Gomez, S.K., Harrison, M.J., Tang, Y., Zhao, P.X., Udvardi, M.K. 2010. Genomic inventory and transcriptional analysis of Medicago truncatula transporters. Plant Physiology 152: 1716–1730.

Berta, G., Fusconi, A., Hooker, J.E. 2002. Arbuscular mycorrhizal modifications to plant root systems: scale, mechanisms and consequences. *In:* Gianinazzi, S., Schuepp, H., Barea, J.M., Haselwandter, K., Eds, Mycorrhizal Technology: from Genes to Bioproducts. Birkhäuser Verlag, Basel-Boston-Berlin, pp 71–85.

Besson-Bard, A., Pugin, A., Wendehenne, D. 2008. New insights into nitric oxide signaling in plants. Annual Review of Plant Biology 59: 21–39.

Biswas, K., Morschhauser, J. 2005. The Mep2p ammonium permease controls nitrogen starvation-induced filamentous growth in *Candida albicans*. Mol Microbiol 56: 649–669.

Bonfante, P., Anca, I.A. 2009. Plants, mycorrhizal fungi, and bacteria: a network of interactions. Annual Review of Microbiology 63: 363–383.

Bonfante, P., Genre, A. 2008. Plants and arbuscular mycorrhizal fungi: an evolutionary-developmental perspective. Trends in Plant Science 13: 492–498.

Bradley, K., Drijber, R.A., Knops, J. 2006. Increased N availability in grassland soils modifies their microbial communities and decreases the abundance of arbuscular mycorrhizal fungi, Vol 38. Elsevier, Oxford, ROYAUME-UNI.

Breuninger, M., Trujillo, C.G., Serrano, E., Fischer, R., Requena, N. 2004. Different nitrogen sources modulate activity but not expression of glutamine synthetase in arbuscular mycorrhizal fungi. Fungal Genetics and Biology 41: 542–552.

Bucher, M. 2007. Functional biology of plant phosphate uptake at root and mycorrhiza interfaces. New Phytologist 173: 11–26.

Bucher, M., Wegmüller, S., Drissner, D. 2009. Chasing the structures of small molecules in arbuscular mycorrhizal signaling. Current Opinion in Plant Biology 12: 500–507.

Burleigh, S.H. 2001. Relative quantitative RT-PCR to study the expression of plant nutrient transporters in arbuscular mycorrhizas. Plant Science 160: 899–904.

Cappellazzo, G., Lanfranco, L., Bonfante, P. 2007. A limiting source of organic nitrogen induces specific transcriptional responses in the extraradical structures of the endomycorrhizal fungus *Glomus intraradices*. Current Genetics 51: 59–70.

Cappellazzo, G., Lanfranco, L., Fitz, M., Wipf, D., Bonfante, P. 2008. Characterization of an amino acid permease from the endomycorrhizal fungus *Glomus mosseae*. Plant Physiology 147: 429–437.

Chalot, M., Blaudez, D., Brun, A. 2006. Ammonia: a candidate for nitrogen transfer at the mycorrhizal interface. Trends in Plant Science 11: 263–266.

Cliquet, J.B., Murray, P.J., Boucaud, J. 1997. Effect of the arbuscular mycorrhizal fungus *Glomus fasciculatum* on the uptake of amino nitrogen by *Lolium perenne*. New Phytologist 137: 345–349.

Couturier, J., Montanini, B., Martin, F., Brun, A., Blaudez, D., Chalot, M. 2007. The expanded family of ammonium transporters in the perennial poplar plant. New Phytologist 174: 137–150.

Cruz, C., Egsgaard, H., Trujillo, C., Ambus, P., Requena, N., Martins-Loucao, M.A., Jakobsen, I. 2007. Enzymatic evidence for the key role of arginine in nitrogen translocation by arbuscular mycorrhizal fungi. Plant Physiology 144: 782–792.

D'Apuzzo, E., Rogato, A., Simon-Rosin, U., El Alaoui, H., Barbulova, A., Betti, M., Dimou, M., Katinakis, P., Marquez, A., Marini, A.M., Udvardi, M.K., Chiurazzi, M 2004. Characterization of three functional high-affinity ammonium transporters in *Lotus japonicus* with differential transcriptional regulation and spatial expression. Plant Physiology 134: 1763–1774.

Delledonne, M., Xia, Y., Dixon, R.A., Lamb, C. 1998. Nitric oxide functions as a signal in plant disease resistance. Nature 394: 585–588.

Egerton-Warburton, L.M., Graham, R.C., Allen, E.B., Allen, M.F. 2001. Reconstruction of the historical changes in mycorrhizal fungal communities under anthropogenic nitrogen deposition. Proceedings. Biological Sciences The Royal Society 268: 2479–2484.

Fitter, A.H. 2006. What is the link between carbon and phosphorus fluxes in arbuscular mycorrhizas? A null hypothesis for symbiotic function. New Phytologist 172: 3–6.

Frenzel, A., Manthey, K., Perlick, A.M., Meyer, F., Pühler, A., Küster, H., Krajinski, F. 2005. Combined transcriptome profiling reveals a novel family of arbuscular mycorrhizal-specific Medicago truncatula lectin genes. Molecular Plant-Microbe Interactions 18: 771–782.

Frey, B., Schuepp, H. 1993. Acquisition of nitrogen by external hyphae of arbuscular mycorrhizal fungi associated with *Zea mays* New Phytologist 124: 221–230.

Garbaye, J. 1994. Helper bacteria—a new dimension to the mycorrhizal symbiosis. New Phytologist 128: 197–210.

George, E., Haussler, K.-U., Vetterlein, D., Gotgus, E., Marschner, H. 1992. Water and nutrient translocation by hyphae of *Glomus mosseae*, Vol 70. National Research Council of Canada, Ottawa, Canada.

Gianinazzi-Pearson, V., Arnould, C., Oufattole, M., Arango, M., Gianinazzi, S. 2000. Differential activation of H^+-ATPase genes by an arbuscular mycorrhizal fungus in root cells of transgenic tobacco. Planta 211: 609–613.

Gianinazzi-Pearson, V., Séjalon-Delmas, N., Genre, A., Jeandroz, S., Bonfante, P. 2008. Plants and arbuscular mycorrhizal fungi: cues and communication in the early steps of symbiotic interactions. Advances in Botanical Research 46: 181–219.

Girlanda, M., Perotto, S., Bonfante, P. 2007. Mycorrhizal fungi: their habitats and nutritional strategies. In Kubicek, C.P., Druzhinina, I.S., Eds., The mycota IV—environmental and microbial relationships, 2nd Ed. Springer, Berlin, Germany, pp. 229–256.

Glass, A.D.M., Erner, Y., Kronzucker, H.J., Schjoerring, J.K., Siddiqi, M.Y., Wang, M.Y. 1997. Ammonium fluxes into plant roots: Energetics, kinetics and regulation. Zeitschrift fur Pflanzenernahrung und Bodenkunde 160: 261–268.

Gomez, S.K., Javot, H., Deewatthanawong, P., Torres-Jerez, I., Tang, Y., Blancaflor, E.B., Udvardi, M.K., Harrison, M.J. 2009. *Medicago truncatula* and *Glomus intraradices* gene expression in cortical cells harboring arbuscules in the arbuscular mycorrhizal symbiosis. BMC Plant Biology 9: 10.

Govindarajulu, M., Pfeffer, P.E., Jin, H., Abubaker, J., Douds, D.D., Allen, J.W., Bucking, H., Lammers, P.J., Shachar-Hill, Y. 2005. Nitrogen transfer in the arbuscular mycorrhizal symbiosis. Nature 435: 819–823.

Guether, M., Balestrini, R., Hannah, M.A., Udvardi, M.K., Bonfante, P. 2009a. Genome-wide reprogramming of regulatory networks, transport, cell wall and membrane biogenesis during arbuscular mycorrhizal symbiosis in *Lotus japonicus* New Phytologist 182: 200–212.

Guether, M., Neuhäuser, B., Balestrini, R., Dynowski, M., Ludewig, U., Bonfante, P. 2009b. A mycorrhizal-specific ammonium transporter from *Lotus japonicus* acquires nitrogen released by arbuscular mycorrhizal fungi. Plant Physiology 150: 73–83.

Güimil, S., Chang, H.S., Zhu, T., Sesma, A., Osbourn, A., Roux, C., Ionnidis, V., Oakeley, E.J., Docquier, M., Descombes, P., Briggs, S.P., Paszkowski, U. 2005. Comparative transcriptomics of rice reveals an ancient pattern of response to microbial colonization. Proceedings of the National Academy of Sciences of the United States of America 102: 8066–8070.

Hampp, R., Wiese, J., Mikolajewski, S., Nehls, U. 1999. Biochemical and molecular aspects of C/N interaction in ectomycorrhizal plants: an update. Springer, Dordrecht, PAYS-BAS.

Harrison, M.J., van Buuren, M.L. 1995. A phosphate transporter from the mycorrhizal fungus *Glomus versiforme*. Nature 378: 626–629.

Hawkins, H.J., Johansen, A., George, E. 2000. Uptake and transport of organic and inorganic nitrogen by arbuscular mycorrhizal fungi. Plant and Soil 226: 275–285.

Helgason, T., Fitter, A.H. 2009. Natural selection and the evolutionary ecology of the arbuscular mycorrhizal fungi (Phylum Glomeromycota). Journal of Experimental Botany 60: 2465–2480.

Hildebrandt, U., Schmelzer, E., Bothe, H. 2002. Expression of nitrate transporter genes in tomato colonized by an arbuscular mycorrhizal fungus. Physiologia Plantarum 115: 125–136.

Hodge, A., Campbell, C.D., Fitter, A.H. 2001. An arbuscular mycorrhizal fungus accelerates decomposition and acquires nitrogen directly from organic material. Nature 413: 297–299.

Hohnjec, N., Vieweg, M.E., Pühler, A., Becker, A., Küster, H. 2005. Overlaps in the transcriptional profiles of *Medicago truncatula* roots inoculated with two different *Glomus* fungi provide insights into the genetic program activated during arbuscular mycorrhiza. Plant Physiology 137: 1283–1301.

Holm, L.M., Jahn, T.P., Moller, A.L., Schjoerring, J.K., Ferri, D., Klaerke, D.A., Zeuthen, T. 2005. NH_3 and NH_4^+ permeability in aquaporin-expressing *Xenopus* oocytes. Pflügers Archiv European Journal of Physiology 450: 415–428.

Hong, Y.Y., Devaiah, S.P., Bahn, S.C., Thamasandra, B.N., Li, M.Y., Welti, R., Wang, X.M. 2009. Phospholipase D epsilon and phosphatidic acid enhance Arabidopsis nitrogen signaling and growth. Plant Journal 58: 376–387.

Jahn, T.P., Moller, A.L., Zeuthen, T., Holm, L.M., Klaerke, D.A., Mohsin, B., Kuhlbrandt, W., Schjoerring, J.K. 2004. Aquaporin homologues in plants and mammals transport ammonia. FEBS Letters 574: 31–36.

Jansa, J., Smith, F.A., Smith, S.E. 2008. Are there benefits of simultaneous root colonization by different arbuscular mycorrhizal fungi? New Phytologist 177: 779–789.

Javot, H., Pumplin, N., Harrison, M.J. 2007. Phosphate in the arbuscular mycorrhizal symbiosis: transport properties and regulatory roles. Plant Cell and Environment 30: 310–322.

Jin, H., Pfeffer, P.E., Douds, D.D., Piotrowski, E., Lammers, P.J., Shachar-Hill, Y. 2005. The uptake, metabolism, transport and transfer of nitrogen in an arbuscular mycorrhizal symbiosis. New Phytologist 168: 687–696.

Johansen, A., Finlay, R.D., Olsson, P.A. 1996. Nitrogen metabolism of external hyphae of the arbuscular mycorrhizal fungus *Glornus intraradices*. New Phytologist 133: 705–712.

Johansen, A., Jakobsen, I., Jensen, E.S. 1992. Hyphal transport of ^{15}N-labelled nitrogen by a vesicular-arbuscular mycorrhizal fungus and its effect on depletion of inorganic soil N. New Phytologist 122: 281–288.

Johansen, A., Jakobsen, I., Jensen, E.S. 1993. External hyphae of vesicular-arbuscular mycorrhizal fungi associated with Trifolium subterraneum L. New Phytologist 124: 61–68.

Jumpponen, A., Trowbridge, J., Mandyam, K., Johnson, L. 2005. Nitrogen enrichment causes minimal changes in arbuscular mycorrhizal colonization but shifts community composition: evidence from rDNA data, Vol 41. Springer, Berlin, ALLEMAGNE.

Krajinski, F., Biela, A., Schubert, D., Gianinazzi-Pearson, V., Kaldenhoff, R., Franken, P. 2000. Arbuscular mycorrhiza development regulates the mRNA abundance of Mtaqp1 encoding a mercury-insensitive aquaporin of *Medicago truncatula* Planta 211: 85–90.

Krajinski, F., Hause, B., Gianinazzi-Pearson, V., Franken, P. 2002. Mtha 1, a plasma membrane H$^+$-ATPase gene from *Medicago truncatula*, shows arbuscule-specific induced expression in mycorrhizal tissue. Plant Biology 4: 754–761.

Küster, H., Hohnjec, N., Krajinski, F., El Yahyaoui, F., Manthey, K., Gouzy, J., Dondrup, M., Meyer, F., Kalinowski, J., Brechenmacher, L., van Tuinen, D., Gianinazzi-Pearson, V., Pühler, A., Gamas, P., Becker, A. 2004. Construction and validation of cDNA-based Mt6k-RIT macro- and microarrays to explore root endosymbioses in the model legume *Medicago truncatula*. Journal of Biotechnology 108: 95–113.

Leigh, J., Hodge, A., Fitter, A.H. 2009. Arbuscular mycorrhizal fungi can transfer substantial amounts of nitrogen to their host plant from organic material. New Phytologist 181: 199–207.

Liu, J., Maldonado-Mendoza, I., Lopez-Meyer, M., Cheung, F., Town, C.D., Harrison, M.J. 2007. Arbuscular mycorrhizal symbiosis is accompanied by local and systemic alterations in gene expression and an increase in disease resistance in the shoots. The Plant Journal 50: 529–544.

Lopez-Pedrosa, A., Gonzalez-Guerrero, M., Valderas, A., Azcon-Aguilar, C., Ferrol, N. 2006. GintAMT1 encodes a functional high-affinity ammonium transporter that is expressed in the extraradical mycelium of *Glomus intraradices* Fungal Genetics and Biology 43: 102–110.

Loque, D., Ludewig, U., Yuan, L., von Wiren, N. 2005. Tonoplast intrinsic proteins AtTIP2;1 and AtTIP2;3 facilitate NH3 transport into the vacuole. Plant Physiology 137: 671–680.

Lorenz, M.C., Heitman, J. 1998. The MEP2 ammonium permease regulates pseudohyphal differentiation in Saccharomyces cerevisiae. The EMBO Journal 17: 1236–1247.

Ludewig, U., Neuhauser, B., Dynowski, M. 2007. Molecular mechanisms of ammonium transport and accumulation in plants. FEBS Letters 581: 2301–2308.

Maldonado-Mendoza, I.E., Dewbre, G.R., Harrison, M.J. 2001. A phosphate transporter gene from the extra-radical mycelium of an arbuscular mycorrhizal fungus *Glomus intraradices* is regulated in response to phosphate in the environment. Molecular Plant-Microbe Interactions 14: 1140–1148.

Manthey, K., Krajinski, F., Hohnjec, N., Firnhaber, C., Puhler, A., Perlick, A.M., Kuster, H. 2004. Transcriptome profiling in root nodules and arbuscular mycorrhiza identifies a collection of novel genes induced during *Medicago truncatula* root endosymbioses. Molecular Plant-Microbe Interactions 17: 1063–1077.

Martin, F., Aerts, A., Ahren, D., Brun, A., Danchin, E.G., Duchaussoy, F., Gibon, J., Kohler, A., Lindquist, E., Pereda, V., Salamov, A., Shapiro, H.J., Wuyts, J., Blaudez, D., Buee, M., Brokstein, P., Canback, B., Cohen, D., Courty, P.E., Coutinho, P.M., Delaruelle, C., Detter, J.C., Deveau, A., DiFazio, S., Duplessis, S., Fraissinet-Tachet, L., Lucic, E., Frey-Klett, P., Fourrey, C., Feussner, I., Gay, G., Grimwood, J., Hoegger, P.J., Jain, P., Kilaru, S., Labbe, J., Lin, Y.C., Legue, V., Le Tacon, F., Marmeisse, R., Melayah, D., Montanini, B., Muratet, M., Nehls, U., Niculita-Hirzel, H., Oudot-Le Secq, M.P., Peter, M., Quesneville, H., Rajashekar, B., Reich, M., Rouhier, N., Schmutz, J., Yin, T., Chalot, M., Henrissat, B., Kues, U., Lucas, S., Van de Peer, Y., Podila, G.K., Polle, A., Pukkila, P.J., Richardson, P.M., Rouze, P., Sanders, I.R., Stajich, J.E., Tunlid, A., Tuskan, G., Grigoriev, I.V. 2008a. The genome of *Laccaria bicolor* provides insights into mycorrhizal symbiosis. Nature 452: 88–92.

Martin, F., Gianinazzi-Pearson, V., Hijri, M., Lammers, P., Requena, N., Sanders, I.R., Shachar-Hill, Y., Shapiro, H., Tuskan, G.A., Young, J.P. 2008b. The long hard road to a completed *Glomus intraradices* genome. New Phytologist 180 (4): 747–750.

Martin, F., Kohler, A., Murat, C., Balestrini, R., Coutinho, P.M., Jaillon, O., Montanini, B., Morin, E., Noel, B., Percudani, R., Porcel, B., Rubini, A., Amicucci, A., Amselem, J., Anthouard, V., Arcioni, S., Artiguenave, F., Aury, J.M., Ballario, P., Bolchi, A., Brenna, A., Brun, A., Buee, M., Cantarel, B., Chevalier, G., Couloux, A., Da Silva, C., Denoeud, F., Duplessis, S., Ghignone, S., Hilselberger, B., Iotti, M., Marcais, B., Mello, A., Miranda, M., Pacioni, G., Quesneville, H., Riccioni, C., Ruotolo, R., Splivallo, R., Stocchi, V., Tisserant, E., Viscomi, A.R., Zambonelli, A., Zampieri, E., Henrissat, B., Lebrun, M.H., Paolocci, F., Bonfante, P., Ottonello, S., Wincker, P. 2010. Perigord black truffle genome uncovers evolutionary origins and mechanisms of symbiosis. Nature 464: 1033–1038.

Maurel, C., Verdoucq, L., Luu, D.T., Santoni, V. 2008. Plant aquaporins: membrane channels with multiple integrated functions. Annual Review of Plant Biology 59: 595–624.

Murphy, P.J., Langridge, P., Smith, S.E. 1997. Cloning plant genes differentially during colonization of roots of *Hordeum vulgare* by the vesicular-arbuscular mycorrhizal fungus *Glomus intraradices* New Phytologist 135: 291–301.

Nagata, M., Hashimoto, M., Murakami, E., Shimoda, Y., Shimoda-Sasakura, F., Kucho, K., Suzuki, A., Abe, M., Higashi, S., Uchiumi, T. 2009. A possible role of class 1 plant hemoglobin at the early stage of legume-rhizobium symbiosis. Plant Signaling and Behavior 4: 202–204.

Nagata, M., Murakami, E., Shimoda, Y., Shimoda-Sasakura, F., Kucho, K., Suzuki, A., Abe, M., Higashi, S., Uchiumi, T. 2008. Expression of a class 1 hemoglobin gene and production of nitric oxide in response to symbiotic and pathogenic bacteria in *Lotus japonicus*. Molecular Plant-Microbe Interactactions 21: 1175–1183.

Neuhäuser, B., Dynowski, M., Ludewig, U. 2009. Channel-like NH3 flux by ammonium transporter AtAMT2. FEBS Letters 583: 2833–2838.

Oláh, B., Briere, C., Becard, G., Denarie, J., Gough, C. 2005. Nod factors and a diffusible factor from arbuscular mycorrhizal fungi stimulate lateral root formation in *Medicago truncatula* via the DMI1/DMI2 signaling pathway. The Plant Journal 44: 195–207.

Olsson, P.A., Burleigh, S.H., van Aarle, I.M. 2005. The influence of external nitrogen on carbon allocation to Glomus intraradices in monoxenic arbuscular mycorrhiza. New Phytologist 168: 677–686.

Parniske, M. 2008. Arbuscular mycorrhiza: the mother of plant root endosymbioses. Nature Review of Microbiology 6: 763–775.

Paungfoo-Lonhienne, C., Lonhienne, T.G.A., Rentsch, D., Robinson, N., Christie, M., Webb, R.I., Gamage, H.K., Carroll, B.J., Schenk, P.M., Schmidt, S. 2008. Plants can use protein as a nitrogen source without assistance from other organisms. Proceedings of the National Academy of Sciences of the United States of America 105: 4524–4529.

Pii, Y., Crimi, M., Cremonese, G., Spena, A., Pandolfini, T. 2007. Auxin and nitric oxide control indeterminate nodule formation. BMC Plant Biology 7:21 doi:10.1186/1471-2229-7-21.

Porras-Alfaro, A., Herrera, J., Natvig, D.O., Sinsabaugh, R.L. 2007. Effect of long-term nitrogen fertilization on mycorrhizal fungi associated with a dominant grass in a semiarid grassland, Vol 296. Springer, Dordrecht, PAYS-BAS.

Requena, N., Breuninger, M., Franken, P., Ocon, A. 2003. Symbiotic status, phosphate, and sucrose regulate the expression of two plasma membrane H^+-ATPase genes from the mycorrhizal fungus *Glomus mosseae*. Plant Physiology 132: 1540–1549.

Ruiz-Lozano, J.M. 2003. Arbuscular mycorrhizal symbiosis and alleviation of osmotic stress. New perspectives for molecular studies. Mycorrhiza 13: 309–317.

Sawers, J.H.R., Yang, S., Gutjahr, C., Paszkowski, U. 2008. The molecular components of nutrient exchange in arbuscular mycorrhizal interaction. *In:* Siddiqui, Akhtar and Futai Eds., Mycorrhizae: Sustainable Agriculture and Forestry. Springer, Dordrecht, The Netherlands, pp. 37–60.

Scheublin, T.R., Sanders, I.R., Keel, C., van der Meer, J.R. 2010. Characterisation of microbial communities colonising the hyphal surfaces of arbuscular mycorrhizal fungi. The ISME Journal 4: 752–763.

Shimoda, Y., Nagata, M., Suzuki, A., Abe, M., Sato, S., Kato, T., Tabata, S., Higashi, S., Uchiumi, T. 2005. Symbiotic rhizobium and nitric oxide induce gene expression of non-symbiotic hemoglobin in Lotus japonicus. Plant Cell Physiology 46: 99–107.

Smith, S.E., Read, D.J. 2008. Mycorrhizal symbiosis, 3rd Ed. Academic Press, London.

Sohlenkamp, C., Wood, C.C., Roeb, G.W., Udvardi, M.K. 2002. Characterization of Arabidopsis AtAMT2, a high-affinity ammonium transporter of the plasma membrane. Plant and Physiology 130: 1788–1796.

Stöhr, C., Stremlau, S. 2006. Formation and possible roles of nitric oxide in plant roots. Journal of Experimental Botany 57: 463–470.

Suenaga, A., Moriya, K., Sonoda, Y., Ikeda, A., Von, W.N., Hayakawa, T., Yamaguchi, J., Yamaya, T. 2003. Constitutive expression of a novel-type ammonium transporter OsAMT2 in rice plants. Plant and Cell Physiology 44: 206–211.

Tamura, K., Dudley, J., Nei, M., Kumar, S. 2007. MEGA4: molecular evolutionary genetics analysis (MEGA) software version 4.0. Molecular Biology and Evolution 24: 1596–1599.

Tian, C., Kasiborski, B., Koul, R., Lammers, P.J., Bücking, H., Shachar-Hill, Y. 2010. Regulation of the nitrogen transfer pathway in the arbuscular mycorrhizal symbiosis: Gene characterization and the coordination of expression with nitrogen flux. Plant Physiology 153: 1175–1187.

Tobar, R., Azcón, R., Barea, J.M. 1994. Improved nitrogen uptake and transport from ^{15}N-labelled nitrate by external hyphae of arbuscular mycorrhiza under water-stressed conditions. New Phytologist 126: 119–122.

Toljander, J.F., Santos-Gonzalez, J.C., Tehler, A., Finlay, R.D. 2008. Community analysis of arbuscular mycorrhizal fungi and bacteria in the maize mycorrhizosphere in a long-term fertilization trial. FEMS Microbiology Ecology 65: 323–338.

Uehlein, N., Fileschi, K., Eckert, M., Bienert, G.P., Bertl, A., Kaldenhoff, R. 2007. Arbuscular mycorrhizal symbiosis and plant aquaporin expression. Phytochemistry 68: 122–129.

Van Baarlen, P., Woltering, E.J., Staats, M., Van Kan, J.A. 2007. Histochemical and genetic analysis of host and non-host interactions of Arabidopsis with three Botrytis species: an important role for cell death control. Molecular Plant Pathology 8: 41–54.

Vidal, E.A., Gutierrez, R.A. 2008. A systems view of nitrogen nutrient and metabolite responses in Arabidopsis. Current Opinion in Plant Biology 11: 521–529.

von Wirén, N., Merrick, M. 2004. Regulation and function of ammonium carriers in bacteria, fungi, and plants. Topics in Current Genetics 9: 95–120.

Walch-Liu, P., Ivanov, I.I., Filleur, S., Gan, Y.B., Remans, T., Forde, B.G. 2006. Nitrogen regulation of root branching. Annals of Botany 97: 875–881.

Weidmann, S., Sanchez, L., Descombin, J., Chatagnier, O., Gianinazzi, S., Gianinazzi-Pearson, V. 2004. Fungal elicitation of signal transduction-related plant genes precedes mycorrhiza establishment and requires the dmi3 gene in *Medicago truncatula*. Molecular Plant-Microbe Interactions 17: 1385–1393.

Yamasaki, H., Sakihama, Y. 2000. Simultaneous production of nitric oxide and peroxynitrite by plant nitrate reductase: in vitro evidence for the NR-dependent formation of active nitrogen species. FEBS Letters 468: 89–92.

Zeidler, D., Zahringer, U., Gerber, I., Dubery, I., Hartung, T., Bors, W., Hutzler, P., Durner, J. 2004. Innate immunity in *Arabidopsis thaliana*: lipopolysaccharides activate nitric oxide synthase (NOS) and induce defense genes. Proceedings of the National Academy of Sciences of the United States of America 101: 15811–15816.

Chapter 4
Ectomycorrhiza and Nitrogen Provision to the Host Tree

Michel Chalot and Claude Plassard

Abbreviations

ABC	ATP-binding cassette
AM	arbuscular mycorrhizal
Amt	ammonium transporter
APC	Amino acid-polyamine-organocations
Ato	ammonium transport outward
DON	dissolved organic N
ECM	ectomycorrhizal
GDH	glutamate dehydrogenase
GS	glutamine synthetase
Mep	methylamine permease
Nir	nitrite reductase
Nr	nitrate reductase
Nrt	nitrate-related transporter
Opt	oligopeptide transporter
Ptr	peptide transporter

In natural ecosystems, the main resources acquired by plants are light and carbon monoxide (CO_2), through photosynthesis in the leaves, and mineral nutrients and water, through root and mycorrhizal uptake. Mycorrhizal interactions established between the root systems of terrestrial plants and hyphae from soilborne fungi are the most ecologically widespread plant symbioses. Arbuscular mycorrhizal (AM) fungi colonize the roots of a host plant intracellularly, whereas ectomycorrhizal (ECM) fungi do not penetrate root cells but rather establish a network between them. AM originated 400 million years ago, coincident with land colonization by plants (Redecker et al., 2000; Heckman et al., 2001). Recent findings and fossil records

Ecological Aspects of Nitrogen Metabolism in Plants, First Edition. Edited by Joe C. Polacco and Christopher D. Todd.
© 2011 by John Wiley & Sons, Inc. Published 2011 by John Wiley & Sons, Inc.

(Hibbett and Matheny, 2009) suggest that the first ECM associations probably involved the Pinaceae. The maximum and minimum ages for the origin of ECM in gymnosperms are therefore set by the split between the lineages leading to gnetophytes and the Pinaceae, which must have occurred by the Permian, about 270 Ma, and the diversification of Pinaceae, which began by the early Cretaceous, ca. 130 Ma.

Symbiotic structures (e.g., mycorrhizas, root nodules) are considered as specialized root structures for nutrient acquisition from nutrient-impoverished soils (Lambers et al., 2008; Smith and Read, 2008). Nutrient transport, namely hyphal absorption from soil solution, nutrient transfer from the fungus to plant, and carbon (C) movement from plant to fungus, is a key feature of mycorrhizal symbioses. Phosphorus (P), nitrogen (N), and carbohydrates are considered to be the main nutrients transferred by the mycorrhizal symbiosis, although transfer of water and trace elements can be very important under certain conditions as well (Smith and Read, 2008). However, differences among mycorrhizal strategies are becoming clearer. Major roles in inorganic P uptake are well established for AM symbiosis, and evidence for involvement in inorganic N nutrition is building. However, AM fungi probably have rather low capacity to release nutrients either from sorbed inorganic forms or from organic combinations; they "scavenge" but do not "mine" the soil. By contrast, ectomycorrhizas play major roles both in uptake of inorganic nutrients from the soil solution and in releasing P and N from organic forms through hydrolysis (Lambers et al., 2008).

Indeed, the expanding ECM mycelium that grows outward from the ECM root (Figure 4.1a) into the surrounding soil (the extraradical mycelium, Figure 4.1a) is a very efficient N scavenger because of its capacity to explore a larger soil volume than do roots alone (Smith and Read, 2008). The ECM symbiosis is indeed crucial for a continuous delivery of plant-available N,

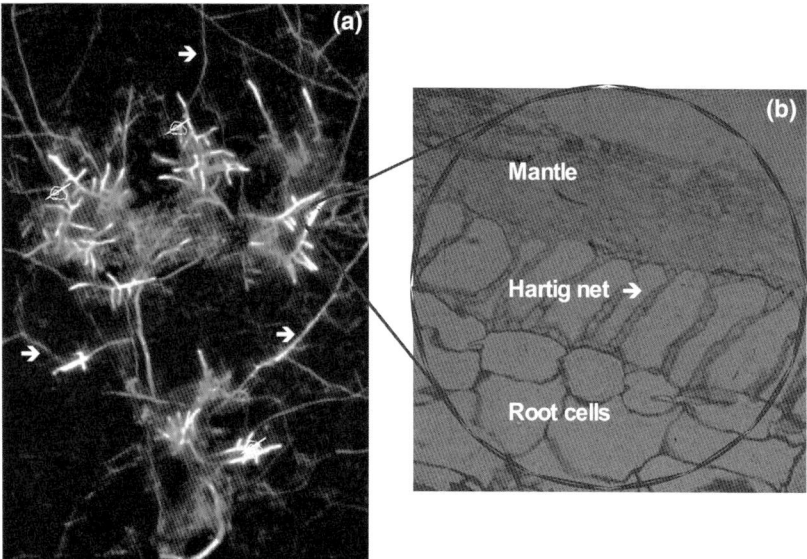

Figure 4.1. Ectomycorrhizal root system from *Betula pendula/Paxillus involutus* association. (a) Intact root system growing on peat. Abundant extraradical mycelia (*arrows*) are expanding from ECM birch roots (*stars*). (b) Cross section of an ectomycorrhiza showing fungal cells forming the mantle, or surrounding the first layer of cortical cells in the Hartig net (*arrow*).

notably because ECM fungi have a demonstrated ability to take up inorganic N and to contribute to depolymerization of N-containing polymers (Chalot and Brun, 1998; Schimel and Bennett 2004; Lambers et al., 2008; Smith and Read, 2008), including N of animal-origin (Klironomos and Hart, 2001). Intimate contact between root and fungal cells within the Hartig net ensures an efficient transfer between the two partners (Figure 4.1b). ECM fungi therefore often enhance resource availability by providing nutrients that are otherwise inaccessible to plant roots (van der Heijden et al., 2008). Pot experiments and field studies have shown that up to 80% of all plant N in boreal forests is derived from ECM fungi (reviewed by van der Heijden et al., 2008).

In addition to the physiological data accumulated on N uptake by the partners and their association, the last decade has witnessed much effort devoted to molecular studies. Molecular and functional characterization of fungal genes has broadened knowledge of N-uptake and assimilation by the mycorrhizae (Chalot et al., 2006; Muller et al., 2007). Studies on a few species of ECM fungi have started to give a good picture of the molecular actors contributing to N uptake and assimilation, but minimal data are available for the endomycorrhizal fungi and even less for the other types of mycorrhizal fungi. *Amanita muscaria*, *Hebeloma cylindrosporum,* and *Tuber borchii* are those ECM fungi for which most is known of expression patterns of genes encoding proteins involved in inorganic and organic N uptake. *H. cylindrosporum* and *A. muscaria* are phylogenetically close (Basidiomycetes, Agaricales); *T. borchii* (an Ascomycetes) is more distant. The strains studied come from various types of soils. The *Tuber* strain comes from a calcareous soil poor in organic matter (high mineralization), the *Hebeloma* strain from a sandy soil (low contents of organic matter and poor mineralization, and the *Amanita* strain comes from a temperate forest (high contents of organic matter and poor mineralization).

Most importantly, the recent sequencing of the genome of *Laccaria bicolor* (Martin et al., 2008) allowed us to identify and present a global view on proteins involved in transport (Lucic et al., 2008) and assimilation of nitrogenous compounds in an ECM fungus. This genomic view has shed light on the symbiotic way of life and the physiology of *L. bicolor* and has allowed us to hypothesize on the forms of N taken up from the soil and transferred to the plant symbiont.

Trees in forest ecosystems from temperate and boreal regions are always associated with ECM fungi, which are believed to play an important role in N uptake and transfer toward the host plant. Since in these forest ecosystems, mineral N fertilization is not often applied, N supply to the plants depends on the N cycle arising from organic matter turnover. As shown in this chapter, ECM fungi can play a role in the numerous steps of N mineralization as well as in uptake of various organic and mineral forms of N whose availability may be highly variable, depending on forest location.

N Availability in Forest Soils

N sources that can be used by plants, their fungal symbionts, and more generally, microorganisms, are either mineral (NH_4^+ and NO_3^-) or organic (mainly small peptides and amino acids). A number of studies have measured the concentrations of these different N sources in soil. Although the absolute concentrations measured in soil solution may vary greatly due to different extraction methods, a number of studies reported that under oceanic and continental climate, nitrate and dissolved organic N (DON) were the two main pools of N constituting up to 90% of total N assayed in soil solution (Jones et al., 2004; Jones and Willett, 2006) with free NH_4^+ and amino acids representing around 10%. With respect to mineral N availability, NO_3^- accumulated in the soil solution, especially in autumn when the forest presents a high degree of fertility (Rothstein 2009) (Figure 4.2). This situation is completely different under boreal and arctic

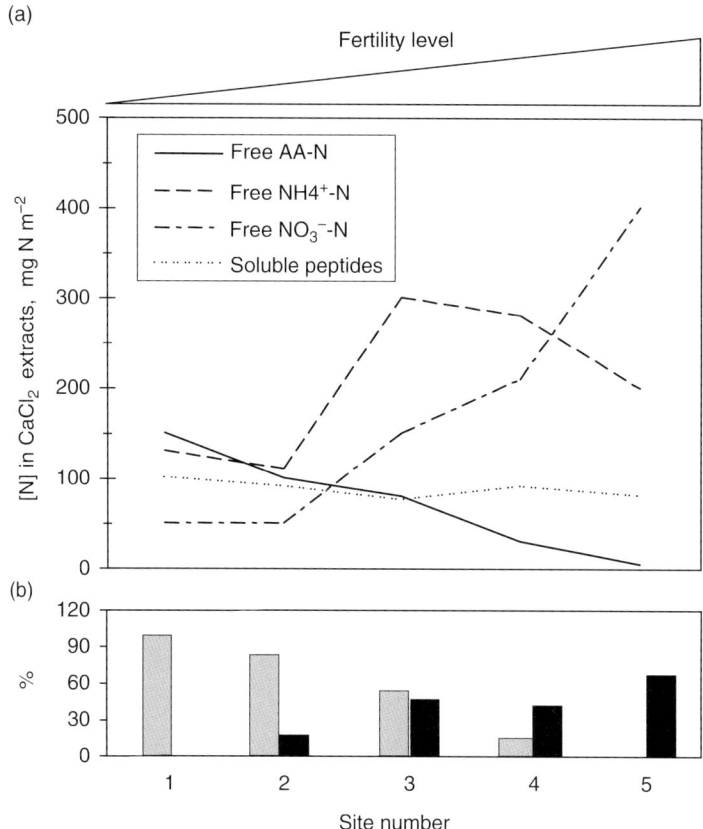

Figure 4.2. (a) Variation of soluble N concentrations assayed in $CaCl_2$ extracts from soils collected in forests of increasing levels of fertility (1 = low fertility, 5 = high fertility). Concentrations given in the figure were the highest ones, and were found in April–May (free amino acids and soluble peptides), June–September (NH_4^+) and September (NO_3^-). (b) Oak (*gray bar*) and maple (*black bar*) as percentages of total inventoried species in the five study sites. Oak (*Q. rubra*, *Q. velutina*, *Q. alba*) percentages and maple (*A. saccharum*, *A. rubrum*) percentages in sites varied inversely and conversely with site fertility, respectively. Redrawn from the data given in Rothstein (2009).

conditions, with DON representing the vast majority of total N (Andersson and Berggren, 2005; Kranabetter et al., 2007; Näsholm et al., 2009), with concentrations of DON typically 10 times higher than those of mineral N. However, even in temperate forests, the proportions of DON can represent a high proportion of mineral N in low fertility sites, as shown by Rothstein (2009) (Figure 4.2). As underlined by Jones and others (2005), such large pools of NO_3^- and DON may occur simply because they are N forms not easily used by plants or microorganisms. If the same argument is made for NH_4^+ and free amino acids, whose concentrations and proportions to total N are always very low, these two N pools should be preferentially taken up. A further complication is that ammonium and free amino acids can be readily adsorbed onto the solid constituents of soil (Barber, 1995; Jones and Hodge, 1999; Qualls and Richardson, 2003; Vieublé Gonod et al., 2006), thus minimizing their concentrations in soil solution.

The rate of flux through N pools is more important than their actual size in soils (Jones et al., 2005). Indeed, using ^{15}N-pool-dilution techniques in short-term experiments has been shown that the inorganic N pools are subjected to high turnover due to microbial use (Kaye and Hart, 1997; Grenon et al., 2005). This was also the case for free amino acids, with half-life expectancy typically ranging from 1–12 h, indicating an extremely fast turnover rate (Jones, 1999; Jones and Kielland, 2002; Jones et al., 2004). Even the taiga forest ecosystems with soil temperature of 10°C or lower showed *in situ* free amino acid pool turnover times of approximately 3–6 h (Kielland et al., 2007). In addition, the measurement of the gross fluxes of amino acid-N over the growing season greatly exceeded the annual vegetation N requirement, suggesting that the pool of amino acids could contribute significantly to the N nutrition of plants (Kielland et al., 2007; Näsholm et al., 2009).

"New" N sources, produced from hydrolysis of complex N sources such as proteins (see Chapter 5), are subjected to competition among several processes that can be physical, such as adsorption on solid phase, and/or biological such as plant or symbiotic fungal uptake or soil microbial utilization. The microbial immobilization of N could be very high, as shown by labeling experiments *in situ*. For example, Clemmensen and others (2008) showed that 90% of an equimolar mixture of ammonium, nitrate, and L-glycine were incorporated into the microbial biomass of two tundra sites. However, these proportions can be much lower in more temperate forest ecosystems, such as those studied by Rothstein (2009) who showed that the microbial compartment contained from 40% (low fertility sites) down to 20% (high fertility sites) of the ^{15}N label, 4 h after addition as ^{15}N-leucine. Conversely, the pool of inorganic N was also rapidly labeled, indicating a fast production of inorganic N *in situ*, especially in high fertility forests (Rothstein, 2009). From these data, it is clear that trees and their ECM fungal symbionts may be supplied with pools of N sources of extremely variable sizes, with a high competition for N capture among the numerous soil organisms including the soil itself.

Organic N Mobilization

Simple organic N compounds, such as small peptides and amino acids, are the preferential forms of N used by plants and their associated fungi. Thus, an appreciable effort has been devoted to examine the capacities of plants to take up organic N (Näsholm et al., 2009). In pioneering work, Read and collaborators (Smith and Read, 1997) demonstrated appreciable growth and N accumulation in young seedlings of ECM *Betula pendula* (Silver Birch) supplied with different organic N sources, in contrast to non-mycorrhizal host plants, which showed restricted capacity to use single amino acids or oligopeptides (Abuzinadah and Read, 1989). The capacity to use organic N, however, may not be restricted to mycorrhizal plants as the non-mycorrhizal plant *Arabidopsis thaliana* and the woody heathland plant *Hakea actites* are able to degrade, take up and assimilate proteins (Paungfoo-Lonhienne et al., 2008). There is now good evidence that dipeptides are also taken up into roots intact, and the ecological relevance of peptides as N sources has now been established for non-mycorrhizal species (Paungfoo-Lonhienne et al., 2009).

Hydrolytic Capacities of ECM Fungi

The first step in the digestion of N-containing organic matter is the cleavage of proteins by proteases. Extracellular protease activities have been measured in ECM fungi (Smith and Read, 1997; Chalot and Brun, 1998; Nehls et al., 2001). In addition, availability of N sources and/or climate conditions appear to select ECM fungal species with proteolytic capacities: Northern boreal forests with raw humus soils had more diverse ECM fungal communities that were more able to develop on proteins supplied as the sole source of N and C than those isolated from southerly locations where N-enrichment occurred (Taylor et al., 2000). Observations by Tibbett and others (1998, 1999) on different strains of the ECM fungus *Hebeloma* also indicated that selection may favor "protein fungi" in arctic and boreal environments; the greatest proteolytic potential was for strains from cold environments, and their proteases had very low thermal optima (0–6°C).

In addition to the mobilization of protein-N by protease-catalyzed release, it would be of great interest to determine whether ECM fungi are also able to take up intact proteins or polypeptides by endocytosis, as recently demonstrated in *Arabidopsis thaliana* and in the woody plant *Hakea actites*, two plants unable to form mycorrhizal associations (Paungfoo-Lonhienne et al., 2008). However, the capacity of trees to take up peptides by endocytosis in the absence of symbionts is highly unlikely as demonstrated in birch (Abuzinadah and Read, 1989). Although the exact contribution of ECM fungi to forest soil protease activity is unknown, it is interesting to note the reported linear relationship between total free amino acids and soil protease activity across successional soils in Alaska differing in their tree composition (Kielland et al., 2007). Remarkably, the lowest levels of protease activity were found in willow and alder forests and the highest levels (12-fold greater) in spruce forests where ECM fungi were the sole fungal species associated with the roots, suggesting a major contribution of ECM to protease activity *in situ*.

Recently, micro-titer plate multiple enzymatic tests were optimized to measure enzyme activities on excised ECM root tips (Courty et al., 2005). Chitinase and leucine aminopeptidase activities were chosen to reflect release of N from organic macromolecules, and were indeed detected in a range of ECM root tips. Remarkably, in the case of the ECM fungus *Tomentella* sp., a shift in distinct activities displayed two ways of N-mobilization from N-containing biopolymers (proteins and chitin): leucine aminopeptidase increase in the organic horizon (Oh), consisting of humic material in advanced stage of decomposition, and chitinase increase in the mineral horizon (Ah), enriched with organic matter. Rineau and Garbaye (2009) showed that the change of ECM community composition due to liming was accompanied by an obvious change of ECM community functioning, estimated as the potential secretion of extracellular enzymes able to degrade organic matter. Among these, leucine aminopeptidase activity was clearly decreased by liming. This was mostly attributed to the reduction of the ECM "specialist" *Xerocomus pruinatus*, the highest contributor to aminopeptidase activity, despite its low abundance.

After protein hydrolysis, a subsequent step in the release of NH_4^+ into the environment would be by deaminase action on amino acids. To our knowledge, there is only one report on the production of free NH_4^+ in the medium by an ECM fungus, *Hebeloma crustuliniforme* (Quoreshi et al., 1995). This net release of NH_4^+ occurred only when the fungus was grown without glucose. The same observation was made on another *Hebeloma* species, *H. cylindrosporum* (Plassard, unpublished data). Deamination of amino acids may be of importance for the fungus under C starvation. Under C-deficient conditions it was recently shown that GDH plays a central role in amino acid breakdown (Miyashita and Good, 2008). However, more work is needed to quantify the importance of this ammonium release *in situ*.

Utilization of Amino Acids

Plassard and others (2000) found that mycorrhizal association between *Pinus pinaster* and *H. cylindrosporum* greatly helps the plant to use L-glutamate. Recently, Clemmensen and others (2008) demonstrated that ECM communities in two tundras (heath and shrub) differed in N-form preferences, with a higher contribution of glycine to total N uptake in the heath tundra, measured by better utilization of ^{15}N-L glycine by *Betula nana*.

The enhancement of amino acid utilization in ECM plants is probably due to fungal ability to use organic N as sole N source, as demonstrated for several ECM fungi (Scheromm et al., 1990; Finlay et al., 1992; Quoreshi et al., 1995; Plassard et al., 2000; Wipf et al., 2002; Guidot et al., 2005). Amino acid uptake was demonstrated for several species of ECM fungi forming ECM tips (Chalot and Brun, 1998; Wallenda and Read, 1999; Boukcim and Plassard, 2003). The apparent Km values were generally lower (hence, higher affinity for amino acids) than those measured in non-mycorrhizal roots, and uptake rates were higher. The differential kinetics may confer on ECM roots or their extraradical hyphae an improved ability to compete with other sinks in soils. However, it should be pointed out that ^{15}N "accumulation" in the root from an organic N source does not imply that it is a good N source. As underlined by Jones and others (2005), root cell walls have negative charges that can adsorb positively charged compounds such as ammonium and the amino acids arginine and lysine, and possibly histidine, depending on pH. Also, it was shown that ECM-free *P. pinaster* plants accumulated root ^{15}N from ^{15}N-labeled glutamate or ^{15}NH$_4^+$ at the same rates over 24 h, although *P. pinaster* plants grew very poorly with L-glutamate without mycorrhizal association (Plassard et al., 2000). This poor growth was due to a very low root to shoot flux of L-glutamate-derived ^{15}N, contrary to ^{15}N originating as ammonium. Thus, to assess accurately the actual utilization of organic N by ECM plants, it is necessary to measure accumulation in the shoots as well as in the roots.

The first ECM amino acid transporter studies were carried out in *A. muscaria* and *H. cylindrosporum*. Gap1, a general amino acid transporter with high affinity for basic amino acids and somewhat lower affinity for neutral and acidic amino acids was identified in *A. muscaria* (Nehls et al., 1999) and in *H. cylindrosporum* (Wipf et al., 2002). All 20 protein amino acids bind to Gap1 as revealed by competition studies indicating a putative broad substrate spectrum, similar to that of the yeast ortholog, ScGap1. In addition, expression of ECM transporters in yeast revealed that they function as high-affinity transporters. Their expression pattern in symbiotic tissues suggests two main functions, namely uptake of amino acid from the soil for nutrition and the prevention of amino acid loss by hyphal leakage in the absence of a suitable N source at a low internal N status.

The genetic potential of *L. bicolor* (ECM fungal model) for modes of amino acid transport was recently addressed (Lucic et al., 2008). Most of the fungal amino acid transporters have been classified into the APC superfamily (Saier et al., 1999). They mediate the transfer of a broad spectrum of amino acids with overlapping specificities. Based on sequence analyses and similarities of putative genes, the APC superfamily was significantly expanded in *L. bicolor* (29 members), compared to the saprophytic (*Coprinopsis cinerea* and *Phanerochaete chrysosporium*, 16 and 22 members, respectively) or parasitic (*Ustilago maydis* and *Cryptococcus neoformans*, 21 and 22 members, respectively) fungi. The expansion of *L. bicolor* APC members could be related to enhanced utilization of organic N (amino acids) and to its dual life style, symbiotic versus saprophytic (Martin et al., 2008).

Utilization of Peptides

Read's group provided the first evidence for peptide utilization by free-living *Laccaria laccata*, *Suillus bovinus*, and *Rhizopogon roseolus* (Abuzinadah and Read, 1986), and by birch plants infected with a range of ECM fungi (Abuzinadah and Read, 1989). In further studies using [^{3}H]-labeled dipeptide, the ECM fungus *H. cylindrosporum* was shown to be able to take up dipeptides and use them as sole N source (Benjdia et al., 2006). Peptide transporters can be grouped into three distinct families based on sequence similarity and mechanism, the Proton-dependent Oligopeptide Transporter (Pot) family, the Oligopeptide Transporter (Opt) family, and the ATP-binding Cassette (ABC) superfamily (Lucic et al., 2008). Two peptide transporter genes (*HcPtr2A* and *HcPtr2B*) were isolated from a *H. cylindrosporum* cDNA library by yeast functional complementation and were shown to mediate dipeptide uptake. Encoded uptake capacities and genetic regulation indicated that HcPtr2A was involved in the high-efficiency peptide uptake under conditions of limited N availability, whereas HcPtr2B was expressed constitutively (Benjdia et al., 2006).

Genome analysis has led to the identification of 10 possible *L. bicolor* Opt orthologs and two orthologs of two Ptrs previously characterized in *H. cylindrosporum* (Benjdia et al., 2006). Interestingly, three *L. bicolor* Opts formed a distinct subgroup with Opts from the two other Basidiomycetes (*P. chrysosporium* and *C. cinerea*) that have known capacities to degrade organic matter. Expression analyses revealed the following regarding the 10 LbOpt genes: (1) four were constitutively expressed, (2) two were specifically highly upregulated in fruiting body mycelium, and (3) two others were upregulated in ECM mycelium. The two LbPtr genes were constitutively expressed in free-living tissues and one of them was highly expressed. These candidate genes may be involved in the constitutive uptake of peptides by mycelium in the free-living condition as well as in ectomycorrhiza (Lucic et al., 2008).

Utilization of Inorganic N

Ammonium Versus Nitrate

Ammonium is a ubiquitous intermediate in N metabolism and one of the major nutrients for plants and microorganisms. Though the soil concentration of the poorly mobile ammonium ion is generally lower than that of nitrate, ammonium is often a preferred N source because of its lower assimilation cost (Marschner, 1995). It is well documented that ECM fungi have a preference for ammonium over nitrate *in vitro* (Rangel-Castro et al., 2002; Guidot et al., 2005) and in the field (Clemmensen et al., 2008), but it is not unusual to find that ECM fungi grow better with nitrate than ammonium (Scheromm et al., 1990; Montanini et al., 2002). In contrast, the ECM fungus *A. muscaria* is not able to grow with nitrate as the sole source of N. The preferential N-source for optimal growth is not only dependent on the species but on the strains studied (Scheromm et al., 1990; Guidot et al., 2005). We can assume that the strains are adapted to (or have been selected by) the available N form in the soil from which they were isolated. These strains must therefore have different ability to take up N. This could help ECM fungi to compete with other soil microbes (bacteria and/or non symbiotic fungi) for the various N sources. However, it must be kept in mind that a harsh competition exists in soil to use available soil N sources. This competition is particularly evident in experiments where a new source of ^{14}N is brought to the soil. Such experiments generally demonstrate that soil microbes (other than ECM fungi) are the first sink for the added N (for example, see Clemmensen et al., 2008).

Early studies conducted on excised and intact ECM root systems demonstrated enhanced ammonium uptake capacity of mycorrhizal-infected plant (Genetet et al., 1984; Rousseau et al., 1994). Based on both ^{14}N labeling and mass balance data, it was demonstrated that hyphal NH_4^+ acquisition contributed 45% of total plant N uptake under N deficiency (Jentschke et al., 2001). The kinetics and energetics of NH_4^+ transport were studied in the ECM fungus *Paxillus involutus* using [^{14}C] methylamine as a labeled ammonium analog (Javelle et al.,

1999). Ammonium transport is mediated by a family of ubiquitous membrane proteins, the Mep/Amt/Rh family, spread throughout all domains of life (Huang and Peng, 2005). Ammonium importers in the fungal plasma membrane could have at least two different functions: uptake of N or prevention of ammonium leakage, consistent with the common presence of multiple ammonium permeases with different kinetic properties within an organism. As in plants, Amt transporters are responsible for NH_4^+ uptake in fungi. Three genes encoding transporters have been cloned from *H. cylindrosporum* (Javelle et al., 2001, 2003b) plus single genes from *T. borchii* (Montanini et al., 2002) and *A. muscaria* (Willmann et al., 2007). HcAmt3 is a low-affinity ammonium transporter, and HcAmt1 and HcAmt2 are high-affinity ammonium transporters/sensors. HcAmt1 is expressed under N-deficiency and NO_3^- feeding and repressed by glutamine. HcAmt3 is highly expressed but not highly regulated.

Most importantly in a symbiotic context, cells growing on arginine-containing media excrete ammonium ions produced by arginine metabolism within the urea cycle. These ammonium ions are taken back up when the high-affinity Amts, HcAmt1, or HcAmt2 are expressed. Additionally, HcAmt1 can fully restore the pseudohyphal growth defect of a *Saccharomyces cerevisiae mep2* mutant—the first evidence that a heterologous member of the Mep/Amt family can complement this dimorphic change defect (Javelle et al., 2003b). It was further proposed that the high-affinity ammonium transporters from mycorrhizal fungi act to sense the environment and induce, via as yet unidentified signal transduction cascade(s), the switch in fungal growth mode observed during mycorrhiza formation (Javelle et al., 2003a). The ability to switch between two morphological forms, a cellular yeast form and a multicellular invasive filamentous form, is a peculiar characteristic of several fungi. This dimorphic change, induced by nutrient starvation (N and/or carbon [C]), is a strategy to improve substrate exploration, and thus nutrient-scavenging capacity.

In Tuber (truffle), TbAmt1 seems to be upregulated only under N deficiency. The upregulation of TbAmt1 after transfer of the fungus to −N solution was slow (a matter of days), but its downregulation after N re-supplementation was fast, occurring in hours. NH_4^+, NO_3^- and glutamine downregulated the transporter (certainly via the glutamine pool as in *H. cylindrosporum*) but proline had no effect (i.e., TbAmt1 remains upregulated), showing that proline was a poor N source as in yeast and *H. cylindrosporum*. In *Amanita*, the gene encoding a high affinity ammonium transporter, AmAmt2, was upregulated in −N and nitrate (a non-usable N source in this strain) solution, also the case for high-affinity HcAmt1. *AmAmt1* expression decreased with increasing concentration of ammonium, with a basal expression reached at 500 μM NH_4^+, as is the case for *TbAmt1*. The expression was further reduced during the formation of ECM roots with the same level of expression in the mantle and the Hartig net. In contrast, expression was high in the extraradical mycelium.

More recent comparative genomics and *in silico* analyses in saprophytic (*P. chrysosporium, C. cinerea*) or parasitic (*U. maydis, C. neoformans*) Basidiomycetes highlighted a remarkable apparent expansion of putative genes encoding Amt transporters in the *L. bicolor* genome (i.e., eight Amt-coding genes) (Lucic et al. 2008). Six *LbAmt* genes were expressed at different levels in the free-living form, the fruiting body, and/or in mycorrhizal mycelium (Figure 4.3). One gene was constitutively expressed in all tissues and did not respond to N starvation, indicating that it could ensure a basal level of ammonium uptake independently of external N status as already demonstrated for the *H. cylindrosporum* ortholog *HcAmt3* (Javelle et al., 2003b). Three other genes were mainly expressed (and one highly so) in ectomycorrhiza, while their expression in mycelia was reduced to a basal level. However, none of the *LbAmts* was specifically expressed in mycorrhizal tissues. Similarly, Willmann and others (2007) showed that in functional ectomycorrhiza, the transcript level of the high affinity ammonium transporter AmAmt2

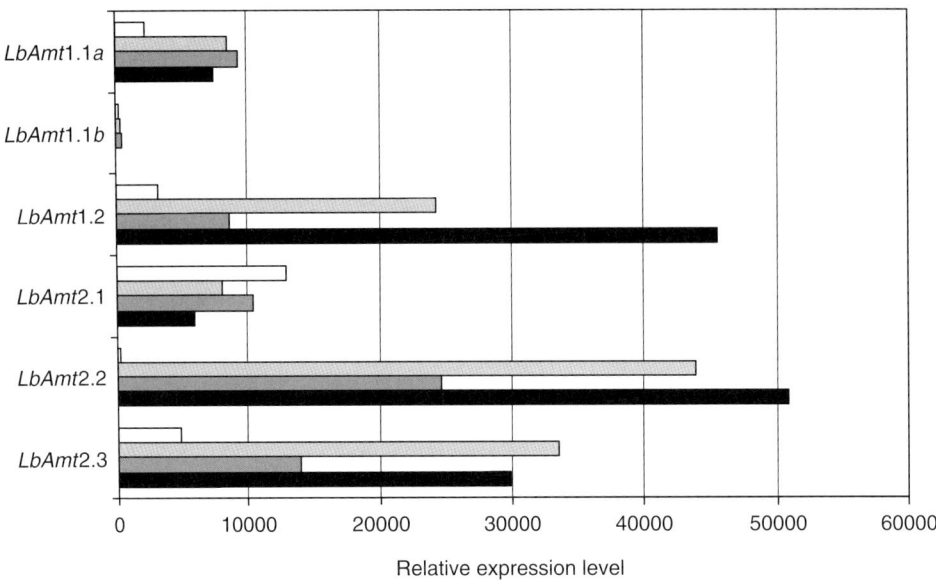

Figure 4.3. Expression levels of *Laccaria bicolor* ammonium transporter genes, as measured by whole-genome expression array, in *L. bicolor* (strain S238) free-living mycelium (*white bar*), *L. bicolor/Pseudotsuga menziesii* mycorrhiza (*light gray bar*), *L. bicolor/Populus tremula* mycorrhiza (*gray bar*) and *L. bicolor/Populus trichocarpa* mycorrhiza (*black bar*). Redrawn from Lucic et al. (2008).

of *A. muscaria* was decreased in both hyphal networks (sheath and Hartig net), while extraradical hyphae retained a strong gene expression. Using laser capture microdissection microscopy to separate Hartig net and fungal sheath mycelium of *L. bicolor* mycorrhiza should allow the expression analyses of Amt transporters in these micro-compartments.

NH_4^+ produced by nitrate reduction or by cellular uptake has to be incorporated in a C skeleton. This can occur through either of two routes: by glutamine synthetase and glutamine oxoglutarate amino-transferase (GS/GOGAT pathway) or by glutamate dehydrogenase and glutamine synthetase (GDH/GS pathway). There have been numerous studies on these enzymes in different fungi (Bedell et al., 1994; Martin et al., 1994; Javelle et al., 2003b; Morel et al., 2006). The GDH/GS pathway seems to predominate in mycorrhizal fungi, but the GS/GOGAT predominates in plants. The function of plant GDH was recently highlighted using a double GDH mutant. The results obtained suggest that GDH plays a central role in amino acid breakdown under C-deficient conditions (Miyashita and Good, 2008). Detection of GDH activity and protein in ECM roots depends on the fungal species (Botton and Dell, 1994). Interestingly, fungal aspartate aminotransferase activity was not detected in ectomycorrhizae whereas the activity of root isoenzymes was stimulated (Botton and Dell, 1994). Taken together, these data indicate that expression of enzymes involved in primary assimilation of ammonium in fungal and plant compartments is highly regulated in the mycorrhizal association.

Nitrate Utilization

Enhanced uptake of nitrate from ECM short roots was observed in the association of *R. roseolus* with *P. pinaster* (Gobert and Plassard, 2002). Again, this positive effect was strongly fungal-species dependent (Plassard et al., 2000). Interestingly, regulation of nitrate uptake seems to be different in the ECM fungus and in the host plant, as shown in the association *R. roseolus* and *P. pinaster* (Gobert and Plassard, 2002). Fungal nitrate uptake was maximal with or without nitrate in the medium, and this was contrary to the nonmycorrhizal host plant in which maximal uptake was reached only after several days of incubation in the presence of nitrate as the sole N-source. The kinetics of nitrate uptake into ECM roots as a function of nitrate availability was identical to that established in the fungus alone, suggesting that in ECM roots, nitrate was taken up mainly by the fungal cells rather than by the plant cells. Such a pattern of uptake could provide the ECM plant a "readier" access to nitrate whose availability fluctuates in the field (Gobert and Plassard, 2002).

Prior to reductive assimilation through the sequential action of nitrate reductase (EC 1.6.6.3) and nitrite reductase (EC 1.6.6.4), nitrate is internalized via an energy-dependent uptake process by specific plasma membrane transporters. A large group of nitrate transporters, from both prokaryotes and eukaryotes, belongs to the Major Facilitator Superfamily and, specifically, to the nitrate/nitrite porter (NNP) family. The best-characterized members of this family in ECM fungi are Nrt2 from *H. cylindrosporum* (Jargeat et al., 2003) and Nrt2 from *T. borchii* (Montanini et al., 2006). They share a similar membrane topology, in which two sets of six transmembrane helices, with N and C termini IN orientations, are connected by a large cytosolic loop. Fungal nitrate transporter genes are usually clustered with Nr and Nir genes and are generally upregulated by N-starvation in NO_3^--grown mycelia and by NO_3^- in NH_4^+-grown mycelia; conversely, gene expression is downregulated by NH_4^+ and glutamine in NO_3^--grown or N-starved mycelia (Montanini et al., 2006).

Gobert and Plassard (2002, 2007) have shown that the ECM fungus *R. roseolus* (Basidiomycetes) displayed only high-affinity NO_3^- uptake kinetics. A single nitrate transporter gene is probably responsible for nitrate uptake in *R. roseolus*, contrary to plants that have several nitrate transporters. Consistent with these data, only a single gene encoding a nitrate transporter was identified in the *L. bicolor* genome (Lucic et al., 2008). This family is exceptionally well conserved with respect to gene number per genome: only one closely related ORF was found in each of five Basidiomycete genomes. Obviously, a strong selection pressure against gene duplication and retention exists for these transporters. Interestingly, all these fungi with a single nitrate permease have multiple ammonium transporters.

Genes encoding nitrate reductase and nitrite reductase were also cloned from *H. cylindrosporum* (Jargeat et al., 2003) and *T. borchii* (Guescini et al., 2003, 2007). These genes were shown to be clustered with *Nrt2* in both fungi. *TbNrt2*, *HcNrt2*, *TbNir1*, and *HcNir1* were upregulated both in the presence of NO_3^- as sole N source and under N starvation, whereas *TbNr1* was only upregulated in the presence of NO_3^- (Jargeat et al., 2000; Guesciniv et al., 2007). All three genes were under ammonium repression in *H. cylindrosporum*, but it appears that *TbNr1* was not repressed by ammonium. *TbNr1* and *TbNrt2* were strongly expressed in the Hartig net and mantle but weakly expressed in the free-living mycelia (Guesciniv et al., 2003; Montanini et al., 2006).

The ability of 106 ECM fungal isolates to grow with nitrate as sole N source was recently assessed (Nygren et al., 2008). All isolates showed some growth, although growth rates differed markedly among taxa. Thirty-five partial sequences for nitrate reductase genes were obtained from the 43 strains tested, and which comprise 31 species and 10 genera, thus demonstrating that the ability to utilize nitrate as an N source is widespread in ECM fungi.

Urea Utilization

Urea, a known N source for many organisms, represents an important fertilizer in forest soils. Since most trees form symbiotic associations with ECM fungi, the capacities of the fungi to take up and assimilate urea influence the efficiency of urea N salvaging by plants. The ECM basidiomycete *P. involutus* is capable of using urea as sole N source by producing 20.9 mg of fresh weight, compared to 67.2 mg and 2.5 mg for equivalent 1 mM N as NH_4^+ and the N-starved condition, respectively (Morel et al., 2008), confirming earlier work by Yamanaka (1999). This suggests that external urea can be used by *P. involutus* although it is not a preferred N source. Molecular characterization of an active urea transporter (PiDur3) was reported in *P. involutus*. PiDur3 expression was upregulated under N deficiency and closely related to the intracellular glutamine pool and urease activity (Morel et al., 2008). Moreover, it was demonstrated that *P. involutus* regulates its intracellular pool of urea by coordinating urea transport and degradation.

Once in the cell, urea is hydrolyzed by the cytoplasmic urease, leading to the formation of two ammonium molecules. This protein has been identified in many microorganisms (Mobley and Hausinger, 1989). For *P. involutus*, an EST was previously identified with sequence homology to *UreG*, a gene encoding an accessory protein indispensable for urease activity in bacteria (Morel et al., 2008) and plants (Freyermuth et al., 2000, Witte et al., 2005).

N Transfer at the Mycorrhizal Interface: A New Face to an Old Question?

The mycorrhizal status relies on the bidirectional transfer of nutrients between the two symbionts, which was first demonstrated by Melin and Nilsson more than 50 years ago (Smith and Read, 1997). In ectomycorrhizae, the bulk of evidence in support of organic N transfer was derived primarily from ^{15}N tracer experiments and enzymatic studies (Smith and Read, 2008). Recently, Philip E. Pfeffer and collaborators (Govindarajulu et al., 2005) reported that beneficial AM fungi transfer substantial amounts of N to their plant hosts, most probably as ammonia (see also Chapter 3). The direct transfer of ammonia in an ECM association was also recently suggested by Uwe Nehls' group (Selle et al., 2005), reinforcing this model. The elucidation of this elegant transfer raised puzzling questions about the mechanisms involved in ammonia transfer and on the N metabolism of the two partners.

However, it is important to keep in mind that mycorrhizae are associations involving plants, fungi, other soil microbes, and, under natural conditions, also N_2-fixing microbes in N_2-fixing plants. It is mycorrhizae and N availability, not N_2 fixation, that plays a vital and decisive role in N redistribution among plants (He, 2004, 2005).

N Transfer in the Field

To our knowledge, very few field studies have addressed the transfer of N between the fungus and the plant. Clemmensen and others (2008) demonstrated that an ^{15}N-labeled equimolar mixture of NH_4^+, NO_3^-, and glycine supplied to the ECM communities grown in an in-growth bag, resulted in ^{15}N detection in *Betula nana* associated with these fungi. Interestingly, the ECM communities clearly discriminated against NO_3^-. Although the percentage of ^{15}N recovery measured in plant tissues was low, no matter the labeled N source (0.2–0.5% after 2 d; 0.3–1.2% after 26 d), *B. nana* accumulated some ^{15}N in the leaves, a strong indication for genuine transfer of N from fungal to host plant cells. In addition, the data of Clemmensen and others (2008) demonstrated that *B. nana* plants did not directly reflect ECM mycelial N uptake, indicating that N uptake by ECM is modulated by the N uptake patterns of both the fungal and plant components of the symbiosis and by competitive interactions in the soil.

In addition to the above study, there are indirect indications of high variability of N transfer among fungal species. Although not specifically designed to assess the N transfer between the mycelium and the host plant, Nara (2005) indicated strong differences among species in aiding young seedlings of *Salix reinii*, connected to the mother plant, to grow in a volcanic desert at Mount Fuji, Japan (Nara, 2005). Eleven native fungal species were used to connect a 1-year-old mother plant with young seedlings of *S. reinii*. After 5 months' growth, young seedlings were harvested and their biomass, N and P contents measured. The results showed that nutrient acquisition and growth of seedlings connected to the mycelial network were improved for most fungal species.

From the data of Nara (2005), we made the following calculations:

- Firstly, we divided the total amount of N and P accumulated in the whole seedlings by the root dry weight to get the accumulation rates of N and P by taking into account the effect of mycorrhizal symbiosis on root growth.
- Secondly, we divided each N and P accumulation rate by that of nonmycorrhizal control plants. Such a ratio will give indication on the possible supplementary N or P accumulation in the host plant through mycelial uptake and/or translocation of N or P from the mother tree.

As shown in Figure 4.4, the effects varied strongly among individual species, with either a ratio <1 only for N (*Laccaria amethystina*, *L. laccata*), suggesting a net transfer of P but not of N; a ratio <1 for N and P (*Hebeloma mesophaeum*), suggesting no net transfer of nutrients for this species; a ratio >1 for N and P (*Hebeloma leucosarx*, *H. pusillum*, *Laccaria murina*, *Russula pectinatoides*, *R. Sororia*, *Scleroderma bovista*), indicating simultaneous transfer of N and P; a ratio slightly >1 for N (*Cenococcum geophilum* and *Inocybe lacera*) suggesting a low effect on N accumulation that is independent of a P effect. These data clearly indicated that interspecific variations in efficiency in N (and P) capture and translocation exist among the fungal species *in situ*.

N Transfer in Microcosms

The demonstration of a net N transfer from fungal cells to root cells was first established by supplying mineral N sources labeled with ^{15}N to the hyphae of an ECM fungus physically separated from the roots of pine seedlings by a barrier (Smith and Read, 1997). The use of mass spectrometry coupled to the specific supply of ^{15}N-labeled substrates to the fungus made it possible to follow the enrichment of individual amino acids all along the way from the extramatrical hyphae to the shoot of the host in ECM plants (Finlay et al., 1988, 1989; Arnebrandt et al, 1993). Finlay and others (1988) showed also that the hyphae of four ECM species associated with *P. sylvestris* seedlings, namely *R. roseolus*, *S. bovinus*, *P. involutus*, and *P. tinctorius*, were able to take up, assimilate, and translocate ^{15}N from $^{15}NH_4^+$. Labeling was measured in the free amino acid fraction extracted from mycelium (above the barrier), mycorrhizal tips, roots, and needles. A feeding period of 73 h with $^{15}NH_4^+$ resulted in high proportions of ^{15}N–labeled free glutamate/glutamine, aspartate/asparagine, and alanine in the mycelium of *R. roseolus*, *S. bovinus*, *P. tinctorius*, and

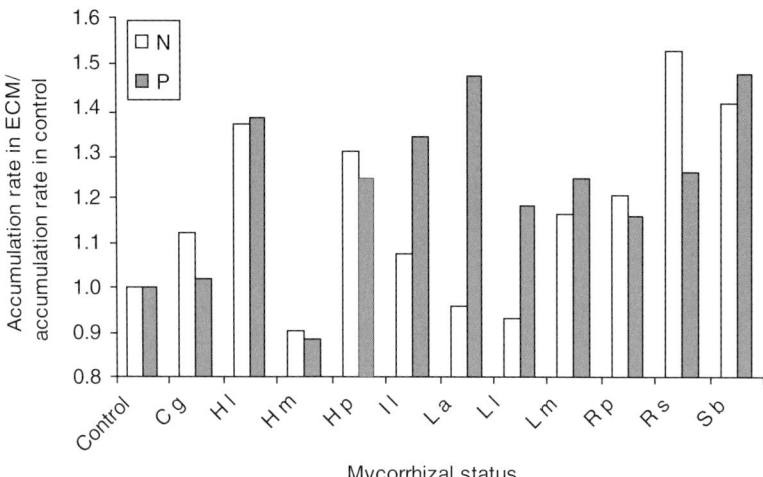

Figure 4.4. Variability of enhancement of N and P accumulation rates in young seedlings of *Salix reinii* connected to mother trees via different native ECM species using data published by Nara (2005). Given are the ratios calculated between accumulation rates of N or P (total amounts of N or P in whole seedlings divided by root dry weight) in ECM plants to that of nonmycorrhizal (control) plant. A ratio <1 indicates a negative effect whereas a ratio >1 indicates a positive effect of ECM symbiosis. The ECM species were Cg: *Cenococcum geophilum*, Hl: *Hebeloma leucosarx*, Hm: *H. mesophaeum*, Hp: *H. Pusillum*, Il: *Inocybe lacera*, La: *Laccaria amethystina*, Ll: *L. Laccata*, Lm: *Laccaria murina*, Rp: *Russula pectinatoides*, R.s: *R. Sororia*, Sb: *Scleroderma bovista*.

P. involutus, although labeled aspartate/asparagine were not found in this last species (Finlay et al., 1988). The same results were found when $^{15}NH_4^+$ or $^{15}NO_3^-$ was supplied to the hyphae of *P. involutus* associated with *Fagus sylvatica* (Finlay et al., 1989). However, high labeling was also found in aspartate/asparagine in the mycelium (Finlay et al., 1989). Other measurements carried out with *P. involutus* associated with two host plants (*Pinus contorta* and *Alnus glutinosa*) showed that citrulline and glutamine were the amino acids with the highest ^{15}N concentrations in the mycelium and in all parts of the system when $^{15}NH_4^+$ was supplied (Arnebrandt et al., 1993). Thus, these data strongly suggested that the form of N translocated throughout the hyphae could be glutamate/glutamine (Chalot and Brun, 1998).

Since these pioneering studies, other ^{15}N sources and different combinations of fungal species and host plant have been examined, but the fungal and root compartments are always separated (for example, Finlay et al., 1989; Arnebrandt et al., 1993; Ek et al., 1996; Ek 1997; Brandes et al., 1998; Jentschke et al., 2001; He et al., 2004, 2005). Generally speaking, a net transfer of ^{15}N from the fungus to the host plant was demonstrated with huge variations in the percentage of N derived from transfer (NDFT) between experiments involving ECM partners (Table 4.1, He et al., 2009). The values varied from 40% (He et al., 2004, 2005) to ≤1% (Ek et al., 1996), depending on the experimental conditions used to supply N to the host plant. The conditions for high NDFT seem to be that (1) the host plant rely fully upon the extraradical hyphae and (2) the hyphae be well supplied with P. Indeed, if there is no P available in the fungal compartment, there is very low NDFT, compared to the values measured with P (1.6 and 11.5, respectively, see Jentschke et al., 2001).

Table 4.1. Transfer of N from one plant to another via a common mycorrhizal network (from He et al., 2009).

Species A	Species B	Mycorrhizal mycelium involved	Element transferred	N transfer (%)	Reference
Alnus glutinosa	*Pinus contorta* (pine)	*Paxillus involutus*	$^{15}NH_4^+$	15.0	Arnebrant et al., 1993
A. incana (alder)	*Pinus sylvestris* (pine)	*P. involutus*	$^{15}NH_4^+$	9.0	Ekblad and Huss-Danell, 1995
Betula pendula (birch)	*Picea abies* (spruce)	*Scleroderma citrinum*	$^{15}NH_4^+$	0.3	Ek et al., 1996
Casuarina cunninghamiana	*C. cunninghamiana*	*Pisolithus tinctorius*	$^{14}NH_4^+$	30.5	Data given in He, 2009
	Eucalyptus maculata	*P. tinctorius*	$^{14}NH_4^+$	32.8	Data given in He, 2009
	E. maculata	*P. tinctorius*	$^{15}NH_4^+$	10.1	He et al., 2004, 2005
	E. maculata	*P. tinctorius*	$^{15}NO_3^-$	5.3	He et al., 2004, 2005
Eucalyptus maculata	*C. cunninghamiana*	*P. tinctorius*	$^{14}NH_4^+$	26.7	Data given in He, 2009
	C. cunninghamiana	*P. tinctorius*	$^{15}NH_4^+$	39.1	He et al., 2004, 2005
	C. cunninghamiana	*P. tinctorius*	$^{15}NO_3^-$	23.6	Data given in He, 2009
	E. maculata	*P. tinctorius*	$^{14}NH_4^+$	10.0	Data given in He, 2009
P. abies (spruce)	*B. pendula* (birch)	*Scleroderma citrinum*	$^{15}NH_4^+$	5.3	Ek et al., 1996

Evidence From $\partial^{15}N$ Data

A useful tool for evaluating N transfer exploits the natural abundance of ^{15}N (expressed as $\partial^{15}N$ values) in soils and plant foliage. Indeed, field and laboratory studies (reviewed in Hobbie and Hobbie, 2008) indicated that fractionation against ^{15}N during the formation of transfer compounds, such as amino acids, led to the transfer of ^{15}N-depleted N to ECM plants and the retention of ^{15}N-enriched N in ECM fungi, with an intrinsic fractionation against ^{15}N of ≈ 10‰). Lindahl and co-workers (2007) found a clear shift in fungal community composition between the surface litter (the L horizon), where the "early" and presumably saprotrophic fungal community dominated, and the underlying F horizon, where the "late" fungal community of mycorrhizal and other presumably root-dependent fungi dominated. This was associated with a simultaneous large increase in the stable isotope ^{15}N, which must be driven by fractionation against the heavier isotope during transfer of N from the soil through mycorrhizal fungi to their host plants (Figure 4.5, Lindahl et al., 2007). During this transfer, ECM fungi became approximately 10-fold ^{15}N enriched over their host plants, which was approximately the observed difference between the litter and the deeper soil layers.

The observed ^{15}N-enrichment of ECM relative to saprophytic sporocarps was attributed to ^{15}N discrimination during the formation and delivery of amino acid N from fungi to plants. The dominance of mycorrhizal fungi in well-degraded litter and humus observed in this study

84 *Plant-Soil Microbe Interactions*

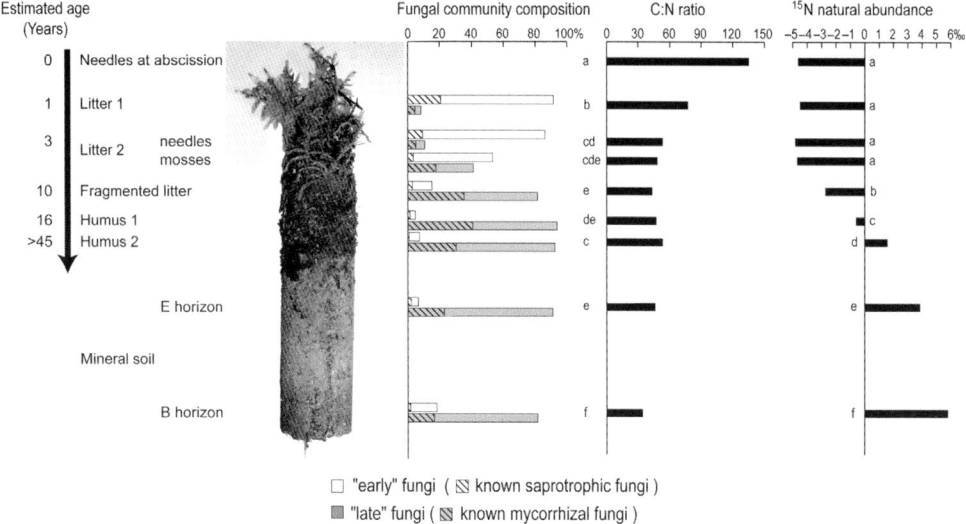

Figure 4.5. Fungal community composition, carbon:nitrogen (C:N) ratio and ^{15}N natural abundance throughout the upper soil profile in a Scandinavian *Pinus sylvestris* forest. Different letters in the diagrams indicate statistically significant differences between horizons in C:N ratios and ^{15}N abundance; the standard error of the mean was <0.3 for ^{15}N natural abundance and <3 for C:N ratio (n = 19–27, for recently abscised needles n = 3). The age of the organic matter is estimated from the average Δ^{14}C of three samples from each horizon (five samples of the litter 2 [needles] fraction) and needle abscission age (3 yr) is subtracted. Community composition data are expressed as the frequency of total observations. "Early" fungi are defined as those occurring with a higher frequency in litter samples compared with older organic matter and mineral soil. "Late" fungi are those occurring with a higher frequency in older organic matter. Reproduced with permission from Lindahl et al. (2007).

supports the hypothesis that mycorrhizal fungi play a significant role in mobilizing N from well-decomposed organic matter in boreal forest soils. This was recently confirmed by a study where compiled analyses of ∂^{15}N and ∂^{13}C isotope values from 813 fungi across 23 sites ranging from boreal tundra to tropical rainforest revealed ^{15}N-enrichment of ECM fungi relative to saprophytic fungi (Mayor et al., 2009). Additional causes of ECM ^{15}N-enrichment relative to saprophytic fungi could include preferential use of ^{15}N-enriched forms of N as well as internal processing within the fungus irrespective of transfer to the host plants. However, field and laboratory observations currently support the process of N delivery to the host plant as having the greatest influence on isotope fractionation. In addition, frequently observed ^{15}N depletion in ECM-associated host plants relative to co-localized non-ECM plants further supports this hypothesis (Hobbie and Hobbie, 2008).

Molecular Data Supporting N Exchanges

Transcript levels of plant and fungal nitrite reductases and fungal ammonium transporter (HcAmt1) were determined in control (uninoculated) roots, extra-radical mycelia, and ECM formed by either wild type or nitrate reductase-deficient fungal strains of *H. cylindrosporum* (Bailly et al., 2007). When the plant-fungal associations were supplied with 1 mM NO_3^-, the plant nitrite reductase was repressed in the association with the two wild-type strains, and fully expressed when associated with the nitrate reductase-deficient strain. Since the nitrite reductase of fungal-free plants was strongly repressed by reduced N and fully expressed by NO_3^-, these data strongly suggest that the wild-type fungus (*H. cylindrosporum*) was able to translocate reduced N to the host plant upon NO_3^- reduction by fungal nitrate and nitrite reductases.

Experimental observations indicated that glutamine (in ECM) and other glutamine-derived amino acids were usually the principal nitrogenous products accumulated at the uptake site during periods of ammonium feeding, providing support for the importance of these amino acids in fungal to plant N transfer—the "traditional view." Rapid assimilation of inorganic N at the uptake site can ensure that internal demands are satisfied prior to transfer to the host plant and can also prevent toxic accumulation of ammonium ions (Finlay et al., 1988, 1989). However, direct transfer of ammonium has been recently hypothesized in ectomycorrhizae (Selle et al., 2005), concordant with the direct transfer of ammonia in AM symbioses (Govindarajulu et al., 2005). Low assimilatory capacities of intraradical fungal cells leading to sustained ammonia transfer is indeed supported by a strong downregulation of the structural gene for ammonium-assimilating glutamine synthetase in ECM hyphae in close contact with root cells (Wright et al., 2005). Previous studies also emphasized both a decrease of the major amino acid pools and a decrease of the activity of fungal enzymes involved in N assimilation during mycorrhizal colonization (Blaudez et al., 1998). However, this model of ammonia transfer implies a lack of NH_4^+ retrieval in the intraradical fungal plasma membrane. Govindarajulu and others (2005) reported a strong expression of a fungal Amt gene in intraradical hyphae and, in principle, Amts can be the path of leakage of ammonia when its internal concentration exceeds that in the medium. However, by the same token, the importance of the fungal Amt family members in the retrieval of ammonium ions cannot be ignored (von Wiren et al., 2000). This is clearly demonstrated by the constant release of ammonia by a yeast mutant defective in all three endogenous Mep (methylammonium permease) genes, a release corrected by expression of *H. cylindrosporum* Amts (Javelle et al., 2003b). If Amts were highly abundant in the plasma membrane of symbiotic hyphae this would result in a futile cycling of inorganic N and probably in an inefficient transfer of N.

The extrusion of ammonia from fungal cells must follow other pathways than those mediated by Amt proteins, either by passive efflux of the deprotonated form or by protein-mediated mechanisms. Alternatively, ammonia could be stored in intracellular vesicles (either by inward diffusion of NH_3 or by active uptake of NH_4^+) and further exported by an exocytotic ammonia excretion mechanism. Such a mechanism would enable fungal cells to maintain a low cytoplasmic ammonia concentration, thus retaining a constant assimilatory capacity and would in turn allow for a sustained export into the plant root cells. This view would also be highly compatible with strong Amt expression in fungal symbiotic cells measured in the AM fungus *G. intraradices* (Govindarajulu et al., 2005). A similar exocytotic model was proposed for the excretion of amino acids by yeast cells (Velasco et al., 2004).

A large array of membrane proteins could fulfill the function of transferring ammonia from the fungal cytosol to the interfacial apoplast (Chalot et al., 2006). Members of the Ato family are integral plasma membrane proteins with six transmembrane domains and hydrophilic N- and C-termini (Augstein et al., 2003). They belong to a "growing" group of highly conserved proteins that are mainly present in prokaryotes and lower eukaryotes (mainly fungi) but not in multicellular eukaryotes. In the former group, many organisms carry several homologues of the Ato proteins. They contain the typical ammonium transporter signature; the hydrophobic transmembrane regions are conserved, but major parts of the hydrophilic N- and C-termini are quite divergent. *S. cerevisiae* encodes three Ato family proteins that appear to function as ammonium/H^+ antiporters, extruding intracellular ammonium, and importing protons (Palkova et al., 2002). These proteins are candidates for the extrusion of ammonia from fungal cells at the symbiotic interface, especially since Ato homologues among mycorrhizal EST databases were shown to be highly expressed in mycorrhizal tissues (Selle et al., 2005). There are two putative Ato homologs (*LbAto1* and *LbAto2*) in the *L. bicolor* genome (Lucic et al., 2008). No expansion was observed for this gene family compared to other basidiomycetes. RT-PCR and microarray analyses revealed a strong expression of *LbAto1* in free-living mycelium, independent of N status, with a 5-fold higher expression in mycorrhiza. *LbAto2* was repressed in N-starved mycelium, and no expression could be detected in Douglas fir mycorrhiza. Although these observations are not compatible with a specific ammonium export role of Ato proteins at the *L. bicolor* symbiotic interface, it cannot be excluded that these proteins directly pump ammonium out of the cells. Alternatively, they may be involved either in transport of other, yet unidentified substrates, or in transfer of a regulatory signal.

Some aquaporins isolated from plants have been implicated in the vacuolar loading of NH_3 (Loque et al., 2005). These proteins would effectively function as a low-affinity version of the Amt mechanism and require an appropriate gradient. Aquaporins in fungal genomes have recently been reviewed, and it appears that overall, the number of aquaporin family members can range from one to five (Pettersson et al., 2005). The function of fungal aquaporin-mediated systems in ammonia transport has, however, not yet been demonstrated.

The transfer of ammonia from the apoplast to the plant cytoplasm requires active transport, and there is a lack of data on ammonia concentrations in the interfacial apoplast. The recent completion of the *Populus trichocarpa*, *Oryza sativa* and *Vitis vinifera* genomes also revealed an unusually high number of genes encoding ammonia transporters of the Amt family, especially from the Amt2 family as compared with the *Arabidopsis* model, a non-mycorrhizal species. Strong expression of plant ammonia transporters of the Amt family in the ECM symbiosis between poplar and *A. muscaria* (Selle et al., 2005) and between poplar and *P. involutus* (Couturier et al., 2007) indicated that a direct transfer of ammonia from the fungus to the plant might occur. These studies were based solely on transcript levels, and there needs to be careful consideration of the function of plant Amt in the re-uptake of fungal-derived ammonia.

Much more attention must also be devoted to non-specific channels such as aquaporins (Loque et al., 2005) or voltage-dependent cation systems (Roberts and Tyerman, 2002), which might also contribute to ammonia import from the interfacial apoplast to the plant cell cytoplasm, where ammonia will be "captured" by the GOGAT/GS enzymes.

Conclusions and Perspectives

The presence of ECM fungi is of pivotal importance for plant productivity in most boreal and temperate forests (van der Heijden et al., 2008). The *in situ* demonstration of extracellular enzyme secretion by ECM fungi demonstrate that they can acquire N from litter through extensive hyphal networks that forage for nutrients (Rineau and Garbaye, 2009). These data, together with the dominance of mycorrhizal fungi over saprotrophic fungi in well-degraded litter and humus observed in recent studies support the hypothesis that mycorrhizal fungi play a significant role in mobilizing N from well-decomposed organic matter (Lindahl et al., 2007; Mayor et al., 2009).

Studies are urgently needed I (1) to identify net N transfer, (2) to determine whether mycorrhizal hyphae play a direct role, (3) to identify the nature of released nutrients or assimilated metabolites into the apoplast interface by the fungus, (4) to determine whether amino acids or inorganic N, or both, are the preferentially transferred N form in ECM symbioses, and

Table 4.2. Number of transporter gene models for N compounds in the genomes of the ECM fungus *Laccaria bicolor* (Martin et al., 2008; Lucic et al., 2008) and the host woody species *Populus trichocarpa* (Tuskan et al., 2006; Couturier et al., 2007), and the characterized orthologs in other ECM or woody species. Nd: not determined. Family names correspond to those described in Saier et al. (1999).

Substrate	Transporter family	*Laccaria bicolor*	Characterized orthologs from other ECM source	*Populus trichocarpa*	Characterized orthologs from other tree source
Nitrate	NNP family	1	HcNrt2 (Jargeat et al., 2003) TbNrt2 (Montanini et al., 2006)	11	none
Ammonium	Amt family	8	AmAmt1 (Willmann et al., 2007). HcAmt1-3 (Javelle et al., 2001, 2003b) TbAmt1 (Montanini et al., 2002)	14	PttAmt1.2 (Selle et al., 2005)
Amino acid	APC family +AAAP family	29	AmGap1 (Nehls et al., 1999) HcGap1 (Wipf et al., 2002).	131	none
Peptide	Opt family	10	None	9	none
	Pot family	2	HcPtr1, HcPtr2 (Benjdia et al., 2006)	90	none
Urea	SSS family	2	PiDur3 (Morel et al., 2008)	nd	none

(5) to study reciprocal C and N transfer. The answer to these key questions will likely arise from a combination of various molecular and biochemical approaches. Some recommended future directions, and perspectives, follow:

- Approaches using ^{15}N-enrichment and/or ^{15}N natural abundance (see discussion by He et al., 2009) and application of fluorescent nanoscale semiconductor technology (Whiteside et al., 2009) and compound-specific gas or liquid chromatography-isotope ratio mass spectrometry (CS-GC- or CS-LC-IRMS) may be combined to trace N fluxes.
- *In silico* analysis of the sequenced genome of symbiotic *L. bicolor* has revealed the presence of at least 128 putative genes encoding putative transport proteins for N-containing compounds (Table 4.2, Lucic et al., 2008). Among them, 118 genes (92%) showed detectable transcript levels and about 30 putative genes belong to gene families in which some members have been characterized. Expression analysis has highlighted some candidate transporters for nutrient uptake or metabolite transfer at the symbiotic interface. However, given the numbers of total potential genes in both *L. bicolor* and the tree host *P. trichocarpa* (see Table 4.2), there is still a long road to go before we get a complete view of transporter function and localization. Sequencing of further fungal (mycorrhizal versus non-mycorrhizal) genomes and comparative genomics will likely prove extremely useful in sorting out key transporter genes for symbiosis.

- The molecular data, combined with direct analysis of apoplast fluid by laser microdissection capture (Balestrini et al., 2007) will be invaluable tools to elucidate transfer between fungal and root cells.

As concluded by He and others (2009), the lack of convincing data underlines the need for creative, careful experimental manipulations.

References

Abuzinadah, R.A., Read, D.J. 1986. The role of proteins in the nitrogen nutrition of ectomycorrhizal plants. I. Utilization of peptides and proteins by ectomycorrhizal fungi. *New Phytologist* 103, 481–493.

Abuzinadah, R.A., Read, D.J. 1989. The role of proteins in the nitrogen nutrition of ectomycorrhizal plants IV. The utilization of peptides by birch (*B. pendula* L.) infected with different mycorrhizal fungi. *New Phytologist* 112, 55–60.

Andersson, P. Berggren, D. 2005. Amino acids, total organic and inorganic nitrogen in forest floor soil solution at low and high nitrogen input. *Water Air Soil Pollution* 162, 369–384.

Arnebrandt, K., Ek, H., Finlay, R., et al. 1993. Nitrogen translocation between *Alnus glutinosa* (L.) Gaertn. Seedlings inoculated with *Frankia* sp. and *Pinus contorta* Doug. Ex Loud seedlings connected by a common ectomycorrhizal mycelium. *New Phytologist* 124, 231–242.

Augstein, A., Barth, K., Gentsch, M., et al. 2003. Characterization, localization and functional analysis of Gpr1p, a protein affecting sensitivity to acetic acid in the yeast *Yarrowia lipolytica*. *Microbiology* 149, 589–600.

Bailly, J., Debaud, J.C., Verner, M.C., et al. 2007. How does a symbiotic fungus modulate expression of its host-plant nitrite reductase? *New Phytologist* 175, 155–165.

Balestrini, R., Gomez-Ariza, J., Lanfranco, L., et al. 2007. Laser microdissection reveals that transcripts for five plant and one fungal phosphate transporter genes are contemporaneously present in arbusculated cells. *Molecular Plant–Microbe Interactions* 20, 1055–1062.

Barber, S.A. 1995. Soil nutrient availability, a mechanistic approach. 2nd ed., John Wiley & Sons, New York.

Bedell, J.P., Garnier, A, Pireaux, J.C., et al. 1994. Study of enzymes involved in nitrogen-metabolism of Douglas-*L. laccata* ectomycorrhizas. *Acta Botanica Gallica* 141, 483–490.

Bendjia, M., Rikirsch, E., Müller, T., et al. 2006. Peptide uptake in the ectomycorrhizal fungus *Hebeloma cylindrosporum*: characterization of two di- and tri-peptide transporters (HcPTR2A and B). *New Phytologist* 17, 401–410.

Blaudez, D. et al. 1998. Structure and function of the ectomycorrhizal association between *Paxillus involutus* (Batsch) Fr. and *B. pendula* (Roth.). II. Modifications of nitrogen and carbon metabolisms during early stages of ectomycorrhizas formation. *New Phytologist* 138, 543–552.

Botton, B., Dell, B. 1994. Expression of glutamate dehydrogenase and aspartate aminotransferase in eucalypt ectomycorrhizas. *New Phytologist* 126, 249–257.

Boukcim, H., Plassard, C. 2003. Juvenile nitrogen uptake capacities and root architecture of two open-pollinated families of *Picea abies*. Effects of nitrogen source and ectomycorrhizal symbiosis. *Journal of Plant Physiology* 160, 1211–1218.

Brandes, B., Godbold, D., Kuhn, A., et al. 1998. Nitrogen and phosphorus acquisition by the mycelium of the ectomycorrhizal fungus *Paxillus involutus* and its effect on host nutrition. *New Phytologist* 140, 735–743.

Chalot, M., Blaudez, D., Brun, A. 2006. Ammonia: a candidate for nitrogen transfer at the mycorrhizal interface. *Trends in Plant Sciences* 11, 263–266.

Chalot, M., Brun, A. 1998. Physiology of organic nitrogen acquisition by ectomycorrhizal fungi and ectomycorrhizas. *FEMS Microbiology Reviews* 22, 21–44.

Clemmensen, K.E., Sorensen, P.L., Michelsen, A., et al. 2008. Site-dependent N uptake from N-form mixtures by arctic plants, soil microbes and ectomycorrhizal fungi. *Oecologia* 155, 771–783.

Courty, P.E., Pritsch, K., Schloter, M., et al. 2005. Activity profiling of ectomycorrhiza communities in two forest soils using multiple enzymatic tests. *New Phytologist* 167, 309–319.

Couturier, J., Montanini, B., Martin, F., et al. 2007. The expanded family of ammonium transporters in the perennial poplar plant. *New Phytologist* 174, 137–150.

Ek, H. 1997. The influence of nitrogen fertilization on the carbon economy of *Paxillus involutus* in ectomycorrhizal association with *B. pendula*. *New Phytologist* 135, 133–142.

Ek, H., Andersson, S., Söderström, B. 1996. Carbon and nitrogen flow in silver birch and Norway spruce connected by a common mycorrhizal mycelium. *Mycorrhiza* 6, 475–467.

Ekblad, A., Huss-Danell, K. 1995. Nitrogen fixation by *Alnus incana* and nitrogen transfer from *A. incana* to *Pinus sylvestris* influenced by macronutrient and ectomycorhiza. *New Phytologist* 131, 453–9.

Finlay, R.D., Ek, H., Odham, G., et al. 1988. Mycelial uptake, translocation and assimilation of nitrogen from ^{15}N-labelled ammonium by *Pinus sylvestris* infected with four different ectomycorrhizal fungi. *New Phytologist* 110, 59–66.

Finlay, R.D., Ek, H., Odham, G., et al. 1989. Uptake, translocation and assimilation of nitrogen from ^{15}N-labelled ammonium and nitrate sources by intact ectomycorrhizal systems of *Fagus sylvatica* infected with *Paxillus involutus*. *New Phytologist* 113, 47–55.

Finlay, R.D., Frostegard, A., Sonnerfeldt, A.M. 1992. Utilization of organic and inorganic nitrogen-sources by ectomycorrhizal fungi in pure culture and in symbiosis with *Pinus contorta* Dougl Ex Loud. *New Phytologist* 120, 105–115.

Freyermuth, S.K., Bacanamwo, M., Polacco, J.C. 2000. The soybean *Eu3* gene encodes a Ni-binding protein necessary for urease activity. *The Plant Journal* 21, 53–60.

Genetet, I., Martin, F., Stewart, G.R. 1984. Nitrogen assimilation in mycorrhizas—Ammonium assimilation in the N-starved ectomycorrhizal fungus *Cenococcum graniforme*. *Plant Physiology* 76, 395–399.

Gobert, A., Plassard, C. 2002. Differential NO_3^- dependent patterns of NO_3^- uptake in *Pinus pinaster*, *Rhizopogon roseolus* and their ectomycorrhizal association. *New Phytologist* 154, 509–516.

Gobert, A., Plassard, C. 2007. Kinetics of NO_3^- net fluxes in *Pinus pinaster*, *Rhizopogon roseolus* and their ectomycorrhizal association, as affected by the presence of NO_3^- and NH_4^+. *Plant, Cell and Environment* 30, 1309–1319.

Govindarajulu, M., Pfeffer, P.E., Jin, H., et al. 2005. Nitrogen transfer in the arbuscular mycorrhizal symbiosis. *Nature* 435, 819–823.

Grenon, F., Bradley, R.L., Jones, M., et al. 2005. Soil factors controlling mineral N uptake by *Picea engelmanii* seedlings: the importance of gross NH_4^+ production rates. *New Phytologist* 165, 791–800.

Guescini, M., Pierleoni, R., Palma, F., et al. 2003. Characterization of the *Tuber borchii* nitrate reductase gene and its role in ectomycorrhizae. *Molecular and General Genomics* 269, 807–816.

Guesciniv M., Zeppa, S., Pierleoni, R., et al. 2007. The expression profile of the *Tuber borchii* nitrite reductase suggests its positive contribution to host plant nitrogen nutrition. *Current Genetics* 51, 31–41.

Guidot, A., Verner, M.C., Debaud, J.C., et al. 2005. Intraspecific variation in use of different organic nitrogen sources by the ectomycorrhizal fungus *Hebeloma cylindrosporum*. *Mycorrhiza* 15, 167–177.

He, X., Critchley, C., Ng, H., et al. 2004. Reciprocal N ($^{15}NH_4^+$ or $^{15}NO_3^-$) transfer between non-N2-fixing *Eucalyptus maculata* and N2-fixing *Casuarina cunninghamiana* linked by the ectomycorrhizal fungus *Pisolithus* sp. *New Phytologist* 163, 629–640.

He, X., Critchley C., Ng, H., et al. 2005. Nodulated N2-fixing *Casuarina cunninghamiana* is the sink for net N transfer from non-N2-fixing *Eucalyptus maculata* via an ectomycorrhizal fungus *Pisolithus* sp. Using $^{15}NH_4^+$ or $^{15}NO_3^-$ supplied as ammonium nitrate. *New Phytologist* 176, 897–712.

He, X., Xy, M., Qiu, G.Y., et al. 2009. Use of ^{15}N stable isotope to quantify nitrogen transfer between mycorrhizal plants. *Journal of Plant Ecology* 2, 107–118.

Heckman, D.S., Geiser, D.M., Eidell, B.R., et al. 2001. Molecular evidence for the early colonization of land by fungi and plants. *Science* 293, 1129–1133.

Hibbett, D.S., Matheny, P.B. 2009. The relative ages of ectomycorrhizal mushrooms and their plant hosts estimated using Bayesian relaxed molecular clock analyses. *BMC Biology* 7, 1–13.

Hobbie, E.A., Hobbie, J.E. 2008. Natural abundance of ^{15}N in nitrogen-limited forests and tundra can estimate nitrogen cycling through mycorrhizal fungi: a review. *Ecosystems* 11, 815–830.

Huang, C.H., Peng, J. 2005. Evolutionary conservation and diversification of Rh family genes and proteins. *Proceedings of the National Academy of Sciences, USA* 102, 15512–15517.

Jargeat, P., Gay, G., Debaud, J.C., et al. 2000. Transcription of a nitrate reductase gene isolated from the symbiotic basidiomycete fungus *Hebeloma cylindrosporum* does not require induction by nitrate. *Molecular and General Genetics* 263, 948–956.

Jargeat, P., Rekangalt, D., Verner, M.C., et al. 2003. Characterisation and expression analysis of a nitrate transporter and nitrite reductase genes, two members of a gene cluster for nitrate assimilation from the symbiotic basidiomycete *Hebeloma cylindrosporum*. *Current Genetic* 43, 199–205.

Javelle, A., Andre, B., Marini, A.M., et al. 2003a. High-affinity ammonium transporters and nitrogen sensing in mycorrhizas. *Trends in Microbiology* 11, 53–55.

Javelle, A., Chalot, M., Soderstrom, B., et al. 1999. Ammonium and methylamine transport by the ectomycorrhizal fungus *Paxillus involutus* and ectomycorrhizas. *FEMS Microbiology Ecology* 30, 355–366.

Javelle, A., Morel, M., Rodriguez-Pastrana, B.R., et al. 2003b. Molecular characterization, function and regulation of ammonium transporters (Amt) and ammonium-metabolizing enzymes (GS, NADP-GDH) in the ectomycorrhizal fungus *Hebeloma cylindrosporum*. *Molecular Microbiology* 47, 411–430.

Javelle, A., Rodriguez-Pastrana, B.R., Jacob, C., et al. 2001. Molecular characterization of two ammonium transporters from the ectomycorrhizal fungus *Hebeloma cylindrosporum*. *FEBS Letters* 505, 393–398.

Jentschke, G., Brandes, B., Kuhn, A., et al. 2001. Interdependence of phosphorus, nitrogen, potassium and magnesium translocation by the ectomycorrhizal fungus *Paxillus involutus*. *New Phytologist* 149, 327–337.

Jones, D.L. 1999. Amino acid biodegradation and its potential effects on organic nitrogen capture by plants. *Soil Biology and Biochemistry* 31, 613–622.

Jones, D.L., Healey, J.R., Willettv V.B., et al. 2005. Dissolved organic nitrogen uptake by plants—an important N uptake pathway? *Soil Biology and Biochemistry* 37, 413–423.

Jones, D.L., Hodge, A. 1999. Biodegradation kinetics and sorption reactions of three differently charged amino acids in soil and their effects on plant organic nitrogen availability. *Soil Biology and Biochemistry* 31, 1331–1342.

Jones, D.L., Kielland, K. 2002. Soil amino acid turnover dominates the nitrogen flux in permafrost-dominated taiga forest soils. *Soil Biology and Biochemistry* 34, 209–219.

Jones, D.L., Shannon, D., Murphy, D.L., et al. 2004. Role of dissolved organic nitrogen (DON) in soil N cycling in grassland soils. *Soil Biology and Biochemistry* 36, 749–756.

Jones, D.L., Willett, V.B. 2006. Experimental evaluation of methods to quantify dissolved organic nitrogen (DON) and dissolved organic carbon (DOC) in soil. *Soil Biology and Biochemistry* 38, 991–999.

Kaye, J.P., Hart, S.C. 1997. Competition for nitrogen between plants and soil microorganisms. *Tree* 12, 139–143.

Kielland, K., McFarland, J.W., Ruess, R.W., et al. 2007. Rapid cycling of organic nitrogen in taiga forest ecosystems. *Ecosystems* 10, 360–368.

Klironomos, J.N., Hart, M.M. 2001. Food-web dynamics. Animal nitrogen swap for plant carbon. *Nature* 410, 651–652.

Kranabetter, J.M, Dawson, C.R., Dunn, D.E. 2007. Indices of dissolved organic nitrogen ammonium and nitrate across productivity gradients of boreal forests. *Soil Biology & Biochemistry* 39, 3147–3158.

Lambers, H., Raven, J.A., Shaver, G.R., et al. 2008. Plant nutrient-acquisition strategies change with soil age. *Trends in Ecology and Evolution* 23, 95–103.

Lindahl, B.D., Ihrmark, K., Boberg, J., et al. 2007. Spatial separation of litter decomposition and mycorrhizal nitrogen uptake in a boreal forest. *New Phytologist* 173, 611–620.

Loque, D., Ludewig, U., Yuan, L. et al. 2005. Tonoplast intrinsic proteins AtTIP2;1 and AtTIP2;3 facilitate NH_3 transport into the vacuole. *Plant Physiology* 137, 671–680.

Lucic, E., Fourrey, C., Kohler, A., et al. 2008. A gene repertoire for nitrogen transporters in *Laccaria bicolor*. *New Phytologist* 180, 343–64.

Marschner, H. 1995. *Mineral Nutrition of Higher Plants*. Academic Press, London.

Martin, F., Cote, R., Canet, D. 1994. NH_4^+ assimilation in the ectomycorrhizal basidiomycete *Laccaria bicolor* (Maire) Orton, a ^{15}N-NMR study. *New Phytologist* 128, 479–485.

Martin, F., Aerts, A., Ahrén, D., et al. 2008. The genome of *Laccaria bicolor* provides insights into mycorrhizal symbiosis. *Nature* 452, 88–92.

Mayor, J.R., Schuur, E.A.G., Henkel, T.W. 2009. Elucidating the nutritional dynamics of fungi using stable isotopes. *Ecology Letters* 12, 171–183.

Mobley, H.L., Hausinger, R.P. 1989. Microbial ureases: significance, regulation, and molecular characterization. *Microbiology Reviews* 53, 85–108.

Montanini, B., Moretto, N., Soragni, E., et al. 2002. A high-affinity ammonium transporter from the mycorrhizal ascomycete *Tuber borchii*. *Fungal Genetic and Biology* 36, 22–34.

Montanini, B., Viscomi, A.R., Bolchi, A., et al. 2006. Functional properties and differential mode of regulation of the nitrate transporter from a plant symbiotic ascomycete. *Biochemistry Journal* 394, 125–134.

Morel, M., Buee, M., Chalot, M., et al. 2006. NADP-dependent glutamate dehydrogenase, a dispensable function in ectomycorrhizal fungi. *New Phytologist* 169, 179–190.

Morel, M., Jacob, C., Fitz, M., et al. 2008. Characterization and regulation of PiDur3, a permease involved in the acquisition of urea by the ectomycorrhizal fungus *Paxillus involutus*. *Fungal Genetic and Biology* 45, 912–21.

Müller, T., Avolio, M., Olivi., M, et al. 2007. Nitrogen transport in the ectomycorrhiza association: The *Hebeloma cylindrosporum-Pinus pinaster* model. *Phytochemistry* 68, 41–51.

Miyashita, Y., Good, A.G. 2008. NAD(H)-dependent glutamate dehydrogenase is essential for the survival of Arabidopsis thaliana during dark-induced C starvation. *Journal of Experimental Botany* 59, 667–80.

Nara, K. 2005. Ectomycorrhizal networks and seedling establishment during early primary succession. *New Phytologist* 169, 169–178.

Näsholm, T., Kielland, K., Ganeteg, U. 2009. Uptake of organic nitrogen by plants. *New Phytologist* 182, 31–48.

Nehls, U., Kleber, R., Wiese, J., et al. 1999. Isolation and characterization of a general amino acid permease from the ectomycorrhizal fungus *Amanita muscaria*. *New Phytologist* 144, 343–349.

Nehls, U., Bock, A., Einig, W., et al. 2001. Excretion of two proteases by the ectomycorrhizal fungus *Amanita muscaria*. *Plant Cell and Environment* 24, 741–747.

Nygren, C.M., Eberhardt, U., Karlsson, M., et al. 2008. Growth on nitrate and occurrence of nitrate reductase-encoding genes in a phylogenetically diverse range of ectomycorrhizal fungi. *New Phytologist* 180, 875–89.

Palkova, Z., Devaux, F., Riccova, M., et al. 2002. Ammonia pulses and metabolic oscillations guide yeast colony development. *Molecular Biology of the Cell* 13, 3901–3914.

Paungfoo-Lonhienne, C., Schenk, P.M., Lonhienne, T.G.A., et al. 2009. Nitrogen affects cluster root formation and expression of putative peptide transporters. *Journal of Experimental Botany* 60, 2665–2676.

Paungfoo-Lonhienne, C., Lonhienne, T.G.A., Rentsch, D., et al. 2008. Plants can use protein as a nitrogen source without assistance from other organisms. *Proceedings of the National Academy of Sciences, USA* 105, 4524–4529.

Pettersson, N., Filipsson, C., Becit, E., et al. 2005. Aquaporins in yeasts and filamentous fungi. *Biology of the Cell* 97, 487–500.

Plassard, C., Bonafos, B., Touraine, B. 2000. Differential effects of mineral and organic N sources, and of ectomycorrhizal infection by *Hebeloma cylindrosporum*, on growth and N utilization in *Pinus pinaster*. *Plant Cell and Environment* 23, 1195–1205.

Qualls, R.G., Richardson, C.J. 2003. Factors controlling concentration, export, and decomposition of dissolved organic nutrients in the Everglades. *Biogeochemistry* 62, 197–229.

Quoreshi, A.M., Ahmad, I., Malloch, D., et al. 1995. Nitrogen metabolism in the ectomycorrhizal fungus Hebeloma crutuliniforme. *New Phytologist* 131, 263–271.

Rangel-Castro, J.I., Danell, E., et al. 2002. Use of different nitrogen sources by the edible ectomycorrhizal mushroom Cantharellus cibarius. *Mycorrhiza* 12, 131–137.

Redecker, D., Kodner, R., Graham, L.E. 2000. Glomalean fungi from the Ordovician. *Science* 289, 1920–1921.

Rineau, F., Garbaye, J. 2009. Does forest liming impact the enzymatic profiles of ectomycorrhizal communities through specialized fungal symbionts? *Mycorrhiza* 19, 493–500.

Roberts, D.M., Tyerman, S.D. 2002. Voltage-dependent cation channels permeable to NH_4^+, K^+, and Ca^{2+} in the symbiosome membrane of the model legume *Lotus japonicus*. *Plant Physiology* 128, 370–378.

Rothstein, D.E. 2009. Soil amino-acid availability across a temperate-forest fertility gradient. *Biogeochemistry* 92, 201–215.

Rousseau, J.V.D., Sylvia, D.M., Fox, A.J. 1994. Contribution of ectomycorrhiza to the potential nutrient-absorbing surface of pine. *New Phytologist* 128, 639–644.

Saier, Jr., M.H., Beatty, J.T., Goffeau, A., et al. 1999. The major facilitator superfamily. *Journal of Molecular Microbiology and Biotechnology* 1, 257–279.

Scheromm, P., Plassard, C., Salsac, L. 1990. Effect of nitrate and ammonium nutrition on the metabolism of the ectomycorrhizal basidiomycete, *Hebeloma cylindrosporum* Romagn. *New Phytologist* 114, 227–234.

Schimel, J.P., Bennett, J. 2004. Nitrogen mineralization: challenges of a changing paradigm. *Ecology* 85, 591–602.

Selle, A., Willmann, M., Grunze, N., et al. 2005. The high-affinity poplar ammonium importer PttAMT1.2 and its role in ectomycorrhizal symbiosis. *New Phytologist* 168, 697–706.

Smith, S.E., Read, D.J. 1997. *Mycorrhizal Symbiosis*, 2nd edition. Academic Press, San Diego, USA.

Smith, S.E., Read, D.J. 2008. *Mycorrhizal Symbiosis*, 3rd edition. Academic Press, San Diego, USA.

Taylor, A.F.S., Martin, F., Read, D.J. 2000. Fungal diversity in ectomycorrhizal communities of Norway spruce (*Picea abies* (L) Karst) and Beech (*Fagus sylvatica* L) in forests along north-south transects in Europe. In: *Carbon Nitrogen Cycling in European Forest Ecosystems* (Ed, Schulze, E.D.). Ecological studies, vol. 142. pp 343–365. Springer-Verlag, Berlin.

Tibbett, M., Sanders, F.E., Cairney, J.W.G., et al. 1999. Temperature regulation of extracellular proteases in ectomycorrhizal fungi (*Hebeloma* spp.) grown in axenic culture. *Mycological Research* 103, 707–714.

Tibbett, M., Sanders, F.E., Minto, S.J., et al. 1998. Utilization of organic nitrogen by ectomycorrhizal fungi (*Hebeloma* spp.) of arctic and temperate regions. *Mycological Research* 102, 1525–1532.

Tuskan, G.A., Difazio, S., Jansson, S., et al. 2006. The genome of black cottonwood, Populus trichocarpa (Torr. & Gray). *Science* 313, 1596–604.

van der Heijden, M.G., Bardgett, R.D., van Straalen, N.M. 2008. The unseen majority: soil microbes as drivers of plant diversity and productivity in terrestrial ecosystems. *Ecology Letters* 11, 296–310.

Velasco, I., Tenreiro, S., Calderon, I.L., et al. 2004. *Saccharomyces cerevisiae* Aqr1 is an internal-membrane transporter involved in excretion of amino acids. *Eukaryotic Cell* 3, 1492–1503.

Vieublé Gonod, L., Jones, D.L., Chenu, C. 2006. Sorption regulates the fate of the amino acids lysine and leucine in soil aggregates. *European Journal of Soil Sciences* 57, 320–329.

von Wiren, N., Gazzarrini, S., Gojon, A., et al. 2000. The molecular physiology of ammonium uptake and retrieval. *Current Opinion Plant Biology* 3, 254–261.

Wallenda, T., Read, D.J. 1999. Kinetics of amino acid uptake by ectomycorrhizal roots. *Plant, Cell Environment* 22, 179–187.

Whiteside, M.D., Treseder, K.K., Atsatt, P.R. 2009. The brighter side of soils: Quantum dots track organic nitrogen through fungi and plants. *Ecology* 90, 100–108.

Willmann, A., Weiss, M., Nehls, U. 2007. Ectomycorrhiza-mediated repression of the high-affinity ammonium importer gene AmAMT2 in *Amanita muscaria*. *Current Genetic* 51, 71–78.

Wipf, D., Bendjia, M., Tegeder, M., et al. 2002. Characterization of a general amino acid permease from *Hebeloma cylindrosporum*. *FEBS Letters* 528, 119–124.

Witte C.-P., Ross, M.G., Romeis, T. 2005. Identification of three urease accessory proteins that are required for urease activation in Arabidopsis. *Plant Physiology* 139, 1155–1162.

Wright, D.P., Johansson, T., Le Quéré, A., et al. 2005. Spatial patterns of gene expression in the extramatrical mycelium and mycorrhizal root tips formed by the ectomycorrhizal fungus *Paxillus involutus* in association with birch (*Betula pendula*) seedlings in soil microcosms. *New Phytologist* 167, 579–596.

Yamanaka, T. 1999. Utilization of inorganic and organic nitrogen in pure cultures by saprotrophic and ectomycorrhizal fungi producing sporophores on urea-treated forest floor. *Mycological Research* 103, 811–816.

Chapter 5
Proteins in the Rhizosphere: Another Example of Plant-Microbe Exchange

Clelia De-la-Peña and Jorge M. Vivanco

Introduction

Nitrogen is a basic macronutrient for plants. It is obtained through root absorption or through associations with microbes such as nodule-forming bacteria or mycorrhizal and ectomycorrhizal fungi. Roots have long been considered to be food and water providers for plants, as well as providing anchorage in the soil. However, it is now appreciated that roots are more than an organic web in the soil. Many morphological and physiological properties of roots such as root vigor, root length/density, and nitrogen transport and metabolism, are very important for nitrogen acquisition and assimilation efficiencies (Garnett et al., 2009). Roots have various roles (Enstone et al., 2002; Hawes et al., 2003; Hodge et al., 2009; Raven and Edwards, 2001), but one of the most important, and least understood, is the release of a wide range of organic and inorganic compounds into the rhizosphere (Badri and Vivanco, 2009; Bais et al., 2004; Bertin et al., 2003; Hinsinger, 1998; Rovira, 1969; Uren, 2007). Root exudation plays important structural, biological, biochemical, and ecological roles in the underground (Bais et al., 2006; Bertin et al., 2003; Currier and Strobel, 1976; Liao et al., 2006; Nóbrega et al., 2005; Parker et al., 2000; Rovira, 1969; Vancura and Hanzlíková, 1972).

The term rhizosphere was defined by Kang and Mills (2004) as the soil closely surrounding the plant roots, although the term has been used since the early 1960s (Rouatt et al., 1960). The rhizosphere is influenced by the flux of mineral nutrients, such as nitrogen, into the roots and accumulation of root and bacterial exudates in the surrounding soil. Organic rhizodeposition consists mainly of border cells, mucilage, carbon-containing compounds, and proteins that are released into the soil (Rovira, 1969; Uren, 2007; Vancura and Hanzlíková, 1972). In the rhizosphere, where there is an abundance of carbon, plants and microorganisms interact very closely and dynamically. Such carbon abundance is one of the reasons why microbes are more abundant in this area than in the bulk soil (Bonkowski et al., 2009; Butler et al., 2003; Haichar, et al. 2008; Morgan et al., 2005). Soil structure is improved by arbuscular mycorrhizal fungi (AM)

Ecological Aspects of Nitrogen Metabolism in Plants, First Edition. Edited by Joe C. Polacco and Christopher D. Todd.
© 2011 by John Wiley & Sons, Inc. Published 2011 by John Wiley & Sons, Inc.

through the secretion of a glycoprotein called glomalin (Purin and Rillig, 2007). This protein has been found to have important environmental implications because once glomalin is detached from the hyphae, it moves into the soil, becoming part of it and contributing to soil structure. Therefore, without glomalin, soil erosion could increase (Haddad and Sarkar, 2003). Purin and Rillig (2007) reported that glomalin was a putative heat shock protein (HSP) homolog that could have dual activity in the soil. One activity could be related to assisting folding of soil proteins, thus protecting them against denaturation; the second activity is related to the soil aggregation properties of glomalin. The authors stated that this second theorized role justifies the energetic cost of glomalin production and secretion, a substantial nitrogen investment by the fungus. Based on these studies, the question is raised as to whether other rhizosphere proteins, in addition to glomalin, could have important role(s). With the advent of new technologies, identification of secreted proteins has revealed important clues about possible function, but the ecological and environmental relevance of many of these macromolecules still remain unresolved (Charmont et al., 2005; Farrar et al., 2003; Hawes, 1990; Wen et al., 2007).

Relatively little is known about the identity or function of proteins secreted by roots, or about the processes involved in protein exudation related to nitrogen investment. There is much work about extracellular proteins possibly involved in the interaction between plants and microbes, and they are listed in Table 5.1. Clearly, however, we still do not know the precise effect on nitrogen cost in the parasitism/pathogenicity relationship. Plants invest "big" in nitrogen resources on extracellular proteins to combat pathogens present in the rhizosphere. Among the extracellular proteins shown to be secreted in the rhizosphere are peroxidases (Barloy-Hubler et al., 2004; De-la-Peña, et al. 2008), phosphatases (Joner et al., 2000; Wasaki et al., 2003), fumarases (Kim and Lee, 2002), esterases (Aparna et al., 2007), proteases (Calmels et al., 1991; De-la-Peña et al., 2008; Godlewski and Adamczyk, 2007; Mastronunzio et al., 2008; Misas-Villamil and van der Hoorn, 2008), chitinases (De-la-Peña et al., 2008; Ovtsyna et al., 2005), glucanases (De-la-Peña et al., 2008; Misas-Villamil and van der Hoorn, 2008), thaumatin-like proteins (Borderies et al., 2004; De-la-Peña et al., 2008), and ribosome-inactivating proteins (Park et al., 2002).

Table 5.1. Extracellular proteins involved in the plant-microbe interaction.

Protein name	Putative function	Plant	Microbe	Reference
Peroxidase	Antifungal activity	*Triticum aestivum*	*Fusarium culmorum*	(1)
	Defense response	*Arabidopsis thaliana*	*Pseudomonas syringae*	(2)
	Nodulation signal	*Medicago sativa*	*Sinorhizobium meliloti*	(2)
	Nodulation signal	*Sesbania rostrata*	*Ralstonia solanacearum*	(3)
Phosphatase	Symbiosis	*Tagetes spp*	*Glamus etonicatum*	(4)
Lipase	Defense	*Arabidopsis thaliana*	*Alternaria brassicicola*	(5)
Protease	Defense	*Medicago sativa*	*Pseudomonas syringae*	(2)
	Fungal resistance	*Lycopersicon esculentum*	*Phytophthora infestans*	(6)
	Exopolysaccharide synthesis and symbiosis	*Medicago sativa*	*Sinorhizobium meliloti*	(7)
Superoxide dismutase	Symbiosis	*Medicago sativa*	*Sinorhizobium meliloti*	(2)
	Signaling	*Arabidopsis thaliana*	*Pseudomonas syringae*	(8)
Lipid transfer proteins	Defense	*Vigna unguiculata*	*Fusarium oxysporum*	(9)
Chitinase	Defense	*Vigna unguiculata*	*Fusarium oxysporum*	(9)
	Nod factor modifier	Pea	*Rhizobium leguminosarum*	(10)
	Nod factor modifier	*Medicago sativa*	*Sinorhizobium meliloti*	(2)

Table 5.1. (cont'd)

Protein name	Putative function	Plant	Microbe	Reference
Glucanase	Defense	*Arabidopsis thaliana*	*Pseudomonas syringae*	(2)
Catalase	Pathogenicity	*Secale cereale*	*Claviceps purpurea*	(11)
	Defense	*Vigna unguiculata*	*Fusarium oxysporum*	(9)
Thaumatin	Symbiosis	*Medicago sativa*	*Sinorhizobium meliloti*	(2)
Lectin	Plant-microbial communication	Wheat	*Azospirillum brasilense*	(12)
	Symbiosis	*Glycine soja*	*Rhizobium japonicum*	(13)
	Symbiosis	*Dolichos biflorus*	*Sinorhizobium meliloti*	(14)
	Symbiosis	*Medicago sativa*	*Sinorhizobium meliloti*	(2)
	Nodulation	*Glycine Max*	*Bradyrhizobium japonicum*	(15)
	Nodulation	*Glycine Max*	*Bradyrhizobium japonicum*	(16)
Polygalacturonase	Pathogenic response	*Arabidopsis thaliana*	*Pseudomonas syringae*	(2)
	Defense response	*Brassica napus*	*Sclerotinia sclerotiorum*	(17)
Urease	Defense	*Glycine max*	*Bacillus pasteurii*	(18)
Flagellin	Signaling	*Arabidopsis thaliana*	*Pseudomonas syringae*	(2)
	Signaling	*Lycopersicon esculentum*	*Pseudomonas syringae*	(19)
	Signaling	*Medicago sativa*	*Sinorhizobium meliloti*	(2)
EF-Tu	Signaling	*Arabidopsis thaliana*	*Pseudomonas syringae*	(2)
	Elicitation	*Arabidopsis thaliana*	*Echerichia Coli*	(20)
Carbamate kinase	Symbiosis	*Medicago sativa*	*Sinorhizobium meliloti*	(2)
Flagellar hook FlgK	Signaling	*Arabidopsis thaliana*	*Pseudomonas syringae*	(2)
Outer Membrane Protein TolC	Exopolysaccharide biosynthesis, antimicrobials resistance, and symbiosis	*Medicago sativa*	*Sinorhizobium meliloti*	(21)
PR-10	Signaling	*Medicago sativa*	*Pseudomonas syringae*	(2)
PR-1	Defense	*Arabidopsis thaliana*	*Pseudomonas syringae*	(2)
Proteinase K	Defense	*Arabidopsis thaliana*	*Pseudomonas syringae*	(?)
Ubiquitin	Cell-cell communication	*Arabidopsis thaliana*	*Pseudomonas syringae*	(8)

Reference List
1. M.P. Aleandri, P. Magro, G. Chilosi, *Plant Pathol* **57**, 1017 (2008).
2. C. De-la-Peña, Z. Lei, B.S. Watson, L.W. Sumner, J.M. Vivanco, *J. Biol. Chem.* **283**, 25247 (2008).
3. J. Den Herder, S. Lievens, S. Rombauts, M. Holsters, S. Goormachtig, *Plant Physiol.* **144**, 717 (2007).
4. T. Ezawa, T. Yoshida, *Soil Sci. Plant Nutr.* **40**, 655 (1994).
5. I.S. Oh et al., *Plant Cell* **17**, 2832 (2005).
6. M.Y. Tian et al., *Plant Physiol.* **143**, 364 (2007).
7. M.L. Summers, L.M. Botero, S.C. Busse, T.R. McDermott, *J. Bacteriol.* **182**, 2551 (2000).
8. F.A.R. Kaffarnik, A.M.E. Jones, J.P. Rathjen, S.C. Peck, *Mol Cell Proteomics* **8**, 145 (2009).
9. F.M. Nóbrega, I.S. Santos, M. Da Cunha, A.O. Carvalho, V.M. Gomes, *Plant Soil* **272**, 223 (2005).
10. A.O. Ovtsyna et al., *Mol Plant Microbe Interact* **13**, 799 (2000).
11. V. Garre, K.B. Tenberge, R. Eising, *Phytopathology* **88**, 744 (1998).
12. L. Antonyuk, N. Evseeva, *Microbiology* **75**, 470 (2006).
13. G. Stacey, A.S. Paau, W.J. Brill, *Plant Physiol.* **66**, 609 (1980).
14. M.E. Etzler et al., *Proc Natl Acad Sci USA* **96**, 5856 (1999).
15. L.J. Halverson, G. Stacey, *Appl. Environ. Microbiol.* **51**, 753 (1986).
16. L.J. Halverson, G. Stacey, *Plant Physiol.* **77**, 621 (1985).
17. D.D. Hegedus et al., *Planta* **228**, 241 (2008).
18. C.R. Carlini, J.C. Polacco, *Crop Sci* **48**, 1665 (2008).
19. G. Felix, J.D. Duran, S. Volko, T. Boller, *Plant J.* **18**, 265 (1999).
20. G. Kunze et al., *Plant Cell* **16**, 3496 (2004).
21. A.M. Cosme et al., *Mol Plant Microbe Interact* **21**, 947 (2008).

Proteins secreted to the rhizosphere are mainly stored in the root apoplast (Buchanan et al., 2000), the collective space outside the plasma membrane (Dietz, 1996). This space includes plant cell walls that usually provide an effective barrier to most microbial pathogens and the environment. Recent work on plant–pathogen interactions has distinguished the complex arsenal of extracellular defenses that are localized in the apoplast (Enstone et al., 2002; Jones and Takemoto, 2004; Veronese et al., 2003; Vorwerk et al., 2004). Therefore, it is imperative to identify bona fide apoplastic proteins and determine their functions. The apoplast has been referred to as a space where nitrogen-fixing endophytes reside and which facilitates the establishment of symbiosis with plants (Dong et al., 1994; Tejera et al., 2006). Tejera and others (2006) suggest that the presence of proteins in the apoplast could regulate the biological activity of endophytes under symbiotic conditions. When a plant establishes a symbiosis (from the Greek συν *syn* "with"; and βιωσισ *biosis* "living") relationship with a bacterium it means that the bacterium as well as the plant can go from parasitism/pathogenicity to commensalism to mutualism (Paszkowski, 2006). However, in this chapter, we will consider symbiosis by the more narrow definition: a relationship from which both the plant and associated bacteria benefit. This means, for instance, that in a symbiotic relationship between leguminous plants and *Rhizobium*, the plant provides the bacteria with carbon and, in return, the bacteria fix nitrogen that plants can take up (Barea et al., 2005).

Although symbiosis has been described as a friendly relationship, before there is any real contact between the plant and the bacteria, checkpoint recognition must first occur (Oldroyd et al., 2005). This recognition starts with plant defense response (Baron and Zambryski, 1995b). In terms of protein secretion, among the proteins best known to be involved in plant defense response and innate immunity are the pathogenesis-related (PR) proteins (De-la-Peña et al., 2008; Fritig et al., 1998; Mitsuhara et al., 2008; Nimchuk et al., 2003; Sels et al., 2008; Wen et al., 2007). The PR proteins thus far identified include chitinases, glucanases, peroxidases, lipid transferases, thaumatin-like proteins, proteinase inhibitors, endoproteinases, ribonuclease-like proteins, defensins, thionins, oxalate oxidases, and others with unknown biochemical roles (De-la-Peña et al., 2008; Ferreira et al., 2007; Sels et al., 2008). The first PR proteins investigated were shown to be present in the leaves of *Samsun NN* and *Xanthi* tobacco (*Nicotiana tabacum* L.) plants during the hypersensitive reaction (HR) against tobacco mosaic virus (TMV) (Carr et al., 1987; van Loon, 1976; van Loon, 1985). More recently, there are reports of PR proteins in root exudates, proteins that are secreted constitutively as well as others under pathogen contact (Basu et al., 2006; De-la-Peña et al., 2008; Nóbrega et al., 2005). The PR proteins exported to the extracellular medium are secreted from conventional secretory pathways as well as from vacuoles (Dore et al., 1991; Hunt and Chrispeels, 1991; Kunze et al., 1998; Surpin and Raikhel, 2004; Vera et al., 1989). Figure 5.1 represents the proposed process of PR protein secretion. When attacked by a pathogen, the plant produces a range of defense-related proteins that are synthesized on the rough endoplasmic reticulum (ER) and then secreted from the cell or deposited in vacuoles. Genes encoding ER resident chaperones, such as the lumenal binding protein (BiP) (Denecke et al., 1991), are also induced under infection by pathogens (Jelitto-Van Dooren et al., 1999). It has been proposed that the induction of BiP expression during plant-pathogen interactions is required as an early response to enhance PR protein synthesis in the rough ER that incites the signal transduction pathway to trigger a rapid defense response (Jelitto-Van Dooren et al., 1999).

The field of plant responses to biotic and abiotic stresses is expanding rapidly; it is not our intention to review this large area of investigation. Instead, this chapter describes what is known about proteins secreted by roots focusing on two examples of plant-microbe interactions (Figure 5.2): the prototypical example of symbiosis in a mutually beneficial association of

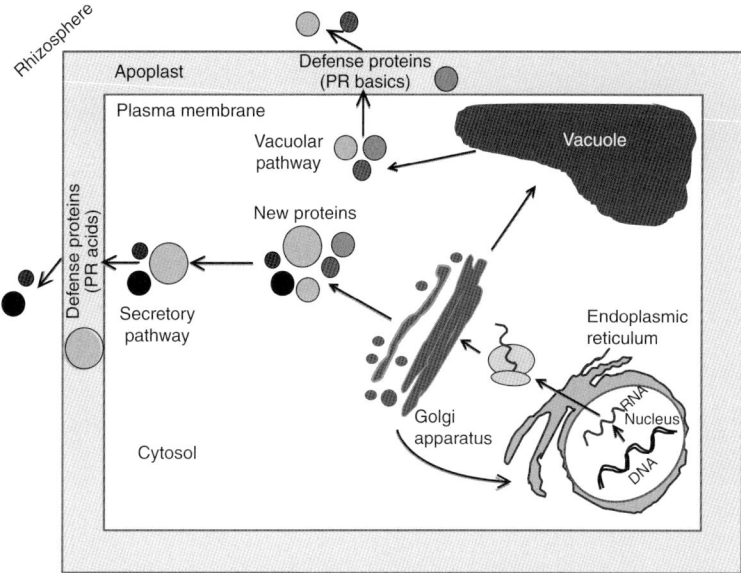

Figure 5.1. A simplified map of the PR protein secretion pathway in higher plants. Proteins are synthesized and processed in the ER. The proteins are then transferred directly to the closely apposed and mobile Golgi apparatus. Some proteins are targeted to the apoplast and secreted into the rhizosphere, and some are kept in the cytosol. However, direct transfer of some proteins to the vacuole may occur. Proteins from vacuolar origin can be secreted to the apoplast and subsequently into the rhizosphere.

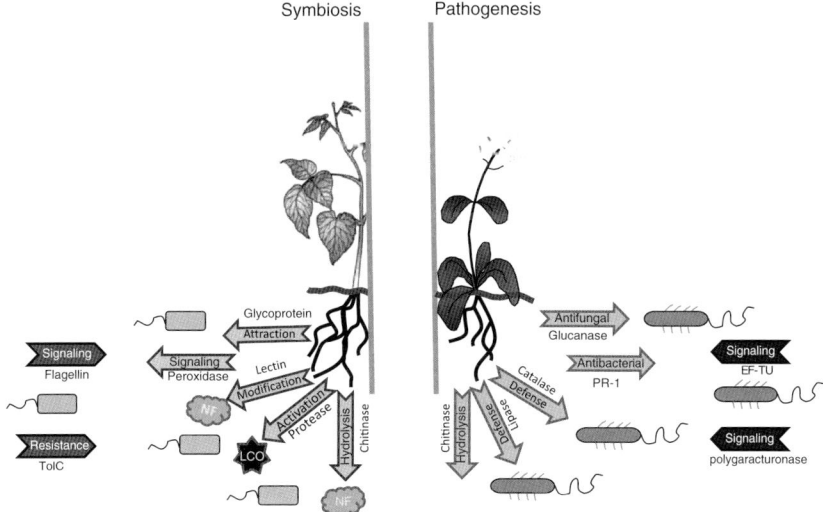

Figure 5.2. Model showing plausible mechanisms of protein exudation in plant-microbe interactions. Proteins of diverse plants and microbes act in the rhizosphere by achieving different activities in legumes (left), *Arabidopsis* (right), rhizobia (left), and *Pseudomonas* (right). Pathogenic bacteria and beneficial symbionts use protein factors to communicate with the host. For instance, polygalacturonases of pathogen origin are harmful to the plant cell wall that contains their substrate. Chitinases and glucanases secreted by plants after pathogen recognition are detrimental to fungal cell walls that contain their substrates. In addition, legumes can secrete proteins that cleave nodulation factors (NF) avoiding the nodulation process. Plant proteases could activate the lipochitooligosaccharides (LCO) that rhizobia secrete inducing the signaling process for nodulation.

bacteria of the family *Rhizobiaceae* with leguminous plants, and the example of the infection of *Arabidopsis thaliana* with pathogenic *Pseudomonas* species that may ultimately kill the infected host plant. Initial cataloguing of proteins secreted into the rhizosphere will help to build models to understand the basic principles of the plant response to intruders versus the response to symbionts. These studies are also leading to an understanding of the pattern of root protein exudation in response to nutrient limitation.

Rhizosphere Proteins Orchestrate Communication Between Plants and Soil Microbes

The chemical nature of most enzymes in soil, particularly due to their size and charge characteristics, is such that they would have to operate close to the point of secretion, and their substrates must be able to diffuse to the enzyme (Uren, 2007) absent major extension of the root network. Proteins secreted into the rhizosphere have been the subject of increasing interest as possible plant defenders and players in the chemotaxis for symbiotic process between *S. meliloti* and alfalfa (Currier and Strobel, 1977; Gramss and Rudeschko, 1998; Nóbrega et al., 2005). The exudation of root proteins, together with other compounds, into the rhizosphere is influenced by the microbial population (Mehta et al., 2008), the availability of such nutrients as nitrogen, and plant competition with other plants and microbes (de Kroon, 2007; Kaye and Hart, 1997; Reynolds et al., 2003). These compounds, metabolites and proteins, are key components in interactions with neighboring plants and microbes (Bais et al., 2004; Bais et al., 2006; Broeckling et al., 2008; De-la-Peña et al., 2008), as well as root-root and even root-insect communication (Bonkowski et al., 2009; Lin et al., 2008; Walker et al., 2003).

Root-exuded plant proteins act as signal molecules to microbes in the rhizosphere (Bais et al., 2006; De-la-Peña et al., 2008; Morgan et al., 2005; Wen et al., 2007). Because microorganisms in the rhizosphere are so dependent on root exudates as a food source, plants largely control the interactions, the abundance of microbes, and processes that exist in the rhizosphere (Kourtev et al., 2003). Furthermore, plants have developed systems for sensing the presence of microorganisms or of microbial molecules (Barea et al., 2005; Morgan et al., 2005; Watt et al., 2006).

Plant cells secrete cocktails of constitutively expressed or inducible proteins into the apoplast. Several studies have described changes in the population of secreted proteins in nutrient deficiency (Marschener, 1998), response to wounding (Kawaoka et al., 1994), fungal infection (Broeckling et al., 2008; Spanu et al., 1989), fungal elicitors (Ndimba et al., 2003) and the defense-related hormone jasmonic acid (Cho et al. 2007; Regvar and Gogala, 1996). Further, plants secrete proteinaceous inhibitors of proteins secreted by the pathogen into the apoplast to aid in colonization (York et al., 2004). It has been recognized that root-secreted proteins play a major role in the defense process against plant pathogens (Carlini and Polacco, 2008; Park et al., 2002). On the other hand, bacterial-secreted proteins have important roles for the successful nodulation in certain rhizobia-legume interactions (Krehenbrink and Downie, 2008). Many plant proteins released into liquid medium remain unexplored (not to mention those released in "more natural" soils), and thus we cannot rule out the possibility of their impact in the rhizosphere for plant-microbe cross talk. De–la-Peña and others (2008) found that the *M. sativa-S. meliloti* interaction caused an increase in the secretion of such plant PR proteins as hydrolases and peroxidases as early as six hours after infection compared with the control plant. In addition, four proteins (a superoxide dismutase, a hypothetical protein SMc02156, a putative glycine betaine-binding ABC transporter protein, and a putative outer membrane lipoprotein, all of bacterial origin, were accumulated 1.5-fold in the exudates of the *M. sativa-S. meliloti* interaction compared with *S. meliloti* alone. The profile of secreted plant proteins when *M. sativa* was inoculated with *P. syringae* DC3000 versus *S. meliloti* was radically different. The profile included high levels of several plant proteins related to defense soon after initial contact with *P. syringae*, proteins that were not secreted in the incompatible interaction with *S. meliloti*.

Although PR proteins were named for their relationship to the pathogenesis response, there are reports of PR proteins found in healthy plants (Grüner and Pfitzner, 1994; Regalado and Ricardo, 1996). The chitinase family, a well-studied group of PR proteins, catalyzes the hydrolysis of the β-1,4-glycosidic bonds linking the *N*-acetyl glucosamine residues of chitin. The chitinases are a structurally diverse group with respect to their physical properties, enzymatic activities and localization (Graham and Sticklen, 1994; Kasprzewska, 2003). It has been found that these enzymes are expressed not only in the presence of a pathogen, but in the presence of nonsymbiotic and symbiotic rhizobia (Salzer et al., 2000; Salzer et al., 2004; Staehelin et al., 1994). We point out that though alfalfa needs the bacteria to fix nitrogen under nitrogen deficiency, it reacts to infection by *Rhizobium* by eliciting a defence mechanism similar to the hypersensitive response (HR) observed in incompatible plant-pathogen interactions (Vasse et al., 1993). Therefore, there must be a common signalling pathway before the plant distinguishes a "malign" from a benign microbe. This detection of microorganisms induces developmental pathways in the plant that culminate in a defence or symbiotic response (Baron and Zambryski, 1995a; Herouart et al., 2002; Mithöfer, 2002). Lohar and others (2006) found that in stage I, the first of four stages in the nodulation process, all the defense/disease/stress responses occur in the plant. This stage is related to root hair swelling, and it happens as early as 1 hour after infection. Therefore, whatever the signal recognition that discriminates between a friend and an enemy, it must happen during the first hour, and whether invading rhizobia are recognized as putative pathogens remains an unresolved question. Many hypotheses about symbiosis and the early defense responses by the host plant have been generated (Lohar et al., 2006; Mithöfer, 2002). However, none of them involved protein secretion as an early signal for symbiotic or pathogenic recognition.

We postulate that during the interaction between the plant and a rhizosphere bacterium, once the plant determines whether the bacterium is pathogenic or symbiotic (or at least non-harmful), the ensuing signalling pathway is different in each case. This signalling process will induce one of two pathways: induced resistance (IR) or systemic acquired resistance (SAR). IR is characterized by increase in the levels of jasmonic acid (JA) (Cooper and Goggin, 2005; van Wees et al., 1999), and SAR by the accumulation of salicylic acid (SA) (Klessig and Malamy, 1994; Sticher et al., 1997). Genes that encode acidic and basic PR proteins are activated by SA- and JA-dependent signalling pathways, respectively (Figure 5.3). Accordingly, De–la-Peña and others (2008) found acidic chitinases accumulated in the rhizosphere in a pathogenic interaction while basic chitinases accumulated in a nonpathogenic interaction.

Signalling and Regulation

At this point the signalling and regulation of proteins secreted in the rhizosphere under environmental influences on plant nitrogen metabolism are not well known. However, the great advance in the discovery of biological roles for proteins with formerly unknown function has helped to understand many dynamics for these macromolecules in the rhizosphere.

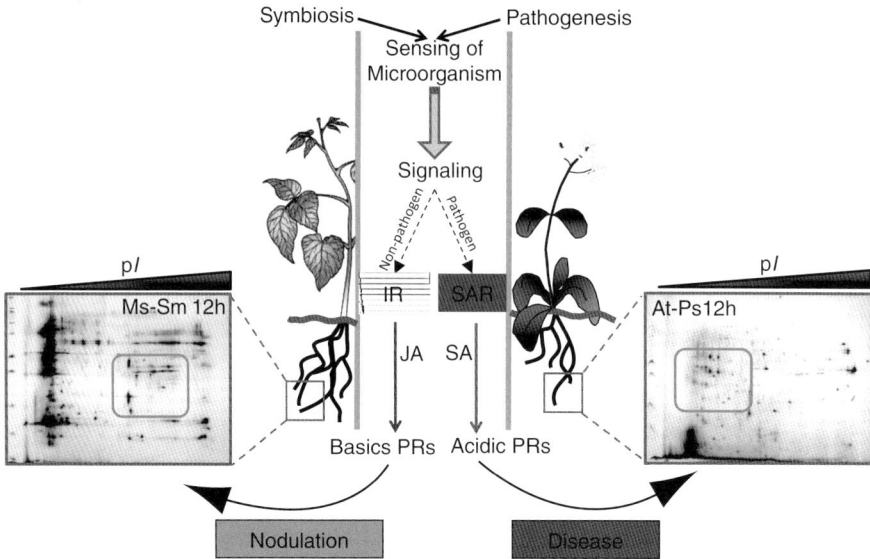

Figure 5.3. Proposed model for the signaling responses mediated by protein secretion in the mutually beneficial association of *Sinorhizobium meliloti* with *Medicago sativa*, and the pathogenic relationship between *Arabidopsis* and *Pseudomonas* species. Here, we describe a few salient features of these signal responses. The first response between plants and microbes is the sensing of the microbe by perceiving signals such as chitin, which unleashes the signaling cascade in the plant. This cascade of signals in the plant will direct the reaction into a pathogenic or nonpathogenic response. Plant defenses targeted by pathogens to promote disease will induce the Systemic Acquired Resistance (SAR) while those induced by Induced Resistance (IR) will be directed to the specific recognition between protective bacteria and the plant. SAR is characterized by an early increase in salicylic acid (SA) and the concomitant activation of genes encoding pathogenesis-related (PR) proteins. The IR, on the other hand, is associated with changes in the expression of jasmonic acid (JA-associated) genes (biosyntheses, signaling, etc.). Genes that encode acidic and basic PR proteins, distinguished by their pI (isoelectric points), are activated by SA- and JA-dependent signaling pathways, respectively.

Most of the principal functions discovered in recent years have been related to defense. Most of the defense-related proteins possess antimicrobial activities *in vitro* through hydrolytic activities on cell walls, contact toxicity, and perhaps an involvement in defense signalling (Blumwald et al., 1998; Côté et al., 1991; de Wit, 2007; van Loon et al., 2006). How the rhizosphere microbial population is influenced by the amount and composition of plant root exudates has been studied for several decades (Rouatt et al., 1960). In general, microbes that inhabit the rhizosphere serve as an intermediary between the plant, which requires soluble inorganic nutrients, and the soil, which contains the necessary nutrients but mostly in complex and inaccessible forms. Rhizosphere microorganisms thus provide a critical link between plant and soil environments. Only microorganisms with appropriate receptors and enzymes can sense and use substrates secreted by plants. For instance, extracellular galactoside found in the rhizosphere of alfalfa (Bringhurst et al., 2001) could have important ecological repercussions (Ankenbauer and Nester, 1990; Nilsson et al., 2005).

Thus, regulation and signalling in the rhizosphere are very important for the establishment of relationships between plants and microbes. For instance, plant-produced/-secreted lectins serve as important signal proteins for symbiosis (Antonyuk and Evseeva, 2006; Stacey et al., 1980), endophytic association between *Azospirillum brasilense* and wheat plants (Antonyuk and Evseeva, 2006), biofilm formation and bacterial attachment to roots (Castellanos et al., 1998; Rodriguez-Navarro et al., 2000), and also for pathogenesis (Yamasaki et al., 2009). Lectins are capable of increasing the attachment, binding, and aggregation of rhizobia at the infection site (Rodriguez-Navarro et al., 2000; Stacey et al., 1980), and act as a factor in plant-microbial communication and a stress response protein (Antonyuk and Evseeva, 2006).

Although *Rhizobium* is considered to be a "good" microbe for the plant, at some stages its interaction with the plant resembles a pathogenic infection (Baron and Zambryski, 1995b). Therefore, the bacterium and the plant need to regulate the plant defense reaction or otherwise the "infection" can abort (Vasse et al., 1993). This defense control is characterized, in part, by necrosis of the bacterial infection threads and limitation of their spread. Also, plant proteins related to defense mechanisms, such as chitinases, tend to accumulate at the sites of "arrested" infection threads. Chitinases, defense proteins mentioned above, have been grouped into three classes. Class I are basic chitinases with an N-terminal cysteine-rich domain; class II chitinases are similar in structure to class I but lack the cysteine-rich domain; and class III are chitinases with conserved sequences different from those of the class I and II enzymes (Shinshi et al., 1990). Chitinase activity, generally at a low level in plants, is induced in response to various stimuli such as ethylene (Boller et al., 1983; Keefe et al., 1990), viral infections (Lawton et al., 1992; Ohme-Takagi et al., 1998; Payne et al., 1990) and various pathogenic microorganisms (Kästner et al., 1998).

The ability of plants to secrete proteins, especially proteases, has been considered an efficient method to increase the levels of free amino acids in the soil as a source of nitrogen (Godlewski and Adamczyk, 2007). Proteases also have been found to have regulatory effects in nodulation. Summers and others (2000) found a Lon protease secreted by *S. meliloti* that is required for nodulation by regulating the synthesis of exopolysaccharides. Although this protein is ubiquitous and conserved among prokaryotes, it has also been found in Arabidopsis, maize, and rice (Barakat et al., 1998; Ostersetzer et al., 2007; Rigas et al., 2009; Su et al., 2006) observations that suggest that plants could use this, or other, proteases to regulate relationships with microorganisms, and that perhaps proteases may have the same effect as their bacterial counterparts. The increasing information about the participation of proteases in nitrogen supply (Godlewski and Adamczyk, 2007; Kingston-Smith et al., 2005) has refocused our attention on the role of many root exudate proteins as important ecological molecules for plant nutrient metabolism.

Acidic phosphatases (APases), fumarases, α-mannosidases, and heat shock proteins are additional proteins indentified among the secreted proteins in several plant culture systems (Kunze et al., 1998; Mita et al., 1997; Petersen and Böttger, 1991; Tarafdar and Marschner, 1994). Root-secreted APases are used by the plant as phosphorous scavengers under phosphorous-limiting conditions (Tomscha et al., 2004). There is some evidence that plants accumulate more nitrogen and biomass when amino acids are the only nitrogen source as opposed to ammonium or nitrate sources (Chapin et al., 1993). Paungfoo-Lonhienne and others (2008) reported that plants can use proteins as a nitrogen source by either of two mechanisms: digestion in the soil with proteases secreted by roots or by taking up the proteins via endocytosis. The same group (Paungfoo-Lonhienne et al., 2009) found that the expression of peptide transporters in the uptake and transport of peptides is regulated in response to nitrogen supply. Furthermore, the uptake of nitrogen in the form of amino acids, with the exception of D-enantiomers, is rapid, and the amino acids are efficiently utilized by the plant (Nasholm et al., 2009). However, the use of organic nitrogen is not restricted to amino acids; dipeptides are also considered to be a nitrogen source and transport form in plants (Komarova et al., 2008; Senwo and Tabatabai, 1998).

Peptides also play crucial roles in defense signalling. Systemin (Pearce and Ryan 2003) and phytosulfokine (PSK) (Matsubayashi and Sakagami, 1996) are two of the most studied peptides. For example, systemin from tomato regulates the synthesis of defensive proteins in plant tissues (McGurl et al., 1992; Pearce et al., 1991). Several groups have isolated peptide signals from plants (Broekaert et al., 1995; García-Olmedo et al., 2001; Lindsey et al., 2002; Matsubayashi and Sakagami, 2006). We point out that while these signalling peptides are generated endogenously, the roles of rhizosphere peptides (generated by microbes or other plants) should not be *a priori* discounted. Indeed, arabinogalactan proteins (AGP), found in root exudates, have not been found to be related to plant nutrition. Rather, they have been considered to be involved in the perception of stimuli-causing defense responses (Mashiguchi et al., 2008).

AGPs are found in the apoplast, culture media, and root exudates (Phillips, 2009; Sanchez et al., 2009; Seifert and Roberts, 2007). Although the precise biological activity of the AGPs is unknown, the molecular composition of AGPs provides the conceptual possibility of involvement in many different signalling processes. For example, AGPs may be enzymatically processed to release oligosaccharide signals, which then could bind to a cell membrane receptor tied into a signal transduction cascade system. Such a model could include extracellular, cell wall, and plasma membrane AGPs as potential substrates.

Peroxidases (POX) may have roles in signalling and regulatory processes in the rhizosphere. POXs are released by roots into the surrounding medium (De-la-Peña et al., 2008). Internally, they catalyze many oxidation-reduction reactions in the presence of H_2O_2 (or organic hydroperoxides) as an oxidizing agent. They are induced by wounding (Kawaoka et al., 1994), pathogen attack (Almagro et al., 2009; Bolwell et al., 2001; Lamb and Dixon, 1997), as well as during plant development (Brownleader et al., 2000). In the plant, POXs may either act on phenylpropanoids in the biosynthesis of lignin or mediate cross-linking of polysaccharides or strengthening cell walls. The distribution of POXs in the cell is spatially differentiated; the most acidic POXs are confined to the cell wall or extracellular sites, whereas the cationic forms are distributed both in the cell wall and the cytoplasmic compartments. The defense signalling of secreted POXs was proposed by Kawano and Muto (2000) who examined a cell suspension culture of tobacco (*N. tabacum* L. cv. Bright Yellow-2, cell line BY-2). They claimed that peroxidase-catalyzed reactions generate the signalling for SA formation, which through the production of active oxygen species increases the cytosolic Ca^{2+} concentration. Elevated cytosolic Ca^{2+}, in turn, results in the induction of PR genes. With respect to root-microbe interactions, one POX gene, *Trprx2*, was expressed only when legume (white clover) roots were infected with either homologous or heterologous *Rhizobia*, but its expression was dramatically reduced with *Pseudomonas* (Crockard et al., 1999).

Extensin is a protein hypothetically secreted into the rhizosphere. Extensins are hydroxyproline-rich cell wall glycoproteins (Shpak et al., 1999) which are synthesized on the ER, glycosylated in the Golgi apparatus, and secreted to the cell walls (Chrispeels, 1976). Bradley and others(1992) have reported that elicitation of soybean results in a rapid, transcription-independent, oxidative cross-linking of cell-wall structural proteins, including extensin-like hydroxyproline-rich glycoproteins. This deposition was concomitant with an increase in cell-wall resistance to enzymatic digestion (Brisson et al., 1994). In the roots of some legumes extensins, named root nodule extensins have been hypothesized to play a role in cell colonization by *Rhizobium* bacteria (Rathbun et al., 2002).

Defense Protein Secretion Involves Sophisticated Cross Talk Between Roots and Pathogens

Root exudates, directly or indirectly, influence the microbial composition and species richness at the soil-plant interface, stimulating different interactions. One of the most studied plant-directed root-microbe interactions involves the secretion and accumulation of antimicrobial compounds in and around the roots (Bais et al., 2004). Wen and others (2007) hypothesize that proteins released during border cell separation may play a vital role in the plant's system of innate immunity.

In a compatible interaction, plant disease develops, and subsequently the host dies. In an incompatible interaction, a resistant host plant establishes a set of different defense mechanisms directed against the pathogen, mechanisms that include the generation and accumulation of reactive oxygen species (ROS) as well as the expression of PR proteins (Benhamou, 1996; Ebel and Mithöfer, 1998). For instance, the role of chitinases as part of the inducible plant defense response is well documented (Boller, 1985; Collinge et al., 1993) and discussed above. Many other secreted proteins contribute to defense responses as well that result in wall reinforcement (Schulze-Lefert, 2004), lipase secretion (Oh et al., 2005), and an apoplastic oxidative burst (Pignocchi and Foyer 2003).

Extracellular enzymes released from soil microorganisms initiate the degradation of high-molecular-weight substrates in plants, such as cellulose, chitin, lignin, etc. As a counterattack many of the different inhibitory compounds in root exudates have both antibacterial and antifungal properties, although the compounds typically work synergistically to achieve general antimicrobial results (Bais et al., 2004). In addition to inhibitory compounds, proteins, such as polygalacturonase-inhibiting proteins (PGIP), which inhibit polygaracturonases (PG) secreted by plant pathogens, have roles in plant defense (Cervone et al., 1989). Furthermore, it has been suggested that in the soil, bacterial urease-induced secretion by plant roots may play a role in rhizosphere relationships, roles that may be distinct from their ureolytic activity (Carlini and Polacco, 2008). Polacco and Holland (1993) had earlier postulated how urease activity, considerable in the seeds of several plants, could play a role in plant defense.

Root-secreted Proteins Involved in Symbiosis

The possible function of proteins in the symbiosis process has been studied over the past decade (Fauvart and Michiels, 2008; Minic et al., 1998; Minic et al., 2000; Xie et al., 1999), though more attention has been paid to secondary metabolite secretion in this process. It is known that plant roots take up low MW nitrogenous compounds including ammonium, nitrate, and amino acids (Rentsch et al., 2007). Paungfoo-Lonhienne and others (2008) showed that the woody heath land plant *Hakea actites* and the herbaceous model plant *Arabidopsis*, which do not form mycorrhizal associations, can use protein as a nitrogen source for growth without assistance from other organisms. Just as proteins were found to be important in N assimilation, also it was found that proteins are important in chemotaxis (Currier and Strobel, 1977), which is one of the earliest essential events in the interaction between plants and microbes (Hawes and Smith, 1989; Manson, 1990). Currier and Strobel (1977) found an ~60 kDa glycoprotein, highly heat stable, produced and exuded by bird's-foot trefoil, that they named trefoil chemotactin, which was able to attract six different strains of rhizobia. Purified trefoil chemotactin attracted more bacteria than the crude root exudates. Furthermore, secreted proteins from alfalfa can also modify the well-known Nod factors from *Rhizobia* (Minic et al., 1998). *Rhizobium, Bradyrhizobium, Mesorhizobium, Sinorhizobium,* and *Azorhizobium* species produce extracellular lipo-oligosaccharidic Nod factors whose chitin-like core structures comprise residues of β-1,4-linked *N*-acetyl-D-glucosamine (Lerouge et al., 1990; Roche et al., 1991a; Roche et al., 1991b; Schultze et al., 1992). Because Nod factors are hydrolyzed by specific chitinases (Roche et al., 1991b), they are believed to act as substrates for chitinase-like enzymes that accumulate in HR cells during abortive infection (Vasse et al., 1993). In the same way, some rhizobia can resist plant defense mechanisms. Some rhizobia, for instance, are resistant to canavanine, which is a vital source of nitrogen in some legume embryos and, as an arginine analog, has defensive properties against herbivores (Rosenthal, 1986). This nonprotein amino acid was found in the exudates of *Glycyrrhiza uralensis* and likely canavanine exudation is a sophisticated mechanism by which legumes can "optimize" the symbiotic events with some rhizobia (Cai et al., 2009). Biochemical studies have already shown that plant defense mechanisms can occur during rhizobia-legume interaction. A localized HR response is one of the mechanisms involved in feedback regulation of the infection process in the alfalfa-*R. meliloti* symbiosis (Vasse et al., 1993). An example of a localized HR is the high levels of H_2O_2 in the roots of pea undergoing symbiotic association (Glyan'ko et al., 2005). This study provides a clue about the role of high levels of peroxidases in the rhizosphere as a possible signalling event between the bacteria and the plant. The increase in peroxidase activity in the root hairs of white clover and pea inoculated by heterologous *R. leguminosarum* nonsymbiont strains has been interpreted as a plant defense response (Salzwedel and Dazzo, 1993). Differential changes in cell wall structure can occur at different stages of nodule ontogeny and suggest that the apoplastic compartment as well as the proteins in it might play a key role in the control of nodulation (De-la-Peña et al., 2008; Vasse et al., 1993).

With respect to the mutualistic AM symbiosis studied in leeks infected with *Glomus mosseae*, chitinase activity increased during early stages of the interaction and was suppressed in later stages (Spanu et al., 1989). Similar results have been obtained in the alfalfa-*G. intraradices* interaction (Volpin et al., 1994). Some antimicrobial proteins such as chitinases and glucanases from root exudates are also effective in repressing the growth of root pathogenic fungi (Nóbrega et al., 2005). However, some root-secreted proteins with antifungal activity reduce the growth of pathogenic fungi whereas they do not interfere with the establishment of symbiosis by the AM fungus (Turrini et al., 2004). Antimicrobial Dm-AMP1 protein in root exudates of *Dahlia merckii* is a clear example of a selective enzyme very beneficial in symbiosis, but it is also a strong antimicrobial protein against some fungi (Turrini et al., 2004). An intriguing aspect of this selective response is that in the case of mutualistic symbioses only acidic PR proteins have been found to be secreted (Vasse et al., 1993). Therefore, we can hypothesize that PR proteins are important signal elements to differentiate between pathogenic and nonpathogenic microbes. The pattern of expression of genes encoding enzymes of the PR family could be crucial in the subsequent development of a mutualistic versus a pathogenic interaction.

Perspectives

We have described here the involvement of proteins in rhizosphere processes. However, it should be noted that most of these protein identifications have been determined under *in vitro* laboratory conditions, and in many cases from nonroot tissues. In some cases, proteins have been identified from other substrates, such as soil but their enzymatic activities have not been determined in these systems. Better systems need to be developed to determine whether these proteins exhibit enzymatic activities in the soil and rhizosphere. Enzymes for the most part need specific pH and co-factor conditions to perform a particular catalysis. Whether these conditions exist in the rhizosphere remains to be determined, and the empirical information developed by *in vitro* systems should provide ammunition to decipher these conditions.

We believe that a likely place to assay enzymatic activities in soils and rhizospheres is a tropical rainforest where the roots of several plant species closely co-interact with a variety of microbes and where tight and efficient degradation of organic matter occurs. In this respect, lignin- and cellulose-degrading activities in these tropical soils should be relatively easy to assay. Finally, the roles of extracellular rhizosphere proteins in plant N nutrition need to be investigated. Indeed, when the plant "makes a choice" to invest valuable nitrogen in root-secreted proteins, the payoff has to be significant, whether in nutritional gain, or in defense.

References

Almagro, L., Gómez Ros, L.V., Belchi-Navarro, S., et al. 2009. Class III peroxidases in plant defence reactions. *Journal of Experimental Botany*, 60 (2), 377–390.

Ankenbauer, R.G., Nester, E.W. 1990. Sugar-mediated induction of *Agrobacterium tumefaciens* virulence genes: structural specificity and activities of monosaccharides. *The Journal of Bacteriology*, 172 (11), 6442–6446.

Antonyuk, L., Evseeva, N. 2006. Wheat lectin as a factor in plant-microbial communication and a stress response protein. *Microbiology*, 75 (4), 470–475.

Aparna, G., Chatterjee, A., Jha, G., et al. 2007. Crystallization and preliminary crystallographic studies of LipA, a secretory lipase/esterase from *Xanthomonas oryzae* pv. *oryzae*. *Acta Crystallographica Section F Structural Biology and Crystallization Communications*, 63, 708–710.

Badri, D.V., Vivanco, J.M. 2009. Regulation and function of root exudates. *Plant, Cell Environment*, 32 (6), 666–681.

Bais, H.P., Park, S.W., Weir, T., et al. 2004. How plants communicate using the underground information superhighway. *Trends Plant Science*, 9 (1), 26–32.

Bais, H.P., Weir, T.L., Perry, L.G., et al. 2006. The role of root exudates in rhizosphere interactions with plants and other organisms. *Annual Review of Plant Biology*, 57 (1), 233–266.

Barakat, S., Pearce, D.A., Sherman, F., et al. 1998. Maize contains a Lon protease gene that can partially complement a yeast *pim1*-deletion mutant. *Plant Molecular Biology*, 37 (1), 141–154.

Barea, J.M., Pozo, M.J., Azcon, R., et al. 2005. Microbial co-operation in the rhizosphere. *Journal of Experimental Botany*, 56 (417), 1761–1778.

Barloy-Hubler, F., Cheron, A., Hellegouarch, A., et al. 2004. Smc01944, a secreted peroxidase induced by oxidative stresses in *Sinorhizobium meliloti* 1021. *Microbiology*, 150 (3), 657–664.

Baron, C., Zambryski, P.C. 1995a. Notes from the underground: Highlights from plant-microbe interactions. *Trends in Biotechnology*, 13 (9), 356–362.

Baron, C., Zambryski, P.C. 1995b. The plant response in pathogenesis, symbiosis, and wounding: Variations on a common theme? *Annual Review of Genetics*, 29 (1), 107–129.

Basu, U., Francis, J.L., Whittal, R.M., et al. 2006. Extracellular proteomes of *Arabidopsis thaliana* and *Brassica napus* roots: analysis and comparison by MudPIT and LC-MS/MS. *Plant Soil*, 286 357–376.

Benhamou, N. 1996. Elicitor-induced plant defence pathways. *Trends in Plant Science*, 1 (7), 233–240.

Bertin, C., Yang, X., Weston, L.A. 2003. The role of root exudates and allelochemicals in the rhizosphere. *Plant and Soil*, 256, 67–83.

Blumwald, E., Aharon, G.S., Lam, H. 1998. Early signal transduction pathways in plant-pathogen interactions. *Trends Plant Science*, 3 (9), 342–346.

Boller, T. 1985. Induction of hydrolases as a defense reaction against pathogens, in *Cellular and Molecular Biology of Plant Stress*, Edited by Key, J.L., Kosuge, T., Alan R. Liss, New York.

Boller, T., Gehri, A., Mauch, F., et al. 1983. Chitinase in bean leaves: induction by ethylene, purification, properties, and possible function. *Planta*, 157, 22–31.

Bolwell, P.P., Page, A., Pišlewska, M., et al. 2001. Pathogenic infection and the oxidative defences in plant apoplast. *Protoplasma*, 217 (1), 20–32.

Bonkowski, M., Villenave, C., Griffiths, B. 2009. Rhizosphere fauna: the functional and structural diversity of intimate interactions of soil fauna with plant roots. *Plant and Soil*, 321 (1), 213–233.

Borderies, G., le Béchec, M., Rossignol, M., et al. 2004. Characterization of proteins secreted during maize microspore culture: arabinogalactan proteins (AGPs) stimulate embryo development. *European Journal of Cell Biology*, 83 (5), 205–212.

Bradley, D.J., Kjellbom, P., Lamb, C.J. 1992. Elicitor- and wound-induced oxidative cross-linking of a proline-rich plant cell wall protein: A novel, rapid defense response. *Cell*, 70 (1), 21–30.

Bringhurst, R.M., Cardon, Z.G., Gage, D.J. 2001. Galactosides in the rhizosphere: utilization by *Sinorhizobium meliloti* and development of a biosensor. *Proceedings of the National Academy of Science USA*, 98 (8), 4540–4545.

Brisson, L.F., Tenhaken, T., Lamb, C. 1994. Function of oxidative cross-Linking of cell wall structural proteins in plant disease resistance. *Plant Cell*, 6, 1703–1712.

Broeckling, C.D., Broz, A.K., Bergelson, J., et al. 2008. Root exudates regulate soil fungal community composition and diversity. *Applied and Environmental Microbiology*, 74 (3), 738–744.

Broekaert, W.F., Terras, F.R.G., Cammue, B.P., et al. 1995. Plant defensins: novel antimicrobial peptides as components of the host defense system. *Plant Physiology*, 108, 1353–1358.

Brownleader, M.D., Hopkins, J., Mobasheri, A., et al. 2000. Role of extensin peroxidase in tomato (*Lycopersicon esculentum* Mill.) seedling growth. *Planta*, 210 (4), 668–676.

Buchanan, B.B., Gruissem, W., Jones, R.L. 2000. *Biochemistry and Molecular Biology of plants*. American Society of Plant Physiologists, Rockville, Maryland.

Butler, J.L., Williams, M.A., Bottomley, P.J., et al. 2003. Microbial community dynamics associated with rhizosphere carbon flow. *Applied and Environmental Microbiology*, 69 (11), 6793–6800.

Cai, T., Cai, W.T., Zhang, J., et al. 2009. Host legume-exuded antimetabolites optimize the symbiotic rhizosphere. *Molecular Microbiology*, 73 (3), 507–517.

Calmels, T.P.G., Martin, F., Durand, H., et al. 1991. Proteolytic events in the processing of secreted proteins in fungi. *Journal of Biotechnology*, 17 (1), 51–66.

Carlini, C.R., Polacco, J.C. 2008. Toxic properties of urease. *Crop Science*, 48 (5), 1665–1672.

Carr, J.P., Dixon, D.C., Nikolau, B.J., et al. 1987. Synthesis and localization of pathogenesis-related proteins in tobacco. *Molecular Cellular Proteomics*, 7 (4), 1580–1583.

Castellanos, T., Ascencio, F., Bashan, Y. 1998. Cell-surface lectins of *Azospirillum* spp. *Current Microbiology*, 36 (4), 241–244.

Cervone, F., Hahn, M.G., De Lorenzo, G., et al. 1989. Host-pathogen interactions: XXXIII. A plant protein converts a fungal pathogenesis factor into an elicitor of plant defense responses. *Plant Physiology*, 90 (2), 542–548.

Chapin, F.S., Moilanen, L., Kielland, K. 1993. Preferential use of organic nitrogen for growth by a non-mycorrhizal arctic sedge. *Nature*, 361 (6408), 150–153.

Charmont, S., Jamet, E., Pont-Lezica, R., et al. 2005. Proteomic analysis of secreted proteins from *Arabidopsis thaliana* seedlings: improved recovery following removal of phenolic compounds. *Phytochemistry*, 66, 453–461.

Cho, K., Agrawal, G.K., Shibato, J., et al. 2007. Survey of differentially expressed proteins and genes in jasmonic acid treated rice seedling shoot and root at the proteomics and transcriptomics levels. *Journal of Proteome Research*, 6 (9), 3581–3603.

Chrispeels, M.J. 1976. Biosynthesis, intracellular transport, and secretion of extracellular macromolecules. *Annual Review of Plant Physiology*, 27 (1), 19–38.

Collinge, D.B., Kragh, K.M., Mikkelsen, J.D., et al. 1993. Plant chitinases. *Plant Journal*, 3 (1), 31–40.

Cooper, W.R., Goggin, F.L. 2005. Effects of jasmonate-induced defenses in tomato on the potato aphid, *Macrosiphum euphorbiae*. *Entomologia Experimentalis et Applicata*, 115 (1), 107–115.

Côté, F., Cutt, J.R., Asselin, A., et al. 1991. Pathogenesis-related acidic b-1,3-glucanase genes of tobacco are regulated by both stress and developmental signals. *Molecular Plant-Microbe Interactions*, 4 (2), 173–181.

Crockard, M.A., Bjourson, A.J., Cooper, J.E. 1999. A new peroxidase cDNA from white clover: Its characterization and expression in root tissue challenged with homologous rhizobia, heterologous rhizobia, or *Pseudomonas syringae*. *Molecular Plant-Microbe Interactions*, 12 (9), 825–828.

Currier, W.W., Strobel, G.A. 1976. Chemotaxis of *Rhizobium spp.* to plant root exudates. *Plant Physiology*, 57 (5), 820–823.

Currier, W.W., Strobel, G.A. 1977. Chemotaxis of *Rhizobium spp.* to a glycoprotein produced by birdsfoot trefoil roots. *Science*, 196 (4288), 434–436.

de Kroon, H. 2007. Ecology– How do roots interact? *Science*, 318 (5856), 1562–1563.

De-la-Peña, C., Lei, Z., Watson, B.S., et al. 2008. Root-microbe communication through protein secretion. *Journal of Biological Chemistry*, 283 (37), 25247–25255.

de Wit, P. 2007. How plants recognize pathogens and defend themselves. *Cellular and Molecular life Sciences*, 64 (21), 2726–2732.

Denecke, J., Goldman, M.H.S., Demolder, J., et al. 1991. The tobacco luminal binding protein is encoded by a multigene family. *Plant Cell*, 3 (9), 1025–1035.

Dietz, K.J. 1996. Function and responses of the leaf apoplast under stress. *Progress in Botany*, 58 221–254.

Dong, Z., Canny, M.J., McCully, M.E., et al. 1994. A nitrogen-fixing endophyte of sugarcane stems (A new role for the apoplast). *Plant Physiology*, 105 (4), 1139–1147.

Dore, I., Legrand, M., Cornelissen, B.J.C., et al. 1991. Subcellular localization of acidic and basic PR proteins in tobacco mosaic virus-infected tobacco. *Archives of Virology*, 120 (1), 97–107.

Ebel, J., Mithöfer, A. 1998. Early events in the elicitation of plant defence. *Planta*, 206 (3), 335–348.

Enstone, D.E., Peterson, C.A., Ma, F.S. 2002. Root endodermis and exodermis: Structure, function, and responses to the environment. *Journal of Plant Growth Regulation*, 21 (4), 335–351.

Farrar, J., Hawes, M., Jones, D., et al. 2003. How roots control the flux of carbon to the rhizosphere. *Ecology*, 84 (4), 827–837.

Fauvart, M., Michiels, J. 2008. Rhizobial secreted proteins as determinants of host specificity in the rhizobium-legume symbiosis. *FEMS Microbiol Letters*, 285, 1–9.

Ferreira, R.B., Monteiro, S., Freitas, R., et al. 2007. The role of plant defence proteins in fungal pathogenesis. *Molecular Plant Pathology*, 8 (5), 677–700.

Fritig, B., Heitz, T., Legrand, M. 1998. Antimicrobial proteins in induced plant defense. *Current Opinion in Immunology*, 10 (1), 16–22.

García-Olmedo, F., Rodríguez-Palenzuela, P., Molina, A., et al. 2001. Antibiotic activities of peptides, hydrogen peroxide and peroxynitrite in plant defence. *FEBS Letters*, 498, 219–222.

Garnett, T., Conn, V., Kaiser, B.N. 2009. Root based approaches to improving nitrogen use efficiency in plants. *Plant, Cell and Environment*, 32 (9), 1272–1283.

Glyan'ko, A.K., Makarova, L.E., Vasil'eva, G.G., et al. 2005. Possible involvement of hydrogen peroxide and salicylic acid in the legume-rhizobium symbiosis. *Biology Bulletin*, 32 (3), 245–249.

Godlewski, M., Adamczyk, B. 2007. The ability of plants to secrete proteases by roots. *Plant Physiology and Biochemistry*, 45 (9), 657–664.

Graham, L.S., Sticklen, M.B. 1994. Plant Chitinases. *Canadian Journal of Botany*, 72, 1057–1083.

Gramss, G., Rudeschko, O. 1998. Activities of oxidoreductase enzymes in tissue extracts and sterile root exudates of three crop plants, and some properties of the peroxidase component. *New Phytologist*, 138 (3), 401–409.

Grüner, R., Pfitzner, U. 1994. The upstream region of the gene for the pathogenesis-related protein 1a from tobacco responds to environmental as well as to developmental signals in transgenic plants. *FEBS Journal*, 220 (1), 247–255.

Haddad, M.J., Sarkar, D. 2003. Glomalin, a newly discovered component of soil organic matter: Part I—Environmental significance. *Environmental Geosciences*, 10 (3), 91–98.

Haichar, F.E., Marol, C., Berge, O., et al. 2008. Plant host habitat and root exudates shape soil bacterial community structure. *ISME J*, 2 (12), 1221–1230.

Hawes, M.C. 1990. Living plant cells released from the root cap: A regulator of microbial populations in the rhizosphere? *Plant Soil*, 129 (1), 19–27.

Hawes, M.C., Bengough, G., Cassab, G., et al. 2003. Root caps and rhizosphere. *Journal of Plant Growth Regulation*, 21 (4), 352–367.

Hawes, M.C., Smith, L.Y. 1989. Requirement for chemotaxis in pathogenicity of *Agrobacterium-tumefaciens* on roots of soil-grown pea-plants. *Journal of Bacteriology*, 171 (10), 5668–5671.

Herouart, D., Baudouin, E., Frendo, P., et al. 2002. Reactive oxygen species, nitric oxide and glutathione: a key role in the establishment of the legume-*Rhizobium* symbiosis? *Plant Physiology and Biochemistry*, 40 (6–8), 619–624.

Hinsinger, P. 1998. How do plants acquire mineral nutrients? Chemical processes involved in the rhyzosphere. *Advances in Agronomy*, 64, 225–265.

Hodge, A., Berta, G., Doussan, C., et al. 2009. Plant root growth, architecture and function. *Plant and Soil*, 321 (1), 153–187.

Hunt, D.C., Chrispeels, M.J. 1991. The signal peptide of a vacuolar protein is necessary and sufficient for the efficient secretion of a cytosolic protein. *Plant Physiology*, 96 (1), 18–25.

Jelitto-Van Dooren, E.P.W.M., Vidal, S., Denecke, J. 1999. Anticipating endoplasmic reticulum stress: A novel early response before pathogenesis-related gene induction. *Plant Cell*, 11 (10), 1935–1944.

Joner, E., van Aarle, I., Vosatka, M. 2000. Phosphatase activity of extra-radical arbuscular mycorrhizal hyphae: A review. *Plant Soil*, 226 (2), 199–210.

Jones, D.A., Takemoto, D. 2004. Plant innate immunity-direct and indirect recognition of general and specific pathogen-associated molecules. *Current Opinion in Immunology*, 16 (1), 48–62.

Kang, S., Mills, A.L. 2004. Soil bacterial community structure changes following disturbance of the overlying plant community. *Soil Science*, 169 (1), 55–65.

Kasprzewska, A. 2003. Plant Chitinases—regulation and function. *Cellular and Molecular Biology Letters*, 8, 809–824.

Kästner, B., Tenhaken, R., Kauss, H. 1998. Chitinase in cucumber hypocotyls is induced by germinating fungal spores and by fungal elicitor in synergism with inducers of acquired resistance. *Plant Journal*, 13 (4), 447–454.

Kawano, T., Muto, S. 2000. Mechanism of peroxidase actions for salicylic acid-induced generation of active oxygen species and an increase in cytosolic calcium in tobacco cell suspension culture. *Journal of Experimental Botany*, 51 (345), 685–693.

Kawaoka, A., Kawamoto, T., Ohta, H., et al. 1994. Wound-induced expression of horseradish peroxidase. *Plant Cell Reports*, 13 (3), 149–154.

Kaye, J.P., Hart, S.C. 1997. Competition for nitrogen between plants and soil microorganisms. *Trends in Ecology and Evolution*, 12 (4), 139–143.

Keefe, D., Hinz, U., Meins, F. 1990. The effect of ethylene on the cell-type-specific and intracellular localization of b-1,3-glucanase and chitinase in tobacco leaves. *Planta*, 182 (1), 43–51.

Kim, S.H., Lee, W.S. 2002. Participation of extracellular fumarase in the utilization of malate in cultured carrot cells. *Plant Cell Reports*, 20 (11), 1087–1092.

Kingston-Smith, A.H., Bollard, A.L., Minchin, F.R. 2005. Stress-induced changes in protease composition are determined by nitrogen supply in non-nodulating white clover. *Journal of Experimental Botany*, 56 (412), 745–753.

Klessig, D.F., Malamy, J. 1994. The salicylic acid signal in plants. *Plant Molecular Biology*, 26 (5), 1439–1458.

Komarova, N.Y., Thor, K., Gubler, A., et al. 2008. AtPTR1 and AtPTR5 transport dipeptides in planta. *Plant Physiology*, 148 (2), 856–869.

Kourtev, P.S., Ehrenfeld, J.G., Haggblom, M. 2003. Experimental analysis of the effect of exotic and native plant species on the structure and function of soil microbial communities. *Soil Biology & Biochemistry*, 35 (7), 895–905.

Krehenbrink, M., Downie, J.A. 2008. Identification of protein secretion systems and novel secreted proteins in *Rhizobium leguminosarum* bv. *viciae*. *BMC Genomics*, 9 (1), 55.

Kunze, I., Kunze, G., Bröker, M., et al. 1998. Evidence for secretion of vacuolar a-mannosidase, class I chitinase, and class I b-1,3-glucanase in suspension cultures of tobacco cells. *Planta*, 205 (1), 92–99.

Lamb, C., Dixon, R.A. 1997. The oxidative burst in plant disease resistance. *Annual Review of Plant Physiology and Plant Molecular Biology*, 48 (1), 251–275.

Lawton, K., Ward, E., Payne, G., et al. 1992. Acidic and basic class III chitinase mRNA accumulation in response to TMV infection of tobacco. *Plant Molecular Biology*, 19 (5), 735–743.

Lerouge, P., Roche, P., Faucher, C., et al. 1990. Symbiotic host-specificity of *Rhizobium meliloti* is determined by a sulphated and acylated glucosamine oligosaccharide signal. *Nature*, 344 (6268), 781–784.

Liao, H., Wan, H., Shaff, J., et al. 2006. Phosphorus and aluminum interactions in soybean in relation to aluminum tolerance. Exudation of specific organic acids from different regions of the intact root system. *Plant Physiology*, 141 (2), 674–684.

Lin, C., Shen, B., Xu, Z., et al. 2008. Characterization of the monoterpene synthase gene tps26, the ortholog of a gene induced by insect herbivory in maize. *Plant Physiology*, 146 (3), 940–951.

Lindsey, K., Casson, S., Chilley, P. 2002. Peptides: new signalling molecules in plants. *Trends Plant Science*, 7 (2), 78–83.

Lohar, D.P., Sharopova, N., Endre, G., et al. 2006. Transcript analysis of early nodulation events in *Medicago truncatula*. *Plant Physiology*, 140 (1), 221–234.

Manson, M.D. 1990. Introduction to bacterial motility and chemotaxis. *Journal of Chemical Ecology*, 16 (1), 107–118.

Marschener, H. 1998. Role of root growth, arbuscular mycorrhiza, and root exudates for the efficiency in nutrient acquisition. *Field Crops Research*, 56 (1–2), 203–207.

Mashiguchi, K., Urakami, E., Hasegawa, M., et al. 2008. Defense-related signaling by interaction of arabinogalactan proteins and b-glucosyl yariv reagent inhibits gibberellin signaling in barley aleurone cells. *Plant and Cell Physiology*, 49 (2), 178–190.

Mastronunzio, J., Tisa, L., Normand, P., et al. 2008. Comparative secretome analysis suggests low plant cell wall degrading capacity in *Frankia* symbionts. *BMC Genomics*, 9 (1), 47.

Matsubayashi, Y., Sakagami, Y. 1996. Phytosulfokine, sulfated peptides that induce the proliferation of single mesophyll cells of *Asparagus officinalis* L. *Proceedings of the National Academy of Sciences USA*, 93 (15), 7623–7627.

Matsubayashi, Y., Sakagami, Y. 2006. Peptide hormones in plants. *Annual Review of Plant Biology*, 57 (1), 649–674.

McGurl, B., Pearce, G., Orozco-Cardenas, M., et al. 1992. Structure, expression, and antisense inhibition of the systemin precursor gene. *Science*, 255 (5051), 1570–1573.

Mehta, A., Magalhães, B.S., Souza, D.S.L., et al. 2008. Rooteomics: The challenge of discovering plant defense-related proteins in roots. *Current Protein & Peptide Science*, 9, 108–116.

Minic, Z., Brown, S., De Kouchkovsky, Y., et al. 1998. Purification and characterization of a novel chitinase-lysozyme, of another chitinase, both hydrolysing *Rhizobium meliloti* Nod factors, and of a pathogenesis-related protein from *Medicago sativa* roots. *Biochemical Journal*, 332 (2), 329–335.

Minic, Z., Leproust-Lecoester, L., Laporte, J., et al. 2000. Proteins isolated from lucerne roots by affinity chromatography with sugars analogous to Nod factor moieties. *Biochemical Journal*, 345 (2), 255–262.

Misas-Villamil, J.C., van der Hoorn, R.A.L. 2008. Enzyme-inhibitor interactions at the plant-pathogen interface. *Current Opinion in Plant Biology*, 11 (4), 380–388.

Mita, G., Nocco, G., Leuci, C., et al. 1997. Secreted heat shock proteins in sunflower suspension cell cultures. *Plant Cell Reports*, 16 (11), 792–796.

Mithöfer, A. 2002. Suppression of plant defence in rhizobia-legume symbiosis. *Trends Plant Science*, 7 (10), 440–444.

Mitsuhara, I., Iwai, T., Seo, S., et al. 2008. Characteristic expression of twelve rice PRI family genes in response to pathogen infection, wounding, and defense-related signal compounds (121/180). *Molecular Genetics and Genomics*, 279 (4), 415–427.

Morgan, J.A.W., Bending, G.D., White, P.J. 2005. Biological costs and benefits to plant-microbe interactions in the rhizosphere. *Journal of Experimental Botany*, 56 (417), 1729–1739.

Nasholm, T., Kielland, K., Ganeteg, U. 2009. Uptake of organic nitrogen by plants. *New Phytologist*, 182 (1), 31–48.

Ndimba, B.K., Chivasa, S., Hamilton, J.M., et al. 2003. Proteomic analysis of changes in the extracellular matrix of Arabidopsis cell suspension cultures induced by fungal elicitors. *Proteomics*, 3, 1047–1059.

Nilsson, M., Rasmussen, U., Bergman, B. 2005. Cyanobacterial chemotaxis to extracts of host and nonhost plants. *FEMS Microbiology Ecology*, 55 (3), 382–390.

Nimchuk, Z., Eulgem, T., Holt III, B.F., et al. 2003. Recognition and response in the plant immune system. *Annual Review of Genetics*, 37 (1), 579–609.

Nóbrega, F.M., Santos, I.S., Da Cunha, M., et al. 2005. Antimicrobial proteins from cowpea root exudates: inhibitory activity against *Fusarium oxysporum* and purification of a chitinase-like protein. *Plant Soil*, 272 (1), 223–232.

Oh, I.S., Park, A.R., Bae, M.S., et al. 2005. Secretome analysis reveals an Arabidopsis lipase involved in defense against *Alternaria brassicicola*. *Plant Cell*, 17 (10), 2832–2847.

Ohme-Takagi, M., Meins, F.J., Shinshi, H. 1998. A tobacco gene encoding a novel basic class II chitinase: a putative ancestor of basic class I and acidic class II chitinase genes. *Molecular and General Genetics*, 259 (5), 511–515.

Oldroyd, G.E.D., Harrison, M.J., Udvardi, M. 2005. Peace talks and trade deals. Keys to long-term harmony in legume-microbe symbioses. *Plant Physiology*, 137 (4), 1205–1210.

Ostersetzer, O., Kato, Y., Adam, Z., et al. 2007. Multiple intracellular locations of Lon protease in arabidopsis: Evidence for the localization of AtLon4 to chloroplasts. *Plant and Cell Physiology*, 48 (6), 881–885.

Ovtsyna, A.O., Dolgikh, E.A., Kilanova, A.S., et al. 2005. Nod factors induce Nod factor cleaving enzymes in pea roots. Genetic and pharmacological approaches indicate different activation mechanisms. *Plant Physiology*, 139 (2), 1051–1064.

Park, S.W., Lawrence, C.B., Linden, J.C., et al. 2002. Isolation and characterization of a novel Ribosome-Inactivating Protein from root cultures of Pokeweed and its mechanism of secretion from roots. *Plant Physiology*, 130 (1), 164–178.

Parker, J.S., Cavell, A.C., Dolan, L., et al. 2000. Genetic interactions during root hair morphogenesis in Arabidopsis. *Plant Cell*, 12 (10), 1961–1974.

Paszkowski, U. 2006. Mutualism and parasitism: the yin and yang of plant symbioses. *Current Opinion in Plant Biology*, 9 (4), 364–370.

Paungfoo-Lonhienne, C., Lonhienne, T.G.A., Rentsch, D., et al. 2008. Plants can use protein as a nitrogen source without assistance from other organisms. *Proceedings of the National Academy of Sciences USA*, 105 (11), 4524–4529.

Paungfoo-Lonhienne, C., Schenk, P.M., Lonhienne, T.G.A., Brackin, R., Meier, S., Rentsch, D., Schmidt, S. 2009. Nitrogen affects cluster root formation and expression of putative peptide transporters. *Journal of Experimental Botany*, 60, 2665–2676.

Payne, G., Ahl, P., Moyer, M., et al. 1990. Isolation of complementary DNA clones encoding pathogenesis-related proteins P and Q, two acidic chitinases from tobacco. *Proceedings of the National Academy of Sciences USA*, 87 (1), 98–102.

Pearce, G., Ryan, C.A. 2003. Systemic signaling in tomato plants for defense against herbivores: Isolation and characterization of three novel defense-signaling glycopeptide hormones coded in a single precursor gene. *Journal of Biological Chemistry*, 278 (32), 30044–30050.

Pearce, G., Strydom, D., Johnson, S., et al. 1991. A polypeptide from tomato leaves induces wound-inducible proteinase inhibitor proteins. *Science*, 253 (5022), 895–897.

Petersen, W., Böttger, M. 1991. Contribution of organic acids to the acidification of the rhizosphere of maize seedlings. *Plant Soil*, 132 (2), 159–163.

Phillips, G.O. 2009. Molecular association and function of arabinogalactan protein complexes from tree exudates. *Structural Chemistry*, 20 (2), 309–315.

Pignocchi, C., Foyer, C.H. 2003. Apoplastic ascorbate metabolism and its role in the regulation of cell signalling. *Current Opinion in Plant Biology*, 6 (4), 379–389.

Polacco, J.C., Holland, M.A. 1993. Roles of urease in plant cells. *International Review of Cytology*, 145, 65–103.

Purin, S., Rillig, M.C. 2007. The arbuscular mycorrhizal fungal protein glomalin: Limitations, progress, and a new hypothesis for its function. *Pedobiologia*, 51 (2), 123–130.

Rathbun, E.A., Naldrett, M.J., Brewin, N.J. 2002. Identification of a family of extensin-like glycoproteins in the lumen of *Rhizobium*-induced infection threads in pea root nodules. *Molecular Plant-Microbe Interactions*, 15 (4), 350–359.

Raven, J.A., Edwards, D. 2001. Roots: evolutionary origins and biogeochemical significance. *Journal of Experimental Botany*, 52 (suppl 1), 381–401.

Regalado, A.P., Ricardo, C.P.P. 1996. Study of the intercellular fluid of healthy *Lupinus albus* organs (presence of a chitinase and a thaumatin-like protein). *Plant Physiology*, 110 (1), 227–232.

Regvar, M., Gogala, N. 1996. Changes in root growth patterns of (*Picea abies*) spruce roots by inoculation with an ectomycorrhizal fungus *Pisolithus tinctorius* and jasmonic acid treatment. *Trees-Structure Function*, 10 (6), 410–414.

Rentsch, D., Schmidt, S., Tegeder, M. 2007. Transporters for uptake and allocation of organic nitrogen compounds in plants. *FEBS Letters*, 581 (12), 2281–2289.

Reynolds, H.L., Packer, A., Bever, J.D., et al. 2003. Grassroots ecology: plant-microbe-soil interactions as drivers of plant community structure and dynamics. *Ecology*, 84 (9), 2281–2291.

Rigas, S., Daras, G., Laxa, M., et al. 2009. Role of Lon1 protease in post-germinative growth and maintenance of mitochondrial function in *Arabidopsis thaliana. New Phytologist*, 181 (3), 588–600.

Roche, P., Debellé, F., Maillet, F., et al. 1991a. Molecular basis of symbiotic host specificity in *Rhizobium meliloti*: *nodH* and *nodPQ* genes encode the sulfation of lipo-oligosaccharide signals. *Cell*, 67 (6), 1131–1143.

Roche, P., Lerouge, P., Ponthus, C., et al. 1991b. Structural determination of bacterial nodulation factors involved in the *Rhizobium meliloti*-alfalfa symbiosis. *Journal of Biological Chemistry*, 266 (17), 10933–10940.

Rodriguez-Navarro, D.N., Dardanelli, M.S., Ruiz-Sainz, J.E. 2000. Attachment of bacteria to the roots of higher plants. *FEMS Microbiology Letters*, 272, 127–136.

Rosenthal, G. 1986. Biochemical insight into insecticidal properties of -Canavanine, a higher plant protective allelochemical. *Journal of Chemical Ecology*, 12 (5), 1145–1156.

Rouatt, J.W., Katznelson, H., Payne, T.M.B. 1960. Statistical evaluation of the rhizosphere effect. *Soil Science Society of America Journal*, 24, 271–273.

Rovira, A.D. 1969. Plant Root Exudates. *Botany Review*, 35, 35–57.

Salzer, P., Bonanomi, A., Beyer, K., et al. 2000. Differential expression of eight chitinase genes in *Medicago truncatula* roots during mycorrhiza formation, nodulation, and pathogen infection. *Molecular Plant-Microbe Interactions*, 13 (7), 763–777.

Salzer, P., Feddermann, N., Wiemken, A., et al. 2004. *Sinorhizobium meliloti*-induced chitinase gene expression in *Medicago truncatula* ecotype R108-1: a comparison between symbiosis-specific class-V and defence-related class-IV chitinases. *Planta*, 219 (4), 626–638.

Salzwedel, J.L., Dazzo, F.B. 1993. pSym *nod* gene influence on elicitation of peroxidase activity from white clover and pea roots by rhizobia and their cell-free supernatants. *Molecular Plant-Microbe Interactions*, 6, 127–134.

Sanchez, A.M.H., Tafur, J.C., Rodriguez-Monroy, M., et al. 2009. Arabinogalactan proteins in plant cell cultures. *Interciencia*, 34 (3), 170–176.

Schultze, M., Quiclet-Sire, B., Kondorosi, E., et al. 1992. *Rhizobium meliloti* produces a family of sulfated lipo-oligosaccharides exhibiting different degrees of plant host specificity. *Proceedings of the National Academy of Sciences USA*, 89 (1), 192–196.

Schulze-Lefert, P. (2004) Knocking on the heaven's wall: pathogenesis of and resistance to biotrophic fungi at the cell wall. *Current Opinion in Plant Biology*, 7 (4), 377–383.

Seifert, G.J., Roberts, K. 2007. The biology of arabinogalactan proteins. *Annual Review of Plant Biology*, 58, 137–161.

Sels, J., Mathys, J., De Coninck, B.M.A., et al. 2008. Plant pathogenesis-related (PR) proteins: A focus on PR peptides. *Plant Physiology and Biochemistry*, 46 (11), 941–950.

Senwo, Z.N., Tabatabai, M.A. 1998. Amino acid composition of soil organic matter. *Biology and Fertility of Soils*, 26 (3), 235–242.

Shinshi, H., Neuhaus, J.M., Ryals, J., et al. 1990. Structure of a tobacco endochitinase gene: evidence that different chitinase genes can arise by transposition of sequences encoding a cysteine-rich domain. *Plant Molecular Biology*, 14 (3), 357–368.

Shpak, E., Leykam, J.F., Kieliszewski, M.J. 1999. Synthetic genes for glycoprotein design and the elucidation of hydroxyproline-O-glycosylation codes. *Proceedings of the National Academy of Sciences USA*, 96 (26), 14736–14741.

Spanu, P., Boller, T., Ludwig, A., et al. 1989. Chitinase in roots of mycorrhizal *Allium porrum*: regulation and localization. *Planta*, 177 (4), 447–455.

Stacey, G., Paau, A.S., Brill, W.J. 1980. Host recognition in the *Rhizobium*-soybean symbiosis. *Plant Physiology*, 66 (4), 609–614.

Staehelin, C., Granado, J., Muller, J., et al. 1994. Perception of *Rhizobium* nodulation factors by tomato cells and inactivation by root chitinases. *Proceedings of the National Academy of Sciences USA*, 91 (6), 2196–2200.

Sticher, L., Mauch-Mani, B., Metraux, A.J. 1997. Systemic Acquired Resistance. *Annual Review of Phytopathology*, 35 (1), 235–270.

Su, W., Lin, C.F., Wu, J.X., et al. 2006. Molecular cloning and expression of a cDNA encoding Lon protease from rice (*Oryza sativa*). *Biotechnology Letters*, 28 (12), 923–927.

Summers, M.L., Botero, L.M., Busse, S.C., et al. 2000. The *Sinorhizobium meliloti* Lon protease is involved in regulating exopolysaccharide synthesis and is required for nodulation of Alfalfa. *The Journal of Bacteriology*, 182 (9), 2551–2558.

Surpin, M., Raikhel, N. 2004. Traffic jams affect plant development and signal transduction. *Nature Reviews Molecular Cell Biology*, 5, 100–109.

Tarafdar, J.C., Marschner, H. 1994. Phosphatase activity in the rhizosphere and hyposphere of VA mycorrhizal wheat supplied with inorganic and organic phosphorus. *Soil Biology & Biochemistry*, 26, 387.

Tejera, N., Ortega, E., Rodes, R., et al. 2006. Nitrogen compounds in the apoplastic sap of sugarcane stem: Some implications in the association with endophytes. *Journal of Plant Physiology*, 163 (1), 80–85.

Tomscha, J.L., Trull, M.C., Deikman, J., et al. 2004. Phosphatase under-producer mutants have altered phosphorus relations. *Plant Physiology*, 135 (1), 334–345.

Turrini, A., Sbrana, C., Pitto, L., et al. 2004. The antifungal Dm-AMP1 protein from Dahlia merckii expressed in *Solanum melongena* is released in root exudates and differentially affects pathogenic fungi and mycorrhizal symbiosis. *New Phytologist*, 163 (2), 393–403.

Uren, N.C. 2007. Types, amounts, and possible functions of compounds released into the rhizosphere by soil-grown plants., in *The Rhizosphere*, Edited by Pinton, R., Varanini, Z., Nanniperi, P. 2nd ed. CRC Press, London.

van Loon, L.C. 1976. Specific soluble leaf proteins in virus-infected tobacco plants are not normal constituents. *Journal of General Virology*, 30, 375–379.

van Loon, L.C. 1985. Pathogenesis-related proteins. *Plant Molecular Biology*, 4, 111–116.

van Loon, L.C., Rep, M., Pieterse, C.M.J. 2006. Significance of inducible defense-related proteins in infected plants. *Annual Review of Phytopathology*, 44 (1), 135–162.

van Wees, S.C.M., Luijendijk, M., Smoorenburg, I., et al. 1999. Rhizobacteria-mediated induced systemic resistance (ISR) in Arabidopsis is not associated with a direct effect on expression of known defense-related genes but stimulates the expression of the jasmonate-inducible gene *Atvsp* upon challenge. *Plant Molecular Biology*, 41 (4), 537–549.

Vancura, V., Hanzlíková, A. 1972. Root exudates of plants. *Plant Soil*, 36 (1), 271–282.

Vasse, J., de Billy, F., Truchet, G. 1993. Abortion of infection during the *Rhizobium meliloti*-alfalfa symbiotic interaction is accompanied by a hypersensitive reaction. *Plant Journal*, 4 (3), 555–566.

Vera, P., Hernandez-Yago, J., Conejero, V. 1989. "Pathogenesis-related" P1 (p 14) protein. Vacuolar and apoplastic localization in leaf tissue from tomato plants infected with citrus exocortis viroid; *in vitro* synthesis and processing. *Journal of General Virology*, 70 (8), 1933–1942.

Veronese, P., Ruiz, M.T., Coca, M.A., et al. 2003. In defense against pathogens. Both plant sentinels and foot soldiers need to know the enemy. *Plant Physiology*, 131 (4), 1580–1590.

Volpin, H., Elkind, Y., Okon, Y., et al. 1994. A vesicular arbuscular mycorrhizal fungus (*Glomus intraradix*) induces a defense response in alfalfa roots. *Plant Physiology*, 104 (2), 683–689.

Vorwerk, S., Somerville, S., Somerville, C. 2004. The role of plant cell wall polysaccharide composition in disease resistance. *Trends Plant Science*, 9 (4), 203–209.

Walker, T.S., Bais, H.P., Grotewold, E., et al. 2003. Root exudation and rhizosphere biology. *Plant Physiology*, 132 (1), 44–51.

Wasaki, J., Yamamura, T., Shinano, T., et al. 2003. Secreted acid phosphatase is expressed in cluster roots of lupin in response to phosphorus deficiency. *Plant Soil*, 248 (1), 129–136.

Watt, M.I.C.H., Silk, W.K., Passioura, J.B. 2006. Rates of root and organism growth, soil conditions, and temporal and spatial development of the rhizosphere. *Annals of Botany*, 97 (5), 839–855.

Wen, F., Van Etten, H.D., Tsaprailis, G., et al. 2007. Extracellular proteins in pea root tip and border cell exudates. *Plant Physiology*, 143, 773–783.

Xie, Z., Staehelin, C., Wiemken, A., et al. 1999. Symbiosis-stimulated chitinase isoenzymes of soybean (*Glycine max* (L.) Merr.). *Journal of Experimental Botany*, 50, (332), 327–333.

Yamasaki, S., Matsumoto, M., Takeuchi, O., et al. 2009. C-type lectin Mincle is an activating receptor for pathogenic fungus, *Malassezia*. *Proceedings of the National Academy of Sciences USA*, 106 (6), 1897–1902.

York, W.S., Qin, Q., Rose, J.K.C. 2004. Proteinaceous inhibitors of *endo*-b-glucanases. *Biochim Biophys Acta (BBA)—Proteins & Proteomics*, 1696 (2), 223–233.

Chapter 6
Actinorhizal Symbioses
Katharina Pawlowski

Introduction

Actinorhizal symbioses occur between gram-positive or gram-variable soil actinomycetes of the genus *Frankia* and a diverse group of dicotyledonous "actinorhizal" plants belonging to 24 genera from eight different families. With the exception of a single species, *Datisca glomerata*, actinorhizal plants are trees or woody shrubs. *Frankia* strains induce the formation of specialized organs, nodules, on the roots of their host plants. In these nodules, the microsymbionts fix dinitrogen while being hosted within nodule cells and export the products of nitrogen fixation to the host plant, thereby rendering it independent of soil nitrogen sources. Meanwhile, the plant provides the intracellular bacteria with carbon sources. Mature actinorhizal nodules are perennial organs consisting of multiple lobes, each of which represents a modified lateral root, without root cap, and with infected cells in the expanded cortex. Although actinorhizal nodules were first described in 1829 (Meyen, 1829), their role in plant nitrogen nutrition was not known before 1895 (Hiltner, 1895), and the actinomycetous nature of the microsymbionts was only recognized in 1932 (Krebber, 1932).

The Microsymbionts

Frankia strains normally grow as a mycelium of vegetative hyphae with a diameter of 0.5–2 µm and limited to extensive branching, and form multilocular sporangia containing nonmotile spores. In contrast to the root nodule symbionts of legumes, rhizobia, *Frankia* strains can fix nitrogen not only in symbiosis, but also in the free-living state. The enzyme catalyzing the reduction of dinitrogen to ammonium, nitrogenase, is highly O_2 sensitive, yet nitrogen fixation requires high amounts of ATP, which is provided by aerobic respiratory processes, necessitating an O_2 protection system for nitrogenase. *Frankia* strains can solve this so-called "O_2

dilemma of nitrogen fixation" by forming special organs, spherical septate vesicles, at the ends of hyphae or on short side hyphae when subjected to nitrogen-limiting aerobic conditions (Figure 6.1A). Vesicles are surrounded by envelopes of multiple layers containing hopanoids, bacterial steroid lipids (Berry et al., 1993). The number of layers comprising the envelopes is correlated with the O_2 tension, indicating that the envelopes act as a gas-diffusion barrier (Meesters et al., 1987; Parsons et al., 1987). When nitrogen limitation takes place under microaerobic conditions, *Frankia* forms nitrogenase in hyphae (Murry et al., 1985). Within nodule cells, the shape, septation, and subcellular localization of *Frankia* vesicles, and whether vesicles are formed at all, depends on the host plant (for reviews, see Silvester et al., 1990; Baker and Mullin, 1992). Hence, within the symbiosis, the plant can direct the differentiation of the intracellular bacteria.

In *Alnus* and *Myrica* symbioses, *Frankia* strains can be classified as spore[+] or spore[−], respectively, on the basis of the presence or absence of sporangia within root nodules (Schwintzer, 1990). Spore[+] strains seem to be more infective than spore[−] strains, but to date, the former have not been successfully maintained in culture.

Several studies have shown that the presence of *Frankia* in soils does not depend on the presence of the host plants (Young et al., 1992; Zimpfer et al., 1997; Maunuksela et al., 1999; for a review, see Dawson, 2008). Infectious *Frankia* can be present in glacial deposits or volcanic soils (Lawrence et al., 1967; Burleigh and Dawson, 1994). However, some actinorhizal genera, particularly *Casuarina* and *Allocasuarina* spp. (Simonet et al., 1999), occasionally are found to be nodulated sparsely, if at all, even in their native habitats (Lawrie et al., 1982) unless compatible *Frankia* strains have been introduced. Host plants release compounds that increase the numbers of infective *Frankia* in soil (Zimpfer et al., 1999; Krumholz et al., 2003); a similar effect could be shown for non-actinorhizal close relatives of host plants (Smolander et al., 1990; Gauthier et al., 2000). *Alnus* leaf litter was shown to increase both growth and infectivity of *Alnus*-infective *Frankia* in soil (Nickel et al., 1999; Nickel 2000). Some effects on infectivity might be indirect, because several soil microbes can synergistically increase *Frankia* nodulation of its host plant (Zimpfer et al., 2003; Solans, 2007).

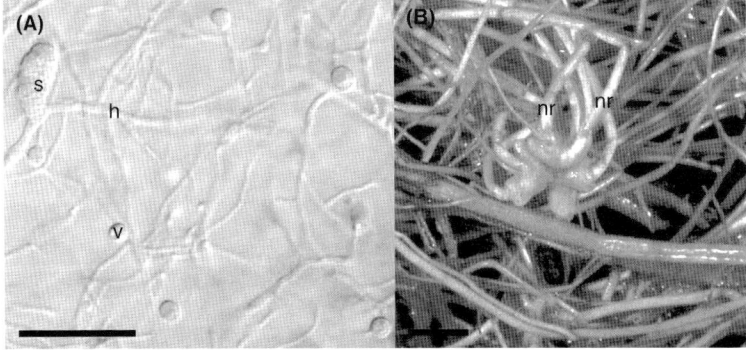

Figure 6.1. Macrosymbiont and microsymbiont. (A) *Frankia* culture grown without a nitrogen source under 21% oxygen. Hyphae (h) and a sporangium (s) can be seen and nitrogen-fixing vesicles (v) have formed. (B) Nodule of *Casuarina glauca* grown in liquid culture, ca. 6 weeks after infection. Nodule roots (nr) grow from the tips of nodule lobes. The size bar in (A) denotes 20 µm, the size bar in (B) 200 µm.

Frankia Host Specificity

Phylogenetic analysis showed that *Frankia* strains can be grouped in three closely related clusters (Normand et al., 1996; Clawson et al., 2004); members of each cluster infect a subset of actinorhizal plants (Figure 6.2). Cluster I strains nodulate plants from the order Fagales of the families Betulaceae, Casuarinaceae, and Myricaceae (Normand et al., 1996). A subclade within cluster I is comprised of the narrow host range "Casuarina strains" that under natural conditions nodulate only *Casuarina* and *Allocasuarina* species from the Casuarinaceae (Benson et al., 2004). Strains from cluster II nodulate plants from the three families in the

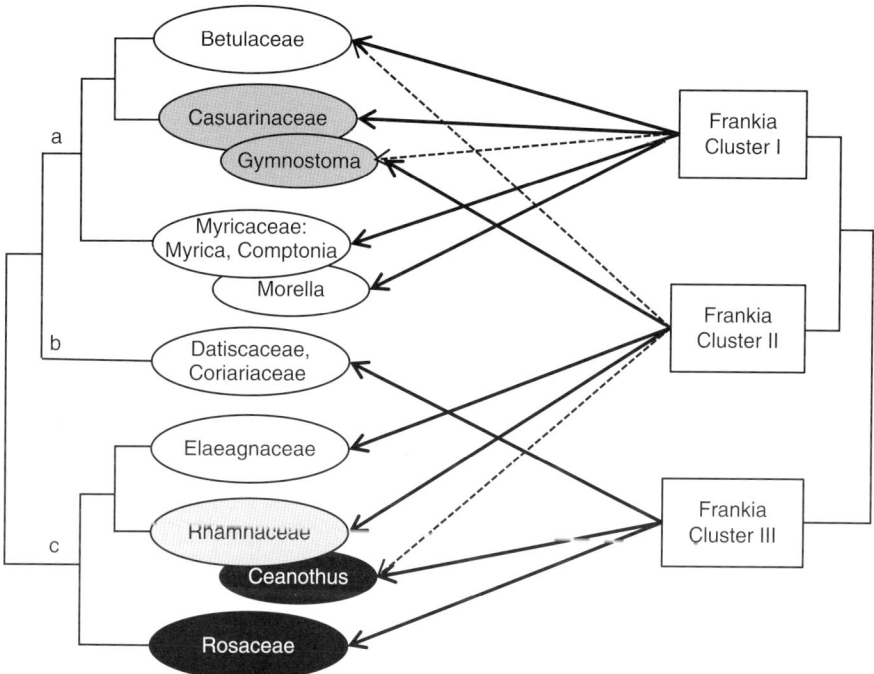

Figure 6.2. Simplified scheme of the relationships between groups of actinorhizal plants and *Frankia* clusters. For actinorhizal plants: a – Fagales; b – Cucurbitales; c – Rosale. *Frankia* clusters are based on Benson et al. (2004). (For *Frankia* strains infecting different Myricaceae genera, see Huguet et al., 2005). Thick arrows connecting *Frankia* clusters with plant families/genera indicate that members of these clusters are commonly associated with the plants. Thin arrows indicate that members of these clusters have been isolated from or detected in an effective or ineffective nodule of a member of the plant genus/family at least once. Some genera (*Gymnostoma, Morella, Ceanothus*) differ in microsymbiont specificity from the rest of the plant family they belong to (the genus *Ceanothus* has an isolated position among actinorhizal Rhamnaceae; the other actinorhizal genera belong to the tribe Colletiae [Benson et al., 2004]). Plant groups that show a particular native geographical distribution are labeled (dark gray, Australia and western Pacific; light gray, South America and southern New Zealand; black, western North America). Actinorhizal Elaeagnaceae and Betulaceae occur on most continents, while Coriariaceae and Datiscaceae are found in temperate zones of both hemispheres. Host specificity exists within the *Frankia* clusters (i.e., not all members of a clade can nodulate all plants associated with that cluster). For instance, all members of cluster I tested can nodulate *Myrica* sp., but only some of them can nodulate either *Alnus* sp. or *Casuarina* sp.

Fagales and from two families in the order Rosales (Elaeagnaceae, Rhamnaceae; Benson et al., 2004). Finally, the strains from cluster III nodulate plants from Rosaceae and from one rhamnaceous genus, *Ceanothus,* and plants from the order Cucurbitales (families Coriariaceae and Datiscaceae; Benson et al., 2004; Vanden Heuvel et al., 2004). All attempts to isolate strains from this cluster and to grow them in culture have failed. Infection studies using soil from beneath nodulated host plants have shown that infective units, presumably spores, persist in soil (Mirza et al., 1994). Cross-inoculation studies using crushed nodules have established that the symbionts from *Dryas* (Rosaceae), *Ceanothus* (Rhamnaceae), *Datisca* (Datiscaceae), and *Coriaria* (Coriariaceae) are in the same cross-inoculation groups (Kohls et al., 1994; Mirza et al., 1994). This cluster occupies a basal position in the phylogeny of *Frankia* (Swensen and Benson, 2008) and is characterized by a low variability among the reported 16S rRNA gene sequences as well as other phylogenetic markers (Vanden Heuvel et al., 2004), in spite of its infecting the broadest range of hosts. The broad host range of cluster III can be interpreted in two ways: (1) that these strains are particularly good at establishing new symbioses or (2) that they represent a remnant of a more ancient symbiosis that was present before the divergence of extant lineages. The second option is consistent with cluster III strains appearing to represent obligate symbionts, a dependency to be expected if the symbiosis were established early in evolution.

Occasionally, attempts to isolate *Frankia* strains from nodules, particularly from nodules induced by cluster III strains, have yielded so-called "*Frankia*-like strains" or "atypical *Frankia* strains," which are phylogenetically related to *Frankia* and form the characteristic multilocular sporangia, but can neither form vesicles nor fix nitrogen nor induce nodules (Hahn et al., 1988; Mirza et al., 1992, 1994; Ramírez-Saad et al., 1998). Detailed analysis showed that these bacteria were present in the nodule periderm (Ramírez-Saad et al., 1998). Since nodules without *Frankia*-like strains were also found (T. Persson and K. Pawlowski, unpublished), such *Frankia*-like strains cannot be involved in the infection process by the cluster III strain.

More recently, the genomes from three *Frankia* strains—two from cluster I and one from cluster II—have been sequenced (Normand et al., 2007); further genome sequence analyses of *Frankia* strains and *Frankia*-like strains are in progress. The sizes of the sequenced genomes varied from 5.43 Mbp for a narrow host range strain from *Casuarina* nodules (*Frankia* sp. strain HFPCcI3) to 7.50 Mbp for a medium host range strain from *Alnus* nodules (*Frankia alni* strain ACN14a) to 9.04 Mbp for a broad host range strain from *Elaeagnus* nodules (*Frankia* sp. strain EAN1pec). This size divergence is the largest yet reported for such closely related soil bacteria (97.8–98.9% 16S rRNA sequence identity; (Normand et al., 2007). Genome sequence analysis showed that the unusual size divergence displayed by the three *Frankia* genomes is due to deletions, duplications, and retentions. These processes led to the reduction of the HFPCcI3 genome, to the expansion of the EAN1pec genome, while the ACN14a genome size remained more or less stable. Combining genomic, ecological, and physiological data led to the hypothesis that the strains from the *Casuarina*-infecting subclade of cluster I evolved to become specialists with reduced genomes (Normand et al., 2007). Although genome reduction is well documented in obligate parasites and symbionts (Moran, 2003), it was not previously reported for facultative symbionts. Given that cluster III strains cannot be cultivated, it will be of interest to see whether their genomes indeed show more reduction than that of the *Casuarina*-infective strain HFPCcI3.

Frankia Carbon and Nitrogen Metabolism

Carbon metabolism has been studied extensively in free-living *Frankia* cultures. *Frankia* strains can be cultured on minimal media and can use a variety of carbon sources, including short chain fatty acids such as propionate and acetate, TCA cycle intermediates, pyruvate, and some sugars (Benson and Silvester, 1993). Propionate can serve as a carbon source for all culturable strains, with the exception of the isolate EAN1pec (Tisa et al., 1983). Some *Frankia* strains have been found to grow on minimal media containing Tween 20 or Tween 80 as a sole carbon source (Lechevalier and Lechevalier, 1990). Several strains can degrade cellulose (Safo-Sampah and Torrey, 1988). Free-living *Frankia* can store carbon in the form of trehalose and glycogen (Lopez et al., 1983, 1984). A variety of organic and inorganic nitrogen sources have been shown to support the growth of free-living *Frankia*, including amino acids, urea, nitrate, ammonium, and dinitrogen (Benson and Silvester, 1993). Ammonium is assimilated via the glutamine synthetase/glutamate synthase (GS/GOGAT) reactions producing glutamate (Benson and Schultz, 1990; Berry et al., 1990).

The Macrosymbionts

Plants able to enter root nodule symbioses—legumes and actinorhizal plants—all belong to a subclade (the so-called nitrogen-fixing clade) of the Fabids, which form a monophyletic part of the Rosids (Zhu et al., 2007). Within the nitrogen-fixing clade, root nodule-forming families or genera are interspersed with nonsymbiotic ones in a pattern suggestive of multiple origins, and multiple losses, of the symbiotic syndrome (Swensen and Benson, 2008). The plant phylogenies suggest that actinorhizal symbioses evolved more than once in evolutionary history, supporting the hypothesis of a genetic predisposition in members of the Rosid I clade that allowed recurrent evolution of root nodule symbioses. However, a detailed analysis of the phylogenetic positions of actinorhizal genera also supports the idea that loss of the symbiotic ability might have been a common feature in evolution. Given that symbiotic species are likely to expend more energy than nonsymbiotic species in the same environment and additionally provide fixed nitrogen to their competitors, it is not surprising that many actinorhizal species do not persist beyond early successional habitats (Côte et al., 1988; Neave et al., 1989; Dawson, 1990). In habitats with sufficient soil nitrogen or under high stress, host plants generally block infection by *Frankia* (Benoit and Berry, 1990). Therefore, the loss of the symbiosis may have been evolutionarily favored in several contexts. Interestingly, the ancient habitats of actinorhizal plants, as indicated by nodule fossils, seem to be similar to those of extant actinorhizal species, indicating that actinorhizal plants may have been nodulated for much of their evolutionary history.

Actinorhizal plants occur on all continents except Antarctica, and are distributed from the arctic to the tropics. They can be found in forest, swamp, riparian, shrub, prairie, and desert ecosystems, yet most of them occur in boreal and temperate ecosystems of both hemispheres. Commonly, they inhabit areas with low soil nitrogen, such as, volcanic deposits, landslides, eroded soils, sand dunes, beaches, and recent glacial deposits. In colder climates, actinorhizal plants become more prevalent and seem to fill the niche dominated by woody legumes in the tropics (Dawson, 1986; Dawson, 2008; chapter by Anderson). Despite actinorhizal symbioses encompassing eight different plant families, actinorhizal trees are far less numerous than nitrogen-fixing legume trees. Nevertheless, their nitrogen fixation has been estimated to range between 2 and 300 kg N ha^{-1} a^{-1}. For comparison, reported maximal rates of N$_2$ fixation in aboveground plant tissues of temperate herbaceous legumes were up to 373 kg N ha^{-1} a^{-1} in red clover (*Trifolium pratense* L.), 545 kg N ha^{-1} a^{-1} in white clover (*T. repens* L.), and 350 kg N ha^{-1} a^{-1} in alfalfa (*Medicago sativa* L.) (Carlsson and Huss-Danell, 2003), while for tropical legume trees, values between 100 and 300 (*Leucaena leucocephala, Gliricidia sepium,* and *Acacia mangium*) and less than 20 kg N ha^{-1} a^{-1} (*Faidherbia albida* and *Acacia senegal*) have been reported (Sanginga et al., 1995). While not as numerous as legumes, actinorhizal plants are important because most represent colonizers able to regenerate poor soils or disturbed sites. Many are pioneer species at early stages of plant succession following disturbances (eruptions, flooding, landslides, fires), such as, *Alnus* sp. in moist environments, *Myrica* sp. on landslides and mining areas, *Dryas drummondi, A. viridis* and *Shepherdia canadensis* in post-glacial successions (Kohls et al., 2003), and *Casuarina* sp., which is used in erosion control, sand dune stabilization, and soil reclamation National Research Council (NRC), 1984, see Chapter 9 on the use of tree legumes to restore degraded tropical areas in Brazil). Some representatives, particularly *Casuarina* sp. and *Alnus* sp., are used to produce timber, firewood, and charcoal. *Casuarina* spp. are also used as components of multipurpose agroforestry plantations (Russo, 2005). *Hippophae rhamnoides* (sea buckthorn; Elaeagnaceae), a salt-tolerant dioecious shrub or small tree with a growth pattern and height that vary with the geographical location, is a "coming plant" (Li and Schroeder, 1996) thanks to its fruits and seed oil. The fruits are rich in vitamins, antioxidants, and trace elements, and the seed oil contains unusually high amounts of essential fatty acids and phytosterols (i.e., it has potential uses in cosmetics and pharmacology) (Beveridge et al., 1999; Li et al., 2007). Several countries, in particular China and Canada, have breeding programs for sea buckthorn (Ruan et al., 2009).

N-limitation is also a feature of many arid ecosystems due to weather conditions that inhibit the mineralization of organic matter; similarly, low-oxygen concentrations and loss of nitrogenous solutes from soil through leaching can lead to nitrogen-limited wetlands. In such habitats, actinorhizal plant species can persist as stable components of plant communities. These species include *Purshia tridentata* (Rosaceae), *Cercocarpus* spp. (Rosaceae), *Cowania mexicana* (Rosaceae), *Shepherdia* spp., *Elaeagnus commutata* and *E. angustifolia* (all Elaeagnaceae), and several *Ceanothus* spp. (Rhamnaceae), which are all native to the rangelands of the western USA (Paschke, 1997). *Ceanothus* spp. are components of the California Floristic Province, particularly in fire-adapted chaparrals and coastal scrubs (Hardig et al., 2000). *Purshia tridentata* (Rosaceae) and *Ceanothus prostrata* form part of the understory in several North American forests (Busse et al., 2007). *P. tridentata* is part of various fire-dominated communities across the western USA (Horning and Cronn, 2009). *Discaria trinervis* (Rhamnaceae) is common along rivers running from the *Nothofagus* forests into the Patagonian semi-deserts and in similar habitats in Chile (Reiche, 1907; Tortosa, 1983).

Nodule Induction and Nodule Structure

Mature actinorhizal nodules are coralloid organs composed of multiple lobes (Figure 6.1B), each of which represents a modified lateral root without root cap, a superficial periderm and infected cells in the expanded cortex as shown in Figure 6.3A/B (Pawlowski and Bisseling, 1996). The activity of the apical meristem leads to the formation of a developmental gradient of infected cortical cells, allowing the distinction of an infection, nitrogen-fixation, and senescence zone in the cortex. In nodules formed on the roots of *Datisca* or *Coriaria* species, the pattern of infected cells is unusual; they form a continuous patch on one side of the acentric stele, not interspersed with uninfected cells as shown in Figure 6.3C (Newcomb and Pankhurst, 1982; Hafeez et al., 1984). In spite of the apical meristem, the growth of individual lobes is limited; additional branch lobes are formed as lateral primordia in the vascular pericycle of the preceding nodule lobe.

In actinorhizal symbioses, two means of infection are known, either intracellularly via root hairs or intercellularly by penetration between epidermal cells. As in legumes, the infection route is determined solely by the host plant (Miller and Baker, 1985; Racette and Torrey, 1989). Members of the Fagales are infected intracellularly, while members of the Rosales are infected intercellularly.

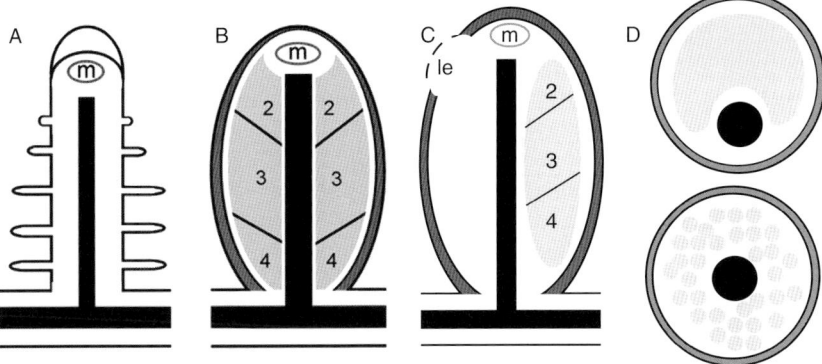

Figure 6.3. Actinorhizal nodule structure. (A) Scheme of a lateral root for comparison. The vascular system is depicted in black. Root cap and root hairs are shown. m, meristem. (B) Scheme of an actinorhizal nodule lobe. Due to the activity of the apical meristem (m), the nodule cortex can be divided into the infection zone (2), where cortical cells become gradually filled with *Frankia* hyphae, and afterward, vesicles differentiate (with the exception of [*Allo-*]*Casuarina* nodules), the fixation zone (3) where *Frankia* fixes dinitrogen in the infected cells, and the senescence zone (4), where *Frankia* hyphae and vesicles are degraded. Nodule lobes are surrounded by a superficial periderm which is shown in dark gray. (C) Longitudinal and cross section of a nodule lobe from *Datisca glomerata*. The infected cells are labeled in light gray. *D. glomerata* nodules can be aerated by lenticels (le) or by antigraviotropically growing nodule roots (not shown) depending on the growth conditions. (D) Distribution of infected cells (gray) in the nodule cortex in nodules of Cucurbitales (top) and other actinorhizal plants (bottom).

During intracellular infection, the first response of the plant to the presence of the microsymbiont is the deformation and branching of growing root hairs (Callaham et al., 1979; Berry et al., 1986) due to factors secreted by the microsymbiont (van Ghelue et al., 1997). The chemical nature of the factors has yet to be determined (Céremonie et al., 1999). In contrast to legumes, root hair deformation can also be induced by nonsymbiotic soil bacteria (Knowlton et al., 1980; Céremonie et al., 1999). Only a few deformed root hairs become infected; a *Frankia* hypha penetrates the root hair cell wall, while the root hair plasma membrane invaginates and an infection thread develops. Within the infection thread, the *Frankia* hypha is embedded within a plant-derived cell-wall-like pectin-rich matrix, the so-called encapsulation (Berry and Torrey, 1983; Berry et al., 1986; Berg, 1990). Concomitantly, cell divisions are induced in the root cortex close to the infected root hair, giving rise to the formation of a so-called prenodule (Callaham et al., 1979). The infection thread-like structures grow toward the prenodule by cell-to-cell passage, and infect some, but not all, prenodule cells by extensive branching within these cells, filling them from the center outward (Schwintzer et al., 1982). *Frankia* starts fixing nitrogen in the infected prenodule cells (Angulo Carmona, 1974; Laplaze et al., 2000). Studies on cell differentiation based on gene expression patterns indicate that, at least in *Casuarina glauca,* prenodules are primitive symbiotic organs that consist of three cell types with unique differentiation features equivalent to their counterparts in mature nodule lobes (Laplaze et al., 2000). Nevertheless, the prenodule is only an intermediate stage in nodule development. While it is developing, the primordium of the nodule lobe is initiated in the root pericycle near the infection site. *Frankia* hyphae in infection threads grow from the prenodule to the nodule primordium, again by cell-to-cell passage, and infect primordium cells.

There is a reluctance to use the term "infection thread" in actinorhizal nodulation. First, there is no actinorhizal equivalent to the infection thread matrix of legumes (Berg, 1999a,b). Actinorhizal "threads" are much thinner than legume infection threads; they enclose a single *Frankia* hypha, which determines their diameter, while legume infection threads tend to contain several contiguous rhizobia. Second, in rhizobial situations where there is no symbiosome formation—in certain legumes (Naisbitt et al., 1992) and in the only nonlegume infected by rhizobia, *Parasponia* sp.—researchers distinguish between "infection threads," which grow from cell to cell, and "fixation threads." In the nonsymbiosome species, fixation threads branch off from infection threads to fill the infected cells wherein the rhizobia develop into bacteroids and symbiotic nitrogen fixation commences (Lancelle and Torrey, 1984). Nevertheless, the term "infection thread" will be used for actinorhiza in this chapter.

As in legumes, stable intracellular accommodation of *Frankia* can only be performed by plant cells formed after the initiation of signal exchange with the microsymbiont. In intracellularly infected actinorhizal plants, infection threads grow by cell-to-cell passage, and infect cells by extensive branching within these cells, filling them from the center outward (Schwintzer et al., 1982). The process has been studied in detail for *Alnus glutinosa* by R.H. Berg (1999a,b) who described infections involving two types of hyphae enclosed in infection threads. He coined the terms "invasive hyphae" and "vegetative hyphae." A plant cell is invaded by the formation of an infection thread containing an invasive hypha sheathed in plant cell wall material and surrounded by the invaginated plasma membrane of the infected cell. Then vegetative hyphae enclosed in infection threads proliferate by branching from this infection thread, filling the cell from the center outward. When the cell has been filled, nitrogen-fixing vesicles differentiate from tips of these vegetative hyphae where nitrogenase is formed (Huss-Danell and Bergman, 1990). In intracellularly infected plants, infected cells often occur in files, since infection threads grow from one cell to the next.

During intercellular infection (Miller and Baker, 1985; Racette and Torrey, 1989), *Frankia* hyphae enter the plant root between epidermal cells and colonize the cortex intercellularly. Although it has been known for several years that *Frankia* strains produce cellulases and that cellulose profiles correlate with host specificity groups (Igual et al., 2001), a recent genome-based secretome analysis revealed that the predicted secretomes of *Frankia* sp. are relatively small and include few hydrolases (Mastronunzio et al., 2008). The lack of secreted hydrolases has been proposed to be typical for microorganisms that form beneficial associations with plants (Nagendran et al., 2009). During the colonization of the cortex by *Frankia* hyphae, epidermal and cortical cells secrete electron-dense material, rich in pectin and proteins, into the intercellular spaces (Liu and Berry, 1991). Infection of nodule primordium cells takes place by the formation of branching infection threads directly in these cells. Infection threads are formed in infected cells and do not grow from one cell to the next (Miller and Baker, 1985; Liu and Berry, 1991).

The infection mechanisms of unculturable strains (cluster III, Figure 6.2), in the families Rosaceae, Datiscaceae, and Coriariaceae, were not analyzed. Although the absence of prenodules and of files of infected cells in the cortex of mature nodules clearly indicates that they are infected intercellularly, analysis of the infection mechanism of the Datiscaceae and Coriariaceae based on the structure of mature nodule lobes is inconclusive. On the one hand, no prenodules are formed, but on the other hand, infection threads grow from one cell to the next (Berg et al., 1999).

Gas Exchange and Oxygen Protection in Nodules

Nodules need gas exchange to obtain N_2 for nitrogen fixation and O_2 to support aerobic respiration. Hence, actinorhizal plants that grow in wet or waterlogged soils have developed mechanisms for gas transport to the nodules. The nodules of these species are surrounded by a periderm that is more or less impermeable to gas. Species of *Casuarina, Gymnostoma, Myrica,* and *Comptonia* form nodule roots at the tips of nodule lobes via redifferentiation of the nodule lobe apical meristems (Figure 6.1B).

These nodule roots contain large air spaces in their cortex and grow upward, functioning as snorkels; their length is negatively correlated with the aeration of the root substrate (Tjepkema, 1978). When *Alnus glutinosa* grows in waterlogged soil, air reaches the nodules *via* thermo-osmotically mediated gas transport from the aerial parts to the roots (Schröder, 1989). In well-drained soils, *A. glutinosa* nodules are aerated via lenticels that interrupt their periderm (Wheeler et al., 1979). Nodules of *Coriaria* and *Datisca* species, in which the infected cells are only found on one side of the acentric stele (Figure 6.3C/D), are surrounded by a dense periderm with suberised cell walls. In *Coriaria* nodules, the periderm is interrupted by a single, large lenticel on the uninfected side of the nodule lobe, limiting the gas-diffusion pathway to the infected cells to the two-cell-layer gap between the internal periderm and the endodermis (Silvester and Harris, 1989). In *Coriaria* nodules, the only oxygen diffusion pathway from the lenticel to the infected cells is the two-cell-layer gap between the endodermis and the internal periderm, which opens the possibility of regulating oxygen diffusion *via* minimal changes in the turgor of these cells. *Coriaria* is unique among actinorhizal plants in that its nodules can perform short-term adaptation to varying external pO_2 levels, as can legume nodules, which also regulate oxygen access to the infected cells via turgor changes (Minchin, 1997). Nodules of *D. glomerata* form nodule roots in liquid culture or water-logged soil and lenticels in well-drained soil.

As described above, although nitrogen fixation by *Frankia* requires aerobic respiration to fulfill its energy demands, the nitrogen-fixing enzyme complex (nitrogenase) is irreversibly denatured by oxygen. As in the free-living state, *Frankia* in most symbioses forms nitrogen-fixing vesicles with multilayered envelopes, thus contributing to the protection of nitrogenase from oxygen. However, in contrast with the free-living state, in symbiosis the shape of the vesicles and their position within the infected cell depends on the host plant species (Baker and Mullin, 1992). For *Alnus* it has been shown that, as in the free-living state, the number of lipid layers of *Frankia* vesicle envelopes in the nodules is correlated with the external oxygen tension, implying that *Frankia* contributes to the protection of nitrogenase from nodule oxygen (Silvester et al., 1988). The fact that the increase in number of layers relative to oxygen tension is lower *in planta* than in the free-living state indicates that the plant must be contributing to oxygen protection as well.

In nodules of *Casuarina*, however, as in legume nodules, the plant seems to be solely responsible for protection of nitrogenase from oxygen. Here, the zone of infected cells contains very few intercellular spaces (Zeng et al., 1989) and the walls of infected cells are impregnated with a very hydrophobic lignin purported to restrict oxygen diffusion (Berg and McDowell, 1988). In the infected cells, *Frankia* does not form vesicles, indicating that nitrogenase is formed in hyphae, which also occurs when *Frankia* is grown in a microaerobic environment (Berg and McDowell, 1987). These observations are consistent with high concentrations of an oxygen-transporting class II hemoglobin in the infected cells (Jacobsen-Lyon et al., 1995). This globin is encoded by a nodule-specific gene, also the situation in legumes. In nodules of *Myrica gale*, microaerobic conditions were also found in infected cells (Zeng and Tjepkema, 1994), and the nodules were shown to contain large amounts of a hemoglobin (Pathirana and Tjepkema, 1995). This hemoglobin, however, was found to represent a class I hemoglobin whose role is ascribed to nitric oxide detoxification, not to oxygen transport (Heckmann et al., 2006). Given that *Frankia* indeed forms vesicles in *M. gale* nodules, it can be concluded that the plant alone does not provide sufficient oxygen protection for nitrogenase.

In nodules of actinorhizal Cucurbitales (*Datisca* and *Coriaria* species), *Frankia* forms finger-shaped vesicles with rather thin envelopes. The vesicles "point at" the central vacuole, forming a hollow sphere around it (Newcomb and Pankhurst, 1982; Hafeez et al., 1984). The vesicles are very tightly packed, leading to low surface exposure, and a continuous blanket of mitochondria is found around their bases (Silvester et al., 1999). This arrangement has led to the proposal that the mitochondrial blanket acts as a gas diffusion barrier and that the short diffusion path for oxygen transport into the vesicles obviates the need for an oxygen transport system, such as hemoglobin in the plant cell. For *D. glomerata*, evidence suggests that a truncated hemoglobin from *Frankia* (trHbO) plays a role in oxygen transport within *Frankia* vesicles (Pawlowski et al., 2007). TrHbO has also been suggested to play a role in the adaptation to low oxygen concentrations during the growth of *Frankia* in culture (Coats et al., 2009).

Taken together, it can be said that, in contrast with legumes, oxygen protection mechanisms in actinorhizal plants can vary depending on the host plant species, and that both macro- and microsymbionts can contribute to oxygen protection. For most actinorhizal symbioses, nodule oxygen protection mechanisms have not yet been examined.

Nodule Metabolism

Nodules are strong carbon sinks, evidenced by many large amyloplasts in uninfected cells in most types of actinorhizal nodules (Newcomb and Wood, 1987; Pawlowski, 2002). Root nodules are provided with photosynthates via the phloem, mostly in the form of sucrose. This sucrose has to support nodule development, *Frankia* nitrogen fixation, and the assimilation of the product of bacterial nitrogen fixation, ammonium. Unloading of sucrose from the phloem in root nodules is followed by catabolism by either cytosolic sucrose synthase or by cytosolic, apoplastic, or vacuolar invertase (van Ghelue et al., 1996; Lalonde et al., 2004).

Enzymatic activity in vesicle clusters and increased vesicle cluster respiration by the combination of malate, glutamate, and NAD indicate that dicarboxylates are the likely carbon source supplied to *Frankia* by its actinorhizal host (reviewed by Huss-Danell, 1997). Malate, and possibly other carbon sources, feed the TCA and glyoxylate cycles and the gluconeogenesis pathway (Benson and Schultz, 1990). Malate, moreover, is the carbon source for amino acid synthesis in *Alnus glutinosa* nodules (McClure et al., 1983). Important additional support for the idea that malate or other dicarboxylates are exported to *Frankia* came from the discovery that a nodule-specific dicarboxylate transporter localizes to the invaginated plant plasma membrane surrounding symbiotic *Frankia* in infected cells of *A. glutinosa* nodules (Jeong et al., 2004). Hence, at least in this symbiosis, *Frankia* can be supplied with dicarboxylates as are rhizobia in legume symbioses (Jeong et al., 2004; Prell and Poole, 2006). *Frankia* can store carbon and energy in the form of glycogen and trehalose in both free-living and symbiotic states (Benson and Eveleigh, 1979; Lopez et al., 1983). Glycogen granules have been observed by transmission electron microscopy in *Frankia* within nodules of *Chamaebatia*, *Myrica*, *Comptonia*, *Coriaria*, and *Elaeagnus* but not in *Discaria* and *Dryas*

(Newcomb and Wood, 1987). Like roots, nodules have a non-light-dependent CO_2-fixing capacity (Huss-Danell, 1990; McClure et al., 1983) proposed to be a back-up process in the event of depletion of carbon skeletons for ammonia assimilation (Valverde and Huss-Danell, 2008). Carbon dioxide fixation has been recorded in different *Alnus* species as well as in *Casuarina montana, Colletia cruciata,* and *Datisca cannabina* (Schubert et al., 1986; McClure et al., 1983; Huss-Danell, 1990). McClure and others (1983) established that CO_2 was incorporated into malate by PEP carboxylase in *A. glutinosa* nodules as occurs in C4/CAM metabolism.

In actinorhizal nodules, N_2 fixation takes place in *Frankia* vesicles. This was confirmed by *in situ* localization of *Frankia nifH* mRNA and immunological detection of nitrogenase (Sasakawa et al., 1988; Huss-Danell and Bergman, 1990; Pawlowski et al., 1994; Valverde and Wall, 2003b). In nodules of *Casuarina* or *Allocasuarina,* where no vesicles are formed, nitrogenase mRNA can be detected in hyphae, as it is in free-living *Frankia* grown under microaerobic conditions (Gherbi et al., 1997).

Based on enzyme activities and NMR studies (Lundberg and Lundquist, 2004), exogenously applied ammonium is assimilated in nodules via GS/GOGAT. Since GS activity and protein levels are low and neither glutamate dehydrogenase nor GOGAT activity can be detected in vesicle clusters isolated from nodules (Blom et al., 1981; Akkermans et al., 1983; Lundquist and Huss-Danell, 1992), it can be assumed that ammonium assimilation either takes place in the hyphae or only in the plant. Plant immunological detection of GS in the cytosol of infected cells in *A. incana* (Lundquist and Huss-Danell, 1992) and high levels of plant cytosolic GS expression in the infected cells (Guan et al., 1996), confirmed the hypothesis that symbiotic *Frankia* bacteria export the fixed nitrogen in the form of ammonium. However, in nodules of *Datisca glomerata*, GS mRNA and protein were not found in the infected cells, but rather in the uninfected cells surrounding the infected cells (Berry et al., 2004). So it seems that in this symbiosis, and probably also in other Cucurbitales nodules, *Frankia* does not export ammonium but an assimilated form of nitrogen. Based on an analysis of nitrogenous solutes in nodules, it was proposed that arginine (or urea) is exported and transported to the uninfected nodule cortical cells, where it is broken down to release ammonium that is re-assimilated in the GS/GOGAT cycle (Berry et al., 2004).

A summary of what is known about enzyme activities in actinorhizal nodules is provided in Figure 6.4. In legumes and actinorhizal plants, the major xylem transport form(s) of fixed nitrogen may comprise up to 70% of total nitrogenous solutes (Valverde and Wall, 2003a). Based on the xylem nitrogen transport forms, actinorhizal plants can be divided into three groups: (1) plants that transport mainly citrulline, (2) plants that transport mainly glutamine and/or asparagine, and (3) plants with a more mixed composition of major xylem compounds. A comparison of the ATP costs of the synthesis of the nitrogen export forms in nodules led to the conclusion that, in decreasing order, allantoin/allantoic acid, arginine, glycine, asparagine, and citrulline are energetically most efficient (Schubert, 1986). The metabolic pathways used in actinorhizal plants for the synthesis of citrulline, glutamine, and asparagine are indicated in Figure 6.4. In *Alnus glutinosa* and in certain legumes, glutamine synthetase transcripts were detected in the nodule vascular system (Guan et al., 1996; Temple et al., 1995; Forde et al., 1989), implying that nitrogenous solutes are degraded and ammonium is re-assimilated in the GS/GOGAT cycle during the transport to the xylem.

The carbon backbone for ammonium assimilation has to be provided by the host, and in legumes, malate seems to be used for this purpose (Vance and Heichel, 1991; Graham and Vance, 2000). That malate seems to be the carbon source provided by actinorhizal plants to

128 Plant-Soil Microbe Interactions

(A)
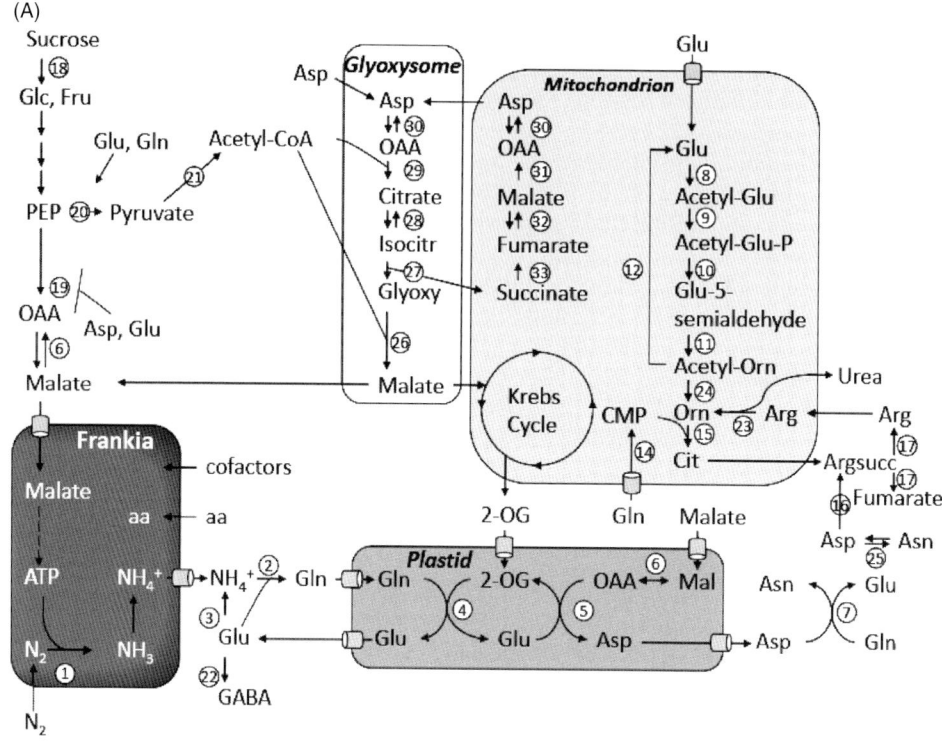

(B)
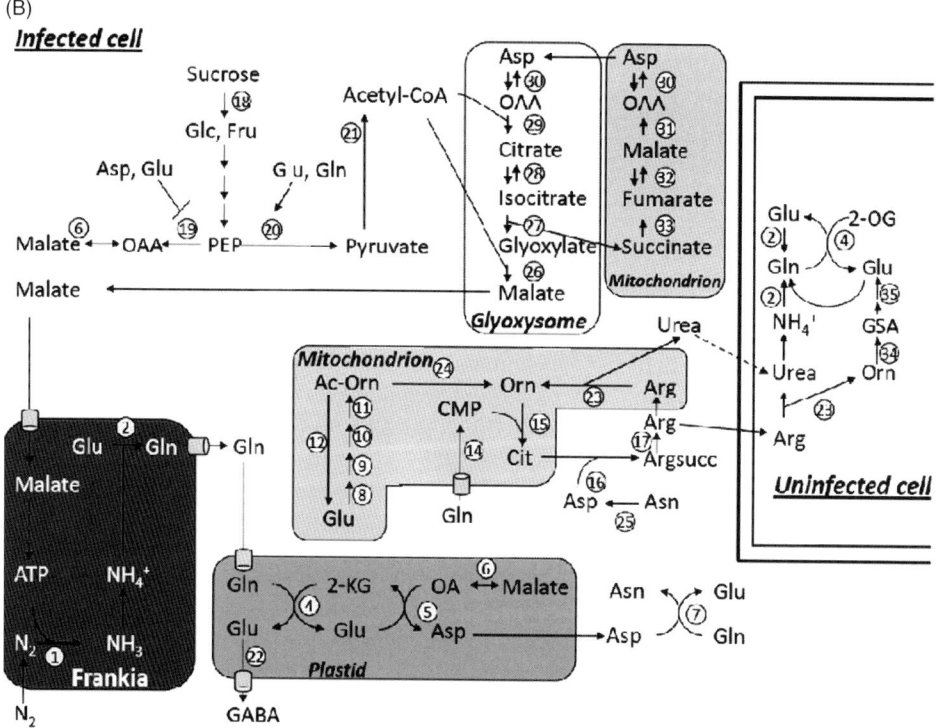

symbiotic *Frankia* (Jeong et al., 2004) leads to the assumption that in actinorhizal plants, as in legumes, the carbon backbones for ammonium assimilation are produced from malate, which can enter the TCA cycle to produce either oxaloacetate for aspartate synthesis or 2-oxoglutarate for the generation of glutamate in the GS/GOGAT cycle.

Outlook

Now that three *Frankia* genomes have been sequenced (Normand et al., 2007) and several more genome sequences are in the pipeline, the stage is set for a serious study of the link between phenotype and genomospecies. This will yield information on the question of co-evolution between micro- and macrosymbionts, and will help us to understand why *Frankia* strains from the cluster with the widest host range (cluster III) as shown in Figure 6.2, infecting host plants present on all continents except Africa and Antarctica, seem to be obligate symbionts. For the distribution of host plants see http://web.uconn.edu/mcbstaff/benson/Frankia/FrankiaHome.htm. In addition, comparative genomics will also be useful to develop markers for screening for important microsymbiont phenotypes such as salt or heavy metal resistance, thus identifying *Frankia* strains that can be used in saline or contaminated soil (Hafeez et al., 1999; Richards et al., 2002; Tani and Sasakawa, 2003).

The growing interest in sustainable agroforestry schemes has led to an increased use of actinorhizal plants, *Alnus* spp. and *Elaeagnus* spp. (see, e.g., Bourke, 1985; Paschke et al., 1989; Bohanek and Groninger, 2005), particularly *Casuarina* spp. in interplanting systems. In India, *C. equisetifolia* has been used successfully to increase the yield from forage grasses in interplanting systems (Kumar, 2006). It is to be hoped that the worldwide interest in energy crops will lead to schemes involving interplanting with nitrogen-fixing species, instead of increased, and sometimes indiscriminate, use of nitrogen fertilizers.

Nitrogen is a plant macronutrient, but the "new" nitrogen cycle can lead to environmental release of damaging waterborne nitrogen species and volatile ones like N_2O. Biological nitrogen fixation does not contribute per se to N_2O emission, which depends on incomplete

Figure 6.4. (*cont'd*) Nodule C and N metabolism. (A) Metabolism in infected cells of most actinorhizal nodules; (B) Metabolism in nodules of *D. glomerata* and probably other actinorhizal Cucurbitales (compartmentalization in uninfected cells is not shown). Enzymes: 1 – Nitrogenase; 2 – glutamine synthetase (GS; Guan et al., 1996; Berry et al., 2004); 3 – glutamate dehydrogenase (Valverde and Wall, 1999); 4 – glutamate oxo-glutarate aminotransferase (GOGAT; Valverde and Wall, 1999); 5 – mino acid transferase (AAT; Valverde and Wall, 1999); 6 – malate dehydrogenase (Valverde and Wall, 1999); 7 – asparagine synthetase (AS) (Baker and Parsons, 1997; Kim et al., 1999; Valverde and Wall, 2003a, 2003b); 8 – glutamate N-acetyl transferase; 9 – N-acetylglutamate 5-phosphotransferase; 10 – N-acetylglutamyl-5-phosphate:NAD(P)+ oxidoreductase; 11 – N2-acetylornithine:2-oxoglutarate aminotransferase; 12 – acetylornithine aminotransferase; 13 – ornithine carbamoylphosphate; 14 – carbamoyl phosphate synthase; 15 – ornithine carbamoyl transferase (Guan et al., 1996; Lundquist and Lundberg, 2004); 16 – argininosuccinate synthase; 17 – argininosuccinate lyase; 18 – sucrose synthase (Guan et al., 1996) or invertase; 19 – phosphoenolpyruvate carboxylase (PEPC; Valverde and Wall, 2003a); 20 – pyruvate kinase; regulation according to Podesta and Plaxton, 1992); 21 – pyruvate dehydrogenase complex; 22 – glutamate decarboxylase; 23 – arginase; 24 – acetylornithinase; 25 – asparagine synthetase; 26 – malate synthase; 27 – isocitrate lyase; 28 – aconitase; 29 – citrate synthase; 30 – aspartate aminotransferase; 31 – malate dehydrogenase; 32 – fumarase; 33 – succinate dehydrogenase; 34 – ornithine delta-aminotransferase; 35 – pyrroline-5-carboxylate dehydrogenase (Delauney and Verma, 1993). Based on Persson and Huss-Danell (2009).

denitrification or nitrification, by the soil microbial community. However, depending on soil composition, moisture, temperature, aeration, and the soil microflora, increased input of N might lead to increased N_2O production. Detailed ecological studies are required to allow predictions as to which biotopes might present risks in this regard.

References

Akkermans, A.D.L., Roelofsen, W., Blom, J., et al. 1983. Utilization of carbon and nitrogen compounds by *Frankia* in synthetic media and in root nodules of *Alnus glutinosa, Hippophae rhamnoides*, and *Datisca cannabina*. *Canadian Journal of Botany* 61, 2793–2800.

Angulo Carmona, A.F. 1974. La formation des nodules fixateurs d'azote chez *Alnus glutinosa* (L.). *Acta Botanica Neerlandica* 23, 257–303.

Baker, A., Parsons, R. 1997. Rapid assimilation of recently fixed N_2 in root nodules of *Myrica gale*. *Physiologia Plantarum* 99, 640–647.

Baker, D., Mullin, B.C. 1992. Actinorhizal symbioses. In: Biological Nitrogen Fixation (Stacey, G., Burris, R.H., Evans, H.J., Eds.), pp. 259–292. New York: Chapman & Hall.

Benoit, L.F., Berry, A.M. 1990. Methods for production and use of actinorhizal plants in forestry, low maintenance landscapes, and revegetation. In Schwintzer, C.R., and Tjepkema, J.D., Eds. *The biology of Frankia and actinorhizal plants,* pp. 281–194. New York: Academic Press.

Benson, D.R., Eveleigh, D.E. 1979. Nitrogen fixing homogenates of *Myrica pensylvanica* (bayberry) non-legume root nodules. *Soil Biology & Biochemistry* 11, 331–334.

Benson, D.R., Schultz, N. 1990. Physiology and biochemistry of *Frankia* in culture. In: The Biology of *Frankia* and Actinorhizal Plants (Schwintzer, C.R., Tjepkema, J.D., Eds.), pp. 107–127. San Diego, CA: Academic Press.

Benson, D.R., Silvester, W.B. 1993. Biology of *Frankia* strains, actinomycete symbionts of actinorhizal plants. *Microbiological Reviews* 57, 293–319.

Benson, D.R., Vanden Heuvel, B.D., Potter, D. 2004. Actinorhizal symbioses: Diversity and biogragraphy. In: Plant Microbiology (Gillings, M., Ed.), pp. 97–127. Oxford: BIOS Scientific Publishers Ltd.

Berg, R.H. 1990. Cellulose and xylans in the interface capsule in symbiotic cells of actinorhizae. *Protoplasma* 159, 35–43.

Berg, R.H. 1999a. *Frankia* forms infection threads. *Canadian Journal of Botany* 77, 1327–1333.

Berg, R.H. 1999b. Cytoplasmic bridge formation in the nodule apex of actinorhizal root nodules. *Canadian Journal of Botany* 77, 1351–1357.

Berg, R.H., Langenstein, B., Silvester, W.B. 1999. Development in the *Datisca-Coriaria* nodule type. *Canadian Journal of Botany* 77, 1334–1350.

Berg, R.H., McDowell, L. 1987. Endophyte differentiation in *Casuarina* actinorhizae. *Protoplasma* 136, 104–117.

Berg, R.H., McDowell, L. 1988. Cytochemistry of the wall of infected cells in *Casuarina* actinorhizae. *Canadian Journal of Botany* 66, 2038–2047.

Berry, A.M., Harriott, O.T., Moreau, R.A., et al. 1993. Hopanoid lipids compose the *Frankia* vesicle envelope, presumptive barrier of oxygen diffusion to nitrogenase. *Proceedings of the National Academy of Sciences of the United States of America* 90, 6091–6094.

Berry, A.M., McIntyre, L., McCully, M.E. 1986. Fine structure of root hair infection leading to nodulation in the *Frankia-Alnus* symbiosis. *Canadian Journal of Botany* 64, 292–305.

Berry, A.M., Murphy, T.M., Okubara, P.A., et al. 2004. Novel expression pattern of cytosolic glutamine synthetase in nitrogen-fixing root nodules of the actinorhizal host, *Datisca glomerata*. *Plant Physiology* 135, 1849–1862.

Berry, A.M., Thayer, J.R., Enderlin, C.S., et al. 1990. Patterns of ^{13}N ammonium uptake and assimilation by *Frankia* HFPArI3. *Archives of Microbiology* 154, 510–513.

Berry, A.M., Torrey, J.G. 1983. Root hair deformation in the infection process of *Alnus rubra*. *Canadian Journal of Botany* 61, 2863–2876.

Beveridge, T., Li, T.S.C., Oomah, B.D., et al. 1999. Sea buckthorn products: manufacture and composition. *Journal of Agricultural and Food Chemistry* 47, 3480–3488.

Blom, J., Roelofsen, W., Akkermans, A.D.L. 1981. Assimilation of nitrogen in root nodules of alder (*Alnus glutinosa*). *New Phytologist* 89, 321–326.

Bohanek, J., Groninger, J. 2005. Productivity of european black alder (*Alnus glutinosa*) interplanted with black walnut (*Juglans nigra*) in Illinois, U.S.A. *Agroforestry Systems* 64, 99–106.

Bourke, R.M. 1985. Food, coffee and casuarina: an agroforestry system from the Papua New Guinea highlands. *Agroforestry Systems* 2, 273–279.

Burleigh, S.H., Dawson, J.O. 1994. Occurrence of *Myrica*-nodulating *Frankia* in Hawaiian volcanic soils. *Plant & Soil* 164, 283–289.

Busse, M.D., Jurgensen, M.F., Page-Dumroese, D.S., et al. 2007. Contribution of actinorhizal shrubs to site fertility in a Northern California mixed pine forest. *Forest Ecology and Management* 244, 68–75.

Callaham, D., Newcomb, W., Torrey, J.G., et al. 1979. Root hair infection in actinomycete-induced root nodule initiation in *Casuasina, Myrica*, and *Comptonia*. *Botanical Gazette* 140 (suppl.), S1–S9.

Carlsson, G., Huss-Danell, K. 2003. Nitrogen fixation in perennial forage legumes in the field. *Plant & Soil* 253, 353–372.

Cérémonie, H., Debelle, F., Fernandez, M.P. 1999. Structural and functional comparison of *Frankia* root hair deforming factor and rhizobia Nod factor. *Canadian Journal of Botany* 77, 1293–1301.

Clawson, M.L., Bourret, A., Benson, D.R. 2004. Assessing the phylogeny of *Frankia*-actinorhizal plant nitrogen-fixing root nodule symbioses with *Frankia* 16S rRNA and glutamine synthetase gene sequences. *Molecular Phylogenetics and Evolution* 31, 131–138.

Coats, V., Schwintzer, C.R., Tjepkema, J.D. 2009. Truncated hemoglobins in *Frankia* Ccl3: effects of nitrogen source, oxygen concentration, and nitric oxide. *Canadian Journal of Microbiology* 55, 867–873.

Côte, B., Carlson, R.W., Dawson, J.O. 1988. Leaf photosynthetic characteristics of seedlings of actinorhizal *Alnus* spp. and *Elaeagnus* spp. *Photosynthesis Research* 16, 211–218.

Dawson, J.O. 1986. Actinorhizal plants: Their use in forestry and agriculture. *Outlook on Agriculture* 15, 202–208.

Dawson, J.O. 1990. Interactions among actinorhizal and associated species. In *The biology of Frankia and actinorhizal plants* (Schwintzer, C.R., Tjepkema, J.D., Eds.), pp. 299–316. New York: Academic Press.

Dawson, J.O. 2008. Ecology of actinorhizal plants. In: Nitrogen-fixing Actinorhizal Symbioses. Nitrogen Fixation: Origins, Applications, and Research Progress, Vol. 6 (Pawlowski, K., Newton, W.E., Eds.), pp. 199–234. New York: Springer.

Delauney, A.J., Verma, D.P.S. 1993. Proline biosynthesis and osmoregulation in plants. *Plant Journal* 4, 215–223.

Forde, B.G., Day, H.M., Turton, J.F., et al. 1989. Two glutamine synthetase genes from *Phaseolus vulgaris* L. display contrasting developmental and spatial patterns of expression in transgenic *Lotus corniculatus* plants. *Plant Cell* 1, 391–401.

Gauthier, D., Jaffre, T., Prin, Y. 2000. Abundance of *Frankia* from *Gymnostoma* spp.in the rhizosphere of *Alphitonia neocaledonica*, a non-nodulated Rhamnaceae endemic to New Caledonia. *European Journal of Biology* 36, 169–175.

Gherbi, H., Duhoux, E., Franche, C., et al. 1997. Cloning of a full-length symbiotic hemoglobin cDNA and in situ localization of the corresponding mRNA in *Casuarina glauca* root nodule. *Physiologia Plantarum* 99, 608–616.

Graham, P.H., Vance, C.P. 2000. Nitrogen fixation in perspective: an overview of research and extension needs. *Field Crops Research* 65, 93–106.

Guan, C., Ribeiro, A., Akkermans, A.D.L., et al. 1996. Glutamine synthetase and acetylornithine aminotransferase in actinorhizal nodules of *Alnus glutinosa*. *Plant Molecular Biology* 32, 1177–1184.

Hafeez, F., Akkermans, A.D.L., Chaudhary, A.H. 1984. Observations on the ultrastructure of *Frankia* sp. in root nodules of *Datisca cannabina* L. *Plant & Soil* 79, 383–402.

Hafeez, F.Y., Hameed, S., Malik, K.A. 1999. *Frankia* and *Rhizobium* strains as inoculum for fast growing trees in saline environment. *Pakistan Journal of Botany* 31, 173–182.

Hahn, D., Starrenburg, M.J.C., Akkermans, A.D.L. 1988. Variable compatibility of cloned *Alnus glutinosa* ecotypes against ineffective *Frankia* strains. *Plant & Soil* 107, 233–243.

Hardig, T.M., Soltis, P.S., Soltis, D.E. 2000. Diversification of the North American shrub genus *Ceanothus* (Rhamnaceae): Conflicting phylogenies from nuclear ribosomal DNA and chloroplast DNA. *American Journal of Botany* 87, 108–123.

Heckmann, A.B., Hebelstrup, K.H., Larsen, K., et al. 2006. A single hemoglobin gene in *Myrica gale* retains both symbiotic and non-symbiotic specificity. *Plant Molecular Biology* 61, 769–779.

Hiltner, L. 1895. Über die Bedeutung der Wurzelknöllchen von *Alnus glutinosa* für die Stickstoffernährung dieser Pflanze. *Landwirtschaftliche Verständnisstudien* 46, 153–161.

Horning, M.E., Cronn, R.C. 2009. Development of variable microsatellite loci and range-wide characterization of nuclear genetic diversity in the important dryland shrub antelope bitterbrush (*Purshia tridentata*). *Journal of Arid Environments* 73, 7–13.

Huguet, V., Gouy, M., Normand, P., et al. 2005. Molecular phylogeny of Myricaceae: a reexamination of host-symbiont specificity. *Molecular Phylogenetics and Evolution* 34, 557–568.

Huss-Danell, K. 1990. The physiology of actinorhizal nodules. In: The Biology of *Frankia* and Actinorhizal Plants (Schwintzer, C.R., Tjepkema, J.D., Eds.), New York: Academic Press, Inc. pp. 129–156.

Huss-Danell, K. 1997. Tansley Review No. 93. Actinorhizal symbioses and their N_2 fixation. *New Phytologist* 136, 375–405.

Huss-Danell, K., Bergman, B. 1990. Nitrogenase in *Frankia* from root nodules of *Alnus incana* (L.) Moench: Immunolocalization of the Fe- and MoFe- proteins during vesicle differentiation. *New Phytologist* 116, 443–455.

Igual, J.M., Velazquez, E., Mateos, P.F., et al. 2001. Cellulase isoenzyme profiles in *Frankia* strains belonging to different cross-inoculation groups. *Plant & Soil* 229, 35–39.

Jacobsen-Lyon, K., Jensen, E. O., Jorgensen, J.-E., et al. 1995. Symbiotic and non-symbiotic hemoglobin genes of *Casuarina glauca*. *Plant Cell* 7, 213–222.

Jeong, J., Suh, S.J., Guan, C., Tsay, Y.-F., et al. 2004. A nodule-specific dicarboxylate transporter from *Alnus glutinosa*. *Plant Physiology* 134, 969–978.

Kim, HB., Lee, S.H., An, C.S. 1999. Isolation and characterization of cDNA synthetase from the root nodules of *Elaeagnus umbellata*. *Plant Science* 149, 85–94.

Knowlton, S., Berry, A., Torrey, J.G. 1980. Evidence that associated soil bacteria may influence root hair infection of actinorhizal plants by *Frankia*. *Canadian Journal of Microbiology* 26, 971–977.

Kohls, S.J., Baker, D.D., van Kessel, C., et al. 2003. An assessment of soil enrichment by actinorhizal N_2 fixation using $\delta^{15}N$ values in a chronosequence of deglaciation at Glacier Bay, Alaska. *Plant & Soil* 254, 11–17.

Kohls, S.J., Thimmapuram, J., Buschena, C.A., et al. 1994. Nodulation patterns of actinorhizal plants in the family Rosaceae. *Plant & Soil* 162, 229–239.

Krebber, O. 1932. Untersuchungen über die Wurzelknöllchen der Erle. *Archives of Microbiology* 2, 588–608.

Krumholz, G.D., Chval, M.S., McBride, M.J., et al. 2003. Germination and physiological properties of *Frankia* spores. In Frankia *symbiosis* (Normand, P., Pawlowski, K., Dawson, J.O., Eds.,) pp. 57–68. Dordrecht, The Netherlands: Kluwer Academic Publishers.

Kumar, B.M. 2006. Agroforestry: the new old paradigm for Asian food security. *Journal of Tropical Agriculture* 44, 1–14.

Lalonde, S., Wipf, D., Frommer, W.B. 2004. Transport mechanisms for organic forms of carbon and nitrogen between source and sink. *Annual Review of Plant Biology* 55, 341–372.

Lancelle, S.A., Torrey, J.G. 1984. Early development of *Rhizobium*-induced root nodules of *Parasponia rigida*. II. Nodule morphogenesis and symbiotic development. *Canadian Journal of Botany* 63, 25–35.

Laplaze, L., Duhoux, E., Franche, C., et al. 2000. Actinorhizal prenodule cells display the same differentiation as the corresponding nodule cells. *Molecular Plant-Microbe Interactions* 13, 107–112.

Lawrence, D.B., Schoenike, R.E., Quispel, A., et al. 1967. The role of *Dryas drummondii* in vegetation development following ice recession at Glacier Bay, Alaska, with special reference to its nitrogen fixation by root nodules. *Journal of Ecology* 55, 793–813.

Lawrie, A.C. 1982. Field nodulation in nine species of *Casuarina* in Victoria. *Australian Journal of Botany* 30, 447–460.

Lechevalier, M.P., Lechevalier, H.A. 1990. Systematics, isolation, culture of *Frankia*. In: The Biology of Frankia and Actinorhizal Plants (Schwintzer, C.R., Tjepkema, J.D. Eds.,) pp. 35–60. San Diego: Academic Press.

Li, T.S.C., Beveridge, T.H.J., Drover, J.C.G. 2007. Phytosterol content of sea buckthorn (*Hippophae rhamnoides* L.) seed oil: Extraction and identification. *Food Chemistry* 101, 1633–1639.

Li, T.S.C., Schroeder, W.R. 1996. Sea buckthorn (*Hippophae rhamnoides* L.): A multipurpose plant. *Horticultural Technology* 6, 370–380.

Liu, Q., Berry, A.M. 1991. The infection process and nodule initiation in the *Frankia-Ceanothus* root nodule symbiosis. *Protoplasma* 163, 82–92.

Lopez, M.F., Fontaine, M.S., Torrey, J.G. 1984. Levels of trehalose and glycogen in *Frankia* sp. HFPArI3 (Actinomycetales). *Canadian Journal of Microbiology* 30, 209–214.

Lopez, M.F., Whaling, C.S., Torrey, J.G. 1983. The polar lipids and free sugars of *Frankia* in culture. *Canadian Journal of Botany* 61, 2834–2842.

Lundberg, P., Lundquist, P.-O. 2004. Primary metabolism in N_2-fixing *Alnus incana-Frankia* symbiotic root nodules studied with ^{15}N and ^{31}P nuclear magnetic resonance spectroscopy. *Planta* 219, 661–672.

Lundquist, P.-O., Huss-Danell, K. 1992. Immunological studies of glutamine synthetase in *Frankia-Alnus incana* symbioses. *FEMS Microbiology Letters* 91, 141–146.

Mastronunzio, J.E., Tisa, L.S., Normand, P., et al. 2008. Comparative secretome analysis suggests low plant cell wall degrading capacity in *Frankia* symbionts. *BMC Genomics* 9, 47.

Maunuksela, L., Zepp, K., Koivula, T. et al. 1999. Analysis of *Frankia* populations in three soils devoid of actinorhizal plants. *FEMS Microbiology Ecology* 28, 11–21.

McClure, P.R., Coker, G.T., Schubert, K.R. 1983. CO_2 fixation in roots and nodules of *Alnus glutinosa*: Role of PEP carboxylase and carbamyl phosphate synthetase in dark CO_2 fixation, citrulline synthesis, and N_2 fixation. *Plant Physiology* 71, 652–657.

Meesters, T.M., Van Vliet, W.M., Akkermans, A.D.L. 1987. Nitrogenase is restricted to the vesicles in *Frankia* strain EAN1pec. *Physiologia Plantarum* 70, 267–271.

Meyen, J. 1829. Über das Hervorwachsen parasitischer Gebilde aus den Wurzeln anderer Pflanzen. *Flora (Jena)* 12, 49–64.

Miller, I.M., Baker, D.D. 1985. The initiation development and structure of root nodules in *Elaeagnus angustifolia* (Elaeagnaceae). *Protoplasma* 128, 107–119.

Minchin, F.R. 1997. Regulation of oxygen diffusion in legume nodules. *Soil Biology & Biochemistry* 29, 881–888.

Mirza, M.S., Hahn, D., Akkermans, A.D.L. 1992. Isolation and characterization of *Frankia* strains from *Coriaria nepalensis*. *Systematic and Applied Microbiology* 15, 289–295.

Mirza, M.S., Hameed, S., Akkermans, A.D. 1994. Genetic diversity of *Datisca cannabina*-compatible *Frankia* strains as determined by sequence analysis of the PCR-amplified 16S rRNA gene. *Applied and Environmental Microbiology* 60, 2371–2376.

Moran, N.A. 2003. Tracing the evolution of gene loss in obligate bacterial symbionts. *Current Opinion in Microbiology* 6, 512–518.

Murry, M.A., Zhang, Z., Torrey, J.G. 1985. Effect of O_2 on vesicle formation, acetylene reduction, and O_2-uptake kinetics in *Frankia* sp. HFPCcI3 isolated from *Casuarina cunninghamiana*. *Canadian Journal of Microbiology* 31, 804–809.

Nagendran, S., Hallen-Adams, H.E., Paper, J.M., et al., 2009. Reduced genomic potential for secreted plant cell-wall degrading enzymes in the ectomycorrhizal fungus *Amanita bisporigera*, based on the secretome of *Trichoderma reesei*. *Fungal Genetics and Biology* 46, 427–435.

Naisbitt, T., James, E.K., Sprent, J.I. 1992. The evolutionary significance of the legume genus *Chamaecrista*, as determined by nodule structure. *New Phytologist* 122, 487–492.

National Research Council. 1984. Casuarinas: Nitrogen-fixing trees for adverse sites. Washington, DC: National Academy Press.

Neave, I.A., Dawson, J.O., DeLucia, E.H. 1989. Autumnal photosynthesis is extended in nitrogen-fixing European black alder when compared with white basswood: Possible adaptive significance. *Canadian Journal of Forestry Research* 19, 12–17.

Newcomb, W., Pankhurst, C.E. 1982. Fine structure of actinorhizal nodules of *Coriaria arborea* (Coriariaceae). *New Zealand Journal of Botany* 20, 93–103.

Newcomb, W.R., Wood, S. 1987 Morphogenesis and fine structure of *Frankia* (Actinomycetales): The microsymbiont of nitrogen-fixing actinorhizal root nodules. *International Review of Cytology* 109, 1–88.

Nickel, A. 2000. Population dynamics of *Frankia* in soil (Ph.D. thesis, Swiss Technical University (ETH), Zürich, Switzerland).

Nickel, A., Hahn, D., Zepp, K. et al. 1999. In situ analysis of introduced *Frankia* populations in root nodules obtained on *Alnus glutinosa* grown under different water availability. *Canadian Journal of Botany* 77, 1231–1238.

Normand, P., Lapierre, P., Tisa, L.S., et al. 2007. Genome characteristics of facultatively symbiotic *Frankia* sp. strains reflect host range and host plant biography. *Genome Research* 17, 7–15.

Normand, P., Orso, S., Cournoyer, B., Jeannin, P., Chapelon, C., Dawson, J., Evtushenko, L., Misra, A. 1996. Molecular phylogeny of the genus *Frankia* and related genera and emendation of the family Frankiaceae. *International Journal of Systematic Bacteriology* 46, 1–9.

Parsons, R., Silvester, W.B., Harris, S. et al. 1987. *Frankia* vesicles provide inducible and absolute oxygen protection for nitrogenase. *Plant Physiology* 83, 728–731.

Paschke, M.W. 1997. Actinorhizal plants in rangelands of the Western United States. *Journal of Range Management* 50, 62–72.

Paschke, M.W., Dawson, J.O., David, M.B. 1989. Soil nitrogen mineralization in plantations of *Juglans nigra* interplanted with actinorhizal *Elaeagnus umbellata* or *Alnus glutinosa*. *Plant and Soil* 118, 33–42.

Pathirana, S.M., Tjepkema, J.D. 1995. Purification of hemoglobin from the actinorhizal root nodules of *Myrica gale* L. *Plant Physiology* 107, 827–831.

Pawlowski, K. 2002. Actinozhizal symbioses. In: Nitrogen Fixation at the Millennium (Leigh, G.J., Ed.), pp. 167–189. Amsterdam: Elsevier Science, Pergamon Press.

Pawlowski, K., Akkermans, A.D.L., van Kammen, A., Bisseling, T. 1994. Expression of *Frankia nif* genes in nodules of *Alnus glutinosa*. *Plant & Soil* 170, 371–376.

Pawlowski, K., Bisseling, T. 1996. Rhizobial and actinorhizal symbioses: What are the shared features? *Plant Cell* 8, 1899–1913.

Pawlowski, K., Jacobsen, K.R., Alloisio, N. et al. 2007. Truncated hemoglobins in actinorhizal nodules of *Datisca glomerata*. *Plant Biology* 9, 776–785.

Persson, T., Huss-Danell, K. 2009. In: Microbiology Monographs: Prokaryotic Symbionts in Plants (Pawlowski, K., Steinbüchel, A., Eds.), pp. 155–178. Berlin, Germany: Springer.

Podestá, F.E., Plaxton, W.C. 1992. Plant cytosolic pyruvate kinase: a kinetic study. *Biochimica et Biophysica Acta* 1160, 213–220.

Prell, J., Poole, P. 2006. Metabolic changes of rhizobia in legume nodules. *Trends in Microbiology* 14, 161–168.

Racette, S., Torrey, J.G. 1989. Root nodule initiation in *Gymnostoma* (Casuarinaceae) and *Shephardia* (Elaeagnaceae) induced by *Frankia* strain HFPGpI1. *Canadian Journal of Botany* 67, 2873–2879.

Ramirez-Saad, H., Janse, J.D., Akkermans, A.D.L. 1998. Root nodules of *Ceanothus caeruleus* contain both the N_2-fixing *Frankia* endophyte and a phylogenetically related Nod⁻/Fix⁻ actinomycete. *Canadian Journal of Microbiology* 44, 140–148.

Reiche, K. 1907. Grundzüge der Pflanzenverbreitung in Chile. Leipzig, Germany: Engelmann.

Richards, J.W., Krumholz, G.D., Chval, M.S., Tisa, L.S. 2002. Heavy metal resistance patterns of *Frankia* strains. *Applied & Environmental Microbiology* 68, 923–927.

Ruan, C.-J., Li, H., Mopper, S. 2009. Characterization and identification of ISSR markers associated with resistance to dried-shrink disease in sea buckthorn. *Molecular Breeding* 24, 255–268.

Russo, R.O. 2005. Nitrogen-fixing trees with actinorhiza in forestry and agroforestry. In: Nitrogen Fixation in Agriculture, Forestry, Ecology, and the Environment. Nitrogen Fixation: Origins, Applications, and Research Progress, Vol. 4, pp. 143–171 New York: Springer.

Safo-Sampah, S., Torrey, J.G. 1988. Polysaccharide-hydrolyzing enzymes of *Frankia* (Actinomycetales). *Plant & Soil* 112, 89–97.

Sanginga, N., Vanlauwe, B., Danso, S.K.A. 1995. Management of biological N_2 fixation in alley cropping systems—estimation and contribution to N balance. Plant & Soil 174, 119–141.

Sasakawa, H., Hiyoshi, T., Sugiyama, T. 1988. Immunogold localization of nitrogenase in root nodules of *Elaeagnus pungens* Thunb. *Plant & Cell Physiology* 29, 1147–1152.

Schröder, P. 1989. Aeration of the root system in *Alnus glutinosa*. *Annales des Sciences Forestières* 46, 592–594.

Schubert, K.R. 1986. Products of biological nitrogen fixation in higher plants: synthesis, transport, and metabolism. *Annual Review of Plant Physiology* 37, 539–574.

Schwintzer, C.R. 1990. Oxygen regulation and hemoglobin. In: The Biology of *Frankia* and Actinorhizal Plants (Schwintzer, C.R., Tjepkema, J.D., Eds.), pp. 177–193. San Diego: Academic Press.

Schwintzer, C.R., Berry, A.M., Disney, L.D. 1982. Seasonal patterns of root nodule growth, endophyte morphology, nitrogenase activity and shoot development in *Myrica gale*. *Canadian Journal of Botany* 60, 746–757.

Silvester, W.B., Harris, S.L. 1989. Nodule structure and nitrogenase activity of *Coriaria arborea* in response to varying oxygen partial pressure. *Plant & Soil* 118, 97–110.

Silvester, W.B., Harris, S.L., Tjepkema, J.D. 1990. Oxygen regulation and hemoglobin. In: The Biology of *Frankia* and Actinorhizal Plants, Schwintzer, (C.R., Tjepkema, J.D., Eds.), pp. 157–176. San Diego: Academic Press.

Silvester, W.B., Langenstein, B., Berg, R.H. 1999. Do mitochondria provide the oxygen diffusion barrier in root nodules of *Coriaria* and *Datisca*? *Canadian Journal of Botany* 77, 1358–1366.

Silvester, W.B., Silvester, J.K., Torrey, J.G. 1988. Adaptation of nitrogenase to varying oxygen tension and the role of the vesicle in root nodules of *Alnus incana* ssp. *rugosa*. *Canadian Journal of Botany* 66, 1772–1779.

Simonet, P., Navarro, E., Rouvier, C. et al. 1999. Co-evolution between *Frankia* populations and host plants in the family Casuarinaceae and consequent patterns of global dispersal. *Environmental Microbiology* 1, 525–533.

Smolander, A., Rönkkö, R., Nurmiaho-Lassila, E.-L. et al. 1990. Growth of *Frankia* in the rhizosphere of *Betula pendula*, a nonhost tree species. *Canadian Journal of Microbiology* 36, 649–656.

Solans, M. 2007. *Discaria trinervis*—*Frankia* symbiosis promotion by saprophytic actinomycetes. *Journal of Basic Microbiology* 47, 243–250.

Swensen, S.M., Benson, D.R. 2008. Evolution of actinorhizal host plants and *Frankia* endosymbionts. In: Nitrogen-fixing Actinorhizal Symbioses. Nitrogen Fixation: Origins, Applications, and Research Progress, Vol. 6, (Pawlowski, K., Newton, W.E., Eds.) pp. 73–104. New York: Springer.

Tani, C., Sasakawa, H. 2003. Salt tolerance of *Casuarina equisetifolia* and *Frankia* Ce strain isolated from the root nodules of *C. equisetifolia*. *Soil Science and Plant Nutrition* 49, 215–222.

Temple, S.J., Heard, J., Ganter, G. 1995. Characterization of a nodule-enhanced glutamine synthetase from alfalfa: nucleotide sequence, *in situ* localization, and transcript analysis. *Molecular Plant-Microbe Interactions* 8, 218–227.

Tisa, L., McBride, M., Ensign, J.C. 1983. Studies of growth and morphology of *Frankia* strains EAN1pec, EuI1c, CpI1, and ACN1ag. *Canadian Journal of Botany* 61, 2768–2773.

Tjepkema, J.D. 1978. The role of oxygen diffusion from the shoots and the nodule roots in nitrogen fixation by root nodules of *Myrica gale*. *Canadian Journal of Botany* 56, 1365–1371.

Tortosa, R.D. 1983. El género *Discaria* (Rhamnaceae) *Boletín de la Sociedad Argentina de Botánica* 22, 301–335.

Valverde, C., Huss-Danell, K. 2008. Carbon and Nitrogen Metabolism in Actinorhizal Nodules. In: Nitrogen-fixing Actinorhizal Symbioses (Pawlowski, K., Newton, W.E., Eds.), pp. 167–198. Dordrecht, The Netherlands: Springer.

Valverde, C., Wall, L.G. 1999. Regulation of nodulation in *Discaria trinervis* (Rhamnaceae)—*Frankia* symbiosis. *Canadian Journal of Botany* 77, 1203–1310.

Valverde, C., Wall, L.G. 2003a. Ammonium assimilation in *Discaria trinervis* root nodules. Regulation of enzyme activities and protein levels by the availability of macronutrients (N, P and C). *Plant & Soil* 254, 139–153.

Valverde, C., Wall, L.G. 2003b. The regulation of nodulation, nitrogen fixation and assimilation under a carbohydrate shortage stress in the *Discaria trinervis*–*Frankia* symbiosis. *Plant & Soil* 254, 155–165.

van Ghelue, M., Lovaas, E., Ringo, E. et al. 1997. Early interactions between *Alnus glutinosa* and *Frankia* strain ArI3—production and specificity of root hair deformation factor(s). *Physiologia Plantarum* 99, 579–587.

van Ghelue, M., Ribeiro, A., Solheim, B. et al. 1996. Sucrose synthase and enolase expression in actinorhizal nodules of *Alnus glutinosa*: comparison with legume nodules. *Molecular & General Genetics* 250, 437–446.

Vance, C.P., Heichel, G.H. 1991. Carbon in N_2 fixation—limitation or exquisite adaptation. *Annual Review of Plant Physiology and Plant Molecular Biology* 42, 373–392.

Vanden Heuvel, B.D., Benson, D.R., Bortiri, E. et al. 2004. Low genetic diversity among *Frankia* spp. strains nodulating sympatric populations of actinorhizal species of Rosaceae, *Ceanothus* (Rhamnaceae) and *Datisca glomerata* (Datiscaceae) west of the Sierra Nevada (California). *Canadian Journal of Microbiology* 50, 989–1000.

Wheeler, C.T., Gordon, E.M., Ching, T.M. 1979. Oxygen relations of the root nodules of *Alnus rubra* Bong. *New Phytologist* 82, 449–457.

Young, D.R., Sande, E., Perters, G.A. 1992. Spatial relationships of *Frankia* and *Myrica cerifera* on a Virginia, U.S.A , barrier island. *Symbiosis* 112, 209–220.

Zeng, S., Tjepkema, J.D. 1994. The wall of the infected cell may be the major diffusion barrier in nodules of *Myrica gale* L. *Soil Biology & Biochemistry* 5, 633–639.

Zeng, S., Tjepkema, J.D., Berg, R.H. 1989. Gas diffusion pathway in nodules of *Casuarina cunninghamiana*. *Plant & Soil* 118, 119–123.

Zhu, X.Y., Chase, M.W., Qiu, Y.L. et al. 2007. Mitochondrial *matR* sequences help to resolve deep phylogenetic relationships in rosids. *BMC Evolutionary Biology* 7, 217.

Zimpfer, J.F., Kaelke, C.M., Smyth, C.A. et al. 2003. *Frankia* inoculation, soil biota, and host tissue amendment influence *Casuarina* nodulation capacity of a tropical soil. *Plant & Soil* 254, 1–10.

Zimpfer, J.F., Kennedy, G.J., Smyth, C.A. et al. 1999. Localization of *Casuarina*—infective *Frankia* near *Casuarina cunninghamiana* trees in Jamaica. *Canadian Journal of Botany* 77, 1248–1256.

Zimpfer, J.F., Smyth, C.A., Dawson, J.O. 1997. The Capacity of Jamaican mine spoils, agricultural and forest soils to nodulate *Myrica Cenfera, Leucaena leucocephala* and *Casuarina Cunninghamiana*. *Physiologia Plantarum* 99, 664–672.

Chapter 7
Two in the Far North: The Alder-*Frankia* Symbiosis, with an Alaskan Case Study

Mike Anderson

Abbreviations

C	Carbon
DNA	Deoxy-ribonucleic acid
EPA	Environmental Protection Agency
G + C	%Guanine + Cytosine content
IGS	Inter-Genic Spacer
N	Nitrogen
nif	Nitrogen-fixation locus
P	Phosphorous
PCR	Polymerase Chain Reaction
RF	Restriction Fragment haplotype
RFLP	Restriction Fragment Length Polymorphism
RNA	Ribonucleic acid
SLW	Specific Leaf Weight = grams of leaf tissue per square meter of dry leaf tissue
SNF	Specific Nitrogen Fixation rate = µmol of N_2 fixed per unit nodule dry mass per hour
Sp–	Spore-negative; *Frankia* strains incapable of forming reproductive spores within host nodules
Sp+	Spore-positive; *Frankia* strains capable of forming reproductive spores within host nodules
ssp	subspecies
subg.	subgenus

Introduction

Root-nodule symbioses, in which plants house nitrogen (N)-fixing bacteria in specialized organs derived from fine roots (the nodules), occur between 10 families of angiosperms and at least 13 genera of bacteria (Sawada et al., 2003). For plants capable of supporting such a symbiosis, the availability of atmospheric N it provides can be a significant ecological benefit, allowing colonization of N-poor soils or alternative N-utilization strategies. Such plants also frequently act as keystone organisms, providing the majority of N entering the N cycle in the ecosystems in which they occur.

As for any interaction between organisms, the outcome (mutualism versus parasitism) and evolutionary trajectory of root-nodule symbioses are subject to modulation both by factors intrinsic to the interaction (e.g., host-symbiont specificity, dispersal mechanisms, etc.) and a variety of extrinsic factors (e.g., availability of light or nutrients, interactions with other organisms, etc.). Variation in such factors can produce very different evolutionary dynamics among local populations, which in turn interact at regional and ultimately global levels to determine macroevolutionary patterns of the interaction (Thompson, 1994, 2005). Because N availability in soils commonly limits plant productivity, factors modulating N-fixing interactions at the local scale can affect local ecosystem function, and the upward cascade of evolutionary effects just described can have broad implications for landscape and regional-scale processes. Many root-nodule-based symbiotic systems are also important crops (e.g., soybean, alfalfa), or have applications in forestry and/or bioremediation (e.g., many species of alder). Detailed characterization of geographic variation in N-fixing symbioses can thus provide insights into the evolutionary ecology of interspecific interactions, into controls over key ecosystem processes, and may have important economic and environmental applications as well.

The symbiosis that occurs between alder (genus *Alnus* Mill. [Betulaceae]) and N-fixing *Frankia* bacteria in Alaskan ecosystems is an excellent subject for such studies, for several reasons. First, the alder-*Frankia* system is widespread and diverse, with a circumpolar distribution in the northern hemisphere providing widely varied environmental contexts and involving several plant species as well as bacterial strains that vary widely in physiology, interaction strategy, and effects on host plants. Alaska provides a microcosm of this global diversity in which alder occurs in habitats ranging from coastal rain forests to arctic tundra at densities ranging from sporadic to abundant. Second, complementing the diversity of regional habitats available in Alaska are: (1) wide differences among local habitats due to variation in landscape position and successional dynamics, and (2) a high degree of habitat redundancy that results from low regional plant diversity and relatively predictable successional pathways following deposition or disturbance (Chapin et al., 2006). Such a combination makes for relatively straightforward habitat replication, allowing methodologically robust testing of ecological hypotheses. Third, the ecological impacts of alder in the region are significant and, for some species, are among the best characterized of any naturally occurring N-fixing system.

This chapter presents recent work conducted across several ecosystems in both arctic and boreal biomes characterizing variation in the alder-*Frankia* interaction, and modulation of N-fixation rates and inputs by some biotic and abiotic factors in boreal forests. Appropriate background information including biology of *Alnus* and *Frankia* genera, evolutionary ecology of the *Alnus-Frankia* interaction, and the ecology of alder in Alaska precedes this discussion.

Background

Alder

Taxonomy, Distribution, and General Biology

Depending on the taxonomic treatment, alder encompasses between 29 and 47 species of trees and shrubs with global distribution throughout the northern hemisphere and south into the Andes (Baker and Schwintzer, 1990; Chen and Li, 2004). Furlow (1979) grouped *Alnus* species into three subgenera—*Alnus*, *Alnobetula*, and *Clethropsis*—a treatment that is largely supported by DNA sequence data (Chen and Li, 2004) (Figure 7.1). The three subgenera have distinct global distribution patterns, with subg. *Alnus* occupying most of Europe and North America and a portion of Andean South America, subg. *Alnobetula* distributed over most of Siberia, northeast Asia and North America, and subg. *Clethropsis* highly disjunct in southern Asia and the United States. Within these ranges, individual species distributions vary from highly cosmopolitan to highly restricted. In the former case is *A. viridis*, various subspecies of which occur continuously across the North American arctic from Alaska to Greenland, south as far as California in the western US and North Carolina in the eastern US, and throughout most of Europe and parts of Asia, as well. An example of a restricted species is *A. maritima*, which consists of small, disjunct populations each representing a distinct subspecies in Delaware, Maryland, and Oklahoma (Furlow, 1979; Schrader and Graves, 2004). Different alder species also vary to a large degree in preferred habitat, with some species such as *A. incana* ssp. *rugosa* restricted to wet habitats (Furlow, 1979), and others such as *A. jorullensis* and *A. viridis* able to occupy relatively dry habitats, including mid-elevation intermittent streams in central Mexico (Furlow, 1979) and rocky slopes in subalpine tundra of boreal Alaska, respectively (personal observation).

Alder species can also vary widely in growth form, both within and among species. *A. viridis*, for example, can occur as a dwarf shrub in arctic tundra habitats, a large shrub in boreal forests, and a small tree in boreal and temperate forests (Viereck and Little, 2007). Several species (e.g., *A. incana*) occur as large (~4 m height) shrubs in boreal and temperate regions, and some species (e.g., *A. glutinosa*, *A. rubra*) are moderately sized trees (McVean, 1953; Viereck and Little, 2007). Both tree and shrub forms can reproduce clonally through formation of multiple stems, and in shrub forms, multiple clumps of stems can be formed by genetically individual plants via lateral stem growth and adventitious rooting (McVean, 1953; Wilson et al., 1985; Harrington, 2006). With respect to sexual reproduction, alders are monoecious, mostly self incompatible, and wind pollinated. The small seeds are also wind dispersed, and bear small wings in some species to aid in dispersal (Furlow, 1979).

N-fixation

All known *Alnus* species form root-nodule symbioses with N-fixing bacteria belonging to the genus *Frankia* (Dawson, 2008). Although such associations can be energetically costly (Lundquist, 2005), the ability to fix atmospheric N also provides several ecological advantages. For example, alders are often among the first plants to colonize N-poor primary seres following glacial retreat (Chapin et al., 1994), floodplain deposition (Walker and Chapin, 1986), or volcanic eruption (Heilman, 1990), and are also important secondary colonists following disturbance such as fire and tree harvest (Harrington, 2006; Mitchell and Ruess, 2009a). The ability of alder to exploit such N-limited habitats is almost certainly related

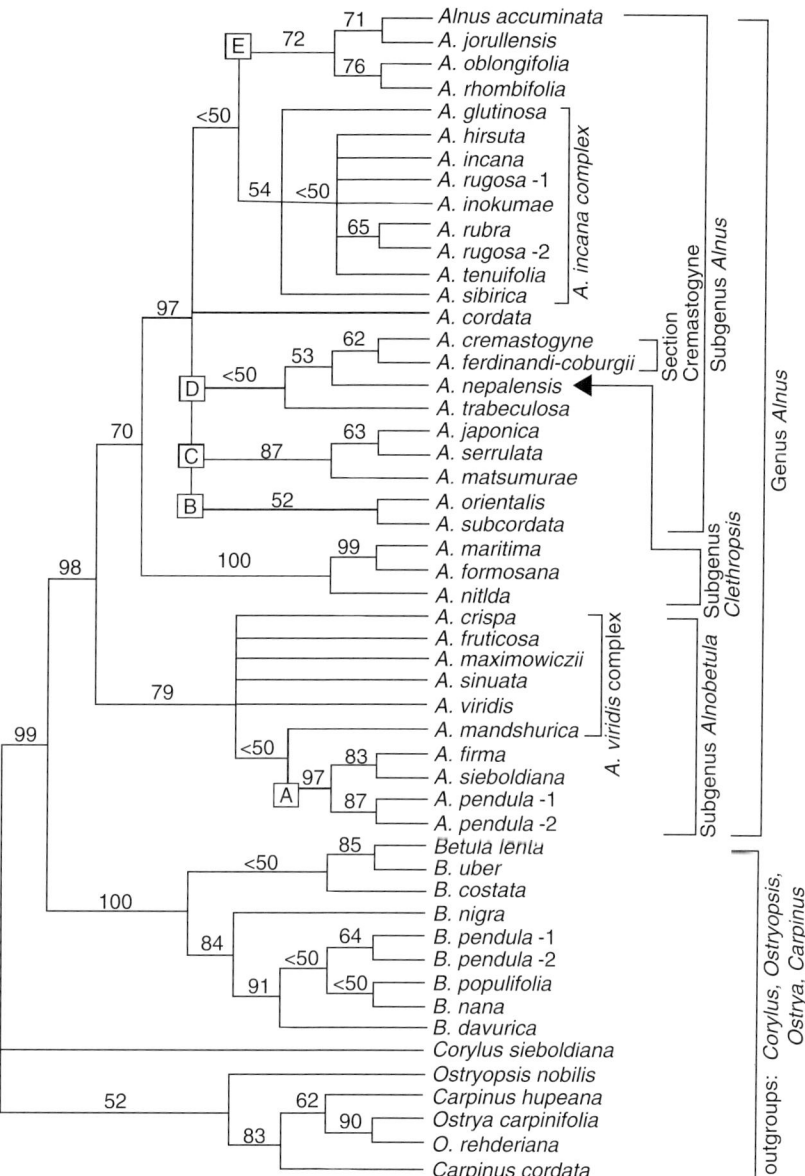

Figure 7.1. Strict consensus of 18 most parsimonious trees of 34 *Alnus* species based on DNA sequence comparison of the ribosomal internally transcribed spacer (ITS) region of the *Alnus* nuclear genome. Numbers above branches are bootstrap percentages. Traditional classifications are shown on the right. Taken from Chen and Li (2004) and reprinted by permission from the University of Chicago Press.

to its relationship with *Frankia*. Alder can persist in the understory throughout succession, and can colonize mineral soil exposed by animal activity or windthrow in mature forests (Wurtz, 1995). The ability of alder to compete in these habitats may be enhanced by an N source independent of soil supplies, which can allow *Alnus* species to supplement available soil N

(Markham and Zekveld, 2007) or, in some species, to drop leaves without N resorption, extending their seasonal growth period relative to non-N-fixing plants (Tateno, 2003). The N-rich leaf litter that results, together with root and nodule turnover in the soil, can enhance N availability in alder-associated soils and plants (e.g., Walker and Chapin, 1986; Chapin et al., 1994), which can have significant ecosystem effects.

Regulation of N-fixation by Alder

The ecological advantages of N-fixation come at a cost to the host plant. N fixation is metabolically expensive in terms of respiration (CO_2 evolution) (Tjepkema and Winship, 1980; Lundquist, 2005), and phosphorous (P) demand (Gentili and Huss-Danell, 2003), and the symbiosis requires tight control to avoid overinvestment in nodules (Simms and Taylor, 2002). Presumably to minimize such costs, *Alnus* species regulate both nodulation and N-fixation rate under varying environmental conditions.

Nodule investment in *Alnus* is regulated at both local and systemic levels; induction of a nodule on a developing root inhibits further nodulation on that root and contributes to a systemic suppression of nodulation over the entire root system (Wall and Huss-Danell, 1997; Wall and Berry, 2008). Both processes are linked to the N and P status of the plant; nodule inhibition in response to N fertilization is well known in *Alnus* (MacConnell and Bond, 1957; Gentili and Huss-Danell, 2003; Laws and Graves, 2005), but systemic inhibition can be countered by addition of P (Gentili and Huss-Danell, 2003). In greenhouse-grown *A. incana* seedlings, the level of nodulation is sensitive to the N:P ratio of the growth medium (Wall, 2000), and in mature *A. incana* ssp. *tenuifolia* plants in Alaskan field sites addition of P results in increased nodulation (Uliassi and Ruess, 2002).

Regulation of symbiotic N fixation in *Alnus* also occurs at the level of N-fixation rate. Availability of soil N has been observed to inhibit N-fixation rate in *Alnus* at both the whole-plant level (e.g., Stewart and Bond, 1961), and per unit nodule biomass (specific-N-fixation [SNF]) (e.g., Gentili and Huss-Danell, 2003; Laws and Graves, 2005). In contrast to its nodulation effects, P does not appear to counteract N inhibition of SNF, which appears to act systemically across the root system (Gentili and Huss-Danell, 2003).

Frankia

Phylogenetics and General Biology

Frankia is an actinomycete, belonging to the high G+C group of gram-positive bacteria. *Frankia* are mostly filamentous in morphology but, unusually for bacteria, form several distinct cell morphologies based on a division of labor among cell types. In culture most of the *Frankia* cell mass consists of filamentous cells (hyphae), while N-fixation is restricted to specialized cells called vesicles, which are generally spherical and have thickened cell membranes for protection of oxygen-labile nitrogenase—the enzyme complex responsible for N-fixation. Reproductive spores, localized in sporangia, are formed freely in culture and occasionally in symbiosis (Benson and Silvester, 1993), from which they appear to enter the soil during nodule senescence (Holman and Schwintzer, 1987).

There is a great deal of evidence to suggest that most symbiotic *Frankia* strains are also capable of independent growth in soil and thus are only facultatively symbiotic with plants. Circumstantial evidence for this includes the ready isolability of numerous strains from nodules on nonspecialized media (Lechevalier and Lechevalier, 1990), the frequent occurrence of symbiotic *Frankia* in soils devoid of any known hosts (e.g., Maunuksela et al., 1999; Batzli et al., 2004), the presence of *Frankia* in decaying wood (Li et al., 1997), and the observation that population size of soil *Frankia* estimated by "trapping" with *Alnus* seedlings is not correlated with estimates based on *Frankia*-specific PCR of DNA from the same soils (Myrold and Huss-Danell, 1994). Direct evidence for saprophytic growth has been obtained using two *Alnus*-infective strains: one isolated from *A. glutinosa* and one from *A. rubra*. Mirza and others (2007) inoculated both strains into two types of media: nonsterile soil with very low organic matter, and sterile mineral medium with no source of C or N. Both strains showed no growth in either medium. However, when each medium was amended with ground leaf litter from *A. glutinosa*, the *A. glutinosa* isolate showed significant temporal increases in filament length, cell number per filament, and RNA content. Interestingly, the *A. rubra* isolate showed significant decreases in the same parameters over the experiment, indicating the effect of C source on *Frankia* growth can be host-strain specific. Subsequent work by these authors has also recently demonstrated growth of several *Frankia* strains on root exudates from birch, and a small group of related strains appear to be capable of growth on birch litter (Mirza et al., 2009a).

In addition to *Alnus*, *Frankia* also forms root-nodule based symbioses with 24 other plant genera from eight families (Benson and Silvester, 1993). Collectively these associations are referred to as "actinorhizal" to distinguish them from the "rhizobial" associations that occur between legume plants (and *Parasponia* of the Ulmaceae) and several genera of N-fixing proteobacteria. Molecular phylogenetic studies indicate that relationships among *Frankia* strains broadly follow host infection patterns. In a broad study using two loci from isolates and nodule DNA extracts from a total of 17 host plant genera, Clawson and others (2004) discerned three clusters within *Frankia* with mostly non-overlapping host infection ranges: clade I, containing strains symbiotic with plants in the Rosaceae, Datiscaceae, and *Ceanothus* spp. (Rhamnnaceae), clade II, symbiotic with the "higher" Hamamelididae (Myricaceae, Casuarinaceae and *Alnus* spp.), and clade III, symbiotic with members of the Elaeagnaceae and most of the actinorhizal Rhamnaceae. Although clade II and clade III appear to be sister groups, very little overlap in infection ranges occurs with respect to *Alnus*-infective strains (e.g., Normand et al., 1996; Welsh et al., 2009). Exceptions include a few related strains from clade III that are able to infect alder to a limited degree (Bosco et al., 1992; Lumini et al., 1996), and a recent report of a clade II sequence derived from an *Elaeagnus angustifolia* nodule (Mirza et al., 2009b).

The Alder-*Frankia* Symbiosis

Frankia Phylogenetics and Host Specificity

Phylogenetic studies of *Frankia* have primarily focused on examining relationships among strains belonging to different host infection groups, so little information has historically been available on relationships within the *Alnus*-infective group. However, recent studies examining large sample collections from *Alnus* species suggest two patterns especially relevant to the symbiosis between alder and *Frankia*: (1) the possible paraphyly of this group, and (2) variability in host specificity among clusters of related sequences.

Phylogenies of the entire *Frankia* genus consistently resolve a single apparent clade containing strains infective on *Alnus*, *Myrica*, and *Casuarina* species (e.g., Normand et al., 1996; Clawson et al., 2004), but such studies typically include only a few representatives from each host genus. Studies of multiple *Alnus*-infective strains occasionally resolve a second group,

Table 7.1. Results of cross-inoculation studies between *Frankia* isolates derived from alder nodules and several alder species from the two major subgenera, as well as *Myrica cerifera* and *Elaeagnus angustifolia*. '+' = positive nodulation, '−' = negative nodulation, '*' = positive, but very low, nodulation. Superscripts indicate studies in which each interaction was observed, and are as follows: [1]Baker, 1987; [2]Du and Baker, 1992; [3]Jiabin et al., 1985; [4]Prat, 1989; [5]Nesme et al., 1985; [6]Hooker and Wheeler, 1987; [7]Weber et al., 1989; [8]Dawson and Sun, 1981; [9]Dillon and Baker, 1982; [10]Maynard, 1980.

Subgenus		Strain, informal	Strain, formal designation	Alnus					Alnobetula	Non-alder hosts	
Species	Subspp			A. glutinosa	A. rubra	A. incana	A. incana ssp. rugosa	A. cremastogyne	A. viridis ssp. crispa	Myrica cerifera	Elaeagnus angustifolia
Alnobetula viridis		AVP3n		+[5]					+[9]	+[1,2]	−[2]
	crispa	Avcl1	DDB01020110	+[2]	+[1,9]	+[7]	+[9]		+[4]		
	crispa	ACN1AG	ULQ0102001007	+[4,5,8,10]	+[4,10]	*[8]	−[9]	+[4]	+[9]		
	sinuata	Avsl2		*[9]							
	sinuata	Avsl3	DDB01360610	+[2]	+[1]					+[1,2]	−[1,2]
	sinuata	Avsl6a	DDB01361410	−[1]	−[1]					−[1]	−[1]
	sinuata	Avsl6b	DDB01361420	+[1]						−[1]	−[1]
	sinuata	54012	DDB01362210	+[1]	+[1]					+[2]	−[2]
	firma	Af2	IMB01040002	+[2]						+[1]	
Alnus incana	rugosa	Airl1	LLR01321							−[1]	−[1]
	rugosa	Airl2	LLR01322	−[1]						−[1]	−[1]
	tenuifolia	R52	LLR013701	−[1]						−[1]	−[1]
	tenuifolia	ATP1d		+[5]							
		54004	DDB01112010	−[1]	−[1]					−[1]	−[1]
		54005	DDB01072110	−[1]	−[1]					−[1]	−[1]
glutinosa		Agc8204	IAE01078204	+[2]				+[3]		+[2]	−[2]
glutinosa		AG10Al	ULF010701001	+[4]	+[4]			+[4]	+[4]		
glutinosa		AGN1g		+[5]							
glutinosa		UGL010703		+[6]	+[6]						
glutinosa		UGL010704		+[6]	+[6]						
glutinosa		UGL010708		+[6]	+[6]						
rubra		Arl4	DDB01310210	+[1,6]	+[1,6]						
rubra		Arl5	DDB01310310	+[1]	+[1,6,9]			+[9]		+[1,2]	−[1]
rhombifolia		Arhl2	DDB01301610	−[1,2]	−[1]					+[1,2]	+[1,2]
cremastogyne		Acc8207	IAE01438207	+[2]				+[3]		+[2]	−[2]
cremastogyne		Acc13						+[3]			
hirsuta		Ahc8201	IAE01098201	+[2]				+[3]		+[2]	−[2]
japonica		Ajc8206						+[3]			
cordata		ACoN24d		+[4]	+[4]			+[4]	+[4]		

which is outgroup to both clade II and clade III *Frankia* (Hahn 2008; Welsh et al., 2009; Mirza et al., 2009b). Among such studies there is a trend toward better resolution of this group with larger samples both in terms of sequence number and sequence length (Hahn et al., 1999; Hahn 2008), and in two recent large data sets, the branch defining this group is statistically significant (Welsh et al., 2009; Mirza et al., 2009b).

Phylograms in such studies also suggest that host specificity among smaller clades within both apparent groups of *Alnus*-infective *Frankia* are highly variable, with all members of some groups derived from nodules of a single host species or subgenus, and other groups made up of samples from plants in multiple subgenera or even different families (Hahn 2008; Welsh et al., 2009). Such a pattern is somewhat at odds with cross-inoculation studies characterizing infectivity of *Frankia* isolates on a range of alder species. Table 7.1 summarizes the results from ten such studies. Collectively, they seem to indicate low levels of host specificity among alder species and *Frankia* strains; although some strains appear to be infective primarily on hosts closely related to their host of origin (e.g., AvsI2), for the most part strains derived from most host species appear to be able to form nodules across a broad range without regard to host phylogeny (e.g., strains derived from both host subgenera are mostly infective on plants of the other subgenus) or geography (e.g., strains derived from *A. glutinosa* and *A. rubra* are mostly compatible with both hosts, despite the former's native Eurasian range and the latter's restricted range within North America). Although such studies may overestimate the breadth of host-symbiont associations in natural environments (Simonet et al., 1999; Huguet et al., 2005), they provide an idealized range of potential associations between both organisms that is likely to be limited mainly by relatively "hard" genetic barriers in each partner. In natural habitats such potential ranges are further subject to a variety of softer barriers.

The data in Table 7.1 are based on host compatibility in very broad terms—the ability of strains to form nodules on a small sample of plants for each host species under greenhouse conditions. Strains with the same host range by this criterion can vary widely in finer-scale compatibility-related traits such as the time required for nodulation of a given host (Nesme et al. 1985), number and/or biomass of nodules formed on a host (Hooker and Wheeler, 1987; Prat, 1989; Weber et al., 1989) and even the ability to nodulate different plant genotypes within a host species (Hahn et al., 1988; van Dijk and Sluimer, 1994). Variation in such traits seems likely to contribute to the restricted realized range of associations in natural versus artificial habitats (others are discussed below), and reflects the complex evolutionary interplay at work between host and symbiont.

Variation in Host Interactions

The interaction between alder and *Frankia* is generally considered a mutualism, which, by definition, means that it is beneficial to both the plant and the bacterium. The plant receives access to atmospheric N, which can provide a competitive edge in some environments, and the bacterium receives access to an exclusive environment, the nodule interior, and a source of C, photosynthate provided by the plant. Under this simple conceptual model, one might expect selection pressure on each partner to increase or at least maintain a constant level of benefit to the other, since this also benefits itself by maintaining or increasing access to the resource provided by the other organism. However, the interaction also carries an inherent cost to each partner in the form of allocation of materials and energy to the symbiotic partner that could otherwise be used to directly support its own growth and reproduction (Bronstein, 2001). This cost results in much more complex evolutionary dynamics in at least two ways: (1) the production of evolutionary pressure on each partner toward "cheating" behavior (i.e., the development of traits that allow it to reap the benefits of the interaction while minimizing or even eliminating the costs, and (2) variation in the cost:benefit ratio of the interaction for both partners with varying biotic and abiotic environmental conditions (Thompson, 1982; Egger and Hibbett, 2004). Such complexity is evident in several respects in the alder-*Frankia* symbiosis, including the following: (1) the spectrum of symbiotic behaviors exhibited by different bacterial strains, (2) varying compatibility of host genotypes within a species with particular bacterial strains, and (3) interacting effects of bacterial strain, host species, and environmental conditions on traits related to host fitness.

Variation in Symbiotic Behavior

Rather than distinct categories of interaction, mutualism and parasitism are now generally thought of as representing opposite ends of a spectrum of symbiotic outcomes mediated by the cost of the interaction to each partner (Bronstein, 1994). A mutualistic outcome occurs when the benefit of the interaction exceeds the cost for both partners; when this occurs for only one partner, the result is a parasitic interaction. One result of the inherent cost of symbiosis is the production of evolutionary pressure to "cheat"—i.e., to utilize resources normally devoted to maintenance of the partnership for an organisms own growth and reproduction. Because a single plant can simultaneously associate with multiple microbial genotypes but not vice-versa and bacteria can complete several generations in a single plant lifetime, this pressure is probably most intense on the microbe (Denison and Kiers, 2004; Kiers and Denison, 2008). In the alder-*Frankia* symbiosis, *Frankia* strains appear to exhibit the full spectrum of symbiotic behaviors, from mutualism to parasitism.

The mutualistic nature of most strains is suggested both by field observations—e.g., the universal occurrence of nodules on alders in the field (Dawson, 2008), the ability of nodulated alders to colonize and maintain high N levels on low N soils, and the high proportion of N derived from fixation of alder species investigated using stable isotope tracer methods (Domenach et al., 1989; Markham and Chanway, 1999)—and greenhouse experiments, which collectively indicate the ability of alder species to meet their entire N requirement in association with a wide range of *Frankia* strains. Studies that have attempted to provide chemical N sources to non-nodulated *A. incana* at the same rate that N was fixed in nodulated plants have found that, despite the C cost to the host plant, nodulated plants with no other N source grew either as well as (in the case of NO_3) or better than (in the case of NH_3) fertilized non-nodulated plants (Sellstedt, 1986; Sellstedt and Huss-Danell, 1986). In the latter case, nodulated alders also had higher N content, and the authors suggest that the plant may be better adapted to deriving N symbiotically than via direct uptake. However, N-fixation rates in both studies were measured as acetylene reduction, which may not be quantitatively reliable under all assay conditions (Anderson et al., 2004).

At the other end of the mutualism-parasitism spectrum are the so-called "ineffective" strains, which induce nodule formation on host roots but do not fix N. This habit appears to occur in at least two ways: (1) an inability of some strains to fix N in symbiosis due to a lack of N fixation genes and/or an inability to form vesicles in symbiosis, and (2) lack of symbiotic compatibility with a particular host species in strains otherwise capable of symbiotic N fixation.

Strains in the former category isolated from *Alnus* nodules that are ineffective on *Alnus* hosts occur in at least two distinct phylogenetic groups: one which clusters with the "atypical" *Frankia* strains (non-N-fixing strains incapable of reinfecting their host of origin), and one which clusters near the group of effective *Frankia* infective on alder (Normand et al., 1996; Wolters et al., 1997).

Both groups form nodules with similar phenotypes, which are distinct from those of effective nodules. Ineffective nodules are typically very small, with bacteria lacking vesicles and producing narrower hyphae than in effective nodules (Lechevalier et al., 1983; Hahn et al., 1988; Wolters et al., 1997). Small nodule size may result from slow nodule growth rather than arrested development; other than the lack of vesicles, bacterial growth in most of these strains appears to be similar to that of effective strains in terms of proportion of nodule cells infected and hyphal density in infected cells (Lechevalier et al., 1983; van Dijk and Sluimer-Stolk, 1990, but see Hahn et al., 1988). Most of these strains appear to lack at least one of the genes necessary for nitrogen fixation (Wolters et al., 1997), and one strain isolated from *A. incana* ssp. *rugosa* exhibited very different physiological characteristics from an effective comparison strain, including higher oxygen tolerance and differences in the ability to utilize various C sources (Lechevalier et al., 1983).

Ineffectivity can also result from incompatibility with particular host species. This has been observed both across host infection groups (e.g., a few strains isolated from *Elaeagnus* spp. nodules form ineffective nodules on some *Alnus* spp.) and also within the *Alnus*-infective group; both crushed nodule inoculum from some *A. incana* nodules and soil from some *A. incana* stands in Finland consistently induce ineffective symbioses with *A. glutinosa* but effective symbioses with *A. incana* (van Dijk et al., 1988; Weber, 1990).

Host Intraspecific Variation in Frankia Compatibility

Intraspecific variation among host genotypes in nodulation with different *Frankia* strains has not been extensively examined within *Alnus*, but studies that have been performed suggest that it may be an important component of the evolutionary dynamics of the symbiosis in some host species. For example, in *A. glutinosa*, the most extensively studied host in this respect, genetic variation in both ability to nodulate and level of nodulation for both effective and ineffective *Frankia* strains has been observed among host seed source families, ecotypes and clones (Hall et al., 1979; Maynard, 1980; Hahn et al., 1988; van Dijk and Sluimer, 1994). In *A. crispa*, a non-nodulating genotype has also been reported (Tremblay et al., 1984). Intraspecific variation in nodulation with different *Frankia* strains has been examined in *A. rubra*, but no evidence for it was found (Monaco et al., 1982).

Effects on Host Fitness

Even among noncheating bacterial genotypes able to nodulate a particular host genotype, considerable difference in mutualistic behavior can occur. Differences among strains in level of nodulation or benefit provided on a given host can interact with similar effects of host genetic variation (both intra- and interspecific) and environmental variation, resulting in considerable ecological and evolutionary complexity in a given interaction, including shifting interaction outcomes among different habitats (Bronstein, 1994), spatial mosaics of host colonization success (Parker, 1999), and increased host-symbiont specificity (Egger and Hibbett, 2004). Because of the importance of fixed N to community structure and ecosystem function, these effects can also ripple out to larger ecological scales.

Numerous studies have investigated the effects of host and strain variation on host performance for different suites of alder species and *Frankia* inocula (both pure isolates and crushed nodules). For a given host species, variation in *Frankia* strain has been consistently observed to affect host growth and N-fixation (Hall et al., 1979; Dawson and Sun, 1981; Dillon and Baker, 1982; Monaco et al., 1982; Sellstedt et al., 1986; Hooker and Wheeler, 1987; Sheppard et al., 1988). Such strain effects can be complex and are not predictable from prior knowledge of host-symbiont associations. For example, the highest host performance is not always obtained with strains originating from a given host species or even genus (Dawson and Sun, 1981; Dillon and Baker, 1982; Prat, 1989; Weber et al., 1989), and there is some evidence that simultaneous inoculation of a single plant with multiple strains can have a synergistic effect on host performance, regardless of symbiont origin relative to the host or even colonization of host nodules by all of the strains (Prat, 1989; Martin et al., 2003). In Martin and others (2003), dual inoculation significantly increased biomass of *A. rubra* seedlings over single inoculation with either strain, but only one of the two strains was detectable in nodules from the dual inoculation treatment. The authors suggest that the undetected strain may have enhanced the speed of nodulation by the detected strain, resulting in more rapid growth in dual inoculated plants over the course of the experiment. Adding further complexity, such strain effects on host performance can also interact with effects of intraspecific host genetic variation (Hall et al., 1979) and environmental variation (Sheppard et al., 1988; Kurdali et al., 1990). Overall, such studies suggest that *Frankia* strains symbiotic with alder define a broad and dynamic mutualism-parasitism spectrum, and that the position along this spectrum for a given strain can be modified by variation in both host genetics and environment.

A special case of these phenomena is provided by the so-called "spore-positive" (Sp+) strains. While most *Frankia* strains sporulate freely in culture (Benson and Silvester, 1993), Sp+ strains are characterized by the ability to form numerous spores within the host nodule, an ability that appears to be generally suppressed in host nodules (Schwintzer, 1990). Considerable circumstantial evidence suggests that this phenotype is a genetically determined property of the bacterium (Schwintzer, 1990; Wheeler et al., 2008). Effects of Sp+ strains on host performance have been well studied and suggest that Sp+ strains generally occupy an intermediate position along the mutualism-parasitism spectrum. Unlike ineffective strains, Sp+ strains retain the ability to fix N, but appear to minimize the cost of mutualism by reallocation of resources from maintenance of the host interaction to their own reproduction (Schwintzer, 1990). Compared to strains that do not sporulate in symbiosis ("spore-negative" or Sp−), Sp+ strains generally result in less growth and lower N-content and fixation rates in host plants (reviewed in Schwintzer, 1990). Sp+ strains also appear to be much more infective on alder host plants than Sp− strains (van Dijk, 1984; Weber, 1990; Markham, 2008), suggesting the Sp+ condition may be a multi-trait alternative strategy based on superior competitivity of Sp+ strains for nodule sites on host roots rather than, and at the expense of, mutualistic behavior. Interestingly however, some evidence suggests that the mutualistic behavior of Sp+ strains may be modulated by both host species and environmental conditions. Markham (2008) found a detectable negative effect on host biomass of Sp+ strains compared to Sp− strains in *A. rubra*, but not in *A. incana* ssp. *rugosa* or *A. viridis* ssp. *crispa*. Kurdali and others (1990) found that when *A. incana* was grown in artificial soil, Sp− strains maintained consistently higher N-fixation rates per unit nodule mass than Sp+ strains across a range of treatments, but that the situation was reversed when plants were grown in natural soil.

Field Associations between Alder and Frankia

As in most symbioses between plant roots and soil microbes, transmission of microbial symbionts from alder parents to offspring is strictly horizontal; germinating alder seedlings form symbiotic associations *de novo* with *Frankia* genotypes encountered in the soil. This is also the case for annual cohorts of nodules produced by perennial alder plants and, in wind-dispersed species such as alder, seedlings are not likely to encounter symbiont genotypes associated with parents. Under this type of transmission, parents have no direct influence over the mutualistic quality of the symbionts utilized by their offspring, and it is thought that evolutionary counter-pressure against the pressure on microbes to cheat is brought about instead by a heritable ability of the plant to withhold resources from "bad" mutualists (host sanctions) and/or selectively allocate resources to "good" mutualists (host choice) (Simms and Taylor, 2002; West et al., 2002). Sanctioning of cheaters is suggested by the small size of ineffective nodules on alder (Hahn et al., 1988; van Dijk and Sluimer, 1994), but the possibility of choice among noncheating *Frankia* strains does not appear to have been examined. Such choice has been observed in N-fixing symbiosis between *Bradyrhizobium* genotypes and the wild legume *Lupinus arboreus* (Simms et al., 2006). Since bacterial densities in nodules can be several orders of magnitude higher than in soil (West et al., 2002), any choice exercised by alder hosts has a strong potential to feed back to soil bacterial populations after nodule senescence.

Field surveys of *Frankia* diversity in alder nodules suggest a degree of influence of the host plant over symbiont populations at the interspecific level; variation in alder host species is generally correlated with genetic structure of *Frankia* assemblages in host nodules (van Dijk et al., 1988; Weber, 1990; Dai et al., 2005; Anderson et al., 2009). However, since different alder species tend to occupy different habitats in most studies, even when they occur in the same region, such structure also correlates with environmental factors in most studies. A few studies have examined sites in which more than one host species occur, and these indicate that different host species can associate with widely different *Frankia* assemblages even in the same soil (Weber, 1986; Anderson et al., 2009).

Within a host species, genetic structure among host populations could result in symbiont structure for alder species that vary intra-specifically in compatibility with specific *Frankia* strains (see the Host Intraspecific Variation in *Frankia* Compatibility section above). Geographic population structure has been reported in such hosts (Bousquet et al., 1987; King and Ferris, 1998), but no studies of whether host structure correlates with structure of *Frankia* assemblages in the field appear to have been conducted.

Field studies of single alder species occurring in different environments commonly report variation with sampling locale in symbiotic *Frankia* structure, whether based on molecular genetic tools (Dai et al., 2004; Huguet et al., 2004; Igual et al., 2006), or proportion of Sp+ nodules (e.g., Weber, 1986; Holman and Schwintzer, 1987; Markham and Chanway, 1998a). Regional-scale studies that have included replicated levels of broad environmental variables have found consistent effects of elevation (Khan et al., 2007) and habitat type (Holman and Schwintzer, 1987) on symbiont structure in *Alnus* spp., and recently such structure was shown to be consistent among replicate sites representing different habitats at the local scale (Anderson et al., 2009).

Alder in Alaska

Distribution, Habits, and Habitats of Alaskan Alder

Three species of *Alnus* are native to Alaska: (1) *A. incana* ssp. *tenuifolia* (also known as [aka] *A. tenuifolia*, thinleaf alder), (2) *A. rubra* (red alder), and (3) two subspecies of *A. viridis*: ssp. *sinuata* (aka *A. sinuata*, Sitka alder) and ssp. *fruticosa* (Siberian alder) (Viereck and Little, 2007). For each, Alaska is the northern and/or westernmost extent of a larger North American range. *A. tenuifolia* is widespread in North America, occurring throughout Canada, the Yukon, and the Pacific Northwest, south nearly to Mexico, and east to the Rockies (eFloras 2008). In Alaska, *A. tenuifolia* occurs across the southern two-thirds of the state, south of the Brooks Range, with the exception of narrow portions of the southeast Alaskan panhandle, the Alaska peninsula, and most offshore islands (see Figure 7.2) (Viereck and Little, 2007). *A. rubra* is mainly a coastal species, occurring along the Pacific coast from central California to the northern extent of the Alaskan panhandle, which is the only portion of the state in which it occurs (Viereck and Little, 2007). Within *A. viridis*, the two Alaskan subspecies have fairly distinct distributions. Subsp. *fruticosa* is primarily northern, occupying arctic portions of western Canada, as well as the entire Yukon and nearly all of Alaska, but occurring only sporadically farther south in the Pacific Northwest (eFloras, 2008). Subsp. *sinuata*, by contrast,

Figure 7.2. Map of Alaska. Reprinted from Chapin et al. (2006) by permission from Oxford University Press. Landmarks mentioned in the text have been added to the original.

occupies large portions of the Pacific Northwest and the southern half of Alaska but is nearly absent north of the Alaska Range (Viereck and Little, 2007; eFloras, 2008).

All Alaskan alders reach tree size except for the Siberian alder, which occurs as a small to large multi-stemmed shrub (Viereck and Little, 2007). However, *A. sinuata* and *A. tenuifolia* only rarely reach tree size in Alaska, and occur more commonly as multi-stemmed shrubs similar to Siberian alder. *A. rubra* occurs as a tree across its range.

All three Alaskan alder species have ruderal growth strategies, with rapid growth and abundant seed production (Viereck and Little, 2007). In *A. rubra*, at least, this strategy also includes a fairly short lifespan (~100 y) (Harrington, 2006), but in shrubby species such as *A. viridis* and *A. tenuifolia*, the lifespan of the plant is difficult to determine due to the continuous production of new stems, which may each be much younger than the parent plant (Wilson et al., 1985). Within its range each species generally reaches its greatest densities in early succession habitats such as fluvial deposits, glacial outwash, or areas disturbed by fire or logging (Harrington, 2006; Viereck and Little, 2007). Despite this early-succession peak, alder presence throughout succession can be significant. On the Tanana River floodplain near Fairbanks, for example, maximum stem density of *A. tenuifolia* occurs 5–10 years after colonization of new substrate (Van Cleve et al., 1971), but this species can continue to make up a significant portion of the shrub layer throughout the successional sequence of canopy dominance by balsam poplar (~50–100 years postsubstrate deposition) and white spruce (~200 years postsubstrate) (Van Cleve and Viereck, 1981). In *A. viridis* ssp. *fruticosa*, Mitchell and Ruess (2009a) even report an increased stem density across an upland secondary sere from postburn through birch dominance to mature white spruce forest. All Alaskan species of alder are also common components of disturbed areas and riparian communities, often forming dense bands along rivers, around lakes, and adjacent to roads and trails (Viereck and Little, 2007). The widespread occurrence of alder in Alaska, together with its symbiotic N-fixing capability, makes it an important organism in Alaskan ecosystems.

Alder in Alaskan Ecosystems

The state of Alaska is large and ecologically diverse, occupying an area of ~1.6×10^2 km^2, or greater than one-fifth that of the contiguous United States, and containing 20 of the 104 ecoregions described for the United States by the US Geological Survey and Environmental Protection Agency (US EPA 2007), distributed in three major biomes: arctic, boreal, and temperate. Alders are common components of Alaskan ecosystems, occurring at important frequency in 15 of the 20 Alaskan ecoregions (Gallant et al., 1995) in all three biomes: *A. rubra* in temperate coastal forests of southeast Alaska, *A. tenuifolia* and *A. sinuata* in boreal and southern coastal regions, and *A. viridis* ssp. *fruticosa* in coastal, boreal, and arctic regions (Viereck and Little, 2007). In arctic regions, *A. viridis* ssp. *fruticosa* is currently undergoing a range expansion concurrent with an expansion of shrub tundra into historically herbaceous tundra that appears to be related to global climate change (Sturm et al., 2001; Tape et al., 2006).

The effects of alder presence on ecosystem structure and function have been studied to varying degrees for each species of *Alnus* (and for each subspecies of *A. viridis*) in all three Alaskan biomes, and in all regions, the effects of alder appear to be significant. In arctic northwest Alaska, the presence of *A. crispa* (syn. *A. viridis* ssp. *fruticosa*) is associated with enhanced N in both soils and plants across multiple ecosystems in habitats that include floodplain terraces, tussock tundra, and valley slopes (Binkley et al., 1994; Rhoades et al., 2001), and in subarctic southwest Alaska both subspecies of *A. viridis* enhance the nutrient concentrations and productivity of alder-associated lakes (Devotta, 2008). Ecosystem effects of

A. rubra have not been intensively studied in the Alaskan portion of its range (Hanley et al., 2006), but the information that exists suggests that the presence of this species enhances understory productivity, diversity, and suitability as wildlife habitat in regional conifer forests (Wipfli et al., 2003; Hanley et al., 2006). Enhancement of soil and plant N has been reported for *A. sinuata* in coastal postglacial areas (Chapin et al., 1994; Kohls et al., 2003) and *A. incana* ssp. *tenuifolia* on river floodplains in interior Alaska (Van Cleve et al., 1971; Walker and Chapin, 1986; Uliassi and Ruess, 2002), in which alder presence is also associated with high N mineralization and turnover rates (Clein and Schimel, 1995). Work with these species, however, suggests that the effects of alder on associated ecosystems are more complex than simple facilitation via enhancement of N availability.

In a series of elegant studies using field observations, field manipulations, and greenhouse experiments, Walker and Chapin (1986) and Chapin and others (1994) demonstrated a range of effects of *A. incana* ssp. *tenuifolia* and *A. sinuata*, respectively, that included both facilitation and inhibition of seedlings of plant species commonly associated with alder in each study area. Both studies demonstrated positive effects of alder on N-status of associated plant species, but in both studies, these facilitative effects on N-status were counterbalanced by inhibitory effects of alder, which included shading and root competition. The net balance of effects appeared to differ between the two systems investigated. In the Chapin and others (1994) study in Glacier Bay on the Alaskan panhandle, the authors conclude that the overall effect of *A. sinuata* on white spruce (the successional climax species in both areas) seedlings is facilitative, while the opposite conclusion is reached by Walker and Chapin (1986) for *A. incana* ssp. *tenuifolia* on the Tanana River floodplain in interior Alaska. The authors suggest this difference may be due to differences in N availability between the two systems, with competition outweighing facilitation where N is less limiting.

Ecology of the Alder-Frankia Interaction in Alaska

This section presents recent and ongoing work investigating sources of variation in three broad aspects of the *Alnus-Frankia* interaction in Alaska: (1) host-endophyte interactions, (2) host physiology, and (3) ecosystem effects. Our ongoing work centers on the hypothesis that alder exercises choice among *Frankia* genotypes and that this choice contributes to the facilitative component of alder effects in some Alaskan ecosystems.

Field Associations between Alder and Frankia in Alaska

Distribution of *Frankia* endophytes has been examined at both local (<15 km) and regional scales (several hundred km) in two of the four host species, *A. tenuifolia* and *A. viridis* ssp. *fruticosa*, in ecosystems from boreal and arctic biomes. As in other geographic areas, composition of symbiotic *Frankia* assemblages in Alaska is correlated with variation in host species and environmental conditions. In Alaska it has been possible to disentangle these effects to an extent, and independent effects of both factors have been observed.

The primary tool used to characterize *Frankia* in these studies is PCR-RFLP of the noncoding bacterial *nif* D-K spacer locus performed on surface-sterilized nodules. We have found that digestion with two restriction enzymes (*Cfo*I and *Hae*III) detects >95% of the sequence variation present in this locus, that this locus is more variable than the 16S-23S IGS spacer, and that both loci yield congruent phylogenies. Across both species and all sampling sites, 16 *nif* D-K RFLP genotypes of *Frankia* have been detected. The following discussion will refer to these genotypes (termed RF1-RF16) when discussing distribution patterns of endophytic *Frankia*.

Frankia assemblages on the two host species differ in composition and phylogeny. In three sites on the Tanana River floodplain with very similar environmental conditions (Table 7.2) and both hosts that are close (≤5 m), very little overlap in composition of symbiont assemblages is evident (Figure 7.3), suggesting a relatively high level of host-symbiont specificity. Such specificity is further suggested by DNA sequence-based phylogenies (both *nif*D-K and IGS loci), in which the RF types most common on *A. tenuifolia* form a single cluster distinct from a second well-defined cluster containing RF types most common on *A. viridis* ssp. *fruticosa* (Anderson, Taylor, and Ruess, unpublished data). The specificity apparent in the two hosts may result from different mechanisms, however; the fact that no *A. viridis* ssp. *fruticosa* nodule has been found to yield any of the dominant RF patterns from *A. tenuifolia* nodules suggests the presence of "hard" genetic barriers, while the detection of both RF8 and RF9 in the latter host, albeit at very low frequency, suggests specificity in this host is mediated by "softer" mechanisms.

Environmental effects on *Frankia* structure also differ between the two hosts. In *A. tenuifolia*, large differences occur between nodule assemblages of plants occupying early (alder canopy) and late (white spruce canopy with alder understory) succession habitats at relatively small (1.5–13.5 km) spatial scales on the Tanana floodplain. *Frankia* structure is consistent among replicate sites representing each habitat, and it appears to be largely consistent year-to-year, particularly in early succession (Figure 7.3). Late succession sites dominated by white spruce (*Picea glauca* [Moench] Voss) support higher richness and evenness of RF types at both the site and individual plant level in this host (Anderson, Taylor, and Ruess, unpublished data). In *A. viridis* ssp. *fruticosa*, by contrast, two genotypes differing by a single base pair in the *nif*D-K spacer (RF8 and RF9) occupy >95% of the nodules examined from sites ranging from the Seward Peninsula to the Brooks Range to the Tanana floodplain (Anderson et al., 2009; Taylor, MacFarland, and Ruess, unpublished data).

The large and consistent environmental effect in *A. tenuifolia* may be due to one or many of the environmental factors differing between these habitats (Table 7.2), and may act at several points in the development of the symbiosis, including dispersal of *Frankia* genotypes to a particular site, survival of *Frankia* in the soil, and/or differential host interactions (e.g., host choice) among bacterial genotypes. For host choice to be important would require, at a minimum, variation in supply and/or demand of plant nutrients among habitats, a host mechanism capable of choosing symbionts based on the plant's varying physiological needs, and variation in relevant parameters among symbionts (e.g., N-fixation rate or unit cost). Two of these three factors have been observed in Alaskan alder.

Host Physiology

Variation Among Host Species and Habitats Environmental variation in host physiological parameters such as nutrient demand, resource allocation among nutrient acquisition options (N-fixation versus root uptake versus mycorrhizae) or nutrient use strategies (e.g., nutrient resorption prior to leaf drop) can modify the outcome of the alder-*Frankia* interaction among plant species and/or habitats by modifying the "value" to the plant of atmospheric N as well as the "cost" of fixed N in terms of other nutrients such as C or P. Such parameters would be expected to interact with any selection mechanism possessed by the plant. Both nutrient demand and use appear to differ between hosts and among habitats in *A. tenuifolia* and *A. viridis* ssp. *fruticosa* in the region of the Tanana River.

Both alder species maintain relatively high leaf N content across succession (Table 7.3), which is similar in the two species by mass, but differ when expressed by area, largely due to

Table 7.2. Site characteristics of successional habitats and mean values (n = 3 representative sites) of selected soil properties in sites in which *Frankia* diversity has been most intensely investigated in Alaska. Soil moisture and temperature data are from Anderson et al. (2009), and the remaining soil variable means were calculated from publicly available data on the Bonanza Creek Long-Term Ecological Research website (www.lter.uaf.edu, Oliver et al. unpublished data). Within a column common superscripts indicate homogeneous subsets (Tukey HSD, $P < 0.05$); no superscript indicates no statistical differences among rows in a column.

The largest differences in most parameters occur between early and late succession floodplain habitats, which corresponds to the largest difference observed in *Frankia* genetic assemblage in nodules of *A. tenuifolia*, the only host species to occur in both of these habitats (see Figure 7.3).

Site characteristics				Soil variables				
Landscape	Stage/sere	Species	Alder crown position	Moisture	Temperature	Organic matter	%N by mass	N:P (mass)	pH
Floodplain	Early Primary	*A. tenuifolia*	Canopy	28.7±0.5^A	9.5±0.3^A	1.6±0.3^A	0.04±0.01^A	0.56±0.11^B	7.4±0.03^A
Floodplain	Late Primary	*A. tenuifolia* + *A. viridis*	Understory	32.7±0.8^B	7.9±0.2^B	11.5±2.5^A	0.21±0.04^B	3.00±0.43^{BC}	5.6±0.09^B
Upland	Early Secondary	*A. viridis*	Canopy	25.7±0.7^A	9.3±0.2^A	6.0±1.1^{AB}	0.13±0.02^{AB}	1.88±0.33^B	5.6±0.26^B
Upland	Late Secondary	*A. viridis*	Understory	26.5±0.6^A	8.8±0.2^A	8.1±0.9^B	0.19±0.02^B	3.96±0.34^A	4.9±0.13^C

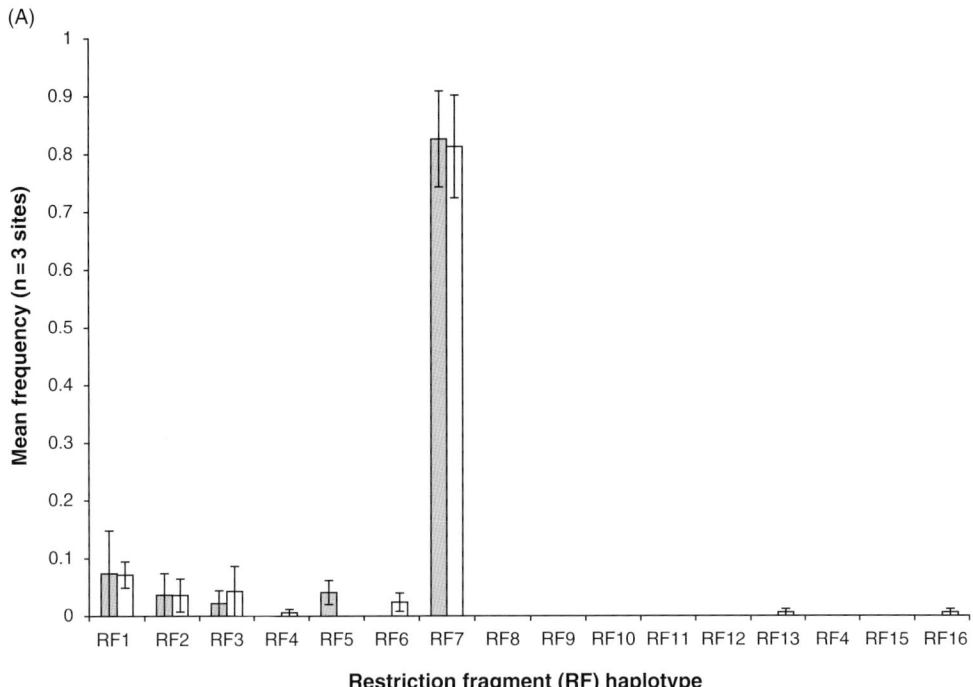

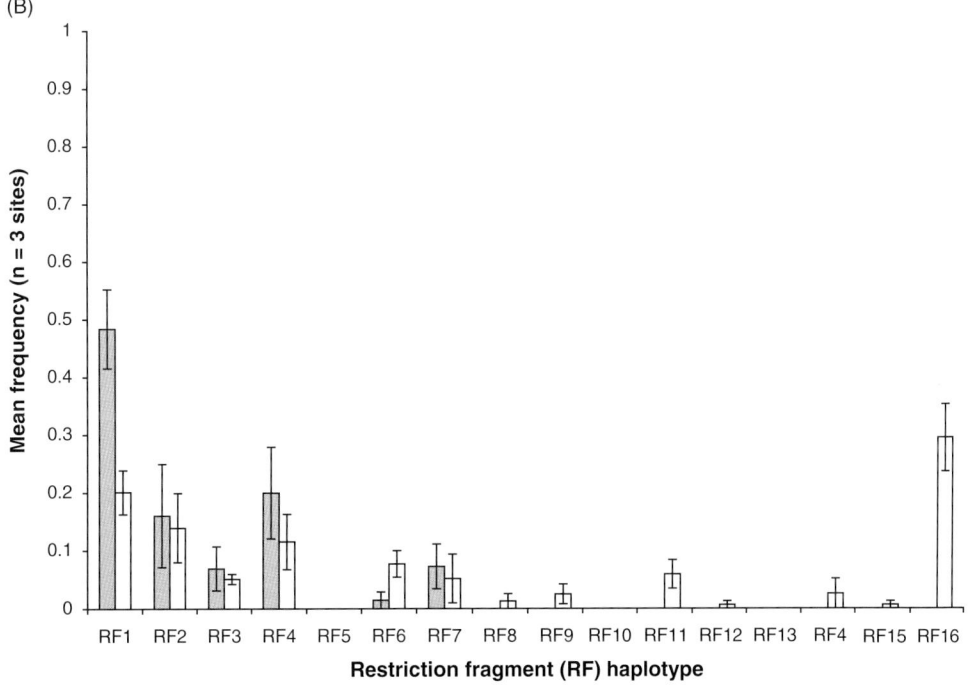

Figure 7.3.

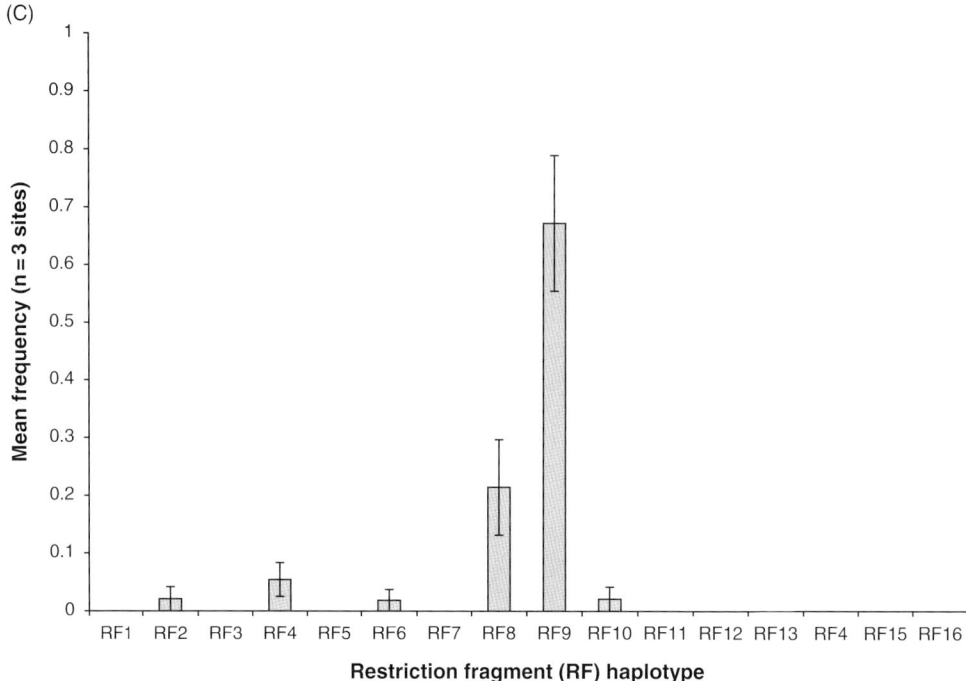

Figure 7.3 (continued). Average frequencies (±1 standard error, n = 3 sites) of *Frankia nif* D-K restriction fragment (RF) genotypes occupying *Alnus tenuifolia* nodules in early (A) and late (B) succession sites, and *Alnus viridis* ssp. *fruticosa* in the same late succession sites (C) on the Tanana River floodplain, interior Alaska. Charcoal bars represent data from a survey conducted in 2002, and open bars represent a survey of the same sites conducted in 2005. 2002 data reused by permission from Springer.

differences in leaf thickness (Table 7.3). In both species the highest leaf N concentrations occur in environments with the greatest light availability, likely in order to maximize photosynthetic rates under high light. In *A. tenuifolia*, this high N demand occurs in an environment with very low soil N, so a large proportion of this N must come from N fixation. Any host choice that occurs would be expected to be most stringent under such conditions, and nodules in these environments are consistently dominated by one *Frankia* genotype—RF7.

Nutrient resorption and interactions between nutrients differ between the two species, and between habitats for *A. tenuifolia*. In *A. viridis* ssp. *fruticosa* in early (alder canopy), mid (paper birch canopy/alder understory), and late succession (white spruce canopy/alder understory) in an upland secondary sere, Mitchell and Ruess (2009b) describe relatively high seasonal N resorption (site means of 19–37% by leaf area) that is on the same order as P resorption (16–33%) in these plants, suggesting a degree of N limitation in this species. By contrast, in *A. tenuifolia* in early (alder) and mid (balsam poplar with alder understory) succession sites on the Tanana floodplain, Uliassi and Ruess (2002) report much lower values of N resorption (7–14%), but much higher P-resorption values (39–51%), suggesting P-limitation rather than N-limitation in this species. Fertilization with P in this study demonstrated that N fixation at the plant level was limited by P availability in early succession, but not in mid-succession.

Table 7.3. Means (n = 3 sites, 30 plants per site) of physiological variables measured on two species of alder in Alaskan successional habitats. Data are from Anderson et al. (2009) and re-used by permission from Springer.

SNF values in field sites are highly variable, and no difference can be detected among sampling blocks. Among habitats leaf N by area parallels light availability for both species (see Table 7.2), and leaf N by both mass and area parallels large differences in *Frankia* genetic assemblage structure in *A. tenuifolia*.

Site characteristics			Physiological variables			
Landscape	Stage	Species	SNF (µmol N$_2$ g noddwt^{-1} h^{-1})	Leaf N (% by mass)	Leaf N by Area (g N/m^2)	SLW (gdwt leaf tissue/m^2)
Floodplain	Early	*A. tenuifolia*	28.3 ± 5.5	2.64 ± 0.05^A	1.59 ± 0.04^A	62.1 ± 1.9^A
	Late		22.4 ± 3.2	2.46 ± 0.03^B	1.23 ± 0.03^B	50.6 ± 1.2^B
Upland	Early	*A. viridis*	30.3 ± 3.3	2.26 ± 0.03^C	1.07 ± 0.03^C	47.9 ± 1.1^B
		ssp. *fruticosa*	33.9 ± 4.5	2.50 ± 0.04^B	1.79 ± 0.03^D	71.7 ± 8.0^C
	Late		34.0 ± 3.9	2.49 ± 0.03^B	1.05 ± 0.02^C	42.4 ± 0.9^D

Differences Among Frankia Genotypes For any host choice mechanism to be effective, variation must exist in relevant physiological parameters among symbiont genotypes. Unfortunately, putatively relevant parameters such as specific N-fixation (SNF) rate or unit cost of N-fixation are highly variable and/or difficult to measure *in situ*. Preliminary evidence for such genotypic differences has nevertheless been observed in Alaskan alder. In the local-scale field survey described above (Anderson et al., 2009), SNF (^{15}N uptake) was measured on all nodules collected, and revealed an overall effect of genetic variation among *Frankia* at the seasonal peak of fixation activity. Recent preliminary data suggests that the unit cost of N fixation may also differ among *Frankia* strains. Simultaneous measurement of ^{15}N uptake and respiration in *A. tenuifolia* nodules examined from three early succession sites on the Tanana floodplain revealed significant correlations between respiration and N fixation rate for two of the three most common bacterial genotypes in the sites (Ruess, unpublished data). The regression slope of the most common genotype among all sites (RF7) was significantly lower than for the other type for which the relationship was significant, suggesting both that genotypes differ in unit cost and that alder preferentially associates with the less expensive type.

Ecosystem Effects of the Alder-Frankia Symbiosis

Factors affecting the alder-*Frankia* symbiosis are likely to modify both the inhibitory ecosystem effects of alder, by altering alder competitivity and its facilitative effects, chiefly by altering N-inputs. However, for the sake of simplicity, the following discussion will focus on the latter.

N-inputs by alder at the ecosystem level depend on the balance between N-fixation and N-uptake in individual alders, and are a product of processes operating at multiple scales—SNF at the scale of individual nodules, total nodule biomass at the plant scale, and plant density and N-release rates from nodules and litter at the ecosystem scale. The variation in the *Alnus-Frankia* symbiosis discussed in the previous sections may affect such inputs directly, or through interactions with other variables. Combined with phylogenetic and/or geographic limitations on the suite of strains available to a host, variation among strains in SNF may directly constrain N-input rates, while similar variation in unit cost of fixation may alter the relationship between alder productivity and N inputs. The latter would also be affected by the ability of alder to minimize the C cost of N-fixation through *Frankia* choice. N availability alters both nodule production and SNF, and other environmental variables such as light, water, and P modify N demand among habitats and, in the case of P, also specifically affect nodule biomass (Uliassi and Ruess, 2002). More complex modulation of N inputs can occur through interactions with other organisms. This section ends with two detailed examples:

1. Mycorrhizae: In addition to symbiosis with *Frankia*, alder forms both arbuscular-mycorrhizal (AM) and ecto-mycorrhizal (EM) interactions with fungi. Considerable variation exists in these interactions in terms of taxonomic and environmental patterns of association and impacts on host performance. Such variation can interact with variation in the alder-*Frankia* symbiosis to alter selection pressures on both partners, and may also modify the ecosystem effects of alder.

Although alder is capable of associating with several AM fungal genera (Gardner, 1986), such associations may not occur at high densities in the field (Helm et al., 1996; Pritsch et al., 1997), or may be limited to early developmental stages of the plant (Arveby and Granhall, 1998). EM associations with alder occur in ~50 distantly related fungal species, many of which are specific to *Alnus* (Bruns et al., 2002; Tedersoo et al., 2009), and a few of which show apparent specificity for particular alder species (Gardner, 1986). Despite this comparatively low overall diversity, EM communities on alder in the field can be quite variable among habitats, particularly across successional seres (Helm et al., 1996; Arveby and Granhall, 1998).

Due to the wide variety of services attributable to mycorrhizal infection—for example, acquisition of water, nitrogen, phosphorous, and various micronutrients, and/or alteration of pathogen susceptibility (Koide, 1991)—such interactions have the potential to modify both the evolutionary dynamics and the ecosystem effects of the alder-*Frankia* association in a given habitat in several opposing ways. For instance, under conditions in which availability of water, P, or other nutrients limits N-fixation, mycorrhizal associations capable of enhancing plant access to such resources could lower the relative cost of N-fixation to the plant and possibly lessen the stringency of any host choice process. Synergistic positive effects of mycorrhizal symbiosis in concert with *Frankia* have been observed in both AM and EM fungal symbionts in greenhouse studies (e.g., Russo, 1989; Yamanaka et al., 2003, 2005), and probably act via enhanced P availability (Mejstrik and Benecke, 1969). Release of limitations on N-fixation by P or other factors would, of course, result in greater N-inputs to associated ecosystems, tipping the balance of alder effects toward facilitation.

On the other hand, mycorrhizal associations with N-mobilizing fungi could provide alders with access to organic soil N. If mycorrhizal-derived N is metabolically cheaper in a given environment than atmospheric N, and alder is capable of selecting between *Frankia* and mycorrhizae, such a process could result either in selection for more efficient *Frankia* strains or in a local breakdown of the coevolutionary process between plant and bacterium. At the point where N-uptake by alder exceeds N-fixation, such a process would also represent a threshold in the ecosystem role of alder between N-source and N-sink.

2. Herbivores and Pathogens: Alders are eaten by a wide variety of herbivores and are host to numerous pathogens. Interactions between alder and both guilds can vary across host species, with abiotic and biotic environmental factors, and across multiple spatial and temporal scales (e.g., Hendrickson et al., 1991; Gange, 1995; Markham and Chanway, 1998b; Mulder et al., 2008). Such variation can be complex and unpredictable. For example, in a

survey of leaf pathogens and five arthropod herbivore guilds on *A. viridis* ssp. *fruticosa* in interior Alaska, Mulder and others (2008) found that damage levels by different guilds were differentially correlated with (1) leaf position along the twig or, (2) distance to the nearest alder or other deciduous tree, and that many of these correlations were significant in 1 year of the study but not the next (several correlations with leaf position), or were significant in both years but the direction of the relationship changed (correlation with tree distance). Additional complexity appears when one considers the interaction of such variation with the alder-*Frankia* symbiosis. For example, increases in foliar N levels in field-grown alders in response to inoculation with *Frankia* can both increase attack rates by insect herbivores and facilitate defenses against them, and such effects differ among herbivores and alder species (Hendrickson et al., 1991, 1993).

The effects of alder herbivory on associated ecosystems may be significant. In *A. tenuifolia*, artificial defoliation of greenhouse grown seedlings ranging from 15–40% leaf area removal resulted in significant reductions of SNF from 33–68%, and one-time removal of 40% leaf area resulted in 73% reduction in SNF after a month of regrowth despite the recovery of biomass to levels indistinguishable from control plants (Ruess et al., 2006). While no such effect on SNF was observed in a recent field study of a current outbreak of stem canker (putatively *Valsa melanodiscus* [anamorph *Cytospera umbrina*]) in *A. tenuifolia* in Alaska, reductions of nodule biomass measured in three Alaskan stands of this species resulted in an estimated 31–38% reduction in N inputs (Ruess et al., 2009).

Conclusion

An emerging paradigm in evolutionary ecology holds that many interspecific interactions shift along a continuum from mutualism to parasitism under varying biotic and abiotic environmental contexts (Bronstein, 1994) and that shifting outcomes among local populations, in addition to genetic variation in each partner, provide the raw material for coevolution of symbiotic partners at the macroevolutionary level (Thompson, 1994, 2005). In the case of N-fixing symbioses, such intrinsic and extrinsic modulators of symbiotic outcomes also have the strong potential to impact ecosystem structure and function, providing an opportunity to study links between evolutionary and ecosystem ecology. The importance of the alder-*Frankia* symbiosis in Alaskan ecosystems, together with the relative simplicity and repetition of such systems across a large geographic area, provides the researcher with an excellent system for such studies and a convenient excuse for doing large amounts of biology in one of the world's wildest and most wondrous areas.

References

Anderson, M.D., Ruess, R.W., Uliassi, D.D., Mitchell, J.S. 2004. Estimating N_2 fixation in two species of *Alnus* in interior Alaska using acetylene reduction and $^{15}N_2$ uptake. *Ecoscience* 11, 102–112.

Anderson, M.D., Ruess, R.W, Myrold, D.D. et al. 2009. Host species and habitat affect nodulation by specific *Frankia* genotypes in two species of *Alnus* in interior Alaska. *Oecologia* 160, 619–630.

Arveby, A.S., Granhall, U. 1998. Occurrence and succession of mycorrhizas in *Alnus incana*. *Swedish Journal of Agricultural Research* 28, 117–127.

Baker, D.D. 1987. Relationships among pure-cultured strains of *Frankia* based on host specificity. *Physiologia Plantarum* 70, 245–248.

Baker, D.D., Schwintzer, C.R. 1990. Introduction. In: *The Biology of Frankia and Actinorhizal Plants*. (Schwintzer, C.R., Tjepkema, J.D., Eds.) Academic Press, Inc., San Diego.

Batzli, J.M., Zimpfer, J.F., Huguet, V. et al. 2004. Distribution and abundance of infective, soilborne *Frankia* and host symbionts *Shepherdia*, *Alnus*, and *Myrica* in a sand dune ecosystem. *Canadian Journal of Botany* 82, 700–709.

Benson, D.R., Silvester, W.B. 1993. Biology of *Frankia* strains, actinomycete symbionts of actinorhizal plants. *Microbiological Reviews* 57, 293–319.

Binkley, D., Stottlemyer, R., Suarez, F. et al. 1994. Soil nitrogen availability in some arctic ecosystems in northwest Alaska: responses to temperature and moisture. *Ecoscience* 1, 64–70.

Bosco, M., Fernandez, M.P., Simonet, P. et al. 1992. Evidence that some *Frankia* sp. strains are able to cross boundaries between *Alnus* and *Elaeagnus* host specificity groups. *Applied and Environmental Microbiology* 58, 1569–1576.

Bousquet, J., Cheliak, W.M., Lalonde, M. 1987. Genetic diversity within and among 11 juvenile populations of green alder (*Alnus crispa*) in Canada. *Physiologia Plantarum* 70, 311–318.

Bronstein, J.L. 1994. Conditional outcomes in mutualistic interactions. *Trends in Ecology and Evolution* 9, 214–217.

Bronstein, J.L. 2001. The costs of mutualism. *American Zoologist* 41, 825–839.

Bruns, T.D., Bidartondo, M.I., Taylor, D.L. 2002. Host specificity in ectomycorrhizal communities: what do the exceptions tell us? *Integrative and Comparative Biology* 42, 352–359.

Chapin III, F.S., Walker, L.R., Fastie, C.L. et al. 1994. Mechanisms of primary succession following deglaciation at Glacier Bay, Alaska. *Ecological Monographs* 64, 149–175.

Chapin III, F.S., Yarie, J., Van Cleve, K. et al. 2006. The conceptual basis of LTER studies in the Alaskan boreal forest. In: *Alaska's Changing Boreal Forest*. (Chapin III, F.S., Osgood, M.W., Van Cleve, K., et al., Eds.) Oxford University Press, New York.

Chen, Z., Li, J. 2004. Phylogenetics and biogeography of *Alnus* (Betulaceae) inferred from sequences of nuclear ribosomal DNS ITS region. *International Journal of Plant Science* 165, 325–335.

Clawson, M.L., Bourret, A., Benson, D.R. 2004. Assessing the phylogeny of *Frankia*-actinorhizal plant nitrogen-fixing root nodule symbioses with *Frankia* 16S rRNA and glutamine synthetase gene sequences. *Molecular Phylogenetics and Evolution* 31, 131–138.

Clein, J.S., Schimel, J.P. 1995. Nitrogen turnover and availability during succession from alder to poplar in Alaskan taiga forests. *Soil Biology and Biochemistry* 27, 743–752.

Dai, Y., Zhang, C., Xiong, Z. et al. 2005. Correlations between the ages of *Alnus* host species and the genetic diversity of associated endosymbiotic *Frankia* strains from nodules. *Science in China Ser. C Life Sciences* 48 Suppl, 76–81.

Dai, Y., He, X.Y., Zhang, C.G. et al. 2004. Characterization of genetic diversity of *Frankia* strains in nodules of *Alnus nepalensis* (D. Don) from the Hengduan Mountains on the basis of PCR-RFLP analysis of the *nifD-nifK* IGS. *Plant and Soil* 267, 207–212.

Dawson, J.O., Sun, S-H. 1981. The effect of isolates from *Comptonia peregrina* and *Alnus crispa* on the growth of *Alnus glutinosa*, *A. cordata*, and *A. incana* clones. *Canadian Journal of Forest Research* 11, 758–762.

Dawson, J.O. 2008. Ecology of actinorhizal plants. In: *Nitrogen Fixation: Origins, Applications, and Research Progress, Vol. 6 Nitrogen-fixing Actinorhizal Symbioses*. (Pawlowski, K., Newton, W.E., Eds.) Springer, Dordrecht, The Netherlands.

Denison, R.F., Kiers, E.T. 2004. Lifestyle alternatives for rhizobia: mutualism, parasitism, and forgoing symbiosis. *FEMS Microbiology Letters* 237, 187–193.

Devotta, D.A. 2008. The influence of *Alnus viridis* on the nutrient availability and productivity of sub-arctic lakes in southwestern Alaska. MS Thesis, University of Illinois at Urbana-Champaign.

Dillon, J.T., Baker D. 1982. Variations in nitrogenase activity among pure-cultured *Frankia* strains tested on actinorhizal plants as an indication of symbiotic compatibility. *New Phytologist* 92, 215–219.

Domenach, A.M., Kurdali, F., Bardin, R. 1989. Estimation of symbiotic dinitrogen fixation in alder forest by the method based on natural ^{15}N abundance. *Plant and Soil* 118, 51–59.

Du, D., Baker, D.D. 1992. Actinorhizal host-specificity of Chinese *Frankia* strains. *Plant and Soil* 144, 113–116.

eFloras. 2008. Published on the Internet http://www.efloras.org [accessed 15 June 2009] Missouri Botanical Garden, St. Louis, MO and Harvard University Herbaria, Cambridge, MA.

Egger, K.N., Hibbett, D.S. 2004. The evolutionary implications of exploitation in mycorrhizas. *Canadian Journal of Botany* 82, 1110–1121.

Furlow, J.J. 1979. The systematics of the American species of *Alnus* (Betulaceae). *Rhodora* 81, 1–121, 151–248.

Gallant, A.L., Binnian, E.F., Omernik, J.M., and Shasby, M.B. 1995. Ecoregions of Alaska: U.S. Geological Survey Professional Paper 1567, Washington DC.

Gange, A.C. 1995. Aphid performance in an alder (*Alnus*) hybrid zone. *Ecology* 76, 2074–2083.

Gardner, I.C. 1986. Mycorrhizae of actinorhizal plants. *MIRCEN Journal of Applied Microbiology and Biotechnology* 2, 147–160.

Gentili, F., Huss-Danell, K. 2003. Local and systemic effects of phosphorous and nitrogen on nodulation and nodule function in *Alnus incana*. *Journal of Experimental Botany* 54, 2757–2767.

Hahn, D., Starrenburg, M.J.C., Akkermans, A.D.L. 1988. Variable compatibility of cloned *Alnus glutinosa* ecotypes against ineffective *Frankia* strains. *Plant and Soil* 107, 233–243.

Hahn, D., Nickel, A., Dawson, J. 1999. Assessing *Frankia* populations in plants and soil using molecular methods. *FEMS Microbiology Ecology* 29, 215–227.

Hahn, D. 2008. Polyphasic taxonomy of the genus *Frankia*. In: *Nitrogen Fixation: Origins, Applications, and Research Progress, Vol. 6 Nitrogen-fixing Actinorhizal Symbioses*. (Pawlowski, K., Newton, W.E., Eds.) Springer, Dordrecht, The Netherlands.

Hall, R.B., McNabb Jr., H.S., Maynard, C.A. et al. 1979. Toward development of optimal *Alnus glutinosa* symbiosis. *Botanical Gazette* 140 (Suppl), S120–S126.

Hanley, T.A., Deal, R.L., Orlikowska, E.H. 2006. Relations between red alder composition and understory vegetation in young mixed forests of southeast Alaska. *Canadian Journal of Forest Research* 36, 738–748.

Harrington, C.A. 2006. Biology and ecology of red alder. In: *Red Alder: A State of Knowledge*. General Technical Report PNW-GTR-669. Portland, OR: U.S. Department of Agriculture, Pacific Northwest Research Station.

Heilman, P.E. 1990. Growth of Douglas-fir and red alder on coal spoils in western Washington. *Soil Science Society of America Journal* 54, 522–527.

Helm, D.J., Allen, E.B., Trappe, J.M. 1996. Mycorrhizal chronosequence near Exit Glacier, Alaska. *Canadian Journal of Botany* 74, 1496–1506.

Hendrickson, O.Q., Fogal, W.H., Burgess, D. 1991. Growth and resistance to herbivory in N_2-fixing alders. *Canadian Journal of Botany* 69, 1919–1926.

Hendrickson, O.Q., Burgess, D., Perinet, P. et al. 1993. Effects of *Frankia* on field performance of *Alnus* clones and seedlings. *Plant and Soil* 150, 295–302.

Holman, R.M., Schwintzer, C.R. 1987. Distribution of spore-positive and spore-negative nodules of *Alnus incana* ssp. *rugosa* in Maine, USA. *Plant and Soil* 104, 103–111.

Hooker, J.E., Wheeler, C.T. 1987. The effectivity of *Frankia* for nodulation and nitrogen fixation in *Alnus rubra* and *A. glutinosa*. *Physiologia Plantarum* 70, 333–341.

Huguet, V., Mergeay, M., Cervantes, E. et al. 2004. Diversity of *Frankia* strains associated to *Myrica gale* in Western Europe: impact of host plant (*Myrica* vs. *Alnus*) and of edaphic factors. *Environmental Microbiology* 6, 1032–1041.

Huguet, V., Gouy, M., Normand, P. et al. 2005. Molecular phylogeny of Myricaceae: a reexamination of host-symbiont specificity. *Molecular Phylogenetics and Evolution* 34, 557–568.

Igual, J.M., Valverde, A., Valázquez, E. et al. 2006. Natural diversity of nodular microsymbionts of *Alnus glutinosa* in the Tormes River basin. *Plant and Soil* 280, 373–383.

Jiabin, H., Zheying, Z., Guanxiong, C. et al. 1985. Host range of *Frankia* endophytes. *Plant and Soil* 87, 61–65.

Khan, A., Myrold, D.D., Misra, A.K. 2007. Distribution of *Frankia* genotypes occupying *Alnus nepalensis* nodules with respect to altitude and soil characteristics in the Sikkim Himalayas. *Physiologia Plantarum* 130, 364–371.

Kiers, E.T., Denison, R.F. 2008. Sanctions, cooperation, and the stability of plant-rhizosphere mutualisms. *Annual Review of Ecology, Evolution, and Systematics* 39, 215–236.

King, R.A., Ferris, C. 1998. Chloroplast DNA phylogeography of *Alnus glutinosa* (L.) Gaertn. *Molecular Ecology* 7, 1151–1161.

Kohls, S.J., Baker, D.D., van Kessel, C. et al. 2003. An assessment of soil enrichment by actinorhizal N_2 fixation using $\delta^{15}N$ values in a chronosequence of deglaciation at Glacier Bay, Alaska. *Plant and Soil* 254, 11–17.

Koide, R.T. 1991. Tansley Review No. 29. Nutrient supply, nutrient demand and plant response to mycorrhizal infection. *New Phytologist* 117, 365–386.

Kurdali, F., Rinaudo, G., Moiroud, A. et al. 1990. Competition for nodulation and $^{15}N_2$-fixation between a Sp+ and a Sp− *Frankia* strain in *Alnus incana*. *Soil Biology and Biochemistry* 22, 57–64.

Laws, M.T., Graves, W.R. 2005. Nitrogen inhibits nodulation and reversibly suppresses nitrogen fixation in nodules of *Alnus maritime*. *Journal of the American Society of Horticultural Science* 130, 496–499.

Lechevalier, M.P., Baker, D., Horrière, F. 1983. Physiology, chemistry, serology, and infectivity of two *Frankia* isolates from *Alnus incana* ssp. *rugosa*. *Canadian Journal of Botany* 61, 2826–2833.

Lechevalier, M.P., Lechevalier, H.A. 1990. Systematics, isolation, and culture of *Frankia*. In: *The Biology of* Frankia *and Actinorhizal Plants* (Schwintzer, C.R., Tjepkema, J.D., Eds.) Academic Press, Inc., San Diego.

Li, C-Y., Crawford, R.H., Chang, T-T. 1997. *Frankia* in decaying fallen trees devoid of actinorhizal hosts and soil. *Microbiological Research* 152, 167–169.

Lumini, E., Bosco, M., Fernandez, M.P. 1996. PCR-RFLP and total DNA homology revealed three related genomic species among broad-host-range *Frankia* strains. *FEMS Microbiology Ecology* 21, 303–311.

Lundquist, P-O. 2005. Carbon cost of nitrogenase activity in *Frankia-Alnus incana* root nodules. *Plant and Soil* 273, 235–244.

MacConnell, J.T., Bond, G. 1957. A comparison of the effect of combined nitrogen on nodulation in non-legumes and legumes. *Plant and Soil* 8, 378–388.

Markham, J.H., Chanway, C.P. 1998a. *Alnus rubra* (Bong.) nodule spore type distribution in southwestern British Columbia. *Plant Ecology* 135, 197–205.

Markham, J.H., Chanway, C.P. 1998b. Response of red alder (*Alnus rubra*) seedlings to a woolly alder sawfly (*Eriocampa ovata*) outbreak. *Canadian Journal of Forest Research* 28, 591–595.

Markham, J.H., Chanway, C.P. 1999 Does past contact reduce the degree of mutualism in the *Alnus rubra—Frankia* symbiosis? *Canadian Journal of Botany* 77, 434–441.

Markham, J.H., Zekveld, C. 2007. Nitrogen fixation makes biomass allocation to roots independent of soil nitrogen supply. *Canadian Journal of Botany* 85, 787–793.

Markham, J.H. 2008. Variability of nitrogen-fixing *Frankia* on *Alnus* species. *Botany* 86, 501–510.

Martin, K.J., Tanaka, Y., Myrold, D.D. 2003. Dual inoculation increases plant growth with *Frankia* on red alder (*Alnus rubra* Bong.) in fumigated nursery beds. *Symbiosis* 34, 253–260.

Maunuksela, L., Zepp, K., Koivula, T. et al. 1999. Analysis of *Frankia* populations in three soils devoid of actinorhizal plants. *FEMS Microbiology Ecology* 28, 11–21.

Maynard, C.A. 1980. Host-symbiont interactions among *Frankia* strains and *Alnus* open-pollinated families. PhD Thesis. Iowa State University, Ames, Iowa.

McVean, D.N. 1953. *Alnus glutinosa* (L.) Gaertn. *Journal of Ecology* 41, 447–466.

Mejstrik, V., Benecke, U. 1969. The ectotrophic mycorrhizas of *Alnus viridis* (Chaix) D.C. and their significance in respect to phosphorous uptake. *New Phytologist* 68, 141–149.

Mirza, B.S., Welsh, A., Hahn, D. 2009a. Growth of *Frankia* strains in leaf litter-amended soil and the rhizosphere of a nonactinorhizal plant. *FEMS Microbiology Ecology* 70, 132–141.

Mirza, B.S., Welsh, A., Rasul, G. et al. 2009b. Variation in *Frankia* populations of the *Elaeagnus* host infection group in nodules of six host plant species after inoculation with soil. *Microbial Ecology* 58, 384–393.

Mirza, B.S., Welsh, A., Hahn, D. 2007. Saprophytic growth of inoculated *Frankia* sp. in soil microcosms. *FEMS Microbiology Ecology* 62, 280–289.

Mitchell, J.S., Ruess, R.W. 2009a. N_2 fixing alder (*Alnus viridis* ssp. *fruticosa*) effects on soil properties across a secondary successional chronosequence in interior Alaska. *Biogeochemistry* 95, 215–229.

Mitchell, J.S., Ruess, R.W. 2009b. Seasonal patterns of climate controls over nitrogen fixation by *Alnus viridis* ssp. *fruticosa* in a secondary successional chronosequence in interior Alaska. *Ecoscience* 16, 241–351.

Monaco, P.A., Ching, K.K., Ching, T. 1982. Host-endophyte effects on biomass production and nitrogen fixation in *Alnus rubra* actinorhizal symbiosis. *Botanical Gazette* 143, 298–303.

Mulder, C.P.H., Roy, B.A., Güsewell, S. 2008. Herbivores and pathogens on *Alnus viridis* subsp. *fruticosa* in interior Alaska: effects of leaf, tree, and neighbor characteristics on damage levels. *Botany* 86, 408–421.

Myrold, D.D., Huss-Danell, K. 1994. Population dynamics of *Alnus*-infective *Frankia* in a forest soil with and without host trees. *Soil Biology and Biochemistry* 26, 533–540.

Nesme, X., Normand, P., Tremblay, F.M. et al. 1985. Nodulation speed of *Frankia* sp. on *Alnus glutinosa*, *Alnus crispa*, and *Myrica gale*. *Canadian Journal of Botany* 63, 1292–1295.

Normand, P., Orso, S., Cournoyer, B. et al. 1996. Molecular phylogeny of the genus *Frankia* and emendation of the family *Frankiaceae*. *International Journal of Systematic Bacteriology* 46, 1–9.

Parker, M.A. 1999. Mutualism in metapopulations of legumes and rhizobia. *The American Naturalist* 153 (Suppl), S48–S60.

Prat, D. 1989. Effects of some pure and mixed *Frankia* strains on seedling growth in different *Alnus* species. *Plant and Soil* 113, 31–38.

Pritsch, K., Munch, J.C., Buscot, F. 1997. Morphological and anatomical characterization of black alder *Alnus glutinosa* (L.) Gaertn. ectomycorrhizas. *Mycorrhiza* 7, 201–216.

Rhoades, C., Oskarsson, H., Binkley, D. et al. 2001. Alder (*Alnus crispa*) effects on soils in ecosystems of the Agashashok River valley, northwest Alaska. *Ecoscience* 8, 89–95.

Ruess, R.W., Anderson, M.D., Mitchell, J.S. et al. 2006. Effects of defoliation on growth and N fixation in *Alnus tenuifolia*: consequences for changing disturbance regimes at high latitudes. *Ecoscience* 13, 404–412.

Ruess, R.W., McFarland, J.W., Trummer, L.M. et al. 2009. Disease-mediated declines in N-fixation inputs by *Alnus tenuifolia* to early-successional floodplains in interior and south-central Alaska. *Ecosystems* 12, 489–502.

Russo, R.O. 1989. Evaluating alder-endophyte (*Alnus acuminata-Frankia*-mycorrhizae) interactions I. Acetylene reduction in seedlings inoculated with *Frankia* strain ArI3 and *Glomus intra-radices*, under three phosphorous levels. *Plant and Soil* 118, 151–155.

Sawada, H., Kuykendall, L.D., Young, J.M. 2003. Changing concepts in the systematics of bacterial nitrogen-fixing legume symbionts. *Journal of General and Applied Microbiology* 49, 155–179.

Schrader, J.A., Graves, W.R. 2004. Systematics of *Alnus maritima* (Seaside alder) resolved by ISSR polymorphisms and morphological characters. *Journal of the American Society for Horticultural Science* 129, 231–236.

Schwintzer, C.R. 1990. Spore-positive and spore-negative nodules. In: *The Biology of Frankia and Actinorhizal Plants*. (Schwintzer, C.R., Tjepkema, J.D., Eds.) Academic Press, Inc., San Diego.

Sellstedt, A. 1986. Nitrogen and carbon utilization in *Alnus incana* fixing N_2 or supplied with NO_3^- at the same rate. *Journal of Experimental Botany* 37, 786–797.

Sellstedt, A., Huss-Danell, K. 1986. Biomass production and nitrogen utilization by *Alnus incana* when grown on N_2 or NH_4^+ made available at the same rate. *Planta* 167, 387–394.

Sellstedt, A., Huss-Danell, K., Ahlqvist, A-S. 1986. Nitrogen fixation and biomass production in symbioses between *Alnus incana* and *Frankia* strains with different hydrogen metabolism. *Physiologia Plantarum* 66, 99–107.

Sheppard, L.J., Hooker, J.E., Wheeler, C.T. et al. 1988. Glasshouse evaluation of the growth of *Alnus rubra* and *Alnus glutinosa* on peat and acid brown earth soils when inoculated with four sources of *Frankia*. *Plant and Soil* 110, 187–198.

Simms, E.L., Taylor, D.L. 2002. Partner choice in nitrogen-fixation mutualisms of legumes and rhizobia. *Integrative and Comparative Biology* 42, 369–380.

Simms, E.L., Taylor, D.L., Povich, J. et al. 2006. An empirical test of partner choice in a wild legume-rhizobium interaction. *Proceedings of the Royal Society B* 273, 77–81.

Simonet, P., Navarro, E., Rouvier, C. et al. 1999. Co-evolution between *Frankia* populations and host plants in the family Casuarinaceae and consequent patterns of global dispersal. *Environmental Microbiology* 1, 525–533.

Stewart, W.D.P., Bond, G. 1961. The effect of ammonium nitrogen on fixation of elemental nitrogen in *Alnus* and *Myrica*. *Plant and Soil* 14, 347–359.

Sturm, M., Racine, C., K. Tape. 2001. Increasing shrub abundance in the Arctic. *Nature* 411, 546.
Tape, K., Sturm, M., Racine, C. 2006. The Evidence for shrub expansion in Northern Alaska and the Pan-Arctic. *Global Change Biology* 12, 686–702.
Tateno, M. 2003. Benefit to N_2-fixing alder of extending growth period at the cost of leaf nitrogen loss without resorption. *Oecologia* 137, 338–343.
Tedersoo, L., Suvi, T., Jairus, T. et al. 2009. Revisiting ectomycorrhizal fungi of the genus *Alnus*: differential host specificity, diversity and determinants of the fungal community. *New Phytologist* 182, 727–735.
Thompson, J.N. 1982. *Interaction and Coevolution*. Wiley New York.
Thompson, J.N. 1994. *The Coevolutionary Process*. University of Chicago Press, Chicago.
Thompson, J.N. 2005. *The Geographic Mosaic Theory of Coevolution*. University of Chicago Press, Chicago.
Tjepkema, J.D., Winship, L.J. 1980. Energy requirement for nitrogen fixation in actinorhizal and legume root nodules. *Science* 209, 279–281.
Tremblay, F.M., Nesme, X., Lalonde, M. 1984. Selection and micropropagation of nodulating and non-nodulating clones of *Alnus crispa* (Ait.) Pursh. *Plant and Soil* 78, 171–179.
Uliassi, D.D., Ruess, R.W. 2002. Limitations to symbiotic nitrogen fixation in primary succession on the Tanana River floodplain. *Ecology* 83, 88–103.
U.S. Environmental Protection Agency. 2007. Level III ecoregions of the continental United States (revision of Omernik, 1987). Corvallis, O: U.S. Environmental Protection Agency—National Health and Environmental Effects Research Laboratory Map M-1, various scales.
Van Cleve, K., Viereck, L.A., Schlentner, R.L. 1971. Accumulation of nitrogen in alder (*Alnus*) ecosystems near Fairbanks, Alaska. *Arctic and Alpine Research* 3, 101–114.
Van Cleve, K., Viereck, L.A. 1981. Forest succession in relation to nutrient cycling in the boreal forest of Alaska. In: *Forest Succession: Concepts and Application*. (West, D.C., Shugart, H.H., Botkin, D.B., Eds.) Springer-Verlag, New York.
van Dijk, C. 1984. Ecological aspects of spore formation in the *Frankia-Alnus* symbiosis. PhD Thesis. State University of Leiden, Leiden, The Netherlands.
van Dijk, C., Sluimer, A., Weber, A. 1988. Host range differentiation of spore-positive and spore-negative strain types of *Frankia* in stands of *Alnus glutinosa* and *Alnus incana* in Finland. *Physiologia Plantarum* 72, 349–358.
van Dijk, C., Sluimer-Stolk, A. 1990. An ineffective strain type of *Frankia* in the soil of natural stands of *Alnus glutinosa* (L.) Gaertner. *Plant and Soil* 127, 107–121.
van Dijk, C., Sluimer, A. 1994. Resistance to an ineffective *Frankia* strain type in *Alnus glutinosa* (L.) Gaertn. *New Phytologist* 128, 497–504.
Viereck, L.A., Little Jr., E.L. 2007. *Alaska Trees and Shrubs,* 2nd ed. University of Alaska Press, Fairbanks, Alaska.
Walker, L.R., Chapin III, F.S. 1986. Physiological controls over seedling growth in primary succession on an Alaskan floodplain. *Ecology* 67, 1508–1523.
Wall, L.G., Huss-Danell, K. 1997. Regulation of nodulation in *Alnus incana-Frankia* symbiosis. *Physiologia Plantarum* 99, 594–600.
Wall, L.G. 2000. The actinorhizal symbiosis. *Journal of Plant Growth Regulation* 19, 167–182.
Wall, L.G., Berry, A.M. 2008. Early interactions, infection and nodulation in actinorhizal symbiosis. In: *Nitrogen Fixation: Origins, Applications, and Research Progress*, Vol. 6 *Nitrogen-fixing Actinorhizal Symbioses*. (Pawlowski, K., Newton, W.E., Eds.) Springer, Dordrecht, The Netherlands.

Weber, A. 1986. Distribution of spore-positive and spore-negative nodules in stands of *Alnus glutinosa* and *Alnus incana* in Finland. *Plant and Soil* 96, 205–213.

Weber, A., Sarsa, M.-L., Sundman, V. 1989. *Frankia-Alnus incana* symbiosis: effect of endophyte on nitrogen fixation and biomass production. *Plant and Soil* 120, 291–297.

Weber, A. 1990. Host specificity and efficiency of nitrogenase activity of *Frankia* strains from *Alnus incana* and *Alnus glutinosa*. *Symbiosis* 8, 47–60.

Welsh, A., Mirza, B.S., Reider, J.P. et al. 2009. Diversity of frankiae in root nodules of *Morella pensylvanica* grown in soils from five continents. *Systematic and Applied Microbiology* 32, 201–210.

West, S.A., Kiers, E.T., Simms, E.L. et al. 2002. Sanctions and mutualism stability: why do rhizobia fix nitrogen? *Proceedings of the Royal Society of London B* 269, 685–694.

Wheeler, C.T., Akkermans, A.D.L., Berry, A.M. 2008. *Frankia* and actinorhizal plants: a historical perspective. In: *Nitrogen Fixation: Origins, Applications, and Research Progress*, Vol. 6 *Nitrogen-fixing Actinorhizal Symbioses*. (Pawlowski, K. Newton, W.E., Eds.) Springer, Dordrecht, The Netherlands.

Wilson, B.F., Patterson III, W.A., O'Keefe, J.F. 1985. Longevity and persistence of alder west of the tree line on the Seward Peninsula, Alaska. *Canadian Journal of Botany* 63, 1870–1875.

Wipfli, M.S., Deal, R.L., Hennon, P.E. et al. 2003. Compatible management of Red Alder-conifer ecosystems in southeastern Alaska. In: *Compatible Forest Management*. (Monserud, R.A., Haynes, R.W., Johnson, A.C., Eds.) U.S. Government Publication. Kluwer Academic, The Netherlands.

Wolters, D.J., van Dijk, C., Zoetendal, E.G. et al. 1997. Phylogenetic characterization of ineffective *Frankia* in *Alnus glutinosa* (L.) Gaertn. nodules from wetland soil inoculants. *Molecular Ecology* 6, 971–981.

Wurtz, T.L. 1995. Understory alder in three boreal forests of Alaska: local distribution and eVects on soil fertility. *Canadian Journal of Forest Research* 25, 987–996.

Yamanaka, T., Li, C.-Y., Bormann, B.T. et al. 2003. Tripartite associations in an alder: effects of *Frankia* and *Alpova diplophloeus* on the growth, nitrogen fixation and mineral nutrition of *Alnus tenuifolia*. *Plant and Soil* 254, 179–186.

Yamanaka, T., Akama, A., Li, C.-Y. et al. 2005. Growth, nitrogen fixation and mineral acquisition of *Alnus sieboldiana* after inoculation of *Frankia* together with *Gigaspora margarita* and *Pseudomonas putida*. *Journal of Forest Research* 10, 21–26.

Chapter 8
The Path of Rhizobia: From a Free-living Soil Bacterium to Root Nodulation

Pedro F. Mateos, Raúl Rivas, Marta Robledo, Encarna Velázquez, Eustoquio Martínez-Molina, and David W. Emerich

Introduction

After water, nitrogen is the nutrient that most often limits crop productivity. This seems to be in conflict with the large amount of nitrogen in various reserves on earth compared to the amount contained in plants. Some 1.9×10^{23} g nitrogen (N) is present in rocks and sediments and another 3.8×10^{21} g is present in the atmosphere as N_2 (Arp, 2000). However, the N in these reserves is not directly available to plants. Nitrate and ammonium, the two forms of N available for most plant and microbes, constitute only about 1.6×10^{16} g N in soils while $1.1 - 1.4 \times 10^{16}$ g N is found in plants (Arp, 2000). Therefore, continuous inputs of fixed N (fertilizer) are required to replenish the inorganic nitrogen available for plant growth. While a relatively small amount of fixed N is produced by lightning, biological and industrial N fixation are the primary mechanisms for introduction of fixed N into the biosphere and hence into the N cycle. Ammonia is produced industrially by the Haber-Bosch process, using an iron-based catalyst, very high pressures of 200 atmospheres, and a temperature of 300°C (Smil, 2001). Synthetic N fertilizers are not completely taken up by crops. Typical estimates are that 10–30% of fertilizer N is lost from croplands due to leaching or conversion to gaseous N forms, making this process economically and ecologically disadvantageous (Arp, 2000). On the other hand, biological N-fixation offers the potential, through symbioses or associations with N_2-fixing bacteria, to meet the N needs of crops without deleterious environmental or economic effects. The reader is directed to Chapter 1 on the new N cycle.

Microorganisms that fix N constitute a heterogeneous taxonomic group. Apart from their being prokaryotes, the only characteristic they share is the presence of an enzymatic complex called nitrogenase, which catalyzes the reduction of molecular N_2 to ammonium. This reaction requires high metabolic energy consumption and the absence of molecular oxygen. Thus, diazotrophic microorganisms have evolved different strategies to create anoxic or nearly

---

Ecological Aspects of Nitrogen Metabolism in Plants, First Edition. Edited by Joe C. Polacco and Christopher D. Todd.
© 2011 by John Wiley & Sons, Inc. Published 2011 by John Wiley & Sons, Inc.

anoxic conditions for nitrogenase. Among these mechanisms are high respiration levels, as in *Azotobacter*, the differentiation of specialized cells, such as the heterocysts of certain filamentous cyanobacteria, and the biosynthesis of oxygen-carrier proteins, such as the leghemoglobin in rhizobia. There are two kinds of N-fixing bacteria, symbiotic and nonsymbiotic. Nonsymbiotic N-fixing bacteria are free-living organisms able to fix N and use it for their own cellular metabolism. The symbiotic species only fix N in symbiosis with host plants, and in most cases, they cannot live independently at the expense of fixed N as the only source of this element. Diazotrophic bacteria–plant symbioses are of critical agronomic and environmental importance, allowing crop and plant production in N-limited soils. The physiological and morphological characteristics of these symbioses vary widely as a function of the microorganism and plant involved. In most cyanobacteria and some proteobacteria, such as *Azospirillum*, symbiosis is carried out outside the plant; however, in other associations, the bacteria are located inside the plant at inter- or intracellular level, as in the case of *Azoarcus*, which is able to penetrate *Oryza sativa* (e.g., Hurek et al., 2002). In other symbioses, the plant creates a specialized structure to host the bacteria, where it will carry out the N fixation. These organs, the nodules, have been described in associations between *Frankia* and actinorhizal plants and between rhizobia and legumes.

In addition to inducing and colonizing root and stem nodules in leguminous plants, rhizobia are also free-living organisms. Rhizobia can invade their plant hosts through colonization of intercellular epidermal spaces, crack entry at emerging lateral roots, or by root hair entry via an infection thread. Unfortunately for agriculture, and for the recovery of N-poor soils in general, these symbioses are very specific. Each of the rhizobia is limited to a small number of plants, not interacting with plants other than the natural hosts. The future use of bacterial N fixation in agricultural and food production depends on our ability to control these symbioses.

Rhizobial Diversity

Legume Rhizobial Sources

Microorganisms able to establish N-fixing symbiosis with legumes were discovered in the nineteenth century; Frank (1889) named the first rhizobial species, which nodulates *Vicia*, as *Rhizobium leguminosarum*. From this date forward, the name rhizobia was applied to all bacteria able to induce the formation of nodules in roots or stems of legumes where bacteria transformed into bacteroids carry out N fixation.

Bergey's Manual of 1974 played a fundamental role in rhizobial taxonomy, and in its 1974 edition, all bacteria able to nodulate legumes were included in a single genus named *Rhizobium* (Frank, 1889), which was included in the Family *Rhizobiaceae* proposed by Conn (1938). Species were distinguished on the basis of a few phenotypic characteristics, such as the morphology of the bacteroids inside the nodules, alkaline or acid production in litmus milk, but mainly based on the legumes they nodulated, according to cross-nodulation experiments (Baldwin and Fred, 1929). Nevertheless, in the case of rhizobia, many of the taxonomic proposals were traditionally made in Bergey's Manual and, in this sense, the first edition of the 1984 Bergey's Manual of Systematic Bacteriology recorded the reclassification of *R. japonicum* into a new genus named *Bradyrhizobium,* proposed to include the slow-growing rhizobial species (Jordan, 1982). The case study of *B. japonicum* nodulation of soybean is covered in the Nodule Formation and Morphology section of this chapter.

Since 1984, slight changes occurred in rhizobial taxonomy, and although the classification of bacteria on the basis of 16S rRNA gene sequences was proposed in the same year (Woese et al., 1984), 16S rRNA sequence was not routinely included in descriptions of rhizobial species until 1991. In that year, this technique was included in the minimal standards for the description of new species of rhizobia and *Agrobacterium* together with DNA-DNA or rRNA-DNA hybridization, RFLP and MLEE analyses and symbiotic characteristics (Graham et al., 1991). In fact, in 1988 the new genus *Azorhizobium* was officially described without its 16S rRNA sequence.

The first description of rhizobial species that included 16S rRNA gene sequences was that of *Rhizobium tropici*, isolated from *Leucaena* nodules in the Americas by Martínez-Romero and others in 1991. Since 1991, 16S rRNA gene sequences have been included in all descriptions or reclassifications of the different taxa from the family *Rhizobiaceae*. In spite of the addition of several legume-nodulating species to the family *Rhizobiaceae*, the number of species ascribed to this family at the end of the last century in comparison with those among other soil-dwelling bacteria was unrealistically low, especially considering that the Leguminosae, with about 19,000 species, form the largest plant family on earth (Polhill et al., 1981). In fact, since 2000 the number of species recognized to nodulate legumes has increased exponentially, and significant changes in the taxonomy of the family *Rhizobiaceae* have been made. The first change was the reclassification of *Agrobacterium* and *Allorhizobium* into the genus *Rhizobium*, proposed by Young and others (2001) in the official journal of bacterial systematics, the International Journal of Systematic and Evolutionary Microbiology (IJSEM). However, we considered this reclassification to be premature, except in the case of the former *Agrobacterium rhizogenes*, which clearly belongs to the genus *Rhizobium* (Willems and Collins, 1993; Yanagi and Yamasato, 1993; Young et al., 2001; Velázquez et al., 2005).

The second change was in the name of the genus *Sinorhizobium*. According to 16S rRNA analysis and the rules of the bacteriological code concerning nomenclature, the name *Ensifer* takes priority over *Sinorhizobium*. These proposals were directly revised by the Judicial Commission of the International Committee on Systematics of Prokaryotes, whose opinion has been recently published (Judicial Commission of the International Committee on Systematics of Prokaryotes, 2008).

The third change was the division of the genera from the family *Rhizobiaceae* into several families and the creation of a new order named *Rhizobiales* proposed in Bergey's Manual from 2005 (Kuykendall, 2005). The new order *Rhizobiales* contains symbiotic, pathogenic and saprophytic bacteria distributed among several families. Most of these names were validated in IJSEM (Validation list No. 107).

A fourth change has been made recently, since in the past few years it has emerged that even though the current bacterial classification based on the 16S rRNA gene allows the description of non-nodulating rhizobial species, 16S rRNA sequence is limited in differentiating closely related rhizobial species (Valverde et al., 2006; Ramírez-Bahena et al., 2008). Thus, several housekeeping genes have been proposed in several groups of bacteria (Gaunt et al., 2001; Maiden, 2006; Valverde et al., 2006; Ramírez-Bahena et al., 2008). In recent years, new schemes for the identification and phylogenetic analysis of bacteria, termed MLSA (MultiLocus Sequence Analysis) and MLST (MultiLocus Sequence Typing) and based on the analysis of several housekeeping genes, have been applied in phylogenetic analyses of specific groups of rhizobia such as *Ensifer* (van Berkum et al., 2006; Martens et al., 2007) and *Bradyrhizobium* (Rivas et al., 2009). The ad hoc committee for the re-evaluation of species definition has suggested that "species should be identifiable by readily available methods (phenotypic, genomic)" and that one approach toward this goal is the determination of a minimum of housekeeping genes (Stackebrandt et al., 2002; Zeigler, 2003).

After all of the changes undergone by rhizobial taxonomy in past decades, the genus *Rhizobium* currently contains 34 species and includes the former genera *Agrobacterium* and *Allorhizobium*. Among other genera, *Sinorhizobium*, currently named *Ensifer*, contains 12 species; *Mesorhizobium*, 20 species; *Phyllobacterium*, 8 species; *Azorhizobium*, 2 species; and *Bradyrhizobium*, 9 species. The complete list of valid species of rhizobia is constantly updated and recorded in the List of Prokaryotic Names with Standing in Nomenclature by Dr. Euzeby (http:// www.bacterio.cict.fr). Considering the large number of legumes hitherto not studied and the few ecosystems analyzed to date, a dramatic increase of the number of rhizobial species can be expected.

Nonlegume Rhizobial Sources

The name rhizobia was derived from the first described legume-nodulating species, *Rhizobium leguminosarum*. From then on, it was long believed (1) that all species of the genus *Rhizobium* were able to nodulate legumes and (2) that nodulation was exclusive to rhizobia. In fact, the identification of these bacteria from sources other than nodules was very difficult until the development of 16S rRNA sequence-based phylogenies and public databases. Until the 1990s, the isolation of rhizobia had to be performed using "trap" plants, and for decades their identification was based on cross-nodulation criteria. Where 16S rRNA-coding sequences were available, it was shown that many species of different genera of rhizobia can nodulate the same legume and, in turn, that different legumes can be nodulated by the same species of rhizobia. Subsequently, the number of rhizobial species known to nodulate legumes has increased dramatically, and it has also been possible to identify species of the classic rhizobial genera in sources other than legume nodules. The description of non-nodulating species of rhizobia has been possible thanks to sequence analysis of 16S rRNA genes and several housekeeping genes such as *recA* and *atpD* (Gaunt et al., 2001).

The first non-nodulating rhizobial species isolated from a non-nodule source was *Bradyrhizobium betae*, described by our research group in 2004 (Rivas et al., 2004). *B. betae* was isolated from tumor-like formations in sugar beet, although it was not the agent responsible for these malformations, and no nodulation or N-fixation genes were found.

Many other species have since been isolated from sources other than nodules, such as *R. daejeonense*, isolated from a cyanide treatment bioreactor, and *R. selenitireducens* also isolated from a bioreactor. Other species, such as *R. alamii*, have been isolated from the plant rhizosphere, while *R. oryzae* is among those isolated from the inner tissues of plant roots. In 2007 we isolated from sawdust of *Populus* a new species (*R. cellulosilyticum*) with marked cellulolytic activity (García-Fraile et al., 2007). In the same year, a species of the genus *Mesorhizobium* was isolated from sources other than legume nodules (i.e., *M. thiogangeticum* isolated from the rhizosphere of legumes) (Ghosh and Roy, 2006).

These results over several years showed that the diversity of rhizobia was higher than had been expected and that these microorganisms are present in very different ecosystems. One of them, *R. oryzae*, is in inner root tissue of nonlegume plants, and some rhizobial species are known as endophytes of different plants. There is currently increasing interest in the study of rhizobial endophytic bacteria, focusing on their potential to promote plant growth (Rosenblueth and Martínez-Romero, 2006; Ryan et al., 2008; Torres et al., 2008). (The reader is directed to Chapter 11 and Chapter 17).

The ability of rhizobia to establish beneficial associations with rice plants has been known for several years, and in 1997 the results of a multinational collaborative study, with the participation of our research group, addressing the beneficial association between *Rhizobium leguminosarum* bv. trifolii and rice were published (Yanni et al., 1997). Since 1997 there have been several studies on the natural association of rice and *R. leguminosarum* strains and their potential for rice growth promotion (Chaintreuil et al., 2000; Yanni et al., 2001; Singh et al., 2006). Currently, rice is the most widely studied plant model for endophytic colonization with different species of rhizobia, and the migration of rhizobial strains from roots to leaves inside rice plants has been reported (Chi et al., 2005). Infection by some rhizobial strains may occur via root hairs located at the emerging lateral roots, followed by extensive spreading throughout the rice plant root (Perrine-Walker et al., 2007). The induction of endophytic colonization of tissue culture-derived rice plants with *Azorhizobium caulinodans* has recently been reported (Senthilkumar et al., 2008).

Other studies have reported the beneficial associations of rhizobia with different field plants rotated with legumes (Gutiérrez-Zamora and Martínez-Romero, 2001; Rosenblueth and Martínez-Romero, 2004). From these studies it was concluded that *R. etli* is a natural endophyte of maize in American soils (Gutiérrez-Zamora and Martínez-Romero, 2001) and that *R. tropici* is also an endophytic bacterium of maize in America (Rosenblueth and Martínez-Romero, 2004). Recently, we found *R. rhizogenes* in endophytic populations from coffee in Cuba (Velázquez et al., 2008).

"Endophytic" rhizobia are also commonly isolated from the nodules of different legumes (Kan et al., 2007); indeed, nonculturable rhizobia have been found together with other endophytic bacteria (Muresu et al., 2008). Many strains of *Rhizobium radiobacter* (formerly *Agrobacterium tumefaciens*), which are unable to produce nodules or tumors, are also frequently recovered from nodules (Mhamdi et al., 2005; Wang et al., 2006; Tiwary et al., 2007).

The identification of rhizobia isolated from sources other than nodules became possible after 2000. These species have great potential as nonlegume plant inoculants. However, we point out that nodulation of nonlegume plants by rhizobial strains was known for several years. Thus, NGR234 is able to nodulate the nonlegume *Parasponia* (Trinick, 1979; Trinick and Hadobas, 1989; Pueppke and Broughton, 1999; Lafay et al., 2006), and the isolation of *Rhizobium* from nodules of the gymnosperm *Podocarpus macrophyllus* has been recently reported (Huang et al., 2007). These findings are also of great interest from the evolutionary point of view of symbiosis, and they may be a window on a more pervasive role of the rhizobia in plant N metabolism on a global scale.

"Non-rhizobia" Able to Induce Nodules in Legumes

A belief that rhizobia are the only bacteria able to initiate nodules in legumes suggested that all nodule-derived colonies lacking the aspect of rhizobia on YMA (Vincent, 1970) solid medium should be discarded, leaving the analysis only to colonies displaying the mucoid aspect of the rhizobia. Colonies with this typical aspect were named *Rhizobium* with no further identification. This situation changed dramatically by 2001, when 16S rRNA gene sequencing was applied to the identification of nodule isolates. In that year the report of two atypical bacteria that nodulate legumes opened the door of legume nodulation to bacteria different from rhizobia. In January, Sy and others (2001) reported that a *Methylobacterium* formed nodules on *Crotalaria*, an African legume. The bacterium, named *M. nodulans* (Jourand et al., 2004), harbors the common nodulation *nodABC* genes and the *nifH* gene encoding the structural nitrogenase enzyme. An even more radical report appeared in a June issue of *Nature*; Moulin and others (2001) showed nodulation of *Mimosa* by *Burkholderia*, a genus belonging to the beta-Proteobacteria. The *Burkholderia* strain is also able to form nodules in the roots of *Macroptilium atropurpureum* and carries nodulation (*nodABC*) genes phylogenetically related to those found in legume symbionts of the class alpha Proteobacteria ("classic" rhizobia), supporting the hypothesis of lateral gene transfer in the rhizosphere, crossing the boundary between the alpha and beta Proteobacteria.

Also in 2001 we analyzed the 16S rDNA of two strains isolated from *Neptunia natans* in India, showing that they belong to the genus *Devosia* from the alpha-Proteobacteria (Rivas et al., 2002). These strains were previously named *Rhizobium neptunii* because of their "rhizobial" colony morphology (Subba-Rao et al., 1995), but 16S rDNA analysis revealed that they identify a new species, designated *D. neptuniae* (Rivas et al., 2003). The strains isolated from *Neptunia* nodules carry *nodD* and *nifH* genes that are closely related to those of *R. tropici* CIAT899^T, suggesting that they were transferred to *D. neptuniae* from *R. tropici*, an American species that nodulates *Leucaena*. Later, this hypothesis was supported by the finding that *R. tropici* nodulated *Neptunia* in America (Zurdo-Piñeiro et al., 2004).

In ensuing years, several nonrhizobial genera from alpha and beta Proteobacteria were reported to be legume endosymbionts. In 2002, in addition to nodulation of *Neptunia* by *Devosia*, the nodulation of *Aeschynomene indica* by *Blastobacter denitrificans* was reported (van Berkum and Eardly, 2002).

Since 2001, several species of beta Proteobacteria have been reported to be legume endosymbionts, such as *Ralstonia taiwanensis* nodulating *Mimosa pudica* and *Mimosa diplotricha* (Chen et al., 2003). In this case, the strain was mistakenly classified as *Ralstonia*, and it has later been reclassified as *Cupriavidus taiwanensis*, a beta Proteobacteria, belonging to the family *Burkholderiaceae* from the order *Burkholderiales* (Vandamme and Coenye, 2004). *C. taiwanensis* carries nodulation genes (*nodBCIJHASUQ*) and one regulatory gene (*nodD*) on the pRalta mega-plasmid. *C. taiwanensis* carries 19 genes, presumably arranged in five operons, covering 25kb adjacent to the *nod* genes, that are involved in nitrogenase synthesis and function (Amadou et al., 2008).

Additionally, several species of *Burkholderia* that nodulate legumes have been described: *B. mimosarum* (Vandamme et al., 2002; Chen et al., 2006), *B. phymatum* (Vandamme et al., 2002), *B. nodosa* (Chen et al., 2007), and *B. sabiae* (Chen et al., 2008), all nodulating *Mimosa*. Although *Burkholderia* mainly nodulates mimosoid legumes, there is evidence that papilionoid legumes such as *Cyclopia* are nodulated by *B. tuberum* (Elliott et al., 2007), and *Dalbergia louvelli* is nodulated by a strain belonging to the *Burkholderia cepacia* complex (Rasolomampianina et al., 2005). It was recently reported that some strains of *Burkholderia* are more competitive than *R. tropici* for the nodulation of *Mimosa* (Elliot et al., 2009).

In 2004, strains from the genus *Ochrobactrum* were found in *Acacia mangium* nodules, but no information about their symbiotic genes was reported (Ngom et al., 2004). In the same year we found that strains from a new species of this genus carrying symbiotic genes closely related to those of rhizobia are able to nodulate *Lupinus* (Trujillo et al., 2005). The strains isolated harbored megaplasmids and the nodulation (*nod*) and N-fixation (*nif*) genes were detected in all of them using *nifH* and *nodD* probes. In 2005, a new species called *Phyllobacterium trifolii*, isolated from *Trifolium pratense* nodules, was also found to be able to nodulate *Lupinus albus*. Although this species harbors symbiotic plasmids in which *nodD* and *nifH* genes are located, it forms ineffective nodules in roots of *Trifolium repens* and *Lupinus albus* (Valverde et al., 2005).

In 2005, we also demonstrated the nodulation of *Phaseolus vulgaris* by two pathogenic strains of the species *Agrobacterium rhizogenes* (currently *Rhizobium rhizogenes*). The *nodD* and *nifH* genes found in these strains, together with the symbiotic plasmids and the presence of tumorigenic or hairy root-inducing plasmids was shown (Velázquez et al., 2005).

In 2007, a second new species of *Ochrobactrum* designated *O. cytisi*, carrying symbiotic genes phylogenetically related to those from rhizobial strains, was isolated from *Cytisus scoparius* nodules in Spain (Zurdo-Piñeiro et al., 2007). In the same year, a new alpha Proteobacteria nodulating *Lupinus* in America and related to the genus *Methylobacterium* was reported (Andam and Parker, 2007). Reports of the presence of nonrhizobia belonging to the alpha and beta Proteobacteria in the nodules of several legumes are increasing rapidly, but even though several studies have reported the presence of gamma Proteobacteria in legume nodules, to date no symbiotic genes have been detected in these strains (Ibáñez et al., 2009). In any case, all of these findings show that legume symbiosis remains poorly understood, that further studies are required, and that perhaps new calculations of global N economy need to take into account these "nonconventional" nodulation processes.

Rhizobial Infection Process

Microbial infection usually means entrance, growth, and multiplication of a microorganism in the body of a host as part of a disease process. Others define infection as the presence of a microorganism in host tissues, whether or not it manifests detectable pathological effects on the host. The latter is the case for rhizobia that have the ability to infect plants, but instead of producing disease, are beneficial for the plant. For this to occur, there must be full coordination between the two partners: the macro- and the micro-symbiont. The establishment of a rhizobia-legume symbiotic relationship follows a series of independent steps that define the so-called symbiotic process: specificity, infectivity, and effectiveness—the result of the properties expressed by the bacterium, the host plant, or both.

Root Hair Entry

In legume symbiosis, bacterial invasion can follow different routes, the best known of which is via root hairs. Rhizobia induce the curling of growing root hairs, become entrapped in the curl, and enter the root hair by local hydrolysis of cell walls and invagination of the plasma membrane. Tip growth toward the base of the root hair results in an intracellular infection thread that proceeds through the cortical cells to reach the nodule primordium, where bacteria are released inside plant cells and differentiate into N-fixing bacteroids. Among other legumes, root hair invasion occurs in clover, pea, bean, soybean, alfalfa, and in the model legumes *Medicago truncatula* and *Lotus japonicus*. The different steps of symbiotic root hair infection are described in the following subsections.

Exchange of Molecular Signals Between the Plant and the Microorganism

Compatible rhizobia and hosts engage in a series of reciprocal signalling events. The principal signals originating from the host and perceived by rhizobia in the soil are derived from the flavonoid family (2-phenyl-1,4-benzopyrone derivatives) of secondary plant metabolites (Gibson et al., 2008). Specific flavonoids released by legume roots serve as chemoattractants for the rhizobial symbiont (Gulash et al., 1984) and also activate the expression of rhizobial *nod* genes. Flavonoids bind bacterial NodD proteins, which are members of the LysR family of transcriptional regulators, and activate these proteins to induce the transcription of specific rhizobial genes (Perret et al., 2000; Barnett and Fisher, 2006). These genes control production of the Nod factors (NF) that consist of a backbone of β-1,4-linked *N*-acetyl-d-glucosamine residues, which can differ in number not only among bacterial species but also within the repertoire of a single species (Perret et al., 2000). NFs are *N*-acylated at the nonreducing end that may also vary among rhizobial species (Perret et al., 2000). These NFs are recognized by a specific LysM domain-containing receptor kinases (Limpens et al., 2003). Besides the LysM receptor kinases, several other components of this NF-induced signalling cascade have been identified. They include the putative cation channel DMI1 (Ané et al., 2004), the leucine-rich- repeat-containing receptor kinase DMI2 (Limpens et al., 2005), and the calcium calmodulin-dependent kinase DMI3 (Levy et al., 2004). The NF receptors trigger a signal-transduction cascade that is essential for the induction of all early symbiotic events, including root hair deformation, pre-infection thread formation, and the induction of cell division in the root cortex that marks the formation of the nodule primordium (Orgambide et al., 1996; Jones et al., 2007). However, the addition of purified compatible NFs to plant roots is not sufficient to cause the formation of tightly curled root hairs (shepherd's crooks), a complete differentiation of the infection thread and mature nodules indicating that the NF is not the only bacterially produced effector required for the symbiont to enter plant tissues and colonize plant cells (Gage, 2004; Jones et al., 2007; Gibson et al., 2008).

Although NF signalling is a nearly universal means of establishing the rhizobia N-fixing symbiosis with compatible legumes, exceptions are emerging. The recent sequencing of some photosynthetic *Bradyrhizobium* strains that form N-fixing nodules on the roots and stems of an aquatic host, *Aeschynomene sensitiva*, revealed that they lack the "common" *nodABC* genes (Giraud et al., 2007). Thus, the host initiates nodule development in a NF-independent manner and instead may respond to the secretion of bacterial purine derivatives with cytokinin-like activity, highlighting the importance that the host hormone balance plays in nodule formation (Gibson et al., 2008).

Root Colonization

All rhizobia are described as motile species (Figure 8.1A) with pronounced chemotactic responses to a wide variety of metabolites likely to be found in soil, in the rhizosphere, and in plant exudates (Yost and Hynes, 2000). In the soil, rhizobia move toward plant roots, and chemotaxis move toward specific compounds exuded by the plant facilitates the colonization of infection sites on the roots of compatible host legumes (Miller et al., 2007).

Rhizobial root colonization is a dynamic, multiphasic process including several nonspecific and host-specific events (Dazzo et al., 1984). The Phase 1 pattern of randomly oriented attachment involves an initial nonhost-specific interaction of bacterial encoded rhicadhesin on individual cells and the root hair tip (Smit ct al., 1992), followed by a more host-specific aggregation of immobilized cells at the root hair tip mediated by secreted, multivalent host lectin (Dazzo and Hubbell, 1975; De Hoff et al., 2009). Cells that have not attached become polarly encapsulated in the external root environment and attach perpendicularly to the sides of the same root hair (Figure 8.1B). This pattern of rhizobial attachment to root hairs (clumps of cells at the root hair tip [Figure 8.1C], and individual polarly attached cells along the shaft of the same root hair) requires the intervention of bacterial proteins and polysaccharides, host lectin, and enzymes that degrade the bacterial polysaccharides; it exhibits host selectivity and is found on approximately 95% of the infected root hairs in the *Rhizobium*-white clover symbiosis (Dazzo and Wopereis, 2000). Phase 1 attachment is distinguished from Phase 2 adhesion. There is a significantly increased force of adhesion of attached cells concurrent with the elaboration of extracellular microfibrils (Figure 8.1D) that increase the degree of contact of the attached bacteria to the root hair surface (Dazzo et al., 1984). Isolated extracellular microfibrils, made *in vitro* by *R. leguminosarum* bv. trifolii, have been shown to consist of microcrystalline cellulose (Napoli et al., 1975). However, the nature of the microfibrils associated with rhizobia

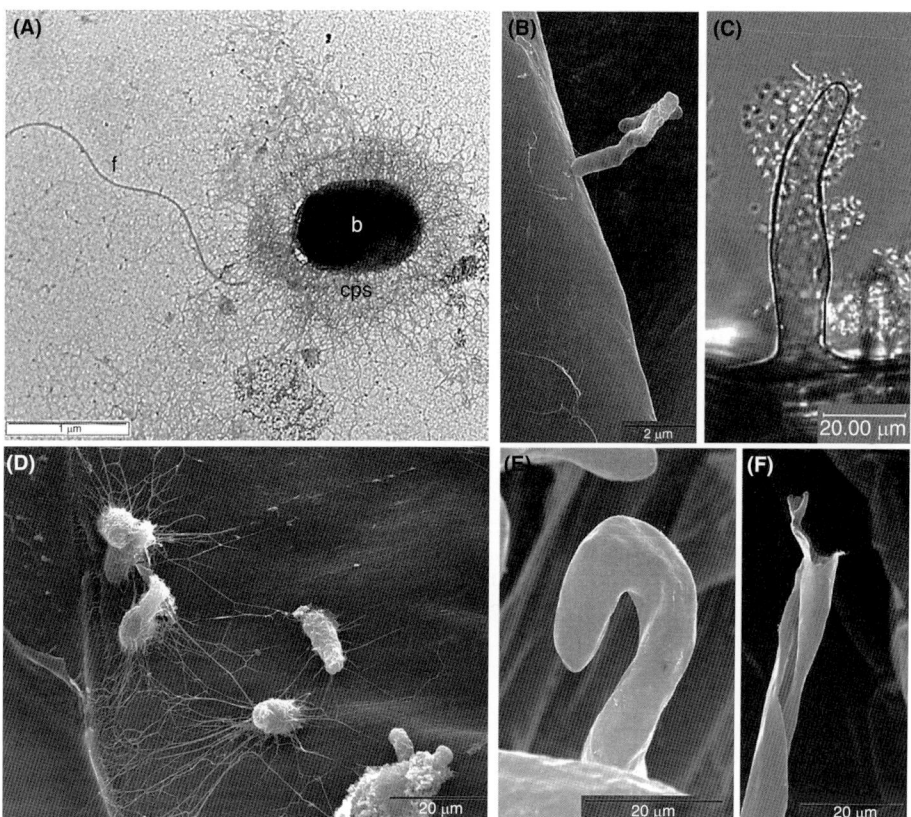

Figure 8.1. Transmission electron micrograph (A), scanning electron micrographs (B, D, E, F), and phase contrast micrograph (C) illustrating primary host infection. (A) Free-living encapsulated and flagelated *Rhizobium leguminosarum* bv. trifolii (b, bacterium; cps, capsule; f, flagellum); (B) individual polarly attached cell along the shaft of the root hair; (C) clumps of cells attached to the root hair tip; (D) root adhesion with associated cellulose microfibrils; (E) deformation of white clover root hair (curling) induced by *Rhizobium leguminosarum* bv. trifolii; (F) Hot (*H*ole *o*n the *t*ip) phenotype induced by purified cellulase CelC2.

firmly attached to the legume root epidermis is more difficult to define. The combined use of scanning electron microscopy, enzyme cytochemistry, and computer-assisted image analysis has provided direct *in situ* evidence of the cellulosic nature of the extracellular microfibrils extending from *R. leguminosarum* bv. trifolii cells colonizing the white clover root epidermis (Mateos et al., 1995).

Rhizobia produce other polysaccharides (capsular, CPS; exopolysaccharides, EPS; lipopolysaccharides, LPS and β 1,2-glucans) that accumulate on the surface of the bacteria (Figure 8.1A). The importance of polysaccharides in bacterial adhesion to the tip of the root hairs and the subsequent development of the infective process seems to be critical because, besides their involvement in the competitiveness of strains, it has been demonstrated that mutants defective in polysaccharide biosynthesis are characterized by low infectivity, a low capacity for nodulation, and in some cases, changes in the host range (Dazzo et al., 1991; Rolfe et al., 1996; Gibson et al., 2008).

Root Hair Invasion

Adhesion is followed by the first microscopically visible reaction of the plant host, which is a deformation and curling of the root hair (Figure 8.1E). Rhizobial NFs elicit highly localized, short-range responses that modulate the differentiation, extension growth dynamics, and crystalline wall architecture of host root hairs (Dazzo et al., 1996). The morphological changes observed in the root hair do not take place in a homogeneous way throughout the root. The most susceptible root hairs are those that have nearly finished growing (root hair zone II). Root hairs that have finished growing (root hair zone III) and root hairs that are actively growing with a strongly polarized internal organization (root hair zone I) are refractory to the deforming activity of NFs (Gage, 2004). The susceptibility of root hairs to deformation in the presence of NF can be modulated by plant hormones such as ethylene, which inhibits NF signal transduction and can influence the degree of root hair deformation and the frequency of productive infections (Oldroyd et al., 2001). This mechanism involves marked curling of the root hair tip to form a typical shepherd's crook. The marked curling of host root hairs is thought to form a pocket that traps the bacteria and creates a suitable microenvironment for a localized degradation of the host cell wall (Figure 8.1F), which permits the bacteria to penetrate it (Robledo et al., 2008). The root hair cytoplasmic membrane becomes invaginated and a new wall of plant origin is deposited between the invading bacteria and the involuted plasma membrane, forming a tubular infection thread (Callaham and Torrey, 1981) that grows down the inside of the root hair and into the body of the epidermal cell.

Infection Thread Development

Infection threads are tubular structures with bacteria embedded in an amorphous lumen matrix surrounded by a more electron-dense fibrillar wall and an infection-thread enclosure membrane that is continuous with the plasmalemma of the infected host cells (Figure 8.2B). The structure of the infection thread wall is similar to that of the root hair. In particular, esterified and unesterified pectins, xyloglucans, and cellulose are likely components of infection thread walls (Rae et al., 1992). The matrix inside the lumen contains material that is normally found in the extracellular matrix of plant cell walls. Consequently, bacteria inside infection threads are topologically "outside" the root hair.

Root hairs and pollen tubes are the best-studied examples of plant cells that elongate through the process of tip growth. Infection threads develop from growing root hairs and are also thought to be tip-growing structures, and therefore are most likely to elongate by using at least some of the machinery involved in root hair growth before infection (Gage, 2004). The tip of the developing infection thread is a site of new wall and membrane synthesis, and has been proposed to involve inversion of the tip growth that is normally exhibited by the root hair, and to be the result of the reorganization of cellular polarity (Gage, 2004). When the infection thread exits the epidermal cell, it fuses with the distal cell wall, and bacteria enter the intercellular space between the epidermal cell and the underlying cell layer. Invagination and tip growth, similar to those seen at the beginning of infection thread growth, recur in the underlying cell, and a thread filled with bacteria is propagated further toward the root interior (van Spronsen et al., 1994). The plant hormone cytokinin and the NF-dependent re-initiation of the cell cycle are involved in directing infection threads to the plant cortex (Jones et al., 2007).

Concomitant with the outward-propagating wave of cell activation, a column or two of cells in the outer cortex also begin to divide. The cytoplasm moves from the cell periphery to a central position, as it normally does during cell division, but in this case the cells usually progress no further through the cell cycle. The cytoplasms in these activated outer cortical

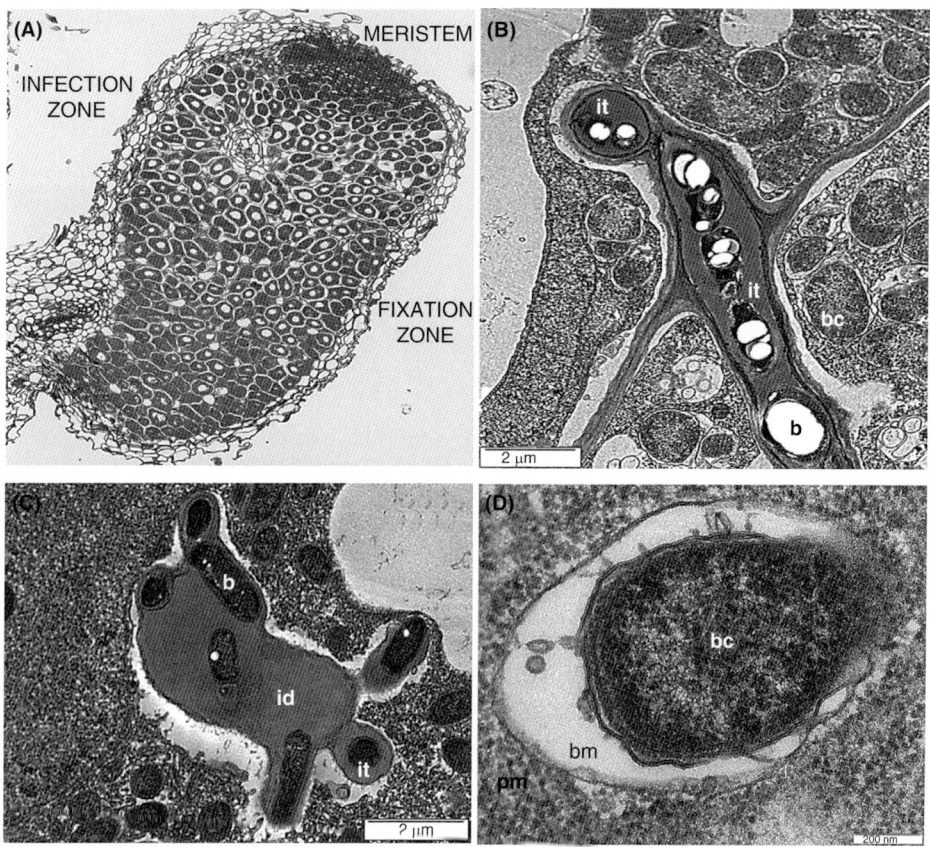

Figure 8.2. Brightfield micrograph (A) and transmission electron micrographs (B, C, D) illustrating secondary host infection. (A) Low magnification view of a longitudinal section of an indeterminate root nodule of white clover, showing the uninfected outer nodule cortex, and various zones in the interior of the invaded nodule; (B) cross- and longitudinal sections of tubular, walled infection threads traversing two adjacent nodule host cells; (C) bacterial release from infection droplets protruding from eroded walls of infection threads within root nodules of white clover; (D) symbiosome containing nitrogen-fixing bacteroid of *Rhizobium leguminosarum* bv. trifolii in a nodule cell of white clover. (b, bacteria; bc, bacteroids; bm, bacteroid membrane; id, infection droplet; it, infection thread; pm, peribacteroid membrane).

cells align with each other, giving rise to columns of cytoplasmic bridges called preinfection threads through which the inwardly growing infection thread propagates (van Brussel et al., 1992). New infection threads are initiated at each cell layer, allowing the bacteria to penetrate deeper layers of root tissue (Jones et al., 2008). Cell walls are degraded in the same direction that the infection thread grows, thus allowing the passage of bacteria from one cell to another (Figure 8.2B). The mechanism is not exactly understood, but it is assumed that pectolytic and cellulolytic enzymes secreted by the host plant, the bacterium, or both symbionts, are responsible for this process. The branching of the thread as it grows through the root and enters the nodule primordium increases the number of sites from which bacteria can exit the thread and enter nodule cells, ensuring that nodule cells, in sufficient number, are infected (Figure 8.2A).

After the thread network has entered the nodule primordium, some of the uninfected, activated cells in the middle cortex become organized into a nodule meristem (Gage, 2004).

In developing indeterminate nodules, infection threads grow through the developing meristematic zone to the underlying cell layers through which the wave of mitotic activity has already passed (Figure 8.2A). These layers of invasion-competent cells have exited mitosis and become polyploid owing to cycles of genomic endoreduplication without cytokinesis (Jones et al., 2007). As the nodule grows, the infection thread network behind the meristem continues to develop, and eventually it forms a highly branched network in a region of the nodule termed the infection zone (Vasse et al., 1990) (Figure 8.2A). The process in determinate nodules, which lack a meristem, is described in the section III "case study" of *B. japonicum* nodulation of soybean.

Bacterial Release

Rhizobia inside the thread undergo cell division, thereby keeping the tubule filled with bacteria (Figure 8.2B). Microscopic analysis of fluorescently tagged bacteria within infection threads indicates that only the bacteria at the tip of the in-growing infection thread are actively dividing (Gage, 2002). When the bacteria reach the infection zone (Figure 8.2A), they must be internalized by a cortical cell, and they must establish a niche within that cell. In a target cell each bacterial cell is endocytosed in an individual, unwalled membrane compartment that originates from the infection thread (Figure 8.2C). The entire unit, consisting of an individual bacterium and the surrounding endocytic membrane, is known as the symbiosome (Brewin, 2004) (Figure 8.2D). Once the bacteria have been engulfed within host cell membranes, they must survive within the symbiosome compartment and differentiate into the N-fixing bacteroid form. Bacteria enclosed within a functional symbiosome membrane have been provided with a low-oxygen environment; they have completed the bacteroid differentiation program, can express the enzymes of the nitrogenase complex, and thus begin to fix N_2 (Fischer, 1994).

The formation of symbiosomes is presumed to represent a major step in the evolution of legume nodule simbiosis. Symbiosome formation does not occur in more primitive legume nodules (e.g., *Andira* spp. and many species belonging to the *Fabaceae* subfamily *Caesalpinoideae*). *Parasponia* spp., the only nonlegume species that can establish a symbiosis with rhizobia, also forms "primitive" nodules. In these species, rhizobia are not released into the nodule host cells but remain in infection threads, called fixation threads, which are enclosed by a cell wall-like structure in which the rhizobia can fix N_2 (Limpens et al., 2005).

Crack Entry

Most temperate legumes are invaded via growing root hairs. However, the invasion of a number of (sub)tropical legumes occurs through crack entry that differs from primary infection via root hairs in that, upon entry, bacteria disseminate intercellularly. There are two types of crack entry depending on the mode of bacterial dissemination within the nodule. In the first, the rhizobia directly invade the cortical cells and dissemination takes place by division of the infected cells without involving infection threads. This is the case of *Arachis, Aeschynomene, Stylosanthes,* and *Chamaecytisus* (Vega-Hernández et al., 2001). In the second, dissemination involves an initial intercellular spreading and later formation of true infection threads that penetrate nodule cells, as is the case for *Sesbania* and *Neptunia* (Subba-Rao et al., 1995). A crack entry has also been described in the infection of the nonlegume *Parasponia* by *Bradyrhizobium*, in which both mechanisms of dissemination appear to take place (Bender et al., 1987).

The simplest mechanism by crack entry occurs in *Arachis hypogaea*. This legume never exhibits structures resembling infection threads in either root hairs or developing nodules. Instead, rhizobia spread entirely intercellularly by way of separating cortical cells at the middle lamellae; they then enter the cells of the nodule primordium by direct uptake from the infection pockets (Chandler, 1978).

Devosia neptuniae occupies a unique ecological niche in aquatic environments by entering into an N_2-fixing root-nodule symbiosis with the aquatic legume, *Neptunia natans*. In its native marsh habitat, *N. natans* does not develop root hairs. The plant floats and develops N-fixing nodules at the emergence of lateral roots on the primary root and on adventitious roots at stem nodes, but not from the stem itself (Subba-Rao et al., 1995). The mode of root infection in this aquatic N_2-fixing symbiosis involves an intercellular route of entry followed by an intracellular route of dissemination within nodule cells. After colonizing the root surface, *D. neptuniae* bacteria enter the primary root cortex through natural wounds caused by the splitting of the epidermis and the emergence of young lateral roots, and then stimulate the early development of nodules at the base of such roots. The bacteria enter the nodule through pockets between separated host cells, then spread deeper into the nodule through a narrower intercellular route, and eventually stimulate the formation of infection threads that penetrate host cells and spread throughout the nodule tissue. Bacteria are released from infection droplets at the unwalled ends of infection threads, where they become enveloped by peribacteroid membrane and are transformed into enlarged bacteroids within symbiosomes (Subba-Rao et al., 1995).

In the semi-aquatic *Sesbania rostrata*, the bacteria enter via root hair curls under nonflooded conditions. However, upon flooding, when root hair growth is prevented and invasion onto accessible root hairs is inhibited, intercellular invasion is recruited. The plant hormone ethylene is involved in these processes (Goormachtig et al., 2004). The bacteria enter via cracks formed by the protrusion of lateral roots and colonize large intercellular spaces called infection pockets. These infection pockets serve as a launching point for intracellular infection threads that grow toward the nodule primordium (Goormachtig et al., 2004). The occurrence of both invasion pathways on the same host plant has allowed the molecular mechanisms governing both processes to be compared in *S. rostrata*. Cortical intercellular invasion at lateral root bases shows less stringent nodulation factor structure requirements and fewer transcriptional changes than epidermal root hair invasion (Capoen et al., 2007).

A Case Study: Determinate Nodule Formation Between *Bradyrhizobium japonicum* and Soybean

We close with an important association, *B. japonicum* and its plant host, soybean. This association has obviated much nitrogen fertilization of a crop that worldwide is one of the major sources of plant protein and oil.

Soybean Nodule Formation and Morphology

B. japonicum forms determinate, or spherical, nodules on soybean. Determinate nodules are developmentally synchronized; that is, they are all predominantly of one developmental form dictated by the age of the nodule. Furthermore, the infection process is controlled in such a way that all of the nodules within the crown of the root are of the same developmental age.

As described elsewhere in this volume (for example, Chapter 15 and Chapter 16), determinate nodules do not have persistent meristems, and the vascular system becomes closed conferring a continuous system of vascular branches. The infection thread elongates into the central cortical region of the root and branches and penetrates meristematic cells depositing bacteria within symbiosomes.

The plant cells differentiate into infected and uninfected cells, which multiply and expand to form a nodule. Infected plant cells are found only in the central portion of the spherically shaped nodules. Each infected cell may contain up to 20,000 bacteroids, which occur in groups of usually 10–20 within a single symbiosome. The bacteroids never come into direct contact with the host cell cytoplasm.

The symbiosome membrane acts as a selective permeability barrier and thus regulates the transfer of metabolites between the infected plant cell and the bacteroids. The flow of metabolites and signal molecules from the plant must pass through the infected plant cell, across the symbiosome membrane, the symbiosome space (the volume outside the bacteroids, yet still inside the symbiosome), the outer bacteroid membrane, the periplasmic space and the inner bacteroid membrane. The flow of metabolites from the bacteroid to the plant would follow the reverse pathway. The transporters at these membrane boundaries are mostly undetermined.

Leghemoglobin, an abundant myoglobin-like protein found in the infected region of the soybean nodule, acts as an oxygen buffer maintaining a low partial pressure of oxygen to permit the functioning of the oxygen-labile nitrogenase component proteins within the bacteroids, while providing sufficient oxygen for respiratory ATP generation.

Interspersed between the infected plant cells are smaller uninfected plant cells occupying about 20% of the central nodule zone (Newcomb, 1981). There are about 1.5 uninfected nodule cells for each infected cell. Uninfected cells convert the fixed nitrogen into ureides (purine degradion products) for transport to the plant shoot.

Metabolic Differentiation of Soybean Root Nodule Cells

The integrated cellular and metabolic differentiation of the plant cells of the soybean nodule can be viewed from several different perspectives. It can be defined transcriptionally or post-translationally (enzymatic and proteomic analyses). In the latter case, plant-encoded proteins that are found only in the root nodules, hence not in any other parts of the host plant, are called nodulins. Nodulins are unique or highly upregulated proteins expressed during symbiosis. Early nodulins are generally defined as being expressed during the infection and invasion process. The "late" nodulins are involved in nodule function and maintenance and are generally defined as coinciding with the beginning of measurable N_2 fixation activity. The two general classes of late nodulins are metabolic nodulins and symbiosome membrane nodulins. Metabolic nodulins include leghemoglobin, which is the primary oxygen regulator of nodules; uricase, which is a key enzyme in the generation of N-rich ureide transport molecules; glutamine synthetase, which catalyzes the first step in ammonium assimilation; and sucrose synthase, which catalyzes the cleavage of sucrose to begin the pathway to produce carbon metabolites for bacteroid energy production. In addition, several enzymes have been detected in root nodules of legumes that differ in their physical, kinetic, and immunological properties from the corresponding root enzymes: phosphoenol pyruvate carboxykinase, choline kinase, xanthine dehydrogenase, purine nucleosidase, and malate dehydrogenase (Franssen et al., 1992). These enzymes may not be true nodulins but rather post-translational modifications of the root enzymes.

Over 100 proteins from the soybean nodule cytosol were identified via mass spectral analysis following separation via 2-D PAGE (Oehrle et al., 2008), annotated with the soybean UniGene database (Mooney and Thelen, 2004). The largest category of proteins was that defined as involved in carbon metabolism (~28%), followed by N-metabolism (~12%), and oxygen protection (~12%). These three categories define the primary metabolic activities of the nodule and, as such, together constitute more than half of all the proteins identified. Surprisingly, among the more abundant proteins were the enzymes that constitute the three-carbon portion of glycolysis. Glyceraldehyde-3-phosphate dehydrogenase, phosphoglycerate kinase, and enolase plus another abundant protein, poorly annotated, but with homology to phosphoglycerate mutase were prominent proteins. This suggests an active three-carbon metabolism within the plant cells.

Lipoxygenase and fatty acid epoxide hydroxylase were both highly expressed proteins. Lipoxygenase activity has been documented in soybean root nodules previously (Mohammadi and Karr, 2003; Junghans et al., 2004), but not the activity of fatty acid epoxide hydroxylase. An active fatty acid epoxide hydrolase provides a metabolic route, alternative to lipoxygenase, to hydroperoxides, which are intermediates in the formation of traumatic acid or jasmonic acid, respectively involved in cell division and senescence.

Comparison of these proteins with those identified as mRNAs in the transcriptome, as reported by Lee and others (2004), revealed congruence of the following major proteins/transcripts: sucrose synthase, coproporphyrinogenase, leghemoglobin, 14-3-3 proteins, peroxidase, and a resistance protein. That these were identified both in the proteome (Oehrle et al., 2008) and in the transcriptome (Lee et al., 2004) demonstrates the critical role these components perform in nodule functioning.

Enzymatic analysis has been based largely on the known or presumed metabolic activities of the nodule. The nodule receives photosynthetically produced carbohydrates, sucrose, and glucose, from the leaves and metabolizes them to organic acids, which are provided to the bacteroids for deriving the energy and low potential electrons needed by nitrogenase. Anthon and Emerich (1990) reported the developmental profiles of carbon metabolic enzymes in soybean nodules during development and concluded that sucrose was metabolized via the sucrose synthase pathway. The glucose and fructose resulting from the sucrose synthase pathway are metabolized to phosphoenol pyruvate via glycolysis. The three-carbon portion of glycolysis has been demonstrated in soybean nodules by proteomic and enzymatic analyses (Oehrle et al., 2008).

Infected and uninfected nodule cell protoplasts have been separated from each other and analyzed for enzymatic activities (Shelp et al., 1983; Suganuma et al., 1987; Kouchi et al., 1988; Copeland et al., 1989). Glycolytic enzymes were found predominately in the outer cortical and uninfected cells. Enzymes of N-assimilation—glutamine synthetase, glutamate synthase, aspartase, and alanine aminotransferase—were several-fold greater in uninfected cells than in infected cells. Ureide metabolism was found primarily in the uninfected cells. Proteomic and metabolite analyses have not been fully exploited to further our understanding of functional activities within specific cell types of the nodule.

Metabolic Differentiation of the Symbiosome Membrane and Symbiosome Space

The symbiosome can be considered a special form of a lytic compartment of the infected plant cell (Mellor, 1989). The symbiosome membrane is produced from the host plant by membrane components from the Golgi and from the endoplasmic reticulum (ER). The proteins identified in soybean symbiosome membranes by biochemical (Werner, 1992) and by proteomic (Panter et al., 2000; Saalbach et al., 2002) analysis include Nodulin 24, Nodulin 26, Nodulin 53, vacuolar ATPase, thiol protease, subtilisin protease, protein disulfide isomerase, heat shock protein 60, and BiP protein. BiP is a widely distributed and highly conserved ER luminal protein that has been implicated in co-translational folding of nascent polypeptides, and in the recognition and disposal of misfolded polypeptides. Panter and others (2000) suggested that the presence of several proteins associated with the symbiosome membrane that function in protein folding, may indicate that symbiosomes import proteins directly from the cytoplasm. This agrees with Simonsen and Rosendahl (1999) who demonstrated that isolated pea symbiosomes can import plant-translated nodule proteins.

Proteins identified in the soybean symbiosome space via biochemical methods include acid phosphatase, α-mannosidase II, proteases, protease inhibitors, α-glucosidase, aspartate aminotransferase, and trehalase (Werner, 1992). Saalbach and others (2002) used proteomic methods to identify 46 proteins in the symbiosome space of pea nodules. Among them were chaperonins, heat shock proteins, protein disulfide isomerase, and ATP synthase components, which would be expected from the reports on the symbiosome membrane.

The combined biochemical and proteomic analyses demonstrate that the symbiosome membrane and symbiosome space are highly active metabolic compartments of the nodule, performing highly specialized functions essential for symbiotic functioning.

Metabolic Differentiation in Bacteroids

In contrast to the transcriptional and post-translational investigations of nodule plant cells, the metabolic differentiation of bacteroids has been studied largely by enzymatic analysis. The primary function of nodules is to reduce atmospheric N_2 to ammonium. Nitrogenase, expressed in the mature bacteroid, catalyzes the MgATP-dependent reduction of N_2 to ammonia:
$N_2 + 8e^- + 16\ MgATP + 8H^+ \rightarrow 2NH_3 + H_2 + 16\ MgADP + P_i$.

Nitrogenase requires a minimum of 16 moles of ATP per mole of N_2, but estimates of the energy needed for the complete nitrogen fixation process are around 40 moles of ATP per mole of N_2, or in terms of carbon, 6 grams of carbon are required for every gram of N reduced (Vance and Heichel, 1991).

Although there are several different types of nitrogenases, rhizobia possess only the molybdenum-containing type (Howard and Rees, 2006). The Mo nitrogenases are composed of two proteins: a Mo-Fe protein and an Fe protein. The MoFe protein is a 220 to 240 kDa tetramer, an $(\alpha\beta)_2$ complex, of the *nif*D (α-subunit) and *nif*K (β-subunit) gene products each of which contains complex metalloclusters. The Fe protein is an ~60 kDa dimer, product of the *nif*H gene, with a single 4Fe–4S cluster located between the subunits. An MgATP binding site is located on each subunit (Howard and Rees, 2006). During catalysis, electrons are delivered one at a time from the Fe protein to the MoFe protein in a reaction coupled to the hydrolysis of 2 MgATP for each electron transferred (Rees and Howard, 2000).

Chang and others (2007) reported the transcriptomic analysis of free-living *B. japonicum* cells cultured on glycerol versus bacteroids. They found 661 genes significantly upregulated and 573 genes significantly downregulated in the bacteroid. Pessi and others (2007) reported the transcriptomic analysis of bacteroids versus peptone-arabinose-grown cultures of *B. japonicum* and found 692 genes significantly upregulated and 2,086 genes significantly downregulated in the bacteroid. Thus, about a third of the genome participates in establishing and maintaining symbiosis. A large number of differentially expressed genes was not in common to both experimental results reflecting differences in experimental conditions and methodologies. As pointed out by Karunakaran and others (2009), a major problem has been determining which medium is the more appropriate for the cultured cell control relative to the bacteroid for which we have only a rudimentary understanding of its metabolism.

Among the genes found to be upregulated in all experiments are those previously identified for nitrogenase and supporting metabolic activities. In addition, there were genes encoding for fatty acid degradation, benzoate degradation and mannitol utilization. Microarray analysis identified nine significantly upregulated genes involved in benzoate metabolism (Chang et al., 2007; Pessi et al., 2007). However, the largest number of proteins and genes in global analyses are classified as unknown or hypothetical, which indicates there is a great deal yet to be learned about the symbiosis.

Proteomics has recently been performed to characterize the bacteroid state further. A partial proteome map consisting of more than 180 proteins obtained from 2-D PAGE showed that the bacteroid expresses a dominant and elaborate protein network for nitrogen and carbon metabolism supported by a selective group of bacteroid transporters (Sarma and Emerich, 2005). However, they seem to lack a defined fatty acid and nucleic acid metabolism. Interestingly, the proteins related to protein synthesis, scaffolding, and degradation were among the most predominant of the bacteroid proteome. In addition, several proteins, relatively abundant, were identified to be involved with cellular detoxification, stress regulation, and signalling communication components. The analysis matched very well with previous biochemical and genetic reports and clearly showed interconnections among several metabolic pathways that meet the needs of the bacteroid. Unfortunately, the primary energy-generating pathways for nitrogenase in *B. japonicum* bacteroids were not obviously displayed in these analyses.

Enzymatic analyses of bacteroids have demonstrated that they are deficient in enzymes of the Embden-Meyerhof, Entner-Doudoroff, and the pentose phosphate pathways (Copeland et al., 1989). Based on the selective permeability of the bacteroid and symbiosome membranes, bacteroids appear to receive only dicarboxylic acids from the plant (Copeland et al., 1989). There is an extensive base of literature documenting the relationship between symbiotic nitrogen fixation and organic acid metabolism (Copeland et al., 1989; Dunn, 1998), but recent analyses of citric acid cycle mutants demonstrate a clear difference in the role of the citric acid cycle between bacteroids from determinant and indeterminant nodules. Phenotypes of *B. japonicum* mutants lacking isocitrate dehydrogenase (Shah and Emerich, 2006), or α-ketoglutarate dehydrogenase (Green and Emerich, 1997a,b), and the bacteroid proteome (Sarma and Emerich, 2005) both suggest that these enzymes are required for proper infection and nodule colonization, but are not needed to supply energy or high-energy electrons for the actual reduction of atmospheric dinitrogen. In contrast, a complete citric acid cycle is essential for bacteroids in indeterminant nodules (McDermott and Kahn, 1992; Warshaw et al., 1997; Mortimer et al., 1999; Dymov et al., 2004).

Prell and Poole (2006) have proposed that amino acid cycles first proposed by Kahn and others (1985) function as nutrient exchange cycles. An example of such an exchange is the supply of glutamate from the plant and bacteroid-derived alanine or aspartate. Neither an amino acid cycle nor a nutrient-exchange system has yet to be conclusively demonstrated (White et al., 2007), but the concept is appealing as nodule nitrogen and carbon metabolism should be inherently integrated. However, the primary nitrogen and carbon metabolism of leguminous nodules are among the least understood interactions of symbiosis.

Summary

Symbiotic associations account for the majority of the estimated 120 million metric tons per year (Tg/Yr) of biological nitrogen fixation in the biosphere (see Chapter 1). Feeding the world's increasing population will exacerbate the environmental effects of fertilizer nitrogen applications. Thus, it becomes imperative to elucidate the molecular and physiological interactions of symbiotic nitrogen fixation to permit its benefits to be broadly applied to agricultural practices. As evident in this chapter, the infection and metabolic processes are understood, but not at a level of sophistication and detail necessary to design directed molecular enhancements of the process in major leguminous crop plants or to extend the process to nonleguminous crops.

Acknowledgments

The authors would like to thank our numerous collaborators and students involved in this research over the years. Funding was provided by Ministerio de Ciencia e Innovación and Junta de Castilla y León from Spain. DWE received support from the United States Department of Agriculture – NRI program.

References

Amadou, C., Pascal, G., Mangenot, S., Glew, M., Bontemps, C., Capela, D., Carrere, S., Cruveiller, S., Dossat, C., Lajus, A., et al. 2008. Genome sequence of the {beta}-rhizobium *Cupriavidus taiwanensis* and comparative genomics of rhizobia. *Genome Research* 18: 1472–1483.

Andam, C.P., Parker, M.A. 2007. Novel alphaproteobacterial root nodule symbiont associated with *Lupinus texensis*. *Applied and Environmental Microbiology* 73: 5687–5691.

Ané, J.M., Kiss, G.B., Riely, B.K., Penmetsa, R.V., Oldroyd, G.E.D., Ayax, C., Lévy, J., Debellé, F., Baek, J.M., Peter Kalo, P., Rosenberg, C., Bruce A. Roe, B.A., Long, S.R., Dénariré, J., Cook, D.R. 2004. *Medicago truncatula DMI1* required for bacterial and fungal symbioses in legumes. *Science* 303: 1364–1367.

Anthon, G.E., Emerich, D.W. 1990. Developmental regulation of enzymes of sucrose and hexose metabolism in effective and ineffective nodules. *Plant Physiology* 92: 346–351.

Arp, D.J. 2000. "The Nitrogen Cycle." *In Prokaryotic Nitrogen Fixation*, edited by Eric W. Triplett, pp. 1–14. Norfolk: Horizon Scientific Press.

Baldwin, I.L., Fred, E.B. 1929. Nomenclature of the root-nodule bacteria of the leguminosae. *Journal of Bacteriology* 17: 141–150.

Barnett, M.J., Fisher, R.F. 2006. Global gene expression in the rhizobial–legume symbiosis. *Symbiosis* 42: 1–24.

Bender, G., Nayder, M., Goydych, W., Rolfe, B. 1987. Early infection events in the nodulation of the nonlegume *Parasponia Andersonii by Bradyrhizobium*. *Plant Science* 51: 285–293.

Brewin, N. J. 2004. Plant cell wall remodelling in the *Rhizobium*–Legume symbiosis. *Critical Reviews in Plant Sciences* 23: 293–316.

Callaham, D., Torrey, J. 1981. The structural basis for infection of root hairs of *Trifolium repens* by *Rhizobium*. *Canadian Journal of Botany* 59: 1647–1664.

Capoen, W., Den Herder, J., Rombauts, S., De Gussem, J., De Keyser, A., Holster, M., Goormachtig, S. 2007. Comparative transcriptome analysis reveals common and specific tags for root hair and crack-entry invasion in *Sesbania rostrata*. *Plant Physiology* 144: 1878–1889.

Casida, L.E.J. 1982. *Ensifer adhaerens* gen nov. sp. nov.: a bacterial predator of bacteria in soil. *International Journal of Systematic Bacteriology* 32: 339–345.

Chaintreuil, C., Giraud, E., Prin, Y., Lorquin, J., Bâ, A., Gillis, M., de Lajudie, P., Dreyfus, B. 2000. Photosynthetic bradyrhizobia are natural endophytes of the African wild rice *Oryza breviligulata*. *Applied and Environmental Microbiology* 66: 5437–5447.

Chandler, M.R. 1978. Some observations on infection of *Arachis hypogea* L. by *Rhizobium*. *Journal of Experimental Botany* 29: 749–755.
Chang, W.-S., Frank, W.L., Cytryn, E., Jeong, S., Joshi, T., Emerich, D.W., Sadowsky, M.J., Xu, D. Stacey, G. 2007. An oligonucleotide microarray for transcriptional profiling of *Bradyrhizobium japonicum*. *Molecular Plant-Microbe Interactions* 20: 1298–1307.
Chen, W.M., James, E.K., Prescott, A.R., Kierans, M., Sprent, J.I. 2003. Nodulation of *Mimosa* spp. by the beta-proteobacterium *Ralstonia taiwanensis*. *Molecular Plant Microbe Interactions* 16: 1051–1061.
Chen, W.M., James, E.K., Coenye, T., Chou, J.H., Barrios, E., de Faria, S.M., Elliott, G.N., Sheu, S.Y., Sprent, J.I., Vandamme, P. 2006. *Burkholderia mimosarum* sp. nov., isolated from root nodules of *Mimosa* spp. from Taiwan and South America. *International Journal of Systematic and Evolutionary Microbiology* 56: 1847–1851.
Chen, W.M., de Faria, S.M., James, E.K., Elliott, G.N., Lin, K.Y., Chou, J.H., Sheu, S.Y., Cnockaert, M., Sprent, J.I., Vandamme, P. 2007. *Burkholderia nodosa* sp. nov., isolated from root nodules of the woody Brazilian legumes *Mimosa bimucronata* and *Mimosa scabrella*. *International Journal of Systematic and Evolutionary Microbiology* 57: 1055–1059.
Chen, W.M., de Faria, S.M., Chou, J.H., James, E.K., Elliott, G.N., Sprent, J.I., Bontemps, C., Young, J.P., Vandamme, P. 2008. *Burkholderia sabiae* sp. nov., isolated from root nodules of *Mimosa caesalpiniifolia*. *International Journal of Systematic and Evolutionary Microbiology* 58: 2174–2179.
Chi, F., Shen, S.H., Cheng, H.P., Jing, Y.X., Yanni, Y.G., Dazzo, F.B. 2005. Ascending migration of endophytic rhizobia, from roots to leaves, inside rice plants and assessment of benefits to rice growth physiology. *Applied and Environmental Microbiology* 71: 7271–7278.
Conn, H.J. 1938. Taxonomic relationships of certain non-sporeforming rods in soil. *Journal of Bacteriology* 36: 320–321.
Copeland, L., Vella, J., Hong, Z. 1989. Enzymes of carbohydrate metabolism in soybean nodules. *Phytochemistry* 28: 57–61.
Dazzo, F., Hubbell, D. 1975. Cross-reactive antigens and lectin as determinants of symbiotic specificity in the *Rhizobium*-clover association. *Applied Microbiology* 30: 1013–1022.
Dazzo, F., Truchet, G., Sherwood, J., Hrabak, E., Abe, M., Pankratz, H.S. 1984. Specific phases of root hair attachment in the *Rhizobium trifolii*-clover symbiosis. *Applied and Environmental Microbiology* 48: 1140–1150.
Dazzo, F.B., Truchet, G.L., Hollingsworth, R.I., Hrabak, E.S., Pankratz, H.S., Philip-Hollingsworth, S., Salzwedel, J.L., Chapman, K., Appenzeller, L., Squartini, A., Gerhold, D., Orgambide, O. 1991. Rhizobium lipopolysaccharide modulates infection thread development in white clover root hairs. *Journal of Bacteriology* 173: 5371–5384.
Dazzo, F.B., Orgambide, G., Philip-Hollingsworth, S., Hollingsworth, R.I., Ninke, K., Salzwedel, J.L. 1996. Modulation of development, growth dynamics, wall crystallinity, and infection thread formation in white clover root hairs by membrane chitolipooligosaccharides from *Rhizobium leguminosarum* bv. *trifolii*. *Journal of Bacteriology* 178: 3621–3627.
Dazzo, F.B., Wopereis, J.L. 2000. "Unraveling the infection process in the *Rhizobium*-Legume symbiosis by microscopy." In: *Prokaryotic Nitrogen Fixation*, edited by Eric W. Triplett, pp. 295–363. Norfolk: Horizon Scientific Press.
De Hoff, P.L., Brill, L.M., Hirsch, A.M. 2009. Plant lectins: the ties that bind in root symbiosis and plant defense. *Molecular Genetics and Genomics* 282: 1–15.
Dunn, M.F. 1998. Tricarboxylic acid cycle and anaplerotic enzymes in rhizobia. *FEMS Micrbiological Reviews* 22: 105–123.

Dymov, S.I., Meek, D.J.J., Steven, B., Driscoll, B.T. 2004. Insertion of transposon Tn5tac1 in the *Sinorhizobium meliloti* malate dehydrogenase (*mdh*) gene results in conditional polar effects on downstream TCA genes. *Molecular Plant-Microbe Interactions* 17: 1318–1327.

Elliott, G.N., Chen, W.M., Bontemps, C., Chou, J.H., Young, J.P., Sprent, J.I., James, E.K. 2007. Nodulation of *Cyclopia* spp. (Leguminosae, Papilionoideae) by *Burkholderia tuberum*. *Annales of Botany (London)* 100: 1403–1411.

Elliott, G.N., Chou, J.H., Chen, W.M., Bloemberg, G.V., Bontemps, C., Martínez-Romero, E., Velázquez, E., Young, J.P., Sprent, J.I., James, E.K. 2009. *Burkholderia* spp. are the most competitive symbionts of *Mimosa*, particularly under N-limited conditions. *Environmental Microbiology* 11: 762–778.

Fischer, H.M. 1994. Genetic regulation of nitrogen fixation in rhizobia. *Microbiology and Molecular Biology Reviews* 58: 352–386.

Frank, B. 1889. Über die Pilzsymbiose der Leguminosen. *Berichte der Deutschen Botanischen Gesellschaft* 7: 332–346.

Franssen, H.J., Nap, J.P., Bisseling, T. 1992. Nodulins in root development. In: *Biological Nitrogen Fixation*, edited by G. Stacey, R.H. Burris, H.J. Evans, pp. 598–624, Chapman & Hall, New York.

Gage, D.J. 2002. Analysis of infection thread development using Gfp- and DsRed-expressing *Sinorhizobium meliloti*. *Journal of Bacteriology* 184: 7042–7046.

Gage, D.J. 2004. Infection and invasion of roots by symbiotic, nitrogen-fixing rhizobia during nodulation of temperate legumes. *Microbiology and Molecular Biology Reviews* 68: 280–230.

García-Fraile, P., Rivas, R., Willems, A., Peix, A., Martens, M., Martínez-Molina, E., Mateos, P.F., Velázquez, E. 2007. *Rhizobium cellulosilyticum* sp. nov., isolated from sawdust of *Populus alba*. *International Journal of Systematic and Evolutionary Microbiology* 57: 844–848.

Gaunt, M.W., Turner, S.L., Rigottier-Gois, L., Lloyd-Macgilp, S.A., Young, J.P. 2001. Phylogenies of *atpD* and *recA* support the small subunit rRNA-based classification of rhizobia. *International Journal of Systematic and Evolutionary Microbiology* 51: 2037–2048.

Ghosh, W., Roy P. 2006. *Mesorhizobium thiogangeticum* sp. nov., a novel sulfur-oxidizing chemolithoautotroph from rhizosphere soil of an Indian tropical leguminous plant. *International Journal of Systematic Bacteriology* 56: 91–7.

Gibson, K.E., Kobayashi, H., Walker, G.C. 2008. Molecular determinants of a symbiotic chronic infection. *Annual Review of Genetics* 42: 413–441.

Giraud, E., Moulin, L., Vallenet, D., Barbe, V., Cytryn, E., Avarre, J.C., Jaubert, M., Simon, D., Cartieaux, F., Prin, Y., Bena, G., Hannibal, L., Fardoux, J., Kojadinovic, M., Vuillet, L., Lajus, A., Cruveiller, S., Rouy, Z., Mangenot, S., Segurens, B., Dossat, C., Franck, W.L., Chang, W.S., Saunders, E., Bruce, D., Richardson, P., Normand, P., Dreyfus, B., Pignol, D., Stacey, G., Emerich, D., Verméglio, A., Médigue, C., Sadowsky, M. 2007. Legumes symbioses: absence of Nod genes in photosynthetic bradyrhizobia. *Science* 316: 1307–1312.

Goormachtig, S., Capoen, W., James, E.K., Holsters, M. 2004. Switch from intracellular to intercellular invasion during water stress-tolerant legume nodulation. *Proceedings of the National Academy of Sciences, USA* 101: 6303–6308.

Graham, P.H., Sadowsky, M.J., Keyser, H.H., Barnet, Y.M., Bradley, R.S., Cooper, J.E., de Ley, D.J., Jarvis, B.D.W., Roslycky, E.B., Strijdom, B.W., Young, J.P.W. 1991. Proposed minimal standards for the description of new genera and species of root- and stem-nodulating bacteria. *International Journal of Systematic Bacteriology* 41: 582–587.

Green, L.S., Emerich, D.W. 1997a. *Bradyrhizobium japonicum* does not require α-ketoglutarate dehydrogenase for growth on succinate and malate. *Journal of Bacteriology* 179: 194–201.

Green, L.S., Emerich, D.W. 1997b. The formation of nitrogen-fixing bacteroids is delayed but not abolished in soybean infected by α-ketoglutarate dehydrogenase-deficient mutant of *Bradyrhizobium japonicum*. *Plant Physiology* 114: 13159–1368.

Gulash, M., Ames, P., Larosiliere, R.C., Bergman, K. 1984. Rhizobia are attracted to localized sites on legume roots. *Applied and Environmental Microbiology* 48: 149–152.

Gutiérrez-Zamora, M.L., Martínez-Romero, E. 2001. Natural endophytic association between *Rhizobium etli* and maize (*Zea mays* L.). *Journal of Biotechnology* 91: 117–126.

Howard, J.B., Rees, D.C. 2006. How many metals does it take to fix N_2? A mechanistic overview of biological nitrogen fixation. *Proceedings of the National Academy of Science USA* 103: 17088–17093.

Huang, B., Lü, C., Wu, B., Fan L. 2007. A rhizobia strain isolated from root nodule of gymnosperm *Podocarpus macrophyllus*. *Science in China Series C Life Science* 50: 228–233.

Hurek, T., Handley, L.L., Reinhold-Hurek, B., Piche, Y. 2002. Azoarcus grass endophytes contribute fixed nitrogen to the plant in an unculturable state. *Molecular Plant Microbe Interactions* 15: 233–242.

Ibáñez, F., Angelini, J., Taurian, T., Tonelli, M.L., Fabra, A. 2009. Endophytic occupation of peanut root nodules by opportunistic Gammaproteobacteria. *Systematic and Applied Microbiology* 32: 49–55.

Jones, K.M., Kobayashi, H., Davies, B.W., Taga, M.E., Walker, G.H. 2007. How rhizobial symbionts invade plants: the *Sinorhizobium-Medicago* model. *Nature Reviews Microbiology* 5: 619–633.

Jones, K.M., Sharopova, N., Lohar, D.P., Zhang, J.Q., VandenBosch, K.A., Walker, G.C. 2008. Differential response of the plant *Medicago truncatula* to its symbiont *Sinorhizobium meliloti* or an exopolysaccharide-deficient mutant. *Proceedings of the National Academy of Sciences, USA* 105: 704–709.

Jordan, D.C. 1982. Transfer of *Rhizobium japonicum* Buchanan 1980 to *Bradyrhizobium* gen. nov., a genus of slow-growing, root nodule bacteria from leguminous plants. *International Journal of Systematic Bacteriology* 32: 136–139.

Jourand, P., Giraud, E., Béna, G., Sy, A., Willems, A., Gillis, M., Dreyfus, B., de Lajudie, P. 2004. *Methylobacterium nodulans* sp. nov., for a group of aerobic, facultatively methylotrophic, legume root-nodule-forming and nitrogen-fixing bacteria. *International Journal of Systematic and Evolutionary Microbiology* 54: 2269–2273.

Judicial Commission of the International Committee on Systematics of Prokaryotes. 2008. The genus name *Sinorhizobium* Chen et al. 1988 is a later synonym of *Ensifer* Casida 1982 and is not conserved over the latter genus name, and the species name '*Sinorhizobium adhaerens*' is not validly published. Opinion 84. *International Journal of Systematic and Evolutionary Microbiology* 58: 1973.

Junghans, T.G., de Almeida-Oliveira, M.G., Moreira, M.A. 2004. Lipoxygenase activities during development of root and nodule of sobyean. *Pesquisa Agropecuaria Brasileira* 39: 625–630.

Kahn, M., Kraus, J. and Sommerville, J.E. 1985. "A model of nutrient exchange in the Rhizobium-legume simbiosis" In: *Nitrogen Fixation Research Progress*, edited by H.J. Evans, P.J., Bottomley, & W.E. Newton. Martinus Nijhoff, Dordrecht, pp. 193–199.

Kan, F.L., Chen, Z.Y., Wang, E.T., Tian, C.F., Sui, X.H., Chen, W.X. 2007. Characterization of symbiotic and endophytic bacteria isolated from root nodules of herbaceous legumes grown in Qinghai-Tibet plateau and in other zones of China. *Archives of Microbiology*, 188: 103–115.

Karunakaran, R., Ramachandran, V.K., Seaman, J.C., East, A.K., Mouhsine, B., Mauchline, T.H., Prell, J., Skeffington, A., Poole, P.S. 2009. Transcriptomic analysis of *Rhizobium leguminosarum* biovar viciae in symbiosis with host plants *Pisum sativum* and *Vicia cracca*. J. Bacteriol. 191: 4002–4014.

Kouchi, H., Fukai, K., Katagin, H., Minamisawa, K., Tajima, S. 1988. Isolation and enzymological characteization of infected and uninfected cell protoplasts from root nodules of *Glycine max*. *Physiologium Plantarum* 73: 327–334.

Kuykendall, L.D. 2005. Order VI. Rhizobiales ord. Nov. In: *Bergey's Manual of Systematic Bacteriology*, second edition, vol. 2 (The Proteobacteria), part C (The Alpha-, Beta-, Delta-, and Epsilonproteobacteria), edited by D.J. Brenner, N.R. Krieg, J.T. Staley and G. M. Garrity, p. 324. Springer, New York.

Lafay, B., Bullier, E., Burdon, J.J. 2006. Bradyrhizobia isolated from root nodules of *Parasponia* (*Ulmaceae*) do not constitute a separate coherent lineage. *International Journal of Systematic and Evolutionary Microbiology* 56: 1013–1018.

Lee, H., Hur, C.-G., Oh, C.J., Ho, B.K., Park, S-Y., An, C.S. 2004. Analysis of the root nodule-enhanced transcriptome in soybean. *Molecules and Cells* 18: 53–62.

Lévy, J., Bres, C., Geurts, R., Chalhoub, B., Kulikova, O., Duc, G., Journet, E.P., Ané, J.M., Lauber, E., Bisseling, T., Dénarié, J., Rosenberg, C., Debellé, F. 2004. A putative Ca^{2+} and calmodulin-dependent protein kinase required for bacterial and fungal symbioses. *Science* 303: 1361–1364.

Limpens, E., Franken, C., Smit, P., Willemse, J., Bisseling, T., Geurts, R. 2003. LysM domain receptor kinases regulating rhizobial nod factor-Induced infection. *Science* 302: 630–633.

Limpens, E., Mirabella, R., Fedorova, E., Franken, C., Franssen, H., Bisseling, T., Geurts, R. 2005. Formation of organelle-like N2-fixing symbiosomes in legume root nodules is controlled by DMI2. *Proceedings of the National Academy of Sciences, USA* 102: 10375–10380.

Maiden, M.C.J. 2006. Multilocus sequence typing of bacteria. *Annual Review Microbiology* 60: 561–588.

Martens, M., Delaere, M., Coopman, R., de Vos, P., Gillis, M., Willems, A. 2007 Multilocus sequence analysis of *Ensifer* and related taxa. *International Journal of Systematic and Evolutionary Microbiology* 57: 489–503.

Martínez-Romero, E., Segovia, L., Mercante, F.M., Franco, A.A., Graham, P., Pardo, M.A. 1991. *Rhizobium tropici*, a novel species nodulating *Phaseolus vulgaris* L. beans and *Leucaena* sp. trees. *International Journal of Systematic Bacteriology* 41: 417–426.

Mateos, P.F., Baker, D., Philip-Hollingsworth, S., Squartini, A., Peruffo, A., Nuti, M., Dazzo, F. B. 1995. Direct *in situ* identification of cellulose microfibrils associated with *Rhizobium leguminosarum* biovar *trifolii* attached to the root epidermis of white clover. *Canadian Journal of Microbiology* 41: 202–207.

McDermott, T.R., Kahn, M.L. 1992. Cloning and mutagenesis of the *Rhizobium meliloti* isocitrate dehydrogenase gene. *Journal of Bacteriology* 174: 4790–4797.

Mellor, R.B. 1989. Bacteroids in the *Rhizobium*-legume symbiosis inhabit a plant internal lytic compartment: Implications for other microbial endosymbioses. *Journal of Experimental Botany* 40: 831–839.

Mhamdi, R., Mrabet, M., Laguerre, G., Tiwari, R., Aouani, M.E. 2005. Colonization of *Phaseolus vulgaris* nodules by *Agrobacterium*-like strains. *Canadian Journal of Microbiology* 51: 105–111.

Miller, L.D., Yost, C.K., Hynes, M.F., Alexandre, G. 2007. The major chemotaxis gene cluster of *Rhizobium leguminosarum* bv. viciae is essential for competitive nodulation. *Molecular Microbiology* 63: 348–362.

Mohammadi, M., Karr, A.L. 2003. Induced lipoxygenases in soybean root nodules. *Plant Science* 164: 471–479.

Mooney, B.P., Thelen, J.J. 2004. High-throughput peptide mass fingerprinting of soybean seed proteins: automated workflow and utility of UniGene expressed sequence tag databases for protein identfication. *Phytochemistry* 65: 1733–1744.

Mortimer, M.W., McDermott,T.R., York, G.M., Walker, G.C., Kahn, M.L. 1999. Citrate synthase mutants of *Sinorhizobium meliloti* are ineffective and have altered cell surface polysaccharides. *Journal of Bacteriology* 181: 7608–7613.

Moulin, L., Munive, A., Dreyfus, B., Boivin-Masson, C. 2001. Nodulation of legumes by members of the beta-subclass of Proteobacteria. *Nature* 411: 948–950. Erratum in: *Nature* 2001 412: 926.

Muresu, R., Polone, E., Sulas, L., Baldan, B., Tondello, A., Delogu, G., Cappuccinelli, P., Alberghini, S., Benhizia, Y., Benhizia, H., Benguedouar, A., Mori, B., Calamassi, R., Dazzo, F.B., Squartini, A. 2008. Coexistence of predominantly nonculturable rhizobia with diverse, endophytic bacterial taxa within nodules of wild legumes. *FEMS Microbiology Ecology* 63: 383–400.

Napoli, C., Dazzo, F., Hubbell, D. 1975. Production of cellulose microfibrils by *Rhizobium*. *Applied Microbiology* 30: 123–131.

Newcomb, E.H. 1981. Nodule morphogenesis and differentiation. *International Review of Cytology Supplement* 13: 247–297.

Ngom, A., Nakagawa, Y., Sawada, H., Tsukahara, J., Wakabayashi, S., Uchiumi, T., Nuntagij, A., Kotepong, S., Suzuki, A., Higashi, S., Abe, M. 2004. A novel symbiotic nitrogen-fixing member of the *Ochrobactrum* clade isolated from root nodules of *Acacia mangium*. *Journal of General and Applied Microbiology* 50: 17–27.

Oehrle, N.W., Sarma, A.D., Waters, J.K., Emerich, D.W. 2008. Proteomic analysis of soybean nodule cytosol. *Phytochemistry* 69: 2426–2438.

Oldroyd, G.E.D., Engstrom E.M., Long, S.R. 2001. Ethylene inhibits the Nod factor signal transduction pathway of *Medicago truncatula*. *Plant Cell* 13: 1835–1849.

Orgambide, G., Philip-Hollingsworth S., Mateos P., Hollingsworth R.I., Dazzo F.B. 1996. Subnanomolar concentrations of membrane chitolipooligosaccharides from *Rhizobium leguminosarum* biovar *trifolii* are fully capable of eliciting symbiosis-related responses on white clover. *Plant Soil* 186: 93–98.

Panter, S., Thomson, R., de Bruxelles, G., Laver, D., Trevaskis, B., Udvardi, M. 2000. Identification of proteomics of novel proteins associated with the peribacteroid membrane of soybean root nodules. *Molecular Plant-Microbe Interactions* 13: 325–333.

Perret, X., Staehelin, C., Broughton, W.J. 2000. Molecular basis of symbiotic promiscuity. *Microbiology and Molecular Biology Reviews* 64: 180–201.

Perrine-Walker, F.M., Prayitno, J., Rolfe, B.G., Weinman, J.J., Hocart, C.H. 2007. Infection process and the interaction of rice roots with rhizobia. *Journal of Experimental Botany* 58: 3343–3350.

Pessi, G., Ahrens, C.H., Rehrauer, H., Lindemann, A., Hauser, F., Fischer, H.-M., Hennecke, H. 2007. Genome-wide transcript analysis of *Bradyrhizobium japonicum* bacteroids in soybean root nodules. *Molecular Plant-Microbe Interactions* 20: 1353–1363.

Polhill, R.M., Raven, P.H., Stirton, C.H. 1981. Evolution and systematics of the Leguminosae. In: *Advances in Legume Systematics* Part 1, edited by R.M. Polhill, P.H. and Raven, pp. 1–26. Royal Botanic Gardens, Kew, UK.

Prell, J., and Poole, P.S. 2006. Metabolic changes of rhizobia in legume nodules. *Trends Microbiology* 14: 161–168.

Pueppke, S.G., Broughton, W.J. 1999. *Rhizobium* sp. strain NGR234 and *R. fredii* USDA257 share exceptionally broad, nested host ranges. *Molecular Plant–Microbe Interactions* 12: 293–318.

Rae, A.L., Bonfante-Fasolo P., Brewin, N.J. 1992. Structure and growth of infection threads in the legume symbiosis with *Rhizobium leguminosarum*. *The Plant Journal* 2: 385–395.

Ramírez-Bahena, M.H., García-Fraile, P., Peix, A., Valverde, A., Rivas, R., Igual, J.M., Mateos, P.F., Martínez-Molina, E., Velázquez, E. 2008. Revision of the taxonomic status of the species *Rhizobium leguminosarum* (Frank 1879) Frank 1889AL, *Rhizobium phaseoli* Dangeard 1926AL and *Rhizobium trifolii* Dangeard 1926AL. *R. trifolii* is a later synonym of *R. leguminosarum*. Reclassification of the strain *R. leguminosarum* DSM 30132 (=NCIMB 11478) as *Rhizobium pisi* sp. nov. *International Journal of Systematic and Evolutionary Microbiology* 58: 2484–2490.

Rasolomampianina, R., Bailly, X., Fetiarison, R. et al. 2005. Nitrogen-fixing nodules from rose wood legume trees (*Dalbergia* spp.) endemic to Madagascar host seven different genera belonging to α- and β-Proteobacteria. *Molecular Ecology* 14: 4135–4146.

Rees, D.C., Howard, J.B. 2000. Nitrogenase: standing at the crossroads. *Current Opinion in Chemical Biology* 4: 559–566.

Rivas, R., Velázquez, E., Willems, A., Vizcaíno, N., Subba-Rao, N.S., Mateos, P.F., Gillis, M., Dazzo, F.B., Martínez-Molina, E. 2002. A new species of *Devosia* that forms a unique nitrogen-fixing root-nodule symbiosis with the aquatic legume *Neptunia natans* (L.f.) Druce. *Applied and Environmental Microbiology* 68: 5217–5222.

Rivas, R., Willems, A., Subba-Rao, N.S., Mateos, P.F., Dazzo, F.B., Kroppenstedt, R.M., Martínez-Molina, E., Gillis, M., Velázquez, E. 2003. Description of *Devosia neptuniae* sp. nov. that nodulates and fixes nitrogen in symbiosis with *Neptunia natans*, an aquatic legume from India. *Systematic and Applied Microbiology* 26: 47–53.

Rivas, R., Willems, A., Palomo, J.L., García-Benavides, P., Mateos, P.F., Martínez-Molina, E., Gillis, M., Velázquez, E. 2004. *Bradyrhizobium betae* sp. nov., isolated from roots of *Beta vulgaris* affected by tumour-like deformations. *International Journal of Systematic and Evolutionary Microbiology* 54: 1271–1275.

Rivas, R., Martens, M., de Lajudie, P., Willems, A. 2009. Multilocus sequence analysis of the genus *Bradyrhizobium*. *Systematic and Applied Microbiology* 32: 101–110.

Robledo, M., Jiménez-Zurdo, J.I., Velázquez, E., Trujillo, M.E., Zurdo-Piñeiro, J.L., Ramírez-Bahena, M.H., Ramos, B., Díaz-Mínguez, J.M., Dazzo, F., Martínez-Molina, E., Mateos, P.F. 2008. *Rhizobium* cellulase CelC2 is essential for primary symbiotic infection of legume host roots. *Proceedings of the National Academy of Sciences, USA* 105: 7064–7069.

Rolfe, B.G., Carlson, R.W., Ridge, R.W., Dazzo, F.B., Mateos, P.F., Pankhurst, C.E. 1996. Defective infection and nodulation of clovers by exopolysaccharide mutants of *Rhizobium leguminosarum* bv. *trifolii*. *Functional Plant Biology* 23: 285–303.

Rosenblueth, M., Martínez-Romero, E. 2004. *Rhizobium etli* maize populations and their competitiveness for root colonization. *Archives of Microbiology* 181: 337–44.

Rosenblueth, M., Martínez-Romero, E. 2006. Bacterial endophytes and their interactions with hosts. *Molecular Plant Microbe Interactions* 19: 827–837.

Ryan, R.P., Germaine, K., Franks, A., Ryan, D.J., Dowling, D.N. 2008. Bacterial endophytes: recent developments and applications. *FEMS Microbiology Letters* 278: 1–9.

Saalbach, G., Erik, P., Weinkoop, S. 2002. Characterisation by proteomics of peribacteroid space and peribaceroid membrane preparations from pea (*Pisum sativum*) symbiosomes. *Proteomics* 2: 325–337.
Sarma, A.D, Emerich, D.W. 2005. Global protein expression pattern of *Bradyrhizobium japonicum* bacteroids: A prelude to functional proteomics. *Proteomics* 5: 4170–4184.
Senthilkumar, M., Madhaiyan, M., Sundaram, S.P., Sangeetha, H., Kannaiyan, S. 2008. Induction of endophytic colonization in rice (*Oryza sativa* L.) tissue culture plants by *Azorhizobium caulinodans*. *Biotechnology Letters* 30: 1477–1487.
Shah, R., D.W. Emerich. 2006. Isocitrate Dehydrogenase of *Bradyrhizobium japonicum* is not required for symbiotic nitrogen fixation with soybean. *Journal of Bacteriology* 188: 7600–7608.
Shelp, B.J., Atkins, C.A., Storer, P.J., Canvin, D.T. 1983. Cellular and subcellular organization of pathways of ammonia assimilation and ureide synthesis in nodules of cowpea (*Vigna unguiculata* L. Walp). *Archives of Biochemistry and Biophysics* 224: 429–441.
Simonsen, A.C.W., L. Rosendahl. 1999. Origin of *de novo* synthesized proteins in the different compartments of pea-*Rhizboium* sp. symbiosomes. *Molecluar Plant-Microbe Interactions* 12: 319–327.
Singh, R.K., Mishra, R.P., Jaiswal, H.K., Kumar, V., Pandey, S.P., Rao, S.B., Annapurna, K. 2006. Isolation and identification of natural endophytic rhizobia from rice (*Oryza sativa* L.) through rDNA PCR-RFLP and sequence analysis. *Current Microbiology* 52: 345–349.
Smil, V. 2001. Enriching the earth: Fritz Haber, Carl Bosch and the transformation of world food production. Cambridge, MA: MIT Press.
Smit, G., Swart, S., Lugtenberg, B., Kijne, J.W. 1992. Molecular mechanisms of attachment of bacteria to plants roots. *Molecular Microbiology* 6: 2897–2903.
Stackebrandt, E., Frederiksen, W., Garrity, G.M., Grimont, P.A., Kämpfer, P., Maiden, M.C., Nesme, X., Rosselló-Mora, R., Swings, J., Trüper, H.G., Vauterin, L., Ward, A.C., Whitman, W.B. 2002. Report of the ad hoc committee for the re-evaluation of the species definition in bacteriology. *International Journal of Systematic and Evolutionary Microbiology* 52: 1043–1047.
Subba-Rao, N.S., Mateos, P.F., Baker, D., Pankrazt, H., Palma, J., Dazzo, F.B., Sprent, J.I. 1995. The unique root-nodule symbiosis between *Rhizobium* and the aquatic legume *Neptunia natans* (L.f.) Druce. *Planta* 196: 311–320.
Suganuma, N., Kitou, M., Yamamoto, Y. 1987. Carbon metabolism in relation to cellular organization of soybean root nodules and respiration of mitochondria aided by leghemoglobin. *Plant and Cell Physiology* 28: 113–122.
Sy, A., Giraud, E., Jourand, P., García, N., Willems, A., de Lajudie, P., Prin, Y., Neyra, M., Gillis, M., Boivin-Masson, C., Dreyfus, B. 2001. Methylotrophic *Methylobacterium* bacteria nodulate and fix nitrogen in symbiosis with legumes. *Journal of Bacteriology* 183: 214–220.
Tiwary, B.N., Prasad, B., Ghosh, A., Kumar, S., Jain, R.K. 2007. Characterization of two novel biovar of *Agrobacterium tumefaciens* isolated from root nodules of *Vicia faba*. *Current Microbiology* 55: 328–333.
Torres, A.R., Araújo, W.L., Cursino, L., Hungria, M., Plotegher, F., Mostasso, F.L., Azevedo, J.L. 2008. Diversity of endophytic enterobacteria associated with different host plants. *Journal of Microbiology* 46: 373–379.
Trinick, M.J. 1979. Structure of nitrogen-fixing nodules formed by *Rhizobium* on roots of *Parasponia andersonii* Planch. *Canadian Journal of Microbiology* 25: 565–578.

Trinick, M.J., Hadobas, P.A. 1989. Competition by *Bradyrhizobium* strains for nodulation of the Nonlegume *Parasponia andersonii*. *Applied and Environmental Microbiology* 55: 1242–1248.

Trujillo, M.E., Willems, A., Abril, A., Planchuelo, A.M., Rivas, R., Ludeña, D., Mateos, P.F., Martínez-Molina, E., Velázquez, E. 2005. Nodulation of *Lupinus albus* by strains of *Ochrobactrum lupini* sp. nov. *Applied and Environmental Microbiology* 71: 1318–327.

Validation List No. 107. 2006. List of new names and new combinations previously effectively, but not validly, published. *International Journal of Systematic and Evolutionary Microbiology* 56: 1–6.

Valverde, A., Velázquez, E., Fernández-Santos, F., Vizcaíno, N., Rivas, R., Mateos, P.F., Martínez-Molina, E., Igual, J.M., Willems, A. 2005. *Phyllobacterium trifolii* sp. nov., nodulating *Trifolium* and *Lupinus* in Spanish soils. *International Journal of Systematic and Evolutionary Microbiology* 55: 1985–1989.

Valverde, A., Igual, J.M., Peix, A., Cervantes, E., Velázquez, E. 2006. *Rhizobium lusitanum* sp. nov. a bacterium that nodulates *Phaseolus vulgaris*. *International Journal of Systematic and Evolutionary Microbiology* 56: 2631–2637.

van Brussel, A.A.N., Bakhuizen, R., van Spronsen, P.C., Spaink, H.P., Tak, T., Lugtenberg, B.J.J., Kijne, J.W. 1992. Induction of pre-infection thread structures in the leguminous host plant by mitogenic lipo-oligosaccharides of *Rhizobium*. *Science* 257: 70–72.

van Berkum, P., Eardly, B.D. 2002. The aquatic budding bacterium *Blastobacter denitrificans* is a nitrogen-fixing symbiont of *Aeschynomene indica*. *Applied and Environmental Microbiology* 68:1132–1136.

van Berkum, P., Elia, P., Eardly, B.D. 2006. Multilocus sequence typing as an approach for population analysis of *Medicago*-nodulating rhizobia. *Journal of Bacteriology* 188: 5570–5577.

van Spronsen, P.C., Bakhuizen, R., van Brussel, A.A., Kijne, J.W. 1994. Cell-wall degradation during infection thread formation by the root-nodule bacterium *Rhizobium leguminosarum* is a 2-step process. *European Journal of Cell Biology* 64: 88–94.

Vance, C.P., Heichel, G.H. 1991. Carbon in N_2 fixation: Limitation or exquisite adaptation. *Annual Reviews of Plant Physiology and Plant Molecular Biology* 42: 373–392.

Vandamme, P., Goris, J., Chen, W.M., de Vos, P., Willems, A. 2002. *Burkholderia tuberum* sp. nov. and *Burkholderia phymatum* sp. nov., nodulate the roots of tropical legumes. *Systematic and Applied Microbiology* 25: 507–512.

Vandamme, P., Coenye, T. 2004. Taxonomy of the genus Cupriavidus: a tale of lost and found. *International Journal of Systematic and Evolutionary Microbiology* 54: 2285–2289.

Vasse, J., de Billy, F., Camut, S., Truchet, G. 1990. Correlation between ultrastructural differentiation of bacteroids and nitrogen fixation in alfalfa nodules. *Journal of Bacteriology* 172: 4295–4306.

Vega-Hernández, M.C., Pérez-Galdona, R., Dazzo, F.B., Jarabo-Lorenzo, A., Alfayate, M.C., León-Barrios, M. 2001. Novel infection process in the indeterminate root nodule symbiosis between *Chamaecytisus proliferus* (tagasaste) and *Bradyrhizobium* sp. *New Phytologist* 150: 707–721.

Velázquez, E., Peix, A., Zurdo-Piñeiro, J.L., Palomo, J.L., Mateos, P.F., Rivas, R., Muñoz-Adelantado, E., Toro, N., García-Benavides, P., Martínez-Molina, E. 2005. The coexistence of symbiosis and pathogenicity-determining genes in *Rhizobium rhizogenes* strains enables them to induce nodules and tumors or hairy roots in plants. *Molecular Plant Microbe Interactions* 18: 1325–1332.

Velázquez, E., Rojas, M., Lorite, M.J., Rivas, R., Zurdo-Piñeiro, J.L., Heydrich, M., Bedmar, E.J. 2008. Genetic diversity of endophytic bacteria which could be find in the apoplastic sap of the medullary parenchym of the stem of healthy sugarcane plants. *Journal of Basic Microbiology* 48: 118–124.

Vincent, J.M. 1970. *A Manual for the Practical Study of Root Nodule Bacteria*. Oxford: Blackwell Scientific.

Wang, L.L., Wang, E.T., Liu, J., Li, Y., Chen, W.X. 2006. Endophytic occupation of root nodules and roots of *Melilotus dentatus* by *Agrobacterium tumefaciens*. *Microbial Ecology* 52: 436–443.

Warshaw, D.L., Wilkinson, A., Mundy, M. Shith, M., Poole, P.S. 1997. Regulation of the TCA cycle and the general amino acaid permease by overflow metabolism in *Rhizobium leguminosarum*. *Microbiology* 143: 2209–2221.

Werner, D. 1992. Physiology of Nitrogen-Fixing Legume Nodules: Compartments and Functions. In *Biological Nitrogen Fixation*, edited by G. Stacey, R.H. Burris, Evans, H.J., pp. 399–431. Chapman & Hall, New York.

White, J., Prell, J., James, E.K., Poole, P. 2007. Nutrient sharing between symbionts. Plant Physiol. 144: 604–614.

Willems, A., Collins, M.D. 1993. Phylogenetic analysis of rhizobia and agrobacteria based on 16S rRNA gene sequences. *International Journal of Systematic Bacteriology* 43: 305–313.

Woese, C.R., Stackebrandt, E., Weisburg, W.G., Paster, B.J., Madigan, M.T., Fowler, V.J., Hahn, C.M., Blanz, P., Gupta, R., Nealson, K.H., Fox, G.E. 1984. The phylogeny of purple bacteria: The alpha subdivision. *Systematic Applied Microbiology* 5: 315–326.

Yanagi, M., Yamasato, K. 1993. Phylogenetic analysis of the family *Rhizobiaceae* and related bacteria by sequencing of 16S rRNA gene using PCR and DNA sequencer. *FEMS Microbiology Letters* 107: 115–20.

Yanni, Y., Rizk, R., Corich, V., Squartini, A., Ninke, K., Philip-Hollingsworth, S., Orgambide, G., deBruijn, F., Stoltzfus, J., Buckley, D., Schmidt, T., Mateos, P., Ladha, J.K., Dazzo, F.B. 1997. Natural endophytic association between *Rhizobium leguminosarum* bv. trifolii and rice roots and assessment of its potential to promote rice growth. *Plant Soil* 194: 99–114.

Yanni, Y., Rizk, R., Abd-El Fattah, F., Squartini, A., Corich, V., de Bruijn, F., Rademaker, J., Maya-Flores, J., Ostrom, P., Vega-Hernández, M., Hollingsworth, R., Martínez-Molina, E., Mateos, P., Velázquez, E., Wopereis, J., Triplett, E., Umali-García, U., Rolfe, B., Ladha, J.K., Hill, J., Dazzo, F.B. 2001. The beneficial plant growth-promoting association of *Rhizobium leguminosarum* bv. *trifolii* with rice roots. *Australian Journal of Plant Physiology* 28: 845–870.

Yost, C.K., Hynes, M.F. 2000. "Rhizobial motility and chemotaxis: Molecular biology and ecological role." In: *Prokaryotic Nitrogen Fixation*, edited by Eric W. Triplett, pp. 237–250. Norfolk: Horizon Scientific Press.

Young, J.M., Kuykendall, L.D., Martínez-Romero, E., Kerr, A., Sawada, H. 2001. A revision of *Rhizobium* Frank 1889, with an emended description of the genus, and the inclusion of all species of *Agrobacterium* Conn 1942 and *Allorhizobium undicola* de Lajudie et al. 1998 as new combinations: *Rhizobium radiobacter*, *R. rhizogenes*, *R. rubi*, *R. undicola* and *R. vitis*. *International Journal of Systematic and Evolutionary Microbiology* 51: 89–103.

Zeigler, D.R. 2003. Gene sequences useful for predicting relatedness of whole genomes in bacteria. *International Journal of Systematic and Evolutionary Microbiology* 53: 1893–1900.

Zurdo-Piñeiro, J.L., Velázquez, E., Lorite, M.J., Brelles-Mariño, G., Schröder, E.C., Bedmar, E.J., Mateos, P.F., Martínez-Molina, E. 2004. Identification of fast-growing rhizobia nodulating tropical legumes from Puerto Rico as *Rhizobium gallicum* and *Rhizobium tropici*. *Systematic and Applied Microbiology* 27: 469–477.

Zurdo-Piñeiro, J.L., Rivas, R., Trujillo, M.E., Vizcaíno, N., Carrasco, J.A., Chamber, M., Palomares, A., Mateos, P.F., Martínez-Molina, E., Velázquez, E. 2007. *Ochrobactrum cytisi* sp. nov., isolated from nodules of *Cytisus scoparius* in Spain. *International Journal of Systematic and Evolutionary Microbiology* 57: 784–788.

Chapter 9
Exploiting Mycorrhizae and Rhizobium Symbioses to Recover Seriously Degraded Soils

Sérgio Miana de Faria, Alexander S. Resende, Orivaldo J. Saggin Júnior, and Robert M. Boddey

Introduction

Since the 1960s and the publication of Paul Ehrlich's book *The Population Bomb* (Ehrlich, 1969), the impression has been given by "futurologists" and others that, even with the intensification of agriculture and the large increases in crop yields induced by the green revolution, there is simply not enough potentially arable land left on the planet to feed the growing human population (Brown, 1995, 1996). A more realistic perspective is now emerging that shows that if farmers, who at present practice extremely low-productivity agriculture in the poor countries of the Third World, were given access to fertilizers, no-till techniques, and markets for their produce, many underdeveloped regions could be self-sufficient in food and even become net food exporters (Thurow and Kilman, 2009).

However, even these more recent reports have entirely ignored the existence of vast regions of well-watered land that over the last few centuries have been degraded and abandoned. In Latin America alone, the United Nations Food and Agriculture Organization (FAO) estimates that 27% of all land area is either severely, or very severely, degraded (Bot et al., 2000). In this same report, the proportion of land in Brazil in these two degradation categories is very similar (28%), and this amounts to a total of 236 million hectares (Mha) or approximately four times the area dedicated to arable crops (including semi-perennial crops such as sugarcane and cassava, and perennial crops such as coffee and cacao). This huge area is already deforested and most receives more than 1,000 mm of annual rainfall, so it theoretically constitutes a vast reserve of potentially arable land. The report of Bot and others (2000) defines severely degraded land as that with "biotic functions largely destroyed; non-reclaimable at farm level" and very severely degraded as "biotic functions fully destroyed, non-reclaimable." Although, the severely degraded category is defined as "non-reclaimable at the farm level," in view of the large breadth of this category, much of the area so classified in Brazil must include degraded areas of planted pastures, which total at least 50 Mha (Zimmer and Euclides, 1997;

Ecological Aspects of Nitrogen Metabolism in Plants, First Edition. Edited by Joe C. Polacco and Christopher D. Todd.
© 2011 by John Wiley & Sons, Inc. Published 2011 by John Wiley & Sons, Inc.

Boddey et al., 2003). Much of this area of degraded pastures is in the central savanna (Cerrado) region of Brazil on mostly gently rolling country, and such areas can be recovered for food crop or sugarcane production using either conventional or no-till techniques along with chemical fertilization. However, the large areas of deforested hillsides in the Atlantic coastal region, especially the more severely degraded areas, need more intensive interventions.

This region is subject to very intense summer rainfall, and in many areas erosion has opened huge gulleys that deepen and lengthen every season. In these gulleys all topsoil is eroded away leaving only subsoil with virtually no organic matter that is extremely low in nutrients for plants. Subsoils are similarly exposed in engineering works, such as road cuttings, land leveling on construction sites, etc. Even in the humid tropical conditions of southeastern Brazil, these subsoils can remain exposed for many years and, while seeds may be spontaneously deposited by wind, water, or animal vectors and various seed species may germinate, they do not thrive, and spontaneous revegetation of the areas does not occur. However, given that nodulating legume trees can fulfill their nitrogen requirement from biological nitrogen fixation (BNF), we were led to test legume trees as useful agents to revegetate seriously degraded soils.

Trees from the genera *Acacia*, *Mimosa*, and *Gliricidia*, among other nitrogen-fixing species, were tried initially, and early investigations soon showed that in these subsoils, there was an almost complete lack of rhizobium bacteria capable of nodulating these species (Franco and Faria, 1997). All of these legume trees are also capable of infection by arbuscular mycorrhizal fungi (AMF), which are known to aid the uptake of soil nutrients, such as phosphorus and low solubility and hence, low mobility. AMF are also known to mitigate water stress. However, as is the case for free-living rhizobium bacteria, propagules (spores) of AMF are almost totally absent in subsoils (Franco and Faria, 1997). This situation led the team at Embrapa Agrobiologia (The Brazilian Agricultural Research Corporation Agrobiology Center), henceforth termed "the team," to integrate inoculation with both selected rhizobium bacteria and AMF for the production of seedlings suitable for the recovery of severely degraded areas.

In this chapter we describe the symbioses of N_2-fixing bacteria of the rhizobium group and AMF, which colonize the roots of fast-growing legume trees, and how seedlings can be produced capable of growing in subsoils almost completely devoid of plant nutrients. Subsequently we describe three case studies where different types of degraded areas were recuperated or revegetated using this technology.

Symbioses of N_2-fixing Bacteria With Species of the Legume Family

The symbioses of nitrogen-fixing bacteria with woody perennial plants are characterized by the formation of nodules. Nodulating species are restricted to families within four component orders of one of the subclades, Eurosideas of the Angiosperms. This classification is based on the gene sequences for ribosomal 18S rDNA and chloroplast *rbcL* and *atpD*. Plant species that form N_2-fixing symbioses with the actinomycete genus *Frankia* (known as actinorhizal symbioses) show the widest diversity of origin, coming from eight different families distributed in the orders Fagales, Cucurbitales, and Rosales (see Chapter 6). Plant species that form symbioses with N_2-fixing microbes other than *Frankia*, are restricted to the families Ulmaceae and Leguminosae, which belong, respectively, to the orders Rosales and Fabales. It should be pointed out that within these orders there are many families lacking species which establish N_2-fixing symbioses. Furthermore, families and genera containing nodulating species may also have genera and species, respectively, which do not nodulate. Further information on the distribution of nodulation and N_2 fixation in the Leguminosae can be found in Sprent (2001, 2009).

Figure 9.1. Spherical determinate nodules of soybean (*Glycine max*). Photo courtesy of Bruno J.R. Alves.

Although some of the actinorhizal associations, especially those with trees such as *Casaurina* spp., have been exploited with success to reclaim degraded soils or prevent soil erosion (e.g., Mailly and Margolis, 1992; Dommergues et al., 1999), the team has concentrated on legume trees as the agents for the recovery of degraded areas.

The family Leguminosae is subdivided into three subfamilies: Mimosoideae, Papilionoideae, and Caesalpinioideae (Polhill et al., 1981, 1994). Woody species are predominant among the approximately 4,000 legume species within the Caesalpinioideae and the Mimosoideae. Although trees are not dominant in the Papiloinoideae there are approximately 4,000 tree species within this large subfamily of ~13,500 species. The occurrence of effective nodulation is much more frequent in legumes of the Papiloinoideae, followed by the Mimosoideae, and in only a small proportion of the legumes in the Caesalpinioideae (Allen and Allen, 1981; Faria et al., 1989; Sprent, 2001, 2009). Different nodule types are found among the wide diversity of legumes, and these are determined by the host plant. The most familiar type is the spherical (determinate growth), which occurs in grain legumes such as soybean (*Glycine max*), cowpea (*Vigna unguiculata*), and French or common bean (*Phaseolus vulgaris*). This type is also found on roots of tree legumes as well as legumes of other growth habit (Figure 9.1). The branched (indeterminate growth) nodule, common among leguminous tree species, can reach much larger size than the spherical type. As in perennials, the nodules are usually also perennial (Figure 9.2).

According to Corby (1981), nodules can be also classified by their predominance in some leguminosae tribes as Lupinoid, Caesalpinoid, Crotalorioid, Aeschynomenoid, and Mucunoid shape of nodules. The most common among the legume trees species are Caesalpinioid (indeterminate) growth nodules. However, we can also find Aeschynomenoid shape in some genera of the tribe Dalbergieae such as *Machaerium*, *Centrolobium*, and *Dalbergia* (see Sprent, 2001). Usually, nodules with indeterminate growth have a much longer life span than nodules with determinate growth.

Figure 9.2. Indeterminate nodules from the tree *Campsiandra comosa* (Caesalpinioideae). Photo by Sérgio M. de Faria.

Host Specificity and Promiscuity of Legumes with Rhizobium

Under natural conditions legumes are nodulated by different strains, species, or even different genera; this diversity can even occur within the same nodule (Giller, 2001). Recent research has increased considerably the known number of genera and species of bacteria that are able to nodulate legumes. Those that produce N_2-fixing symbioses are generally termed "effectively nodulating." At first all bacteria that effectively nodulated legumes were included in one genus, named *Rhizobium*. Many strains that nodulated legumes of tropical origin were found to grow slowly on solid medium and the genus subsequently was divided in two, *Rhizobium* and *Bradyrhizobium*, for the fast- and slow-growing strains, respectively (Jordan, 1982). With the advent of molecular-based phylogeny, the number of genera and species has grown rapidly.

Until 2001, all bacteria that effectively nodulated legumes were found to belong to the order Rhizobiales of the α-Proteobacteria. At present these α-Proteobacteria are recognized to be in 9 or 10 genera spread among approximately 60 species (Sprent, 2009). The numbers of genera and species cannot be given with precision because (1) some of the classifications are under dispute, and, (2) continuing investigations make it inevitable that the numbers of genera and species will increase.

In a more surprising development, Moulin and others (2001) isolated β-Proteobacteria from the nodules of a leguminous tropical tree, *Machaerium lunatum*. After a flurry of competing research activity, it was found that nodulators from several other legume species, mostly species of the genus *Mimosa*, belonged to the β-Proteobacteria, and many of the bacterial partners have been classified as members of the genus *Burkholderia* (e.g., Chen et al., 2005; Elliot et al., 2007). There are many ongoing studies on these symbioses, and according to Sprent (2009), there are likely to be several new nodulating species from the β-Proteobacteria designated in the near future.

With regard to nomenclature, a consensus is being reached where all bacteria capable of forming effective N_2-fixing nodules on legume plants are known as "rhizobium" (no initial upper case "r" and no italics, plural rhizobia) or "of the rhizobium group." The groups from the α- and β-Proteobacteria are known as α- and β-rhizobia, respectively (Sprent, personal communication).

Different bacterial strains and species vary widely in the efficiency of BNF depending on which host plant they nodulate. However, there does not seem to be any relationship between efficiency of N_2 fixation and the phylogeny of the strain. The grower's selection of strains of N_2-fixing bacteria compatible with a plant species, selection based on their high efficiency and competitivity, and their adaptation to the environmental conditions experienced at planting, can have very significant benefits for the BNF input and growth of the host plant. In general, knowledge of the phylogeny of the rhizobium strain is not helpful in the prediction of the efficiency of N_2 fixation (Giller, 2001), but strains isolated from a given plant species are much more likely to nodulate that same species efficiently.

In experiments on legume host plant specificity realized at Embrapa Agrobiologia in the 1960s, it was observed that certain tribes of legumes were very specific with respect to rhizobium strain, while others were very promiscuous in relation to these bacteria (Campelo and Döbereiner, 1969). Plant species in the tribe Mimoseae form nodules only with bacteria isolated from the same genus, while species of the tribe Phaseoleae nodulated efficiently with a majority of isolates from any legume (Faria et al., 1999; Faria, unpublished results). Furthermore, it was observed that some rhizobia were able to nodulate efficiently with various legume species from other tribes or even other subfamilies. In summary, there is a range, from promiscuity to specificity, with reference to both the plant species and the rhizobium, in the formation of effective nodules (Faria et al., 1999; Faria, unpublished results).

In the case of tree species, the majority of responses obtained with inoculation of rhizobium were based on observations in the greenhouse or in a nursery for seedling production. Fewer studies have shown a significant effect after the plants had been transferred to the field for a few years (Balieiro et al., 2004, 2007). One reason, of course, is the long time required to obtain significant growth responses, unless the soil is extremely poor in nutrients. However, in areas of mine waste or completely decapitated soils, inoculation effects can be very significant, even spectacular. Positive results are often recorded for the rate of survival of the tree seedlings in the field (Costa et al., 2004).

Selection of Rhizobium Strains for High BNF Efficiency

The selection of symbiotically-efficient rhizobium strains for species of economic importance has been one of the principal targets of research in the area of biological N_2 fixation. Apart from high BNF efficiency, these strains should have high competitivity with less efficient strains for infection sites and the capacity to survive under different edaphoclimatic conditions.

Surveys of the nodulation of legume species in Brazil have been made in the Amazon (Moreira et al., 1992; Faria et al., unpublished data) and the Atlantic Forest (Faria et al., 1984, 1987, 1989) regions. As a result, the nodulation status of hundreds of species has been recorded, many for the first time. Not all results were positive, since many tree species were found to be incapable of forming nodules with any available rhizobium strain. These results and many others have been summarized by Sprent (2001, 2009).

At Embrapa Agrobiologia, strains were isolated from nodules according to Somasegaran and Hoben (1994). Briefly, bacteria are isolated after surface sterilization of the nodules, which are then squashed and plated on YMA medium. The Embrapa Agrobiologia rhizobium collection has >15,000 isolates. However, only ~3,000 have so far been registered in the online database. For further information see http://www.cnpab.embrapa.br/.

For woody legume species, the collection of rhizobia contains about 5,500 strains. These strains have been selected for legume species with potential use for revegetation of degraded land. To date the laboratory has recommended efficient nitrogen-fixing bacteria for 87 legume species.

The selection of these bacteria is made in several steps: First, they are inoculated onto the legume species under sterile conditions in "Leonard jars" (Somasegaran and Hoben, 1994). These greenhouse experiments, which from 3 to 6 months, aim to determine whether the isolate is really a rhizobium strain and gives an indication of potential efficiency in BNF. Subsequently, experiments are performed in pots of (nonsterilized) soil. In this case the objective is to determine whether the strain selected in the first step is competitive with the native strains present in the soil. Finally, further tests may be performed under field conditions to confirm the greenhouse results. The selection of rhizobium strains is very important for inoculant preparation, because in a considerable number of cases there is a high specificity between bacteria and host plant. There are groups in the Leguminosae family that are nodulated preferentially by rhizobia isolated from the same group. This is the case for the Mimoseae tribe. On the other hand, there are groups that show a wide promiscuity toward the rhizobium strain, which is the case of the Phaseoleae and Milletieae tribes (Faria et al., 1999; Faria, unpublished results).

Specificity in the Symbioses of Arbuscular Mycorrhizal Fungi (AMF) With Higher Plants

AMF of the phylum Glomeromycota form symbioses with roots of higher plants—symbioses which are more widespread geographically and phylogenetically than any other plant-microbe association. These symbioses have very significant benefits for the host plant and for nutrient cycling in ecosystems. Surveys indicate that 80% of plant families form some type of mycorrhizal symbiosis (Moreira et al., 2006). Arbuscular mycorrhizae are of more ancient evolutionary origin than other mycorrhizal or N_2-fixing symbioses and are more widely occurring in the tropics than other mycorrhizae (Brundrett, 2002). The AMF function as an extension of plant root systems increasing their potential to absorb nutrients, principally those of lower soil mobility such as phosphorus. They also aid in the extraction of water from soils under conditions of low availability (hydric stress). The AMF can also have synergistic effects on BNF in tripartite symbioses of plant/fungus/bacteria (Bethlenfalvay et al., 1982, 1989). Apart from these benefits, AMF play a most important ecological role, especially in forest fragments where they alter the competitive relationships among plants (Janos, 1996; Flores-Aylas et al., 2003). The AMF are obligate biotrophs, that is, they cannot complete their life cycle unless they infect plants. Until recently it was generally believed that there was little specificity in the association of AMF with higher plants. The apparent absence of specificity for arbuscular mycorrhizae was basically established from the observation that one AM fungal species could colonize different species or even families of plants, suggesting that a small diversity of fungi could colonize a great diversity of plants (Smith and Read, 1997).

Recent molecular studies suggest that the diversity of AMF is much greater than heretofore suspected (Robinson-Boyer et al., 2009) and that locally adapted ecotypes are frequently described in the literature (Schubert and Hayman, 1986; Louis and Lim, 1988; Saggin Júnior and Siqueira, 1995). Recent studies have demonstrated the existence of host specificity or selectivity in the arbuscular mycorrhizal symbioses in grassland communities (Vandenkoornhuyse et al., 2002, 2003). In the case of tropical forest species, few studies have been made but it is now becoming apparent that there can be a considerable degree of specificity between host plant and AMF (Siqueira and Saggin Júnior, 2001).

Pouyu-Rojas and others (2006) studied several diverse forest species in combination with different isolates of AMFs and found that the different symbioses indicated a certain degree of specificity. Knowledge of selective, or preferential, plant-AMF associations is of considerable importance for the rehabilitation of low fertility tropical soils, conditions in which mycorrhizae are considered essential for revegetation (Jasper et al., 1992). Some AM fungi were capable of colonizing and forming an efficient symbiosis with a large number of host plants, while other AMF isolates were very restricted with respect to the host species (Pouyu-Rojas et al., 2006). It was likewise found that plants can range from "extremely promiscuous" to very restrictive with respect to the AM fungi that can colonize their roots.

Glomus clarum and *Gigaspora margarita* are AM fungi that can efficiently colonize a large number of plant species. They have played an important role in the recuperation of degraded areas (RDA) as they have been considered to be able to colonize most plants that facilitate revegetation (Rocha et al., 2006). At Embrapa Agrobiologia these two AM fungi (*G. clarum* + *G. margarita*) have been used to inoculate the seedlings used for the RDA program. Among the widely used species are those of the genus *Acacia* (*A. mangium*, *A. holosericea* and *A. auriculiformis*). These trees of Australian origin have been utilized by others in revegetation of degraded areas (Deans et al., 2003; Midgley and Turnbull, 2003) in Australia and Senegal. However, in comparison with other tree species used in the Embrapa RDA program, the results of inoculation with AM fungi were not very satisfactory, resulting in slow-growing seedlings and increased cost. Thus, Angelini (2008) evaluated the symbioses of these three species of *Acacia* inoculated with different isolates from the AMF collection at Embrapa Agrobiologia. In this study, the three tree species showed very similar preferences for the establishment of efficient symbioses with the same AMFs: *Acaulospora morrowiae*, *Scutellospora calospora*, *S. gilmorei*, and *S. heterogama*. The two species customarily used for inoculation, *Glomus clarum* and *Gigaspora margarita,* showed poor efficiency for these species of *Acacia* (Figure 9.3).

Those AM fungi showing the highest symbiotic efficiency had >75% colonization rates. *Acacia mangium* showed the greatest response to mycorrhizal inoculation, followed by *A. holosericea* and *A. auriculiformis*. The results suggest that the specificity may be host plant-associated, since all three *Acacia* species increased yield very significantly when *Glomus clarum* and *Gigaspora margarita* were substituted by *Scutellospora calospora*, *S. gilmorei*, *S. heterogama*, or *Acaulospora morrowiae*.

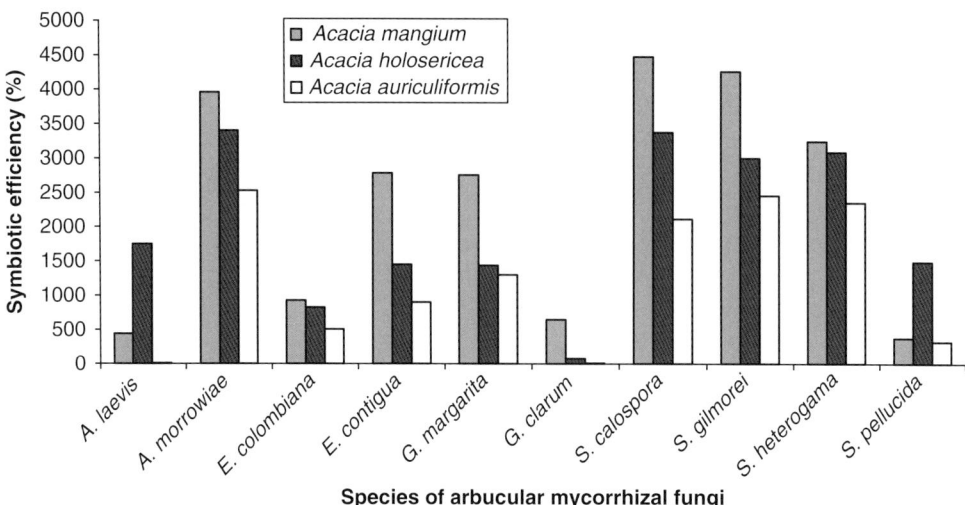

Figure 9.3. Growth response of three Acacia species to inoculation with 10 different species of arbuscular mycorrhizal fungi. Plants were grown in 700 ml of autoclaved soil (pH 5.2) and inoculated with the strains of rhizobium recommended by Embrapa Agrobiologia. Symbiotic efficiency was measured as a percentage of growth of the non-inoculated control plants.

Preparation of Tree Seedlings for the Recovery of Degraded Areas

The success of the Embrapa Agrobiologia RDA program is mainly due to the selection of fast-growing legume trees and their specific mycorrhizal and rhizobial symbiotic partners. A great deal of experience has been obtained with many legume trees suitable for different climates and soils that vary in salinity, acidity, etc. These species, classified by climatic region and soil restriction, can be found in the Technical Bulletin of Faria and others, 2010. (http://www.cnpab.embrapa.br/publicacoes/download/cot127.pdf)

For most trees employed in RDA, it is necessary to scarify the seeds to break dormancy and induce germination. This is normally performed with either hot water or concentrated sulfuric acid, and the time of immersion differs among species. The seedlings are produced in a mixture of sandy and clay soils and organic compost in the ratio of 1:1:1 (v/v) contained in polystyrene planting boxes (e.g., Plantagil®, Plantagil Comercial Agrícola, São Paulo, SP) with typically 72 pockets (seedlings) per box, or in plastic tubes or plastic bags.

The mycorrhizal fungal inoculant is prepared as follows: the mycorrhizal fungi (currently *Glomus clarum* and *Gigaspora margarita*) are multiplied in pots (20 L) using *Brachiaria decumbens* as a trap plant. The mycorrhizal inoculant is a mixture of soil, *Brachiaria* roots, and fungal mycelia/spores. The inoculant (approximately 1 g) is inserted in the hole made for the seed immediately prior to planting. At planting the seeds are treated with standard rhizobium inoculant using sterile peat as a base (Somasegaran and Hoben, 1994). Seeds are wetted and mixed with the peat-based inoculant to coat them—the objective is to make the seedling roots pass through both inoculants, thus making direct contact with them.

The seedlings are left to grow in a nursery for 3 to 4 months, depending on the species and the season. Once ready, seedlings are transported to the field. Correction of soil pH is not usually necessary because these tree species withstand very acid soil and/or substrates. However, soil fertility analyses are always made. Holes spaced at about 30 cm × 30 cm × 30 cm are made in the area. For planting, 250 g of rock phosphate and 10 g of fritted trace elements, and usually 1 or 2 L of green manure or chicken dung (when available) are added because most of the subsoils and other substrates are almost completely devoid of phosphorus and micronutrients. Planting is carried out at the start of the rainy season where possible, although in dry areas it may be possible to plant if irrigation is available.

Case Studies

Recuperation of Erosion Gulleys at a Rural Site

When Brazil was first "discovered" by the Portuguese in 1500 AD, the Atlantic coastal region was covered by a dense tropical forest that stretched from latitude 5° to 30° south and occupied approximately 100 Mha. Today it is estimated that less than 8% of this forest remains. The deforestation started with the extraction of the red dye from the "pau brasil" tree after which the country is named, and continued through the cycle of sugarcane, coffee, and timber extraction and, eventually, in the last century, large areas were felled for charcoal production for iron founding. This sad history has been related in detail by Dean (1995).

A large proportion of this deforested land, especially in the states of Rio de Janeiro, Minas Gerais, and Espirito Santo is in mountainous areas. After deforestation and/or coffee cropping, the hillsides became colonized principally by grasses such as green panic (*Panicum maximum*) or molasses grass (*Melinis minutiflora*). Owing to the almost continuous presence of grazing animals and regular burning (usually intentional, to provoke new growth for forage), woody species have never returned to these hills. As mentioned in the Introduction, the heavy summer rainfall often opens up erosion gulleys that gradually expand. No complete survey of the area covered by such erosion gulleys appears to have been made but, to give an example, along a 70-km stretch of the valley of the river Paraíba do Sul (the source of much of Rio's drinking water) in the interior of the State of Rio de Janeiro, 160 erosion gulleys have been counted (Machado et al., 2006).

In this case study, we report the results of the use of the fast-growing legume tree (FGLT) technology on the environmental recovery of one such erosion gulley with a mean depth of 10 m and a volume of approximately 10,000 m^3. The gulley before intervention is pictured in Figure 9.4. It is estimated that during the "growth" (erosion) of the gulley, the equivalent of 2,000 truckloads of sediment were washed out and deposited in the areas below and in the stream and rivers of the watershed.

The intervention started in 2000. Initially, narrow terraces were formed at the upper and lower ends of the gulley, and walls of bamboo and rubber tires were formed in the central streambed within the gulley in order to trap sediments (Figure 9.5.) Following this engineering, the legume tree seedlings, all inoculated with strains of rhizobium and AMF selected at Embrapa Agrobiologia, were planted with a spacing of 2 × 2 m along the gulley and into holes cut into the walls. The success of the technology was evaluated by measurements on tree growth and the output of sediments collected in sediment tanks.

The best results were obtained with the species *Acacia mangium, Mimosa artemisiana, M. caesalpiniifolia*, and *Pseudosamanea guachapele*. After 170 days the mean heights (cm) for these species were *A. mangium* 53, *Mimosa artemisiana* 38, *M. caesalpiniifolia* 67, and *Pseudosamanea guachapele* 67. The species *Acacia auriculiformes, A. angustissima, Albizia lebbek, Enterolobium contortisiliquum,* and *Samanea saman* showed low indices of survival,

204 *Plant-Soil Microbe Interactions*

Figure 9.4. The erosion gulley at the start of the recovery program in 2000. Note the rubber tires and bamboo barriers installed to slow down stream flow. The white plastic netting hanging from the sides of the gulley have pockets to hold seedlings.

Figure 9.5. Sediment trap at the bottom of the erosion gulley where no recuperation had been attempted. The image shows the scene shortly after a 30-mm rainfall in 30 minutes. The sediment traps had to be redesigned and rebuilt! Photo by Eduardo F.C. Campello.

sometimes because of their lower resistance to drought, or their position in the gulley where water was not retained, or because they suffered from attack by leaf-cutting ants.

Sediment tanks were installed in three adjacent gulleys of similar size, one which was left without a recovery operation, one where the recovery operation started 2 years before, and the third 6 years before. From the no-intervention gulley 52.5, 20.5, and 29.0 Mg of sediment at the outlet was collected in three evaluations during the rainy season from November 2005 to February 2006: the first over a period of 17 days (134 mm rainfall), the second over the next 24 days (134 mm rainfall), and the final evaluation over a period of 8 days (123 mm rainfall). The gulley where the recuperation had started 2 years before produced quantities of sediment of 1.0, 1.7, and 1.8 Mg over the three measurement periods. The sediment output from the gulley where the recuperation had started 7 years before was too low to be measured, although it is possible that very fine particles did not have time to settle in the sediment tanks before washout.

With respect to cost of this operation, a total of R$10,904 (one $US = ~R$1.70) was spent to recover one of the gullies. The largest proportion of the cost (64%) was for labor followed by the cost of the 4,000 seedlings (20%, or R$0.55 per seedling) and transport, and the other costs were for materials such as fencing posts and wire, rock phosphate and fritted trace elements (FTE), manure, and insecticide.

Rehabilitation of a Decapitated Hillside in the Atlantic Forest Region

The objective of this study was to examine the effect of soil rehabilitation using leguminous nitrogen-fixing trees on the recovery of nutrient cycling processes and soil C and N stocks 13 years after planting. The area was located in the town of Angra dos Reis, along the western coast of the State of Rio de Janeiro, 23° 02′ 30″ S and 44° 11′ 30″ W, 100–200 m above sea level, within the limits of the Atlantic Forest biome. According to Köppen's method, the climate of Angra dos Reis is classified as Af, moist tropical forest. Average annual precipitation and temperature are 2,300 mm and 22.5°C, respectively, without a distinct dry season (Chada et al., 2004). The area is steeply sloping and the soil is classified as a Ferralsol under the FAO soil classification system (Oxisol, United States Department of Agriculture [USDA] Soil Taxonomy).

In 1991, when the area was dominated by grass vegetation, the topsoil was removed from the site and used for the foundations of a shopping mall. Exposure of the subsoil to rainfall led to severe erosion, which, after a short period of time, resulted in the formation of erosion gullies. The area was restored by planting seedlings of *Acacia mangium, A. auriculiformis, Enterolobium contortisiliquum, Gliricidia sepium, Leucaena leucocephala, Mimosa caesalpiniifolia,* and *Paraserianthes falcataria,* all of which were inoculated with selected rhizobia and arbuscular mycorrhizal fungi as described in the Preparation of Tree Seedlings for the Recovery of Degraded Areas section. The recovered area was approximately 1 hectare (ha) in size. To avoid further erosion, when the seedlings were planted, bamboo stems were anchored crosswise in the erosion channels or gullies to slow down the rainwater running down the slope.

In September 2004 soil samples were collected from the two reference areas and the area rehabilitated with legume trees. One of these reference areas was a fragment of native forest (Atlantic Forest), with few signs of human presence or disturbance, while the other consisted of 2 ha of deforested land. These three areas are located close to each other on the same hillside. The deforested area (where no intervention was performed) was spontaneously overgrown with green panic (*Panicum maximum*). In contrast to the rehabilitated area, the topsoil of this control area was not completely removed. Soil samples were collected from the following depth intervals: 0–5, 5–10, 10–20, 20–30, 30–40, and 40–60 cm. The samples were collected from three replicate trenches excavated to a total depth of 70 cm and spaced 4 m apart. Soil bulk density was estimated from samples collected using volumetric rings at each depth interval. Four soil bulk density samples were used. Additional samples were taken from each depth interval for chemical analyses, which were performed on triplicate subsamples.

To obtain bulk density, samples were dried at 105°C for >72 h and weighed. Samples for analysis of texture, nutrient, and N and C concentration were air-dried and sieved to 2 mm. Soil samples were analyzed for the fertility parameters: pH, exchangeable Al, Ca, and Mg, available P and K (Mehlich-1 extractant) using standard methods (Embrapa, 1997). Total N was determined using semi-micro Kjeldahl digestion (Urquiaga et al., 1992). Total C content of finely ground (<100 mesh) samples was measured using a LECO CHN-600 analyzer (Tarré et al., 2001).

The soil C and N stocks were calculated from the C and N concentrations measured at each depth interval multiplied by the respective bulk density and the thickness values of the corresponding soil layer. To avoid overestimates of C and N in compacted soils, stocks were corrected for differences in soil mass to 60 cm depth using the procedure of Veldkamp (1994).

The amount of standing litter on the soil surface was determined on recovered and native samples collected during the dry season (September 2004) and rainy season (March 2005). Samples taken using a steel quadrant frame (0.25 m^2), were dried at 65°C and weighed. (Full details are given in Macedo et al., 2008.)

The soil C and N concentrations of the recuperated area were higher than those of the deforested area (Table 9.1), and both were similar to the C and N values of the native forest soil Macedo and others (2008). The total soil C stocks of the 0–30 cm depth range were similar among the three areas, but the C stocks to a depth of 60 cm in the deforested area were only 60% of those under the native forest (Table 9.1). Assuming that the C stock of the deforested area was equivalent to the C stock of the recovered area prior to planting the legume trees, it may be concluded that the soil C stock increased by 23 Mg ha^{-1} in 13 years or a mean of over 1.7 Mg C ha^{-1} year^{-1}.

The litter stocks of the recovered and native forest areas were statistically similar for rainy and dry seasons and ranged from 5.0 to 6.7 Mg ha^{-1} (Table 9.2), indicating that the net aerial primary productivity of the area planted to legume trees was at least equal to that of the native forest. These results are similar to those reported by Vital and others (2004) for a steady state forest (6.2 Mg ha^{-1}), although somewhat lower than the values observed by Arato and others (2003) in a 9- to 10-year old agroforestry system established on degraded land (8.7 Mg ha^{-1}).

Soil C and N were restored within a few years after planting of legume trees in symbiotic association with nitrogen-fixing bacteria and arbuscular mycorrhizal fungi. Other studies have shown the increase of soil C and N during forest development (Brown and Lugo, 1990; Gleason and Tilman, 1990; Feldpaush et al., 2004) but not in a situation where the soil had been decapitated.

Soil N increase is very important in degraded land rehabilitation projects, since, according to Francis and Read (1994), it enhances the capacity of the system to support a more complex community. The N increase was directly related to C incorporation, as indicated by the strong correlation of soil C and N in all areas in this study (r = 0.78, P<0.0001, n = 50). Because of their ability to fix nitrogen, legume species have been used as an N source in a several tropical agroecosystems, including pastures (Fisher et al., 1994; Tarré et al., 2001), no-till fields (Sisti et al., 2004; Boddey et al., 2010), tree plantations (Resh et al., 2002; Balieiro et al., 2008), and agroforestry (Handayanto et al., 1995). In these diverse systems

Table 9.1. C and N stocks (0–60 cm) whole soil profile of recovered, native forest and deforested areas in Angra dos Reis, RJ, Brazil. (n = 3)

Depth (cm)	Recovered	Native forest	Deforested
		C stock (Mg ha^{-1})	
0–5^{ns}	10.9 +/– 0.2	12.0 +/– 1.5	7.1 +/– 0.3
5–10^{ns}	10.6 +/– 1.0	10.9 +/– 1.0	7.2 +/– 0.1
10–20^{ns}	19.7 +/– 1.2	21.7 +/– 2.5	13.3 +/– 0.5
20–30^{ns}	15.9 +/– 2.4	16.8 +/– 4.3	11.0 +/– 2.4
30–40^{ns}	14.4 +/– 1.9	17.9 +/– 4.8	10.1 +/– 3.0
40–60^{ns}	21.5 +/– 2.7	34.1 +/– 14.4	19.8 +/– 6.7
0–30^{ns}	54.8 +/– 2.1	58.3 +/– 7.7	35.4 +/– 1.7
0–60	88.1ab +/– 0.4	107.7a +/– 22.1	65.1b +/– 11.2
Depths (cm)		N stock* (Mg ha^{-1})	
0–5^{ns}	0.94 +/– 0.08	1.17 +/– 0.07	0.59 +/– 0.10
5–10^{ns}	0.92 +/– 0.09	1.00 +/– 0.06	0.52 +/– 0.04
10–20	1.67a +/– 0.15	1.70a +/– 0.12	0.99b +/– 0.07
20–30	1.41ab +/– 0.14	1.57a +/– 0.18	0.94b +/– 0.08
30–40^{ns}	0.98 +/– 0.10	1.27 +/– 0.21	0.97 +/– 0.02
40–60	1.75b +/– 0.12	2.54a +/– 0.56	1.96ab +/– 0.09
0–30	5.0a +/– 0.4	5.4a +/– 0.2	3.0b +/– 0.1
0–60	7.7ab +/– 0.3	9.1a +/– 0.9	6.0b +/– 0.1

Numbers followed by the same letter are not significantly different according to the Bonferroni "t" Test (P<0.05). +/– = Standard error. ns = not significant.
*Data Transformed by Ln (x)
Mg = mega grams

Table 9.2. Total nutrient content of different litter fractions on native forest, recovered and deforested areas in Angra dos Reis, RJ, Brazil, collected during the dry and rainy season. (n = 5)

	Leaves		Stem		Decomposed		Total	
Season:	Dry	Rainy	Dry	Rainy	Dry	Rainy	Dry	Rainy
Area				kg ha^{-1}				
Native forest	1278Aa	855Aa	1419Aa	1088Aa	2312Ab	3809Aa	5009Aa	5753Aa
Recovered area	1725Aa	543Ab	1463Aa	1935Aa	2765Ab	4308Aa	5954Aa	6786Aa
CV(%)		44.97		12.02*		29.47		37.34

A different superscript letter indicates a significant difference between the areas in the same season according to the Bonferroni "t" Test P < 0.05. A subscript letter indicates a significant difference between the seasons relative to the same area according to the Bonferroni "t" Test P<0.05.
*Data transformed by Ln (x)

soil N content and soil organic matter (SOM) stocks were found to increase. Organic matter in tropical soils plays a crucially important role in the formation and maintenance of soil structure and fertility as well as nutrient and water availability (Bayer et al., 2001; Craswell and Lefroy, 2001; Six et al., 2002).

Revegetation of Iron Mining Waste

According to Brazilian law any mining site can be exploited provided that it is revegetated as close to the original as possible, or the mining license is not ceded to mining enterprise. For mining companies, the legislation is quite rigorous.

The "waste" from iron mining is defined as part of surface soil and subsoil between the vegetation down to the haematite level. Haematite is the economically exploitable layer and is often mined at depths of 6 to 10 m but sometimes even down to 20 m. The overlying "waste" material is generally composed of a mixture of laterite and filite (iron and aluminum oxides). This material is removed to give access to the iron ore and is deposited in embankments.

The original practice of the mining company in this case study was to plant the embankments with a mixture of two herbaceous legumes, *Crotalaria* spp. and *Cajanus cajan* (neither inoculated with rhizobium), and two grasses, signal grass (*Brachiaria decumbens*) and molasses grass (*Melinus minutiflora*). This system was shown not to be sustainable. After a few years, it results in a complete dominance by molasses grass, which is not only very prone to fire in the dry season but also does not allow natural vegetative succession to take place (Chada et al., 2004). The Embrapa FGLT technology has been applied since 1992 in the Mariana-Ouro Preto districts of Minas Gerais State (Southeast Brazil) where approximately 25 ha per year are revegetated.

Fertility analyses of the material at this site revealed that it had extremely low phosphorus content, no detectable organic matter, traces of potassium and pH ranging from 4.0 to 5.5. In the first tests to evaluate native nitrogen-fixing legume species, we did not have a very wide range of choices nor were many seeds available. Therefore, non-native species were also tested. Fifteen species were tested in this area (Figure 9.6) including nitrogen-fixing legumes, non nitrogen-fixing legumes, and nonlegume species. Seedlings of all 15 species were prepared in a nursery at the mining site. Seeds of the nitrogen-fixing species were inoculated both with rhizobium strains specific for each species and with mycorrhizal fungi (*G. clarum* and *Gl. margarita*) as described in section titled Preparation of Tree Seedlings for the Recovery of Degraded Areas. Plants were left to grow for 4 months in a nursery bed at the mine site. The trees were seeded in January (rainy season).

The strategy was to mix nitrogen-fixing species with species attractive to fauna (e.g., birds and bats) that would ultimately bring in native species not introduced initially.

The seedlings were planted in a triangular array with a distance between them of 1.5 m. Each planting hole in the iron mining waste had received 100 g lime, 200 g rock phosphate, 10 g fritted trace elements (containing Mo, Zn, and B) and 1 L green manure. Control of leaf-cutting ants was performed 1 month before planting using traps containing poison (Mirex). Tree heights were measured 7, 14, and 24 months after planting, and the results after 2 years are shown in Figure 9.6. *Acacia angustissima*, *Mimosa artemisiana*, and *Enterolobium contortisiliquum* (all nitrogen-fixing legumes) reached more than 2 m in height over this period. Some species were not adapted to these conditions, and disappeared, including *Parkinsonia aculeta* (non nitrogen-fixing legume), mulberry (*Morus* sp., non-legume species), and *Mimosa bimucronata* (a nitrogen-fixing legume). After this time, the canopy of plants completely covered the embankments of the iron-mining waste. Colonization by non-planted species was observed as a result of seed dispersal by wind, birds, and bats.

The National Forestry Institute no longer allows planting of such exotic species as Acacias, in the first few years. In the more recent interventions, native nitrogen-fixing legume trees and shrubs have been used together with plant species attractive to the avifauna with equal success.

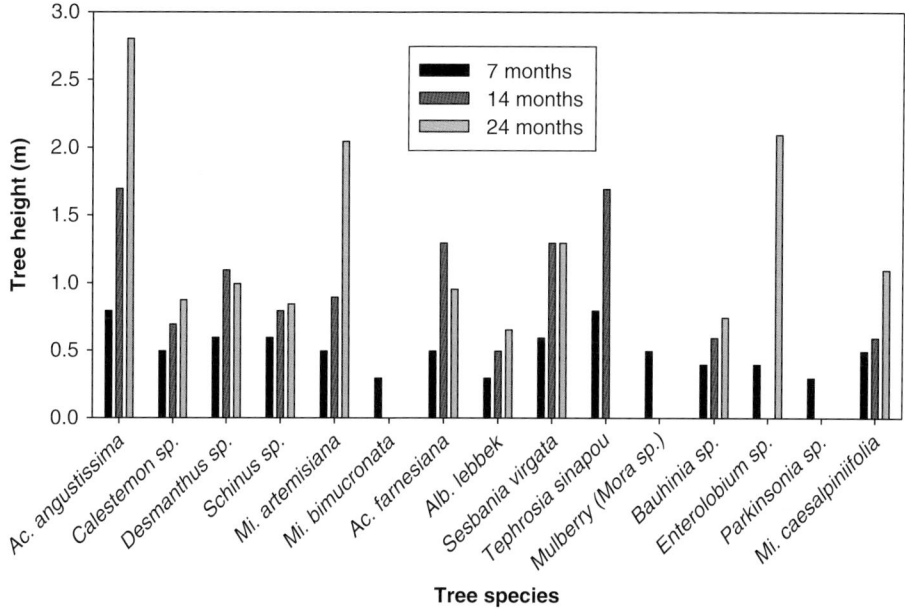

Figure 9.6. Growth of trees (height) on embankments of mine waste 7, 14, and 24 months after planting.

This technology developed by Embrapa Agrobiologia has been successfully applied at other sites in Brazil such as wastes/tailings from bauxite, gold, and nickel mines.

Conclusions

The use of fast-growing legume trees for the recuperation or revegetation of degraded areas or mine wastes developed by the RDA team at Embrapa Agrobiologia has met with great success. As mentioned in the Introduction, areas covered by severely degraded land and erosion gulleys in Brazil are extremely large, and the older strategy of using grasses or herbaceous species rarely resulted in recolonization of the areas by native species, even after many decades. However, fast-growing legume trees inoculated with carefully selected rhizobium strains and AM fungi can result in rapid plant growth. Recent results of the Embrapa Agrobiologia RDA team indicate that an effort should be made to select more effective AM fungi, as recent results are showing that the use of just two species (*Glomus clarum* and *Gigaspora margarita*) is inadequate for some tree species such as those of the genus *Acacia*, and more work on the suitable matching of legume tree species with AMF species is necessary.

The RDA program has proven to have multiple applications in areas with different problems. In all but the most arid regions of Brazil, the revegetation of subsoils that have been exposed by erosion or by land leveling/moving for roads and construction, will result in the areas being totally covered by trees in less than 24 months. These fast-growing legume trees, whether native or exotic, are necessarily pioneer species and if the areas are close to reserves of native forest, spontaneous colonization of secondary species can occur via the agency of wind, birds, bats, and other fauna. When forested areas are distant, secondary native species need to be planted subsequently. In both cases the succession gradually leads to a restoration of the forest composition.

Aide and others (2000) studied the natural recovery at 64 sites in four regions of Puerto Rico where pastures had been abandoned 5 to 75 years before. They showed that basal area, aboveground biomass and species richness recovered to the same levels as those measured in old-growth forests within 40 years. The plant species composition was, however, very different from that of the old growth forests. This recovery of species richness would obviously be much more rapid with an active intervention such as the Embrapa Agrobiologia RDA program. Holl and others (2000) showed that in Costa Rica, the process of recovery could be speeded up if woody species were planted in the abandoned areas. These were not mycorrhizal-nodulated legume trees.

The environmental services of such degraded area recovery programs are diverse and highly valuable. The recovery of plant diversity is of both direct and indirect value, in that it necessarily provides habitats for fauna, whether they are endangered species or not. The elimination or mitigation of the negative effects of erosion, such as the reduction of the sediment load reaching rivers and reservoirs and the increase in infiltration and reduced flooding potential are also highly beneficial consequences. The data show that not only can the trees themselves act as a sink for carbon dioxide and be used to substitute fossil fuel or ameliorate the destruction of native forest, but the trees can also promote the accumulation of N and C in the soil as SOM (Resh et al., 2002; Macedo et al., 2008, Table 9.1). SOM accumulation is of great intrinsic value in degraded soils in that virtually all physical, chemical, and biological soil properties favorable for agricultural production or forestry are improved by its increase. Apart from this, soil C accumulation is synonymous with atmospheric CO_2 sequestration, considered by many as a strategy to mitigate global warming (e.g., Lal, 2004).

Here there should be a word of caution: forests of the humid and semi-humid tropics are known to be the natural ecosystems having the highest emissions of nitrous oxide (N_2O), an extremely potent greenhouse gas (GHG). Estimated emissions from such forest ecosystems can range from 1 to 12 kg N_2O-N h^{-1} yr^{-1}. For example, Silver and others (2005) and Maddock and others (2001) estimated annual emissions of approximately 3 kg N_2O-N ha^{-1} yr^{-1} in the Atlantic forest region of Rio de Janeiro State.

On a molar basis, N_2O absorbs approximately 300 times the amount of infrared radiation as CO_2. This means that the emission of 1 kg of N as N_2O is the equivalent to the emission of 471 kg of CO_2 or 150 kg of C as CO_2. If the N_2O emissions under the legume trees were the same as that under the native Atlantic Forest in southeastern Brazil (~3 kg N ha^{-1} yr^{-1}), then as long as the plant biomass and soil accumulated more than 450 kg C ha^{-1} yr^{-1}, the use of these trees for the revegetation of degraded areas will promote net mitigation of GHG emissions. Studies are under way at Embrapa Agrobiologia to assess the N_2O emissions specifically from the areas under FGLTs.

Although no long-term studies have yet been made, there seems no reason why through such recuperation of soils, very large areas in Brazil and many other "degraded" regions of the tropics could not return to be used for forestry, productive pastures, and even food production. One alternative to increase a cropping area without further deforestation is to intensify crop production on fertile areas, and another alternative is to recuperate the vast areas that have been abandoned over the last century or so. This latter alternative is much more likely to result in the mitigation of greenhouse gas emissions, and those of reactive N, than would be incurred by the increased use of fertilizers and other agrochemicals.

References

Aide, T.M., Zimmerman, J.K., Pascarella, J.B., et al. 2000. Forest regeneration in a chronosequence of tropical abandoned pastures: Implications for restoration ecology. *Restoration Ecology*, 8, 328–338.

Allen, E.K., Allen, O.N. 1981. The leguminosae: A source book of characteristics uses and nodulation. University of Wisconsin Press, Wisconsin.

Angelini, G.A.R. 2008. Seleção de fungos micorrízicos arbusculares e ectomicorrízicos para simbioses eficientes com leguminosas arbóreas do gênero Acacia. MSc Thesis. Universidade Federal Rural do Rio de Janeiro, Seropédica, RJ, Brazil.

Arato, H.D., Martins, S.V., Ferrari, S.H.S. 2003. Produção e decomposição de serapilheira em um sistema agroflorestal implantado para a recuperação de área degradada em Viçosa-MG. *Revista Árvore*, 27, 715–721.

Balieiro, F.C., Franco, A.A., Pereira, M.G., et al. 2004. Dinâmica da serapilheira e transferência de nitrogênio ao solo em plantios de *Pseudosamanea guachapele* e *Eucalyptus grandis*. *Pesquisa Agropecuária Brasileira*, 39, 597–601.

Balieiro, F.C., Franco, A.A., Dias, L.E., et al. 2007. Evaluation of the throughfall and stemflow nutrient contents in mixed and pure plantations of *Acacia mangium*, *Pseudosamenea guachapele* and *Eucalyptus grandis*. *Revista Árvore*, 31, 339–346.

Balieiro, F.C., Pereira, M.G., Alves, B.J.R., et al. 2008. Soil carbon and nitrogen in pasture soil reforested with eucalyptus and guachapele. *Revista Brasileira da Ciência do Solo*, 32, 1253–1260.

Bayer, C., Martin-Neto, L., Mielniczuk, J., et al. 2001. Changes in soil organic matter fractions under subtropical no-till cropping systems. *Soil Science Society of America Journal*, 65, 1473–1478.

Bethlenfalvay, G.J., Pacovsky, R.S., Bayne, H.G., 1982. Interactions between nitrogen-fixation, mycorrhizal colonization, and host-plant growth in the *Phaseolus-Rhizobium-glomus* symbiosis. *Plant Physiology*, 70, 446–450.

Bethlenfalvay, G.J., Franson, R.L., Brown, M.S., et al. 1989. The *Glycine-Glomus-Bradyrhizobium* symbiosis. IX. Nutritional, morphological and physiological responses of nodulated soybean to geographic isolates of the mycorrhizal fungus *Glomus mosseae*. *Plant Physiology*, 76, 226–232.

Boddey, R.M., Xavier, D.F., Alves, B.J.R., et al. 2010. Brazilian agriculture: The transition to sustainability. *Journal of Crop Production*, 9, 593–621.

Boddey, R.M., Jantalia, C.P., Conceição, P.C., et al. 2010. Carbon accumulation at depth in Ferralsols under zero-till subtropical agriculture. *Global Change Biology*, 16, 784–795.

Bot, A.J., Nachterhaele, R.O., Young, A. 2000. *Land Resource Potential and Constraints at Regional and Country Levels*. Land and Water Development Division, Food and Agriculture Organization of the United Nations, Rome.

Brown, L.R. 1995. *Who Will Feed China*. W.W. Norton & Co. Inc. New York.

Brown, L.R. 1996. *Tough Choices: Facing the Challenge of Food Scarcity*. W.W. Norton & Co. Inc. New York.

Brown, S., Lugo, A.E. 1990. Tropical secondary forests. *Journal of Tropical Ecology*, 6, 1–32.

Brundrett, M.C. 2002. Tansley Review: Coevolution of roots and mycorrhizas of land plants. *New Phytologist*, 154, 275–305.

Campelo, A.B., Döbereiner, J. 1969. Estudo sobre a inoculação cruzada de algumas leguminosas florestais. *Pesquisa Agropecuária Brasileira*, 4, 67–72.

Chada, S. de S., Campello, E.F.C., de Faria, S.M. 2004. Sucessão vegetal em uma encosta reflorestada com leguminosas arbóreas em Angra dos Reis, RJ. *Revista Árvore*, 28, 801–809.

Chen, W-M., Faria, S.M. de. et al. 2005. Proof that *Burkholderia* strains form effective symbioses with legumes: a study of novel *Mimosa*-nodulating strains from South America. *Applied and Enviromental Microbiology*, 71, 7461–7471.

Corby, H.D.L. 1981. The systematic value of leguminous root nodules. In: *Advances In Legume Systematics*. (Polhill, R.M., Raven, P.H., Eds.) pp. 657–669, Royal Botanic Gardens, Kew, London.

Costa, G.S, Franco, A.A., Damasceno R.N., et al. 2004. Aporte de nutrientes pela serapilheira em uma área degradada e revegetada com leguminosas arbóreas. *Revista Brasileira de Ciência do Solo*, 28, 919–927.

Craswell, E.T., Lefroy, R.D.B. 2001. The role and function of organic matter in tropical soils. *Nutrient Cycling in Agroecosystems*, 61, 7–18.

Dean, W. 1995. With Broadax and Firebrand: The Destruction of the Brazilian Atlantic Forest. University of California Press, Berkeley, CA.

Deans, J.D., Diagne, O., Nizinski, J., et al. 2003. Comparative growth, biomass production, nutrient use and soil amelioration by nitrogen-fixing tree species in semi-arid Senegal. *Forest Ecology and Management*, 176, 253–264.

Dommergues, Y., Duhoux, E., Diem, H.G. 1999. *Les Arbres Fixateurs D'Azote*. Editions Espaces, Montpellier, France.

Elliot, G.N., Chen, W.-M., Chou, J.-H., et al. 2007. *Burkholderia phymatum* is a highly effective nitrogen-fixing symbiont of *Mimosa* spp. and fixes nitrogen *ex planta*. *New Phytologist*, 173, 168–180.

EMBRAPA. 1997. *Manual de métodos de análises de solo*. Embrapa Solos, Rio de Janeiro.

Erhlich, P. 1969. *Population Bomb* 1st Edition. Mass Market Paperback, New York.

Faria, S.M. de, Jesus, R.M., Menandro, M.S., et al. 1984. New nodulating legume trees from South East Brazil. *New Phytologist*, 98, 317–328.

Faria, S.M. de, Lewis, G.P., Sprent, J.I., et al. 1989. Occurrence of nodulation in the Leguminosae. *New Phytologist*, 111, 607–619.

Faria, S.M. de, Lima, H.C., Mucci, E.S.F. et al. 1987. Nodulation of legume trees from South East Brazil. *Plant and Soil*, 99, 347–356.

Faria, S.M. de, Lima, H.C., Oliveira, F.F., et al. 1999. Nodulação em espécies florestais, especificidade hospedeira e implicações na sistemática de Leguminosae morfologia dos nódulos em florestais. *Revista Brasileira de Ciência do Solo*, 23, 667–686.

Faria, S.M. de, Campello, E.F.C. Xavier, D.F., et al. 2010. Multi-Purpose Fast-Growing Legumes Trees for Smallholders in the Tropics and Sub-Tropics: Firewood, fencing and fodder. Comunicado Técnico No. 127, Embrapa Agrobiologia, Seropédica, RJ, Brazil.

Feldpausch, T.R., Rondon, M.A., Fernandes, E.S., et al. 2004. Carbon and nutrient accumulation in secondary forests regenerating on pastures in central Amazônia. *Ecological Applications*, 14, S164–S176.

Fisher, M.J., Rao, I.M., Ayarza, M.A., et al. 1994. Carbon storage by introduced deep-rooted grasses in the South American savannas. *Nature*, 371, 236–238.

Flores-Aylas, W.W., Saggin Júnior, O.J., Siqueira, J.O., et al. 2003. Efeito de *Glomus etunicatum* e fósforo no crescimento inicial de espécies arbóreas em semeadura direta. *Pesquisa Agropecuária Brasileira*, 38, 257–266.

Francis, R., Read, D.J. 1994. The contributions of mycorrhizal fungi to the determination of plant community structure. *Plant and Soil*, 159, 11–25.

Franco, A.A., Faria, S.M. de. 1997. The contribution of N_2-fixing tree legumes to land reclamation and sustainability in the tropics. *Soil Biology and Biochemistry*, 29, 897–903.

Giller, K.E. 2001. *Nitrogen fixation in tropical cropping systems*. CAB Publishing, Wallingford, Oxon, UK.

Gleason, S.K., Tilman, D. 1990. Allocation and transient dynamics of sucession on poor soils. *Ecology*, 71, 1144–1155.

Handayanto, E., Cadisch, G., Giller, K.E. 1995. Manipulation of quality and mineralization of tropical legume tree prunings by varying nitrogen supply. *Plant and Soil*, 176, 149–160.

Holl, K.D., Loik, M.E., Lin, E.H.V., et al. 2000. Tropical montane forest restoration in Costa Rica: Overcoming barriers to dispersal and establishment. *Restoration Ecology*, 8, 339–349.

Janos, D.P. 1996. Mycorrhizas, succession and rehabilitation of deforested lands in the humid tropics. In: Fungi and Environmental Change. (Frankland, J.C., Gadd, G.M., Eds.), p. 1–18. Cambridge, University Press, Cambridge, UK.

Jasper, D.A., Abbott, L.K., Robson, A.D. 1992. Soil disturbance in native ecosystems—The decline and recovery of infectivity of VA mycorrhizal fungi. In: *Mycorrhizas in Ecosystems* pp. 151–155.(Read, D.J., Lewis, D.H., Fitter, A.H. et al., Eds.) CAB International, Wallingford, UK.

Jordan, D.C. 1982. Transfer of *Rhizobium japonicum* Buchanan (1980) to *Bradyrhizobium* gen. nov., a genus of slow-growing, root nodule bacteria from leguminous plants. *International Journal of Systematic Bacteriology*, 32, 136–139.

Lal, R. 2004. Soil carbon sequestration impacts on global climate change and food security. *Science*, 34, 1623–1627.

Louis, I., Lim, G. 1988. Differential response in growth and mycorrhizal colonisation of soybean to inoculation with two isolates of *Glomus clarum* in soils of different P availability. *Plant and Soil*, 112, 37–43.

Macedo, M.O., Resende, A.S., Garcia, P.C., et al. 2008. Changes in soil C and N stocks and nutrient dynamics 13 years after recovery of degraded land using leguminous nitrogen-fixing trees. *Forest Ecology and Management*, 255, 1516–1524.

Machado, R.L., Resende, A.S. de, Campello, E.F.C., et al. 2006. *Recuperação de voçorocas em áreas rurais*. Sistemas de Produção, No. 1, Embrapa Agrobiologia, Seropédica, RJ, Brazil.

Maddock, J.E.L., dos Santos, M.B.P., Prata, K.R. (2001). Nitrous oxide emission from soil of the Mata Atlantica, Rio de Janeiro state, Brazil. *Journal of Geophysical Research*, 106, 23055–23060.

Mailly, D., Margolis, H.A. 1992. Forest floor and mineral soil development in *Casuarina equisetifolia* plantations on the coastal sand dunes of Senegal. *Forest Ecology and Management*, 55, 259–278.

Midgley, S.J., Turnbull, J.W. 2003. Domestication and use of Australian Acacias: an overview. *Australian Systematic Botany*, 16, 89–102.

Moreira, F.M.S., Silva, M.F., de Faria, S.M. 1992. Occurrence of nodulation in legumes species in the Amazon region of Brazil. *New Phytologist*, 121, 563–570.

Moreira, F.M.S., Siqueira, J.O., Brussard, L. 2006. *Soil Biodiversity in Amazonian and Other Brazilian Ecosystems*. CAB International, Wallingford, UK.

Moulin, L., Munive, A., Dreyfus, B., et al. 2001. Nodulation of legumes by members of the β-subclass of proteobacteria. *Nature*, 411, 948–950.

Polhill, R.M., Raven, P.H., Stirton, C.H. 1981. Evolution and systematics of the Leguminosae. In: *Advances In Legume Systematics*. (Polhill, R.M., Raven, P.H., Eds.) pp. 126. Royal Botanic Gardens, Kew, London.

Polhill, R.M. 1994. Classificação of the Leguminosae. In: *Phytochemical Dictionary 1*. (Bisby, F.A., Buckingham, J., Harborne, J.B., Eds.) pp. xxxv–lvii. Chapman & Hall, London.

Pouyu-Rojas, E., Siqueira, J.O., Santos, J.D.G. 2006. Symbiotic compatibility of arbuscular mycorrhizal fungi with tropical tree species. *Revista Brasileira da Ciência do Solo*, 30, 413–424.

Resh, S.C., Binkley, D., Parrotta, J.A. 2002. Greater soil carbon sequestration under nitrogen-fixing trees compared with *Eucalyptus* species. *Ecosystems*, 5, 217–231.

Robinson-Boyer, L., Grzyb, I., Jeffries, P. 2009. Shifting the balance from qualitative to quantitative analysis of arbuscular mycorrhizal communities in field soils. *Fungal Ecology*, 2, 1–9.

Rocha, F.S., Saggin Júnior, O.J., Silva, E.M.R., et al. 2006. Dependência e resposta de mudas de cedro a fungos micorrízicos arbusculares. *Pesquisa Agropecuária Brasileira*, 41, 77–84.

Saggin Júnior, O.J., Siqueira, J.O. 1995. Avaliação da eficiência simbiótica de fungos endomicorrízicos para o cafeeiro. *Revista Brasileira da Ciência do Solo*, 19, 221–228.

Schubert, A., Hayman, D.S. 1986. Plant growth responses to vesicular-arbuscular mycorrhiza. XVI. effectiveness of different endophytes at different levels of soil phosphate. *New Phytologist*, 103, 79–90.

Siqueira, J.O., Saggin Júnior, O.J. 2001. Dependency on arbuscular mycorrhizal fungi and responsiveness of some Brazilian native woody species. *Mycorrhiza*, 11, 245–255.

Silver, W., Thompson, A.W., McGroddy, M.E., Varner, R.K., Dias, J.D., Silva, H., Cril, P.M., Keller, M. (2005). Fine root dynamics and trace gas fluxes in two lowland tropical forest soils. *Global Change Biology*, 11, 290–306.

Sisti, C.P.J., Santos, H.P.de, Kochhann, R.A., et al. 2004. Change in carbon and nitrogen stocks in soil under 13 years of conventional or zero tillage in southern Brazil. *Soil and Tillage Research*, 76, 39–58.

Six, J., Conant, R.T., Paul, E.A., et al. 2002. Stabilization mechanisms of soil organic matter: Implications for C-saturation of soils. *Plant and Soil*, 241, 155–176.

Smith, E.S., Read, J.D. 1997. *Mycorrhizal Symbiosis*. 2 ed. Academic Press, New York.

Somasegaran, P., Hoben, H.J. 1994. *Handbook for Rhizobia: Methods in Legume-Rhizobium Technology*. Springer Verlag, Berlin.

Sprent, J.I. 2001. Nodulation in Legumes. Kew Publishing, London.

Sprent, J.I. 2009. *Legume Nodulation: a Global Perspective*. John Wiley & Sons, Chichester, UK.

Tarré, R.M., Macedo, R., Cantarutti, R.B., et al. 2001. The effect of the presence of a forage legume on nitrogen and carbon levels in soils under *Brachiaria* pastures in the Atlantic forest region of the South of Bahia, Brazil. *Plant and Soil*, 234, 15–26.

Thurow, R., Kilman, S. 2009. *ENOUGH: Why the World's Poorest Starve in and Age of Plenty*. BBS Public Affairs, New York.

Urquiaga, S., Cruz, K.H.S., Boddey, R.M. 1992. Contribution of nitrogen fixation to sugar cane: Nitrogen-15 and nitrogen-balance estimates. *Soil Science Society of America Journal*, 56, 105–114.

Vandenkoornhuyse, P., Husband, R., Daniell, T.J., et al. 2002. Arbuscular mycorrhizal community composition associated with two plant species in a grassland ecosystem. *Molecular Ecology*, 11, 1555–1564.

Vandenkoornhuyse, P., Ridgway, K.P., Watson, I.J., et al. 2003. Co-existing grass species have distinctive arbuscular mycorrhizal communities. *Molecular Ecology*, 12, 3085–3095.

Veldkamp, E. 1994. Organic carbon turnover in three tropical soils under pasture after deforestation. *Soil Science Society of America Journal*, 58, 175–180.

Vital, A.R.T., Guerrini, T.A., Franken, W.K., et al. 2004. Produção de serapilheira e ciclagem de nutrientes de uma floresta estacional semidecidual em zona ripária. *Revista Árvore*, 28, 793–800.

Zimmer, A.H., Euclides Filho, K.P. 1997. As pastagens e a pecuária de corte brasileira. In: Proceedings of an International Symposium on Animal Production under Grazing, pp. 349–379. Departamento de Zootecnia da Universidade Federal de Viçosa. Viçosa, MG.

Section 3
Epi- and Endo-Phytic Microbes

Chapter 10
Nitrogen: Give and Take from Phylloplane Microbes
Mark A. Holland

Introduction

Imagine this vast landscape: an uneven terrain of undulating mounds, steep crags and deep crevasses, wild swings in temperature that can shift from subfreezing to scalding hot within a few hours, extended periods of drought punctuated by torrential rain and flooding, and intense solar radiation from which there is no escape. From the perspective of a microbial colonist, this is the surface of a leaf (Hirano and Upper, 2000; Suslow, 2002).

Leaf surfaces comprise an enormous habitat. It has been estimated that the phylloplane may extend to 4×10^8 km² globally (Morris and Kinkle, 2002). To the microbes inhabiting this surface, it is operationally even larger given its uneven topography. Of course, the phylloplane is not a uniform environment, but even at its most inhospitable, it is home to a microbial flora adapted to life in its extremes. Many of the same genera (and even the same species) of microbe are found as regular inhabitants of leaf surfaces worldwide, and their relationships with their plant hosts range from pathogenic to parasitic, commensal to mutualistic. Other bacteria are only occasional or accidental visitors to leaf surfaces. All together, the phylloplane assemblage on a single plant may include dozens of bacterial genera and species (Beattie and Lindow, 1999). Sometimes the same bacterial species can be found on the leaf surface, embedded in or beneath the leaf cuticle, and on internal leaf surfaces, bringing to question whether "phylloplane," "phyllosphere," and "leaf" are functionally distinct. In this chapter, "phylloplane" is applied *sensu latu* to describe the leaf habitat.

The variability of niches on leaf surfaces and the diversity of microbial life found there would make generalizations about nitrogen economy and resource allocation difficult, were it not for the fact that so much remains to be learned about life in the phylloplane. Questions abound. On the microbial side of the equation, what is the nature of the microbial load and the microbial demand for nitrogen? For the microbes living on leaf surfaces, what sources and

Ecological Aspects of Nitrogen Metabolism in Plants, First Edition. Edited by Joe C. Polacco and Christopher D. Todd.
© 2011 by John Wiley & Sons, Inc. Published 2011 by John Wiley & Sons, Inc.

forms of nitrogen are available? What evidence is there that microbial growth is limited by the availability of nitrogen or other nutrients? On the plant side, how do microbial activities influence plant nitrogen metabolism, both positively and negatively?

What is the nature of the microbial load and the microbial demand for nitrogen?

Do phylloplane microbes command a large quantity of nitrogen? Is nitrogen likely to be limited for them? Do their needs compromise plant resources?

In some rainforest ecosystems, microbial life on the surfaces of leaves has been reported to form a layer 50 μm thick (Tukey, 1971). Population densities for phylloplane bacteria are reported to range up to 10^8 colony-forming units (cfu) /cm^2 of leaf; fungal populations being somewhat smaller (Morris and Kinkle, 2002). Some "back of the envelope" calculations suggest the scope of the nitrogen demands of such a population. If we take the protein content of a typical bacterial cell to be approximately 0.1 pg, the protein content of the bacterial population at 10^8 cfu/cm^2 of leaf is roughly equivalent to 10 μg/cm^2 of leaf. Using a standard conversion factor for total N to protein of 6.25 (Van Gelder, 1981), this is equivalent to 1.6 μg N/cm^2 of leaf tissue tied up in bacterial cells. If an average plant leaf is 2–5% protein (wet weight) and we assume that an average plant leaf weighs 150 g/m^2 (see, for example, Prado and DeMoraes, 1997) then leaf tissue weighs approximately 15 mg/cm^2. At 3.75% protein, the leaf contains 0.56 mg protein/cm^2. Converting protein to N, the leaf contains 90 mg N/cm^2. So at maximal bacterial load, the bacterial population contains only about 2% as much nitrogen as its plant host tissue. Although these calculations are admittedly crude, it seems unlikely that the bacterial demand for nitrogen has a large negative impact on the nitrogen economy of the plant.

What is the nutrient stream or pool that feeds phylloplane microbes?

Microbes on leaf surfaces must get all of their nitrogen (1) from their plant hosts, (2) from the atmosphere, or (3) from foliar fertilizers of some kind. Discussion of these potential N sources follows:

1. Nitrogen Resources from the Plant Host: Plant leaves, generally considered to be relatively water-tight, have been found to leach a variety of organic and inorganic compounds, as much as 6% of their dry weight in the case of one study on *Phaseolus* (Tukey, 1971). Leaf leachates include all the essential minerals, macro- and micro-nutrients, sugars and sugar alcohols, pectins, all amino acids, other organic acids, growth regulators, alkaloids, phenolics, CO_2, acetone, terpenoids, aldehydes, alcohols, sulfides, and vitamins (Tukey, 1971; Godfrey, 1976; Whipps et al., 2008). Sattelmacher and others (1998) wrote, "…so far it is not possible to explain the precise path by which ions are leached from the leaf apoplast, but one may speculate that stomates or hydathodes and trichomes are of significance." Hydathodes and trichomes have been implicated in the secretion of various metabolites, including glutamine, by other authors as well (Cutter, 1976; Pilot et al., 2004). Indeed, phylloplane microbes are often found clustered about glandular and nonglandular trichomes and leaf hairs as well as around the stomata, in the junctions between adjacent cells and near leaf veins (Omer, 2004; DeCosta et al., 2006; Baldotto and Olivares, 2008).

   In one study, the diffusion of water across plant cuticles was shown to be enhanced up to 50% by phylloplane bacteria of genera *Pseudomonas, Stenotrophomonas, Achromobacter, Xanthomona*, and *Corynebacterium* (Schreiber et al., 2005). Other studies demonstrated nutrient leakage associated with the indole acetic acid secreted by microbial epiphytes (Brandl and Lindow, 1998; Manulis et al., 1998; Beattie and Lindow, 1999). Although the physiological mechanism(s) for this increase in leaf permeability has not been studied specifically, one may speculate that IAA-induced modifications to cell wall are involved. Significantly, Brandl and Lindow (1998) showed a connection between low water potential and upregulation of bacterial IAA synthesis genes.

The immediate source of leachate is the apoplast, a plant compartment sometimes taken for granted as the nonliving "box" that surrounds the plant cell. But the apoplast of the leaf represents the end of the transpiration stream (Canny 1990), contiguous with the walls of the xylem and receiving the contents of xylem for distribution throughout the leaf (Sattelmacher, 2001). The fate of materials arriving at the leaf through the xylem, whether they remain in the apoplast or move into the symplast, depends on the metabolic demands of the plant cells and the balance between osmotic and hydrostatic pressures within the leaf tissue (Westgate and Steudle, 1985). Analysis of the nitrogen contents of the apoplast of *Brassica napus* revealed a "highly dynamic NH4+ pool" (Neilsen and Schjoerring, 1998) with ammonium concentrations as high as 1 mM (Husted and Schjoerring, 1995). In the apoplast of sugarcane grown either under a high nitrogen fertilization regime or unfertilized, or inoculated with *Gluconacetobacter diazotrophicus*, Tejera and others (2006) found amino acids (70–90% of N) and protein (20–30% of N) as well as inorganic forms: ammonium, nitrate and nitrite (<20% of N). In the apoplastic sap of the three cultivars they tested, total amino acid concentration averaged approximately 1 mg/mL.

Some of the nitrogen found in the apoplast has its origin in the transpiration stream, some comes from exchange with the symplast (presumably, in the case of ammonium, to control cytoplasmic pH), and some of it comes from the atmosphere.

2. Nitrogen Resources from the Atmosphere: Leaves can function as either a source or a sink for ammonia from the atmosphere, depending on the physiological status of apoplast tissue and on the age of the leaf (Husted and Schjoerring, 1996; Mattson et al., 1998; Mattson and Schjoerring, 2003; Hayashi et al., 2008). The size of the nitrogen pool in apoplast, combined with the physical parameters (temperature, humidity, light intensity, pH) that influence stomatal conductance contribute to a "stomatal NH_3 compensation point," which can be used to predict the flux of atmospheric ammonia into and out of the leaf (Herrmann et al., 2002). In addition to uptake by the stomata, ammonia from the atmosphere can also enter the leaf nitrogen pool by dissolution in leaf surface water and subsequent equilibration with the apoplastic sap (Flechard et al., 1999). Ammonia uptake by this route must have an immediate impact on surface-dwelling microbes. Various atmospheric oxides of nitrogen can also be taken up by leaf tissue. Pollutants like N_2O are readily absorbed through the stomata and can be used as a nitrogen source by the plant and presumably by microbial symbionts as well (Wellburn, 1990; Hufton et al., 1996). Recent recognition of a "new N cycle" that includes a link converting nitrite and ammonium to dinitrogen gas through the activities of the anammox bacteria may also be of significance here (see Chapter 1).

3. Nitrogen Resources from Foliar Fertilization: The application of nitrogen-containing fertilizers directly to leaf surfaces has obvious implications for the nutrition of the phylloplane community. The application of organic nitrogen to leaves of turfgrass, for example, was shown to have a direct stimulatory effect on phylloplane yeast populations (Nix-Stohr et al., 2008). Foliar fertilization increased the levels of essentially all nutrients in the leaves and heads of broccoli, and it increased the chlorophyll content of leaf tissue (Yildirim et al., 2007). Foliar applications of urea to *Rhododendron* resulted in measurable and long-lasting changes in nutrient uptake and accumulation (Scagel et al., 2008). In both the broccoli and *Rhododendron* studies, changes in the patterns of

resource allocation and/or increases in chlorophyll content of treated tissues suggest a possible role for microbial symbionts mediated by their production of plant growth regulators. Controlled studies that assess the contributions of phylloplane microbes to plant resource allocation have not been carried out.

What evidence is there that microbial growth is limited by the availability of nitrogen or other nutrients?

Data regarding nutrient–imposed limitations to microbial growth in the phylloplane are themselves limited and conflicting. Mercier and Lindow (2000) suggested that the growth of bacterial populations on bean (*Phaseolus*), corn, cucumber, pea, tobacco, and tomato is limited by the availability of a carbon source. Their results show that the amounts of free sugars found on the leaves of different plant species were correlated directly with the extent of the bacterial populations developed. Brandl and Amundson (2008) found a correlation between both carbon and nitrogen content of lettuce leaf exudates on leaf surfaces and population growth of phylloplane bacteria. However, they also demonstrated that the addition of ammonium nitrate, but not glucose, to leaves stimulated bacterial growth, suggesting that nitrogen is a limiting factor for the population. Phylloplane yeast populations responded with growth to organic nitrogen, but not to inorganic nitrogen or carbohydrates or other nutrients (Nix-Stohr et al., 2008). In extracts of pesticide-treated leaves of maize, amino acids were reduced compared to carbohydrates, and bacterial populations were correlated with the size of the nitrogen pool (Annapurna and Rao, 1982). To confuse the issue further, Yadav and others (2005) found no correlation between bacterial population size and either soluble sugar content or nitrogen content in the leaves of various Mediterranean perennials. The only parameters that seemed to limit bacterial populations in their study group were water and phosphorus content followed by thickness of the adaxial epidermis. Wilson and Lindow (1994) demonstrated that the ability of pairs of bacterial species to establish large populations on *Phaseolus* leaves was inversely correlated with the overlap in their nutrient use profiles, suggesting that competition for nutrients is a limiting factor to bacterial growth.

How do microbial activities influence plant nitrogen metabolism both positively and negatively?

It seems that the nitrogen demands of phylloplane bacteria are at best a minor drain on plant resources. It may even be the case that the nutritional needs of the microbial population are met by material that simply leaches through the cuticle of the leaf or volatilizes through the stomata whether or not bacteria are present. But if the phylloplane symbionts are not compromising plant metabolism as nitrogen parasites, do they actually have a positive influence on nitrogen economy?

One place where microbial contributions to nitrogen metabolism are clear is in epiphytic tank bromeliads. These plants, with a vase-like tank formed from overlapping leaves, trap standing pools of water. A study of nutritional strategies in these plants by Inselbacher and others (2007) demonstrated that nutrients are taken up prominently by the submerged surfaces of their tank leaves. Ammonium is a preferred nitrogen source, although some urea is also taken up, but the activities of microbes in the tank are responsible for the mineralization of essentially all of the nitrogen that enters the tank water. Similarly, microbes may contribute to the mineralization of nitrogen in insectivorous plants like the pitcher plants (Cutter, 1976).

A less obvious, but no less interesting, case was demonstrated by Papen and others (2002). Their study revealed the presence of aerobic chemolithoautotrophic ammonia oxidizing and chemolithoautotrophic nitrite oxidizing bacteria in the apoplasts of spruce needles in a German

Table 10.1. Recent references to nitrogen fixation by microbes inhabiting the leaves of plants.

Plant host	Microbe	Reference
Maize	*Azospirillum lipoferum*, *A. amazonense*, *Burkholderia* sp.	Miyauchi et al., 2008
Mosses in the arctic	Cyanobacteria	Solheim et al., 2004
Sugarcane, coffee, corn, and rice	Acetobacteriaceae	Saravanan et al., 2008
Sugarcane	*Gluconacetobacter diazotrophicus* and *Herbaspirillum* sp.	Suman et al., 2008
Rice	*Burkholderia kururiensis*	Mattos et al. 2008
Rice	*Klebsiella* sp.	Razi and Sen, 1996
Rice	*Pantoea agglomerans*	Feng et al., 2006
Rice	*Pantoea*, *Methylobacterium*, *Azospirillum*, *Herbaspirillum*, *Burkholderia*, and *Rhizobium*	Mano and Morisaki, 2008
Tomato	*Burkholderia* spp.	Caballero-Mellado et al., 2007
Typha australis	*Klebsiella oxytoca*	Jha and Kumar, 2007
Wheat	*Azospirillum brasilense*	Van Dommelen et al., 2009
Sweet potato	Unidentified bacteria	Terakado-Tonooka et al., 2008
Tillandsia	*Pseudomonas stutzeri*	Puente and Bashan, 1993
Cotton	*Bacillus macerans*	Vigneshwaran and Natarajan, 2003
Tropical legumes	*Methylobacterium nodulans*	Sy et al., 2001; Jourand et al., 2004; Renier et al., 2008; Madhaiyan et al., 2009
Tropical epiphytes	Cyanobacteria and gamma-proteobacteriaceae	Furnkranz et al., 2008
Nonlegumes	*Rhizobium* and relatives	Bhattacharjee et al., 2008

forest exposed to high levels of atmospheric pollution. The activities of the bacteria in the needles were sufficiently high to feed nitrogen to the plant. Interestingly, the forest soil of their study site, which might have been expected to harbor high numbers of nitrifiers, was shown to have very low populations of these bacteria. Annual variations in numbers of the nitrifiers in soil led to the conclusion that leaf litter containing an inoculum of nitrifying bacteria actually seeds the soil annually. This effect on the soil flora recalls the statement of Tukey (1971) that, "Not only is the plant a product of the soil, but the soil is a product of the plant." Phylloplane microbes certainly enter the soil community through leaf drop and probably contribute to the decomposition of leaf litter. Leachates from the leaves of plants must also add to the nutrient pool in the soil.

The apoplast contains a high concentration of organic resources, but may also have a relatively low partial pressure of oxygen. This makes the apoplast an apt environment for nitrogen-fixing bacteria (Hecht-Buchholz, 1998; Sattelmacher et al., 1998). There is an enormous body of literature devoted to microbial nitrogen fixation, and the topic is addressed in other chapters in this volume. Not to duplicate that material, Table 10.1 summarizes some recent references to nitrogen fixation by leaf-associated bacteria that illustrate the extent of this symbiosis. General reviews of the subject are found in Kneip and others (2007) and Roesch and others (2008).

Phylloplane microbes may also exert a substantial effect on nitrogen allocation in their plant host through their production of plant growth regulators. Sakakibara and others (2006) stated, "(there are) multiple signaling routes that communicate internal and external nitrogen status. One route depends on nitrate itself and one uses cytokinin as a messenger." Taiz and Zeiger (2006) discuss cytokinin-induced nutrient mobilization, a phenomenon in which nutrients are preferentially transported to and accumulate in cytokinin-treated tissue. Walch-Liu and others (2005), Cline and others (2006), Tamaki and Mercier (2007), and Hirose and others (2007) show this effect on long-distance nitrogen transport. In what seems to be the converse of this cytokinin effect, Peng and others (2008) showed that cytokinin levels rise in response to foliar application of nitrate through upregulation of a cytokinin biosynthesis gene. Cytokinins are known to be produced by a number of plant-associated microbes including *Agrobacterium, Methylobacterium*, and *Pantoea* (see for example Koenig et al., 2002; Omer et al., 2004a,c). The growth regulators produced by bacteria have been demonstrated to have a measurable effect on plant metabolism (Butler 2001; Arkhipova et al., 2005; Abanda-Nkpwatt et al., 2006; Madhaiyan et al., 2006). It may be profitable to ask whether the cytokinins produced on the leaf by microbial symbionts signal a demand for nitrogen resources from root.

Cytokinins are known to play a signalling role in nodulation by nitrogen-fixing bacteria. It has even been suggested that cytokinins may be a ubiquitous signal in the establishment of all plant microbe symbioses (Frugier et al., 2008). In fact, cytokinin production by phylloplane bacteria is well established and will be further discussed in the *Methylobacterium* case study that follows this section.

Auxin, in the form of indole acetic acid (IAA), has already been mentioned here as a signal that enhances nutrient leakage through the leaf cuticle. Brandl and Lindow (1997) demonstrated that IAA synthesis is upregulated in *Erwinia herbicola* when it is growing on plants, but not when it is in culture. Like cytokinin, auxin has been implicated as an integrative signal of nitrogen status, produced in shoots, and communicated to roots (Guo et al., 2005). Auxin has been demonstrated to transfer from epiphytic bacteria on maize to the coleoptile (Libbert and Silhengst, 2006).

Perhaps the current record holder at contributing to plant nitrogen metabolism is *Pantoea agglomerans* YS19. This bacterium not only fixes nitrogen but has been demonstrated to produce indole acetic acid, abscisic acid, gibberellic acid, and cytokinin. Foliar application of the organism to rice results in changes in resource allocation patterns (Feng et al., 2006) and stimulates plant growth.

Methylobacterium spp. as Phylloplane Symbionts: A Case Study

Phylloplane symbionts of the genus *Methylobacterium*, often referred to in the literature as PPFM (pink-pigmented facultatively methylotrophic) bacteria, illustrate many of the ideas presented here.

PPFMs are widely distributed symbionts of plants. They have been isolated from the leaves, roots, and seeds of more than 100 different plant species and from the soil in which plants are growing (Corpe and Basile, 1982; Corpe and Rheem, 1989). They are seed-transmitted, so presumably plants start out at germination with an inoculum of the bacteria and maintain a population of them throughout their lives, although PPFM colonists may periodically arrive via blown dust, etc., on plant leaf surfaces (Omer, 2004c). The bacteria have not been shown to be pathogens on any of their hosts. Populations have been shown to change during the growing season (Omer et al., 2004c), and on at least some species of plant, they are the single most abundant species of symbiont (Hirano and Upper, 1992).

Despite their abundance, most plant scientists seem never to have heard of them or to consider their activities important to plants. This was certainly true for this author personally. Working with the soybean urease enzymes, we came to appreciate the contribution of PPFM bacteria only when genetic ablation of both the plant isozymes failed to remove urease activity from certain plant tissues (Holland and Polacco, 1992). PPFM bacteria on soybean have been shown to consume ureides transported from nitrogen-fixing nodules of tropical legumes like soybean (reviewed by Polacco and Holland, 1993). The bacteria also consume urea, and there is evidence that urease-negative plant tissues *in vitro* are able to grow on urea as the sole nitrogen source when supported by urease-positive PPFM bacteria (Stebbins et al., 1991). This observation indicates that the bacterial urease generates sufficient ammonia to meet bacterial metabolic demands as well as those of their host. Low levels of urease enzyme are found ubiquitously throughout the plant kingdom. An interesting possibility is that in at least some cases, this enzyme is the product of phylloplane microbes rather than of the plant itself. Possibly this is why Cambui and others (2009) found large quantities of urease associated with cell wall in a tank bromeliad?

The PPFMs produce the plant growth regulators cytokinin (Butler et al., 2000; Butler, 2001; Koenig et al., 2002) and auxin (Ivanova et al., 2001; Omer et al., 2004b; Abanda-Nkpwatt et al., 2006), and there is convincing evidence that these growth regulators are physiologically meaningful to the plant host. The size of the PPFM population is highly significantly correlated both with the growth of the host and with the concentration of cytokinin extractable from host tissue (Butler et al., 2000; Butler, 2001). Significantly, the cytokinins produced by PPFMs are trans-Zeatin and its riboside (Koenig et al., 2002). These are the same cytokinins suggested by Sakakibara and others (2006) as long-distance signals for nitrogen status from root to shoot

Another example of PPFM contribution to nitrogen metabolism comes from experiments with PPFM mutants that overproduce the amino acid methionine. Overproducers were isolated in culture from a mutagenized population by their resistance to ethionine and their phenotype confirmed biochemically. The mutant was applied to soybean leaves, and the beans produced on treated plants were assayed for bound methionine in seed protein. When compared to controls, the seed proteins of treated plants contained as much as 15% more methionine (Witzig and Holland, 1998). These experiments suggest that the activities of the PPFMs can alter host nitrogen metabolism and improve the nutritional quality of their host as a crop plant. Similar experiments were performed with vitamin B12 overproducing mutants and resulted in the accumulation of nanogram quantities of the vitamin in lettuce leaf tissue (Witzig and Holland, 1998, 1999). Although higher plants are not thought to require B12, the results are significant for two reasons:

- First, Basile and others (1985) demonstrated that PPFM bacteria support the growth of liverworts in culture through the production of B12, which is required by some nitrogen-fixing organisms (Cowles et al., 1969).
- Second, the results in lettuce are a second example using the phylloplane PPFMs to alter the nutritional qualities of plant tissue.

Certain species of the PPFM bacteria are also known to nodulate plants and fix nitrogen (Sy et al., 2001; Jourand et al., 2004; Lee et al., 2006; Abanda-Npkwatt et al., 2006).

A Note on Nitrogenous Waste

In General Biology class, we all learn that metabolism produces nitrogenous waste. Although that general principle certainly applies to our own (and other animal) metabolism, the situation is far from clear in plants and bacteria. Halle (2002) suggests that plants store their nitrogenous wastes in the form of alkaloids, glucosinolates, and other nasty molecules—compounds that do double duty as antimicrobials. Perhaps some of the nitrogenous components of leachate from leaf should also be considered as waste. Generally speaking, bacteria do seem to have difficulties with plant secondary metabolites. However, it is relatively easy to find exceptions in the literature, such as bacteria that use them as a carbon source (see for example Santana et al., 2002). Is it legitimate to cast phylloplane microbes in the role of waste managers? Earlier, it was mentioned that phylloplane microbes are frequently found associated with trichomes. Interestingly, the trichomes are among favored sites for the secretion of these putatively antimicrobial compounds (Shepherd et al., 2006).

The nitrogenous wastes of bacteria are similarly elusive. Among the nitrifiers, nitrite, nitrate, and other oxides could be called nitrogen waste products. When these organisms inhabit leaves, of course, the plant host may be beneficiary to such waste as a usable nitrogen source. Ammonia is sometimes also identified as a waste product of bacteria. As a waste deposited on leaves, it too proves beneficial to the plant.

A Note on Pathogen Versus Nonpathogen

In this review, a distinction between pathogens and nonpathogens as phylloplane symbionts has not been made. Although there certainly may be a difference where the fate of the plant is concerned, the plant response to both pathogen and nonpathogen appears to employ similar biochemistry, at least at the outset of the relationship (Soto et al., 2009). The same plant resources seem to be available to both pathogens and nonpathogens alike. Even the signals that establish the symbioses of nonpathogens may be shared with pathogens. A notable example of shared signals and pathogenic exploitation for nitrogenous resources is given by *Agrobacterium*, which engineers infected plant tissue in such a way that it manufactures opines, which is produced and secreted to the benefit of the bacterium alone (Yuan et al., 2008).

Summary

Phylloplane microbes are at once common and unknown, easily found but invisible in their activity. The data summarized here hint at significant contributions to plant metabolism and ecology from the microbes, but much remains to be learned. Questions range beyond those specifically raised in this review. How do we tease apart the activities of plant and microbe? What signals establish the symbioses and what distinguishes pathogen from nonpathogen? How is the bacterial community on a leaf structured and how do phylloplane microbes interact with one another? What are the relationships among phylloplane, rhizosphere and soil microbes? Finally, how much of what we now know as plant metabolism is really microbial in origin?

References

Abanda-Nkpwatt, D., Musch, M., Tschiersch, J., Boettner, M., Schwab, W. 2006. Molecular interaction between *Methylobacterium extorquens* and seedlings: growth promotion, methanol consumption, and localization of the methanol emission site. J Exper Bot 57(15): 4025–4032.

Annapurna, Y., Rao, P.R. 1982. Influence of foliar application of pesticides on leaf extracts and phylloplane microflora of corn. Folia Microbiol 27(4): 250–255.

Arkhipova, T.N., Veselov, S.U., Melentiev, A.I., Martynenko, E.V., Kudayarova, G.R. 2005. Ability of bacterium *Bacillus subtilis* to produce cytokinins and to influence the growth and endogenous hormone content of lettuce plants. Plant Soil 272: 201–209.

Basile, D.V., Basile, M.R., Li, Q.Y., Corpe, W.A. 1985. Vitamin B12-stimulated growth and development of *Jungermannia leiantha* Grolle and *Gymnocolea inflata* (Huds.) Dum. (Hepaticae). Bryologist: 88(2): 77–81.

Baldotto, L.E.B., Olivares, F.L. 2008. Phylloepiphytic interaction between bacteria and different plant species in a tropical agricultural system. Can J Microbiol 54: 918–931.

Beattie, G.A., Lindow, S.E. 1999. Bacterial colonization of leaves: A spectrum of strategies. Phytopathology 89: 353–359.

Bhattacharjee, R.B., Singh, A., Mukhopadhyay, S.N. 2008. Use of nitrogen-fixing bacteria as biofertiliser for non-legumes: prospects and challenges. Appl Microbiol Biotechnol 80: 199–209.

Brandl, M.T., Amundson, R. 2008. Leaf age as a risk factor in contamination of lettuce with *Escherichia coli* O157:H7 and *Salmonella enterica*. Appl Environ Micro 74(8): 2298–2306.

Brandl, M.T., Lindow, S.E. 1997. Environmental signals modulate expression of an indole-3-acetic acid biosynthesis gene in *Erwinia herbicola*. MPMI 10(4): 499–505.

Brandl, M.T., Lindow, S.E. 1998. Contribution of indole-3-acetic acid production to the epiphytic fitness of *Erwinia herbicola*. Appl Envirn Micro 64: 3256–3263.

Butler, H.S.K. 2001. Contribution of PPFM (*Methylobacterium mesophilicum*), a bacterial symbiont, to cytokinin content and biomass accumulation in soybean [*Glycine max* (L) Merr.] seedlings. Masters Thesis. University of Maryland Eastern Shore, Princess Anne, Maryland.

Butler, H.S.K., Dadsen, R., Holland, M.A. 2000. Evidence that trans-zeatin riboside produced by a microbial symbiont is physiologically meaningful to its host plant. (abstract available at: http://www.ryconusa.com/aspp2000/public/P43/0604.html).

Caballero-Mellado, J., Onofre-Lemus, J., Estrada-de los Santos, P., Martinez-Aguilar, L. 2007. The tomato rhizosphere, an environment rich in nitrogen-fixing *Burkholderia* species with capabilities of interest for agriculture and bioremediation. Appl Environ Microbiol 73(16): 5308–5319.

Cambui, C.A., Gaspar, M., Mercier, H. 2009. Detection of urease in the cell wall and membranes of bromeliad species. Physiol Plantarum 136(1): 86–93.

Canny, M.J. 1990. What becomes of the transpiration stream? New Phytol 114: 341–368.

Cline, M.G., Thangavelu, M., Dong-Il, K. 2006. A possible role of cytokinin in long-distance signaling in the promotion of sylleptic branching in hybrid poplar. J Plant Physiol 163: 684–688.

Corpe, W.A., Basile, D.V. 1982. Methanol-utilizing bacteria associated with green plants. Dev Indust Microbiol 32: 483–493.

Corpe, W.A., Rheem, S. 1989. Ecology of the methylotrophic bacteria living on leaf surfaces. FEMS Microbiol Ecol 62: 243–250.

Cowles, J.R., Evans, H.J., Sterling, A.R. 1969. B$_{12}$ coenzyme-dependent ribonucleotide reductase in *Rhizobium* species and the effects of cobalt deficiency on the activity of the enzyme. J Bacteriol 97(3): 1460–1465.

Cutter, E.G. 1976. Aspects of the structure and development of the aerial surfaces of higher plants. In: Dickinson, C.H., Preece, T.F., Eds. *Microbiology of Aerial Plant Surfaces*. Academic Press, NY. ISBN 0-12-215050-3.

DeCosta, D.M., Rathnayake, R.M.P.S., DeCosta, W.A.J.M., Kumari, W.M.D., Dissanayake, D.M.N. 2006. Variation of phyllosphere microflora of different rice varieties in Sri Lanka and its relationship to leaf anatomical and physiological characters. J Agron Crop Sci 192: 209–220.

Feng, Y., Shen, D., Song, W. 2006. Rice endophyte *Pantoea agglomerans* YS19 promotes host plant growth and affects allocations of host photosynthates. J Appl Micro 100: 938–945.

Flechard, C., Fowler, D., Sutton, M.A., Cape, J.N. 1999. A dynamic chemical model of bi-directional ammonia exchange between semi-natural vegetation and the atmosphere. Quart J Royal Meteorological Soc 125(559): 2611–2641.

Frugier, F., Kosuta, S., Murray, J., Crespi, M., Szczyglowski, K. 2008. Cytokinin: secret agent of symbiosis. Trends Plant Sci 13(3): 115–120.

Furnkranz, M., Wanek, W., Richter, A., Abell, G., Rasche, F. 2008. Nitrogen fixation by phyllosphere bacteria associated with higher plants and their colonizing epiphytes of a tropical lowland rainforest of Costa Rica. ISME J 2: 561–570.

Godfrey, B.E.S. 1976. Leachates from aerial parts of plants and their relation to plant surface microbial populations. In: Dickinson, C.H., Preece, T.F., Eds. *Microbiology of Aerial Plant Surfaces*. Academic Press, New York. ISBN 0-12-215050-3.

Guo, Y., Chen, F., Zhang, F., Mi, G. 2005. Auxin transport from shoot to root is involved in the response of lateral root growth to localized supply of nitrate in maize. Plant Sci 169: 894–900.

Halle, F. 2002. *In Praise of Plants*. Timber Press, Portland, OR. ISBN 0-88192-550-0.

Hayashi, K., Hiradate, S., Ishikawa, S., Nouchi, I. 2008. Ammonia exchange between rice leaf blades and the atmosphere: effect of broadcast urea and changes in xylem sap and leaf apoplastic ammonium concentrations. Soil Sci Plant Nutrition 54(5): 807–818.

Hecht-Buchholz, C. 1998. The apoplast-habitat of endophytic dintrogen-fixing bacteria and their significance for the nutrition of non-leguminous plants. Z Pflanz Bodenkunde 161: 509–520.

Herrmann, B., Mattsson, M., Fuhrer, J., Schoerring, J.K. 2002. Leaf-atmosphere NH$_3$ exchange of white clover (*Trifolium repens* L.) in relation to mineral N nutrition and symbiotic N$_2$ fixation. J Exp Bot 53(366): 139–146.

Hirano, S.S., Upper, C.D 1992. Bacterial Community Dynamics. In: *Microbial Ecology of Leaves*. pp. 271–294. Andrews, J.H., Hirano, S.S., Eds. Springer-Verlag, New York. ISBN 0387975795.

Hirano, S.S., Upper, C.D. 2000. Bacteria in the leaf ecosystem with emphasis on *Pseudomonas syringae*—a pathogen, ice nucleus and epiphyte. Microbiol Mol Biol Rev 64(3): 624–653.

Hirose, N., Takei, K., Kuroha, T., Kamada-Nobusada, T., Hayashi, H., Sakakibara, H. 2007. Regulation of cytokinin biosynthesis, compartmentation and translocation. J Exp Bot 59(1): 75–83.

Holland, M.A., Polacco, J.C. 1992. Urease-null and hydrogenase-null phenotypes of a plylloplane bacterium reveal altered nickel metabolism in two soybean mutants. Plant Physiol 98: 942–948.

Hufton, C.A., Besfoed, R.T., Wellburn, A.R. 1996. Effects of NO (+NO2) pollution on growth, nitrate reductase activities and associated proteincontents in glasshouse lettuce grown hydroponically in winter with CO2 enrichment. New Phytol 133: 495–501.

Husted, S., Schjoerring, J.K. 1995. Apoplastic pH and ammonium concentration in leaves of *Brassica napus* L. Plant Physiol 109: 1453–1460.

Husted, S., Schjoerring, J.K. 1996. Ammonia flux between oil seed rape plants and the atmosphere in response to leaf temperature, light intensity and air humidity—interactions with leaf conductance and apoplastic NH4+ and H+ concentrations. Plant Physiol 112: 67–74.

Inselbacher, E., Cambui, C.A., Richter, A., Stange, C.F., Mercier, H., Wanek, W. 2007. Microbial activities and foliar uptake of nitrogen in the epiphytic bromeliad *Vriesea gigantea*. New Phytol 175: 311–320.

Ivanova, E.G., Doronina, N.V., Trotsenko, Y.A. 2001. Aerobic methylobacteria are capable of synthesizing auxins. Microbiol 70(4): 392–397.

Jha, P.N., Kumar, A. 2007. Endophytic colonization of *Typha australis* by a plant growth-promoting bacterium *Klebsiella oxytosa* strain GR-3. J Appl Micro 103: 1311–1320.

Jourand, P., Giraud, E., Bena, G., Sy, A., Willems, A., Gillis, M., Dreyfus, B., deLajudie, P. 2004. *Methylobacterium nodulans* sp.nov., for a group of aerobic, facultatively methylotrophic, legume root-nodule-forming and nitrogen-fixing bacteria. Int J Syst Evol Microbiol 54(6): 2269–2273.

Kneip, C., Lockhart, P., Voss, C., Maier, U.-G. 2007. Nitrogen fixation in eukaryotes—New models for symbiosis. BMC Evolut Biol 7: 55.

Koenig, R.L., Morris, R.O., Polacco, J.C. 2002. tRNA is the source of low-level trans-zeatin production in *Methylobacterium spp*. J Bact 184(7): 1832–1842.

Lee, H.S., Madhaiyan, M., Kim, C.W., Choi, S.J., Chung, K.Y., Sa, T.M. 2006. Physiological enhancement of early growth of rice seedlings (*Oryza sativa* L.) by production of phytohormone of N_2-fixing methylotrophic isolates. Biol Fertil Soils 42: 402–408.

Libbert, E., Silhengst, P. 2006. Interactions between plants and epiphytic bacteria regarding their auxin metabolism: VII. Transfer of 14C indoleacetic acid from epiphytic bacteria to corn coleoptiles. Physiol Plantarum 23(3): 480–487.

Madhaiyan, M., Poonguzhali, S., Sundaram, S.P., Sa, Tongmin. 2006. A new insight into foliar applied methanol influencing phylloplane methylotrophic dynamics and growth promotion of cotton (*Gossipium hirsutum* L.) and sugarcane (*Saccharum officinarum* L.). Environ Exp Bot 57: 168–176.

Madhaiyan, M., Poonguzhali, S., Senthilkumar, M., Sundaram, S.P., Sa, T. 2009. Nodulation and plant-growth promotion by methylotropic bacteria isolated from tropical legumes. Microbiol Res 164: 114–120.

Mano, H., Morisaki, H. 2008. Endophytic bacteria in the rice plant. Microbes Environ 23(2): 109–117.

Manulis, S., Haviv-Chesner, A., Brandl, M.T., Lindow, S.E. 1998. Differential involvement of indole-3-acetic acid biosynthetic pathways in pathogenicity and epiphytic fitness of *Erwinia herbicola* pv. *Gypsophilae*. Mol Plant Microbe Interact 11: 634–642.

Mattos, K.A., Padua, V.L.M., Romeiro, A., Hallack, L.F., Neves, B.C., Ulisses, T.M.U., Barros, C.F., Todeschini, A.R., Previato, J.O., Mendonca-Previato, L. 2008. Endophytic colonization of rice (*Oryza sativa* L.) by the diazotrophic bacterium *Burkholderia kururiensis* and its ability to enhance plant growth. Anais Acad Brasileira Ciencias 80(3): 477–493.

Mattson, M., Husted, S. Schjoerring, J.K. 1998. Influence of nitrogen nutrition and metabolism on ammonia volatilization in plants. Nutr Cycling Agroecosyst 51: 35–40.

Mattson, M., Schjoerring, J.K. 2003. Senescence-induced changes in apoplastic and bulk tissue ammonia concentrations of ryegrass leaves. New Phytol 160(3): 489–499.

Mercier, J., Lindow, S.E. 2000. Role of leaf surface sugars in colonization of plants by bacterial epiphytes. Appl Environ Micro 66(1): 369–374.

Miyauchi, M.Y.H., Lima, D.S., Nogueira, M.A., Lovato, G.M., Murate, L.S., Cruz, M.F., Ferreira, W.Z., Andrade, G. 2008. Interactions between diazotrophic bacteria and mycorrhizal fungus in maize genotypes. Sci Agric (Piracicaba, Braz.) 65(5): 525–531.

Morris, C.E., Kinkle, L.L. 2002. Fifty years of phyllosphere microbiology: Significant contributions to research in other fields. In: Lindow, S.E., Hecht-Poinar, E.I., Elliott, V.J. 2002. *Phyllosphere Microbiology*. APS Press, St. Paul, MN ISBN 0-89054-286-4.

Neilsen, K.H., Schjoerring, J.K. 1998. Regulation of NH_4^+ concentration in leaves of oil seed rape. Plant Physiol 118: 1361–1368.

Nix-Stohr, S., Burpee, L.L., Buck, J.W. 2008. The influence of exogenous nutrients on the abundance of yeasts on the phylloplane of turfgrass. Microb Ecol 55(1): 15–20.

Omer, Z.S. 2004. Bacterial-plant associations with special focus on pink-pigmented facultative methylotrophic bacteria (PPFMs). Doctoral dissertation. Swedish University of Agricultural Sciences, Uppsala, Sweden. ISSN 1401–6249, ISBN 91-576-6473-0.

Omer, Z.S., Bjorkman, P.O., Nicander, B., Tillberg. E., Gerhardson, B. 2004a. 5-Deoxyisopentenyladenosine and other cytokinins in culture filtrates of the bacterium *Pantoea agglomerans*. Physiol Plantarum 121(3): 439–447.

Omer, Z.S., Tombolini, R., Broberg, A., Gerhardson, B. 2004b. Indole-3-acetic acid production by pink-pigmented facultative methylotrophic bacteria. Plant Growth Regul 43(1): 93–96.

Omer, Z.S., Tombolini, R., Gerhardson, B. 2004c. Plant colonization by pink-pigmented facultative methylotrophic bacteria (PPFMs). FEMS Microbiol Ecol 47: 319–326.

Papen, H., Gessler, A., Zumbusch, E., Rennenberg, H. 2002. Chemolithoautotrophic nitrifiers in the phyllosphere of a spruce ecosystem receiving high atmospheric nitrogen input. Current Microbiol 44: 56–60.

Peng, J., Peng, F., Zhu, C., Wei, S. 2008. Molecular cloning of a putative gene encoding isopentenyltransferase from pigyitiancha (*Malus hupehensis*) and characterization of its response to nitrate. Tree Physiol 28: 899–904.

Pilot, G., Stransky, H., Bushey, D.F., Pratelli, R., Ludewig, U., Wingate, V.P.M., Frommer, W.B. 2004. Overexpression of glutamine dumper1 leads to hypersecretion of glutamine from hydathodes of Arabidopsis leaves. Plant Cell 16(7): 1827–1840.

Polacco, J.C., Holland, M.A. 1993. Roles of urease in plant cells. In: Jeon, K.W., Jarvik, J., Eds. *International Review of Cytology: A Survey of Cell Biology*, Vol. 145, pp. 65–103. Academic Press, Inc., San Diego, CA. ISBN 0-12-364548-4.

Prado, C.H.B.A., DeMoraes, J.A.P.V. 1997. Photosynthetic capacity and specific leaf mass in twenty woody species of Cerrado vegetation under field conditions. Photosynthetica 33(1): 103–112.

Puente, M., Bashan, Y. 1993. The desert epiphyte *Tillandsia recurvata* habours the nitrogen-fixing bacterium *Pseudomonas stutzeri*. Can J Bot 72: 406–408.

Razi, S.S., Sen, S.P. 1996. Amelioration of water stress effects on wetland rice by urea-N, plant growth regulators and foliar spray of a diazotrophic bacterium *Klebsiella sp*. Biol Fertil Soils 23: 454–458.

Renier, A., Jourand, P., Rapior, S., Poinsot, V., Sy, A., Dreyfus, B., Moulin, L. 2008. Symbiotic properties of *Methylobacterium nodulans* ORS 2060: A classic process for an atypical symbiont. Soil Biol Biochem 40: 1404–1412.

Roesch, L.F.W., Camargo, F.A.O., Bento, F.M., Triplett, E.W. 2008. Biodiversity of diazotropic bacteria within the soil, root, and stem of field-grown maize. Plant Soil 302: 91–104.
Sakakibara, H., Takei, K., Hirose, N. 2006. Interactions between nitrogen and cytokinin in the regulation of metabolism and development. Trends Plant Sci 11(9) 440–448.
Santana, F.M.C., Pinto, T., Fialho, A.M., Sa-Correia, I., Empis, J.M.A. 2002. Bacterial removal of quinolizidine alkaloids and other carbon sources from a *Lupinus albus* aqueous extract. J. Agric. Food Chem 50: 2318–2323.
Saravanan, V.S., Madhaiyan, M., Osborne, J., Thangaraju, M., Sa, T.M. 2008. Ecological occurance of *Gluconacetobacter diazotrophicus* and nitrogen-fixing *Acetobacteriaceae* members: their possible role in plant growth promotion. Microbial Ecol 55: 130–140.
Sattelmacher, B. 2001. The apoplast and its significance for plant mineral nutrition. New Phytol 149: 167–192.
Sattelmacher, B., Muhling, K.H., Pennewiss, K. 1998. The apoplast—its significance for the nutrition of higher plants. Z Pflanz Bodenkunde 161(5): 485–498.
Scagel, C.F., Bi, G.H., Fuchigami, L.H., Regan, R.P. 2008. Rate of nitrogen application during the growing season and spraying plants with urea in the autumn alters uptake of other nutrients by deciduous and evergreen container-grown Rhododendron cultivars. Hortscience 43(5) 1569–1579.
Schreiber, L., Krimm, U., Knoll, D., Sayed, M., Auling, G., Kroppenstedt, R.M. 2005. Plant-microbe interations: identification of epiphytic bacteria and their ability to alter leaf surface permeability. New Phytol 166(2): 589–594.
Shepherd, R.W., Wagner, G.J. 2006. Phylloplane proteins: emerging defenses at the aerial frontline? Trends Plant Sci 12(2): 51–56.
Solheim, B., Wiggen, H., Roberg, S., Spaink, H.P. 2004. Associations between arctic cyanobacteria and mosses. Symbiosis 37: 169–187.
Soto, M.J., Dominguez-Ferreras, A., Perez-Mendoza, D., Sanjuan, J., Olivares, J. 2009. Mutualism vs. pathogenesis: the give-and-take in plant-bacteria interactions. Cellular Microbiol 11(3): 381–388.
Stebbins, N., Holland, M.A., Cianzio, S.R., Polacco, J.C. 1991. Genetic tests of the roles of the embryonic ureases of soybean. Plant Physiol 97: 1004–1010.
Suman, A., Shrivastava, A.K., Gaur, A., Singh, P., Singh, J., Yadav, R.L. 2008. Nitrogen use efficiency of sugarcane in relation to its BNF potential and population of endophytic diazotrophs at different N levels. Plant Growth Regul 54: 1–11.
Suslow, T.V. 2002. Production practices affecting the potential for persistent contamination of plants by microbial foodborne pathogens. In: Lindow, S.E., Hecht-Poinar, E.I., Elliott, V.J. 2002. *Phyllosphere Microbiology*, APS Press, St. Paul, MN ISBN 0-89054-286-4.
Sy, A., Giraud, E., Jourand, P., Garcia, N., Williams, A., DeLajudie, P., Prin, Y., Neyra, M., Gillis, M., Masson-Boivin, C., Dreyfus, B. 2001. Methylotrophic *Methylobacterium* bacteria nodulate and fix nitrogen in symbiosis with legumes. J Bact 183: 214–220.
Taiz, L., Zeiger, E. 2006. *Plant Physiology*, 4th ed. Sinauer Assoc., Inc., Sunderland, MA. ISBN 0-87893-856-7.
Tamaki, V., Mercier, H. 2007. Cytokinins and auxin communicate nitrogen availability as long-distance signal molecules in pineapple (*Ananas comosus*). J Plant Physiol 164: 1543–1547.
Tejera, N., Ortega, E., Rodes, R., Lluch, C. 2006. Nitrogen compounds in the apoplastic sap of sugarcane stem: Some implications in the association with endophytes. J Plant Physiol 163: 80–85.

Terakado-Tonooka, J., Ohwaki, Y., Yamakawa, H., Tanaka, F., Yoneyama, T., Fujihara, S. 2008. Expressed *nifH* genes of endophytic bacteria detected in field-grown sweet potatoes (*Ipomoea batatas* L.). Microb Environ 23(1): 89–93.

Tukey, H.B. Jr. 1971. Leaching Substances from Plants. In: Preece, T.F., Dickinson, C.H., Eds. *Ecology of Leaf Surface Microorganisms*. Academic Press, London. ISBN 0125639503.

Van Dommelen, A., Croonenborghs, A., Spaepen, S., Vanderleyden, J. 2009. Wheat growth promotion through inoculation with an ammonium-excreting mutant. Biol Fertil Soils 45: 549–553.

Van Gelder, W.M.J. 1981. Conversion factor from nitrogen to protein for potato tuber protein. Potato Res 24(4): 423–425.

Vigneshwaran, N., Natarajan, T. 2003. Nitrogen fixation by phylloplane bacterium, *Bacillus macerans*, in cotton. Indian J Microbiol 43(2): 107–109.

Walch-Liu, P., Filleur, S., Gan, Y., Forde, B.G. 2005. Signalling mechanisms integrating root and shoot responses to changes in the nitrogen supply. Photosynth Res 83: 239–250.

Wellburn, A.R. 1990. Why are atmospheric oxides of nitrogen usually phytotoxic and not alternative fertilizers? New Phytol 115: 395–429.

Westgate, M.E., Steudle, E. 1985. Water transport in the midrib tissue of maize leaves. Direct measurement of the propagation of changes in cell turgor across a plant tissue. Plant Physiol 78: 183–191.

Whipps, J.M., Hand, P., Pink, D., Bending, G.D. 2008. Phyllosphere microbiology with special reference to diversity and plant genotype. J Appl Microbiol 105: 1744–1755.

Wilson, M., Lindow, S.E. 1994. Coexistence among epiphytic bacterial populations mediated through nutritional resource partitioning. Appl Environ Micro 60(12): 4468–4477.

Witzig, S.B., Holland, M.A. 1998. A microbial symbiont used to alter the nutritional quality of soybean and other plants. Presented poster at the 8th Gatlinburg Symposium, Molecular and Cellular Biology of the Soybean Biennial Meeting, Knoxville, TN.

Witzig, S.B., Holland, M.A. 1999. A microbial symbiont shown to alter the nutritional quality of plants. Poster presented at the American Society of Plant Physiologists Annual Meeting, Baltimore, MD.

Yadav, R.K.P., Karamanoli, K., Vokou, D. 2005. Bacterial colonization of the phyllosphere of Mediterranean perennial species as influenced by leaf structural and chemical features. Microbial Ecol 50: 185–196.

Yildirim, E., Guvenc, I., Turan, M., Karatas, A. 2007. Effect of foliar urea application on quality, growth, mineral uptake and yield of broccoli (*Brassica oleracea*, L. var. italica). Plant Soil Environ 53(3): 120–128.

Yuan, Z.C., Haudecoeur, E., Faure, D., Kerr, K.F., Nester, E.W. 2008. Comparative transcriptome analysis of *Agrobacterium tumefaciens* in response to plant signal salicylic acid, indole-3-acetic acid, and γ-amino butyric acid reveals signaling cross-talk and *Agrobacterium* plant co-evolution. Cellular Microbiol 10(11): 2339–2354.

Chapter 11
N_2-Fixing Endophytes of Grasses and Cereals

Veronica Massena Reis, Jos Vanderleyden, and Stijn Spaepen

Introduction

An endophyte is an endosymbiont, often a bacterium or fungus, which lives within a plant for at least part of its life without causing apparent disease. Vascular plants have been considered to be autonomous organisms. Nevertheless, it is well recognized that symbiosis is a common and fundamental condition of plants (Sapp, 2004). Current research suggests that all plants in natural ecosystems are symbiotic with bacteria and/or fungi on their leaf and root surfaces, rhizosphere and internal tissues that influence their performance (Vandenkoornhuyse et al., 2002). However, only a few plants have ever been thoroughly studied relative to their endophytic bacteria. Consequently, the opportunity to find new endophytic microorganisms among the diversity of plants in different ecosystems is considerable (Ryan et al., 2008). Historically, the search for endophytic bacteria has been biased toward bacteria that are able to reduce molecular nitrogen to ammonia, the so-called nitrogen-fixing bacteria or diazotrophs. These bacteria will be the subject of this chapter. Currently the genomes of a limited number of endophytic bacteria, such as *Azoarcus* BH72 (Krause et al., 2006), *Pseudomonas stutzeri* A1501 (Yan et al., 2008), and *Klebsiella pneumoniae* 342 (Fouts et al., 2008) have been sequenced, and others are in progress, such as *Herbaspirillum seropedicae* and *Gluconacetobacter diazotrophicus*, all of which are diazotrophic bacteria. Comparative analysis of the genomes of the three sequenced strains did not reveal any genes/systems specific for endophytes. *Azoarcus* BH72 was described as a disarmed plant pathogen (Krause et al., 2006). The three strains can metabolize a broad range of carbohydrates for which a number of metabolic pathways and transport systems are encoded in their genomes. Other common features of endophytes (but also other plant-associated bacteria)

Ecological Aspects of Nitrogen Metabolism in Plants, First Edition. Edited by Joe C. Polacco and Christopher D. Todd.
© 2011 by John Wiley & Sons, Inc. Published 2011 by John Wiley & Sons, Inc.

are synthesis of exo- and lipopolysaccharides, the presence of a type IV pilus, flagella and chemotaxis systems (Krause et al., 2006; Yan et al., 2008; Fouts et al., 2008).

We will try to address questions as to how these endophytes enter and are maintained in the plant, how the plant controls the invasion, whether the endophytic relationship is necessary for both organisms, whether and how the partners benefit from this relationship, and how the best plant-bacterium interactions in nature can be selected for agriculture utilization. In an agricultural sense these questions are prerequisites to understanding the impact of N_2-fixing endophytes on crop N economy. Globally, it is important to understand plant-endophyte relationship with respect to overall plant N metabolism, a topic also covered in Chapter 10.

Bacterial Endophytes: Occurrence, Distribution, and Plant Localization

Bacterial endophytes are bacteria that occur inside a plant (endo = inside; phyte = plant) and have been described for more than 120 years (Hardoim et al., 2008). Endophytes colonize intercellular spaces and vascular tissues of plants but do not inhabit living vegetative cells (James et al., 1994; Reinhold-Hurek and Hurek, 1998b). Although Pirtillä and colleagues (2004) and others have described an internal colonization of plant tissue, the authors did not use ultrastructural criteria to prove it. It is necessary to visualize the internal structure of the vegetative cell to confirm colonization of a living cell. This ecological niche is the same as observed for bacterial phytopathogens. Initially, endophytic bacteria were regarded as latent pathogens or as contaminants from incomplete plant surface sterilization (Berg et al., 2005). Later on, various reports demonstrated that bacterial endophytes are able to promote plant growth and health. These effects include the nitrogen status of the plant, since part of the endophyte population consists of nitrogen-fixing bacteria. A limited number of studies have tried to isolate and identify all the culturable bacterial endophytes of a given plant, and it has been estimated that around 10% of the total bacterial endophytic population consists of diazotrophs (Fuentes-Ramirez et al., 1999; Seghers et al., 2004). Baldani and others (1997) proposed a division of the diazotrophs localized inside plant tissue of several grasses into facultative and obligatory endophytes depending on their capacity to survive also outside the plant (mainly in the soil).

Isolating a bacterial strain from plant tissue, following surface sterilization, is not a sufficient criterion to recognize a "true" endophyte. To identify endophytes, two steps are required, as described by Reinhold-Hurek and others (1998a, b). The first step is recovery of the endophyte from the plant. Lodewyckx and others (2002) described some methods for isolation of endophytes from different plant tissues. The type of sterilization affects the recovery of an endophyte, as exemplified by the use of chloramine T. This compound is commonly used for root surface sterilization but is not effective for aerial plant tissues because these surfaces possess many compounds that protect plants from desiccation. The duration of the sterilization process can also affect the recovery of the real internal population. An excessively long treatment time can kill the bacteria localized inside the plant, and a shorter time may only kill the bacteria that are loosely attached to the plant surface (Döbereiner et al., 1995). The second step involves inoculating plants with the isolated (and characterized) strain and monitoring the internal plant colonization of the inoculant strain (Koch's postulate). This can be achieved by using labeled bacteria (expressing reporter genes

Table 11.1. Occurrence of *Gluconacetobacter diazotrophicus* in sugarcane varieties.

Variety	Number of fields sampled	Plant part	Origin of the plant sample	
			pre-germinated	Direct from the field
SP 80-3280	1	Root	10^5	N.D.
		Aerial part	N.D.	N.D.
SP 80-1842	1	Root	N.D.	N.D.
		Aerial part	N.D.	N.D.
SP 80-4445	1	Root	10^3	N.D.
		Aerial part	N.D.	N.D.
SP 80-1816	2	Root	10^7	N.D.
		Aerial part	10^3	N.D.
SP 80-1842	2	Root	10^3	N.D.
		Aerial part	10^3	N.D.
RB 855563	2	Root	10^5	N.D.
		Aerial part	10^3	N.D.

Maximum dilution of isolates identified as *G. diazotrophicus* isolated from different sugarcane varieties obtained directly from the field. These are compared to samples collected from stem cuttings "pre-germinated" in sand.
Sample locations: 1. Usina São José - Pirassununga (SP), 2. Usina São João - Araras (SP). The numbers refer to colony forming units (CFU) per gram fresh weight of plant material after surface sterilization. N.D., not detected (below the detection limit). Values are the mean of two samples of each plant part analyzed.

such as *lacZ* and *gusA*) or by using molecular detection techniques based on strain-specific antibodies or nucleic acid probes. Additional data of the ecology of the species can reinforce the criterion of endophytism.

G. diazotrophicus and *Azoarcus* sp. strain BH72 species are well-characterized diazotrophic endophytic bacteria documented by several publications (James et al., 1994, 1997; Reinhold et al., 2007). Neither species survives in soil. They are not able to grow on many of the organic compounds found in the soil, and both species are transferred by plant material to the next generation. Even insects can be involved in transmission and colonization of the flowers. Consequently seeds can carry the bacteria for the next generation (transmission from seed to seed).

The plant growth stage influences the population of the endophytic diazotrophs. *G. diazotrophicus*, a bacterial species commonly found in association with sugarcane, offers a noteworthy example of the dynamics of the endophyte population. In a survey of plants taken from the field, only 3 out of 24 varieties sampled (13%) tested positive for *G. diazotrophicus*. However, when stem cuttings of these plants were cultured for at least 30 days, more than half of the varieties tested positive for *G. diazotrophicus* (Table 11.1) It was shown that the titer of the endophytic natural population gradually increased during the first 30 days after explanting the stem cuttings on a sterilized substrate. Moreover, it was demonstrated that the population change is linked to the outgrowth of the stem cuttings, whereby the sucrose content of the stem is metabolized and transformed to several other sugars such as glucose and fructose, which are used by the bacteria as preferential carbon sources. Moreover, it also was shown that the isolates obtained from the roots differ from the isolates obtained using the aerial plant part (Figure 11.1)

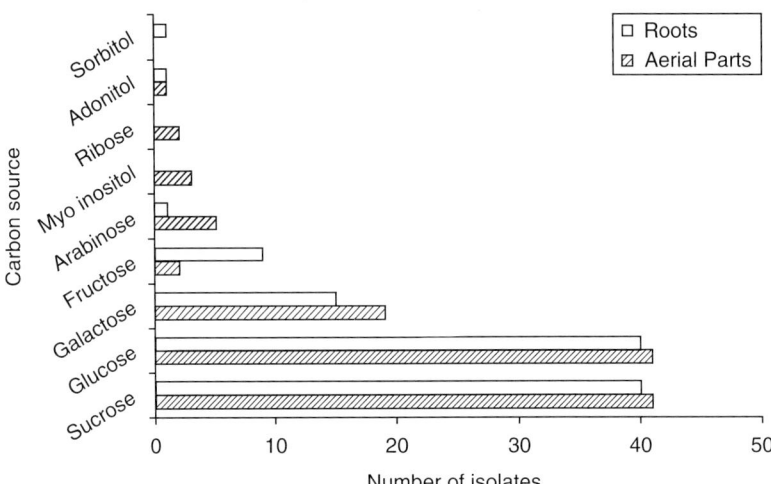

Figure 11.1. Utilization of carbon sources by different isolates and type strains of *G. diazotrophicus*. Samples were obtained from roots (*white bars*) and aerial parts (*grey bars*) of different sugarcane varieties maintained in a germplasm bank (Guedes et al., 2008).

Proposed Mode of Colonization

Plant Colonization

The rhizosphere is the pool of the highest bacterial diversity associated with plants. The endophytic bacterial communities are usually less complex in structure (Hardoim et al., 2008) and are often a subset of the rhizosphere population. It is hypothesized that endophytes originate from the outside environment and have the capability to enter the plant. Obligatory endophytes might have lost, during their evolution, the capability to survive outside the plant. Specific traits most likely contribute to the capability to enter the plant and survive in the intercellular spaces and these traits would be retained. This follows from observations of Schloter and Hartmann (1998) testing two closely related *Azospirillum brasilense* strains, Sp7 and Sp245. Only the latter was able to colonize substantially the inner part of wheat roots, although it is not known which genes are important for this behavior (Schloter and Hartmann, 1998; Rothballer et al., 2003). Mavingui and others (1992) also found that some *Paenibacillus polymyxa* strains can colonize the rhizoplane (root surface), and these strains differ from those recovered from the rhizosphere. Similar results were obtained by Rosenblueth and Martínez-Romero (2004) who tested *Rhizobium etli* strains found in the inner tissue of maize roots and *Phaseolus vulgaris* nodules. Whether *R. etli* fixes nitrogen inside the plant is still under debate.

The first step in bacterial colonization of the plant root is adhesion to the root surface. Several bacteria produce fibrils that allow the bacteria to attach to the plant surface. The presence of pili can also contribute to this process as shown for *Azoarcus* sp. type IV pili (Dorr et al., 1998; Böhm et al., 2007). The process of root adhesion and colonization has been intensively studied for *Azospirillum* species. Bacterial attachment to roots is a biphasic process as demonstrated by *in vitro* binding assays. In a first phase, cells adsorb to the root cells in a weak reversible manner, and the involvement of the polar flagellum in this phase could be demonstrated (Michiels et al., 1991; Croes et al., 1993). In the second, or anchoring phase, bacterial aggregates are formed and irreversibly attached to the roots by the bacterial production of extracellular polysaccharides (Michiels et al., 1991; Steenhoudt and Vanderleyden, 2000).

Although the adhesion is not well understood, it is obvious that anchoring on a preferential location can contribute to the exchange of molecular signals between the bacteria and the host. This interplay of signals is well characterized in the legume rhizobia symbiosis (Cooper, 2007), but the players, involved in the communication in other plant-bacterium interactions, are not known. Interesting candidates as signal molecules between bacteria and plants are molecules related to quorum sensing (QS). QS is a mechanism that induces changes in gene expression in a population density-dependent manner by the accumulation of bacterial signal molecules. The best-documented system in Gram-negative bacteria is the *N*-acyl-homoserine lactone (AHL) system, and it is involved in symbiosis, virulence, motility, biofilm maturation, and survival (Hooshangi and Bentley, 2008). Eukaryotes (both mammalian and plant) can sense and respond to QS molecules (through a variety of different physiological functions, such as hormone signalling, primary and secondary metabolism, and plant defense [Mathesius et al., 2003; Bauer and Mathesius, 2004; Hughes and Sperandio, 2008]). In addition, AHL-producing rhizobacteria can trigger systemic resistance in plants by inducing salicylic acid-dependent and ethylene/jasmonic acid-dependent defense pathways (Schuhegger et al., 2006). Since plants can secrete QS-mimicking compounds, which can interfere with bacteria, they can counteract pathogen virulence or initiate symbiosis (Teplitski et al., 2004; Pierson and Pierson, 2007).

Both the plant and bacterial partner control the interaction. For example, *Herbaspirillum rubrisubalbicans* causes foliar mottled red stripe disease in a single sugar cane variety (variety B4362). In other varieties, symptoms are not visible after foliar inoculation (Olivares et al., 1997). The same cultivar-dependent outcome could be shown for *G. diazotrophicus* (James et al., 1994, 1997). Dong and others (2003a, b) reported that two different strains of *Klebsiella* differed in the invasion process. One strain is so aggressive that a single cell was sufficient to colonize a plant; however, colonization differs with the cultivar of the tested plant (as shown for *Medicago sativa, M. truncatula, Arabidopsis thaliana, Triticum aestivum*, and *Oryza sativa*).

Endophytic Colonization

In plant tissues, bacterial endophytes may originate from the rhizosphere, seeds, the phylloplane, and vegetative (nonreproductive) material used for propagation (McInroy and Kloepper, 1995; Raaijmakers et al., 1995). Entry of these bacteria into the plant can be through sites of emergence of lateral roots, other wounds, natural openings (including lenticels and stomata and hydathodes), and germinating radicles (Hallmann et al., 1997). The zones of lateral root emergence and of root elongation and differentiation are well-characterized openings for several endophytes as described by James and others for *G. diazotrophicus* and *H. seropedicae* colonization of sugarcane (James et al., 1994, 1997) and by Dong and others (2003a, b) for *Klebsiella* sp. strain Kp342 colonizing *Medicago* and wheat. Bacteria can also use the loose cells of the root cap of the root tip for entering the plant. Once inside, bacteria multiply and may colonize several niches but only outside the plant cell. The apoplast is the space where bacteria live inside the plant, and spreading to other plant parts occurs via the vascular system (James et al., 1994; Sprent and James, 1995; Reinhold-Hurek and Hurek, 1998a; James et al., 2001).

Some aspects of the infection process of diazotrophic bacteria were reported for sorghum and sugarcane, inoculated with *H. seropedicae* and *H. rubrisubalbicans* (James et al., 1997; Olivares et al., 1997). In these studies, a strong interaction of plant genotype and the species inoculated was demonstrated. *H. rubrisubalbicans* is a mild phytopathogenic agent causing mottled stripe disease in some susceptible varieties of sugarcane (Pimentel et al., 1991). When a suspension of this bacterium was injected into the leaves of the susceptible variety B-4362, *H. rubrisubalbicans* completely blocked some of the xylem vessels and colonized the intercellular space of the mesophyll cells. At this particular location, the presence of the nitrogenase in the center of the micro-colonies was observed using an antiserum against the FeMoCo subunit of the enzyme (James et al., 1997). In sorghum, the metaxylem was also colonized by this bacterium, and the nitrogenase antigen was also found to be associated with it (Olivares et al., 1997). In the genotype resistant to mottled stripe disease, SP 701143, the bacterium also colonized the xylem but formed clusters of 10 to 20 cells encapsulated by membranes, probably of plant origin. The methodology applied could not demonstrate whether the cells were fixing nitrogen.

Monitoring Colonization

Colonization of sugarcane by diazotrophs has been studied by conventional microscopy. Recently, cryotechniques have allowed the ultrastructural investigation of the interaction between sugarcane plantlets and *H. seropedicae* (Silva et al., 2003). Microscopic observation showed consistent differences between cryo- and conventional preparations, especially related to the appearance of the bacterial cell and the type of the endophytic attachment to the host cell wall. *H. seropedicae* was found to be localized as a single cell or a micro-colony inside the apoplast of leaves. Bacteria were also found inside apparently dead vascular parenchyma cells. Root cortex apoplast showed bacteria surrounded by an amorphous matrix of unknown composition, as well as adhering to the plant cell wall by a stalk of electron-dense material (Silva et al., 2003). Similar results were observed by Gyaneshwar and others (2001). In both roots and leaves, the bacterial cytoplasm was shown to contain numerous poly-β-hydroxybutyrate granules. At adhesion sites both bacteria and plant cell walls showed altered aspects and bacterial protrusions could be observed (Silva et al., 2003).

Active penetration of the bacteria into the plant tissue is supported by the presence of cellulytic and pectinolytic enzymes produced by several endophytes such as *Azoarcus* (Hurek et al., 1994) and *Azospirillum irakense* (Khammas and Kaiser, 1991; Bekri et al., 1999), but very little is known about the regulation of these enzymes. A direct infection of the aerial plant part is also possible, using the natural openings of the stomata and to a minor extent via broken trichomes. Several authors described the presence of diazotrophs on the leaf mesophyll dispersed into the spaces between plant cells and even colonizing the stomata (Olivares et al., 1997; Reinhold-Hurek and Hurek, 1998a; Elbeltagy et al., 2001; Gyaneshwar et al., 2001).

Several plant-endophyte interactions have been monitored using molecular tagging methods. These techniques detect microorganisms and estimate their number *in situ* on the plant surfaces and *in planta*. Colonization patterns of wheat by *A. brasilense* Sp245 were monitored using a *gusA*-expressing derivative. Initially, bacteria were found mainly on root hairs and at sites of lateral root emergence. By coupling the *nifH* promoter to the *gusA* gene, the sites of nitrogen fixation could be observed. BNF was only found in N-free growth medium, and both oxygen and carbon availability were limiting factors for nitrogen fixation under associative conditions (Vande Broek et al., 1993). James and others (2002) used a *gusA*-marked *H. seropedicae* Z67 strain to follow the colonization of rice seedlings. The cracks at lateral root emergence were

Table 11.2. Mechanisms involved in plant growth promotion by endophytic diazotrophs.

Mechanism	Plant	Reference
BNF	Sugarcane, rice, wheat	James et al., 1994, 2001; Iniguez et al., 2004; Hurek et al., 2002
Phytohormones	Sugarcane	Spaepen et al., 2007a
Hormone transport and biosynthesis	Sugarcane	Nogueira et al., 2001; Vargas et al., 2003
Plant defense genes	*Medicago truncatula*, *Arabidopsis thaliana*, wheat	Iniguez et al., 2005
Bacterial extracellular components	*Medicago sativa* and wheat; Rice	Iniguez et al., 2005; Böhm et al., 2007

used by the bacteria to enter and colonize the inner plant tissue reaching the aerenchyma and cortical cells, with a few penetrating the stele and the vascular tissue. The vascular xylem vessels in leaves and stems were also colonized. A fluorescent *in situ* hybridization (FISH) analysis based on rRNA-targeted oligonucleotide probes confirmed that in micropropagated sugarcane inoculated with a mixture of five different endophytic diazotrophs, all inoculated species (*G. diazotrophicus, H. seropedicae, H. rubrisubalbicans, Azospirillum amazonense,* and *Burkholderia tropica*) reached the endophytic habitat of micropropagated sugarcane plantlets through active infection of the root cap and emerging zone of secondary roots, although with different efficiencies due to apparently different competitiveness for colonization. A putative antagonistic effect between inoculated *H. seropedicae* and *H. rubrisubalbicans* was observed, suggesting that the specific *H. seropedicae* strain that was used outcompeted the *H. rubrisubalbicans* strain. With respect to the time course of root colonization by both strains, the detection of *H. seropedicae* occurs just 6 h after inoculation, suggesting that this species is one of the most aggressive in colonizing sugarcane plantlets (Oliveira et al., 2009).

Progress in confocal laser scanning microscopy, in combination with the use of (auto-) fluorescent proteins (FP) to tag bacteria, has enabled the visualization of the colonization process and pattern (Bloemberg and Lugtenberg, 2001; Bloemberg, 2007) in different associations such as the *Herbaspirillum* sp. – rice, *Herbaspirillum frisingense* – *Miscanthus,* and *Azoarcus* BH72 : rice interactions (Egener et al., 1998; Elbeltagy et al., 2001; Rothballer et al., 2008). In the latter association, the *gfp* gene was fused to the *nifH* promoter, allowing visualization of promoter activity *in planta*. *nifH* expression was observed at sites of lateral root emergency and above the root tips at the zone of elongation and differentiation. These sites are zones of primary colonization and infection. Fluorescence could also be observed inside plant cells (Egener et al., 1998). The use of FPs could differentiate two closely related *A. brasilense* strains Sp245 and Sp7. *A. brasilense* Sp245 is able to penetrate root tissue, indicating an endophytic behavior of this strain under certain conditions, while strain Sp7 was not detected in plant tissue. These results were also confirmed by FISH analysis (Rothballer et al., 2003).

Mechanisms of Action

The involved mechanisms in plant growth promotion by endophytes can be divided into direct and indirect modes of action (Table 11.2). The first direct mechanism studied was improved N availability as a possible result of biological nitrogen fixation (BNF). Other direct mechanisms are phytohormone production, production of volatile organic compounds (VOC), phosphate solubilization, and 1-aminocyclopropane-1-carboxylate (ACC) deaminase activity. Endophytes can also stimulate plant growth indirectly by protecting the plant against phytopathogens. Most of these mechanisms are well characterized in plant growth-promoting rhizobacteria (PGPR) and seem to be important also for diazotrophic bacterial endophytes, although not as well documented.

Direct Mechanisms

Most PGPR are able to fix atmospheric N_2 into ammonium in pure culture when an appropriate energy source is available in combination with the optimal temperature, pH, and O_2 concentration (Elmerich, 2007). The contribution of BNF to the plant-growth promoting capacities of endophytes is variable and depends on several parameters such as the plant species/genotype, the bacterial strain, and the environment (Boddey et al., 1995; Engelhard et al., 2000; Iniguez et al., 2004). The importance of BNF was demonstrated by inoculation experiments with nitrogen-fixing-negative mutants (Nif$^-$) in combination with ^{15}N applications. From the limited studies reported, it appears that BNF is usually not involved in plant growth promotion since Nif$^-$ mutants are still able to stimulate plant growth. However, for *Azoarcus* sp. BH72, it was demonstrated that inoculation of Kallar grass plantlets with its Nif$^-$ derivative (*nifK* mutant) resulted in significantly decreased plant dry weight and lowered nitrogen accumulation. In addition, abundant *nifH* transcripts could be retrieved from inoculated plants in contrast to non-inoculated plants or plants inoculated with Nif$^-$ mutants. These experiments indicate that in the case of the Kallar grass – *Azoarcus* association, fixed N is transferred to the plant. The mode of transfer is not known, but it may be a direct transfer or an indirect process via death and mineralization of the bacteria (Hurek et al., 2002). BNF was also demonstrated in wheat inoculated with *Klebsiella pneumonia* 342. Inoculation resulted in increased dry weight, chlorophyll content, and total N concentration. In addition, ^{15}N was diluted in the plant tissue as a result of inoculation, and the contribution of BNF was only found for one wheat variety. Using a nitrogen-fixing-negative mutant, these effects were not observed upon inoculation (Iniguez et al., 2004).

Most PGPR have a broad spectrum of N and C metabolism, mainly adapted to their environment. N metabolism is especially important in relation to the plant because it can discriminate the form of nitrogen available in the soil. Microbial conversions are necessary in the overall N cycle (see Chapter 1). The process of denitrification is a capacity of many PGPR (Steenhoudt and Vanderleyden, 2000). In this process, nitrate is reduced to nitrite, which can mimic the effects of indole-3-acetic acid (IAA) on plants (see below) (Zimmer et al., 1988). Nitrite itself can be converted to N-oxides (NO, N_2O) or NH_3. NO is a signal molecule that participates in several processes in plants, such as IAA-dependent root development and growth (see Chapter 17). In *A. brasilense* it was shown that periplasmic nitrate reductase-dependent NO production contributes to the plant growth-promoting effect (Molina-Favero et al., 2008).

The production of phytohormones by PGPR has been reported for many decades. These hormonal compounds influence physiological processes in the plant at low concentrations. There are five major classes of phytohormones: auxins, gibberellins, cytokinins, abscisic acid, and ethylene. However, more phytohormones are currently known, and new ones are still being discovered, such as strigolactone, brassinosteroids, and peptide hormones (Santner et al., 2009; Santner and Estelle, 2009). Here we discuss endophyte production of auxin (IAA) and the modulation of ethylene by endophytic bacteria.

The class of auxins has been extensively studied, with IAA as the best-documented member (Davies, 2004; Woodward and Bartel, 2005; Santner et al., 2009; Santner and Estelle, 2009). The production of IAA has been reported for many bacteria. It is assumed that over 80% of the bacteria isolated from the rhizosphere are capable of synthesizing IAA (Patten and Glick, 1996; Khalid et al., 2004). For endophytic bacteria, there are no estimates, but IAA production is a general feature of many of these bacteria. In several (plant-associated) bacteria, such as *Azospirillum* species, the production of auxins has been shown to contribute to their plant growth promoting capacities by use of mutants with altered auxin production (overproducers or knock-out mutants) in inoculation experiments. Inoculation of wheat seedlings with increasing concentrations of wild-type *A. brasilense* Sp245 resulted in a decrease in root length and increase in root hair formation. In a mutant producing only 10% wild-type IAA level, this effect could only be observed at higher bacterial concentrations (Dobbelaere et al., 1999). Using recombinant strains, with a constitutive or plant-inducible promoter driving the key gene involved in IAA biosynthesis resulting in a higher IAA production, the observed root hair promotion was more pronounced than in inoculations with the wild-type strain (Spaepen et al., 2008).

As in plants, several IAA biosynthetic pathways have been described in bacteria, mainly starting from the precursor tryptophan (Costacurta and Vanderleyden, 1995; Patten and Glick, 1996; Spaepen et al., 2007a). The two best-characterized pathways use the intermediates indole-3-acetamide (IAM) or indole-3-pyruvate (IPyA). The first pathway, described primarily in plant pathogens and in some rhizobial strains, converts tryptophan to indole-3-acetamide by tryptophan monooxygenase (encoded by *iaaM*). An amide hydrolase (encoded by *iaaH*), converts IAM to IAA (Sekine et al., 1989; Clark et al., 1993; Morris, 1995; Manulis et al., 1998). The IPyA pathway has been characterized in many beneficial bacteria and consists of three steps: tryptophan is transaminated to IPyA by an aminotransferase. In the next and rate-limiting step, IPyA is converted to indole-3-acetaldehyde (IAAld) by the key enzyme IPyA decarboxylase. In the third step, IAAld is oxidized to IAA by a dehydrogenase (Costacurta and Vanderleyden, 1995; Patten and Glick, 1996). Many genes/enzymes involved in the key step (catalyzed by IPyA decarboxylases) have been characterized (Koga et al., 1991; Costacurta et al., 1994; Brandl and Lindow, 1996; Patten and Glick, 2002; Schütz et al., 2003), and recently this group of enzymes was classified into two subgroups based on a phylogenetic and biochemical analyses (Spaepen et al., 2007b). Mutants in genes involved in IAA biosynthetic pathways are still capable of producing IAA, indicating that multiple pathways are present in a single organism.

The production of auxins has been reported for many endophytes, but a mechanistic proof, using, for example knock-out mutants, is still lacking. Since endophytes are located inside the plant, IAA production can have a dramatic direct effect on the plant IAA pool and gradient (Hardoim et al., 2008). In a screening assay, the majority of isolated endophytes (mainly *Pseudomonas* species) from field-grown *Solanum nigrum* were capable of producing IAA. Although the beneficial effects of these strains can be very host-dependent, they are mainly determined by the bacteria (Hoa Long et al., 2008).

The plant hormone ethylene influences many aspects of plant development, including senescence. However, if the ethylene concentration is too high, plant growth is inhibited (Mattoo and Suttle, 1991; Abeles et al., 1992). Therefore, lowering the ethylene level can induce plant growth and this has been attributed to some bacteria that produce the enzyme ACC deaminase (encoded by *acdS*). ACC is the ethylene precursor and ACC deaminase converts it to α-ketobutyrate and ammonia. These products can be used by the bacterium as nitrogen and carbon source. In a model for plant growth promotion by ACC deaminase activity, proposed by Glick and others (1998), ACC is exuded by plant roots and subsequently metabolized by bacteria, leading to more ACC exudation. This leads to a decrease in ACC concentration in the plant and consequently a lower ethylene production due to the lack of precursor. In addition, IAA (contributed by bacteria) stimulates the synthesis of ACC (at the level of its biosynthetic enzyme ACC synthase), resulting in a higher ACC exudation. This depletion of the internal ACC pool by exudation and bacterial degradation ultimately leads to a reduced ethylene level inside the plant, elevating the inhibitory effect of ethylene and increasing plant growth (Chae and Kieber, 2005; Glick et al., 2007).

A recently emerging field in the mechanisms of plant growth promotion by bacteria focuses on the production of VOCs. In *Bacillus subtilis*, bacterial production of acetoin and 2,3-butanediol was shown to induce plant growth, and the use of a knock-out mutant in the biosynthetic pathway demonstrated the contribution of VOCs to plant growth promotion (Ryu et al., 2003). Later, VOC production was also detected in *Pseudomonas chlororaphis* and *Bacillus amyloliquefaciens* (Han et al., 2006; Cho et al., 2008). The production of acetoin (and 2,3-butanediol) is induced under low atmospheric oxygen partial pressure to provide an alternative electron sink for the regeneration of needed NAD+ (acetoin is reduced to 2,3-butanediol with NADH as e-donor). Besides growth promotion, VOCs can also induce systemic resistance (ISR) (see below) in plants, priming the plant against subsequent infections by pathogens (Ryu et al., 2004) and triggering plant growth promotion by regulating auxin homeostasis and cell expansion, thereby inducing an optimized coordination between root and leaf development (Zhang et al., 2007).

Indirect Mechanisms

The most important mechanism for indirect plant growth promotion is biocontrol, defined as the reduction of diseases by 'natural' mechanisms. Several types of biocontrol can be described. Competition for nutrients and niche is a characteristic of all bacteria competing around and in the root and leads to the exclusion of pathogens. Iron competition involves the production of low molecular weight iron-chelating compounds (siderophores), which bind very efficiently Fe^{3+}, the abundant form of iron. Loaded siderophores are transported back into the cells, thus making iron available for the bacteria. It has also been suggested that plants can take up bacterial siderophores via specialized transporters, thus contributing to plant growth promotion, especially in soils with low iron availability (Marschner and Romheld, 1994; Whipps, 2001; Robin et al., 2008). Plants themselves also produce siderophores, and their production is increased under iron deficiency, leading to complex interactions between plants and microorganisms in relation to iron dynamics (Lemanceau et al., 2009).

Several bacterial species are capable of producing secondary metabolites that act as antibiotic compounds. The best characterized are phenazines, 2,4-diacetylphloroglucinol, pyoluteorin, pyrrolnitrin, cyclic (lipo-)peptides, hydrogen cyanide, and hydrogen peroxide although the mode of action is still unknown for most molecules. These compounds were mainly identified in pseudomonads and bacilli, and their biosynthetic pathways and regulation have been described in detail (for reviews see Cook et al., 1995; Raaijmakers et al., 2002; Haas and Keel, 2003; Chin-A-Woeng et al., 2003; De Weert and Bloemberg, 2006; Weller, 2007; Dubuis et al., 2007). Biocontrol activity has not been established for endophytic bacteria and therefore, its contribution to their plant growth promoting capacities is unknown.

Most rhizobacteria are able to induce systemic resistance in plants, resulting in an enhanced protection against a broad spectrum of phytopathogens. The molecular determinants from bacteria that induce this resistance are mainly known as microbe-associated molecular patterns

(MAMP), such as flagellin and lipopolysaccharides. However, siderophores, antibiotics, AHLs, and 2,3-butanediol can also be considered to be MAMPs (van Loon, 2007). Upon recognition of MAMPs by the plant, signalling networks via the plant hormones salicylic acid, jasmonic acid and ethylene are triggered, priming for enhanced defense. After a pathogen attack, defense responses are accelerated (in comparison to nonprimed plants), resulting in increased resistance against plant pathogens (Van Wees et al., 2008).

Some enzymatic activities also contribute to the suppression of phytopathogens, exemplified by chitinolytic activity. Several endophytic *Streptomyces* sp. produce chitinases that inhibit fungal growth by hydrolyzing chitin, a structural compound of the cell wall of some classes of fungi. However, *in planta* experiments are needed to confirm the chitinase contribution to the biocontrol effect (Quecine et al., 2008).

Genes Related to Colonization and Plant Response

Although endophytic bacterial interactions are considered to provide a potential benefit to plants, it remains to be elucidated how plants recognize endophytes as such and not as phytopathogens. The initial steps in colonization by neutral, beneficial, and phytopathogenic bacteria are identical, suggesting that they use common mechanisms for entry. Once inside the plant, neutral or beneficial bacteria most likely either suppress or evade the plant defense system or do not elicit the plant defense system (Sikorski et al., 1999).

It is generally accepted that in bacteria-plant interactions, the plant is in control of the interaction. Several studies evaluating the culturable bacterial population of endophytes (about 10^5 to 10^6 CFU per g fresh tissue) indicate that the endophytes do not achieve the high numbers of a pathogenic interaction (about 10^{10} CFU per g fresh tissue) (Lodewyckx et al., 2002). Plant control over endophyte titer is well described for sugarcane (Fuentes-Ramırez et al., 1993; dos Reis et al., 2000; Nogueira et al., 2001; Vargas et al., 2003; Vinagre et al., 2006).

In accordance with plant control, microarray analysis indicated that four *R*-genes and a gene involved in the biosynthesis of salicylic acid are coordinately upregulated in the process of sugarcane colonization by *Herbaspirillum* spp. and *G. diazotrophicus* (Rocha et al., 2007). These *R*-genes control the recognition of a plant pathogen (Ellis et al., 2000).

A preliminary assay demonstrated that *G. diazotrophicus* is able to control *X. albilineans*, causative agent of a foliar disease in sugarcane. By cDNA-AFLP analysis, some plant genes (using leaf tissue) involved in biocontrol activity were identified (Arencibia et al., 2006). These results indicate that inoculation stimulates genes involved in plant defense such as genes controlling the ethylene defense pathway. This pathway is activated when sugarcane micropropagated plants are inoculated with endophytic bacteria (Nogueira et al., 2001; Vargas et al., 2003; Cavalcante et al., 2007).

It is believed that certain endophytic bacteria elicit induced systemic resistance, discussed below, and that it can be effective against different types of pathogens (Bakker et al., 2007; van Loon, 2007). This mechanism is poorly understood but can explain why bacteria can live inside the plant and act as a beneficial organism for plant growth promotion.

Endophytic bacteria can induce plant defense pathways, and these pathways can determine the degree of bacterial colonization. Endophytic colonization is decreased by addition of ethylene, a signal molecule for ISR. In addition, endophytic colonization is hampered by plant defense responses, induced by bacterial extracellular components (MAMP) (Iniguez et al., 2005). After priming of defense pathways by endophytic bacteria, defense genes are more highly upregulated upon pathogen attack, compared with non-endophyte-treated plants, leading to better resistance (Conn et al., 2008).

The signalling pathways mediating the association between endophytic bacteria and plants has not yet been established, but in the case of sugarcane – *Gluconacetobacter/ Herbaspirillum* interactions, some plant proteins have been characterized, including a receptor kinase SHR5 and proteins of the ethylene signalling pathway (Vinagre et al., 2006; Cavalcante et al., 2007). We point out that the above examples are for diazotrophs, and other examples for commensal bacteria have been reported (e.g., Gourion et al., 2003).

Contribution to Plant Growth—A Role for N-fixation?

Internal colonization can provide a protected environment for the endophyte, with decreased competition for carbon sources and less oxygen present to inhibit bacterial nitrogenase activity. Metabolic exchanges between bacterial and plant cells are facilitated in this contact. The presence of nonpathogenic bacteria can control the disease spread of some pathogens (see discussion above). These factors can be divided into direct and indirect effects, and the sum of them is visualized by the increment in biomass and nitrogen accumulation.

Several publications show the contribution of BNF to plant nitrogen assimilation. BNF associated with the culture of sugarcane is estimated to supply up to 60% of the total accumulated nitrogen (Urquiaga et al., 1992; Boddey et al., 2001). Recently, Montañez and others (2009) measured the nitrogen-fixing capacity of a range of commercial cultivars of maize (*Zea mays* L.) evaluated by the ^{15}N isotope-dilution method. BNF expressed as percent nitrogen derived from air (NDFA) ranged from 12 to 33%, regardless of nitrogen fertilization.

However, in several cases the nitrogen source affected BNF. Medeiros and others (2006) evaluated the effect of the addition of increasing amounts of two sources of mineral nitrogen (ammonium sulphate and calcium nitrate) on the population of *G. diazotrophicus*, on nitrogenase activity (measured by acetylene reduction assay –[ARA]), and on accumulation of N by two sugarcane hybrids, SP 701143 and SP 792312. The results showed that both varieties differed in the form of nitrogen they prefer for uptake from the soil. The acetylene reduction activity (as a measure of nitrogenase activity) was inhibited in both varieties, especially in variety SP 792312, in the presence of either of two nitrogen sources, ammonium sulphate and calcium nitrate (Figure 11.2). In both varieties, the addition of increased doses of ammonium and nitrate inhibited the population of *G. diazotrophicus*.

Suman and others (2008) observed that the natural population of *G. diazotrophicus* in subtropical Indian sugarcane varieties was increased at a medium level of N-fertilizer applied compared to the control level (without added nitrogen) and a fully fertilized condition, whereas the *Herbaspirillum* population increased gradually according the amount N fertilizer. ARA was positively correlated with the *G. diazotrophicus* population in the rhizosphere and root, whereas it had a poor correlation with the *Herbaspirillum* population. These data are in agreement with those reported by Fuentes-Ramirez and others (1993). They suggested that the association between *G. diazotrophicus* and sugarcane in Mexico can be severely limited when nitrogen fertilizers are applied in large quantities.

Similarly, in India, in sugarcane plantations fertilized with 300 kg N ha^{-1}, the percentages of *G. diazotrophicus* isolates obtained were much lower compared to those measured under fertilization with 120 kg N ha^{-1}, although it should be noted that the *Herbaspirillum* population was unaffected (Muthukumarasamy et al., 1999).

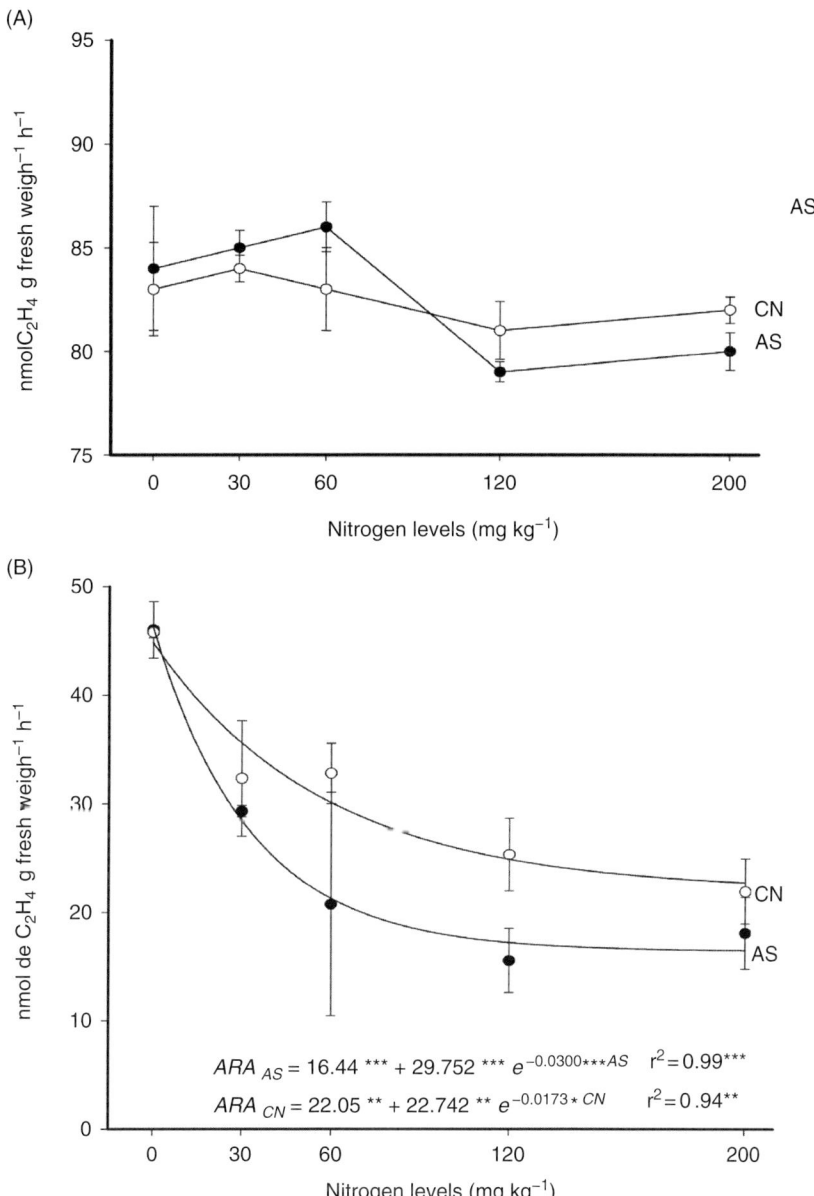

Figure 11.2. Acetylene reduction activity of *G. diazotrophicus* as a function of different amounts of N applied (in mg N per kg substrate). Ammonium sulfate (AS, *closed symbols*) and calcium nitrate (CN, *open symbols*), at 30 DAP (a) and 60 DAP (b) in stem cuttings of sugarcane (SP 701143 variety). Values are the average of seven repetitions. Error bars represent standard error of the means. ***, **, *, ns = significant at 0.1%, 1%, 5%, and not significant in Student T test, respectively (Medeiros et al., 2006).

Application and Commercial Products

During the last 25 years several companies in different countries have developed and commercialized bacterial products that are used for different purposes in plant growth. Many *Azospirillum* species have been widely applied by industry. In some cases, co-inoculations are used to improve root development in order to induce more nodules in legumes or to improve the uptake of water and other nutrients (such as nitrogen and phosphorus) (Casanovas et al., 2003; Creus et al., 2004; Cassan et al., 2009). It is also well known that the plant genotype plays a significant role (see above). Therefore, it is important to include the role of beneficial bacteria in the breeding aspects of plants in order to find the optimal variety-bacterium interaction.

Inoculation experiments with endophytes have been published from studies in several countries using different crops such as rice (Chi et al., 2005; Pedraza et al., 2009), sugarcane (Oliveira et al., 2002; Oliveira et al., 2009), wheat (Diaz-Zorita and Fernandez-Canigia, 2009), and maize (Miyauchi et al., 2008). Contributions of BNF have been calculated using several techniques such as N balance, $^{15}N_2$ gas incorporation, ^{15}N enrichment or delta ^{15}N (Boddey, 1987; Boddey et al., 1995). However, it is not possible to discriminate the contribution of internal diazotrophic bacteria from that of external applied bacteria.

Several groups have tried to use the conserved *nif* genes (such as *nifH*) as a probe to correlate the signal of nitrogen-fixing bacteria population in plants to the shifts in environmental factors such as nitrogen level (Diallo et al., 2008) and water content (Hai et al., 2009). This has led to the accumulation of considerable evidence on how diazotrophs can individually or in a community infect and colonize several plants, but has not answered the question of which microorganisms are responsible for the N_2-fixing activity detected under natural conditions or even where it happens and when.

Concluding Remarks

Although endophytes have been studied for more than 50 years and their isolation from internal plant tissues and some mechanisms regarding infection and plant growth promotion have been described, many aspects of the endophyte-plant interaction still need to be elucidated. Nitrogen transfer by nitrogenase activity of several diazotrophs that colonize plants is still controversial, mainly due to the lack of adequate methods to measure this effect. Most studies evaluate the total nitrogen accumulated in the plant in comparison to control plants. Even in this case, reliable estimates are difficult since control plants also obtain some nitrogen from associated nitrogen-fixing bacteria. In general, it can be stated that new imaging technologies with higher resolution will also benefit the field of plant-endophyte interactions, as is the case today in the study of bacterial symbiosis in animals and humans.

The diversity of endophytic bacteria is high, illustrated by several examples in this chapter. They colonize diverse plant species without causing disease and even protect plants from pathogens. The repertoire of their functions still needs to be further explored in order to manage the microbial communities favoring plant colonization and to increase the potential of these bacteria. This will allow decreasing the dependency on chemical control agents against phytopathogens. In line with several themes of this book, N-fertilization may also be obviated, thus ameliorating many of the negative effects of anthropogenic inputs into the modern N cycle (see Chapter 1). Thus, the contribution of endophyte research will certainly have economic and environmental impacts for agriculture.

Acknowledgments

The authors thank European Union FP6 programme for financial support (RHIBAC project FP6-FOOD-CT-2006-036297). S.S. is a recipient of a postdoctoral fellowship grant from the Research Foundation-Flanders (FWO Vlaanderen).

References

Abeles, F.B., Morgan, P.W., Saltveit, M.E. 1992. *Ethylene in Plant Biology.* Academic Press, San Diego, CA.

Arencibia, A., Vinagre, F., Estevez, Y., Bernal, A., Perez, J., Cavalcante, J., Santana, I., Hemerly, A.S. 2006. *Gluconacetobacter diazotrophicus* elicidate a sugarcane defense response against a pathogenic bacteria *Xanthomonas albilineans*. *Plant Signaling and Behavior*, 1 (5)265–273.

Bakker, P.A.H.M., Pieterse, C.M.J., van Loon, L.C. 2007. Induced systemic resistance by fluorescent *Pseudomonas* spp. *Phytopathology*, 97 (2), 239–243.

Baldani, J.I., Caruso, L., Baldani, V.L.D., Goi, S.R., Döbereiner, J. 1997. Recent advances in BNF with non-legume plants. *Soil Biology & Biochemistry*, 29 (5–6), 911–922.

Bauer, W.D., Mathesius, U. 2004. Plant responses to bacterial quorum sensing signals. *Current Opinion in Plant Biology*, 7 (4), 429–433.

Bekri, M.A., Desair, J., Keijers, V., Proost, P., Searle-Van Leeuwen, M., Vanderleyden, J., Vande Broek, A. 1999. *Azospirillum irakense* produces a novel type of pectate lyase. *Journal of Bacteriology*, 181 (8), 2440–2447.

Berg, G., Eberl, L., Hartmann, A. 2005. The rhizosphere as a reservoir for opportunistic human pathogenic bacteria. *Environmental Microbiology*, 7 (11), 1673–1685.

Bloemberg, G.V. 2007. Microscopic analysis of plant-bacterium interactions using auto fluorescent proteins. *European Journal of Plant Pathology*, 119 (3), 301–309.

Bloemberg, G.V., Lugtenberg, B.J. 2001. Molecular basis of plant growth promotion and biocontrol by rhizobacteria. *Current Opinion in Plant Biology*, 4, 343–350.

Boddey, R.M. 1987. Methods for quantification of nitrogen fixation associated with Gramineae. *CRC Critical Reviews in Plant Sciences*, 6 (3), 209–266.

Boddey, R.M., de Oliveira, O.C., Urquiaga, S., Reis, V.M., de Olivares, F.L., Baldani, V.L.D., Döbereiner, J. 1995. Biological nitrogen fixation associated with sugar cane and rice—contributions and prospects for improvement. *Plant and Soil*, 174 (1–2), 195–209.

Boddey, R.M., Polidoro, J.C., Resende, A.S., Alves, B.J.R. Urquiaga, S. 2001. Use of the N-15 natural abundance technique for the quantification of the contribution of N_2 fixation to sugar cane and other grasses. *Australian Journal of Plant Physiology*, 28 (9), 889–895.

Böhm, M., Hurek, T., Reinhold-Hurek, B. 2007. Twitching motility is essential for endophytic rice colonization by the N_2-fixing endophyte *Azoarcus* sp. strain BH72. *Molecular Plant-Microbe Interactions*, 20 (5), 526–533.

Brandl, M.T., Lindow, S.E. 1996. Cloning and characterization of a locus encoding an indolepyruvate decarboxylase involved in indole-3-acetic acid synthesis in *Erwinia herbicola*. *Applied and Environmental Microbiology*, 62, 4121–4128.

Casanovas, E.M., Barassi, C.A., Andrade, F.H., Sueldo, R.J. 2003. *Azospirillum*-inoculated maize plant responses to irrigation restraints imposed during flowering. *Cereal Research Communications*, 31 (3–4), 395–402.

Cassan, F., Perrig, D., Sgroy, V., Masciarelli, O., Penna, C., Luna, V. 2009. *Azospirillum brasilense* Az39 and *Bradyrhizobium japonicum* E109, inoculated singly or in combination, promote seed germination and early seedling growth in corn (*Zea mays* L.) and soybean (*Glycine max* L.). *European Journal of Soil Biology*, 45 (1), 28–35.

Cavalcante, J.J.V., Vargas, C., Nogueira, E.M., Vinagre, F., Schwarcz, K., Baldani, J.I., Ferreira, P.C.G., Hemerly, A.S. 2007. Members of the ethylene signalling pathway are regulated in sugarcane during the association with nitrogen-fixing endophytic bacteria. *Journal of Experimental Botany*, 58 (3), 673–686.

Chae, H.S., Kieber, J.J. 2005. Eto Brute? Role of ACS turnover in regulating ethylene biosynthesis. *Trends in Plant Science*, 10 (6), 291–296.

Chi, F., Shen, S.H., Cheng, H.P., Jing, Y.X., Yanni, Y.G., Dazzo, F.B. 2005. Ascending migration of endophytic rhizobia, from roots to leaves, inside rice plants and assessment of benefits to rice growth physiology. *Applied and Environmental Microbiology*, 71 (11), 7271–7278.

Chin-A-Woeng, T.F.C., Bloemberg, G.V., Lugtenberg, B.J.J. 2003. Phenazines and their role in biocontrol by *Pseudomonas* bacteria. *New Phytologist*, 157 (3), 503–523.

Cho, S.M., Kang, B.R., Han, S.H., Anderson, A.J., Park, J.Y., Lee, Y.H., Cho, B.H., Yang, K.Y., Ryu, C.M., Kirn, Y.C. 2008. 2R,3R-butanediol, a bacterial volatile produced by *Pseudomonas chlororaphis* O6, is involved in induction of systemic tolerance to drought in *Arabidopsis thaliana*. *Molecular Plant-Microbe Interactions*, 21 (8), 1067–1075.

Clark, E., Manulis, S., Ophir, Y., Barash, I., Gafni, Y. 1993. Cloning and characterization of *iaaM* and *iaaH* from *Erwinia herbicola* pathovar *gypsophilae*. *Phytopathology*, 83 (2), 234–240.

Conn, V.M., Walker, A.R., Franco, C.M.M. 2008. Endophytic actinobacteria induce defense pathways in *Arabidopsis thaliana*. *Molecular Plant-Microbe Interactions*, 21 (2), 208–218.

Cook, R.J., Thomashow, L.S., Weller, D.M., Fujimoto, D., Mazzola, M., Bangera, G., Kim, D.S. 1995. Molecular mechanisms of defense by rhizobacteria against root disease. *Proceedings of the National Academy of Sciences of the United States of America*, 92, 4197–4201.

Cooper, J.E. 2007. Early interactions between legumes and rhizobia: disclosing complexity in a molecular dialogue. *Journal of Applied Microbiology*, 103 (5), 1355–1365.

Costacurta, A., Keijers, V., Vanderleyden, J. 1994. Molecular cloning and sequence analysis of an *Azospirillum brasilense* indole-3-pyruvate decarboxylase gene. *Molecular and General Genetics*, 243, 463–472.

Costacurta, A. Vanderleyden, J. 1995. Synthesis of phytohormones by plant-associated bacteria. *Critical Reviews in Microbiology*, 21, 1–18.

Creus, C.M., Sueldo, R.J., Barassi, C.A. 2004. Water relations and yield in *Azospirillum*-inoculated wheat exposed to drought in the field. *Canadian Journal of Botany-Revue Canadienne de Botanique*, 82 (2), 273–281.

Croes, C.L., Moens, S., Van Bastelaere, E., Vanderleyden, J., Michiels, K.W. 1993. The polar flagellum mediates *Azospirillum brasilense* adsorption to wheat roots. *Journal of General Microbiology*, 139, 2261–2269.

Davies, P.J. 2004. *Plant Hormones: Biosynthesis, Signal Transduction and Action*, 3rd edn. Springer, Dordrecht, The Netherlands.

De Weert, S., Bloemberg, G.V. 2006. Rhizosphere competence and the role of root colonization in biocontrol. In: *Plant-associated Bacteria*, (Gnanamanickam, S.S., Ed.), pp. 317–333 Springer, Dordrecht, The Netherlands.

Diallo, M.D., Reinhold-Hurek, B., Hurek, T. 2008. Evaluation of PCR primers for universal *nifH* gene targeting and for assessment of transcribed *nifH* pools in roots of *Oryza longistaminata* with and without low nitrogen input. *FEMS Microbiology Ecology*, 65 (2), 220–228.

Diaz-Zorita, M., Fernandez-Canigia, M.V. 2009. Field performance of a liquid formulation of *Azospirillum brasilense* on dryland wheat productivity. *European Journal of Soil Biology*, 45 (1), 3–11.

Dobbelaere, S., Croonenborghs, A., Thys, A., Vande Broek, A., Vanderleyden, J. 1999. Phytostimulatory effect of *Azospirillum brasilense* wild type and mutant strains altered in IAA production on wheat. *Plant and Soil*, 212, 155–164.

Döbereiner, J., Urquiaga, S., Boddey, R.M. 1995. Alternatives for nitrogen nutrition of crops in tropical agriculture. *Fertilizer Research*, 42 (1–3), 339–346.

Dong, Y.M., Iniguez, A.L., Ahmer, B.M.M., Triplett, E.W. 2003a. Kinetics and strain specificity of rhizosphere and endophytic colonization by enteric bacteria on seedlings of *Medicago sativa* and *Medicago truncatula*. *Applied and Environmental Microbiology*, 69 (3), 1783–1790.

Dong, Y.M., Iniguez, A.L., Triplett, E.W. 2003b. Quantitative assessments of the host range and strain specificity of endophytic colonization by *Klebsiella pneumoniae* 342. *Plant and Soil*, 257 (1), 49–59.

Dorr, J., Hurek, T., Reinhold-Hurek, B. 1998. Type IV pili are involved in plant-microbe and fungus-microbe interactions. *Molecular Microbiology*, 30 (1), 7–17.

dos Reis, F.B., da Silva, L.G., Reis, V.M., Döbereiner, J. 2000. Occurrence of diazotrophic bacteria in different sugar cane genotypes. *Pesquisa Agropecuaria Brasileira*, 35 (5), 985–994.

Dubuis, C., Keel, C., Haas, D. 2007. Dialogues of root-colonizing biocontrol pseudomonads. *European Journal of Plant Pathology*, 119 (3), 311–328.

Egener, T., Hurek, T., Reinhold-Hurek, B. 1998. Use of green fluorescent protein to detect expression of *nif* genes of *Azoarcus* sp. BH72, a grass-associated diazotroph, on rice roots. *Molecular Plant-Microbe Interactions*, 11 (1), 71–75.

Elbeltagy, A., Nishioka, K., Sato, T., Suzuki, H., Ye, B., Hamada, T., Isawa, T., Mitsui, H., Minamisawa, K. 2001. Endophytic colonization and in planta nitrogen fixation by a *Herbaspirillum* sp isolated from wild rice species. *Applied and Environmental Microbiology*, 67 (11), 5285–5293.

Ellis, J., Dodds, P., Pryor, T. 2000. Structure, function and evolution of plant disease resistance genes. *Current Opinion in Plant Biology*, 3 (4), 278–284.

Elmerich, C. 2007. Historical perspective: From bacterization to endophytes. In: *Associative and Endophytic Nitrogen-fixing Bacteria and Cyanobacterial Associations*, Vol. 5, (Elmerich, C., Newton, W.E., Eds.), pp. 1–20. Springer, Dordrecht, The Netherlands.

Engelhard, M., Hurek, T., Reinhold-Hurek, B. 2000. Preferential occurrence of diazotrophic endophytes, *Azoarcus* spp., in wild rice species and land races of *Oryza sativa* in comparison with modern races. *Environmental Microbiology*, 2 (2), 131–141.

Fouts, D.E., Tyler, H.L., Deboy, R.T., Daugherty, S., Ren, Q.H., Badger, J.H., Durkin, A.S., Huot, H., Shrivastava, S., Kothari, S., Dodson, R.J., Mohamoud, Y., Khouri, H., Roesch, L.F.W., Krogfelt, K.A., Struve, C., Triplett, E.W., Methe, B.A. 2008. Complete genome sequence of the N_2-fixing broad host range endophyte *Klebsiella pneumoniae* 342 and virulence predictions verified in mice. *PLoS Genetics*, 4 (7).

Fuentes-Ramirez, L.E., Caballero-Mellado, J., Sepulveda, J., Martinez-Romero, E. 1999. Colonization of sugarcane by *Acetobacter diazotrophicus* is inhibited by high N-fertilization. *FEMS Microbiology Ecology*, 29 (2), 117–128.

Fuentes-Ramirez, L.E., Jimenez-Salgado, T., Abarca-Ocampo, I.R., Caballero-Mellado, J. 1993. *Acetobacter diazotrophicus*, an indoleacetic acid producing bacterium isolated from sugarcane cultivars of Mexico. *Plant and Soil*, 154 (2), 145–150.

Glick, B.R., Penrose, D.M., Li, J.P. 1998. A model for the lowering of plant ethylene concentrations by plant growth-promoting bacteria. *Journal of Theoretical Biology*, 190 (1), 63–68.

Glick, B.R., Todorovic, B., Czarny, J., Cheng, Z.Y., Duan, J., McConkey, B. 2007. Promotion of plant growth by bacterial ACC deaminase. *Critical Reviews in Plant Sciences*, 26 (5–6), 227–242.

Gourion, B., Rossignol, M., Vorholt, J.A. 2003. A proteomic study of Methylobacterium extorquens reveals a response regulator essential for epiphytic growth. *Proceedings of the National Academy of Sciences of the United States of America*, 103, 13186–13191.

Guedes, H.V., dos Santos, S.T., Perin, L., Teixeira, K.R.D., Reis, V.M., Baldani, J.I. 2008. Polyphasic characterization of *Gluconacetobacter diazotrophicus* isolates obtained from different sugarcane varieties. *Brazilian Journal of Microbiology*, 39 (4), 718–723.

Gyaneshwar, P., James, E.K., Mathan, N., Reddy, P.M., Reinhold-Hurek, B., Ladha, J.K. 2001. Endophytic colonization of rice by a diazotrophic strain of *Serratia marcescens*. *Journal of Bacteriology*, 183 (8), 2634–2645.

Haas, D., Keel, C. 2003. Regulation of antibiotic production in root-colonizing *Pseudomonas* spp. and relevance for biological control of plant disease. *Annual Review of Phytopathology*, 41, 117–153.

Hai, B., Diallo, N.H., Sall, S., Haesler, F., Schauss, K., Bonzi, M., Assigbetse, K., Chotte, J.L., Munch, J.C., Schloter, M. 2009. Quantification of key genes steering the microbial nitrogen cycle in the rhizosphere of sorghum cultivars in tropical agroecosystems. *Applied and Environmental Microbiology*, 75 (15), 4993–5000.

Hallmann, J., Quadt-Hallmann, A., Mahaffee, W.F., Kloepper, J.W. 1997. Bacterial endophytes in agricultural crops. *Canadian Journal of Microbiology*, 43 (10), 895–914.

Han, S.H., Lee, S.J., Moon, J.H., Park, K.H., Yang, K.Y., Cho, B.H., Kim, K.Y., Kim, Y.W., Lee, M.C., Anderson, A.J., Kim, Y.C. 2006. GacS-dependent production of 2R, 3R-butanediol by *Pseudomonas chlororaphis* O6 is a major determinant for eliciting systemic resistance against *Erwinia carotovora* but not against *Pseudomonas syringae* pv. *tabaci* in tobacco. *Molecular Plant-Microbe Interactions*, 19 (8), 924–930.

Hardoim, P.R., Van Overbeek, L.S., Van Elsas, J.D. 2008. Properties of bacterial endophytes and their proposed role in plant growth. *Trends in Microbiology*, 16 (10), 463–471.

Hoa Long, H., Schmidt, D.D., Baldwin, I.T. 2008. Native bacterial endophytes promote host growth in a species-specific manner; phytohormone manipulations do not result in common growth responses. *PLoS One*, 3 (7), e2702.

Hooshangi, S., Bentley, W.E. 2008. From unicellular properties to multicellular behavior: bacteria quorum sensing circuitry and applications. *Current Opinion in Biotechnology*, 19 (6), 550–555.

Hughes, D.T., Sperandio, V. 2008. Inter-kingdom signalling: communication between bacteria and their hosts. *Nature Reviews Microbiology*, 6 (2), 111–120.

Hurek, T., Handley, L.L., Reinhold-Hurek, B., Piché, Y. 2002. *Azoarcus* grass endophytes contribute fixed nitrogen to the plant in an unculturable state. *Molecular Plant-Microbe Interactions*, 15 (3), 233–242.

Hurek, T., Reinhold-Hurek, B., Van Montagu, M., Kellenberger, E. 1994. Root colonization and systemic spreading of *Azoarcus* sp strain BH72 in grasses. *Journal of Bacteriology*, 176 (7), 1913–1923.

Iniguez, A.L., Dong, Y.M., Carter, H.D., Ahmer, B.M.M., Stone, J.M., Triplett, E.W. 2005. Regulation of enteric endophytic bacterial colonization by plant defenses. *Molecular Plant-Microbe Interactions*, 18 (2), 169–178.

Iniguez, A.L., Dong, Y.M., Triplett, E.W. 2004. Nitrogen fixation in wheat provided by *Klebsiella pneumoniae* 342. *Molecular Plant-Microbe Interactions*, 17 (10), 1078–1085.

James, E.K., Gyaneshwar, P., Mathan, N., Barraquio, Q.L., Reddy, P.M., Iannetta, P.P.M., Olivares, F.L., Ladha, J.K. 2002. Infection and colonization of rice seedlings by the plant growth-promoting bacterium *Herbaspirillum seropedicae* Z67. *Molecular Plant-Microbe Interactions*, 15 (9), 894–906.

James, E.K., Olivares, F.L., Baldani, J.I., Döbereiner, J. 1997. *Herbaspirillum*, an endophytic diazotroph colonizing vascular tissue in leaves of *Sorghum bicolor* L Moench. *Journal of Experimental Botany*, 48 (308), 785–797.

James, E.K., Olivares, F.L., de Oliveira, A.L.M., dos Reis, F.B., da Silva, L.G., Reis, V.M. 2001. Further observations on the interaction between sugar cane and *Gluconacetobacter diazotrophicus* under laboratory and greenhouse conditions. *Journal of Experimental Botany*, 52 (357), 747–760.

James, E.K., Reis, V.M., Olivares, F.L., Baldani, J.I., Döbereiner, J. 1994. Infection of sugar cane by the nitrogen-fixing bacterium *Acetobacter diazotrophicus*. *Journal of Experimental Botany*, 45 (275), 757–766.

Khalid, A., Tahir, S., Arshad, M., Zahir, Z.A. 2004. Relative efficiency of rhizobacteria for auxin biosynthesis in rhizosphere and non-rhizosphere soils. *Australian Journal of Soil Research*, 42 (8), 921–926.

Khammas, K.M., Kaiser, P. 1991. Characterization of a pectinolytic activity in *Azospirillum irakense*. *Plant and Soil*, 137 (1), 75–79.

Koga, J., Adachi, T., Hidaka, H. 1991. Molecular cloning of the gene for indolepyruvate decarboxylase from *Enterobacter cloacae*. *Molecular and General Genetics*, 226, 10–16.

Krause, A., Ramakumar, A., Bartels, D., Battistoni, F., Bekel, T., Boch, J., Bohm, M., Friedrich, F., Hurek, T., Krause, L., Linke, B., McHardy, A.C., Sarkar, A., Schneiker, S., Syed, A.A., Thauer, R., Vorholter, F.J., Weidner, S., Puhler, A., Reinhold-Hurek, B., Kaiser, O., Goesmann, A. 2006. Complete genome of the mutualistic, N_2-fixing grass endophyte *Azoarcus* sp. strain BH72. *Nature Biotechnology*, 24 (11), 1385–1391.

Lemanceau, P., Bauer, P., Kraemer, S., Briat, J.-F. 2009. Iron dynamics in the rhizosphere as a case study for analyzing interactions between soils, plants and microbes. *Plant and Soil*, 321 (1–2), 513–535.

Lodewyckx, C., Vangronsveld, J., Porteous, F., Moore, E.R.B., Taghavi, S., Mezgeay, M., van der Lelie, D. 2002. Endophytic bacteria and their potential applications. *Critical Reviews in Plant Sciences*, 21 (6), 583–606.

Manulis, S., Haviv-Chesner, A., Brandl, M.T., Lindow, S.E., Barash, I. 1998. Differential involvement of indole-3-acetic acid biosynthetic pathways in pathogenicity and epiphytic fitness of *Erwinia herbicola* pv. *gypsophilae*. *Molecular Plant-Microbe Interactions*, 11 (7), 634–642.

Marschner, H., Romheld, V. 1994. Strategies of plants for acquisition of iron. *Plant and Soil*, 165 (2), 261–274.

Mathesius, U., Mulders, S., Gao, M., Teplitski, M., Caetano-Anolles, G., Rolfe, B.G., Bauer, W.D. 2003. Extensive and specific responses of a eukaryote to bacterial quorum-sensing signals. *Proceedings of the National Academy of Sciences of the United States of America*, 100, 1444–1449.

Mattoo, A.K., Suttle, J.C. 1991. *The Plant Hormone Ethylene*. CRC Press, Baco Raton.

Mavingui, P., Laguerre, G., Berge, O., Heulin, T. 1992. Genetic and phenotypic diversity of *Bacillus polymyxa* in soil and in the wheat rhizosphere. *Applied and Environmental Microbiology*, 58 (6), 1894–1903.

McInroy, J.A., Kloepper, J.W. 1995. Survey of indigenous bacterial endophytes from cotton and sweet corn. *Plant and Soil*, 173 (2), 337–342.

Medeiros, A.F.A., Polidoro, J.C., Reis, V.M. 2006. Nitrogen source effect on *Gluconacetobacter diazotrophicus* colonization of sugarcane (*Saccharum* spp.). *Plant and Soil*, 279 (1–2), 141–152.

Michiels, K.W., Croes, C.L., Vanderleyden, J. 1991. 2 different modes of attachment of *Azospirillum brasilense* Sp7 to wheat roots. *Journal of General Microbiology*, 137 2241–2246.

Miyauchi, M.Y.H., Lima, D.S., Nogueira, M.A., Lovato, G.M., Murate, L.S., Cruz, M.F., Ferreira, J.M., Zangaro, W., Andrade, G. 2008. Interactions between diazotrophic bacteria and mycorrhizal fungus in maize genotypes. *Scientia Agricola*, 65 (5), 525–531.

Molina-Favero, C., Creus, C.M., Simontacchi, M., Puntarulo, S., Lamattina, L. 2008. Aerobic nitric oxide production by *Azospirillum brasilense* Sp245 and its influence on root architecture in tomato. *Molecular Plant-Microbe Interactions*, 21 (7), 1001–1009.

Montañez, A., Abreu, C., Gill, P.R., Hardarson, G., Sicardi, M. 2009. Biological nitrogen fixation in maize (*Zea mays* L.) by N-15 isotope-dilution and identification of associated culturable diazotrophs. *Biology and Fertility of Soils*, 45 (3), 253–263.

Morris, R.O. 1995. Genes specifying auxin and cytokinin biosynthesis in prokaryotes. In: *Plant Hormones*, (Davies, P.J., Ed.), 2nd edn, pp. 318–339. Kluwer Academic Publishers, Dordrecht, The Netherlands.

Muthukumarasamy, R., Revathi, G., Lakshminarasimhan, C. 1999. Influence of N fertilisation on the isolation of *Acetobacter diazotrophicus* and *Herbaspirillum* spp. from Indian sugarcane varieties. *Biology and Fertility of Soils*, 29 (2), 157–164.

Nogueira, E.D., Vinagre, F., Masuda, H.P., Vargas, C., de Padua, V.L.M., da Silva, F.R., dos Santos, R.V., Baldani, J.I., Cavalcanti, P., Ferreira, G., Hemerly, A.S. 2001. Expression of sugarcane genes induced by inoculation with *Gluconacetobacter diazotrophicus* and *Herbaspirillum rubrisubalbicans*. *Genetics and Molecular Biology*, 24 (1–4), 199–206.

Olivares, F.L., James, E.K., Baldani, J.I., Döbereiner, J. 1997. Infection of mottled stripe disease-susceptible and resistant sugar cane varieties by the endophytic diazotroph *Herbaspirillum*. *New Phytologist*, 135 (4), 723–737.

Oliveira, A.L.M., Stoffels, M., Schmid, M., Reis, V.M., Baldani, J.I., Hartmann, A. 2009. Colonization of sugarcane plantlets by mixed inoculations with diazotrophic bacteria. *European Journal of Soil Biology*, 45 (1), 106–113.

Oliveira, A.L.M., Urquiaga, S., Döbereiner, J., Baldani, J.I. 2002. The effect of inoculating endophytic N_2-fixing bacteria on micropropagated sugarcane plants. *Plant and Soil*, 242 (2), 205–215.

Patten, C.L., Glick, B.R. 1996. Bacterial biosynthesis of indole-3-acetic acid. *Canadian Journal of Microbiology*, 42, 207–220.

Patten, C.L., Glick, B.R. 2002. Role of *Pseudomonas putida* indoleacetic acid in development of the host plant root system. *Applied and Environmental Microbiology*, 68 3795–3801.

Pedraza, R.O., Bellone, C.H., de Bellone, S., Sorte, P.M.B., Teixeira, K.R.D. 2009. *Azospirillum* inoculation and nitrogen fertilization effect on grain yield and on the diversity of endophytic bacteria in the phyllosphere of rice rainfed crop. *European Journal of Soil Biology*, 45 (1), 36–43.

Pierson, L.S., Pierson, E.A. 2007. Roles of diffusible signals in communication among plant-associated bacteria. *Phytopathology*, 97 (2), 227–232.
Pimentel, J.P., Olivares, F., Pitard, R.M., Urquiaga, S., Akiba, F., Döbereiner, J. 1991. Dinitrogen fixation and infection of grass leaves by *Pseudomonas rubrisubalbicans* and *Herbaspirillum seropedicae*. *Plant and Soil*, 137 (1), 61–65.
Pirttilä, A.M., Joensuu, P., Pospiech, H., Jalonen, J., Hohtola, A. 2004. Bud endophytes of Scots pine produce adenine derivatives and other compounds that affect morphology and mitigate browning of callus cultures. *Physiologia Plantarum*, 121, 305–312.
Quecine, M.C., Araujo, W.L., Marcon, J., Gai, C.S., Azevedo, J.L., Pizzirani-Kleiner, A.A. 2008. Chitinolytic activity of endophytic *Streptomyces* and potential for biocontrol. *Letters in Applied Microbiology*, 47 (6), 486–491.
Raaijmakers, J.M., Vandersluis, I., Koster, M., Bakker, P.A.H.M., Weisbeek, P.J., Schippers, B. 1995. Utilization of heterologous siderophores and rhizosphere competence of fluorescent *Pseudomonas* spp. *Canadian Journal of Microbiology*, 41 (2), 126–135.
Raaijmakers, J.M., Vlami, M., de Souza, J.T. 2002. Antibiotic production by bacterial biocontrol agents. *Antonie Van Leeuwenhoek International Journal of General and Molecular Microbiology*, 81 (1–4), 537–547.
Reinhold, B., Krause, A., Boehm, M., Friedrich, F., Gemmer, S., Miche, L., Oetjen, J., Sarkar, A., Hurek, T. 2007. Life inside cereals: genomic insights into the lifestyle of the mutualistic, N_2-fixing endophyte *Azoarcus* sp strain BH72. *FEBS Journal*, 274, 15.
Reinhold-Hurek, B., Hurek, T. 1998a. Interactions of gramineous plants with *Azoarcus* spp. and other diazotrophs: Identification, localization, and perspectives to study their function. *Critical Reviews in Plant Sciences*, 17 (1), 29–54.
Reinhold-Hurek, B., Hurek, T. 1998b. Life in grasses: diazotrophic endophytes. *Trends in Microbiology*, 6 (4), 139–144.
Robin, A., Vansuyt, G., Hinsinger, P., Meyer, J.M., Briat, J.F., Lemanceau, P. 2008. Iron dynamics in the rhizosphere: Consequences for plant health and nutrition. *Advances in Agronomy*, 99, 183–225.
Rocha, F.R., Papini-Terzi, F.S., Nishiyama, M.Y., Vencio, R.Z.N., Vicentini, R., Duarte, R.D.C., de Rosa, V.E., Vinagre, F., Barsalobres, C., Medeiros, A.H., Rodrigues, F.A., Ulian, E.C., Zingaretti, S.M., Galbiatti, J.A., Almeida, R.S., Figueira, A.V.O., Hemerly, A.S., Silva-Filho, M.C., Menossi, M., Souza, G.M. 2007. Signal transduction-related responses to phytohormones and environmental challenges in sugarcane. *BMC Genomics*, 8: 71.
Rosenblueth, M., Martinez-Romero, E. 2004. *Rhizobium etli* maize populations and their competitiveness for root colonization. *Archives of Microbiology*, 181 (5), 337–344.
Rothballer, M., Eckert, B., Schmid, M., Fekete, A., Schloter, M., Lehner, A., Pollmann, S., Hartmann, A. 2008. Endophytic root colonization of gramineous plants by *Herbaspirillum frisingense*. *FEMS Microbiology Ecology*, 66 (1), 85–95.
Rothballer, M., Schmid, M., Hartmann, A. 2003. In situ localization and PGPR-effect of *Azospirillum brasilense* strains colonizing roots of different wheat varieties. *Symbiosis*, 34 (3), 261–279.
Ryan, R.P., Germaine, K., Franks, A., Ryan, D.J., Dowling, D.N. 2008. Bacterial endophytes: recent developments and applications. *FEMS Microbiology Letters*, 278 (1), 1–9.
Ryu, C.M., Farag, M.A., Hu, C.H., Reddy, M.S., Kloepper, J.W., Pare, P.W. 2004. Bacterial volatiles induce systemic resistance in *Arabidopsis*. *Plant Physiology*, 134 (3), 1017–1026.
Ryu, C.M., Farag, M.A., Hu, C.H., Reddy, M.S., Wei, H.X., Pare, P.W., Kloepper, J.W. 2003. Bacterial volatiles promote growth in *Arabidopsis*. *Proceedings of the National Academy of Sciences of the United States of America*, 100, 4927–4932.

Santner, A., Calderon-Villalobos, L.I.A., Estelle, M. 2009. Plant hormones are versatile chemical regulators of plant growth. *Nature Chemical Biology*, 5 (5), 301–307.

Santner, A., Estelle, M. 2009. Recent advances and emerging trends in plant hormone signalling. *Nature*, 459 (7250), 1071–1078.

Sapp, J. 2004. The dynamics of symbiosis: an historical overview. *Canadian Journal of Botany-Revue Canadienne de Botanique*, 82 (8), 1046–1056.

Schloter, M., Hartmann, A. 1998. Endophytic and surface colonization of wheat roots (*Triticum aestivum*) by different *Azospirillum brasilense* strains studied with strain-specific monoclonal antibodies. *Symbiosis*, 25 (1–3), 159–179.

Schuhegger, R., Ihring, A., Gantner, S., Bahnweg, G., Knappe, C., Vogg, G., Hutzler, P., Schmid, M., Van Breusegem, F., Eberl, L., Hartmann, A.. Langebartels, C. 2006. Induction of systemic resistance in tomato by N-acyl-L-homoserine lactone-producing rhizosphere bacteria. *Plant Cell and Environment*, 29 (5), 909–918.

Schütz, A., Golbik, R., Tittmann, K., Svergun, D.I., Koch, M.H., Hübner, G., König, S. 2003. Studies on structure-function relationships of indolepyruvate decarboxylase from *Enterobacter cloacae*, a key enzyme of the indole acetic acid pathway. *European Journal of Biochemistry*, 270 2322–2331.

Seghers, D., Wittebolle, L., Top, E.M., Verstraete, W., Siciliano, S.D. 2004. Impact of agricultural practices on the *Zea mays* L. endophytic community. *Applied and Environmental Microbiology*, 70 (3), 1475–1482.

Sekine, M., Watanabe, K., Syono, K. 1989. Molecular cloning of a gene for indole-3-acetamide hydrolase from *Bradyrhizobium japonicum. Journal of Bacteriology*, 171 (3), 1718–1724.

Sikorski, M.M., Biesiadka, J., Kasperska, A.E., Kopcinska, J., Lotocka, B., Golinowski, W., Legocki, A.B. 1999. Expression of genes encoding PR10 class patho genesis-related proteins is inhibited in yellow lupine root nodules. *Plant Science*, 149 (2), 125–137.

Silva, L.G., Miguens, F.C., Olivares, F.L. 2003. *Herbaspirillum seropedicae* and sugarcane endophytic interaction investigated by using high pressure freezing electron microscopy. *Brazilian Journal of Microbiology*, 24 (S1), 69–71.

Spaepen, S., Dobbelaere, S., Croonenborghs, A., Vanderleyden, J. 2008. Effects of *Azospirillum brasilense* indole-3-acetic acid production on inoculated wheat plants. *Plant and Soil*, 312 (1–2), 15–23.

Spaepen, S., Vanderleyden, J., Remans, R. 2007a. Indole-3-acetic acid in microbial and microorganism-plant signaling. *FEMS Microbiology Reviews*, 31 (4), 425–448.

Spaepen, S., Versées, W., Gocke, D., Pohl, M., Steyaert, J., Vanderleyden, J. 2007b. Characterization of phenylpyruvate decarboxylase, involved in auxin production of *Azospirillum brasilense. Journal of Bacteriology*, 189 (21), 7626–7633.

Sprent, J.I., James, E.K. 1995. N_2 fixation by endophytic bacteria: questions of entry and operation. In: *Azospirillum VI and Related Microorganisms* (Fendrik, I., del Gallo, M., Vanderleyden, J., de Zamaroczy, M., Eds.), pp. 15–30. Springer-Verlag, Berlin.

Steenhoudt, O., Vanderleyden, J. 2000. *Azospirillum*, a free-living nitrogen-fixing bacterium closely associated with grasses: genetic, biochemical and ecological aspects. *FEMS Microbiology Reviews*, 24, 487–506.

Suman, A., Shrivastava, A.K., Gaur, A., Singh, P., Singh, J., Yadav, R.L. 2008. Nitrogen use efficiency of sugarcane in relation to its BNF potential and population of endophytic diazotrophs at different N levels. *Plant Growth Regulation*, 54 (1), 1–11.

Teplitski, M., Chen, H.C., Rajamani, S., Gao, M.S., Merighi, M., Sayre, R.T., Robinson, J.B., Rolfe, B.G., Bauer, W.D. 2004. *Chlamydomonas reinhardtii* secretes compounds that mimic bacterial signals and interfere with quorum sensing regulation in bacteria. *Plant Physiology*, 134 (1), 137–146.

Urquiaga, S., Cruz, K.H.S., Boddey, R.M. 1992. Contribution of nitrogen fixation to sugar cane—N-15 and nitrogen balance estimates. *Soil Science Society of America Journal*, 56 (1), 105–114.

van Loon, L.C. 2007. Plant responses to plant growth-promoting rhizobacteria. *European Journal of Plant Pathology*, 119 (3), 243–254.

Van Wees, S.C.M., Van der Ent, S., Pieterse, C.M.J. 2008. Plant immune responses triggered by beneficial microbes. *Current Opinion in Plant Biology*, 11 (4), 443–448.

Vande Broek, A., Michiels, J., Vangool, A., Vanderleyden, J. 1993. Spatial-temporal colonization patterns of *Azospirillum brasilense* on the wheat root surface and expression of the bacterial *nifH* gene during association. *Molecular Plant-Microbe Interactions*, 6 (5), 592–600.

Vandenkoornhuyse, P., Baldauf, S.L., Leyval, C., Straczek, J., Young, J.P.W. 2002. Evolution—Extensive fungal diversity in plant roots. *Science*, 295 (5562), 2051.

Vargas, C., de Padua, V.L.M., Nogueira, E.D., Vinagre, F., Masuda, H.P., da Silva, F.R., Baldani, J.I., Ferreira, P.C.G., Hemerly, A.S. 2003. Signaling pathways mediating the association between sugarcane and endophytic diazotrophic bacteria: A genomic approach. *Symbiosis*, 35 (1–3), 159–180.

Vinagre, F., Vargas, C., Schwarcz, K., Cavalcante, J., Nogueira, E.M., Baldani, J.I., Ferreira, P.C.G., Hemerly, A.S. 2006. SHR5: a novel plant receptor kinase involved in plant-N_2-fixing endophytic bacteria association. *Journal of Experimental Botany*, 57 (3), 559–569.

Weller, D.M. 2007. *Pseudomonas* biocontrol agents of soilborne pathogens: Looking back over 30 years. *Phytopathology*, 97 (2), 250–256.

Whipps, J.M. 2001. Microbial interactions and biocontrol in the rhizosphere. *Journal of Experimental Botany*, 52 (Spec Issue), 487–511.

Woodward, A.W., Bartel, B. 2005. Auxin: Regulation, action, and interaction. *Annals of Botany*, 95 (5), 707–735.

Yan, Y.L., Yang, J., Dou, Y.T., Chen, M., Ping, S.Z., Peng, J.P., Lu, W., Zhang, W., Yao, Z.Y., Li, H.Q., Liu, W., He, S., Geng, L.Z., Zhang, X.B., Yang, F., Yu, H.Y., Zhan, Y.H., Li, D.H., Lin, Z.L., Wang, Y.P., Elmerich, C., Lin, M., Jin, Q. 2008. Nitrogen fixation island and rhizosphere competence traits in the genome of root-associated *Pseudomonas stutzeri* A1501. *Proceedings of the National Academy of Sciences of the United States of America*, 105 (21), 7564–7569.

Zhang, H., Kim, M.S., Krishnamachari, V., Payton, P., Sun, Y., Grimson, M., Farag, M.A., Ryu, C.M., Allen, R., Melo, I.S., Pare, P.W. 2007. Rhizobacterial volatile emissions regulate auxin homeostasis and cell expansion in *Arabidopsis*. *Planta*, 226 (4), 839–851.

Zimmer, W., Roeben, K., Bothe, H. 1988. An alternative explanation for plant-growth promotion by bacteria of the genus *Azospirillum*. *Planta*, 176 (3), 333–342.

Section 4
Arthropods

Chapter 12
Effects of Insect Herbivores on the Nitrogen Economy of Plants

Leiling Tao and Mark D. Hunter

Introduction

As the authors of previous chapters in this volume have stressed, nitrogen (N) availability limits the growth and fitness of many plant species, leading to a wide variety of adaptations to secure N from the environment. From an ecological perspective, N availability often places a fundamental limitation on primary production and therefore energy flow through terrestrial ecosystems (Aber and Melillo, 2001). In short, the energy (as organic carbon, C) required to fuel trophic interactions and support the web of life can be limited by the availability of N. As a result, the accumulation of herbivore biomass, including that of the insect herbivores discussed here, is tightly linked to the N economy of plants (Speight et al., 2008).

In addition to its impacts on herbivores through limitations on productivity, N availability influences herbivores because of its relative scarcity compared with carbon. Plant tissues contain lower concentrations of N than do insect tissues, and gaining sufficient N from plant material, with its relatively high C:N ratio, is a fundamental problem for insect herbivores (Mattson, 1980; Fagan et al., 2002). The stoichiometric mismatch between plants and herbivores has received considerable attention and is a major driver of both population (Hillebrand et al., 2009) and ecosystem processes (Elser et al., 1996, 2000). For example, stoichiometric differences between insects and plants are reflected in the low assimilation and growth efficiencies (g gain per g ingested) of insect phytophages compared with those of predatory insects (2–38% for phytophages versus 38–51% for predators) (Speight et al., 2008).

Because N limits their growth and fitness, we might expect insect herbivores to evolve mechanisms for the efficient extraction of N from plant tissue, while plants respond with adaptations to protect their N from consumption. Indeed the coevolutionary arms race between plants and insects (Darwin, 1859; Ehrlich and Raven, 1964; Thompson and Cunningham, 2002) may be viewed in part as a struggle over access to limiting nutrients such as N. In turn,

Ecological Aspects of Nitrogen Metabolism in Plants, First Edition. Edited by Joe C. Polacco and Christopher D. Todd.
© 2011 by John Wiley & Sons, Inc. Published 2011 by John Wiley & Sons, Inc.

the nutrient allocation strategies of plants, in response to herbivory, may be viewed as evolutionary adaptations to protect limiting resources from consumption; they are an *evolutionary effect* of insect herbivores on the N economy of plants (Frost and Hunter, 2008a). However, plants inevitably lose some of their tissue N to insect (and other) herbivores, and that "lost" N then passes through various biotic and abiotic pools in the ecosystem; this flux of N through the ecosystem represents an *ecological effect* of insect herbivores on the N economy of plants (Hunter, 2001a).

In this chapter, we review some evolutionary and ecological effects of insect herbivores on the N economy of plants. Under evolutionary effects, we focus on strategies of nutrient allocation in the presence and absence of herbivores, and under ecological effects, we focus on the fate of N "lost" to herbivores as it moves through pools in the ecosystem. We conclude with some suggestions of areas for future research.

Patterns of N and C Allocation Following Herbivory

Background

Plants are not passive recipients of herbivore damage. Rather, herbivory induces a wide range of responses in plants, including increases in the expression of physical, chemical, and indirect defenses, and the reallocation of resources to and from sites of damage (Green and Ryan, 1972; Baldwin and Schultz, 1983; Van der Meijden et al., 1988; Karban and Baldwin, 1997). The evolution of plant resistance mechanisms against insect herbivores remains a very active area of research (Stamp, 2003; Speight et al., 2008) and is discussed in Chapter 13. Here, we focus on damage-induced changes in N allocation in plants, to consider how insect damage influences patterns of N allocation among processes (what the plant is doing) and organs (where those processes take place). By necessity, we will also comment on patterns of C allocation, because the uptake and allocation of C and N are intimately linked.

Wound-induced reallocation of resources underlies plant "tolerance" to herbivory, the capacity of plants to maintain their fitness through growth and reproduction after suffering herbivore damage (Van der Meijden et al., 1988; Rosenthal and Kotanen, 1994). Plants control resource allocation mainly by transporting and transferring carbon and nitrogen through vascular connections between sources and sinks (Wardlaw, 1990). Phenotypic plasticity in resource partitioning can maximize plant growth and fitness under variable environments (Bazzaz and Grace, 1997) including the unpredictable densities of insect herbivores (Rosenthal and Kotanen, 1994). Levels of defoliation may be unpredictable, and almost all plants will suffer some degree of herbivory during their lives so that selection pressure is maintained for plastic rather than fixed strategies of allocation. We might therefore predict that most plant species will exhibit some capacity for adaptive reallocation of resources following herbivory.

However, wound-induced changes in nutrient allocation occur under significant constraints. Principally, insects often consume the organs that plants use for resource uptake (leaves and roots), reducing overall resource levels for future allocation and constraining the ability of plants to tolerate damage. This observation leads to one critical issue in studies of wound-induced changes in resource allocation by plants; we often do not know whether observed changes in N and C partitioning following damage result from reallocation of "old" resources, accumulated before the damage took place, or allocation of "new" resources, accumulated after damage began. Labeling with stable isotopes (Frost and Hunter, 2008a) and the analysis of structural and storage compounds (Lindroth et al., 2007) can help to separate the two. It is generally assumed that stored resources provide materials for short-term regrowth (e.g., several days after defoliation) whereas newly gained resources are more important for long-term tolerance. However, data are sorely lacking in this area, and we believe that distinguishing between these processes should be a principal focus of future research.

The plasticity to reassign the allocation of new resources, and to reallocate old resources, should provide plants with many options following damage. For example, they may increase allocation of new resources to damaged organs or transfer old resources away from herbivore attack (Henkes et al., 2008), use stored resources for regrowth (Palacio et al., 2008), release resources into the soil to stimulate microbial activity (Holland et al., 1996), or modify C output to symbionts (Saravesi et al., 2008). In the remainder of this section, we follow Stowe and others (2000) and characterize patterns of plant resource allocation following damage by (1) relative allocation to different processes (e.g. storage, growth, reproduction and defense); (2) relative allocation to different organs (e.g. shoot, root, different tillers); and (3) the timing of allocation. We note that relatively few studies measure accurately the fitness consequences of changes in N allocation following insect damage, and so whether such changes represent true adaptations or are simply the unavoidable consequences of insect damage remains to be evaluated fully.

Constrained or Flexible Allocation Responses?

Historically, patterns of resource allocation within individual plants have been seen as the consequences of source-sink relationships, with the relative strengths of different sinks determining the accumulation of resources among organs (Ho, 1988). In tandem with this view is the assumption that there exists some "optimal" allocation strategy among plant organs, based on plant life history, ontogeny, and the abiotic environment, which maximizes plant growth and fitness (Thornley, 1972). By consuming sinks and/or sources, herbivores will alter source-sink relationships and modify subsequent allocation patterns. Additionally, we might predict that plants will actively replace lost tissue in order to reestablish the "optimal" ratios of resources among organs that were in place before herbivore attack. For example, imagine an optimal root N:shoot N ratio for a given plant species growing in a particular environment. The most simple view would predict that, following shoot herbivory, N will be allocated actively to shoot tissue; shoots now represent strong N sinks, and there is a need to reestablish the optimal (preherbivory) root N:shoot N ratio.

However, this simple view fails to recognize that the threat of future herbivory itself represents a significant environmental factor that should influence resource allocation during regrowth. For example, plants can strategically move resources away from sites of herbivore attack (Babst et al., 2005), either as an escape strategy or for future growth (Karban and Baldwin, 1997). In at least some plant species, the expression of SNF1-related kinase genes allows plants actively to allocate resources from shoots to roots against concentration gradients after shoot herbivory (Schwachtje et al., 2006). The ability to move resources against simple source-sink flows should provide plants with significant flexibility in allocation responses to insect herbivory. The key to understanding wound-induced patterns of resource allocation depends upon our ability to recognize the adaptive value of different allocation strategies and the physiological, morphological, and resource constraints that operate upon them (Orians, 2005). In other words, we need to focus equally both on the potential value of allocation strategies and on any constraints that act on their expression.

Allocation of N and C after Damage to Aboveground Tissues

Concentrations of the N-rich compounds RuBisCO and chlorophyll determine in part rates of photosynthesis and C gain by plants (Aber and Melillo, 2001). The dual pressures of protecting photosynthetic N while maintaining C gain likely constrain wound-induced changes in N allocation in many plant species (Karban and Baldwin, 1997). Simply put, without continued allocation of N to photosynthetic enzymes, C gain will cease. There are additional ecological pressures that favor the maintenance of leaf area and canopy volume in the face of herbivore pressure, including competition for the light environment with other plant individuals (Edwards and Wratten, 1983; Wratten et al., 1990). Rapid canopy re-establishment following defoliation enables plants to maintain a competitive advantage over individuals with less tolerance to herbivory. Regrowth mechanisms include retention of meristems, remobilization of N and C reserves, and sharing of resources among connected plant parts (Cullen et al., 2006). The ability of plants to reprioritize N and C allocation between roots and shoots in response to defoliation aboveground may be an important determinant of tolerance (Briske et al., 1996).

Reallocation of nutrients to regrowing shoots at the expense of roots has been reported in numerous plant species, and many studies suggest that resource reallocation from belowground to aboveground can contribute to grazing tolerance. For example, in a comparison among three grass species with different grazing tolerances, Guitian and Bardgett (2000) reported that two species tolerant of herbivory, *Festuca rubra* and *Cynosurus cristatus*, increase resource allocation to shoots at the expense of roots after herbivore damage. In contrast, the grazing-sensitive species *Anthoxanthum odoratum* increases the relative allocation of resources belowground. Similarly, in a congeneric comparison between two semiarid *Agropyron* bunchgrasses, Caldwell and others (1981) reported that *A. desertorum*, the grazing-tolerant species, exhibits higher flexibility in resource allocation following herbivore damage aboveground. Specifically, it has a greater capacity to curtail root system growth in favor of regrowing tillers in comparison with *A. spicatum*, the herbivore-sensitive species.

If many plants favor shoot growth at the expense of root growth after defoliation, leading to lower root:shoot ratios, we might predict that the N concentration in leaves and shoots would decrease following herbivore attack because more shoots (sinks) are competing for the N supplied by a smaller number of roots (sources). However, the pattern is quite often the opposite, especially in herbaceous plants. Repeated defoliation of plants often results in higher N concentrations in aboveground tissues (Green and Detling, 2000; Hiernaux and Turner, 1996; Raillard and Svoboda, 1999), at least until some defoliation threshold, beyond which N concentration declines again (Green and Detling, 2000). Such data have been used to support the "grazing optimization" hypothesis, which states that plants grazed at intermediate pressures exhibit the highest rates of production and foliar quality (McNaughton, 1979).

How do plants accumulate shoot N despite lower allocation of resources to roots after defoliation? Some authors believe that the mechanisms are mainly sink-driven (Hiernaux and Turner, 1996; Wardlaw, 1990). The differentiation of crown meristems following clipping, and subsequent increases in photosynthetic rates, could increase N sink strength in shoots. This could lead to higher N uptake by roots even though root growth is reduced by damage. Indeed, studies show that defoliation can increase root N uptake per unit biomass by 20–60% (Jaramillo and Detling, 1988; Ruess et al., 1983; Thomas et al., 2008). This increase in uptake activity per unit root mass may be achieved by higher rates of root respiration (Ruess et al., 1983), although further work is required to confirm this.

However, there is growing evidence that increases in N uptake following insect damage may be mediated by more than simple sink strength. Morphological flexibility (root diameter, branching, and specific root length) of roots might contribute to increased N uptake after defoliation. For example, the diameter of seminal and first order roots and overall specific root length increase after defoliation of *Agropyron desertorum* (Arredondo and Johnson, 1998). This means that *A. desertorum* has a thicker main root and longer/thinner fine roots following damage. Greater specific root length of fine roots likely enhances N uptake from soil. Coarse root morphology may also respond to shoot damage (Anderson et al., 2007), increasing subsequent rates of N uptake. Of course, if plants can increase N uptake after herbivore damage, it begs the question of why plants fail to maximize their N uptake before defoliation occurs. Presumably, tradeoffs in the use of additional resources required for N uptake (C and other nutrients) constrain N uptake below maximal levels. However, such tradeoffs remain poorly understood and worthy of further study.

Not all plants can increase N uptake after defoliation (Millard et al., 2001; Lovett and Tobiessen, 1993; Fahnestock and Detling, 1999). In such cases, stored N can play an important role in reallocation after damage. For example, in trees and some perennial grasses, N storage in root and stem can represent as much as 60% of the total N used during the following growth year, and stored N plays an essential part in plant regrowth following damage (Heilmeier et al., 1986). Reallocation of stored N from roots, stems, and old leaves to new leaves after defoliation has been observed in studies using stable isotopes (Thomas et al., 2008; Millard et al., 2001).

We argued above that plants would gain flexibility in response to herbivore damage if they were able to move N and C resources against simple chemical gradients. In response to herbivore damage aboveground, plants may benefit from transferring resources belowground for safe storage and future reallocation (Karban and Baldwin, 1997). For example, oaks that are defoliated by caterpillars preferentially allocate newly acquired N to stem and taproot storage and newly acquired C to foliage (Frost and Hunter, 2008a). The result is a wound-induced increase in oak foliar C:N ratio and a decline in the palatability of foliage for subsequent herbivore populations (Hunter and Schultz, 1995; Frost and Hunter, 2007). In further support of the idea that insect damage can stimulate reallocation of resources against concentration gradients and into storage, jasmonic acid treatment, which induces defense signalling pathways in many plants (Liechti and Farmer, 2002), can also increase bulb formation (Nojiri et al., 1992) and initiate potato tuberization (Koda et al., 1992). However, interactions between defense signals and other plant hormones that play key roles in sink formation (e.g., cytokinins) can be antagonistic (Sarkar et al., 2006). Effects of defoliation on the expression of nondefense hormones are not well characterized and should be a fruitful area for future study.

For some plants, the maintenance of extensive root systems is an important contributor to tolerance. Wound-induced resource allocation to roots can increase root growth and respiration, increase support of rhizosphere mutualists (by root exudation, below), and thereby increase nutrient uptake to compensate for N lost to herbivory (Olson and Wallander, 1999). A key point here is that defoliation by herbivores aboveground may have a greater fitness cost to plants in terms of N losses than C losses (Reynolds et al., 2000; Hunter, 2001a; Hunter et al., 2003), particularly in ecosystems where N is strongly limiting and harder to replace than C. In such environments, replacing lost N may be best accomplished by wound-induced allocation of resources belowground following insect damage. Increased root growth will, of course, allow plants to forage for soil resources more extensively. Additionally, plant roots engage in rhizodeposition, the secretion of compounds into the soil that directly surrounds plant roots (the "rhizosphere," see Chapter 5). Compounds secreted during rhizodeposition include ions, free oxygen and water, enzymes, mucilage, and many carbon-containing primary and secondary metabolites (Bais et al., 2006). Between 10 and 40% of carbon allocated belowground goes to rhizodeposition in various plant-soil systems (Cheng and Gershenson, 2007), making simple root biomass measurements a huge underestimate of resource allocation by plants belowground. Rhizodeposition is very important in mediating various plant-microbe interactions, including positive interactions with mycorrhizal fungi, N-fixing bacteria, and saprophytic microbes (Bais et al., 2006) and maintaining rhizodeposition in the face of herbivory may be a priority for plants (Frost and Hunter, 2008a). In other words, plants can actively control microbial productivity and nutrient acquisition processes (Paterson, 2003) (Chapters 2–8) and may do so in response to defoliation.

In a study with maize, Holland and others (1996) used pulse-labeled $^{14}CO_2$ to track the allocation of newly gained carbon to root exudation following defoliation aboveground. They reported a positive correlation between damage intensity and soluble root exudates and root-soil respiration within 24 hours of defoliation (Holland et al., 1996). Similarly, some temperate grass species increase exudation of newly acquired C from roots by 1.5-fold after defoliation (Hamilton and Frank, 2001). However, not all wound-induced increases in root exudation rely on newly acquired C. In a steady-state labeling experiment with *Festuca rubra*, Paterson and others (2005) found that the source of increased exudation following damage was predefoliation assimilate. The use of stored resources for damage-induced increases in root exudation may explain two observations. First, in the face of repeated defoliation, some plants decrease root exudation, apparently because of depleted C and N stores in the roots. For example, in ryegrass and clover, continuous defoliation reduces root carbon exudation by 14 and 4%, respectively (Mawdsley and Bardgett, 1997). Second, most wound-induced increases in root exudation are observed within 24–72 hours of defoliation but then decline over time (Hamilton and Frank, 2001).

In addition to changing the amount of root exudation, defoliation can also change its chemical composition (Clayton et al., 2008) with potential effects on rhizosphere symbionts. Any shift in the quality or quantity of flow into the rhizosphere can affect microbial community composition, subsequent rates of N cycling, and soil organic matter dynamics (Frost and Hunter, 2004, 2008a). Increased exudation of soluble compounds can result in increased microbial biomass (Holland, 1995) and higher bacterial:fungal ratios in the rhizosphere (Mawdsley and Bardgett, 1997), which can increase N cycling rates and create a positive feedback between plant and soil. For example, aboveground defoliation of *Poa pratensis* increases net N mineralization in the rhizosphere and consequently, ammonium and nitrate concentrations. These in turn have a positive effect on plant shoot nitrogen content and net photosynthetic rate (Hamilton and Frank, 2001). Likewise, intact *Phleum pratense* plants, raised in sieved soil that has previously hosted defoliated *Phleum pratense* plants for 4 weeks, have shoot N concentrations that are 10% higher than those of plants grown in "clean" soils (Mikola et al., 2005). Presumably, exudation by defoliated plants increases the availability of N in soils, emphasizing the importance of plant-soil feedbacks in the N economy of plants that are under attack by herbivores. N conversions by rhizosphere bacteria are discussed in more detail in Chapter 17. Overall, N losses to herbivory aboveground can stimulate changes in allocation patterns in plants that favor the replacement of lost N from belowground sources.

In addition to saprophytic bacteria and fungi in the rhizosphere, mycorrhizal fungi can take up a substantial proportion of plant C fixed by photosynthesis (Smith and Read, 2008). Indeed, wound-induced allocation of C to mycorrhizae may come at the expense of allocation to rhizosphere saprotrophs. In *Pisum sativum*, increases in rhizosphere microbial biomass induced by defoliation are negated in the presence of mycorrhizal fungi (Wamberg et al., 2003). Whether plants actively control wound-induced C allocation between mycorrhizal fungi and rhizosphere microbes is currently unclear. Overall, patterns of abundance of mycorrhizal fungi following defoliation aboveground are highly variable, with reports of increases, decreases, and no change in fungal biomass following defoliation (Gehring and Whitham, 1994; Saikkonen et al., 1999; Wearn and Gange, 2007). Much additional work needs to be done in this area to measure direct changes in C allocation to mycorrhizal fungi (not just fungal biomass) across a broad range of damage intensities and grazing types.

In their attempts to replace the N lost to herbivores, plants may compete intensely for soil N after insect defoliation aboveground. Competition for belowground resources requires a healthy root system and some plants, like *Centaurea maculosa*, may compromise the root health of their competitors. *Centaurea* plants exude the allelopathic chemicals (±)-catechin, which can have deleterious effects on native plant roots (Callaway and Ridenour, 2004). Under insect herbivory, catechin exudation increases by 4- to 5-fold, presumably conferring a competitive advantage on *C. maculosa* (Thelen et al., 2005).

Allocation of N and C after Damage to Belowground Tissues

In many ecosystems, more than 50% of the carbon captured during photosynthesis (net primary production [NPP]) is allocated belowground (Hendrick and Pregitzer, 1992). For example, the area of fine roots in forests generally exceeds that of leaf area aboveground (Jackson et al., 1997) and, in sugar maple, *Acer saccharum* Marshall, forests, the NPP of fine roots can exceed 8,000 kg ha^{-1} year^{-1}, nearly 2-fold that of foliage NPP (Hendrick and Pregitzer, 1993). Given the wide distribution and high abundance of root-feeding organisms in soil (Andersen, 1987), it is extraordinary how little we know about the influence of root-feeding insects on the physiology, ecology, and evolution of land plants (Hunter, 2008). Published papers on the effects of root-feeders on plants comprise no more than 2% of the total literature on plant-herbivore interactions (Hunter, 2001b). Yet roots and their herbivores are obviously important. In addition to their role in plant attachment and support, roots are primary water- and nutrient-absorbing tissues, principal sites of exchange with rhizosphere mutualists, and critical storage organs. In other words, they play vital roles in the N economy of plants, and damage by root herbivores has the potential to influence plant N economy and reduce plant fitness (Johnson and Murray, 2008).

Effects of root-feeding insects on N and C allocation among plant organs are poorly understood. Fine roots, the major sites of exchange with the soil environment, vary markedly in their nutrient concentrations. In perennial plants, the N content of fine roots increases on gradients of increasing soil nitrate availability (Hendricks et al., 2000), suggesting a strong role of environmental heterogeneity in determining root N content and defense. In a study of forest trees, root N concentrations ranged from 0.85–3.09% across nine tree species, with gymnosperm roots averaging lower N concentrations than those of angiosperms (Pregitzer et al., 2002). These values of root N concentrations are slightly lower than, but overlapping with, values of foliar N concentrations (Speight et al., 2008). Given that root N concentrations can be high, and that their biomass belowground is enormous, the loss of fine roots may impose significant N costs on plants. So how much root biomass and root N is lost to herbivores?

There is no question that losses from roots can be substantial. In some coniferous forests, root nematode densities range from 1,250,000–15,000,000 m^{-2} (Sohlenius, 1980). Such densities simply cannot be supported without significant damage to root systems. In peach orchards, soil insecticide application increases the longevity of fine roots by up to 125 days (Wells et al., 2002) while application of insecticide and fungicide to fine roots of sugar maples increases their fine root longevity by 500 days (Eissenstat et al., 2000) (although, in the experiment by

Eissenstat and others, effects of fungicide on mycorrhizae are not separated from effects on root pathogens). In some South Carolina forests, fine root biomass is 2-fold higher in the absence than in the presence of root-feeding insects (Stevens and Jones, 2006).

What do these densities of herbivores and levels of root loss mean to the N economy of plants? We can make some indirect estimates for particular systems. For example, cicadas are root xylem feeders and consume an average of 1.3 liters of xylem nymph^{-1} year^{-1} (Ausmus et al., 1978). Using published data on the amino acid and ureide content of the xylem of 60 eastern tree species (Barnes, 1963), Hunter (2008) estimated recently the root N consumed by cicadas on an annual basis. Each cicada consumes an average of 4–20 mg N per year in 1.3 liters of xylem (mostly water). Given that *Magicicada* densities can reach 3.75 million ha^{-1} (Dybas and Davis, 1962), cicadas can cost their tree hosts between 15 and 75 kg of nitrogen ha^{-1} y^{-1} in 4,875,000 l of water. Cicadas do not assimilate all of the xylem N that they take in and, as with other herbivores (Frost and Hunter, 2004, 2007), much of it will be excreted, recycled, and become available to tree roots quite rapidly (see discussion below). More profound N losses to the soil system may come at peak cicada emergence, which will move between 4.25 and 8.5 kg N ha^{-1} and between 1,200 and 2,400 l water ha^{-1} from belowground to aboveground (Hunter, 2008). These estimates agree well with those from a riparian forest in Kansas, where nitrogen flux during cicada emergence was estimated to average 6.3 kg N ha^{-1}, peaking at 30 kg N ha^{-1} at the highest densities (Whiles et al., 2001). By any measure, this represents a substantial impact on the N economy of plants, and we might expect that periodical cicadas will impose chronic losses to the growth and fitness of forest trees during their long nymphal period belowground. Of course, as we describe below, some of the N lost from plants to cicadas may be returned to soils by N cycling processes including the decomposition of cicada cadavers (Yang, 2004, 2008).

How do plants respond to the N losses imposed by belowground herbivores? Perhaps not surprisingly, roots generally exhibit a reduction in the rate of N uptake per unit mass following damage (Vaast et al., 1998). At the same time, export of N from root tissues to aboveground tissues can decline, particularly following damage by root-knot nematodes, potentially a form of parasite-modified host behavior. The result is the accumulation of N in the parasitized root tissues (Burgeson, 1966) and a decline in shoot N content (Mohanty et al., 1990). However, severe root damage can also cause a water "stress response" that impairs protein metabolism and amino acid synthesis in plants. As a result, there can be increases in the hydrolysis of existing proteins to free amino acids, leading to higher mobile N availability in shoots (Johnson et al., 2008). This response may explain the observation that damage by root herbivores can sometimes favor the performance of phloem-feeding insects above ground (Masters et al., 1993).

Evidence suggests that plants can tolerate moderate levels of root damage without significant changes in resource allocation but that the replacement of root tissue following significant damage often occurs at the expense of allocation to aboveground shoots. For example, in tomatoes, removal of 25% of lateral roots does not affect rates of new shoot or root growth (Belsky, 1986). However, excision of 40% of the lateral roots reduces shoot growth by 20% without any effect on new root growth. Removal of all lateral tomato roots reduces subsequent root growth by 54% (Belsky, 1986). The maintenance of root growth rate after insect attack relies in part on preferential allocation of newly acquired resources. For example, the proportion of newly acquired ^{14}C allocated to the roots of root-grazed *Bouteloua gracilis* plants significantly exceeds that allocated to the roots of controls. One consequence is the maintenance of predamage rates of root growth but a significant decline in shoot growth rates (Detling et al., 1980). Sometimes, root herbivory can stimulate the growth of new root tillers, leading to a change in root architecture, but not in root total biomass (Treonis et al., 2007).

It may be premature to conclude that the preferential allocation of resources to roots at the expense of shoots is the most common response to root herbivory; there are still too few data to be sure. The opposite pattern can occur, and there may be a significant advantage to the translocation of N and soluble C out of belowground tissues following damage: reallocation of N and C aboveground reduces the energy content of the remaining root tissue and reduces its digestibility to herbivores (Andersen, 1987). For example, wireworm damage to the roots of *Plantago lanceolata* decreases root growth rates by 13% but increases shoot growth rates (Wurst and van der Putten, 2007). In fact, roots may have considerable flexibility in the export of resources under root herbivory. Following jasmonic acid treatment to individual roots of barley, 16% of carbohydrates were transferred to neighboring roots of the same plant within 2 hours (Henkes et al., 2008).

Plants respond differentially to attack by different species or guilds of insect herbivores aboveground (Hunter, 2000), and we should also expect variation in the effects of different belowground herbivore guilds on plant performance. Brown and Gange (1990) define four functional groups of root herbivorous insects: those that feed on the root collar, on the central vascular tissues, on the root cortex, or externally on the root. The four different functional guilds damage root function in different ways; thus, we should expect to see different induced responses by plants. In *Centaurea maculosa*, root feeding by the moth *Agapeta zoegana* has no effect on shoot or root mass, but feeding by the weevil *Cyphocleomus achates* on the same plant reduces shoot biomass (Steinger and Muller-Scharer, 1992). Similar patterns have been found in *Plantago lanceolata*, where feeding by wireworms causes increased shoot growth at the expense of root compensation, but feeding by nematodes has no effect (Wurst and van der Putten, 2007). We stress that future studies should examine the effects of herbivore identity on patterns of wound-induced resource reallocation and compare them with the relative importance of other ecological variables.

Ecological Consequences of N Loss from Plants Following Herbivory

Background

In the preceding section we argued that, because of deleterious effects on plant fitness, N losses from plants caused by insect herbivores should select for adaptive wound-induced changes in resource allocation. Here, we change our focus to the fate of N that is lost from plants during herbivory and its movement through the ecosystem. Such ecological consequences for the N economy of plants can be dramatic; we described earlier the vast quantities of N that are mobilized from plants as the result of cicada activities (Yang, 2004; Hunter, 2008; Yang, 2008). However, much lower insect herbivore densities than those typically associated with periodical cicada populations can still result in significant mobilization of N in ecosystems (Hunter et al., 2003). Whether the N lost to herbivores is tightly conserved and recycled within the ecosystem (Lovett et al., 2002) or instead lost in export from the ecosystem (Reynolds et al., 2000) has profound consequences for the N economy of plants. Nutrient conservation and internal recycling will make at least some of the N lost to herbivores available once more to plants (Frost and Hunter, 2004, 2007), mitigating in part the deleterious effects of insects on plant N economy.

The influence of herbivores on nutrient cycling and primary production has been studied for several decades (Chew, 1974; Mattson and Addy, 1975), with interest in this research area accelerating remarkably in recent years (Madritch et al., 2007; Meehan and Lindroth, 2007; Dunham, 2008; Haase et al., 2008; Hamilton et al., 2008; Kurz et al., 2008). In the remainder of this section, we focus on the mechanisms by which insect herbivores influence N cycling in ecosystems and the consequences for the N economy of plants.

Herbivore Effects on N Cycling in Ecosystems

In previous work, we have established seven broad mechanisms by which the activity of insect herbivores can cause changes in N cycling in ecosystems (Hunter, 2001a). Key among these are the deposition of fecal material (frass) and leaf fragments/prematurely abscised leaves (so-called "greenfall" or "orts") on the forest floor. The mobilization of N from plant tissue to the soil that is caused by insect herbivores can be substantial. For example, in deciduous forests, where defoliation acts before trees resorb nutrients prior to leaf senescence, nutrients returned to soils by these routes can exceed that in leaf litter (Fogal and Slansky, 1985; Grace, 1986), and can double overall rates of N return from plants to soil (Hollinger, 1986).

Frass and greenfall are generally labile substrates for fungal and bacterial decomposers (Choudhury, 1988; Lovett and Ruesink, 1995; van der Wal et al., 2004), which are subsequently grazed by soil and litter microbivores (mites, collembola, nematodes). Microbivores are, in turn, consumed by their own natural enemies (predacious insects, spiders, etc.) (Schowalter and Sabin, 1991; Reynolds et al., 2003). In other words, canopy herbivores tend to stimulate soil food webs (Hunter, 2001a). In our studies, microbial and faunal activity on herbivore-derived frass and greenfall increases rates of leaf litter decay, soil respiration, and N mineralization (Reynolds and Hunter, 2001; Hunter et al., 2003; Frost and Hunter 2007, 2008b); although plants lose N to insect herbivores, the N that is lost tends to stimulate biological activity in soils that will make some of that "lost" N available once more to plants.

The results of our studies are consistent with other research showing increases in rates of N (and C) cycling following insect damage in forest systems (Swank et al., 1981; Seastedt and Crossley, 1984; Fonte and Schowalter, 2005; Stadler et al., 2006). For example, in a recent study, endemic herbivore densities contributed up to 33 and 58% of soil N and P inputs, respectively, of that in annual foliar senescence (Domisch et al., 2009). We stress that experiments and sampling demonstrate that typical endemic densities of insects on plants influence N dynamics in soils, explaining up to 62% of the variation in soil N availability (Hunter et al., 2003; Frost and Hunter, 2004). This means that the insects that feed on plants typically have a substantial influence on the availability of N in soils and the subsequent N economy of plants. We believe that, in many systems, it is simply unrealistic to study N availability in soils without considering the role of insect herbivores in influencing that availability.

In addition to solid inputs such as frass, greenfall, and insect cadavers (Yang, 2004, 2008), defoliation by insects changes the N content of precipitation as it passes through plant canopies. Known as "throughfall," precipitation can leach nutrients from damaged leaves and dissolve frass from foliage (Hunter, 2001a). We have demonstrated that increases in soil N availability and the activities of fungal- and bacterial-feeding nematodes are associated with experimental additions of throughfall and variation in natural throughfall (Reynolds and Hunter, 2001; Hunter et al., 2003; Reynolds et al., 2003). Additional mechanisms by which insect herbivores can influence N cycling in ecosystems include wound-induced increases in foliar phenolics (Findlay et al., 1996; Frost and Hunter, 2008b), root mortality (Ruess et al., 1998), and community-wide changes in the relative abundance of plant species or genotypes that vary in their litter quality (Pastor et al., 1993; Uriarte, 2000). Herbivore-mediated changes in plant community composition influence litter quality, subsequent rates of N mineralization, and the availability of N to plants (Cornwell et al., 2008). Additionally, changes in plant community composition caused by insect herbivores will affect the utilization of soil N by the new community (Kielland and Bryant, 1998). As discussed above, herbivory may influence root exudates or interactions between roots and their symbionts (Bardgett et al., 1998; Frost and Hunter, 2004, 2008a) both of which are known to influence N dynamics. Finally, herbivores can influence the structure of plant canopies, and the cover that they provide, with concomitant changes in light availability, soil temperature, and moisture. The changes in soil microclimate that result from herbivore activity can alter the cycling of N (Mulder, 1999). Similarly, herbivore-induced changes in light availability may influence litter quality through effects on leaf chemistry (Hunter and Forkner, 1999) or plant productivity and diversity (van der Wal et al., 2000).

The effects of herbivores on N and C dynamics that we have reported in our own work (Reynolds and Hunter, 2001, Hunter et al., 2003, Reynolds et al., 2003) are analogous to the "fast cycle" (McNaughton et al., 1988) or the "acceleration hypothesis" (Ritchie et al., 1998). In essence, the acceleration hypothesis predicts that the activities of insect herbivores will increase the rate at which nutrients, including N, cycle through ecosystems. However, herbivore activity can also decrease rates of N cycling in some ecosystems (Ritchie et al., 1998; Belovsky and Slade, 2000; Bradford et al., 2008). For example, wound-induced increases in recalcitrant compounds such as tannins and lignin, or herbivore-mediated increases in the representation of resistant plant genotypes, might act to decelerate nutrient cycling (Findlay et al., 1996; Uriarte, 2000; Chapman et al., 2006; Kay et al., 2008) analogous to McNaughton and others' (1988) "slow cycle." Both fast- and slow-cycle effects, and acceleration or deceleration of cycles, may occur in the same system but on different time scales (Bradford et al., 2008). For example, herbivore activity may increase rates of N cycling and primary production in the short-term (e.g., through frass inputs) while causing both to decline in the long-term (e.g., through natural selection for resistant plants) (Uriarte, 2000).

It should be clear from the preceding discussion that insect herbivores can inflict substantial N losses on plants and also influence the rate at which soil N is available for plant uptake. How much of the N that is lost to herbivores in the short term is re-assimilated by plants in the long-term? In some sense, because the earth is a closed system, all matter will cycle through all compartments of all ecosystems over very long (millennial or greater) time scales (Odum, 1953). Perhaps more relevant to our present discussion is to ask whether individual plants ever recover any of the N that they lose to insect herbivores and, if so, how much? This is a very difficult question to answer, in part because of the difficulty in tracing N accurately among compartments in natural ecosystems. We have attempted to address the question using mesocosm experiments in which plants and herbivores are manipulated under seminatural conditions (Frost and Hunter, 2004).

We added ^{15}N-enriched insect frass to a series of replicated red oak, *Quercus rubra*, mesocosms that had been damaged experimentally by insect herbivores. We then traced the frass N over a period of 2 years to measure its form in the soil, its assimilation by plants, and its loss through export from the mesocosms (Frost and Hunter, 2007) (Figure 12.1). Some frass N was lost from the mesocosms in both organic and inorganic forms within 1 week of application, reflecting permanent losses to the trees. However, within 1 month, some frass N had been acquired by the oaks and had enriched the foliage in ^{15}N. Remarkably, insect herbivores feeding in the mesocosms also assimilated some frass N within the first year (Figure 12.1). In other words, N had cycled from oak foliage, through a first generation of insects, into insect frass, into the soil, back into trees, and into a second generation of insect herbivores, all within a single growing season (Frost and Hunter, 2007). Such rapid recycling of N is strong evidence

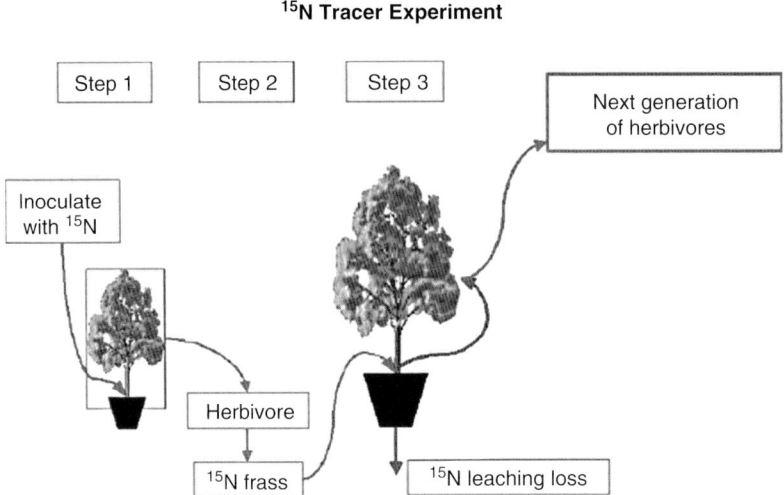

Figure 12.1. Experimental design used to trace nitrogen from foliage (Step 1), into insect herbivores and their frass (Step 2), back into the soil (Step 3), and into a second generation of insect herbivores within the same growing season. Courtesy of Chris Frost.

for the acceleration hypothesis (Ritchie et al., 1998), and suggests that reuptake of lost N can compensate in part for N losses imposed on plants by insect herbivores.

However, plants were unable to compensate fully for N lost to insect herbivores. In the second year of our experiment, herbivore damage from the previous year lowered total leaf N contents and the concentration of ^{15}N recovered in the foliage. Interestingly, colonization of oak trees by insects also declined in the year following defoliation, reflecting perhaps the requirements of insect herbivores for high N foliage (Mattson, 1980; Fagan et al., 2002). There is evidence to suggest that initial losses caused by insects to the N economy of plants may reduce subsequent N losses to insects because of declines in the palatability of foliage to herbivores (Baltensweiler et al., 1977; Haukioja et al., 1985; Hunter and West, 1990; Hunter and Schultz, 1995; Frost and Hunter, 2008b).

Overall, trees recovered a little over 16% of the N during a 2-year period that they lost initially to insect herbivores (11% in foliage, 5% in roots, 0.25% in stem). About 44% of frass N was recovered from the soil, perhaps representing a storage pool that may be accessed by plants over the longer term; leaching losses of N that can result from herbivory (Reynolds et al., 2000) appear to be buffered to some degree by storage in soil pools (Christenson et al., 2002; Frost and Hunter, 2007).

Nitrogen Deposition and the Effects of Insects on the N Economy of Plants

In the previous sections, we have described both evolutionary and ecological effects of insect herbivores on the N economy of plants. With significant increases in the deposition of N from the atmosphere caused by human activities (Vitousek et al., 1997; Galloway et al., 2004;

Chapter 1), we might ask what impacts insects will have on plant N economy in a eutrophic world. One simple prediction is that atmospheric N deposition should benefit N-limited insect herbivores, increasing their abundance and subsequent levels of defoliation (Erelli et al., 1998). In other words, the impacts of insect herbivores on the N economy of plants may well increase under N deposition. As expected, N deposition often leads to increases in foliar N concentrations and plant biomass (Aber et al., 2003, Throop and Lerdau, 2004, Tylianakis et al., 2008). In at least some systems, there are subsequent increases in the abundances, per capita rates of increase, and carrying capacities of insect herbivores (Kerslake et al., 1998; Hartley et al., 2003; Throop and Lerdau, 2004; Jones et al., 2008; Zehnder and Hunter, 2008). However, atmospheric N deposition may change the quality or quantity of plant defenses or alter amino acid profiles, both of which can reduce herbivore performance (Throop and Lerdau, 2004). Moreover, insects on high N plants may simply eat less per individual (Speight et al., 2008), with no overall change in defoliation levels. Additionally, N deposition can affect insect herbivores by altering plant community composition (Stevens et al., 2004) as well as the abundance and activity of natural enemies (Forkner and Hunter, 2000; Strengbom et al., 2005; Schmitz, 2008; Tylianakis et al., 2008). In reality, then, N deposition may have positive, negative, or neutral effects on insect herbivore population dynamics (Ritchie, 2000; Throop and Lerdau, 2004; Sudderth et al., 2005; Throop, 2005; Cleland et al., 2006; Cuesta et al., 2008; Lau et al., 2008). Although synthesis papers suggest that a majority of systems exhibit increases in herbivore abundance and defoliation under N deposition (Throop and Lerdau, 2004; Tylianakis et al., 2008), there remain relatively few studies of the effects of N deposition *per se* on insect herbivores from which to draw generalizations.

Conclusions

Insect herbivores are in the business of turning plant tissue into insect biomass and are often limited in their capacity to do so by the availability of N (Zehnder and Hunter, 2008). Natural selection has favored adaptations by insects to gain N from plants (Speight et al., 2008) and adaptations by plants to protect that N (discussed above and in Chapter 13). The inevitable N losses that occur during herbivory (Hunter, 2001a) may be mitigated in part by rapid cycling of N back to plants and longer-term N storage in soils (Frost and Hunter, 2007). However, there is still much to learn about the influences of insect herbivores on the N economy of plants, and we suggest some important areas of future research below.

First, we have assumed above that wound-induced changes in resource allocation are adaptive for plants. However, almost no studies measure fitness variation among plants in response to their allocation patterns following herbivory. We recommend that genetic variation in wound-induced changes in resource allocation be exploited to explore the fitness consequences of such changes. Genetic modification of plants may also provide insight into the fitness consequences of resource reallocation. Second, isotopic techniques are now readily available to measure the reallocation of old resources and the allocation of new resources by plants following attack by herbivores. We need many more studies of this type if we are to recognize broad-scale patterns among the available data. Currently, we can report examples of plants that vary in their strategies, but with no real understanding of why such variation exists. Predictable patterns based on plant taxonomy, growth form, life history, ecology, or ontogeny may emerge as we accumulate studies that explore in detail resource allocation following damage. Third, allocation strategies are not free to evolve without constraints. We need a better understanding of how source-sink relationships, physiology, and morphology constrain wound-induced changes in resource allocation by plants. Finally, an understanding of ecological contingency is necessary if we want to move beyond laboratory studies to a holistic view of plant responses to insect damage in natural systems. How does nutrient limitation, water, or light availability influence plant responses? How do plant responses differ with the identity and ecology of the attacker? Given that the majority of plants are attacked simultaneously by aboveground and belowground herbivores, how do these forces interact to influence resource allocation by plants? A focus on questions such as these will improve our understanding of how insect herbivores influence the N economy of plants.

Glossary

Fine root: Nonperennial lateral root tissue with a diameter less than a defined size (usually either <2 mm or <0.5 mm, depending on the author).
Frass: The feces of insect herbivores.
Guild: A group of organisms (here plants or insects) that make a living in a similar way. For insects, guilds are generally defined by feeding strategy (chewing, sucking, galling, etc.) whereas for plants they represent functional groups (e.g., nitrogen fixers, C3 plants, etc.).
Mesocosm experiments: Experiments of intermediate scale, generally larger than those that can be accomplished in the laboratory, yet smaller than whole-ecosystem manipulations.
Ort: A scrap of living plant material dropped by an insect herbivore.
Rhizodeposition: The secretion of plant-derived material into the soil by plant roots.

Acknowledgements

This work was supported by the Department of Ecology and Evolutionary Biology at the University of Michigan (LT) and The National Science Foundation (Grant DEB 0814340 to MDH). We gratefully acknowledge their support.

References

Aber, J.D., Melillo, J.M. 2001. *Terrestrial Ecosystems*. Harcourt Academic Press, San Diego.
Aber, J.D., Goodale, C.L., Ollinger, S., Smith, M.L., Magill, A.H., Martin, M.E., Hallett, R.A., Stoddard, J.L. 2003. Is nitrogen deposition altering the nitrogen status of northeastern forests? *Bioscience*, 53:375–389.
Andersen, D.C. 1987. Below-ground herbivory in natural communities: a review emphasizing fossorial animals. *Quarterly Review of Biology*, 62:261–286.
Anderson, T.M., Starmer, W.T., Thorne, M. 2007. Bimodal root diameter distributions in Serengeti grasses exhibit plasticity in response to defoliation and soil texture: implications for nitrogen uptake. *Functional Ecology*, 21:50–60.
Arredondo, J.T., Johnson, D.A. 1998. Clipping effects on root architecture and morphology of 3 range grasses. *Journal of Range Management*, 51:207–213.
Ausmus, B.S., Ferris, J.M., Reichle, D.E., Williams, E.C. 1978. Role of belowground herbivores in mesic forest root dynamics. *Pedobiologia*, 18:289–295.

Babst, B.A., Ferrieri, R.A., Gray, D.W., Lerday, M., Schlyer, D.J., Schueller, M., Thorpe, M.R., Orians, C.M. 2005. Jasmonic acid induces rapid changes in carbon transport and partitioning in *Populus*. *New Phytologist*, 167:63–72.

Bais, H.P., Weir, T.L., Perry, L.G., Gilroy, S., Vivanco, J.M. 2006. The role of root exudates in rhizosphere interactions with plants and other organisms. *Annual Review of Plant Biology*, 57:233–266.

Baldwin, I.T., Schultz, J.C. 1983. Rapid changes in tree leaf chemistry induced by damage: evidence for communication between plants. *Science*, 221:277–279.

Baltensweiler, W., Benz, G., Bovey, P., Delucchi P. 1977. Dynamics of larch bud moth populations. *Annual Review of Entomology*, 22:79–100.

Bardgett, R.D., Wardle, D.A., Yeates, G.W. 1998. Linking above-ground and below-ground interactions: how plant responses to foliar herbivory influence soil organisms. *Soil Biology and Biochemistry*, 30:1867–1868.

Barnes, R.L. 1963. Organic nitrogen compounds in tree xylem sap. *Forest Science*, 9: 98–102.

Bazzaz, F.A., Grace, J. 1997. *Plant Resource Allocation*. Academic Press, San Diego, CA.

Belovsky, G.E., Slade, J.B. 2000. Insect herbivory accelerates nutrient cycling and increases plant production. *Proceedings of the National Academy of Sciences*, 97:14412–14417.

Belsky, A.J. 1986. Does herbivory benefit plants? A review of the evidence. *American Naturalist*, 127:870–892.

Bradford, M.A., Gancos, T., Frost, C.J. 2008. Slow-cycle effects of foliar herbivory alter the nitrogen acquisition and population size of Collembola. *Soil Biology & Biochemistry*, 40:1253–1258.

Briske, D.D., Boutton, T.W., Wang, Z. 1996. Contribution of flexible allocation priorities to herbivory tolerance in C_4 perennial grasses: an evaluation with ^{13}C labeling. *Oecologia*, 105:151–159.

Brown, V.K., Gange, A.C. 1990. Insect herbivory belowground. *Advances in Ecological Research*, 20:1–58.

Burgeson, G.B. 1966. Mobilization of minerals to the infection site of root-knot nematodes. *Phytopathology*, 56:1287–1289.

Caldwell, M.M., Richards, J.H., Johnson, D.A., Nowak, R.S., Dzurec, R.S. 1981. Coping with herbivory: photosynthetic capacity and resource allocation in two semiarid *Agropyron* bunchgrasses. *Oecologia*, 50:14–24.

Callaway, R.M., Ridenour, W.M. 2004. Novel weapons: invasive success and the evolution of increased competitive ability. *Frontiers in Ecology and the Environment*, 2:436–443.

Chapman, S.K., Schweitzer, J.A., Whitham, T.G. 2006. Herbivory differentially alters plant litter dynamics of evergreen and deciduous trees. *Oikos*, 114:566–574.

Cheng, W., Gershenson, A. 2007. Carbon fluxes in the rhizosphere. Pp 31–56. In: *The Rhizosphere: An Ecological Perspective*. Cardon, Z.G., Whitbeck, J.L., Eds. Academic Press, San Diego.

Chew, R.M. 1974. Consumers as regulators of ecosystems: an alternative to energetics. *Ohio Journal of Science* 74:359–370.

Choudhury, D. 1988. Herbivore induced changes in leaf litter resource quality: a neglected aspect of herbivory in ecosystem dynamics. *Oikos*, 51:389–393.

Christenson, L.C., Lovett, G.M., Mitchell, M.J., Groffman, P.G. 2002. The fate of nitrogen in gypsy moth frass deposited to an oak forest floor. *Oecologia*, 131:444–452.

Clayton, S.J., Read, D.B., Murray, P.J., Gregory, P.J. 2008. Exudation of alcohol and aldehyde sugars from roots of defoliated *Lolium perenne* L. grown under sterile conditions. *Journal of Chemical Ecology*, 34:1411–1421.

Cleland, E.E., Peters, H.A., Mooney, H.A., Field, C.B. 2006. Gastropod herbivory in response to elevated CO2 and N addition impacts plant community composition. *Ecology*, 87:686–694.

Cornwell, W.K., Cornelissen, J.H.C., Amatangelo, K., Dorrepaal, E., Eviner, V.T., Godoy, O., Hobbie, S.E., Hoorens, B., Kurokawa, H., Perez-Harguindeguy, N., Quested, H.M., Santiago, L.S., Wardle, D.A., Wright, I.J., Aerts, R., Allison, S.D., van Bodegom, P., Brovkin, V., Chatain, A., Callaghan, T.V., Diaz, S., Garnier, E., Gurvich, D.E., Kazakou, E., Klein, J.A., Read, J., Reich, P.B., Soudzilovskaia, N.A., Vaieretti, M.V., Westoby M. 2008. Plant species traits are the predominant control on litter decomposition rates within biomes worldwide. *Ecology Letters*, 11:1065–1071.

Cuesta, D., Taboada, A., Calvo, L., Salgado, J.M. 2008. Short- and medium-term effects of experimental nitrogen fertilization on arthropods associated with *Calluna vulgaris* heathlands in north-west Spain. *Environmental Pollution*, 152:394–402.

Cullen, B.R., Chapman, D.F., Quigley, P.E. 2006. Comparative defoliation tolerance of temperate perennial grasses. *Grass and Forage Science*, 61:405–412.

Darwin, C. 1859. *The Origin of Species*. John Murray, London.

Detling, J.K., Winn, D.T., Procter-Gregg, C., Painter, E.L. 1980. Effects of simulated grazing by below-ground herbivores on growth, CO_2 exchange, and carbon allocation patterns of *Bouteloua gracilis*. *Journal of Applied Ecology*, 17:771–778.

Domisch, T., Finer, L., Neuvonen, S., Niemela, P., Risch, A.C., Kilpelainen, J., Ohashi, M., Jurgensen, M.F. 2009. Foraging activity and dietary spectrum of wood ants (*Formica rufa* group) and their role in nutrient fluxes in boreal forests. *Ecological Entomology*, 34:369–377.

Dunham, A.E. 2008. Above and below ground impacts of terrestrial mammals and birds in a tropical forest. *Oikos*, 117:571–579.

Dybas, H.S., Davis, D.D. 1962. Population census of 17-year periodical cicadas (Homoptera: Cicadidae, *Magicicada*). *Ecology*, 43:432.

Edwards, P.J., Wratten, S.D. 1983. Wound-induced defenses in plants and their consequences for patterns in insect grazing. *Oecologia*, 59:88–93.

Ehrlich, P.R., Raven, P.H. 1964. Butterflies and plants: a study in coevolution. *Evolution*, 18:586–608.

Eissenstat, D.M., Wells, C.E., Yanai, R.D., Whitbeck. J.L. 2000. Building roots in a changing environment: implications for root longevity. *New Phytologist*, 147:33–42.

Elser, J.J., Dobberfuhl, D.R., MacKay, N.A., Schampel, J.H. 1996. Organism size, life history, and N:P stoichiometry. *Bioscience*, 46:674–684.

Elser, J.J., Sterner, R.W., Gorokhova, E., Fagan, W.F., Markow, T.A., Cotner, J.B., Harrison, J.F., Hobbie, S.E., Odell, G.M., Weider, L.J. 2000. Biological stoichiometry from genes to ecosystems. *Ecology Letters*, 3:540–550.

Erelli, M.C., Ayres, M.P., Eaton, G.K. 1998. Altitudinal patterns in host suitability for forest insects. *Oecologia*, 117:133–142.

Fagan, W.F., Siemann, E., Mitter, C., Denno, R.F., Huberty, A.F., Woods, H.A., Elser, J.J. 2002. Nitrogen in insects: implications for trophic complexity and species diversification. *American Naturalist*, 160:784–802.

Fahnestock, J.T., Detling, J.K. 1999. Plant responses to defoliation and resource supplementation in the Pryor Mountains. *Journal of Range Management*, 52:263–270.

Findlay, S., Carriero, M., Krischik, V., Jones, C.C. 1996. Effects of damage to living plants on leaf litter quality. *Ecological Applications*, 6:269–275.

Fogal, W.H., Slansky, F., Jr. 1985. Contribution of feeding by European pine sawfly larvae to litter production and element flux in Scots pine plantations. *Canadian Journal of Forest Research*, 15:484–487.

Fonte, S.J., Schowalter, T.D. 2005. The influence of a neotropical herbivore (*Lamponius portoricensis*) on nutrient cycling and soil processes. *Oecologia*, 146:423–431.

Forkner, R.E., Hunter, M.D. 2000. What goes up must come down? Nutrient addition and predation pressure on oak herbivores. *Ecology*, 81:1588–1600.

Frost, C.J., Hunter, M.D. 2004. Insect canopy herbivory and frass deposition affect soil nutrient dynamics and export in oak mesocosms. *Ecology*, 85:3335–3347.

Frost, C.J., Hunter, M.D. 2007. Recycling of nitrogen in herbivore feces: plant recovery, herbivore assimilation, soil retention, and leaching losses. *Oecologia*, 151:42–53.

Frost, C.J., Hunter. M.D. 2008a. Herbivore-induced shifts in carbon and nitrogen allocation in red oak seedlings. *New Phytologist*, 178:835–845.

Frost, C.J., Hunter, M.D. 2008b. Insect herbivores and their frass affect *Quercus rubra* leaf quality and initial stages of subsequent litter decomposition. *Oikos*, 117:13–22.

Galloway, J.N., Dentener, F.J., Capone, D.G., et al. 2004. Nitrogen cycles; past, present and future. *Biogeochemistry*, 70:153–226.

Gehring, C.A., Whitham, T.G. 1994. Interactions between aboveground herbivores and the mycorrhizal mutualists of plants. *Trends in Ecology & Evolution*, 9:251–255.

Grace, J.R. 1986. The influence of gypsy moth on the composition and nutrient content of litter fall in a Pennsylvania oak forest. *Forest Science*, 32:855–870.

Green, T.R., Ryan, C.A. 1972. Wound-induced proteinase inhibitor in plant leaves: a possible defense mechanism against insects. *Science*, 175:776–777.

Green, R.A., Detling, J.K. 2000. Defoliation-induced enhancement of total aboveground nitrogen yield of grasses. *Oikos*, 91:280–284.

Guitian, R., Bardgett, R.D. 2000. Plant and soil microbial responses to defoliation in temperate semi-natural grassland. *Plant and Soil*, 220:271–277.

Haase, J., Brandl, R., Scheu, S., Schadler, M. 2008. Above- and Belowground Interactions Are Mediated by Nutrient Availability. *Ecology*, 89:3072–3081.

Hamilton, E.W., Frank, D.A. 2001. Can plants stimulate soil microbes and their own nutrient supply? Evidence from a grazing tolerant grass. *Ecology*, 82:2397–2402.

Hartley, S.E., Gardner, S.M., Mitchell, R.J. 2003. Indirect effects of grazing and nutrient addition on the hemipteran community of heather moorlands. Journal of Applied Ecology 40:793–803.

Haukioja, E., Niemela, P., Siren, S. 1985. Foliage phenols and nitrogen in relation to growth, insect damage, and ability to recover after defoliation, in the mountain birch *Betula pubescens* ssp *tortuosa. Oecologia*, 65:214–222.

Heilmeier, H., Schulze, E.D., Whale, D.M. 1986. Carbon and nitrogen partitioning in the biennial monocarp *Arctium tomentosum* Mill. *Oecologia*, 70:466–474.

Hendrick, R.L., Pregitzer, K.S. 1992. The demography of fine roots in a northern hardwood forest. *Ecology*, 73:1094–1104.

Hendrick, R.L., Pregitzer, K.S. 1993. Patterns of fine root mortality in 2 sugar maple forests. *Nature*, 361:59–61.

Hendricks, J.J., Aber, J.D., Nadelhoffer, K.J., Hallett, R.D. 2000. Nitrogen controls on fine root substrate quality in temperate forest ecosystems. *Ecosystems*, 3:57–69.

Henkes, G.J., Thorpe, M.R., Minchin, P.H., Schurr, U., Rose, U.S.R. 2008. Jasmonic acid treatment to part of the root system is consistent with simulated leaf herbivory, diverting recently assimilated carbon towards untreated roots within an hour. *Plant, Cell and Environment*, 31:1229–1236.

Hiernaux, P., Turner, M.D. 1996. The effect of clipping on growth and nutrient uptake of Sahelian annual rangelands. *Journal of Applied Ecology*, 33:387–399.

Hillebrand, H., Borer, E.T., Bracken, M.E.S., Cardinale, B.J., Cebrian, J., Cleland, E.E., Elser, J.J., Gruner, D.S., Harpole, W.S., Ngai, J.T., Sandin, S., Seabloom, E.W., Shurin, J.B., Smith, J.E., Smith, M.D. 2009. Herbivore metabolism and stoichiometry each constrain herbivory at different organizational scales across ecosystems. *Ecology Letters*, 12:516–527.

Ho, L.C. 1988. Metabolism and compartmentation of imported sugars in sink organs in relation to sink strength. *Annual Review of Plant Physiology and Plant Molecular Biology*, 39:355–378.

Holland, J.N. 1995. Effects of aboveground herbivory on soil microbial biomass in conventional and no-tillage agroecosystems. *Applied Soil Ecology*, 2:275–279.

Holland, J.N., Cheng, W., Crossley, D.A., Jr. 1996. Herbivore-induced changes in plant carbon allocation: assessment of below-ground C fluxes using carbon-14. *Oecologia*, 107:87–94.

Hollinger, D.Y. 1986. Herbivory and the cycling of nitrogen and phosphorus in isolated California oak trees. *Oecologia*, 70:291–297.

Hunter, M.D. 2000. Mixed signals and cross-talk: Interactions between plants, insect herbivores, and plant pathogens. *Agricultural and Forest Entomology*, 2:155–160.

Hunter, M.D. 2001a. Insect population dynamics meets ecosystem ecology: Effects of herbivory on soil nutrient dynamics. *Agricultural and Forest Entomology*, 3:77–84.

Hunter, M.D. 2001b. Out of sight, out of mind: The impacts of root-feeding insects in natural and managed systems. *Agricultural and Forest Entomology*, 3:3–10

Hunter, M.D. 2008. Root herbivory in forest ecosystems. In: *Root Feeders, an Ecosystem Perspective* (Johnson, S.N., Murray, P.J., Eds.), pp. 68–95. CAB Biosciences, Ascot, UK.

Hunter, M.D., Forkner, R.E. 1999. Hurricane damage influences foliar polyphenolics and subsequent herbivory on surviving trees. *Ecology*, 80:2676–2682.

Hunter, M.D., Linnen, C.R., Reynolds, B.C. 2003. Effects of endemic densities of canopy herbivores on nutrient dynamics along a gradient in elevation in the southern Appalachians. *Pedobiologia*, 47:231–244.

Hunter, M.D., Schultz, J.C. 1995. Fertilization mitigates chemical induction and herbivore responses within damaged oak trees. *Ecology*, 76:1226–1232.

Hunter, M.D., West, C. 1990. Variation in the effects of spring defoliation on the late season phytophagous insects of *Quercus robur*. In: *Population Dynamics of Forest Insects* (Watt, A.D., Leather, S.R., Hunter, M.D., Kidd, N.A.C., Eds.). Intercept, Andover.

Jackson, R.B., Mooney, H.A., Schulze, E.D. 1997. A global budget for fine root biomass, surface area, and nutrient contents. *Proceedings of the National Academy of Sciences of the United States of America*, 94:7362–7366.

Jaramillo, V.J., Detling, J.K. 1988. Grazing history, defoliation, and competition: effects on shortgrass production and nitrogen accumulation. *Ecology*, 69:1599–1608.

Johnson, S.N., Murray, P.J. 2008. *Root Feeders, an Ecosystem Perspective*. CAB Biosciences, Acot, UK.

Johnson, S.N., Bezemer, T.M., Jones, T.H. 2008. Linking aboveground and belowground herbivory, pp. 153–170. In: Johnson, S.N., Murray, P.J., Eds. *Root Feeders, an Ecosystem Perspective*. CAB Biosciences, Acot, UK.

Jones, M.E., Paine, T.D., Fenn, M.E. 2008. The effect of nitrogen additions on oak foliage and herbivore communities at sites with high and low atmospheric pollution. *Environmental Pollution*, 151:434–442.

Karban, R., Baldwin, I.T. 1997. *Induced Responses to Herbivory*. University of Chicago Press, Chicago.

Kay, A.D., Mankowski, J., Hobbie, S.E. 2008. Long-term burning interacts with herbivory to slow decomposition. *Ecology*, 89:1188–1194.

Kerslake, J.E., Woodin, S.J., Hartley, S.E. 1998. Effects of carbon dioxide and nitrogen enrichment on a plant-insect interaction: the quality of *Calluna vulgaris*, as a host for *Operophtera brumata*. *New Phytologist*, 140:43–53.

Kielland, K., Bryant, J.P. 1998. Moose herbivory in taiga: effects on biogeochemistry and vegetation dynamics in primary succession. *Oikos*, 82:377–383.

Koda, Y., Kikuta, Y., Kithara, T., Nishi, T., Mori, K. 1992. Comparisons of various biological activities of sterioisomers of methyl jasmonate. *Phytochemistry*, 31:1111–1114.

Kurz, W.A., Dymond, C.C., Stinson, G., Rampley, G.J., Neilson, E.T., Carroll, A.L., Ebata, T., Safranyik, L. 2008. Mountain pine beetle and forest carbon feedback to climate change. *Nature*, 452:987–990.

Lau, J.A., Strengbom, J., Stone, L.R., Reich, P.B., Tiffin, P. 2008. Direct and indirect effects of CO2, nitrogen, and community diversity on plant-enemy interactions. *Ecology*, 89:226–236.

Liechti, R., Farmer, E.E. 2002. The jasmonate pathway. *Science*, 296:1649–1650.

Lindroth, R.L., Donaldson, J.R., Stevens, M.T. 2007. Browse quality in quaking aspen (*Populus tremuloides*): effects of genotype, nutrient, defoliation, and coppicing. *Journal of Chemical Ecology*, 33:1049–1064.

Lovett, G.M., Ruesink, A.E. 1995. Carbon and nitrogen mineralization from decomposing gypsy moth frass. *Oecologia*, 104:133–138.

Lovett, G.M., Christenson, L.M., Groffman, P.M., Jones, C.G., Hart, J.E., Mitchell, M.J. 2002. Insect defoliation and nitrogen cycling in forests. *Bioscience*, 52:335–341.

Lovett, G.M., Tobiessen, P. 1993. Carbon and nitrogen assimilation in red oaks (*Quercus rubra* L.) subject to defoliation and nitrogen stress. *Tree Physiology*, 12:259–269.

Madritch, M.D., Donaldson, J.R., Lindroth, R.L. 2007. Canopy herbivory can mediate the influence of plant genotype on soil processes through frass deposition. *Soil Biology & Biochemistry*, 39:1192–1201.

Masters, G.J., Brown, V.K., Gange, A.C. 1993. Plant mediated interactions between above- and below-ground insect herbivores. *Oikos*, 66:148–151.

Mattson, W.J., Addy, N.D. 1975. Phytophagous insects as regulators of forest production. *Science*, 190:515–521.

Mattson, W.J., Jr. 1980. Herbivory in relation to plant nitrogen content. *Annual Review of Ecology and Systematics*, 11:119–161.

Mawdsley, J.L., Bardgett, R.D. 1997. Continuous defoliation of perennial ryegrass (*Lolium perenne*) and white clover (*Trifolium repens*) and associated changes in the composition and activity of the microbial population of an upland grassland soil. *Biology and Fertility of Soils*, 24:52–58.

McNaughton, S.J. 1979. Grazing as an optimization process: grass–ungulate relationships in the Serengeti. *American Naturalist*, 113:691–703.

McNaughton, S.J., Ruess, R.W., Seagle, S.W. 1988. Large mammals and process dynamics in African ecosystems: Herbivorous mammals affect primary productivity and regulate recycling balances. *Bioscience*, 38:794–800.

Meehan, T.D., Lindroth, R.L. 2007. Modeling nitrogen flux by larval insect herbivores from a temperate hardwood forest. *Oecologia*, 153:833–843.

Mikola, J., Ilmarinen, K., Nieminen, M., Vestberg, M. 2005. Long-term soil feedback on plant N allocation in defoliated grassland miniecosystems. *Soil Biology and Biochemistry*, 37:899–904.

Millard, P., Hester, A., Wendler, R., Baillie, G. 2001. Interspecific defoliation responses of trees depend on sites of winter nitrogen storage. *Functional Ecology*, 13:535–543.

Mohanty, K.C., Das, S.N., Patnaik, P.R., Sahoo, B. 1990. Chemical changes in horse gram plants infected with root knot nematode, *Meloidogyne incognita*. *Comparative Physiology and Ecology*, 15:100–102.

Mulder, C.P.H. 1999. Vertebrate herbivores and plants in the Arctic and subarctic: effects on individuals populations, communities and ecosystems. *Perspectives in Plant Ecology, Evolution and Systematics*, 2:29–55.

Nojiri, H., Yamane, H., Seto, H., Yamaguchi, I., Murofushi, N., Yoshihara, T., Shibaoka, H. 1992. Qualitative and quantitative analysis of endogeous jasmonic acid in bulbing and non-bulbing onion plants. *Plant Cell Physiology*, 33:1225–1231.

Odum, E.P. 1953. *Fundamentals of Ecology*. W.B. Saunders Company, Philadelphia.

Olson, B.E., R.T. Wallander. 1999. Carbon allocation in *Euphorbia esula* and neighbours after defoliation. *Canadian Journal of Botany*, 77:1641–1647.

Orians, C.M. 2005. Herbivores, vascular pathways and systemic induction: facts and artifacts. *Journal of Chemical Ecology*, 32:2231–2242.

Palacio, S., Hester, A.J., Maestro, M., Millard, P. 2008. Browsed *Betula pubescens* trees are not carbon-limited. *Functional Ecology*, 22:808–815.

Pastor, J., Dewey, B., Naiman, R.J., McInnes, P.F., Cohen, Y. 1993. Moose browsing and soil fertility in the boreal forests of Isle-Royale National Park. *Ecology*, 74:467–480.

Paterson, E. 2003. Importance of rhizodeposition in the coupling of plant and microbial productivity. *European Journal of Soil Science*, 54:741–750.

Paterson, E., Thornton, B., Midwood, A.J., Sim, A. 2005. Defoliation alters the relative contributions of recent and nonrecent assimilate to root exudation from *Festuca rubra*. *Plant, Cell and Environment*, 28:1525–1533.

Pregitzer, K.S., DeForest, J.L., Burton, A.J., Allen, M.F., Ruess, R.W., Hendrick, R.L. 2002. Fine root architecture of nine North American trees. *Ecological Monographs*, 72:293–309.

Raillard, M.C., Svoboda, J. 1999. Exact growth and increase nitrogen compensation by the arctic sedge *Carex aquatilis* var. stans after simulated grazing. *Arctic, Antarctic and Alpine Research*, 31:21–26.

Reynolds, B.C., Hunter, M.D. 2001. Responses of soil respiration, soil nutrients, and litter decomposition to inputs from canopy herbivores. *Soil Biology & Biochemistry*, 33:1641–1652.

Reynolds, B.C., Crossley, D.A., Hunter, M.D. 2003. Response of soil invertebrates to forest canopy inputs along a productivity gradient. *Pedobiologia*, 47:127–139.

Reynolds, B.C., Hunter, M.D., Crossley, D.A., Jr. 2000. Effects of canopy herbivory on nutrient cycling in a northern hardwood forest in western North Carolina. *Selbyana*, 21:74–78.

Ritchie, M.E. 2000. Nitrogen limitation and trophic vs. abiotic influences on insect herbivores in a temperate grassland. *Ecology*, 81:1601–1612.

Ritchie, M.E., Tilman, D., Knops, J.M.H. 1998. Herbivore effects on plant and nitrogen dynamics in oak savanna. *Ecology*, 79:165–177.

Rosenthal, J.P., Kotanen, P.M. 1994. Terrestrial plant tolerance to herbivory. *Trends in Ecology and Evolution*, 9:145–148.

Ruess, R.W., Hendrick, R.L., Bryant, J.P. 1998. Regulation of fine root dynamics by mammalian browsers in early successional Alaskan taiga forests. *Ecology*, 79:2706–2720.

Ruess, R.W., McNaughton, S.J., Coughenour, M.B. 1983. The effects of clipping, nitrogen source and nitrogen concentration on the growth responses and nitrogen uptake of an east African sedge. *Oecologia*, 59:253–261.

Saikkonen, K., Ahonen-Jonnarth, U., Markkola, A.M., Helander, M., Tuomi, J., Roitto, M., Ranta, H. 1999. Defoliation and mycorrhizal symbiosis: a functional balance between carbon sources and below-ground sinks. *Ecology Letters*, 2:19–26.

Saravesi, K., Markkola, A., Rautio, P., Roitto, M., Tuomi, J. 2008. Defoliation causes parallel temporal responses in a host tree and its fungal symbionts. *Oecologia*, 156:117–123.

Sarkar, D., Pandey, S.K., Sharma, S. 2006. Cytokinins antagonize the jasmonates action on the regulation of potato (*Solanum tuberosum*) tuber formation in vitro. *Plant Cell Tissue and Organ Culture*, 87:285–295.

Schmitz, O.J. 2008. Herbivory from individuals to ecosystems. *Annual Review of Ecology Evolution and Systematics*, 39:133–152.

Schowalter, T.D., Sabin, T.E. 1991. Litter microarthropod responses to canopy herbivory, season and decomposition in litterbags in a regenerating conifer ecosystem in Western Oregon. *Biology and Fertility of Soils*, 11:93–96.

Schwachtje, J., Minchin, P.E.H., Jahnke, S., Dongens, J.T., Schittko, U., Baldwin, I.T. 2006. SNF1-related kinases allow plant to tolerate herbivory by allocating carbon to roots. *Proceedings of National Academy of Sciences of the United States of America* 103:12935–12940.

Seastedt, T.R., Crossley, D.A., Jr. 1984. The influence of arthropods on ecosystems. *BioScience*, 34:157–161.

Smith, S.E., Read, D.J. 2008. *Mycorrhizal Symbiosis*. 3rd edn. Academic Press, San Diego.

Sohlenius, B. 1980. Abundance, biomass and contribution to energy flow by soil nematodes in terrestrial ecosystems. *Oikos*, 34:186–194.

Speight, M.R., Hunter, M.D., Watt, A.D. 2008. *The Ecology of Insects: Concepts and Applications*. 2nd edn, Blackwells Scientific, Oxford.

Stadler, B., Muller, T., Orwig, D. 2006. The ecology of energy and nutrient fluxes in hemlock forests invaded by hemlock woolly adelgid. *Ecology*, 87:1792–1804.

Stamp, N. 2003. Out of the quagmire of plant defense hypotheses. *Quarterly Review of Biology*, 78:23–55.

Steinger, T., Muller-Scharer, H. 1992. Physiological and growth responses of *Centaurea maculosa* (Asteraceae) to root herbivory under varying levels of interspecific plant competition and soil nitrogen availability. *Oecologia*, 91:1432–1939.

Stevens, C.J., Dise, N.B., Mountford, J.O., Gowing, D.J. 2004. Impact of nitrogen deposition on the species richness of grasslands. *Science*, 303:1876–1879.

Stevens, G.N., Jones, R.H. 2006. Patterns in soil fertility and root herbivory interact to influence fine-root dynamics. *Ecology*, 87:616–624.

Stowe, K.A., Marquis, R.J., Hochwender, C.G., Simms, E.L. 2000. The evolutionary ecology of tolerance to consumer damage. *Annual Review of Ecology and Systematics*, 31:565–595.

Strengbom, J., Witzell, J., Nordin, A., Ericson, L. 2005. Do multitrophic interactions override N fertilization effects on *Operophtera* larvae? *Oecologia*, 143:241–250.

Sudderth, E.A., Stinson, K.A., Bazzaz, F.A. 2005. Host-specific aphid population responses to elevated CO2 and increased N availability. *Global Change Biology*, 11:1997–2008.

Swank, W.T., Waide, J.B., Crossley, D.A., Todd, R.L. 1981. Insect defoliation enhances nitrate export from forest ecosystems. *Oecologia*, 51:297–299.

Thelen, G.C., Vivanco, J.M., Newingham, B., Good, W., Bais, H.P., Landres, P., Caesar, A., Callaway, R.M. 2005. Insect herbivory stimulates allelopathic exudation by an invasive plant and the suppression of natives. *Ecology Letters*, 8:209–217.
Thomas, M.M., Millard, P., Watt, M.S., Turnbull, M.H., Peltzer, D., Whitehead, D. 2008. The impact of defoliation on nitrogen translocation patterns in the woody invasive plant, *Buddleia davidii*. *Functional Plant Biology*, 35:462–469.
Thompson, J.N., Cunningham, B.M. 2002. Geographic Structure and Dynamics of Coevolutionary Selection. *Nature*, 417:735–738.
Thornley, J.H.M. 1972. *Mathematical Models in Plant Physiology*. Academic Press, San Diego.
Throop, H.L. 2005. Nitrogen deposition and herbivory affect biomass production and allocation in an annual plant. *Oikos*, 111:91–100.
Throop, H.L., Lerdau, M.T. 2004. Effects of nitrogen deposition on insect herbivory: *Implications for community and ecosystem processes*. *Ecosystems*, 7:109–133.
Treonis, A.M., Cook, R., Dawson, L., Grayston, S.J., Mizen, T. 2007. Effects of a plant parasitic nematode (*Heterodera trifolii*) on clover roots and soil microbial communities. *Biology and Fertility of Soils*, 43:541–548.
Tylianakis, J.M., Didham, R.K., Bascompte, J., Wardle, D.A. 2008. Global change and species interactions in terrestrial ecosystems. *Ecology Letters,* 11:1351–1363.
Uriarte, M. 2000. Interactions between goldenrod (*Solidago altissima* L.) and its insect herbivore (*Trirhabda virgata*) over the course of succession. *Oecologia*, 122:521–528.
Vaast, P., Caswell-Chen, E.P., Zasoski, R.J. 1998. Effects of two endoparasitic nematodes (*Pratylenchus coffeae* and *Meloidogyne konaensis*) on ammonium and nitrate uptake by Arabica coffee (*Coffea arabica* L.). *Applied Soil Ecology*, 10:171–178.
Van der Meijden, E., Wijn, M., Verkaar, H.J. 1988. Defence and regrowth, alternative plant strategies in the struggle against herbivores. *Oikos*, 51:355–363.
van der Wal, R., Bardgett, R.D., Harrison, K.A., Stien, A. 2004. Vertebrate herbivores and ecosystem control: cascading effects of faeces on tundra ecosystems. *Ecography*, 27:242–252.
van der Wal, R., van Wijnen, H., van Wieren, S., Beucher, O., Bos, D. 2000. On facilitation between herbivores: How Brent Geese profit from brown hares. *Ecology*, 81:969–980.
Vitousek, P.M., Aber, J.D., Howarth, R.W., Likens, G.E., Matson, P.A., Schindler, D.W., Schlesinger, W.H., Tilman, D.G. 1997. Human alteration of the global nitrogen cycle: Sources and consequences. *Ecological Applications*, 7:737–750.
Wamberg, C., Christensen, S., Jakobsen, I. 2003. Interaction between foliar-feeding insects, mycorrhizal fungi, and rhizosphere protozoa on pea plants. *Pedobiologia*, 47:281–287.
Wardlaw, I.F. 1990. The control of carbon partitioning in plants. *New Phytologist*, 116:341–381.
Wearn, J.A., Gange, A.C. 2007. Above-ground herbivory causes rapid and sustained changes in mycorrhizal colonization of grasses. *Oecologia*, 153:959–971.
Wells, C.E., Glenn, D.M., Eissenstat, D.M. 2002. Soil insects alter fine root demography in peach (*Prunus persica*). *Plant Cell and Environment*, 25:431–439.
Whiles, M.R., Callaham, M.A., Meyer, C.K., Brock, B.L., Charlton, R.E. 2001. Emergence of periodical cicadas (*Magicicada cassini*) from a Kansas riparian forest: Densities, biomass and nitrogen flux. *American Midland Naturalist*, 145:176–187.
Wratten, S.D., Edwards, P.J., Barker, A.M. 1990. Consequences of rapid feeding-induced changes in trees for the plant and the insect: individuals and populations, pp. 137–145 in

Watt, A.D., Leather, S.R., Hunter, M.D., Kidd, N.A.C., Eds. *Population Dynamics of Forest Insects*. Intercept, Andover, England.

Wurst, S., van der Putten, W.H. 2007. Root herbivore identity matters in plant-mediated interactions between root and shoot herbivores. *Basic and Applied Ecology*, 8:491–499.

Yang, L.H. 2004. Periodical cicadas as resource pulses in North American forests. *Science*, 306:1565–1567.

Yang, L.H. 2008. Pulses of dead periodical cicadas increase herbivory of American bellflowers. *Ecology*, 89:1497–1502.

Zehnder, C.B., Hunter, M.D. 2008. Effects of nitrogen deposition on the interaction between an aphid and its host plant. *Ecological Entomology*, 33:24–30.

Chapter 13
Plant Defense Proteins That Inhibit Insect Peptidases

Carlos Peres Silva and Richard Ian Samuels

Introduction

Herbivorous insects face a variety of challenges when considering their "chosen" lifestyle. The major challenge is dealing with the nutritional discrepancy that exists between the available plant nutrients and the needs of the insects for normal growth and reproduction. A comparison of the chemical composition of plants and insects shows that these organisms differ greatly. Plants are usually carbohydrate-rich, while insects are protein-rich. Among all of the chemical needs, obtaining nitrogen is a critical problem for insects that live on plants. For most insects that rely on plants as a food source, the available forms of nitrogen determine the nutritional value of their diets. In fact, plants and animals (insects included) differ in the elemental composition at the genome (Acquisti et al., 2009) and proteome (Elser et al., 2006) levels. Plants not only differ from other living organisms in the ratio of carbohydrates to proteins, but also tend to "favor" amino acids and nitrogen bases with lower total nitrogen content.

Further complicating the situation for herbivorous insects, plants possess chemical defenses that deploy primary gene products capable of inhibiting the digestion of proteins by the insects. This fascinating defense system, that involves both constitutive and induced systemic production of proteins that inhibit plant protein breakdown, is the subject of this chapter.

Occurrence, Classification, and Functions of Plant Peptidase Inhibitors

Plant peptidase inhibitors (PI) are a highly diverse group of proteins capable of inactivating or lowering the activity of peptidases, the enzymes involved in the cleavage of peptide bonds. PIs form stable complexes with target peptidases, preventing access to the enzyme active site and generally acting as tight-binding, reversible inhibitors (Laskowsky and Kato, 1980; Rawlings et al., 2008). PIs are common in the plant kingdom. In fact, PIs have been found in all plant species so far investigated, occurring in reproductive or vegetative tissues. They can be constitutive or induced upon attack or abiotic stress. These inhibitors are active toward four, of the six known mechanistic classes of peptidases (Rawlings et al., 2008). We use the term peptidase, a synonym of protease or proteolytic enzymes, as a general term for enzymes that hydrolyze peptide bonds. The classification of peptidases depends on the active amino acid in the reactive center. Peptidases have traditionally been classified based on the mechanistic catalytic classes as aspartic, cysteine, glutamic, metallo, serine, and threonine peptidases (Rawlings and Barrett, 1993; Rawlings et al., 2008).

Peptidases are also classified into different families and clans on the basis of the similarity of their primary and tertiary structures, which are used to establish evolutionary relationships (Rawlings and Barrett, 1993). In this chapter we use the information available at the MEROPS database (http://merops.sanger.ac.uk) (release 8.4) (Rawlings et al., 2008).

Proteinaceous inhibitors were traditionally described after the peptidases used as models to assay their inhibitory activity. In the past, the protocols to assay PIs normally used conditions suitable for mammalian enzymes, mainly trypsin and chymotrypsin, and the PIs were therefore classified as trypsin or chymotrypsin inhibitors. When more information was obtained it became obvious that some of the PIs were functionally and phylogenetically related, thus PI families were described. PIs that are active against the main four mechanistic classes of peptidases (aspartic, cysteine, metallo, and serine) have been described in plants. Inhibitors of cysteine and serine peptidases seem to be universal throughout the plant kingdom and have been extensively characterized since their discovery, while detection and characterization of aspartic and metallo-peptidase inhibitors are relatively rare (Ryan, 1990; Xavier-Filho and Campos, 1989).

Besides the original classification based on the mechanistic classes of the inhibited peptidases, more recent classification of PIs is also based on the relationship between amino acid sequences and/or on the tertiary structure of the inhibitory domain or unit (Rawlings et al., 2008). An inhibitory domain here is defined as the amino acid sequence of the inhibitor that contains the reactive site, or in other words, the sequence that is responsible for the inhibitory activity, excluding any part that is not directly involved in the recognition of the peptidase. PIs containing a single inhibitor domain are termed simple inhibitors, and those that contain two or more inhibitory domains or inhibitory units in the same molecule are termed compound inhibitors. Compound inhibitors with domains from the same family are known as homotypic inhibitors. However, they can also possess a combination of domains from different families, and are thus called heterotypic inhibitors.

As in the case of peptidases that they inhibit, PIs are classified into different families and clans on the basis of structural and evolutionary relationships, amino acid sequences, and the three-dimensional structures of the inhibitor domains. Families of PIs share significant similarities in their primary sequences; a clan is formed by families that are recognizably homologous (Rawlings et al., 2008).

Currently, from a total number of 79 known peptidase inhibitor families, 19 of these are of plant origin (MEROPS database release 8.4, http://merops.sanger.ac.uk). Table 13.1 summarizes some of the data on plant PIs. Based solely on the tertiary structure of different plant PIs, it is possible to separate 17 of these 19 families into individual clans (Table 13.2). This data suggests that most of the PIs are therefore not closely related. Two families, I71 and I73, have not yet been assigned to any clan.

Table 13.1. Families of plant peptidase inhibitors.

Inhibitor family	Inhibitor common name	Peptidases inhibited	Occurrence in plant species (species and number of species)	Plants species with complete genome sequences but lacking inhibitor homologues
I1	Kazal family	S1, S8	Populus trichocarpa	Arabidopsis thaliana; Oryza sativa; Vitis vinifera
I2	Kunitz-A family	S1	Chlamydomonas reinhardtii	Arabidopsis thaliana; Oryza sativa; Vitis vinifera
I3	Kunitz-P family	S1, A1, C1	87	
I4	Serpin	S1, S8, C1, C14	11	Vitis vinifera
I6	Cereal trypsin/α-amylase inhibitors	S1	12	Arabidopsis thaliana; Vitis vinifera
I7	Squash inhibitors	S1	15	Arabidopsis thaliana; Oryza sativa; Vitis vinifera
I12	Bowman-Birk	S1, S3	63	Arabidopsis thaliana; Vitis vinifera
I13	Potato inhibitor 1	S1, S8	33	
I18	Mustard trypsin inhibitor	S1	5	Oryza sativa; Vitis vinifera
I20	Potato inhibitor 2	S1, S8	34	Arabidopsis thaliana; Oryza sativa; Vitis vinifera
I25	Cystatin	C1, C13, S8, M12	53	Vitis vinifera
I29	Cytotoxic T-lymphocyte antigen-2 alpha	C1	50	
I37	Potato carboxypeptidase inhibitor	M14	Solanum lycopersicum and S. tuberosum	Arabidopsis thaliana; Oryza sativa; Vitis vinifera
I39	Alpha-2-macroglobulin	The majority of endopeptidases regardless of family or catalytic type	Populus trichocarpa	Arabidopsis thaliana; Oryza sativa; Vitis vinifera
I51	Serine carboxypeptidase Y inhibitor	S10	36	
I63	Metallopeptidase pappalysin-1 inhibitor	M43	Populus trichocarpa	Arabidopsis thaliana; Oryza sativa; Vitis vinifera
I67	Bromein	C1, S1	Ananas comusus	Arabidopsis thaliana; Oryza sativa; Vitis vinifera
I71	Falstatin	C1, C2	Populus trichocarpa	Arabidopsis thaliana; Oryza sativa; Vitis vinifera
I73	Veronica trypsin inhibitor	C1	Veronica hederifolia	Arabidopsis thaliana; Oryza sativa; Vitis vinifera

Sources: (http://merops.sanger.ac.uk) (release 8.4).

Table 13.2. Clans of plant peptidase inhibitors.

Clan	Family type inhibitor	Plant families
IA	Ovomucoid inhibitor	I1, I20
IB	Aprotinin	I2
IC	Soybean Kunitz trypsin inhibitor	I3
ID	α-1-peptidase inhibitor	I4
IE	Trypsin inhibitor MCTI-1	I7, I37
IF	Bowman-Birk inhibitor	I12, I67
IG	Eglin c	I13
IH	Cystatin A	I25
IJ	Ragi seed trypsin/α-amylase inhibitor	I6
IL	α-2-macroglobulin	I39
JB	Pro-eosinophil major basic protein	I63
JD	Mustard trypsin inhibitor-2	I18
JE	Serine carboxypeptidase Y inhibitor	I51
JF	Cytotoxic T-lymphocyte antigen-2 α	I29
I-		I71, I73

Sources: (http://merops.sanger.ac.uk) (release 8.4).

The following description is an attempt to give a short comprehensive presentation of some of the most frequently found families of plant PIs. The structure of each family of PIs is briefly described, as is also their classification in the MEROPS database, but emphasis was given to the description of their role in plant defense against insects.

Family I3 (Kunitz-P Inhibitors)

There are two unrelated families classified as Kunitz inhibitors, after M. Kunitz, who first described the members of these families (Kunitz, 1945; Rawlings et al., 2008). Both families I2 and I3 are found in plants. The inhibitors of family I3 are widespread in the plant kingdom and have been described in legumes, cereals, and solanaceous species. The majority of family I3 inhibitors are active against serine peptidases of S1 family (mostly active against trypsin, chymotrypsin, and elastase), but a minority of its members (subfamily I3A) can also inhibit other peptidases, including the aspartic peptidase cathepsin D (family A1) and the cysteine peptidases of the C1 family (Rawlings et al., 2008). Kunitz-type inhibitors from plants are in general single-chain molecules with low cysteine content, but they contain one or two intra-chain disulfide bridges and a single reactive site. The molecular mass range of family I3 members is 18–24 kDa, and they form 1:1 tight complex with target peptidases according to the "Laskowski mechanism" (Laskowsky and Kato, 1980). It has been suggested that Kunitz inhibitors evolved through ancestral gene duplication with the region of the active site largely conserved and a rapid diversification caused by mutations in other regions giving rise to multi-gene families of iso-inhibitors (Christeller, 2005; Habu et al., 1997; Major and Constabel, 2008).

Inhibitors of family I3 are inducible upon wounding by insects as described, for example, in poplar trees *Populus trichocarpa* and *Populus deltoids* (Major and Constabel, 2008) and in *Passiflora* f. *edulis flavicarpa* (passion fruit) (Botelho-Júnior et al., 2008). Some recent examples of genes that were used to transform plants causing antibiosis against insects are the inhibitor from *Cicer arietinum* (chickpea) versus larvae of *Helicoverpa armigera* (podborer) (Srinivasan et al., 2005), and the chymotrypsin inhibitor from *Psophocarpus tetragonolobus* (winged bean) against larval *H. armigera* (Telang, 2009).

Family I4 (Serpins)

Members of the serpin (serine peptidase inhibitors) family (I4) are present throughout the entire plant kingdom. The discovery of serpins in plants was associated with studies on the occurrence of allergens in beer. A protein of 43 kDa, which was first named Protein Z and found in the barley grain, was later identified as a serpin (Hejgaard et al., 1985).

Serpins possess a reactive site containing a trap or bait sequence that recognizes the target peptidase. Following the formation of the inhibitor-enzyme complex, the peptide bond adjacent to the P1 position is cleaved. The cleavage of this peptide bond in the reactive site of the inhibitor triggers a rapid conformational change. Catalysis does not proceed beyond the formation of the acyl-enzyme complex, with the serine of peptidase catalytic triad covalently attached to P1 (the position in the N-terminus side immediately adjacent to the scissile bond) in the serpin. Most serpins inhibit peptidases through the kinetic trap described above, which is effectively irreversible. Trapping reactions irreversible because the unmodified inhibitor is not released. Due to this mechanism, serpins are known as "suicide" inhibitors or "molecular mouse traps" (Roberts and Hejgaard, 2008; Ye et al., 2001).

The formation of an acyl-enzyme complex is essential to serpin activity. In spite of the fact that plant serpins are active against serine peptidases of the S1 and S8 families, some serpins are also active against cysteine peptidases of the C1 and C14 families (Table 13.1). As serine and cysteine peptidases form acyl-enzymes complexes, the serpin-sensitive serine peptidases are also inhibited through the same "mouse trap" mechanism (Roberts and Hejgaard, 2008).

Serpins have been purified and characterized from 11 plant species, but the serpin gene is absent from the genome of *V. vinifera* (Table 13.1). These inhibitors were described from grains in the Triticeae family, as well as in pollen and from phloem exudates of *Cucurbita maxima* (Wieczorek et al., 1985). Despite the large distribution of serpin genes among plants, there is little information about their functions in plant physiology. There are no obvious targets for these inhibitors in plants, since plants appear to lack genes encoding peptidases of the S1 family such as chymotrypsin, trypsin or elastase, which would be the expected targets for these serpins. It has been suggested that serpins found in the phloem sap could be involved in protection from proteolysis of signalling proteins or peptides that travel between different parts of the plant or that they might interfere with the immune system of insects (Law et al., 2006; Yoo et al., 2000). Functions in defense against herbivores through peptidase inhibition have been suggested for serpins found in high concentrations in seeds (Hejgaard et al., 2005; Law et al., 2006; Roberts and Hejgaard, 2008). Induction of serpins in response to insect attack was reported in the phloem sap of pumpkin (*C. maxima*) when challenged by the aphid *Myzus persicae* (Yoo et al., 2000). Heterologous expression of a *Manduca sexta* (Lepidoptera) serpin in alfalfa (*Medicago sativa*) and tobacco (*Nicotiana tabacum*) showed anti-metabolic effects on a range of insect species (Thomas et al., 1995; Thomas et al., 1994).

Family I6 (Cereal Bifunctional Trypsin/α-amylase Inhibitors)

Members of family I6 have been isolated from seeds, where their relatively high quantities permit us to consider them as storage proteins, with the additional function of inhibiting both serine peptidases from the S1 family (through the Laskowski mechanism) and α-amylases. They were first recognized as bifunctional inhibitors in 1983 after analysis of the complete sequences of inhibitors purified from seeds of barley (Odani et al., 1983) and ragi (finger millet, *Eleusine coracana*) (Campos and Richardson, 1983). These inhibitors are rich in cysteine residues, exhibit five disulfide bonds, and are glycosylated with molecular masses of approximately 13 kDa. They are known from 12 plant species (Table 13.1), but they are absent from the genome of *A. thaliana* and *V. vinifera*. Their tertiary structures are more similar to those of 2S storage albumins and lipid-transfer proteins than to other peptidase inhibitors (Ryan, 1990; Rawlings et al., 2008).

The transgenic expression of a barley bifunctional inhibitor in wheat resulted in a high accumulation in the seeds and reduced weight gain and survival of the early instar of the Angoumois grain moth (*Sitotroga cerealella*), but no significant effect was observed in leaf-feeding insects (Altpeter et al., 1999).

Family I12 (Bowman-Birk Inhibitors)

On the basis of sequence homology, the Bowman-Birk inhibitors (BBI) are classified in the I12 family, forming a very important and widespread group of plant serine PIs. The first BBI was isolated from soybean (*Glycine max*) by D.E. Bowman in 1946 (Bowman, 1946) and further characterized by Y. Birk and colleagues (1963). BBIs purified from dicotyledonous plants consist of single polypeptide chains with molecular masses around 8 kDa containing a conserved and characteristic group of 14 cysteine residues involved in intra-chain disulfide bridges. They have two homologous domains each bearing a separate reactive site that can react independently and simultaneously with two equal or different serine peptidases, and for this reason they are known as double-headed inhibitors. The specificity of each reactive site in BBIs is determined by the identity of the amino acid at position P1. In contrast, monocotyledonous plants have two different types of BBIs, a group consisting of 8 kDa single-headed and another group of 16 kDa double-headed BBIs. It has been suggested that the 16 kDa double-headed BBIs evolved from one 8 kDa single-headed ancestral inhibitor via internal gene duplication and fusion (Mello et al., 2003). Alternatively, it has been suggested, based on the intra molecular sequence homology, that one of the reactive sites of the 8 kDa double-headed BBI from monocotyledonous plants became nonfunctional, resulting in the 8 kDa single-headed BBIs (Mello et al., 2003; Qi et al., 2005). The pioneer work on this inhibitor family has been carried out on the double-headed inhibitors, and it is now clear that the I12 family of inhibitor molecules can present one, two, four, five, and even more inhibitor units (Qi et al., 2005).

Currently, sequences of more than 100 plant BBIs are available at the MEROPS database. BBIs are generally found in seeds of legumes and cereals and in the grass family Poaceae, where they may function as part of the constitutive arsenal of defense proteins, but they are also wound-inducible, as observed in maize (*Zea mays*) (Eckelkamp et al., 1993). An example of a transgenic plant encoding a BBI gene, showing anti-metabolic effects against an insect pest, is sugarcane expressing BBI developed with the aim of controlling the sugarcane stalk borer, *Diatraea saccharalis* (Falco and Silva-Filho, 2003).

Family I13 (Potato Inhibitor I)

The first member of this family was isolated from potato tubers and described by Ryan and Balls in 1962. Inhibitors from family I13 are generally monomeric, have molecular masses of approximately 8 kDa, and are extremely stable. They generally lack disulfide bonds. However, some of the inhibitors from cucurbit and potato tubers do have a single disulfide bridge. The members of the I13 family are inhibitors of serine peptidases, mostly from the S1 family (for example, chymotrypsin). However, a few have activity against peptidases of the S8 family (for example, subtilisin), and they follow the classic Laskowski mechanism of inhibition. Members of this family are known from 33 species of plants so far and are also present in all species for which complete genomes have been published (Table 13.1). Inhibitors of family I13 have been found in potato tubers, in aerial parts as in fruits and leaves, and even in phloem saps (Singh et al., 2009). They have been implicated in plant defense against herbivores as part of the constitutive arsenal, as well being inducible upon wounding by insects or abiotic stresses (Lee et al., 1986; Singh et al., 2009).

Family I20 (Potato Inhibitor II)

The I20 family of inhibitors has only been reported from species of the Solanaceae. Family I20 contains compound inhibitors with multiple repeat inhibitory units varying from one to eight (Kong and Ranganathan, 2008). They are inhibitors of serine peptidases from the S1 and S8 families (Rawlings et al., 2008).

At present, members of the I20 family are known from 34 species but are absent from the genomes of *A. thaliana*, *O. sativa* and *V. vinifera* (Table 13.1). The first inhibitor of this family was isolated from potato tubers, but they can be found in aerial parts such as leaves, flowers, fruits, and phloem sap, both as constitutive and induced in response to mechanical injury (Tamhane et al., 2007). The potato tuber I20 inhibitor is one of the most intensively studied at the protein level. Several transgenic plants have demonstrated the potential of these inhibitors to interfere with development of pest insects, for example, in *Nicotiana atteanuata* versus the tobacco hornworm (*M. sexta*) and the mirid *Tupiocoris notatus* (Zavala et al., 2004).

Family I25, Subfamily I25B (Phytocystatins)

Cystatins are mainly inhibitors of cysteine peptidases. In plants, cystatins are known as phytocystatins, included in family I25, subfamily I25B and grouped in clan IH (Tables 13.1 and 13.2). The majority of phytocystatins inhibit cysteine peptidases from the C1 family (papain-like), but some can act on peptidases of the C13 family (legumain-like) (Mazumdar-Leighton and Broadway, 2001). The majority of the phytocystatins are devoid of disulfide bonds and glycosylation regions. Their molecular masses generally range between 12 and 16 kDa, but some phytocystatins show an extension at the carboxy-terminal that is involved in the inhibition of peptidases of the C13 family (Margis-Pinheiro et al., 2008; Martinez and Diaz, 2008; Rawlings et al., 2008).

There are several reports on the activity of phytocystatins against insect digestive cysteine peptidases (reviews in Arai et al., 2002; Haq et al., 2004). It has also been reported that phytocystatins are part of the constitutive defensive arsenal of plants but that they can also be induced in response to insect attack (Girard et al., 2007). Several transgenic plants encoding phytocystatins have been documented to show antibiosis against insects, for example, in the case of tomato (*Solanum lycopersicum*) versus the Colorado potato beetle (*Leptinotarsa decemlineata*) (Goulet et al., 2008).

Family I37 (Potato Carboxypeptidase Inhibitor)

The inhibitors of family I37 have been reported only from two species of the Solanaceae family, *Solanum lycopersicum* and *Solanum tuberosum*. The first inhibitor of this small family was isolated by Rancour and Ryan in 1968 (Rancour and Ryan, 1968) and was shown to inhibit the metallopeptidase carboxypeptidase B, a member of the M14 family (Rawlings et al., 2008). Members of this plant inhibitor family are capable of inhibiting both carboxypeptidases A and B and are also known as PCI (potato carboxypeptidase inhibitors). PCIs are small proteins, 39 amino acid residues in length, and their tertiary structures are stabilized by three disulfide bonds forming a structure known as the "cysteine knot" or "T-knot" (Rees and Lipscomb, 1982). This structure is found in other peptidase inhibitor families grouped in the 1E clan (Table 13.2). Based on studies of the crystal complex of PCI and carboxypeptidase A, it is known that the four C-terminal amino acid residues of PCI bind onto the active site cleft of the enzyme in a substrate-like manner. The glycine-39 residue splits off from the C-terminal position, and its binding to the S1' subsite is crucial in the inhibition process (Rees and Lipscomb, 1982). Members of the I37 family are also induced in plants as a response to wounding by insects or by mechanical damage (Villanueva et al., 1998).

The description of other minor plant PIs (those known from less than 10 plant species) will be discussed in the following sections concerning their activities as plant anti-insect defensive compounds and the ways that insects deal with these inhibitors.

General Considerations

The "MEROPS" families of PIs cannot always be reconciled with classical nomenclature based solely on the mechanistic classes of the peptidases that they inhibit, since a number of families contain inhibitors that are active against peptidases of different mechanistic classes (Rawlings et al., 2008). Classical examples of these are the plant PIs grouped into the I13 family, the so-called Kunitz-type PIs, and the I14 family of serpin inhibitors. The former family contains PIs that generally inhibit serine peptidases (family S1), but they also include inhibitors of aspartic (family A1) and cysteine peptidases (family C1) (Heibges et al., 2003), and the second family mostly contains inhibitors of serine peptidases but also contains inhibitors of some cysteine peptidases (Roberts and Hejgaard, 2008).

Inhibitors of serine peptidases are predominant in the plant kingdom (Table 13.1). In fact, of all known plant PI families, there are only two exceptions to this rule: the phytocystatin subfamily (I25B) that contains cysteine-peptidase inhibitors, and the potato metallocarboxypeptidase inhibitor family (I37) that includes inhibitors of carboxypeptidases A and B. It is no surprise that serine peptidase inhibitors are so abundant in the plant kingdom, considering that serine peptidases are the major digestive proteolytic enzymes in herbivorous animals, including herbivorous insects, which represent an important herbivore pressure (Lopes et al., 2006; Lopes et al., 2009; Terra and Ferreira, 2005).

Plant PIs are usually found as isoform products of multi-gene families. The occurrence of iso-inhibitor genes among plant species and the details of how those genes are expressed are still far from complete, but recent studies are bringing to light important new issues (Jander and Howe, 2008). Many plant PIs are multifunctional proteins, but here we will only explore their roles as insect anti-nutritional compounds, their potential use in pest control strategies, and how insects adapt to either natural (endogenous) or transgene-encoded PIs. Several functions are usually attributed to PIs in plant tissues, such as storage proteins, regulators of endogenous peptidases, protection from unwanted endogenous or exogenous proteolysis, radio-protection, tolerance to abiotic factors such as cold, heat, drought and salinity stresses, component of signal cascades, regulators of apoptosis, and part of the plant arsenal of defenses against herbivores and pathogens (Huang, 2007;

Mosolov et al., 2001; Mosolov and Valueva, 2005; Shan et al., 2008; Soloman et al., 1999; Srinivasan et al., 2009).

Plant PIs as Defensive Compounds Against Phytophagous Insects

Results of many studies have highlighted the importance of PIs as defensive agents against herbivores and pathogens. The following arguments support this hypothesis: (1) The high concentrations of PIs in certain tissues of many plant species exceed the quantity necessary to control unwanted endogenous or exogenous proteolytic activities; (2) Digestive peptidases of all mechanistic classes from insects can be inhibited *in vitro* and *in vivo* by plant PIs; (3) Plants have evolved a sophisticated mechanism to induce inhibitors of different families in response to wounding caused by insect attack; (4) Herbivorous insects have evolved mechanisms to adapt to the presence of PIs by regulating the pattern of gut digestive peptidases; (5) Transgenic plants expressing PIs confer resistance against some insect species; (6) The overall abundance, as well as the specificities, of cysteine and serine plant PIs parallels the occurrence of cysteine and serine digestive peptidases in most herbivorous insects.

The first conclusive experimental results indicating that plant PIs possess anti-nutritional activity to insects date from the 1950s, when Lipke, Fraenkel, and Liener (1954) reported that a protein fraction from seeds of *G. max* (soybean) inhibited growth of larval *Tribolium confusum* (Lipke et al, 1954). They also demonstrated that the fraction was capable of inhibiting the proteolytic activity of larval extracts *in vitro* (Lipke et al., 1954). The use of artificial diets to study the effects of plant PIs on insects became progressively more common. Soon it became clear that different inhibitors had a range of effects on different insect species. The incorporation of a purified Bowman-Birk inhibitor from soybean in artificial diet was capable of inhibiting the growth of *T. castaneum*, but a Kunitz inhibitor was much less effective (Birk, 1985). Comparing the effects of two purified inhibitors on the development of the European corn borer, *Ostrinia nubilalis*, Steffens and colleagues(1978) demonstrated that the Kunitz trypsin inhibitor from soybean was effective in inhibiting growth and delaying pupation, but it did not prevent completion of the life cycle, whereas a trypsin inhibitor from corn had no effect on growth or metamorphosis of the larvae (Staswick and Tiryaki, 2004). Broadway and Duffey demonstrated in 1986 that purified soybean trypsin inhibitor (Kunitz, I3) and potato inhibitor 2 (I20) had detrimental effects on larval *Heliothis zea* (now *Helicoverpa zea*) and *Spodoptera exigua* (Broadway and Duffey, 1986).

Following these early discoveries, there have been many examples of both *in vitro* and *in vivo* inhibition of insect digestive peptidases by plant PIs, and this number is still growing. Important reviews on this issue have been published by Ryan (1990), Jongsma and Boulter (1997), and Browse and Howe (2008).

Induction of PIs in Response to Insect Attack

Convincing evidence that plant PIs are important defensive macromolecules against insect attack was furnished by the seminal work of Green and Ryan published in 1972. They were the first to demonstrate that plants produce and accumulate, in response to mechanical injury or insect wounding, multiple inhibitory proteins that reduce the digestibility and nutritional quality of their leaves.

Firstly, it was observed that wounding of the leaves of potato or tomato plants by larval or adult Colorado potato beetle (*L. decemlineata*), induced a rapid accumulation of serine peptidase inhibitors throughout the plants' aerial tissues. The systemic production of PIs was also achieved by mechanical damage to the leaves. A remarkable observation was that the accumulation of the inhibitors throughout the plant from the point of injury increased rapidly, within a few hours (Green and Ryan, 1972). The administration of extracts from wounded leaves to excisions on undamaged leaves could also induce the synthesis of PIs, suggesting the existence of a mobile signalling molecule, initially called PIIF (Proteinase Inhibitor Inducing Factor) (Ryan, 1974). At that time, in the late 1970s and early 1980s, several reports pointed to the involvement of fragments of plant cell walls, especially oligogalacturonides (OGA), released during attacks on the plants by herbivores and pathogens, in the induction of defensive compounds, among them PIs (Bishop et al., 1981; Ryan et al., 1981). For some time, these compounds were considered as putative PIIFs. However, the inability of these compounds to be transported away from the site of injury suggested that they were unlikely to be the long-distance molecules of the defense response (Baydoun and Fry, 1985). The OGAs were probably among the signals produced at the sites of attacks, where they help produce localized defensive chemicals.

In 1990, a lipid-derived molecule, methyl jasmonate, was hypothesized to be a key component of the defensive signalling pathway triggered by herbivore or pathogen attack (Farmer and Ryan, 1990). Later it was proposed that linolenic acid was released into the cytoplasm from the plant plasma membrane and converted through the octadecanoid pathway to jasmonic acid or closely related derivatives such as methyl jasmonate or jasmonate-isoleucine (Farmer and Ryan, 1992; Howe and Jander, 2008). Currently, the plant signalling compounds collectively known as jasmonates are considered ubiquitous molecules involved in the recognition of tissue damage and subsequent induction of an arsenal of defensive compounds against herbivorous insects and other invaders (Bostock, 2005; Howe and Jander, 2008; Chung and Howe, 2009; Maffei et al., 2007; Ryan, 1990).

As indicated above, the first inducible plant PIs described by the pioneering work of Green and Ryan (1972) were serine peptidase inhibitors of the families I13 and I20 (potato inhibitors I and II). Subsequently, inducible inhibitors of four mechanistic classes of peptidases have been described as a result of insect attack or mechanical damage (Habib and Fazili, 2007; Jongsma and Bolter, 1997; Ryan, 1990).

Graham and Ryan (1981) showed that mechanical wounding of young potato plants induced the expression of a carboxypeptidase inhibitor. In this case, the accumulation of PCI was shown to parallel the wound-induced expression of the serine peptidase inhibitors I and II. Eckelkamp and collaborators (1993) and Cordero and colleagues (1994) demonstrated that a Bowman-Birk inhibitor gene (family I12) was systemically induced in maize, a monocot. In 1992, Hildmann and colleagues (1992) reported the induction of genes encoding aspartic and cysteine PIs upon wounding in potato plants (Zhao et al., 1996). Yoo and colleagues demonstrated that the increase in serpin concentration in the phloem of *C. maxima* was correlated with the mortality of the aphid *Myzus persicae* (Yoo et al., 2000).

More recently, the identification of genes induced in response to insect attack has been carried out by the use of molecular approaches such as microarrays. The identification of genes differentially expressed in plants challenged by insects or different elicitors has been carried out in tomato (*L. esculentum*) in response to the Colorado potato beetle (*L. decemlineata*) (Lawrence et al., 2008), in *N. attenuata* in response to *M. sexta* (Voelckel and Baldwin, 2004), in *Arabidopsis* in response to the specialist *Pieris rapae* and to the generalist *Spodoptera littoralis* (Bodenhausen and Reymond, 2007), and in sorghum (*Sorghum bicolor*) in response to the greenbug aphid (*Schizaphis graminum*) (Zhu-Salzman et al., 2004).

The present scenario reveals that insect attack (either as phloem feeders, in the case of aphids, chewing insects, or cell-content feeders such as thrips) tends to cause wounding and thus induce expression of genes that are involved in direct defenses such as production of PIs, polyphenol oxidases, threonine deaminase, healing macromolecules and chitinases, as well as genes involved in indirect defenses, such as the release of volatile compounds that attract natural enemies (predators or parasitoids) of the feeding herbivores (see review in Howe and Jander, 2008). Conversely, housekeeping genes or genes responsible for photosynthesis are repressed by insect attack. There is a great diversity of signals capable of mediating the induction of jasmonate, salicylic acid, and ethylene as a wound response, and it is clear now that there is cross talk among the major signalling pathways generating a signalling network capable of discriminating and dealing with abiotic stresses or phytophagous insect attack (Bostock, 2005; Howe and Jander, 2008). The molecules in the oral secretions and patterns of wounding may determine the type of responses by the plants to the feeding insect (Halitschke et al., 2001; Peiffer and Felton, 2009).

Different works have demonstrated the importance of fatty acid-amino acid conjugates in eliciting the wound response in plants. Following the recognition of the insect oral components and elicitors from the injured cell walls, the primary signal is transmitted by electrical signals as forerunners (Rhodes et al., 1996; Zimmermann et al., 2009) and by a chemical signal transduction pathway that involves calcium ion fluxes, protein phosphorylation, degradation of repressors, jasmonate conjugate synthesis, differential regulation of transcription factors, and subsequent gene expression or repression (AbuQamar et al., 2008; Bari and Jones, 2009; Howe and Jander, 2008; Bostock, 2005; Chung and Howe, 2009; Erb et al., 2008; Katsir et al., 2008; Kazan and Manners, 2008; Maffei et al., 2007; Mithöfer and Boland, 2008; Zheng and Dicke, 2008).

Because of space limitations, not all relevant papers and reviews on this topic could be cited. However, Figure 13.1 summarizes the main events involved in the signal transduction pathway trigged by insect wounding.

The Effects of Plant PIs on Gut Physiology of Insects

The protective function of plant PIs is attributed to the formation of tight or irreversible complexes with target digestive peptidases blocking dietary protein breakdown. This inhibition could ultimately cause reduction in the availability of amino acids necessary for growth, development, and reproduction. As nitrogen-based compounds have a major impact on the performance of plant-feeding insects, any factor decreasing the acquisition of nitrogen from the diet is a threat to the animal.

Broadway and Duffey (1986) were among the first to study insect midgut physiology in response to the ingestion of plant PIs. They initially suggested that growth inhibition effects caused by plant PIs on insects were not only due to direct inhibition of digestive peptidases but that inhibition could also induce hyper-production of peptidases in an attempt to overcome the anti-nutritional effects (Broadway and Duffey, 1986). This massive production of constitutive (or novel) peptidases would then enhance the loss of amino acids to the detriment of the synthesis of other proteins. They fed larvae of *H. zea* and *S. exigua* on two inhibitors, the soybean trypsin inhibitor and the potato inhibitor II, capable of inhibiting trypsin and chymotrypsin. The authors observed growth inhibition of the animals and an increase in proteolytic activity against

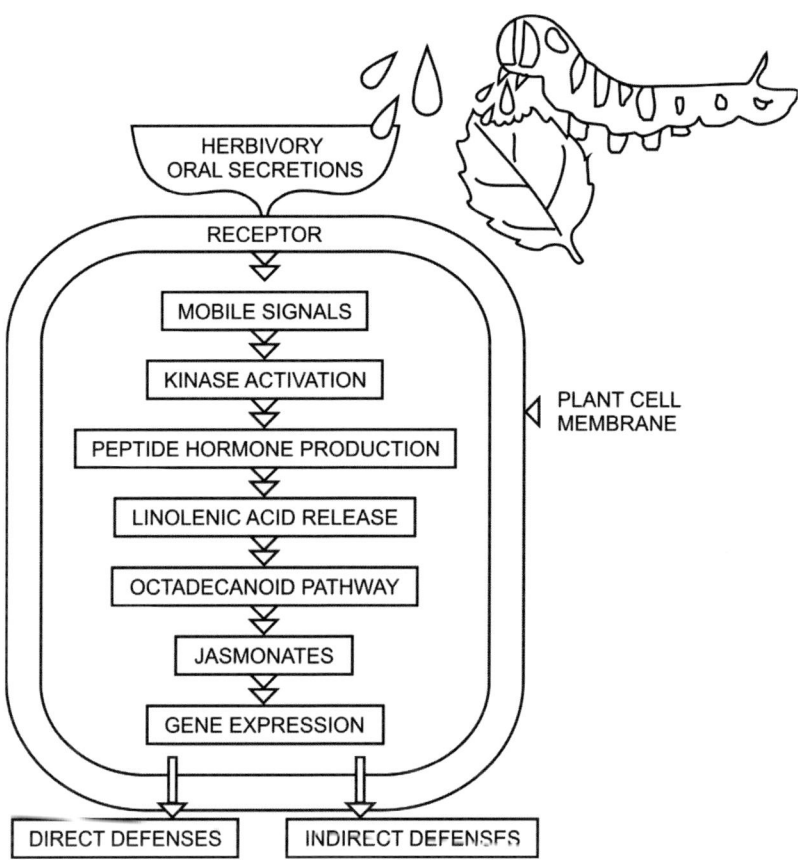

Figure 13.1. Schematic representation of the signaling pathway trigged by insect wounding. Mobile signals are released at the site of wounding following insect attack and initiate the defensive responses. Plants react by the induction of local and systemic regulation of gene expression. Current evidence suggests that insect oral secretions can be recognized and initiate a specific defense response, but several signals are common for different stresses and there is cross talk between abiotic and biotic defense responses. For more details see the text and references therein. Obs: The signal pathway is not confined to one cell but for simplicity is represented as such.

a synthetic substrate. As the use of methionine supplements reversed the negative effects of the inhibitors, the authors also concluded that the hyper-production of peptidases lead to the "loss" of available sulfur-containing amino acids (Broadway and Duffey, 1986). Recalculation of the original data, however, showed that the hypothesis of hyper-production of digestive peptidases was not correct (Broadway, 1996; Jongsma and Bolter, 1997). The increase of proteolytic activity was due to the induction of novel, and insensitive peptidases and the inhibitor-sensitive enzymes were largely replaced by the insensitive ones. Later investigations proved that a lack of amino acids as a result of hyper-production of peptidases was not the cause of growth inhibition (Broadway, 1996; Broadway, 1997; Jongsma and Bolter, 1997).

The effects caused by inhibition of digestive peptidases have been expressed using different fitness parameters, such as reduced feeding, extended retention time of food in the gut, reduced weight gain, reduction of adult emergence, reduced fecundity, reduced egg hatching, mortality, among others (Broadway, 1997; Jongsma and Bolter, 1997; Ryan, 1990). Even though the direct inhibition of target peptidases is the primary cause of the anti-nutritional activity of plant PIs, there are many other factors involved in the success or failure of a certain PI in affecting an insect species. The following parameters may be considered in the assessment of PIs as insect resistance factors: PI concentration, the affinity of the inhibitor for the target peptidase (measured as K_i), the stability of the PI in the physico-chemical conditions of the insect gut, the presence of different peptidases (some of them not inhibited by the PI), the modulation of gene expression of different peptidases in response to ingestion of the PI, and interactions with other phytochemicals (Johnson and Felton, 1996; Jongsma and Bolter, 1997; Ryan, 1990).

The concentration of PIs can be different among plant species, among genotypes within each species, among different tissues in the same individual, and in the same tissue as an adaptation to abiotic and biotic stresses. PI concentration ranges of 100–200 μM in storage tissues and 10–50 μM in leaves (following induction by insect feeding) have been reported. The dominant peptidases concentrations in insect guts range from 10–30 μM, implying that PI concentrations found in nature and in transgenic leaves are capable of complete enzyme inhibition (Jongsma and Bolter, 1997).

The affinity of PIs varies with different peptidases. The lower the K_i, the lower the PI concentration necessary to inhibit a fixed concentration of the enzyme. Knowledge of the affinity of a certain PI to a digestive peptidase is crucial information when deciding if the inhibitor could be a good candidate for insect pest control. To illustrate the importance of PI affinity, two wound-inducible phytocystatins (I25 family) from soybean were compared to a constitutive form as inhibitors of the digestive cysteine peptidases from the western corn rootworm (*Diabrotica virgifera*) and the beetle *L. decemlineata* (Zhao et al., 1996). The K_is against papain (C1 family) of the inducible forms were 500 to 1,000 times lower than the K_i of the constitutive form; indeed, the inducible forms had substantially higher inhibitory activity against the gut peptidases of both insect species (Zhao et al., 1996).

In addition to a selection pressure to modify PI quantities, plants may also be under selection pressure to adjust PI affinity to the digestive peptidases of the attacking insects. Another strategy to achieve efficient inhibition of target peptidases is the evolution of multidomain or multimeric inhibitors, leading to more inhibition per N invested in PI (Christeller, 2005; Jongsma and Bolter, 1997). Homotypic compound inhibitors against serine and cysteine peptidases are the rule among plants and there are other combinations of heterotypic compound inhibitors with domains against serine and aspartic or cysteine peptidases or serine and α-amylases (Campos and Richardson, 1983; Mares et al., 1991; Odani et al., 1983).

Several research groups are attempting to use a combination of different plant defensive proteins to improve efficiency and long-term control of insect pests. Such combinations could be mixtures of plant PIs, transgenic expression of fusion protein constructs, or of two or more potentially toxic proteins. The combination of PIs from non-host plants with endogenous host plant inhibitors proved to be much more efficient in inhibiting growth of larval *H. armigera* than the use of isolated inhibitors (Harsulkar et al., 1999). Alternatively, several PI fusion proteins have been devised. Abdeen and collaborators (2005) reported tomato leaf-specific over-expression of potato inhibitor II and carboxypeptidase inhibitor and their effects on larvae of two insects, the lepidopteran *Heliothis obsoleta* and the dipteran *Liriomyza trifolii* (Abdeen et al., 2005). These workers obtained a relatively high expression of the inhibitors (>1% of the total soluble proteins) and observed an increased resistance to both insects. In another study, the hybrid inhibitor combining the activities of a phytocystatin and a carboxypeptidase inhibitor showed high inhibitory activity against the cysteine enzymes of the Colorado potato beetle *L. decemlineata* (Benchabane et al., 2008).

A recombinant fusion protein containing a phytocystatin domain plus the lectin from *Griffonia simplicifolia* retained both peptidase and lectin (specific carbohydrate-binding) activities and showed synergistic delay in the development of the bruchid beetle *Callosobruchus maculatus*, whereas the dietary administration of a mixture of the separate proteins showed only an additive effect (Stratmann and Ryan, 1997; Zhu-Salzman et al., 2003). A synergistic effect against larval *C. maculatus* was also observed with the combination of a phytocystatin and pepstatin A, a specific inhibitor of aspartic peptidases (Amirhusin et al., 2007). The combination of the cystatin and pepstatin A with a soybean Kunitz inhibitor protected the latter against degradation by the bruchid digestive cysteine and aspartic peptidases, causing a synergistic effect (Amirhusin et al., 2007). This synergistic effect probably involves the protection of a bean α-amylase inhibitor that would normally be detoxified by a minor serine peptidase of the bruchid larva (Silva et al., 2001a; Silva et al., 2001b; Zhu-Salzman et al., 2003).

A synergistic effect was also reported for the combination of the phytocystatin soyacystatin N (scN) and a wheat α-amylase inhibitor against larval *C. maculatus* developing inside artificial seeds containing both inhibitors. The delay in larval development was more pronounced than when the individual inhibitors were used separately, probably because the cysteine peptidase inhibitor protects the α-amylase inhibitor from proteolytic degradation (Amirhusin et al., 2004). The combination of soybean trypsin inhibitor with the insecticidal protein from the soil bacterium *Bacillus thuringiensis* (Bt toxin), the most successful protein used in transgenic crops to date, caused protection against proteolysis of the toxin in the gut of larval *H. armigera*, resulting in a synergistic effect reflected by higher insect mortality (Zhang et al., 2000).

Interactions of plant PIs with other phytochemicals have the potential to modify the toxic effect of the inhibitor. These phytochemicals can be either constitutive or induced in the plant in response to insect attack. Indeed, most plants produce a complex mixture of potentially toxic substances (including alkaloids, glucosinolates, glycosides, tannins, anti-nutritional proteins, etc.) that can act as deterrents, anti-feedants, or toxins. The combination of different activities results in a synergistic and efficient form of plant defense. Because some insects can compensate for the presence of PIs in their diets by increasing consumption, the production of PIs by plants may be a liability. This could be avoided by the combined activity of secondary substances with PIs. In *N. attenuata*, insect wounding causes a rapid increase in leaf concentrations of serine peptidase inhibitors and the alkaloid nicotine. The generalist lepidopteran *S. exigua* is able to adapt to serine inhibitors by inducing insensitive peptidases and by compensatory feeding behavior, resulting in an enhanced performance triggered by the over-expression of the potato inhibitor II (Jongsma et al., 1995). However, the presence of nicotine *in planta* results in a strong synergism with the PI, because the alkaloid acts as an anti-feedant toxin, preventing the compensatory feeding adaptation and favoring the inhibitory activity (Steppuhn and Baldwin, 2007).

The combined and coordinated action of multiple plant toxins targeting several nutritional vulnerabilities in insects and other herbivores seem to be the rule rather than the exception (Dyer et al., 2003). A major vulnerability is the capacity to acquire essential amino acids (Awmack and Leather, 2002; Schoonhoven et al., 1998). The digestion of dietary proteins and absorption of free amino acids in insects are carried out by a combination of enzymes and other proteins working in concert. Each step could be targeted by different plant defense compounds and not solely by the action of PIs. In tomato plants, besides the induction of PIs, several other proteins are induced, among them are the enzymes arginase and threonine deaminase. One of the isoforms of threonine deaminase is induced by insect wounding (and jasmonic acid) and seems to be involved in the degradation of threonine in the midgut, an essential amino acid for herbivorous insects like the specialist *M. sexta* and the generalist *Trichoplusia ni* (Chen et al., 2007; Chen et al., 2005). The combined actions of induced PIs and enzymes such as arginase (that degrade the amino acid arginine), threonine deaminase, and polyphenol oxidase (that can modify dietary proteins, decreasing their digestibility) may potentiate the defense responses of challenged plants (Chen et al., 2007; Howe and Jander, 2008).

Another factor to be evaluated in the use of plant PIs as defensive agents against insects is the physical-chemical milieu prevalent in the digestive tract. Parameters such as pH, redox potential, metabolites, and even the microbiota can have an important role in the capacity of the insect to adapt or detoxify ingested plant PIs and other defensive compounds (Johnson and Felton, 1996; Visôtto et al., 2009). The stability of PIs toward mixtures of peptidases with different mechanistic catalysis, a situation frequently found in the insect midgut, is crucial to the success of the inhibitor. Many insects can inactivate the ingested plant PIs by degradation via digestive non-inhibited peptidases (Girard et al., 1998; Giri et al., 1998; Yang et al., 2009). As proteins with biological activity, plant PIs are dependent on pH and redox potential to have full activity, because these parameters have an influence on the native conformation of the inhibitors (Johnson and Felton, 1996).

Besides the influence of physical and chemical parameters in the insect gut on the activities of plant PIs, the quantity and quality of dietary proteins are important in the establishment of anti-nutritional effects (Awmack and Leather, 2002). The protein quality, in terms of digestibility and amino acid composition greatly alters the toxicity of soybean trypsin inhibitor and potato inhibitor II to *H. zea* and *S. exigua* (Broadway and Duffey, 1986), and to the black field cricket *Teleogryllus commodus* (Burgess et al., 1994). Concentrations as low as 0.1% (w/v) of the cowpea trypsin inhibitor, soybean trypsin inhibitor, wheat germ trypsin inhibitor, and potato inhibitors I and II caused high levels of cricket mortality. However, in all cases the levels of casein in the diets significantly influenced the efficacy of the inhibitors. The higher the concentration of dietary casein, the lower the efficiency of the inhibitors in reducing cricket growth (Burgess et al., 1994). The authors also showed that potato inhibitor II and soybean trypsin inhibitors were particularly effective against the cricket and that there was a synergistic effect with the combination of the potato inhibitor I and the wheat germ trypsin inhibitor, together as efficient as potato inhibitor II (Burgess et al., 1994). Another comparison between suboptimal and rich protein diets and the effects of plant PIs on the development and mortality of insects was carried out with the larval sugar cane stalk borer *Diatraea saccharalis* and the soybean inhibitors of the I3 family (Kunitz) and the I12 family (Bowman-Birk). It was demonstrated that despite the effects of both inhibitors on midgut enzymes, the effectiveness of PIs was directly influenced by the quality of protein in the diet. When the PIs were incorporated into artificial diets, the negative effects on larval fitness were significantly higher for larvae fed on the poorer diet than on the protein-rich diet (Pompermayer et al., 2003).

In summary, the constitutive or induced post-ingestive direct defenses of plants against insects frequently result in synergistic interactions between different PIs, amino acid degrading enzymes, oxidases, and secondary substances (Amirhusin et al., 2007; Burgess et al., 1994; Howe and Jander, 2008; Jongsma and Bolter, 1997). In terms of insect control, the combination with other potentially toxic substances in addition to plant PIs may increase efficiency when compared to individual defense molecules, due to the capacity of insects to adapt to plant defensive compounds.

The effects of the ingestion of plant PIs on the gut physiology of insects have also been studied by the use of state-of-the-art techniques, such as proteomics of midgut contents and feces after ingestion of plant PIs or determinations of differentially expressed genes in response to the inhibitors present in the diet (Chi et al., 2009; De Vos et al., 2005; Delp et al., 2009; Ferry et al., 2004; Liu et al., 2004; Vogel et al., 2007).

Subtractive hybridization and cDNA microarrays showed that when challenged by the incorporation of soyacystatin N in artificial diets, the second larval instar of the southern corn rootworm (*Diabrotica undecimpunctata howardi*) responded by differential over-expression of 29 genes in the midgut, among them, cysteine and aspartic peptidases and also a peritrophin gene (Liu et al., 2004). The data suggested that the enhanced expression of constitutive cathepsin L- and B-like peptidases and the induction of peritrophin were involved in the adaptive response, compensating for the inhibition of the pre-existing peptidases and by strengthening the peritrophic membrane. It was also observed that several genes encoding proteins associated with metabolism and development were down-regulated, reflecting, according to the authors, a reallocation of resources to prioritize the counter-defense response (Liu et al., 2004).

The effects of soyacystatin N were also investigated using larvae of the cowpea weevil *C. maculatus* as a model (Chi et al., 2009). This involved the production of a library containing 20,352 cDNAs from the midgut of larvae fed a control diet and diets containing anti-nutritional compounds both chronically (sublethal scN dose throughout larval development) and acutely (fourth instar larvae fed on a high concentration of scN for 24 h). The transcript responses to dietary scN were analyzed by cDNA microarray followed by quantitative (real-time) RT-PCR. It was observed that 1,756 cDNAs were induced or down-regulated in at least one larval instar fed chronic or acute scN doses. Overall, hydrolases involved in digestion of protein and carbohydrates and proteins associated with lipid mobilization were upregulated by scN, whereas housekeeping genes, such as those encoding structural proteins and stress-related proteins, other than the anti-nutritional effect triggered by the PI, were largely down-regulated (Chi et al., 2009). Induction of both scN-sensitive cathepsin L-like and scN-insensitive cathepsin B-like peptidases was also observed. The authors also suggested that the induction of a midgut aquaporin gene is related to the necessity of an enhanced secretion of water in the luminal contents to facilitate the transit of poorly digested food. These authors also observed that to compensate for the inhibited protein digestion, the insect compromised its ability to cope with other stresses.

A midgut proteome approach was used to study the response of larval *Drosophila melanogaster* to the ingestion of a Bowman-Birk inhibitor (Li et al., 2007). In spite of a delay in larval development, there was no difference in mortality of inhibitor-fed larvae compared to the controls. Analysis using two-dimensional gel-electrophoresis showed that nine proteins were differentially expressed in larvae fed the PI, which were associated with protein degradation, transport and lipid metabolism.

There is a growing number of proteomic studies of the fate and effects of ingested plant PIs and other plant defensive proteins on insect gut physiology. Proteomic analysis of the midgut lumen and of the fecal bolus of insects fed on plants provides interesting information concerning the stability of defensive proteins in the adverse environment of the insect midgut. The findings to date show that plant proteins have a wide range of stability during the passage through the insect gut, and that the bulk of dietary proteins are degraded, but some potentially toxic proteins remain hyper-stable and active following the passage through the gut, being enriched in the feces (Chen et al., 2005, 2007; Schilmiller and Howe, 2005).

Transgenic Plants Over-expressing PIs

Since the pioneer work of Hilder and collaborators (1987), many papers have reported the production of transgenic plants expressing PIs. Particular efforts have been directed to produce effective control of insect pests (Haq et al., 2004; Mosolov et al., 2001). The majority of economically important insect species belong to the orders Coleoptera and Lepidoptera, which rely on cysteine and serine peptidases (mostly members of families S1 and C1) to digest plant proteins (Lopes et al., 2009; Terra and Ferreira, 2005). Accordingly, genes encoding PIs active against these mechanistic classes of peptidases have generally been used in the development of transgenic plants (Haq et al., 2004; Howe and Jander, 2008; Jouanin et al., 1998; Ryan, 1990; Velkov et al., 2005).

Several studies have questioned how PIs in transgenic plants can affect the natural community (beneficial predators, parasitoids, nontarget herbivores, the local microbiota etc.). Assessments of tri-trophic interactions have shown conflicting results, likely due to differences in the species and type of inhibitor. The success of the ecto-parasitoid *Eulophus pennicornis* was reduced when wasp larvae fed on the host (*Laconobia oleracea*) reared on potato leaves expressing cowpea (*V. unguiculata*) trypsin inhibitor (Bowman-Birk, family I12) (Bell et al., 2001). Conversely, a rice phytocystatin expressed in transgenic potato did not affect survival, growth or development of the two-spotted stink bug *Perillus bioculatus* preying on the herbivore *L. decemlineata* feeding on transgenic plants (Bouchard et al., 2003). In this case, the predator adapted its digestive proteolytic machinery to the presence of the plant PI ingested by its prey. Álvarez-Alfageme and colleagues (2007) also reported that another predatory bug, *Podisus maculiventris*, was not significantly affected when fed on *L. decemlineata* or *S. litotallis* reared on transgenic tobacco expressing a barley cystatin (Álvarez-Alfageme et al., 2007). Ferry and colleagues (2005) investigated the effect of the mustard trypsin inhibitor-2 (I18 family) over-expressed in oilseed rape (*Brassica napus*), on the predatory carabid beetle *Pterostichus madidus* feeding on the susceptible lepidopteran *Plutella xylostella* (Ferry et al., 2005). They also found that the predator was able to adapt to the ingestion of the inhibitor present in its prey by altering the composition of its digestive peptidases, and thus overcoming the deleterious effects of exposure to the inhibitor. The effects on longterm dietary exposure of another carabid (*Ctenognathus novaezelandiae*) to a prey species (*Spodoptera litura*) feeding on transgenic tobacco expressing a serine trypsin inhibitor were described by Burgess and colleagues (2008). The authors reported that there was no significant effect on survival or fecundity, and only minor or transient effects on the beetles were observed.

In general, it appears that large-scale use of transgenic plants expressing plant PIs does not result in significant negative effects on ecosystems (O'Callaghan et al., 2005; Velkov et al., 2005, for reviews). Overall, it seems that predatory or parasitoid insects can adapt their digestive metabolism to the presence of plant PIs ingested by their herbivorous prey or hosts. Nevertheless, additional studies are still required to assess the risks, mainly in field trials to gain more information on the multi-trophic interactions involving insect-resistant transgenic plants over-expressing broad spectrum PIs (Lövei et al., 2009; Melo-Martén and Meghani, 2008; Velkov et al., 2005). However, insect adaptation to transgenic plants expressing PIs has become a major obstacle in the successful use of this strategy for pest management. In the next section we discuss the strategies insects employ to overcome the effects of plant PIs.

Adaptation of Insects to Plant PIs

Over the last 20 years several crop plants have been transformed with genes that express plant PIs. Nevertheless, resistance against pest insects has been limited, and to date there are no commercial plant species expressing heterologous PIs. The lack of protection, or transient protection at best, of plants expressing PI transgenes is attributed to the capacity of herbivorous insects to adapt to the ingested PIs. Since the pivotal work of Jongsma and others (1995) demonstrating that larvae of *S. exigua* adapt to transgenic tobacco transformed with the potato inhibitor II gene by replacing sensitive peptidases with insensitive ones (Jongsma et al., 1995), numerous papers have reported that insects employ different strategies to suppress the antinutritional effects of plant PIs. Basically, insects adapt to dietary PIs by over-production of pre-existing (constitutive) peptidases (Bonadé-Bottino et al., 1999; Brioshi et al., 2007); selective *de novo* expression of insensitive peptidases (Bolter and Jongsma, 1995; Bown et al., 1997; Cloutier et al., 2000; Jongsma et al., 1995; Mazumdar-Leighton and Broadway, 2001; Zhu-Salzman et al., 2003); or by proteolysis of the ingested PI (Girard et al., 1998; Giri et al., 1998; Yang et al., 2009). Furthermore, some of the physiological adaptations also involve overconsumption of plant material to compensate for the presence of inhibitors (Cloutier et al., 2000; Winterer and Bergelson, 2001).

Despite the discovery that insects adapt to the ingestion of plant PIs, the possible use of transgenic plants expressing PIs is still a promising strategy for pest control, especially as the regulatory mechanisms of counter defense genes of insects challenged by plant PIs are largely unknown. It is probable that the insensitive peptidases are recruited from a repertoire of digestive peptidases available from multigene families found in the insect genome. Support for this hypothesis has been published by several groups (Bown et al., 2004; Brioshi et al., 2007; Broadway, 1997; Chi et al., 2009; Jongsma et al., 1995; Mazumdar-Leighton and Broadway, 2001; Rispe et al., 2008). It has also been demonstrated that the expression and secretion of insensitive peptidases are transcriptionally regulated and that this process is triggered by the ingestion of PIs in a dose and time-dependent manner (Bown et al., 2004; Broadway, 1997). Figure 13.2 summarizes the events that can be observed when insects ingest plant PIs.

The multiple regulatory mechanisms that permit insects to evade the toxic effects of plant PIs were demonstrated by the researchers who first transformed plants for resistance to insect pests. This adaptive capacity imposes a serious obstacle to the development of PI-based resistant plants, and it continues to be a serious challenge in the understanding of plant-insect interactions. Some important questions need to be addressed: What is the mechanism involved in the recognition of the PI in the insect gut? Are receptors or monitor peptide-like molecules analogous to those found in mammals involved in the recognition process? How do insect midgut tissues control gene expression resulting in down-regulation of sensitive and up-regulation of insensitive peptidases? How did this adaptive capacity evolve?

It is hoped that the answers to these questions will permit advances in the development of biorational techniques of crop protection. Researchers would be able to choose an effective combination of defensive proteins based on the knowledge of the capacity of the insect to counter plant defenses. Once the complete repertory of peptidases multi-gene families and adaptive mechanisms have been determined, this information will be important for selection of defensive proteins to be used in plant genetic engineering.

Concluding Remarks

The co-evolutionary arms race between plants and herbivorous insects has resulted in sophisticated plant defense mechanisms and the development of counter-defenses by the insects. Among the defensive compounds in plants, PIs play an important role, probably due to their nature as primary gene products, which permit the modification of the digestive peptidases of the attackers. Both defense and counter-defense are energetically expensive. Under insect attack, plants reallocate resources for gene expression and induction of several defensive

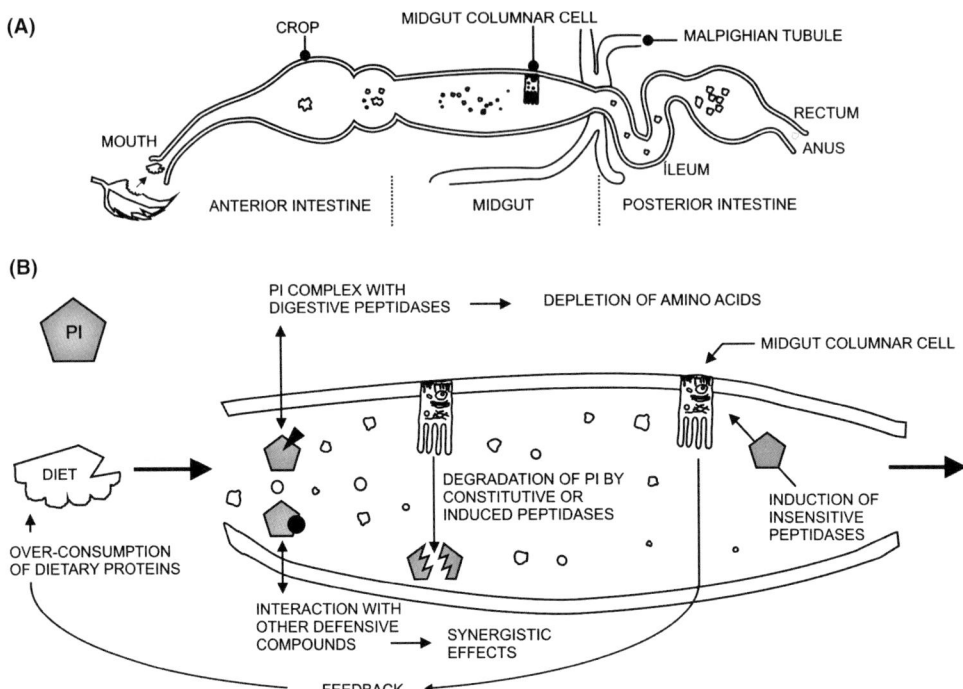

Figure 13.2. A, schematic representation of the insect digestive system, showing arrangement of the different compartments; and B, details of the interactions between Plant Protein Inhibitors (PI) and insect peptidases in the insect midgut. The peritrophic membrane is omitted in order to simplify the figure.

compounds, some of which act synergistically with PIs. To be successful, the counter-defenses of insects against these dietary threats also involve a high-energy cost. Enhanced expression of insensitive peptidases represents an important strategy of the insects to deal with constitutive and induced plant PIs. Consequently, both plants and herbivorous insects can survive under suboptimal conditions as an adaptive strategy. With the advances of proteomic, genomic and metabolomic techniques, combined with the increasing number of complete genomes of plants and insects, there is a great expectation that more details will be revealed concerning the intriguing relationships between plant and insects.

References

Abdeen, A., Virgos, A., Olivella, E., Villanueva, J., Aviles, S., Gabarra, R., Prat, S. 2005. Multiple insect resistance in transgenic tomato plants over-expressing two families of plant proteinase inhibitors. *Plant Molecular Biology* 57: 189–202.

AbuQamar, S., Chai, M., Luo, H., Song, F. Mengiste, T. 2008. Tomato protein kinase 1b mediates signaling of plant responses to necrotrophic fungi and insect herbivory. *The Plant Cell* 20: 1964–1983.

Acquisti, C., Elser, J.J., Kumar, S. 2009. Ecological nitrogen limitation shapes the DNA composition of plant genomes. *Molecular Biology and Evolution* 26: 953–956.

Altpeter, F., Díaz, I., McAuslane, H., Gaddour, K., Carbonero, P., Vasil, I.K. 1999. Increased insect resistance in transgenic wheat stably expressing trypsin inhibitor CMe. *Molecular Breeding* 5: 53–63.

Álvarez-Alfageme, F., Martínez, M., Pascual-Ruiz, S., Castañera, P., Diaz, I., Ortego, F. 2007. Effects of potato plants expressing a barley cystatin on the predatory bug *Podisus maculiventris* via herbivorous prey feeding on the plant. *Transgenic Research* 16: 1–13.

Amirhusin, B., Shade, R.E., Koiwa, H., Hasegawa, P.M., Bressan, R.A., Murdock, L.L., Zhu-Salzman, K. 2007. Protease inhibitors from several classes work synergistically against *Callosobruchus maculatus*. *Journal of Insect Physiology* 53: 734–740.

Amirhusin, B., Shade, R.E., Koiwa, H., Hasegawa, P.M., Bressan, R.A., Murdock, L.L., Zhu-Salzman, K. 2004. Soyacystatin N inhibits proteolysis of wheat α-amylase inhibitor and potentiates toxicity against cowpea weevil. *Journal of Economic Entomology* 97: 2095–2100.

Arai, S., Matsumoto, I., Emori, Y., Abe, K. 2002. Plant seed cystatins and their target enzymes of endogenous and exogenous origin. *Journal of Agricultural and Food Chemistry* 50: 6612–6617.

Awmack, S.C., Leather, S.R. 2002. Host plant quality and fecundity in herbivorous insects. *Annual Review of Entomology* 47: 817–844.

Bari, R., Jones, J.D.G. 2009. Role of plant hormones in plant defense responses. *Plant Molecular Biology* 69: 473–488.

Baydoun, E.H., Fry, S.C. 1985. The immobility of pectic substances in injured tomato leaves and its bearing on the identity of the wound hormone. *Planta* 165: 269–276.

Bell, H.A., Fitches, E.C., Down, R.E., Ford, L., Marris, G.C., Edwards, J.P., Gatehouse, J.A., Gatehouse, A.M.R. 2001. Effect of dietary cowpea trypsin inhibitor (CpTI) on the growth and development of the tomato moth *Lacanobia oleracea* (Lepidoptera: Noctuidae) and on the success of the gregarious ectoparasitoid *Eulophus pennicornis* (Hymenoptera : Eulophidae). *Pest Management Science* 57: 57–65.

Benchabane, M., Goulet, M., Dallaire, C., Côté, P., Michaud, D. 2008. Hybrid protease inhibitors for pest and pathogen control—a functional cost for the fusion partners? *Plant Physiology and Biochemistry* 46: 701–708.

Birk, Y. 1985. The Bowman-Birk inhibitor. Trypsin and chymotrypsin inhibitor from soybeans. *International Journal of Peptide and Protein Research* 25: 113–131.

Birk, Y., Gertler, A., Khalef, S. 1963. A pure trypsin inhibitor from soya beans. *Biochemical Journal* 87: 281–284.

Bishop, P.D., Makus, D.J., Pearce, G., Ryan, C.A. 1981. Proteinase inhibitor-inducing factor activity in tomato leaves resides in oligosaccharides enzymatically released from cell walls. *Proceedings of the National Academy of Science USA* 78: 3536–3540.

Bodenhausen, N., Reymond, P. 2007. Signaling pathways controlling induced resistance to insect herbivores in *Arabidopsis*. *Molecular Plant-Microbe Interactions* 20: 1406–1420.

Bolter, C.J., Jongsma, M.A. 1995. Colorado potato beetles (*Leptinotarsa decemlineata*) adapt to proteinase inhibitors induced in potato leaves by methyl jasmonate. *Journal of Insect Physiology* 41: 1071–1078.

Bonadé-Bottino, M., Lerin, J., Zaccomer, B., Jouanin, L. 1999. Physiological adaptation explains the insensitivity of *Baris coerulescens* to transgenic oilseed rape expressing oryzacystatin I. *Insect Biochemistry and Molecular Biology* 29: 131–138.

Bostock, R.M. 2005. Signal crosstalk and induced resistance: Straddling the line between cost and benefit. *Annual Review of Phytopathology* 43: 545–580.

Botelho-Júnior, S., Siqueira-Júnior, C.L., Jardim, B.C., Machado, O.L.T., Neves-Ferreira, A.G.C., Perales, J., Jacinto, T. 2008. Trypsin inhibitors in passion fruit (*Passiflora* f. *edulis flavicarpa*) leaves: Accumulation in response to methyl jasmonate, mechanical wounding, and herbivory. *Journal of Agricultural and Food Chemistry* 56: 9404–9409.

Bouchard, É., Cloutier, C., Michaud, D. 2003. Oryzacystatin I expressed in transgenic potato induces digestive compensation in an insect natural predator via its herbivorous prey feeding on the plant. *Molecular Ecology* 12: 2439–2446.

Bowman, D.E. 1946. Differentiation of soy bean anti-tryptic factors. *Proceedings of the Society for Experimental Biology and Medicine* 63: 547–550.

Bown, D.P., Wilkinson, H.S., Gatehouse, J.A. 1997. Differentially regulated inhibitor-sensitive and insensitive protease genes from the phytophagous insect pest, *Helicoverpa armigera*, are members of complex multigene families. *Insect Biochemistry and Molecular Biology* 27: 625–638.

Bown, D.P., Wilkinson, H.S., Gatehouse, J.A. 2004. Regulation of expression of genes encoding digestive proteases in the gut of a polyphagous lepidopteran larva in response to dietary protease inhibitors. *Physiological Entomology* 29: 278–290.

Brioshi, D., Naldani, L.D., Bengstan, M.H., Sogayar, M.C., Moura, D.S., Silva-Filho, M.C. 2007. General up regulation of *Spodoptera frugiperda* trypsins and chymotrypsins allows its adaptation to soybean proteinase inhibitor. *Insect Biochemistry and Molecular Biology* 37: 1283–1290.

Broadway, R.M. 1996. Dietary proteinase inhibitors alter compliment of midgut proteases. *Archives of Insect Biochemistry and Physiology* 32: 39–53.

Broadway, R.M. 1997. Dietary regulation of serine proteinases that are resistant to serine proteinase inhibitors. *Journal of Insect Physiology* 43: 855–874.

Broadway, R.M., Duffey, S.S. 1986. Plant proteinase inhibitors: mechanism of action and effect on the growth and digestive physiology of larval *Heliothis zea* and *Spodoptera exigua*. *Journal of Insect Physiology* 32: 827–833.

Browse, J., Howe, G.A. 2008. New weapons and a rapid response against insect attack. *Plant Physiology* 146: 832–838.

Burgess, E.P.J., Main, C.A., Stevens, P.S., Christeller, J.T., Gatehouse, A.M.R., Laing, W.A. 1994. Effects of protease inhibitor concentration and combinations on the survival, growth and gut enzyme activities of the black field cricket, *Teleogryllus commodus*. *Journal of Insect Physiology* 40: 803–811.

Burgess, E.P.J., Philip, B.A., Christeller, J.T., Page, N.E.M., Marshall, R.K., Wohlers, M.W. 2008. Tri-trophic effects of transgenic insect-resistant tobacco expressing a protease inhibitor or a biotin-binding protein on adults of the predatory carabid beetle *Ctenognathus novaezelandiae*. *Journal of Insect Physiology* 54: 518–528.

Campos, F.A.P., Richardson, M. 1983. The complete amino acid sequence of bifunctional α-amylase/trypsin inhibitor from seeds of ragi (Indian finger millet, *Eleusine coracana* Gaertn.). *FEBS Letters* 152: 300–304.

Chen, H., Gonzales-Vigil, E., Wilkerson, C.G., Howe, G.A. 2007. Stability of plant defense proteins in the gut of insect herbivores. *Plant Physiology* 143: 1954–1967.

Chen, H., Wilkerson, C.G., Kuchar, J.A., Phinney, B.S., Howe, G.A. 2005. Jasmonate-inducible plant enzymes degrade essential amino acids in the herbivore midgut. *Proceedings of the National Academy of Science USA* 102: 19237–19242.

Chi, Y.H., Salzman, R. Balfe, A., S., Ahn, J.-E., Sun, W., Moon, J., Yun, D.-J., Lee, S. Y., Higgins, T.J.V., Pittendrigh, B., Murdock, L.L., Zhu-Salzman, K. 2009. Cowpea bruchid midgut transcriptome response to a soybean cystatin—costs and benefits of counter-defence. *Insect Molecular Biology* 18: 97–110.

Christeller, J.T. 2005. Evolutionary mechanisms acting on proteinase inhibitor variability. *FEBS Journal* 272: 5710–5722.

Chung, H.S., Howe, G.A. 2009. A critical role for the TIFY motif in repression of jasmonate signaling by a stabilized splice variant of the JASMONATE ZIM-domain protein JAZ10 in *Arabidopsis*. *The Plant Cell* 21: 131–145.

Cloutier, C., Jean, C., Fournier, M., Yelle, S., Michaud, D. 2000. Adult Colorado potato beetles, *Leptinotarsa decemlineata* compensate for nutritional stress on Oryzacystatin I—Transgenic potato plants by hypertrophic behavior and overproduction of insensitive proteases. *Archives of Insect Biochemistry and Physiology* 44: 69–81.

Cordero, M.J., Raventos, D., San Segundo, B. 1994. Expression of a maize proteinase inhibitor gene is induced in response to wounding and fungal infection: systemic wound-response of a monocot gene. *Plant Journal* 6: 141–150.

De Vos, M., Van Oosten, V.R., Van Poecke, R.M.P., Van Pelt, J.A., Pozo, M.J., Mueller, M.J., Buchala, A.J., Metraux, J.P., Van Loon, L.C., Dicke, M., Pieterse, C.M.J. 2005. Signal signature and transcriptome changes of *Arabidopsis* during pathogen and insect attack. *Molecular Plant-Microbe Interaction* 18: 923–937.

Delp, G., Gradin, T., Ahman, I. Jonsson, L.M.V. 2009. Microarray analysis of the interaction between the aphid *Rhopalosiphum padi* and host plants reveals both differences and similarities between susceptible and partially resistant barley lines. *Molecular Genetics and Genomics* 281: 233–248.

Dyer, L.A., Dodson, C.D., Stireman, J.O., Tobler, M.A., Smilanich, A.M., Fincher, R.M., Letourneau, D.K. 2003. Synergistic effects of three *Piper* amides on generalist and specialist herbivores. *Journal of Chemical Ecology* 29: 2499–2514.

Eckelkamp, C., Ehmann, B., Schopfer, P. 1993. Wound-induced systemic accumulation of a transcript coding for a Bowman-Birk trypsin inhibitor-related protein in maize (*Zea mays* L.) seedlings. *FEBS Letters* 323: 73–76.

Elser, J.J., Fagan, W.F., Subramanian, S., Kumar, S. 2006. Signature of ecological resource availability in the animal and plant proteomes. *Molecular Biology and Evolution* 23: 1946–1951.

Erb, M., Ton, J., Degenhardt, J., Turlings, T.C.J. 2008. Interactions between arthropod-induced above ground and below ground defenses in plants. *Plant Physiology* 146: 867–874.

Falco, M.C., Silva-Filho, M.C. 2003. Expression of soybean proteinase inhibitors in transgenic sugarcane plants: Effects on natural defense against *Diatraea saccharalis*. *Plant Physiology and Biochemistry* 41: 761–766.

Farmer, E.E., Ryan, C.A. 1990. Interplant communication: airborne methyl jasmonate induces synthesis of proteinase inhibitors in plant leaves *Proc. Natl. Acad. Sci. USA* 87: 7713–7716.

Farmer, E.E., Ryan, C.A. 1992. Octadecanoid precursors of jasmonic acid activate the synthesis of wound-inducible proteinase inhibitors. *The Plant Cell* 4: 129–134.

Ferry, N., Edwards, M.G., Gatehouse, J.A. Gatehouse, A.M.R. 2004. Plant–insect interactions: molecular approaches to insect resistance. *Current Opinion in Biotechnology* 15: 155–161.

Ferry, N., Jouanin, L., Ceci, L.R., Mulligan, E.A., Emami, K., Gatehouse, J.A., Gatehouse, A.M.R. 2005. Impact of oilseed rape expressing the insecticidal serine protease inhibitor, mustard trypsin inhibitor-2 on the beneficial predator *Pterostichus madidus*. *Molecular Ecology* 14: 337–349.

Girard, C., Le Métayer, M., Bonadé-Bottino, M., Pham-Delégue, M., Jouanin, L. 1998. High level of resistance to proteinase inhibitors may be conferred by proteolytic cleavage in beetle larvae. *Insect Biochemistry and Molecular Biology* 28: 229–237.

Girard, C., Rivard, D., Kiggundu, A., Kunert, K., Gleddie, S.C., Cloutier, C., Michaud, D. 2007. A multicomponent, elicitor-inducible cystatin complex in tomato, *Solanum lycopersicum*. *New Phytologist* 173: 841–851.

Giri, A.P., Harsulkar, A.M., Deshpande, V.V., Sainani, M.N., Gupta, V.S., Ranjekar, P.K. 1998. Chickpea defensive proteinase inhibitors can be inactivated by podborer gut proteinases. *Plant Physiology* 116: 393–401.

Goulet, M.-C., Dallaire, C., Vaillancourt, L.-P., Khalf, M., Badri, A.M., Preradov, A., Duceppe, M.-O., Goulet, C., Cloutier, C., Michaud, D. 2008. Tailoring the specificity of a plant cystatin toward herbivorous insect digestive cysteine proteases by single mutations at positively selected amino acid sites. *Plant Physiology* 146: 1010–1019.

Graham, J.S., Ryan, C.A. 1981. Accumulation of a metallo-carboxypeptidase inhibitor in leaves of wounded potato plants. *Biochemical and Biophysical Research Communications* 101: 1164–1170.

Green, T.R., Ryan, C.A. 1972. Wound-induced proteinase inhibitor in plant leaves. A possible defense mechanism against insects. *Science* 175: 776–777.

Habib, H., Fazili, K.M. 2007. Plant protease inhibitors: a defense strategy in plants. *Biotechnology and Molecular Biology Review* 2: 68–85.

Habu, Y., Peyachoknagul, S., Sakata, Y., Fukusawa, K. 1997. Evolution of a multigene family that encodes the Kunitz chymotrypsin inhibitor in winged bean: a possible intermediate in the generation of a new gene with a distinct pattern of expression. *Molecular & General Genetics* 254: 73–80.

Halitschke, R., Schittko, U., Pohnert, G., Boland, W., Baldwin, I.T. 2001. Molecular interactions between the specialist herbivore *Manduca sexta* (Lepidoptera, Sphingidae) and its natural host *Nicotiana attenuata*. III. Fatty acid-amino acid conjugates in herbivore oral secretions are necessary and sufficient for herbivore-specific plant responses. *Plant Physiology* 125: 711–717.

Haq, S.K., Atif, S.M., Khan, R.H. 2004. Protein proteinase inhibitor genes in combat against insects, pests, and pathogens: natural and engineered phytoprotection. *Archives of Biochemistry and Biophysics* 431: 145–159.

Harsulkar, A.M., Giri, A.P., Patankar, A.G., Gupta, V.S., Sainani, M.N., Rajankar, P.K., Deshpande, V.V. 1999. Successive use of non-host plant proteinase inhibitors required for effective inhibition of gut proteinases and larval growth of *Helicoverpa armigera*. *Plant Physiology* 121: 497–506.

Heibges, A., Salamini, F., Gebhardt, C. 2003. Functional comparison of homologous members of three groups of Kunitz-type enzyme inhibitors from potato tubers (*Solanum tuberosum* L.). *Molecular Genetics and Genomics* 269: 535–541.

Hejgaard, J., Laing, W.A., Marttila, S., Gleave, A.P., Roberts, T.H. 2005. Serpins in fruit and vegetative tissues of apple (*Malus domestica*): expression of four serpins with distinct reactive centres and characterisation of a major inhibitory seed form, MdZ1b. *Functional Plant Biology* 32: 517–527.

Hejgaard, J., Rasmussen, S.K., Brandt, A., Svendsen, I. 1985. Sequence homology between barley endosperm protein Z and protease inhibitors of the alpha-1-antitrypsin family. *FEBS Letters* 180: 89–94.

Hilder, V.A., Gatehouse, A.M.R., Sheerman, S.E., Barker, R.F., Boulter, D. 1987. A novel mechanism of insect resistance engineered into tobacco. *Nature* 333: 160–163.

Hildmann, T., Ebneth, M., Cortth, H.P., Serrano, J.J.S., Willmitzer, L., Prat, S. 1992 General roles of abscisic and jasmonic acids in gene activation as a result of mechanical wounding. *The Plant Cell* 4: 1157–1170.

Howe, G.A., Jander, G. 2008. Plant immunity to insect herbivores. *Annual Review of Plant Biology* 59: 41–66.

Huang, Y., Xiao, B., Xiong, L. 2007. Characterization of a stress responsive protease inhibitor gene with positive effect in improving drought resistance in rice. *Planta* 226: 73–85.

Jander, G., Howe, G.A. 2008. Plant interactions with arthropod herbivores: state of the field. *Plant Physiology* 146: 801–803.

Johnson, K.S., Felton, G.W. 1996. Potential influence of midgut pH and redox potential on protein utilization in insect herbivores. *Archives of Insect Biochemistry and Physiology* 32: 85–105.

Jongsma, M.A., Bakker, P.L., Peters, J., Bosch D., Stiekema, W.J. 1995. Adaptation of *Spodoptera exigua* larvae to plant proteinase inhibitors by induction of gut proteinase activity insensitive to inhibition. *Proceedings of the National Academy of Science USA* 92: 8041–8045.

Jongsma, M.A., Bolter, C. 1997. The adaptation of insects to plant protease inhibitors. *Journal of Insect Physiology* 43: 885–895.

Jouanin, L., Bonade-Bottino, M., Girard, C., Morrot, G., Giband, M. 1998. Transgenic plants for insect resistance. *Plant Science* 131: 1–11.

Katsir, L., Chung, H.S., Koo, A.J.K., Howe, G.A. 2008. Jasmonate signaling: a conserved mechanism of hormone sensing. *Current Opinion in Plant Biology* 11: 428–435.

Kazan, K., Manners, J.M. 2008. Jasmonate signaling: toward an integrated view. *Plant Physiology* 146: 1459–1468.

Kong, L., Ranganathan, S. 2008. Tandem duplication, circular permutation, molecular adaptation: how *Solanaceae* resist pests *via* inhibitors. *BMC Bioinformatics* 9(Suppl 1): S22.

Kunitz, M. 1945. Crystallization of a trypsin inhibitor from soybean. *Science* 101: 668–669.

Laskowsky, M., Kato, I., 1980. Protein inhibitors of proteinases. *Annual Review of Biochemistry* 49: 593–626.

Law, R.H.P., Zhang, Q.W., McGowan, S., Buckle, A.M., Silverman, G.A., Wong, W., Rosado, C.J., Langendorf, C.G., Pike, R.N., Bird, P.I., Whisstock, J.C. 2006. An overview of the serpin superfamily. *Genome Biology* 7: 216.

Lawrence, S.D., Novak, N.G., Ju, C.J.T., Cooke, J.E.K. 2008. Potato, *Solanum tuberosum*, defense against Colorado potato beetle, *Leptinotarsa decemlineata* (Say): microarray gene expression profiling of potato by Colorado potato beetle regurgitant treatment of wounded leaves. *Journal of Chemical Ecology* 34: 1013–1025.

Lee, J.S., Brown, W.E., Graham, J.S., Pearce, G., Fox, E.A., Dreher, T.W. Ahern, K.G., Pearson, G.D., Ryan, C.A. 1986. Molecular characterization and phylogenetic studies of a wound-inducible proteinase inhibitor I gene in *Lycopersicon* species. *Proceedings of the National Academy of Science USA* 83: 7277–7281.

Li, H.-M., Margam, V., Muir, W.M., Murdock, L.L., Pittendrigh, B.R. 2007. Changes in *Drosophila melanogaster* midgut proteins in response to dietary Bowman–Birk inhibitor. *Insect Molecular Biology* 16: 539–549.

Lipke, H., Fraenkel, G.S., Liener, I.E. 1954. Effect of soybean inhibitors on growth of *Tribolium confusum*. *Journal of Agricultural and Food Chemistry* 2: 410–414.

Liu, Y.L., Salzman, R.A., Pankiw, T. Zhu-Salzman, K. 2004. Transcriptional regulation in southern corn rootworm larvae challenged by soyacystatin N. *Insect Biochemistry and Molecular Biology* 34: 1069–1077.

Lopes, A.R., Juliano, M.A., Marana, S.R., Juliano, L., Terra, W.R. 2006. Substrate specificity of insect trypsins and the role of their subsites in catalysis. *Insect Biochemistry and Molecular Biology* 36: 130–140.

Lopes, A.R., Saro, P.M., Terra, W.R. 2009. Insect chymotrypsins: chloromethyl ketone inactivation and substrate specificity relative to possible coevolutional adaptation of insects and plants. *Archives of Insect Biochemistry and Physiology* 70: 188–203.

Lövei, G.L., Andow, D.A., Arpaia, S. 2009. Transgenic insecticidal crops and natural enemies: a detailed review of laboratory studies. *Environmental Entomology* 38: 293–306.

Maffei, M.E., Mithöfer, A., Boland, W. 2007. Insects feeding on plants: rapid signals and responses preceding the induction of phytochemical release. *Phytochemistry* 68: 2946–2959.

Major, I.T., Constabel, C.P. 2008. Functional analysis of the Kunitz trypsin inhibitor family in poplar reveals biochemical diversity and multiplicity in defense against herbivores. *Plant Physiology* 146: 888–903.

Mares, M., Fusek, M., Kostka, V., Baudys, M. 1991. Cathepsin D inhibitor from potato tubers (*Solanum tuberosum* L.). *Advances in Experimental Medicine and Biology* 306: 349–353.

Margis-Pinheiro, M., Zolet, A.C.T., Loss, G., Pasquali, G., Margis, R. 2008. Molecular evolution and diversification of plant cysteine proteinase inhibitors: new insights after the poplar genome. *Molecular Phylogenetics and Evolution* 49: 349–355.

Martinez, M., Diaz, I. 2008. The origin and evolution of plant cystatins and their target cysteine proteinases indicate a complex functional relationship. *BMC Evolutionary Biology* 8: 198.

Mazumdar-Leighton, S., Broadway, R. 2001. Transcriptional induction of digestive midgut trypsin in larval *Agrotis ipsilon* and *Helicoverpa zea* feeding on the soybean trypsin inhibitor. *Insect Biochemistry and Molecular Biology* 31: 645–657.

Mello, M.O., Tanaka, A.S., Silva Filho, M.C. 2003. Molecular evolution of Bowman–Birk type proteinase inhibitors in flowering plants. *Molecular Phylogenetics and Evolution* 27: 103–112.

Melo-Martén, I., Meghani, Z. 2008. Beyond risk—A more realistic risk-benefit analysis of agricultural biotechnologies. *EMBO Reports* 9: 302–306.

Mithöfer, A., Boland, W. 2008. Recognition of herbivory-associated molecular patterns. *Plant Physiology* 146: 825–831.

Mosolov, V.V., Grigoréva, L.I., Valueva, T.A. 2001. Involvement of proteolytic enzymes and their inhibitors in plant protection. *Applied Biochemistry and Microbiology* 37: 115–123.

Mosolov, V.V., Valueva, T.A., 2005. Proteinase inhibitors and their function in plants: a review. *Applied Biochemistry and Microbiology* 41: 227–246.

O'Callaghan, M., Glare, T.R., Burgess, E.P.J., Malone, L.A. 2005. Effects of plants genetically modified for insect resistance on nontarget organisms. *Annual Review of Entomology* 50: 271–292.

Odani, S., Koide, T., Ono, T. 1983. The complete amino acid sequence of barley trypsin inhibitor. *Journal of Biological Chemistry* 258: 7998–8003.

Pauchet, Y., Muck, A., Svatos, A., Heckel, D.G., Preiss, S. 2008. Mapping the larval midgut lumen proteome of *Helicoverpa armigera*, a generalist herbivorous insect. *Journal of Proteome Research* 7: 1629–1639.

Peiffer, M., Felton, G.W. 2009. Do caterpillars secrete 'oral secretions'? *Journal of Chemical Ecology* 35: 326–335.

Pompermayer, P., Falco, M.C., Parra J.R., P., Silva-Filho, M.C. 2003. Coupling diet quality and Bowman-Birk and Kunitz-type soybean proteinase inhibitor effectiveness to *Diatraea saccharalis* development and mortality. *Entomologia Experimentalis et Applicata* 109: 217–224.

Qi, R.F., Song, Z.W., Chi, C.W. 2005. Structural features and molecular evolution of Bowman-Birk protease inhibitors and their potential application. *Acta Biochimica et Biophysica Sinica* 37: 283–292.

Rancour, J.M., Ryan, C.A. 1968. Isolation of a carboxypeptidase B inhibitor from potatoes. *Archives of Biochemistry and Biophysics* 125: 380–383.

Rawlings, N.D., Barrett, A.J. 1993. Evolutionary families of peptidases. *Biochemical Journal* 290: 205–218.

Rawlings, N.D., Morton, F.R., Kok, C.Y., Kong, J. Barrett, A.J. 2008. *MEROPS*: the peptidase database. *Nucleic Acids Research* 36: D320–D325.

Rees, D.C., Lipscomb, W.N. 1982. Refined crystal structure of the potato inhibitor complex of carboxypeptidase A at 2.5Ä resolution. *Journal of Molecular Biology* 160: 475–498.

Reymond, P., Bodenhausen, N., Van Poecke, R.M.P., Krishnamurthy, V., Dicke, M., Farmer, E.E. 2004. A conserved transcript pattern in response to a specialist and a generalist herbivore. *The Plant Cell* 16: 3132–3147.

Rhodes, J.D., Thain, J.F., Wildon, D.C. 1996. The pathway for systemic electrical signal conduction in the wounded tomato plant. *Planta* 200: 50–57.

Rispe, C., Kutsukake, M., Doublet, V., Hudaverdian, S., Legeai, F., Simon, J.C., Tagu, D., Fukatsu, T. 2008. Large gene family expansion and variable selective pressures for cathepsin B in aphids. *Molecular Biology and Evolution* 25: 5–17.

Roberts, T.H., Hejgaard, J. 2008. Serpins in plants and green algae. *Functional and Integrative Genomics* 8: 1–27.

Ryan, C.A. 1974. Assay and biochemical properties of the proteinase inhibitor inducing factor, a wound hormone. *Plant Physiology* 54: 328–332.

Ryan, C.A. 1990. Protease inhibitors in plants: genes for improving defense against insects and pathogens. *Annual Review of Phytopathology* 28: 25–49.

Ryan, C.A., Balls, A.K. 1962. An inhibitor of chymotrypsin from *Solanum tuberosum* and its behavior toward trypsin. *Proceedings of the National Academy of Science USA* 48: 1839–1844.

Ryan, C.A., Bishop, P., Pearce, G., Darvill, A.G., McNeil, M., Albersheim, P. 1981. A sycamore cell wall polysaccharide and a chemically related tomato leaf polysaccharide possess similar proteinase inhibitor-inducing activities. *Plant Physiology* 68: 616–618.

Scheer, J., Ryan, C.A. 2002. The systemin receptor SR160 from *Lycopersicon peruvianum* is a member of the LRR receptor kinase family. *Proceedings of the National Academy of Science USA* 99: 9585–9590.

Schilmiller, A.L., Howe, G.A. 2005. Systemic signaling in the wound response. *Current Opinion in Plant Biology* 8: 369–377.

Schoonhoven, L.M., Jermy, T., van Loon, J.J.A. 1998. *Insect-plant Biology. From physiology to evolution*. 1st Ed., London: Chapman & Hall Publishing.

Shan, L., Li, C., Chen, F., Zhao, S., Xia, G. 2008. A Bowman-Birk type protease inhibitor is involved in the tolerance to salt stress in wheat. *Plant, Cell and Environment* 31: 1128–1137.

Silva, C.P., Terra, W.R., de Sá, M.F.G., Isejima, E.M., Bifano, T.D., Almeida, J.S. 2001a. Induction of digestive α-amylases in larvae of *Zabrotes subfasciatus* (Coleoptera: Bruchidae) in response to ingestion of the common bean α-amylase inhibitor 1. *Journal of Insect Physiology* 47: 1283–1290.

Silva, C.P., Terra, W.R., Lima, R.M. 2001b. Differences in midgut serine proteinases from larvae of the bruchid beetles *Callosobruchus maculatus* and *Zabrotes subfasciatus*. *Archives of Insect Biochemistry and Physiology* 47: 18–28.

Singh, A., Sahi, C., Grover, A. 2009. Chymotrypsin protease inhibitor gene family in rice: genomic organization and evidence for the presence of a bidirectional promoter shared between two chymotrypsin protease inhibitor genes. *Gene* 428: 9–19.

Solomon, M., Belenghi, B., Delledonne, M., Menachem, E., Levine, A. 1999. The involvement of cysteine proteases and protease inhibitor genes in the regulation of programmed cell death in plants. *The Plant Cell* 11: 431–443.

Srinivasan, A., Giri, A.P., Harsulkar, A.M., Gatehouse, J.A., Gupta, V.S. 2005. A Kunitz trypsin inhibitor from chickpea (*Cicer arietinum* L.) that exerts anti-metabolic effect on podborer (*Helicoverpa armigera*) larvae. *Plant Molecular Biology* 57: 359–374.

Srinivasan, T., Kumar, K.R.R., Kirti P.B. 2009. Constitutive expression of a trypsin protease inhibitor confers multiple stress tolerance in transgenic tobacco. *Plant Cell Physiology* 50: 541–553.

Staswick, P.E., Tiryaki, I. 2004. The oxylipin signal jasmonic acid is activated by an enzyme that conjugates it to isoleucine in *Arabidopsis*. *The Plant Cell* 16: 2117–2127.

Steffens, R., Fox, F.R., Kassell, B. 1978. Effect of trypsin inhibitors on growth and metamorphosis of corn borer larvae, *Ostrinia nubilalis* (Hübner). *Journal of Agricultural and Food Chemistry* 26: 170–174.

Steppuhn, A., Baldwin, I.T. 2007. Resistance management in a native plant: nicotine prevents herbivores from compensating for plant protease inhibitors. *Ecology Letters* 10: 499–511.

Stratmann, J.W., Ryan, C.A. 1997. Myelin basic protein kinase activity in tomato leaves is induced systemically by wounding and increases in response to systemin and oligosaccharide elicitors. *Proceedings of the National Academy of Science USA* 94: 11085–11089.

Tamhane, V.A., Giri, A.P., Sainani, M.N., Gupta, V.S. 2007. Diverse forms of Pin-II family proteinase inhibitors from *Capsicum annuum* adversely affect the growth and development of *Helicoverpa armigera*. *Gene* 403: 29–38.

Telang, M.A., Giri, A.P., Pyati, P.S., Gupta, V.S., Tegeder, M., Franceschi, V.R. 2009. Winged bean chymotrypsin inhibitors retard growth of *Helicoverpa armigera*. *Gene* 431: 80–85.

Terra, W.R., Ferreira, C. 2005. Biochemistry of digestion. In: *Comprehensive Molecular Insect Science*. Gilbert, L.I., Latrou, K., Gill, S.S., Eds. v. 4, pp. 171–224, Oxford: Elsevier.

Thomas, J.C., Adams, D.G., Keppenne, V.D., Wasmann, C.C., Brown, J.K., Kanost, M.R., Bohnert, H.J. 1995. *Manduca sexta* encoded protease inhibitors expressed in *Nicotiana tabacum* provide protection against insects. *Plant Physiology and Biochemistry* 33: 611–614.

Thomas, J.C., Wasmann, C.C., Echt, C., Dunn, R.L., Bohnert, H.J., McCoy, T.J. 1994. Introduction and expression of an insect proteinase inhibitor in alfalfa (*Medicago sativa* L.). *Plant Cell Reports* 14: 31–36.

Velkov, V.V., Medvinsky, A.B., Sokolov, M.S., Marchenko, A.I. 2005. Will transgenic plants adversely affect the environment? *Journal of Biosciences* 30: 515–548.

Villanueva, J., Canals, F., Prat, S., Ludevid, D., Querol, E., Avileès, F.X. 1998. Characterization of the wound-induced metallocarboxypeptidase inhibitor from potato. *FEBS Letters* 440: 175–182.

Visôtto, L.E., Oliveira, M.G.A., Guedes, R.N.C., Ribon, A.O.B., Good-God, P.I.V. 2009. Contribution of gut bacteria to digestion and development of the velvet bean caterpillar, *Anticarsia gemmatalis*. *Journal of Insect Physiology* 55: 185–191.

Voelckel, C., Baldwin, I.T. 2004. Generalist and specialist lepidopteran larvae elicit different transcriptional responses in *Nicotiana attenuata*, which correlate with larval FAC profiles. *Ecology Letters* 7: 770–775.

Vogel, H., Kroymann, J., Mitchell-Olds, T. 2007. Different transcript patterns in response to specialist and generalist herbivores in the wild *Arabidopsis* relative *Boechera divaricarpa*. *PLoS ONE* 2(10): e1081. doi:10.1371/journal.pone.0001081.

Wieczorek, M., Otlewski, J., Cook, J., Parks, K., Leluk, J., Wilimowska-Pelc, A., Polanowski, A., Wilusz, T., Laskowski Jr., M. 1985. The squash family of serine proteinase inhibitors. Amino acid sequences and association equilibrium constants of inhibitors from squash, summer squash, zucchini, and cucumber seeds. *Biochemical and Biophysical Research Communications* 126: 646–652.

Winterer, J., Bergelson, J. 2001. Diamondback moth compensatory consumption of protease inhibitor-transformed plants. *Molecular Ecology* 10: 1069–1074.

Xavier-Filho, J., Campos, F.A.P. 1989. Proteinase inhibitors. In: *Toxicants of Plant Origin*. Cheeke, P.R., Ed., v. 3, pp 1–27. London: CCR Press.

Yang, L., Fang, Z., Dicke, M., van Loon, J.J.A., Jongsma, M.A. 2009. The diamondback moth, *Plutella xylostella*, specifically inactivates Mustard Trypsin Inhibitor 2 (MTI2) to overcome host plant defence. *Insect Biochemistry and Molecular Biology* 39: 55–61.

Ye, S., Cech, A.L., Belmares, R., Bergstrom, R.C., Tong, Y.R., Corey, D.R., Kanost, M.R., Goldsmith, E.J. 2001. The structure of a Michaelis serpin–protease complex. *Nature Structural Biology* 8: 979–983.

Yoo, B.C., Aoki, K., Xiang, Y., Campbell, L.R., Hull, R.J., Xoconostle-Cazares, B., Monzer, J. Lee, J.Y., Ullman, D.E., Lucas, W.J. 2000. Characterization of *Cucurbita maxima* phloem serpin-1 (CmPS-1)—a developmentally regulated elastase inhibitor. *Journal of Biological Chemistry* 275: 35122–35128.

Zavala, J.A., Patankar, A.G., Gase, K., Hui, D., Baldwin, I.T. 2004. Manipulation of endogenous trypsin proteinase inhibitor production in *Nicotiana attenuata* demonstrates their function as antiherbivore defenses. *Plant Physiology* 134: 1181–1190.

Zhang, J., Wang, C., Qin, J. 2000. The interactions between soybean trypsin inhibitor and δ-endotoxin of *Bacillus thuringiensis* in *Helicoverpa armigera*. *Journal of Invertebrate Pathology* 75: 259–266.

Zhao, Y., Botella, M.A., Subramanian, L., Niu, X., Nielsen, S.S., Bressan, R.A., Hasegawa, P.M., 1996. Two wound-inducible soybean cysteine proteinase inhibitors have greater insect digestive proteinase inhibitory activities than a constitutive homolog. *Plant Physiology* 111: 1299–1306.

Zheng, S., Dicke, M. 2008. Ecological genomics of plant-insect interactions: from gene to community. *Plant Physiology* 146: 812–817.

Zhu-Salzman, K., Ahn, J.E., Salzman, R.A., Koiwa, H., Shade, R.E., Balfe, S. 2003. Fusion of a soybean cysteine protease inhibitor and a legume lectin enhances anti-insect activity synergistically. *Agricultural and Forest Entomology* 5: 317–323.

Zhu-Salzman, K., Salzman, R.A., Ahn, J.E., Koiwa, H. 2004. Transcriptional regulation of sorghum defence determinants against a phloem-feeding aphid. *Plant Physiology* 134: 420–431.

Zimmermann, M.R., Maischak, H., Mithöfer, A., Boland, W., Felle, H.H. 2009. System potentials, a novel electrical long-distance apoplastic signal in plants, induced by wounding. *Plant Physiology* 149: 1593–1600.

Chapter 14
Nutrient Acquisition and Concentration by Ant Symbionts: The Incidence and Importance of Biological Interactions to Plant Nutrition

Cynthia L. Sagers

Overview

The rationale of this chapter is to try to reach a clearer understanding of what is known about nutrient provisioning by insects to their plant hosts. Provisioning, particularly by ants, has arisen repeatedly throughout the history of the angiosperms, and functional traits related to nutrient acquisition are sometimes so odd and elaborate that they appear to be adaptations to ant symbiosis. What evidence exists for plant use of ant-provided nutrients, and have morphologies and behaviors evolved in response to ant symbiosis? My suspicions have been confirmed in that we know very little about how, when, or where ants provision their hosts, or what the ecological or evolutionary implications of these behaviors have been. What is known is reviewed here.

Introduction

Plants have evolved remarkable strategies for the more efficient capture of nitrogen and other limiting nutrients. Among these is the propensity of plants to manipulate the foraging behaviors of other organisms to their own benefit. These other organisms range in diversity and complexity from microbial endosymbionts to free-living animals, all of which concentrate nutrients in the environment and make them available to their host. As a rare, spectacular example, when the herb *Gunnera* L. (Gunneraceae) is grown on nutrient-poor medium, cyanobacteria colonize its stems and fix atmospheric nitrogen (Bonnett, 1990). Because new sources of nitrogen are available via symbiosis, *Gunnera* hosts can colonize and survive in nitrogen-poor habitats (Chiu et al., 2005). Some free-living insects, in particular, ants, appear to behave in a broadly similar fashion by collecting and depositing waste and debris around plant tissues.

Ecological Aspects of Nitrogen Metabolism in Plants, First Edition. Edited by Joe C. Polacco and Christopher D. Todd.
© 2011 by John Wiley & Sons, Inc. Published 2011 by John Wiley & Sons, Inc.

Presumably, plants that can capitalize on this newly available resource will be at a competitive advantage or be better able to persist in marginal environments.

Close association with ants has been documented in more than 400 plant species, the most conspicuous of which are the specialized ant-plants, or myrmecophytes, common in lowland tropical forests. Among myrmecophytes, ants frequently exchange nutrients and other services with their hosts for housing or food. Elaborate examples of specialized trading partnerships and biological markets involving nutrient exchange are discussed in this chapter and, although I broach the topic, few studies have explicitly examined nutrient provisioning from the perspective of population biology or evolutionary history. When waste and debris are made available for assimilation, plants that associate with ants may benefit. If host populations are genetically variable for traits related to nutrient assimilation, and if plants benefit from such traits, plants may evolve morphologies, physiologies, and behaviors to capitalize on ant-provided nutrients. Are the remarkable relationships that plants form with insects in nature adaptations evolved by natural selection, or are they merely the manifestation of opportunistic foraging? Have ant-plants evolved to optimize nutritional gains of ant association or are they merely taking advantage of a rich and available resource?

The primary objective of this chapter is to explore the ecology and evolution of plant association with insects as a means of acquiring limiting nutrients, especially N. Specifically this chapter will:

1. Review evidence for the uptake of ant-provided nutrients.
2. Discuss mechanisms of nutrient concentration and assimilation.
3. Introduce ecological factors that may influence the importance of ant-provided nutrients.
4. Evaluate evidence that ant associations are evolutionary adaptations for nutrient concentration and provisioning.

Evidence for Uptake of Ant-provided Nutrients

Ants concentrate nutrients in the environment (Pêtal, 1998; Wagner and Jones, 2006). Predatory, long-distance foraging and fungus-tending behaviors result in high concentrations of nutrients and crude organic matter at ant nests where workers nourish the brood in closed, often subterranean, chambers (Sternberg et al., 2007; Cammeraat and Risch, 2008). Nutrient-rich islands around ant nests are most evident in grasslands (Dauber and Wolters, 2008) and arid systems (Wagner et al., 2004) where ant mounds are a dominant feature and where nutrient enrichment can easily be detected (although effects are seen in systems as diverse as spruce forests (Stadler et al., 2006), tropical wet forests (Sternberg et al., 2007), and permanently disturbed roadsides (Farji-Brener and Ghermandi, 2009)). Burrowing and tunneling behaviors of ground-nesting ants cause further changes in soil bulk density, porosity, and texture (Czerwinski et al., 1971; Dlussky, 1981; Eldridge, 1993; Wang et al., 1996; Dean et al., 1997; Levan and Stone, 1983;), which lead to increased decomposition rates (Culver and Beattie, 1983; Pêtal and Kusinska, 1994) and improved soil fertility (Dean and Yeaton, 1993; McGinley et al., 1994; Pêtal, 1998; Lenoir et al. 2001; Ginzburg et al. 2008; Paris et al., 2008). Because of their impact on their surroundings, ants have been termed ecosystem engineers (Folgarait, 1998; Cammeraat and Risch, 2008; Dauber and Wolters, 2008) and have been credited with a number of beneficial ecosystem processes (Folgarait, 1998; Stadler et al., 2006). For example, Ginzburg and others (2008) conclude that harvester ants (*Messor* spp.) in the

Negev desert substantially increase soil fertility as a result of increased microbial biomass and microbial activity on ant nest-modified soils.

Evidence that plants take up nutrients provided by ants comes largely from isotope tracer studies of highly specialized ant plant associations. Specialized ant-plants, or myrmecophytes, offer ants housing (domatia) in the form of hollow stems, folded leaves, tangles of trichomes, sheathing petioles, or other hollowed structures such as leaves or stipules, thorns, and tubers (Beattie, 1985). Beccari (1884–1886) was the first to suggest that debris and waste concentrated by ants within domatia may be a source of nutrients for the plant. The first direct evidence of nutrient translocation from ant-provided debris involved the use of radioactive tracers and autoradiographs or Geiger counters to track the movement of radiolabeled nutrients into host tissues. Benzing and others (1976), working with Central American *Tillandsia* spp. (Bromeliaceae), found that soluble ^{45}Ca was translocated from basal bulbs of the plant to new leaf tissues. In a highly specialized ant-plant, Huxley (1978) confirmed with radiolabels an earlier suggestion by Janzen (1974) that two epiphytic genera, *Hydnophytum* Jack and *Myrmecodia* Jack (Rubiaceae), use ant-provided waste. She fed honey water containing ^{32}P-phosphate, ^{35}S-sulfate, or ^{35}S-methionine to specialist symbiotic *Iridomyrmex* ants and detected the tracers in new plant tissues. Rickson (1979) went one step further to show that ants in this system introduce non-host sources of nutrients to their hosts by feeding the ants irradiated flies and tracing the radiolabel to host plant tissues. Gay (1993a) showed that uptake of ant-borne radiolabel is not limited to higher plants but also occurs in an epiphytic fern (*Lecanopteris* Reinw.).

Presumably, it was these early radioactive tracer studies that led Thompson (1981) to conclude that epiphytic, but not geophytic, ant-plants used ant debris as a nutrient source. A contemporary suggestion (Risch et al., 1977), that shrubs in the neotropical genus *Piper* (Piperaceae) are fed by ants, was confirmed with stable isotope tracers only later (Fischer et al., 2002). In the years after Thompson (1981), stable isotope technologies were popularized in the field of ecology, and a succession of studies gave further evidence that not only do plants feed ants (Fisher et al., 1990; Sagers et al., 2000; Tillberg, 2004), but ants provide nutrients to their hosts (Cabrera and Jaffe, 1994; Treseder et al., 1995; Sagers et al., 2000; Fischer et al., 2002; Solano and Dejean, 2004; Watkins et al., 2008). Each of these examples involves plant species that possess highly modified, hollow structures in which ants dwell and from which nutrients can be absorbed. Moreover, more recent examples include a diverse group of terrestrial plants (neotropical trees, treelets, and shrubs), as well as epiphytes (vines, orchids, and ferns). Radioactive and stable isotopic tracer studies involving a group of taxonomically distant, tropical myrmecophytes provide clear evidence of the widespread adoption of ant refuse as a source of nutrients. The history of these discoveries is summarized in Table 14.1.

Tracer studies provide convincing evidence of myrmecotrophy (plants fed by ants), but the extent to which nutritional supplements from ants benefit the plant remains uncertain. To a population biologist, benefits are measured as fitness gains, or a boost in total lifetime reproduction. A growing number of studies have documented changes in plant fitness or fitness correlates with ant association (Huxley, 1978; Vasconcelos and Casimiro, 1997; Schupp, 1986; Heil et al., 2001), but few have tested explicitly whether measurable benefits are due to nutritional supplements provided by ant symbionts. One rare exception is an elegant study by Wagner (1997) of the fitness effects of *Formica purpilosa* (Hymenoptera: Formicidae) ground nests on the growth and reproduction of *Acacia constricta* Bentham (Fabaceae) in the deserts of the southwestern U.S. She used stable isotopes to confirm the movement of nutrients from *F. purpilosa* ant mounds to leaf tissues of nearby *Acacia* trees. She then observed that trees

Table 14.1. Evidence for nutrient provisioning by ants.

Plant order	Plant family	Host plant	Ant symbiont	Evidence for provisioning	Source
Gentianales	Asclepiadaceae	Dischidia major	Philidris	Stable isotopes	Treseder et al., 1995
	Rubiaceae	Myrmecodia cf. tuberosa	Diverse	[^{32}P] phosphate, [^{35}S)] sulfate or [^{35}S] methionine	Merrill, 1945; Huxley, 1978, 1986
Asparagales	Orchidaceae	Schomburgkia tibicinis	Camponotus, Crematogaster, Ectatomma	^{14}C-labelled glucose in honey	Rico-Gray et al., 1989
		Myrmecophila christinae		^{14}C-labeled glucose in honey	Rico-Gray et al., 1989
Poales	Bromeliaceae	Tillandsia caput-medusae	Diverse	^{45}Ca	Benzing, 1970
Myrtales	Melastomataceae	Tococa guianensis	Azteca	irradiated argenine	Cabrera and Jaffe, 1994?
		Maieta guianensis	Pheidole	^{15}N	Solano and Dejean, 2004
Piperales	Piperaceae	Piper fimbriulatum	Pheidole	^{15}N-labelled glycine	Fischer et al., 2003
		Piper obliquum	Pheidole	^{15}N-labelled glycine	Fischer et al., 2003
Rosales	Cecropiaceae	Cecropia peltata	Azteca	^{15}N	Sagers et al., 2000
Lamiales	Gesneriaceae	Codonanthe crassifolia	Crematogaster	Seedling growth	Kleinfeldt, 1978
Pteridophyte	Polypodiaceae	Lecanopteris sp.	Camponotus, Crematogaster, Iridomyrmex	^{14}C-labelled glucose, ^{86}Rb, ^{32}P	Gay, 1993a,b; Gay and Hensen, 1992
		Antrophylum lanceolatum	Pheidole	^{15}N, ^{13}C	Watkins et al., 2008

growing on *F. puprilosa* nests produced nearly twice as many seeds as plants without ant nests, presumably due to nutritional supplements from ant associates.

Fitness benefits have profound implications for the ecology and evolution of ant-plant systems, but do myrmecophytes benefit substantially from ant-provided nutrition? Fischer and others (2002) argue not. They confirmed a report by Risch and others (1977) that shrubs in the genus *Piper* assimilate and translocate ant-provided nutrients. ^{15}N-labelled glycine in sugar water solution was taken up from hollowed stems of *Piper* spp. at a rate of 10 mg day^{-1}, which is equivalent to about 1% of the aboveground demand. However, because *Pheidole* symbionts receive much of their food from their hosts, they argued that assimilation of ant debris constitutes little more than nitrogen recycling and that the small nutritional benefit provided by ants is unlikely to lend a substantial fitness advantage to the host plant. On the contrary, it could be argued that a small advantage, even on the order of 1%, may provide a competitive edge in extreme or marginal environments. Nitrogen acquisition is expensive; a minimum of 30 carbons is required for the assimilation of a single nitrogen atom, and up to 30% of a plant's total energy may be committed to ammonia-N uptake and assimilation (Schubert, 1986). Plants in the understory of tropical wet forests may receive as little as 1% full sun during the course of the day (Chazdon and Fetcher, 1984), and most understory plants are operating at compensation point (Baltzer and Thomas, 2007). *Piper* shrubs, by reducing the loss or recycling even small amounts of carbon or nitrogen, may gain an advantage, especially if there are few costs in doing so. Additional benefits may accrue if the forms of nitrogen made available to plants by ants differ in the physiological costs of uptake and assimilation. If, for example, the energetic costs of assimilation of amino-, protein, nitrate, or ammonium-N or NO_2-N differ among tissues, availability and assimilation of one or another form may reduce the total costs of nitrogen metabolism. Finally, as Beattie (1985) has argued, ants provide benefits beyond nutrient provisioning, such as gleaning leaves of epiphylls that may limit photosynthesis, ridding the plant of herbivores and their eggs, and trimming encroaching vegetation. By his accounting, nutrient provisioning may simply subsidize the overall costs of feeding and housing costly ant mutualists. To determine conclusively the fitness effects of nutrient provisioning by ants would require additional study that incorporates both ant exclusion and nutritional supplements. As it stands, further research on the evolutionary ecology of ant provisioning is clearly merited.

Plants appear to take advantage of nutrients made available by the behaviors of ants. In desert systems, where nutrients are limiting and ants dominate the landscape, free-living ants benefit the plants associated with them (Rissing, 1986; Wagner, 1997). The case of myrmecophytes is especially compelling, however, as an exchange of nutrients may contribute to mutual fitness benefits for ant and plant partners. If functional traits express genetic variability, mutual benefits for ant and plant partners could lead to specialization and coevolution. Beattie (1985) estimates there are more than 400 species of specialized ant-plants in nearly 20 plant families, largely in the tropics. This number is likely to be conservative as cryptic associations, such as between neotropical *Ocotea atirrensis* Mez. and Donn. Sm. (Lauraceae) shrubs and their timid, obligate symbiont, *Myrmelachista flavocotea* (Hymenoptera: Formicidae), may number in the tens or hundreds but are likely to escape study (Stout, 1979; McNett et al., 2010). Nutrient exchange may be an important part of each of these interactions, yet the nutritional relationships of ant-plants are understood in only a small number of the most highly specialized ant-plant species pairs. Even less is known of the contributions of third party partners in the interaction such as sap-feeding insects, bacteria, or fungi. Where it has been examined, more often than not, plants assimilate nutrients provided by ants. Whether nutrient provisioning translates to fitness benefits for the host is simply unknown. Additional long-term manipulative studies will provide a clearer understanding of the significance of ant provisioning to the physiology, ecology, and coevolutionary dynamics of these captivating systems.

Mechanisms of Nutrient Concentration and Assimilation

Speculation that colony waste might be beneficial to plants began more than a century ago (Schimper, 1888; Karsten, 1895), but isotopic tracer studies now leave little doubt that plants assimilate ant-provided nutrients. The forms of nitrogen made available to plants are not well known, however, nor is the relative importance of this nutrient source to plant growth and reproduction. Nitrogen assimilation in terrestrial higher plants most commonly begins with ammonium and/or nitrate present in the soil solution, where it is taken up by the root and either assimilated by a series of enzymatic steps, or transported as nitrate and assimilated remotely. Mechanisms of nitrogen uptake are most clearly understood in the root, but most of the plant body (leaves, stems, and roots) appears capable of nitrogen uptake. Hutchinson and others (1972), for example, claimed that up to 10% of a plant's nitrogen could be taken up from the atmosphere by the shoot. Although it is likely that the metabolism of ant-provided nitrogen is similar to that taken up from the soil solution, details, such as the dominant form of nitrogen taken up, uptake efficiencies by different plant parts, or the importance of these nutrients to plant growth and reproduction, have yet to be evaluated in the context of evolutionary ecology.

Among myrmecophytes, ants can be found nesting in modified leaves and stems. Specialized domatia have been fashioned from pseudobulbs, thorns, leaf blades, petioles, stipules, tubers, rhizomes, and primary stems and branches. Debris piles made up of dead ants, insect skeletons, frass, and carton are frequently packed into these closed chambers where fungi and bacteria decompose the waste into a thick paste of rotting sludge (Bailey, 1924; Janzen, 1972; Risch et al., 1977; Huxley, 1986). The amount of debris build-up can be substantial, and often maggots and grubs can be found feeding in this sludge. In hollow-stemmed *Cecropia* trees, specialized *Azteca* ants colonize successive stem chambers, leaving behind piles of colony waste (Beattie, 1989). As a result, debris builds up in a series of closed chambers, which, if gases or solutions can be transported into the plant body, may provide a reliable and renewable nutritional supplement to the plant. Nutrients trapped within ant domatia can conceivably move into tissues and supplement plant nutrition, but a number of questions remain:

1. Are plant surfaces adjacent to ant waste capable of absorbing essential nutrients?
2. What form of nitrogen is taken up?
3. Have plant tissues become specialized for the assimilation of ant-provided nutrients?

Each of these questions is addressed separately for three parts of the plant body, roots, leaves, and stems, and examples of each are discussed.

Nitrogen Uptake in Roots

Our understanding of nitrogen uptake in plants comes largely from studies of the root. Roots are highly specialized for the uptake and transport of nutrients, although they are rarely modified as ant domatia. (A rare exception is *Pachycentria* (Melastomataceae) (Beattie, 1985)). Ammonia, uric acid, amino acids, and proteins that accumulate in ground nests of ants may become available to plants that are rooted there or growing nearby. The traditional perspective of nitrogen assimilation by roots holds that organic N is decomposed by the action of fungi and bacteria to ammonia, then further modified to NH_4^+ or NO_3^- before uptake. Ammonium and nitrate enter the root via carriers in the root hair and are either fixed in the root (assimilated into C compounds), or transported as NO_3^- and fixed in aerial portions of the plant. The ratio of NO_3^- to NH_4^+ uptake at the root hair is largely influenced by environmental factors that include the pH of the soil, rooting zone, temperature, and rooting depth (Bergback and Borg, 1989; Henry and Raper, 1989; Vessey et al., 1990). Uptake of ammonium-N is a complex, highly regulated, and energetically costly process. Until recently it was believed that terrestrial plants depend absolutely on mineral N for metabolism, but now it is broadly accepted that plants assimilate a number of forms of nitrogen, including amino acids (Raab et al., 1996; Näsholm et al., 1998) and proteins (Paungfoo-Lonhienne et al., 2008). At present, we know little about the uptake of alternate forms of nitrogen, how uptake efficiency differs among nitrogen species and between plant tissues, or what ecological factors, other than soil pH, water content, and other physical factors, influence the availability of each nitrogen species to plants. (see Chapter 4 on plant N-uptake facilitated by ubiquitous ectomycorrhizal fungi.)

Specializations by Which Plants Take Up Nutrients Through the Root

Colony Waste and Ground-nesting Organisms Nutrients provided by ground dwelling ants are known to influence plant growth, fitness, and even the structure of plant communities in a broad range of ecosystems including arid lands (Wagner, 1997; Wagner and Jones, 2006) and tropical forests (Farji-Brener and Medina, 2000; Sternberg et al., 2007). Plant associations with ground-nesting ant species appear largely opportunistic, although Rissing (1986) reported that some plant species of the U.S. desert southwest were found only on mounds of seed-harvesting ants. Whether a restricted distribution such as this is sufficient evidence of ecological specialization or evolutionary adaptation remains open to question.

Ant Gardens Ant gardens were first described by Ule (1901) who observed clusters of epiphytic plants on ant nests in tropical forest canopies. Ants construct aerial nests of accumulated organic material, waste, and debris, which are tacked together with exudates from ant salivary glands. This "carton" is used by ants as nest material or to construct protected passageways on plant stems. A small number of plant species in the Araceae, Gesneriaceae and in the genus *Peperomia* (Piperaceae) are found only in ant gardens and may be dispersed by their ant symbionts (Madison, 1979). Their seeds germinate and take root in the aerial carton, which strengthens and adds durability to the ant nest (Longino, 1987). In turn, epiphytes produce more seeds when rooted in the carton than when growing away from ant nests (Kleinfeldt, 1978). Because both partners benefit from the association, the interaction may evolve into a specialized symbiosis if functional traits involved in the association are heritable. However, epiphytic plants appear neither structurally nor physiologically adapted to life on ant carton, nor has evolutionary specialization of epiphytic species to ant nests been fully substantiated. Rather, these plants may simply be taking advantage of an abundant and rich resource found in tropical forest canopies (Longino, 1987). There is clearly a need for experimentation to determine the specialized nature of these associations.

Adventitious Roots Plants forage opportunistically aboveground by forming adventitious roots, roots that develop from tissues other than the root vascular cambium. In some instances, adventitious roots form within ant domatia where resources are concentrated and readily available. One spectacular example is the case of *Dischidia major* (Vahl) Merr. (Asclepiadaceae), an epiphytic vine of Malaysia, that produces sac-like leaves frequently inhabited by ants (Figure 14.1). Adventitious roots grow into the sac through a small pore at the base of the petiole and proliferate where ant debris has accumulated. Stable isotope evidence suggests that plants receive carbon and nitrogen supplements from their symbionts by way of the domatia (Treseder et al., 1995).

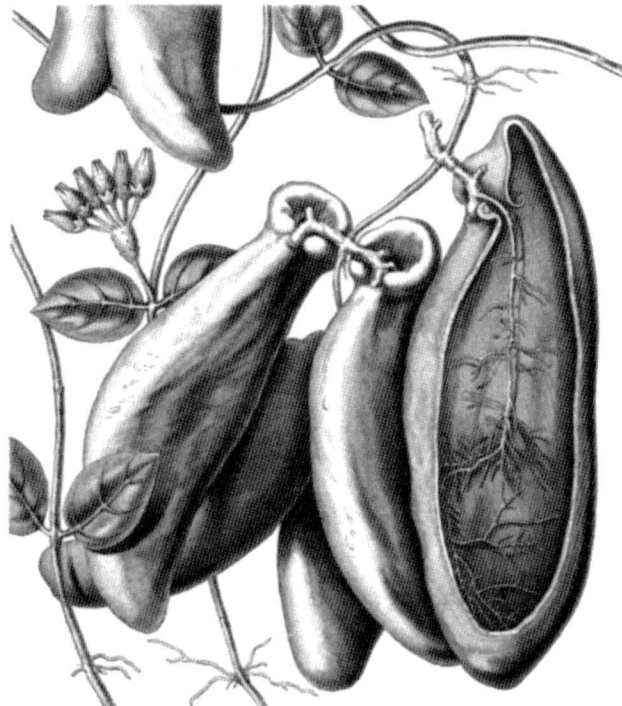

Figure 14.1. Line drawing of *Dischidia major* (Vahl) Merr. (Asclepiadaceae) illustrating adventitious root growth in an ant domatium.

Dischidia is a fascinating case of morphological modification to capitalize on symbiosis with ants, but because the origin of adventitious roots likely predates the origin of ants, the formation of adventitious roots may be merely a generalized plant response to nutrient availability rather than an adaptation to ant symbiosis. Once again, whether these traits are adaptations is unknown and requires further investigation.

Roots of terrestrial plants are efficient at nutrient uptake, and there appear to be no particularly specialized physiologies or morphologies of roots to be found among ant-plants. Moreover, roots are rarely modified as ant domatia (Beattie, 1985). The remainder of the plant body is a different story.

Nitrogen Uptake in Leaves

Leaves are frequently modified as domatia and show a remarkable capacity to assimilate nitrogen. At least three forms of nitrogen are known to enter leaves via the stomata: NH_3, NH_4^+, and NO_2. Atmospheric NH_3 enters the leaves of higher plants almost exclusively through the stomata and is dissolved in water of the leaf interior to form NH_4^+ (Hutchinson et al., 1972; Van Hove et al., 1987a; Krupa, 2003). NH_3 taken up by the leaf rapidly reaches equilibrium with NH_4^+ in the water film of the mesophyll tissue where NH_4^+ strongly dominates NH_3 and is assimilated. In bean leaves, NH_3 flux increases linearly with NH_3 concentration, meaning there is little mesophyll resistance to NH_3 uptake (Van Hove et al., 1987b). As long as ambient NH_3 concentrations exceed the mesophyll concentration, NH_3 uptake by the leaf will occur. Because the cuticle is nearly impermeable to NH_3 (Van Hove et al., 1987b, 1989), NH_3 moves into the leaf by way of the stomata. Uptake of NH_3, therefore, depends directly on stomatal conductance and indirectly on stomatal response to microclimate (e.g., radiation, air temperature, and plant water status).

NH_4^+ accumulates on leaf surfaces by both dry and wet deposition (Gmur et al., 1983; Van Hove et al., 1989; Bobbink et al., 1992). Fine particles and gases enter the leaf through stomata, in the same manner as NH_3 (Chevone et al., 1986). The question of how NH_4^+ enters the mesophyll has not yet been fully answered, but Wilson (1992) believes it is likely to occur by diffusion. (see Chapter 10 for a discussion on nutrient exchanges on the leaf surface.)

NO_2 uptake in leaves has been clearly demonstrated in a diverse group of plant species. Uptake is correlated with stomatal conductance and is presumably via stomata (Porter et al., 1972; Okano et al., 1986; Ammann et al., 1995; Sparks et al., 2001). Moreover, uptake is likely to occur by diffusion, which depends upon ambient NO_2 concentration and stomatal conductance to NO_2 transport (Klepper, 1979; Johansson, 1989; Theone et al., 1991; Weber and Rennenberg, 1996). It has been proposed that NO_2 dissolves in the leaf apoplast to form NO_2^- and NO_3^- (Zeevaart, 1976; Wellburn, 1990; Gessler and Rennenberg, 1998; Sparks, 2009). Both forms are transported into the cell where NO_3^- is rapidly reduced to NO_2^- by the enzyme nitrate reductase (Stulen and ter Steege, 1995). NO_2^- is then transported into the chloroplast and reduced to NH_4^+ by nitrite reductase and eventually assimilated via the GS-GOGAT cycle (Lea et al., 1994; Sparks, 2009). Plant emission of NO_2 at low atmospheric concentrations has also been documented, suggesting an NO_2 compensation point (Johansson, 1989; Rondon et al., 1993; Weber and Rennenberg, 1996; Wildt et al., 1997). Beyond these observations, the physiological controls and environmental influences of NO_2 uptake and loss by leaves are not well known (Sparks et al., 2001; Sparks, 2009).

Specializations By Which Plants Take Up Nutrients Through Leaves

Leaf Lamina Leaf domatia frequently form by curling up or down, expanding, or folding the leaf blade. The leaves of many melastomes (Melastomataceae: *Tococa, Maieta, Myrmidone, Clidemia, Conostegia, Ossaea, Henriettella, Allomaieta,* and *Blakea*) have been modified by invagination and expansion of the abaxial surface of the leaf blade to make a pair of pouches that rest on the leaf petiole. Domatia may also form by curling up or under the margin of the leaf blade. For example, leaf domatia of *Duroia* and *Remijia* (Rubiaceae) result from curling up (*Duroia*) or under (*Remijia*) the leaf margin on the edge of the petiolar insertion. In this case, and in the case of leaf pouches, leaf epidermis lines the domatium. Because N-gases move into the leaf via stomata, stomata within the domatia provide a direct route for nitrogen uptake. (see the discussion on tank bromeliads in Chapter 10.)

Petiole Domatia occasionally form by folding of the leaf petiole. *Piper* (Piperaceae), a species-rich genus of the neotropics, includes three specialized ant-plants that all form domatia this way. The appressed margins of the petiole and the sheathing leaf bases of *Piper* form tubular structures that are lined with epidermis (Risch et al., 1977; Svoma and Morawetz, 1992; Fischer et al., 2003). Specialized *Pheidole bicornis* (Formicidae: Mymicinae) ants, as well as a suite of facultative ant associates, establish colonies in these domatia and subsequently form galleries throughout the *Piper* stem. Labeling studies show co nvincingly that nutrients introduced to the chamber are rapidly translocated to new plant tissues (Fischer et al., 2003). One possible route of uptake is via stomata or lenticels of the sheathing petiole (Fischer et al., 2003), but another route is absorption from the plant stem, which is also inhabited by *Pheidole*.

Trichomes The leaf surface is often modified with uni-, or multicellular trichomes, extensions of the epidermis. Benzing and others (1976, 1991) found that the trichomes of aerial bromeliads are specialized for nitrogen uptake. Trichomes line the expanded leaf bases of *Tillandsia* where organic matter accumulates through the action of ant associates. Benzing and others (1976) confirmed with autoradiographs that radioactive calcium introduced to the leaf bases was taken up from these structures and translocated to expanding leaf tissues. Leaf domatia are found in a number of plant families including the Bromeliaceae, Fabaceae, Melastomataceae, Chrysobalanaceae, and Moraceae, and in each case the domatium is lined with epidermis. Stomata are a direct route for the uptake and assimilation of atmospheric nitrogen by leaves. Leaves take up at least three nitrogen species: NO_2, NH_3, and NH_4^+, and although the mechanism of uptake is thought to be by diffusion, the details of uptake and assimilation have yet to be fully documented. One suspected route of nitrogen uptake is via stomata or lenticels, but this has not yet been demonstrated directly. Additional modifications, such as specialization of leaf trichomes, are known, but the morphological specialization and physiological details of uptake have received little attention. Each of these processes assumes that all organic matter is broken down to an inorganic form to be assimilated, implying further cooperation with a microbial partner.

Nitrogen Uptake in Stems

Plant stems have been highly modified as ant domatia and are capable of assimilating gases and solutes. Early studies by Billings and Godfrey (1967) found that hollow stems of a non-ant plant, *Mertensia ciliata* (James ex Torr.) G. Don. (Boraginaceae), fixed CO_2 respired by root and stems cells. Osmond and others (1987) made a similar observation in the desert ephemeral *Eriogonum inflatum* Torr. and Fém. (Polygonaceae). Janzen (1974) first tested the idea with a myrmecophyte by tracing the absorption of India ink from the ant chambers of *Hydnophytum* to the parenchyma of the plant's hollowed tuber. He reasoned that if NO_3^- or NH_4^+ can cross the cell wall, it will be transported across the plasma membrane and be assimilated by the same pathways as those for nitrate and ammonium transported from the roots.

Where hollow spaces in the plant stem are lined with moist parenchyma cells it is easy to imagine uptake, translocation, and assimilation of ant-provided nutrients (Raven, 1996; Rickson, 1979; Fischer et al., 2003; McNett et al., 2010). Often, however, hollowed stems are lined with horny schlerynchyma that appears impenetrable to gases and solutes, as is the case in the ant-plants *Cordia* (Boraginaceae), *Cecropia*, (Moraceae), and *Macaranga* (Euphorbiaceae). In the case of *Cecropia*, the pith is surrounded by a thin layer of extremely hard schlerynchyma. The cambium, and therefore all woody growth, is external to this capsule of horny tissue (Bailey, 1922; Longino, 1991). The route to cambial tissues may be via ants themselves as ants often chew away at the stem lining (Beattie, 1989). This may be the case for *Cecropia*, as there is little question that these trees assimilate ant-provided nutrients (Sagers et al., 2000).

Stems have been variously modified into domatia from thorns (Janzen, 1966, 1969), tubers (Huxley, 1991), pseudobulbs (Rico-Gray et al., 1989), rhizomes (Gay and Hensen, 1992), and stem tips (Bailey, 1924), yet few details about nutrient uptake from these structures are known.

Of the more than 400 described myrmecophytes the importance of nutrient exchange has been quantified in one and investigated in only a handful. The plant body, and particularly the shoot, has been highly modified to accommodate ant colonies, and many of these structures appear capable of nutrient uptake. Structural modifications may promote nitrogen assimilation by making new sources of nitrogen available to the host or reducing the loss of nitrogen from the plant. Ants make nitrogen available in the form of uric acid (Krupka and Towers, 1959), amino acids and proteins (McKee, 1962; Benzing, 1970). Moreover, particulate matter, solid waste, and insect debris may be slowly made available to the plant following microbial decomposition (Benzing, 1970). Nitrogen uptake by plant cells is likely via carrier systems in the root, through stomata in leaves and by diffusion into parenchyma in plant stems. Plant parts are likely to differ in their affinities for various nitrogen species, but much of this research, particularly for NO_2 and organic nitrogen, has only just begun (Näsholm et al., 1998, 2009). Because the availability of ant-provided nutrients depends on the behaviors of animals, the relative importance of ant-provided nutrients will vary among environments, among species of ant associate, and among the forms of nitrogen made available.

Ecological Factors That Influence the Importance or Quality of Ant-provided Nutrients

Plants assimilate nitrogen provided by ants, but precisely what factors affect uptake and how this benefit varies spatially or temporally are unknown. Nitrogen uptake is in part a function of the ambient concentration of plant-available nitrogen (Evans, 2001). Therefore, we may find the sources of variability are those factors that affect either the quality or quantity of ant waste. We know that larger ant colonies provide relatively more nitrogen than small colonies, for example. Moreover, if ant specialists rely more heavily on their host for food than generalists, as suggested by Schemske (1983), then specialist ant species may introduce fewer new sources of nitrogen to the plant. Moreover, species pairings shift over ecological gradients as both ant and plant demonstrate habitat preferences. Because ant-plant associations vary geographically, there may be strong ecological influences on the relative importance of ant-provided nutrients. The following discussion, largely speculative, is of ecological factors that may influence the importance of ant provisioning.

Most of the nitrogen made available by myrmecotrophy is probably recycled from the plant itself (Fischer et al., 2003). Frass that accumulates in ant domatia is largely metabolized sap, nectar, or food bodies processed by ants but produced by the host. Although the degree of recycling is likely to vary, there is no evidence in any case that ant and plant rely entirely on each other for food. Complete reliance would constitute a closed system, one in which nutrients are recycled between partners and no new nutrients are introduced. Nitrogen recycling within a closed system drives isotopic discrimination between partners toward zero. However, in all studies that have considered to be the diets of specialized plant-ants, ants are enriched in nitrogen isotopic composition relative to their hosts (Sagers et al., 2000; Tillberg, 2004; Tillberg et al., 2006; Trimble and Sagers, 2004; Fiedler et al., 2007). That is, ant-plant systems are open, and both partners use sources of nutrients outside of the association. Less specialized ants, ants that feed on sources other than their hosts, are more likely to introduce new sources of nitrogen to the plant. Isotopic enrichment may shift further away from the host as novel items are added to the ant diet, as has been found along ecological gradients (Trimble and Sagers, 2004) and among species pairs (Fisher et al., 1990; Tillberg et al., 2004; Tillberg et al., 2006). For example, Tillberg (2004) found large differences (up to 4.5±) in nitrogen isotopic composition among ant symbionts of *Cordia alliadora* (Ruiz & Pav.) Oken (Boraginaceae) (Figure 14.2). Presumably, "predatory ants," those with the highest discrimination values, are more likely to introduce new nitrogen to their host (Tillberg, 2004). The importance of ant provisioning in finding new sources or nutrients, then, is likely to vary among species of ant symbionts and among habitats. Systematic study of variation in isotopic discrimination is rare, however, and interspecific variability in nutrient provisioning has yet to be fully explored.

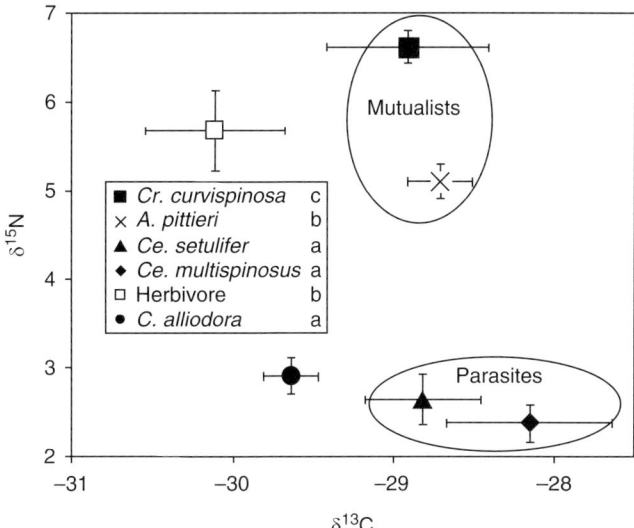

Figure 14.2. $\delta^{13}C$ by $\delta^{15}N$ of ant symbionts of *Cordia alliadora*. Two species, *Azteca pittieri* and *Crematogaster curvispinosa*, are classified by Tillberg as mutualists, while two others, *Cephalotes setulifer* and *C. multispinosus*, are classified as parasites based on their carbon and nitrogen isotopic composition. Potential diet sources, *C. alliadora* and plant herbivores, are shown for comparison. Different letters in the figure legend indicate significant differences among species pairs in K nearest neighbors test (Tillberg, 2004).

Habitat preferences appear to be the dominant driver of ant-plant species associations (Davidson and Fisher, 1991; Yu and Davidson, 1997; Michelangeli, 2003). For example, Davidson and Fisher (1991) found that *Azteca* and *Cecropia* species pairings differed predictably among forest environments. Because plant and ant species pairs shift across their distributions (Longino, 1989; Koptur, 1991) and because ant symbionts may differ in the amount of new nitrogen they bring to the host, we expect the benefits of ant provisioning may vary spatially and temporally (Thompson, 1988; Beattie, 1991; Koptur, 1991; Rico-Gray and Castro, 1996; Rico-Gray et al., 1998). As yet, there are few data to support this supposition.

Shifts in isotopic enrichment between ant and host along complex gradients signal shifts in ant diet. One study that explicitly measured geographic variation in isotopic enrichment of ant to host found it (Trimble and Sagers, 2004). If ant-host associations are influenced by environment, then the amount or quality of nitrogen provided by symbionts is likely to vary as well (Trimble and Sagers, 2004). Because ecology influences the nature of ant-plant associations, the relative importance of myrmecotrophy may vary geographically. However, there are few data available to inform this discussion.

Evidence That Ant Associations Are Evolutionary Adaptations for Nutrient Concentration and Provisioning

Have ant-plants evolved adaptations for nutrient acquisition? An adaptation can be recognized as any morphology, physiology, behavior, or reproductive quality that is characteristic of a population, has arisen by natural selection, and that confers a reproductive advantage to its bearer (Endler, 1986). Several strategies have emerged to investigate whether a character is an adaptation, but it has been argued that population studies provide the most compelling evidence (Endler, 1986). Adopting this perspective, characters are likely to be adaptations if the following are true: (1) they are heritable (i.e., genetic), (2) they confer a fitness advantage, and (3) they are characteristic of a population.

Agronomic studies provide ample evidence that crop varieties differ in nitrogen assimilation rate and nitrogen use efficiency, but most of these studies focus on uptake by roots from the soil solution. In a rare study of uptake by other plant tissues, Ashraf and others (2003) demonstrated that foliar uptake of labeled N differed among varieties of rice (*Oryza sativa* (Poaceae). Evidence for genetic variation in foliar nitrogen uptake argues for the potential for this trait to evolve. In natural systems, and particularly among myrmecophytes, data for heritable variation in foliar uptake or uptake by the plant shoot are rare. One study clearly demonstrates interspecific differences in myrmecotrophy among three species of tropical myrmecophytes that host the same ant species (Solano and Dejean, 2004). To the best of this author's knowledge, however, there are no population studies of genetic variability in functional traits related to the uptake of ant-provided nutrients.

Cost-benefit analysis of functional traits provides additional evidence for adaptation. Where a heritable trait proves beneficial in terms of energy efficiency, plant growth, or ultimately, reproduction, it may evolve by natural selection. Exclusion studies that examine host benefit in the absence and presence of an ant symbiont have repeatedly demonstrated the benefits of ant symbiosis (Vasconcelos and Casimiro, 1997; Schupp, 1986; Wagner, 1997). For example, Schupp (1986) showed that *Cecropia* saplings experienced significantly less herbivory when growing with ant symbionts than without. Benefits in the presence and costs in the absence of ant symbionts suggest that a character has evolved as an adaptation to ant association, given that the character is heritable. Although there is ample evidence that ant symbionts benefit their hosts, benefits are frequently attributed to functions other than nutrient provisioning, such as anti-herbivore defense, pruning away encroaching vegetation, or cleaning leaves of colonizing epiphylls. Genetic variation in fitness benefits of ant provisioning have yet to be examined in any myrmecophytic system.

Morphological specialization provides further evidence that characters have evolved as adaptations. These are the traits that "…just so excite the imagination" (Darwin, 1859). It is tempting to invoke evolution by natural selection to explain the amazing finger-like protuberances that develop within the leaf pouches of *Maieta* (Belin-Depoux and Bastien, 2002; Solano and Dejean, 2004) or the rough and smooth chambers of *Myrmecodia* tubers: smooth chambers in which ant colonies reside, rough chambers where debris piles accumulate. These knobs, protuberances, and projections are believed to increase the surface area of ant-inhabited chambers and increase the rate of absorption of ant-provided nutrients (Solano and Dejean, 2004). At present, population biology studies of these systems have not yet been done; so whether these traits have adaptive significance remains speculative. A final bit of evidence that might be used to argue adaptation for ant provisioning is that plants that use ant-provided nutrients have radiated following the adoption of ant symbionts. A slight physiological advantage over the long term may allow niche expansion and the colonization of marginal environments, as suggested by Thompson (1981). For example, *Hydnophytum* is found growing at the edge of the earth on red mangrove stems amid salt spray, intense light, and in the virtual absence of soil (Huxley, 1978). This and other rubiaceous genera have diversified into more than 40 species of specialized myrmecophytes, all with a similar growth form, all with similar ant associates, all in marginal environments, and all presumably absorbing ant-provided nutrients. Rapid speciation following the development of an evolutionary innovation, such as ant association,

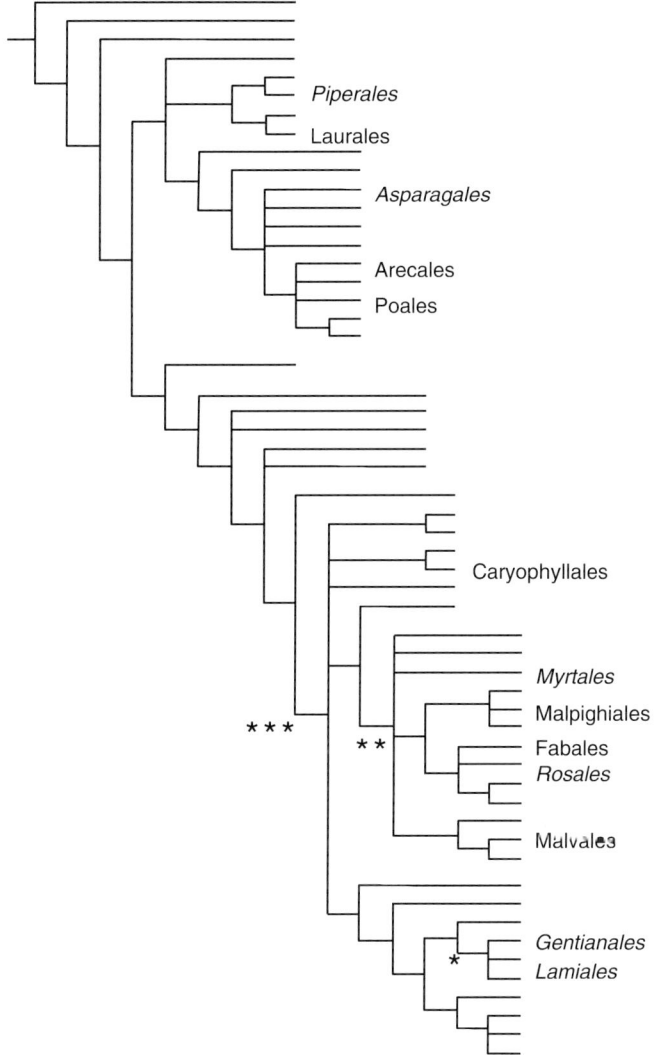

Figure 14.3. Phylogenetic distribution of specialized ant-plants within a flowering plant phylogeny. Nominal plant families contain at least one species of specialized ant-plant. Italics indicates evidence of ant provisioning. Phylogeny adapted from The Angiosperm Phylogeny Group (2003). Positions of three families in this phylogeny are uncertain. Their relative positions are indicated by asterisks: Boraginaceae(*), Vitaceae (**), and Dilleneaceae (***). Modified from Angiosperm Group (2002).

has been interpreted as adaptive evolution to novel environments. To date, phylogenetic analysis of ant provisioning has not yet been attempted in this group, or any others and therefore the question of adaptive radiation following ant association remains open.

Ant plants and evidence of ant-fed plants are scattered throughout the evolutionary history of the flowering plants (Figure 14.3). Considering the widely scattered distribution of ant-plants among angiosperm orders, it is clear that the ant-plant habit has arisen multiple times and that traits linked to ant association have arisen independently. Moreover, modern associations appear to have arisen by shifts of ant species onto pre-adapted hosts, rather than by cospeciation via coevolution (Davidson and McKey, 1993). For example, recent evidence suggests that the ant habit originated two to four times and was lost one to three times in the paleotropical genus *Macaranga* (Davies et al., 2001), although this interpretation has not been entirely supported (Itino et al., 2002; Bänfer et al., 2006). On the ant side, Ward (1991) estimates that within the ant sub-family Pseudomyrmicinae host plant specialization has arisen 12 times on three distinctive plant groups. Because the ant-habit is dispersed throughout the angiosperm phylogeny, it is unlikely that specialized domatia are homologous. Rather, structures that house ants have arisen repeatedly, and presumably independently, multiple times in the evolution of higher plants. All vegetative structures have been modified to form ant domatia. However, the mechanisms by which nutrients enter the plant and are subsequently assimilated are constrained, so pathways of nutrient uptake are most likely convergent among plant lineages. In any case, the evolutionary histories of ant-plant association have been hypothesized for only a few groups, and at present little more can be said regarding the evolution of ant provisioning.

Conclusions

Because work on ant provisioning of plants has focused on only a few specialized ant-plant associations, the ideas presented here are largely speculative. We do know, however, that isotope approaches have proven tremendously useful in the study of nutrient translocation and assimilation. Isotopic tracers have verified the phenomena of uptake, translocation, and assimilation and, in one case, made possible an estimate of the rate of uptake of ant-provided nutrients. The stable isotope work, in combination with studies of plant growth, provide convincing evidence that nutrient provisioning by ants can result in substantial ecological benefits to the host. Moreover, the broad distribution of ant provisioning throughout the evolutionary history of higher plants suggests frequent and independent origins of the ant habit. This review has revealed a wealth of compelling research problems surrounding the mechanisms of nutrient uptake and the ecological and evolutionary significance of ant provisioning. First and foremost is to develop a better understanding of the mechanisms and pathways of nitrogen assimilation from plant shoots. Addressing questions, such as what forms of nitrogen are taken up, the efficiency of uptake, and the pathways by which nitrogen moves out of domatia, are fundamental to understanding the evolution of ant provisioning. A second overarching question is whether nutrient assimilation shown by isotopic tracer studies translates into whole plant benefits such as greater likelihood of survival or reproduction. Are plants with ants better able to survive in marginal habitats? From an evolutionary perspective, it is essential to understand the relative importance of ant-provided nutrients, much like Fischer and others (2002) have done. Their approach, to estimate the proportion of daily nitrogen demand that is satisfied by ant provisioning, must be translated into fitness units to be meaningful in an ecological or evolutionary context. Once this approach is established, we can begin to address questions of the evolutionary ecology of ant provisioning: Do the net benefits of nutrient provisioning differ among ant symbionts? Among habitats? Is this variation heritable? A research framework grounded in evolutionary ecology will reveal patterns of ecological variation in nutrient provisioning that is the raw material of evolutionary change (Thompson, 1981). Finally, when we understand ant provisioning at the population level, we may begin to examine in a phylogenetic context the evolution of relevant plant traits and its significance in the coevolution of ant-plant associations.

As summarized by Thompson (1981), nutrient provisioning by ant symbionts appears far more commonly in tropical than temperate systems and among species with a perennial life history. Specialized associations appear not to be limited to nutrient-poor sites, however, but rather more generally to resource-poor sites such as the low-light environment of tropical forest understory. Moreover, the ant habit is not limited to epiphytes, but can be found in many life forms. Much remains to be done before we understand the evolution of ant provisioning, its benefits to plant fitness, or its influence on the structure of plant communities. When these studies are undertaken, they should include a careful look at temporal and spatial variability to understand more clearly the ecological and evolutionary influence of nutrient provisioning by ants.

References

Ammann, M., Von Ballmoos, P., Stalder, M., Suter, M., Brunold, C. 1995. Uptake and assimilation of atmospheric NO_2-N by spruce needles (*Picea abies*): a field study. Water Air Soil Pollution 85: 1497–1502.

Angiosperm Phylogeny Group. 2002. An update of the angiosperm Phylogeny Group classification for the orders and families of flowering plants: APG II. Botanical Journal of the Linnean Society 144: 399–436.

Ashraf, M., Mahmood, T., Azram, F. 2003. Translocation and recovery of ^{15}N-labelled N vderived from foliar uptake of $^{15}NH_3$ by rice (*Oryza sativa* L.) cultivars. Biology and Fertility of Soils 38:257–260.

Bailey, I.W. 1922. Notes on neotropical ant-plants. I. *Cecropia angulata*, sp. Nov. Botanical Gazette 74: 369–91.

Bailey, I.W. 1924. Notes on neotropical ant-plants. III. *Cordia nodosa* Lam. Botanical Gazette 77: 32–49.

Baltzer, J.L., Thomas, S.C. 2007. Determinants of whole-plant light requirements in Bornean rain forest tree saplings. Journal of Ecology 95: 1208–1221.

Bänfer, G., Moog, U., Fiala, B., Mohamed, M.A., Weising, K., Blattner, F.R. 2006. A chloroplast genealogy of myrmecophytic *Macaranga* species (Euphorbiaceae) in Southeast Asia reveals, hybridization, vicariance and long-distance dispersals. Molecular Ecology 15: 4409–4424.

Beattie, A.J. 1985. The Evolutionary Ecology of Ant-Plant Mutualisms. Cambridge University Press, Cambridge, England. p 182.

Beattie, A.J. 1989. Myrmecotrophy: plants fed by ants. Trends in Ecology and Evolution 4: 172–176.

Beattie, A.J. 1991. Problems outstanding in ant-plant interaction research. In: Huxley, C.R., Cutler, D.F., Eds. Ant-Plant Interactions. Oxford University Press, Oxford, U.K., pp. 559–576.

Beccari, O. 1884–86. Piante Ospitatrici. Malesia vol. 2. Bibografia del istituto Sordo Muti: Genoa.

Belin-Depoux, M., Bastien, D. 2002. Regards sur la myrmécophilie en Guyane française: Les dispositifs d'absorption de *Maieta guianensis* et la triple association *Philodendron*-foumis-*Aluerodes*. Acta Botanica Gallica 149:299–318.

Benzing, D.H. 1970. An investigation of two bromeliad myrmecophytes: *Tillandsia butzii* Mez., *T. caput-medusae* E.Morren and their ants. Bulletin of the Torrey Botanical Club 97:109–115.

Benzing, D.H., Henderson, K., Kessel, B., Sulak, J. 1976. The absorptive capacities of Bromeliad trichomes. American Journal of Botany 63:1009–1014.

Benzing, D.H. 1991. Myrmecotrophy: origins, operation, and importance. In: Huxley, C.R., Cutler, D.F., Eds. Ant–Plant Interactions. Oxford University Press, Oxford England, pp. 353–373.

Bergback, B., Borg, L. 1989. Effect of acidification on uptake of rubidium, potassium and ammonium ions by spruce. Journal of Plant Nutrition. 12:1473–1482.

Billings, W.D., Godfrey, P.J. 1967. Photosynthetic utilization of internal carbon dioxide by hollow-stemmed plants. Science 158:121–123.

Bobbink, R., Heil, G.W., Raessen, M.B.A.G. 1992. Atmospheric deposition and canopy exchange processes in heathland ecosystems. Environmental Pollution 75:29–37.

Bonnett, H.T 1990. The *Nostoc-Gunnera* association. In: Rai, A.N., Ed. Handbook of Symbiotic Cyanobacterium, Ed. 1, CRC Press, Boca Raton, FL, pp. 161–171.

Cabrera, M., Jaffe, K. 1994. A trophic mutualism between the myrmecophytic Melastomataceae *Tococa guianensis* Aublet and an *Azteca* ant species. Ecotropicos 7:1–10.

Cammeraat, E.L.H., Risch, A.C. 2008. The impact of ants on mineral soil properties and processes at different spatial scales. Journal of Applied Entomology 132:285–294.

Chazdon, R.L., Fetcher, N. 1984. Photosynthetic light environments in a lowland tropical rain forest in Costa Rica. Journal of Ecology 72:553–564.

Chevone, B.I., Herzfeld, D.E., Krupa, S.V., Chappelka, A.H. 1986. Direct effects of atmospheric sulfate deposition on vegetation. The Journal of the Air & Waste Management Association 363:813–815.

Chiu, W.-L., Peters, G.A., Levieille, G., Still, P.C., Cousins, S., Osborne, B., Elhai, J. 2005. Nitrogen deprivation stimulates symbiotic gland development in *Gunnera manicata*. Plant Physiology 139:224–230.

Culver, D.C., Beattie, A.J. 1983. Effects of ant mounds on soil chemistry and vegetation patterns in a Colorado montane meadow. Ecology 64:485–492.

Czerwinski, A., Jakubczyk, H., Petêl, J. 1971. The influence of ants of the genus *Myrmica* on the physico-chemical and microbiological properties of soil within the compass of ant hills in Strzeleckie Meadows. Polish Journal of Soil Science 3:51–58.

Darwin, C.1859. The Origin of Species by Means of Natural Selection: or, the Preservation of Favored Races in the Struggle for Life. John Murray, London.

Dauber, J., Wolters, V. 2008. Microbial activity and functional diversity in the mounds of three different ant species. Soil Biology and Biochemistry 32:93–99.

Davidson, D.W., Fisher, B.L. 1991. Symbiosis of ants with *Cecropia* as a function of light regime. In: Huxley, C.R., Cutler, D.F., Eds. Ant-Plant Interactions. Oxford University Press, Oxford, UK.

Davidson, D.W., McKey, D. 1993. The evolutionary ecology symbiotic ant-plant relationships. Journal of Hymenopteran Research 2:13–83.

Davies, S.J., Lum, S.K.Y., Chan, R., Wang, L.K. 2001. Evolution of myrmecophytism in wester Mealesian Macaraga (Euphrbiaceae). Evolution 55:1542–1559.

Dean, W.R.J., Yeaton, R.I. 1993. The effects of harvester ant *Messor capensis* nest-mounds on the physical and chemical properties of soils in the southern Karoo, South Africa. Journal of Arid Environments 25:249–260.

Dean, W.R.J., Milton, S.J., Klotz, S. 1997. The role of ant nest-mounds in maintaining small-scale patchiness in dry grasslands in Central Germany. Biodiversity and Conservation 6:1293–1307.

Dlussky, G.M. 1981. Ants of Deserts. Nauka Press, Russia.

Eldridge, D.J. 1993. Effects of ants on sandy soils in semi-arid eastern Australia—local distribution of nest entrances and their effect on infiltration of water. Australian Journal of Soil Research 31:508–518.

Endler, J.A. 1986. Natural Selection in the Wild. Princeton University Press, Princeton, NJ.
Evans, R.D. 2001. Physiological mechanisms influencing plant nitrogen isotope composition. Trends in Plant Science 6:121–126.
Farji-Brener, A.G., Medina, C.A. 2000. The importance of where to dump the refuse: seed banks and fine roots in nests of the leaf-cutting ants *Atta cephalotes* and *A. colombica*. Biotropica 32:120–126.
Farji-Brener, A.G., Ghermandi, L. 2009. Leaf-cutting ant nests near roads increase fitness of exotic plant species in natural protected areas. Proceedings of the Royal Society of London. Series B 275:1431–1440.
Fiedler, K., Kuhlmann, F., Schlick-Steiner, B.C., Steiner, F.M., Gebauer, G. 2007. Stable N-isotope signatures of central European ants: assessing positions in a trophic gradient. Insectes Sociaux 54:393–402.
Fischer, R.C., Richter, A., Wanek, W., Mayer, V. 2002. Plants feed ants: food bodies of myrmecophytic *Piper* and their significance for the interaction with *Pheidole bicornis* ants. Oecologia 133:186–192.
Fischer, R.C., Wanek, W., Richter, A., Mayer, V. 2003. Do ants feed plants? A ^{15}N labeling study of nitrogen fluxes from ants to plants in the mutualism of *Pheidole* and *Piper*. Journal of Ecology 91:126–134.
Fisher, B.L., Sternberg, L.S.L., Price, D. 1990. Variation in the use of orchid extrafloral nectar by ants. Oecologia 83:263–266.
Folgarait, P.J. 1998. Ant biodiversity and its relationship to ecosystem functioning: a review. Biodiversity and Conservation 7:1221–1244.
Gay, H. 1993a. Animal-fed plants: an investigation in to the uptake of ant-derived nutrients by the far-eastern epiphytic fern *Lecanopteris* Reinw. (Polypodiaceae). Biological Journal of the Linnean Society 50:221–233.
Gay, H 1993b. Rhizome structure and evolution in the ant-associated epiphytic fern *Lecanopteris* Reinw. (Polypodiaceae). Botanical Journal of the Linnean Society 113: 135–160.
Gay, H., Hensen, R. 1992. Ant specificity and behaviour in mutualisms with epiphytes: the case of *Lecanopteris* (Polypodiaceae). Biological Journal of the Linnean Society 47: 261–284.
Gessler, A., Rennenberg, H. 1998. Atmospheric ammonia: mechanisms of uptake and impacts on N metabolism of plants. In: De Kok, L.J., Studen, I., Eds. Responses of Plant Metabolism to Air Pollution and Global Change. Backhuys, Leiden, pp. 81–94.
Ginzburg, O., Whitford, W.G., Steinberger, Y. 2008. Effects of harvester ant (*Messor* spp.) activity on soil properties and microbial communities in a Negev Desert ecosystem. Biology and Fertility of Soils 45:165–173.
Gmur, N.R., Evans, L.S., Cunningham, E.A. 1983. Effects of ammonium sulfate aerosols on vegetation. II. Mode of entry and responses of vegetation. Atmospheric Environment 17:715–721.
Heil, M., Fiala, B., Maschwitz, U., Linsenmair, K.E. 2001. On benefits of indirect defence: short- and long-term studies of antiherbivore protection via mutualistic ants. Oecologia 126:395–403.
Henry, L.T., Raper Jr., C.D. 1989. Effects of root-zone acidity of utilization of nitrate and ammonium in tobacco plants. Journal of Plant Nutrition 12:811–826.
Hutchinson, G.L., Millington, R.J., Peters, D.B. 1972. Atmospheric ammonia: absorption by plant leaves. Science 175:771–772.
Huxley, C.R. 1978. The ant-plants *Myrmecodia* and *Hydnophytum* (Rubiaceae), and the relationships between their morphology, ant occupants, physiology and ecology. New Phytologist 80:231–268.

Huxley, C.R. 1986. Evolution of benevolent ant-plant relationships. In: Huxley, C.R., Cutler, D.F., Eds. Ant–Plant Interactions. Oxford University Press, Oxford England. pp. 257–282.

Huxley, C.R. 1991. The tuberous epiphytes of the Rubiaceae I. The Hydnophytinaceae. Blumea 36:1–20.

Itino, T., Davies, S.J., Tada, H., Hieda, Y., Inoguchi, M., Itioka, T., Yamane, S., Inoue, T., 2002. Cospeciation of ants and plants. Ecological Research 16:787–793.

Janzen, D.H. 1966. Coevolution of mutualism between ants and Acacias in Central America. Evolution 20:249–275.

Janzen, D.H 1969. Allelopathy by myrmecophytes: the ant *Azteca* as an allelopathic agent of *Cecropia*. Ecology 50:147–153.

Janzen, D.H. 1972. Protection of *Barteria* (Passifloraceae) by *Pachysima* ants (Pseudomyrmecinae) in a Nigerian rain forest. Ecology 53:885–892.

Janzen, D.H. 1974. Epiphytic myrmecophytes in Sarawak: mutualism through the feeding of plants by ants. Biotropica 6:237–239.

Johansson, C. 1989. Pine forest: a negligible sink for atmospheric NO_x in rural Sweden. Tellus 39:395–412.

Karsten, G. 1895. Morphologische und biologische Untersuchungen uber einige Epiphytenformen der Molukken. Ann Jardin Bot Buitenzorg, XII, pp. 117–195, Pls. XIII–XIX.

Kleinfeldt, S.E. 1978. Ant-gardens: the interaction of *Codonanthe crassifolia* (Gesneriaceae) and *Crematogaster longispina* (Formicidae). Ecology 59:449–456.

Klepper, L.A. 1979. Nicric oxide (NO) and nitrogen dioxide emissions from herbicide-treated soybean plants. Atmospheric Environment 13:57–542.

Koptur, S. 1991. Extrafloral nectaries of herbs and trees: modeling the interaction with ants and parasitoids. In: Huxley, C.R., Cutler, D.F., Eds. Ant-Plant Interactions. Oxford University Press, Oxford, U.K. pp. 213–230.

Krupka, R.M., Towers, G.H.N. 1959. Studies of the metabolic relations in wheat. Canadian Journal of Botany 37:539–545.

Krupa, S.V. 2003. Effects of atmospheric ammonia (NH_3) on terrestrial vegetation: a review. Environmental Pollution 124:179–221.

Lea, P.J., Wolfenden, J., Wellburn, A.R. 1994. Influence of air pollutants upon nitrogen metabolism. In: Alscher, R., Wellburn, A.R., Eds. Plant Responses to Gaseous Environments. Chapman & Hall, London, pp. 279–299.

Lenoir, L., Persson, T., Bengtsson, J. 2001. Wood ant nests as potential hot spots for carbon and nitrogen mineralization. Biology and Fertility of Soils 34:235–240.

Levan, M.A., Stone, E.L. 1983. Soil modifications by colonies of black meadows ants in a New York old field. Soil Science Society of America Journal 47:1192–1196.

Longino, J.T. 1987. Ants provide substrate for epiphytes. Selbyana 9:100–103.

Longino, J.T. 1989. Geographic variation and community structure in an ant-plant mutualism: *Azteca* and *Cecropia* in Costa Rica. Biotropica 21:126–132.

Longino, J.T. 1991. *Azteca* ants in *Cecropia* trees: taxonomy, colony structure, and behavior. In: Huxley, C.R., Cutler, D.F., Eds.. Ant-Plant Interactions. Oxford University Press, Oxford. pp. 271–288.

Madison, M. 1979. Additional observations on ant-gardens in Amazonas. Selbyana 5:107–115.

McGinley, M.A., Dhillion, S.S., Neumann, J.C. 1994. Environmental heterogeneity and seedling establishment: ant-plant-microbe interactions. Functional Ecology 8:607–615.

McKee, H.S. 1962. Nitrogen Metabolism in Plants. Clarendon Press, Oxford.

McNett, K., Longino, J., Barriga, P., Vargas, O., Phillips, K., and Sagers, C.L. 2010. Stable isotope investigation of a cryptic ant-plant association: *Mymelachista flavocotea* (Hymenoptera: Formicidae) and *Ocotea* spp. (Lauraceae). Insectes Sociaux 57:67–72.

Merrill, E.D. 1945. Plant Life of the Pacific World. New York, Macmillan.

Michelangeli, F.A. 2003. Ant protection against herbivory in three species of *Tococa* (Melastomataceae) occupying different environments. Biotropica 35:181–188.

Näsholm, T., Ekblad, A., Nordin, A., Glesler, R., Höberg, M., and Höberg, P. 1998. Boreal forest plants take up organic nitrogen. Nature 392:914–916.

Näsholm, T., Kielland, K., Ganeteg, U. 2009. Uptake of organic nitrogen by plants. New Phytologist 182:31–48.

Okano, K., Fukuzawa, T., Tazaki, T. Totsuka, T. 1986. N dilution method for estimating the absorption of atmospheric NO_2 by plants. New Phytologist 102:73–84.

Osmond, C.B., Smith, S.D., Gui-Ying, B., Sharkey, T.D. 1987. Stem photosynthesis in a desert ephemeral, *Eriogonum inflatum*: characterization of leaf and stem CO_2 fixation and H_2O vapor exchange under controlled conditions. Oecologia 72:542–549.

Paris, C.I., Polo, M.G., Garbagnoli, C., Martinez, P., de Ferre, G.S., Folgarait, P.J. 2008. Litter decomposition and soil organisms within and outside of *Camponotus punctualtus* nests in sown pasture in northeastern Argentina. Applied Soil Ecology 40:271–282.

Paungfoo-Lonhienne C., Lonhienne, T.G.A., Rentsch, D., Robinson, N., Christie, C.M., Webb, R.I., Garnage, H.K., Carroll, B.J., Schenk, P.M., Schmidt, S. 2008. Plants can use protein as a nitrogen source without assistance from other organisms. Proceedings of the National Academy of Sciences, USA 105:4524–4529.

Pêtal, J. 1998. The influence of ants on carbon and nitrogen mineralization in drained fen soils. Applied Soil Ecology 9:271–275.

Pêtal, J., Kusinska, A. 1994. Fractional composition of organic matter in the soil of anthills and of the environment of meadows. Pedobiologia 38:493–501.

Porter, L.K., Viets, Jr., F.G., Hutchinson, G.L. 1972. Air containing nitrogen-15 ammonia: foliar absorption by corn seedlings. Science 175:759–761.

Raab, T.K., Lipson, D.A., Monson, R.K. 1996. Non-mycorrhizal uptake of amino acids by roots of the alpine sedge *Lobresia myosuroides*: implications of the alpine nitrogen cycle. Oecologia 108:488–494.

Raven, J.A. 1996. Into the voids: the distribution, function, development and maintenance of gas spaces in plants. Annals of Botany 78:137–142.

Rickson, F.R. 1979. Absorption of animal tissue breakdown products into a plant stem-the feeding of a plant by ants. American Journal of Botany 66:87–90.

Rico-Gray, V., Barber, J.T., Thein, L.B., Ellgaard, E.G., Toney, J.J. 1989. An unusual animal-plant interaction: feeding of *Schomburgkia tibicinis* (Orchidaceae) by ants. American Journal of Botany 76:603–608.

Rico-Gray, V., Castro, G. 1996. Effect of an ant-aphid interaction on the reproductive fitness of *Paullinia fuscensens* (Sapindaceae). Southwestern Naturalist 41:434–440.

Rico-Gray, V., Palacios-Rios, M., Garcia-Franco, J.G., MacKay, W.P. 1998. Richness and seasonal variation of ant-plant associations mediated by plant-derived food resources in the semiarid Zapotitlan Valley, Mexico. American Midland Naturalist 140:21–26.

Risch, S., McClure, M., Vandermeer, J., Waltz, S. 1977. Mutualism between three species of tropical piper (Piperaceae) and their ant inhabitants. American Midland Naturalist 98:433–444.

Rissing, S.W. 1986. Indirect effects of granivory by harvester ants: plant species composition and reproductive increase near ant nests. Oecologia 68:231–234.

Rondon, A., Johansson, C., Granat, L. 1993. Dry deposition of nitrogen dioxide and ozone to coniferous forest. Journal of Geophycial Research 98:5159–5172.

Sagers, C.L., Ginger, S.M., Evans, R.D. 2000. Carbon and nitrogen isotopes trace nutrient exchange in an ant-plant mutualism. Oecologia 123:582–586.

Schimper, A.F.W. 1888. Die Wechselbeziehungen zwischen Pflanzen und Ameisen im tropischen Amerika. Botan. Mitt. aus den Tropen, Jena, Heft 1, pp. 1–95.

Schemske, D.W. 1983. Limits to specialization and coevolution in plant-animal mutualism. In: Nitecki, M.H., Eds. Coevolution. University of Chicago Press, Chicago, pp. 67–109.

Schubert, K.R. 1986. Products of biological nitrogen fixation in higher plants: synthesis, transport, and metabolism. Annual Review of Plant Physiology 37:539–574.

Schupp, E.W. 1986. *Azteca* protection of *Cecropia*: ant occupation benefits juvenile trees. Oecologia 70:379–386.

Solano, P., Dejean, A. 2004. Ant-fed plants: comparison between three geophytic myrmecophytes. Biological Journal of the Linnean Society 83:433–439.

Sparks, J., Monson, R.K., Sparks, K.L., Lerdau, M. 2001. Leaf uptake of nitrogen dioxide (NO_2) in a tropical wet forest: implications for tropospheric chemistry. Oecologia 127:214–221.

Sparks, J. 2009. Ecological ramifications of direct foliar uptake of nitrogen. Oecologia 159:1–13.

Stadler, B., Schramm, A., Kalbitz, K. 2006. Ant-mediated effects on spruce litter decomposition, solution chemistry, and microbial activity. Soil Biology and Biochemistry 38:561–572.

Sternberg, L. da S.L., Pinzon, M.C., Moreira, M.Z., Mountinho, P., Rojas, E.I., Herre, E.A. 2007. Plants use macronutrients accumulated in leaf-cutting ant nests. Proceedings of the Royal Society Series B: Biological Sciences 274:315–321.

Stout, J.1979. An association of an ant, a mealy bug, and an understory tree from a Costa Rican rainforest. Biotropica 11:309–311.

Stulen, I., ter Steege, M. 1995. Light and nitrogen assimilation. In Srivastava, H.S., Singh, R.P., Eds. Nitrogen nutrition in higher plants. Associated Publishing, New Delhi, India. pp. 367–384.

Svoma, E., Morawetz, W. 1992. Drüsenhaare, Emergenzen und Blattdomatien bei der Ameisenpflanze *Tococa occidentalis*. Botanishce Jarhbuch 114:185–200.

Theone, B., Schroder, P., Papen, H., Egger, A., Rennenberg, H. 1991. Absorption of atmospheric NO_2 by spruce (*Picea abies* L. Karst) trees. I. NO_2 influx and its correlation with nitrate reduction. New Phytologist 117:575–585.

Thompson, J.N. 1981. Reversed animal-plant interactions: the evolution of insectivorous and ant-fed plants. Biological Journal of the Linnean Society 26:147–155.

Thompson, J.N. 1988. Variation in interspecific interactions. Annual Review of Ecology and Systematics 19:65–85.

Tillberg, C.V. 2004. Friend or foe? A behavioral and stable isotopic investigation of an ant-plant symbiosis. Oecologia 140:506–515.

Tillberg, C.V., McCarthy, D.P., Dolezal, A.G., Suarez, A.V. 2006. Measuring the trophic ecology of ants using stable isotopes. Insectes Sociaux 53:65–69.

Treseder, K.K., Davidson, D.W., Ehleringer, J.R. 1995. Absorption of ant-provided carbon dioxide and nitrogen by a tropical epiphyte. Nature 375:137–139.

Trimble, S.T., Sagers, C.L. 2004. Differential host use in two highly specialized ant-plant associations: evidence from stable isotopes. Oecologia 138:74–82.

Ule, E. 1901. Ameisengärten in Amazonas-gebiet. Botanische Jarhbuch 30:45–51.

Van Hove, L.W.A., Koops, A.J., Adema, E.H., Vredenberg, W.J., Pieters, G.A. 1987a. Analysis of the uptake of atmospheric ammonia by leaves of *Phaseolus vulgaris* L. Atmospheric Environment 21:1759–1763.

Van Hove, L.W.A., Adema, E.H., Vredenberg, W.J. 1987b. The uptake of atmospheric ammonia by leaves. In: Mathy, P., Ed. Air Pollution and Ecosystems, Proceedings of the International Symposium, Grenoble, France, 18–22 May 1987. Reidel, Dordrecht, pp. 734–742.

Van Hove, L.W.A., Adema, E.H., Vredenberg, W.J., Pieters, G.A. 1989. A study of the adsorption of NH_3 and SO_2 on leaf surfaces. Atmospheric Environment 23:1479–1486.

Vasconcelos, H.L., Casimiro, A.B. 1997. Influence of *Azteca alfari* ants on the exploitation of *Cecropia* trees by a leaf-cutting ant. Biotropica 29:84–92.

Vessey, J.K., Henry, L.T., Chaillou, S., Raper, Jr., C.D. 1990. Root-zone acidity affects relative uptake of nitrate and ammonium from mixed nitrogen sources. Journal of Plant Nutrition 13:95–116.

Wagner, D. 1997. The influence of ant nests on *Acacia* seed production, herbivory and soil nutrients. Journal of Ecology 85:83–93.

Wagner, D., Jones, J.B., Gordon, D.M. 2004. Development of harvester ant colonies alters soil chemistry. Soil Biology and Biochemistry 36:797–804.

Wagner, D., Jones, J.B. 2006. The impact of harvester ants on decomposition, N mineralization, litter quality, and the availability of N to plants in the Mojave Desert. Soil Biology and Biochemistry 38:2593–2601.

Wang, D., Lowery, B., McSweeney, K., Norman, J.M. 1996. Spatial and temporal patterns of ant burrow openings as affected by soil properties and agricultural practices. Pedobiologia 40:201–211.

Ward, P.S. 1991. Phylogenetic analysis of pseudomyrmecine ants associated with domatia-bearing plants. In: Huxley, C.R., Cutler, D.F., Eds. Ant-Plant Interactions. Oxford University Press. pp. 335–352.

Watkins, J.E., Jr., Cardelás, C.L., Mack, M.C. 2008. Ants mediate nitrogen relations of an epiphytic fern. New Phytologist 180:5–8.

Weber, P., Rennenberg, H. 1996. Dependency of nitrogen dioxide (NO_2) fluxes to wheat (*Triticum aestivum* L.) leaves on NO_2 concentration, light intensity, temperature and relative humidity determined from controlled dynamic chamber environments. Atmospheric Environment 30:3001–3009.

Wellburn, A.R. 1990. Why are atmospheric oxides of nitrogen usually phytotoxic and not alternative fertilizers? New Phytologist 115:395–429.

Wildt, J., Kley, D., Rockel, A., Rockel, P., Segschnieder, H.J. 1997. Emission of NO from several higher plant species. Journal of Geophysical Research 102:5919–5927.

Wilson, E.J. 1992. Foliar uptake and release of inorganic nitrogen compounds in *Pinus sylvestris* L. and *Picea abies* (L.) Karst. New Phytologist 120:407–416.

Yu, D.W., Davidson, D.W. 1997. Experimental studies of species-specificity in *Cecropia*-ant relationships. Ecological Monographs 67:273–294.

Zeevaart, A.J. 1976. Some effects of fumigating plants for short periods with NO_2. Environmental Pollution 11:97–108.

Section 5
Environmental Signalling in N Acquisition

Chapter 15
The Functions of Flavonoids in Legume-Rhizobia Interactions

Oliver Yu and Yechun Wang

Summary

Flavonoids are a group of phenolic secondary metabolites found in all higher plants. Some flavonoids are important free radical scavengers protecting plants against ultraviolet (UV) irradiation, and others are pigments and co-pigments that attract pollinators. Legumes also synthesize a unique group of flavonoid compounds called isoflavones, which are the dominant flavonoid compounds in their roots and seeds. Advances in molecular biology and genetics have revealed much about isoflavone biosynthesis and biological functions in leguminous species. Here, we outline our latest understanding of the unique role of flavonoids in legume nitrogen acquisition, focusing mainly on their complex functions in plant-symbiont interactions. In addition to serving as external rhizobium attractants and nodulation gene inducers, flavonoids play distinct roles in formation and function of both determinate and indeterminate nodules. In summary, flavonoids are found to play vital, complex, and previously unknown tasks in many aspects of plant growth and development. Understanding how plants respond to N-deficiency begs an understanding of the roles of flavonoids in signaling and development.

Overview

The phenylpropanoid pathway is one of the major pathways of secondary metabolism in higher plants. Many phytochemists oppose the term "secondary metabolism" because many so-called "secondary" metabolic pathways produce compounds indispensable for plant growth and development. The phenylpropanoid pathway is one such example. It synthesizes at least 20% of plant biomass in the form of various lignins, which are components of secondary cell walls in all monocot and dicot plants (Weng et al., 2008). Lignins provide the rigidity and

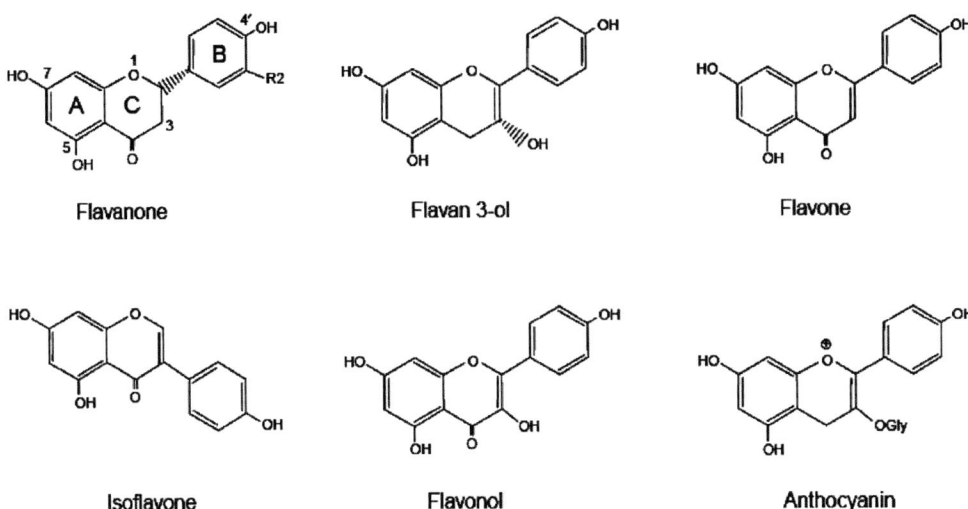

Figure 15.1. The structures of representative phenylpropanoid compounds. The carbon numbering is marked on flavanones only.

hydrophobicity of the secondary cell walls, and thus are essential for the terrestrial lifestyle of plants. The phenylpropanoid pathway also produces other phenolic compounds that are important for plant survival, such as anthocyanins and proanthocyaninidins (Yu and Jez, 2008). Anthocyanins are the major pigments of many flowers. Proanthocyanindins, responsible for the brownish color of many seed coats, are polymers of flavan-3-ols. Both compounds play essential roles in plant-animal interactions, fostering pollination and seed dispersal by animals of different sizes. In addition to pigments, another major group of phenylpropanoid compounds found in all higher plants is the flavonoids (Figure 15.1) whose functions are the focus of this chapter. These compounds all bear a core structure of a C6-C3-C6 phenyl-benzopyran ring (Dixon, 2005). Flavonoids have been proven to function in abiotic and biotic stress responses. For example, UV light irradiation induces rapid accumulation of flavonoids in both leaves and roots, suggesting that flavonoids are an endogenous plant "sunscreen." In agreement, mutants that lack flavonoid biosynthesis are more sensitive to UV irradiation (Franke et al., 2002). Various flavonoids with UV absorbance ranging roughly from 240–320 nm may reduce the intensity of damaging UV-A and UV-B light reaching important cellular organelles. More importantly, the multiple hydroxyl groups attached to the basic flavone rings can combat free radicals and reactive oxygen species generated by UV-irradiation (Korkina, 2007). For biotic stress responses, pathogen invasions normally cause enhanced flavonoid biosynthesis (Graham et al., 2008). Some of the flavonoid compounds are well-known phytoalexins, the "antibiotics" that inhibit the growth and propagation of invading pathogens, though the molecular bases of phytoalexin effects on pathogens have not been clearly characterized. The biotic and abiotic stress responses of flavonoids have been investigated for many years and reviewed in detail elsewhere (Dixon et al., 1996; Korkina, 2007; Ververidis et al., 2007).

Flavonoids are important players in the nitrogen cycle of the biosphere. For the Fabaceae, which can fix atmospheric nitrogen by interacting with symbiotic bacteria, flavonoids play prominent roles in both defense against pathogenic microbes and promotion of interactions with symbiotic microbes. Legumes have evolved two seemly contradictory functions for the same group of compounds: In mutants impaired in flavonoid biosynthesis, neither defense responses nor symbiotic responses were as effective as in wild type. In many legumes, isoflavones are the dominant form of flavonoid compounds. The phenyl ring is attached to the C3 of the benzopyran ring in all isoflavones, instead of C2 in the flavones. Isoflavones and their derivatives are collectively named isoflavonoid compounds, which represent the major phytoalexins in the Fabaceae family. In general, more complex isoflavonoids, such as coumestrols and ptreocarpans, are more potent antibacterial and antifungal compounds than simple isoflavones. However, both simple and complex isoflavonoids are important in the defense against pathogens. For mutualistic interactions, flavonoids and isoflavonoids are active participants in rhizobia bacterial signaling during symbiotic processes. We review here both defense and symbiosis functions of flavonoid compounds, but primarily focus on the latest discoveries in symbiosis, demonstrating the unique and essential functions of flavonoid and isoflavonoid compounds in legume-*Rhizobium* interactions, and hence on an important strategy for plant nitrogen acquisition.

The Phenylpropanoid Biosynthetic Pathway

The phenylpropanoid pathway leading to the flavonoid core structure has been well characterized and extensively reviewed (Winkel-Shirley, 2001; Ferrer et al., 2008; Yu and Jez, 2008). Indeed, this pathway is one of the best-studied natural product pathways in plants. The catalytic mechanisms, kinetics of key enzymes, transcriptional and post-transcriptional regulations, and tissue-specific and subcellular localizations of the pathway components have been investigated for decades. However, important to remember, is that surprising discoveries are still being made within this pathway, broadening our understanding of natural product biosynthesis in all areas. Figure 15.2 outlines flavonoid biosynthesis. The amino acid phenylalanine is typically the starting substrate of the entire pathway. Phenylalanine is synthesized in the chloroplast by the shikimate pathway (in photosynthetic tissues). Phenylalanine ammonia lyase (PAL) catalyzes the deamination of phenylalanine to cinnamic acid (Appert et al., 1994). Cinnamate 4-hydroxylase (C4H) then converts cinnamic acid to *p*-coumaric acid (Bell-Lelong et al., 1997). This compound is activated by ATP-dependent 4-coumaroyl-CoA ligase (4CL) resulting in the *p*-coumaroyl-CoA thioester (Lee and Douglas, 1996). A major pathway branch point occurs here. In cells that produce lignin, the majority of the *p*-coumaroyl-CoA is conjugated with shikimate by hydroxycinnamoyl-CoA:shikimate/quinate hydroxycinnamoyl transferase (HCT), which initiates G and S type monolignol biosynthesis (Wagner et al., 2007). These monolignols eventually will be polymerized to lignins that enhance secondary cell walls.

In flavonoid-producing cells, *p*-coumaroyl-CoA is condensed with three molecules of malonyl-CoA to form naringenin chalcone (Ryder et al., 1987). This reaction is catalyzed by chalcone synthase (CHS), a ubiquitous enzyme in higher plants. CHS is the entry-point enzyme leading to myriad flavonoid compounds, and not surprisingly, it is regulated by many environmental and developmental signals (Faktor et al., 1996). The CHS product naringenin chalcone and many other chalcones are unstable under acidic conditions. A chalcone isomerase (CHI) converts chalcones into flavanones (van Tunen et al., 1988). There are two types of CHI in legumes (Kimura et al., 2001; Ralston et al., 2005). One closely resembles CHIs of nonleguminous plants; it has a strict substrate-specificity toward naringenin chalcone. The other CHI is prevalent in plants that produce both 5-hydroxyflavanones and 5-deoxyflavanones. This so-called Type II CHI catalyzes cyclization of the benzopyran ring from both naringeninchalcone and 5′-deoxychalcones. In addition, the production of 5′-deoxychalcones and

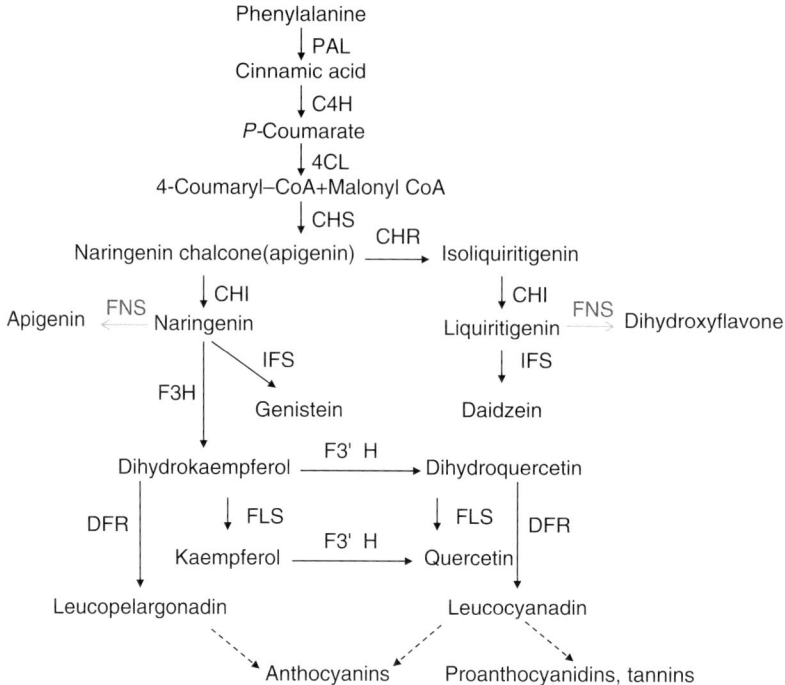

Figure 15.2. A schematic diagram of the main phenylpropanoid pathway. The enzyme abbreviations are introduced in the text. Genistein and daidzein are isoflavonoids made mostly in legumes. Arrows do not necessarily signify single enzyme-catalyzed steps.

5-deoxyflavanones requires another enzyme that is primarily distributed within the Fabaceae family, the chalcone reductase (CHR). This enzyme removes a hydroxyl group during p-coumaroyl-CoA condensation (Ballance and Dixon, 1995).

Flavanones are the precursors of many downstream flavonoid compounds. For example, flavonoid 3-hydroxylase (F3H) can convert flavanones into dihydroflavonols (Charrier et al., 1995), which in turn can be converted to either leucoanthocyanindins by dihydroflavonol reductase (DFR) (Beld et al., 1989), or to flavonols by flavonol synthase (FLS) (Holton et al., 1993). Flavanones can also be converted to flavones directly by flavone synthase (FNS) (Akashi et al., 1998). In legumes, FNS belongs to the CYP93B family of cytochrome P450 monooxygenase. Leucoanthocyanindins (or flavan-3,4-diols) are the precursors of proanthocyanindins and anthocyanins, conversions that require additional modifications of the core structure by several oxido-reductases and conjugation enzymes.

In addition to flavanols (above), isoflavonoids are also made from flavanones. Isoflavone synthase (IFS), found in many plants of the Fabaceae, is a cytochrome P450 enzyme in the CYP93C family. IFS converts flavanones to isoflavanones by catalyzing an aryl migration, which first breaks and then reestablishes a C-C bond between the benzopyran and the phenyl group (Akashi et al., 1999; Steele et al., 1999; Jung et al., 2000). This reaction is followed by an isoflavanone 2, 3-dehydration reaction catalyzed by isoflavanone dehydratase (IDH) (Akashi et al., 2005). After these reactions, the isoflavones genistein and daidzein can be produced from the flavanones naringenin and liquiritigenin, respectively. In most legume species, isoflavones are further modified to produce complex isoflavonoid compounds by methylation, hydroxylation, conjugation, and other reactions.

The 12 enzymes mentioned above constitute the core and major branches of the phenylpropanoid pathway. All of these enzymes have been isolated initially through traditional biochemical and genetic approaches. Since mutation in flavonoid production often affects flower or seed color, identification of these mutants is relatively easy. Mutants of Arabidopsis, and especially maize, with defects in pigment accumulation were isolated many years ago, and characterizations of these mutants have led to cloning of many structural enzymes. Traditional biochemical genetic approaches, based on mutant analyses and enzyme purification, continue to bear fruit. Over the past 2 years, newly identified phenylpropanoid pathway enzymes include pterocarpan 4-dimethylallyltransferase (G4DT) from soybean (Akashi et al., 2009), phaselic acid synthase (PAS) from red clover (Sullivan, 2009), flavone-C-glucosyltransferase (C-GT) from rice (Brazier-Hicks et al., 2009), and the TT12 transporter from Arabidopsis (Zhao and Dixon, 2009). Discovery of these genes has opened new frontiers in phenylpropanoid pathway research.

New Structural Genes

G4DT catalyzes the key prenylation step in the biosynthesis of glyceollins (Akashi et al., 2009). The identification of this enzyme followed "classical" approaches, and it had been known for many years to be associated with plastids. A recently identified cDNA for homogentisate phytyltransferase (in tocopherol biosynthesis) was used to probe induced cDNAs of soybean seedlings that accumulated significant amount of glyceollins in response to pathogen signals. One homolog with a plastid targeting sequence showed strong dimethylallyl diphosphate: (6aS,11aS)-3,9,6a-trihydroxypterocarpan [(2)-glycinol] 4-dimethylallyltransferase activity. Genetic analysis later confirmed this enzyme as G4DT.

Identification of PAS demonstrated once again that homologs of the phenylpropanoid pathway genes might have different activities at different steps of the pathway (Sullivan, 2009). There are two HCT homologs in red clover (*Trifolium pretense*). One is 75% identical to known shikimate-HCT enzymes mentioned above in monolignol biosynthesis; the other, HCT2, is only 34% identical. Enzymatic analysis showed that HCT2 carried out transfer of caffeoyl or *p*-coumaroyl moieties from a CoA-thioester to malate, instead of shikimate, thus making phaselic acid.

Most intracellular flavonoids are conjugated with one or more molecules of glucose by O-glucosyltransferase (O-GT). These conjugates are the major storage forms that allow transport and sequestering of flavonoid compound into vacuoles. However, C-glycosylation has also been identified in a few plant species. C-GT was identified by enzymatic activity screening of candidate genes (Brazier-Hicks et al., 2009). Interestingly, kinetic analysis, and intermediates discovered in the *in vitro* reaction, suggested a unique reaction mechanism that required generations of dibenzoylmethane tautomers, followed by dehydration to form a mixture of 6C- and 8C-glucosyl derivatives, agreeing perfectly with the distributions of C-glycosides found in plant cells. The functional identification of TT12 took advantage of over-expression of the Myb-like transcription factor TT2 in both *Arabidopsis* and *Medicago truncatula*. In *TT2* over-expressing *M. truncatula*, multidrug and toxin extrusion 1 protein (MATE1) was shown to transport epicatechin 3′-O-glucoside preferentially, not anthocyanin conjugates (Zhao and Dixon, 2009). The over-expressed MtMATE1 ortholog in Arabidopsis, TT12, had a similar activity. This transporter is essential for proanthocyanidin accumulation.

Transcriptional Regulation of the Phenylpropanoid Pathway

Transcription factors that coordinately regulate different steps of the pathway have been discovered following similar mutant analyses that identified structural enzymes (Memelink, 2005). The maize C1 factor, one of the first plant transcription factors to be cloned (Cone et al., 1986), belongs to a family of R2R3-Myb-like transcription factors. Mutations in C1 cause the maize kernel to lose the red pigment of the aleurone layer. It was later learned that many of the transcription factors that regulate the phenylpropanoid pathway are Myb-like transcription factors, which bind to a conserved *cis*-element located in the promoters of many structural genes (Martin and Paz-Ares, 1997). In Arabidopsis and other plants studied, a cascade of transcription factors activates expression of the Myb-like transcription factors, which eventually activate the expression of the pathway enzymes.

Importantly, R2R3-Myb-like transcription factors not only activate the phenylpropanoid biosynthetic pathway, but some members act to inhibit this pathway. For example, Myb308 and Myb330 of *Antirrhinum* strongly inhibit both flavonoid and lignin production (Tamagnone et al., 1998). More recently, some single domain Myb transcription factors, the R3-Myb-like proteins, were found to suppress phenylpropanoid biosynthesis. Arabidopsis MybL2 is one such example (Dubos et al., 2008).

In many cases, the Myb-like transcription factors require helper proteins (i.e., co-transcription factors) to control tissue-specific expression. A group of basic Helix-Loop-Helix (bHLH) – type transcription factors has been shown to bind C1 and other phenylpropanoid-specific Myb-like transcription factors (Rabinowicz and Grotewold, 2000). Strikingly, the maize bHLH transcription factor (R gene) that, with the C1 gene, co-activates the phenylpropanoid pathway was shown to modify histone acetylation by recruiting an EMSY-related factor (Hernandez et al., 2007; Zhao et al., 2008; Morohashi and Grotewold, 2009). In turn, the Myb-bHLH complex had been previously shown to be downstream of a WD40-type transcription factor TTG1 (Zhao et al., 2008). Taken together, the energy-consuming phenylpropanoid pathway is understandably regulated by an extensive signal transduction network that responds to various environmental and developmental signals.

Flavonoids as Nod Operon Inducers

Symbiotic nodules are root lateral organs that are the result of interactions between leguminous plants and nitrogen-fixing rhizobial bacteria. (We do not discuss nodulation by gram-positive *Frankia* on 'actinorhizal' nonleguminous plants; see Chapter 6.) Nitrogen is a major plant nutrient and a main component of chemical fertilizers. Worldwide, the production of nitrogenous fertilizers consumes approximately 5% of global natural gas and 2% of all energy produced (Alper and Stephanopoulos, 2009; Wen et al., 2009). Symbiotic nodule development in cultivated legumes obviates much use of chemical fertilizers. U.S. farmers used ~12 million tons of nitrogen in 2006 costing on average $12 billion, a cost that could be significantly reduced through greater reliance on biological nitrogen fixation. Biological fixation is more "environmentally friendly," reducing energy inputs into artificial fertilizer and their excessive applications to crops. Nodules resulting from legume-rhizobial interactions account for the majority of biological nitrogen fixation (Alper and Stephanopoulos, 2009; Wen et al., 2009).

The Nodulation Process

Members of the α-proteobacteria (e.g., *Rhizobium, Sinorhizobium, Bradyrhizobium,* and *Methylobacterium*) and the β-proteobacteria (e.g., *Burkholderia*) infect the roots of leguminous plants and establish a nitrogen-fixing symbiosis (Downie and Walker, 1999; Ferguson and Mathesius, 2003; Markmann and Parniske, 2009). This process, initiated at the root hair, requires the close coordination of host and symbiont functions, achieved by the recognition and exchange of diffusible chemical signals (Magori and Kawaguchi, 2009; Martinez-Romero, 2009). Among them are plant-secreted flavonoids and isoflavonoids, which are recognized by the compatible bacteria resulting in induction of bacterial nodulation (*nod*) genes. Their expression leads to the synthesis and excretion of a specific lipo-chitin nodulation signal (Nod factor) from the bacteria, which activates many of the early events in root hair infection. The infection process occurs by bacteria entering the plant via the root epidermis and inducing the reprogramming of root cortical cell development and formation of a nodule. In the best-studied cases, infection initiates through root hairs, which subsequently curl. The bacteria become enclosed within the root hair curl where the plant cell wall is degraded, the cell membrane is invaginated, and an intracellular, tubular structure (infection thread) is formed. Within the infection thread, the bacteria enter the root hair cell and eventually ramify into the root cortex. Before the infection thread reaches the base of the root hair cell, the root cortical cells are induced to de-differentiate, activating their cell cycle and cell division to form the nodule primordium (Sprent, 2008). When the infection thread reaches the cells of the developing primordium, the bacteria are released into cells via endocytosis. Inside a plant cell, the bacteria are enclosed in vacuole-like structures (symbiosomes) where they differentiate into bacteroids. It is within these symbiosomes that the bacteria convert N_2 to NH_4^+.

Nod Operon Induction

As outlined above, flavonoids are the first set of signals that trigger the complicated chemical signal dialog between the host and its symbiotic partner. During the 1980s, various *Rhizobium nod-lacZ* reporters facilitated bioassays of different flavonoids and other factors that regulate this cross-kingdom dialog (Peters et al., 1986; Fisher and Long, 1992). It soon became clear that host-specific flavonoids also determined the specificity of the symbiotic partner. For example, *Sinorhizobium meliloti,* that nodulates *Medicago sativa,* responds to the flavone luteolin (Peters et al., 1986), while *Bradyrhizobium japonicum,* that nodulates soybean, responds to the isoflavone daidzein (Banfalvi et al., 1988). Each legume host produces a unique mixture of root exudates that serve as specific chemo-attractants and Nod gene inducers of their own symbionts (Barbour et al., 1991). In *in vitro*-cultured bacteria, the *Rhizobium* nod operon is suppressed and no Nod factors are synthesized. Upon adding the corresponding flavonoid compound(s) to the growth media, Nod factor biosynthesis is activated, the composition and molecular mass distribution of extracellular polysaccharides is altered, and the growth rate of the bacteria is significantly increased. Since flavonoids are also phytoalexins that inhibit bacterial growth, the co-evolution of legume and rhizobia suggests flavonoids play interesting roles in this complex process. Currently, many commercial rhizobial inoculants contain stimulating flavonoid compounds to increase nodulation efficiencies.

The control of *nod* gene expression in response to plant flavonoids appears to be regulated by at least two independent pathways. One involves the NodD protein and its homologs found in many rhizobia (Hirsch and LaRue, 1997). The other is more specific to *Bradyrhizobium* and requires NodV and NodW proteins (Loh et al., 1997). Both regulatory pathways act upon the signal of flavonoid compounds. NodD is a member of the LysR family of transcription factors, which includes proteins such as *Escherichia coli* LysR and *Salmonella typhimurium* MetR (Bellato et al., 1996). The NodD protein of *Rhizobium sp.* NGR234 was demonstrated to bind flavonoid (Kobayashi et al., 2004). Upon binding flavonoids, the activated NodD binds the conserved nod box sequences present upstream of the respective nod operons, and kicks off Nod factor biosynthesis. In a *Bradyrhizobium* mutant lacking NodD1, nodulation is significantly reduced, but the mutant is still capable of nodulating host plants. In this case, the secondary regulating factors, NodV and NodW, transmit the isoflavone signal to the transcription mechanism in the absence of NodD via a series of phosphorylation steps (Loh et al., 1997). As an interesting side note, flavonoids are not necessary plant signal compounds in induction of the arbuscular mycorrhizal symbiosis. Root exudates depleted of flavonoids and mutant plants without flavonoid accumulation can both promote normal mycorrhizal fungal growth (see Chapter 16). However, flavonoid regulation has been implicated to be important for nematode infection, and flavonoid biosynthesis is indeed induced at the site of infection (Grunewald et al., 2009; Mathesius, 2009; Wasson et al., 2009).

For many years since the discovery of flavonoids as Nod operon inducers, the primary function of these flavonoid compounds was assumed to be external (i.e., flavonoids function extracellularly in the rhizosphere where they interact with symbiotic bacteria). However, recent results on flavonoid effects on auxin transport are now changing this concept.

Flavonoid as Auxin Transport Inhibitors

Root lateral organs such as symbiotic nodules and lateral roots are unique in that they arise from nonmeristematic differentiated cells, whereas initiation of lateral organs in the shoot occurs from founder cells in the meristem. The sequence of events in the initiation of root nodules has many parallels with that of lateral roots (Ferguson and Mathesius, 2003). Correlation between the number of lateral roots and nodules has been observed in many legume species (Carroll et al., 1985; Nishimura et al., 2002). Since many of the downstream genes activated during nodule primordia development were also found to be activated during lateral root development, it has been postulated that nodulation might have originated from mechanisms pre-existing for lateral root formation (Sprent, 2007). However, the site of initiation of legume nodules is generally different from that of lateral roots; legume nodules arise from cortex cells whereas lateral roots arise from the pericycle cells. Exceptions have been observed in peanut, pea, and red clover (McIver et al., 1989), where nodules might arise directly from pericycle cells. Additionally, after initiation, the two lateral organs have clear differences in their development. A conspicuous example is the presence of central vasculatures in lateral roots as opposed to the peripheral vasculatures in nodules. However, it is clear that *de novo* organogenesis induced by rhizobial infection requires significant changes in the hormonal status of the root tissue, particularly the auxin/cytokinin ratio. It is well documented that alterations in this ratio precede initiation of the cell cycle and cell de-differentiation during lateral organ development.

Auxin Role in Root Lateral Organ Development

The plant hormone auxin has been clearly shown to play an important role in the initiation and development of lateral roots (Casimiro et al., 2003). Auxin promotes lateral root initiation and growth by activating the initial pericycle cell divisions (Fukaki and Tasaka, 2009). The dominant auxin species in plants is indole acetic acid (IAA), and mutants that over-produce it, such as *sur1/alf1/rty1*, *sur2*, and *yucca*, produce more lateral roots (Boerjan et al., 1995; Celenza et al., 1995; Zhao et al., 2001), while mutants with reduced auxin-sensitivity, such as *axr1* and *axr4*, produce fewer lateral roots (Hobbie and Estelle, 1995; Hobbie et al., 2000). Additionally, roots deprived of endogenous auxin, by growth in the presence of the auxin transport inhibitor 1-naphthylphthalamic acid (NPA), fail to initiate lateral root primordial. Supplementation with exogenous auxin (IAA or synthetic forms) can restore lateral root primordia initiation underscoring the role of auxin in lateral root initiation (Casimiro et al., 2001; Marchant et al., 2002).

Auxin Signaling

Auxin-induced lateral root initiation results from regulated gene expression upon perception and transduction of the auxin signal (Vierstra, 2009). Auxin is perceived by a set of transport inhibitor response 1 F-box proteins (TIR1) in Arabidopsis (Robert and Friml, 2009). Recessive mutations in TIR1 (or other auxin receptor F-box proteins) led to reduced sensitivity to auxin. Molecular and biochemical characterization of TIR1 revealed that it directly binds auxin and activates auxin-responsive gene expression by causing disassociation of ARF-Aux/IAA protein complexes that repress auxin-responsive (ARF) gene expression (Vanneste and Friml, 2009). ARFs are transcriptional activators and repressors that bind with specificity to promoters of early auxin response genes. Aux/IAA proteins are short-lived nuclear proteins encoded in general by early auxin response genes. The model is that Aux/IAA proteins prevent transactivation by ARF transcriptional activators by forming a protein complex with ARFs at low auxin concentrations. As auxin concentrations increase, TIR1 bound to auxin mediates degradation of Aux/IAA proteins through the ubiquitin pathway, thus releasing the ARFs for activation of auxin-responsive promoters. TIR1 acts as part of an SCF-type ubiquitin ligase complex. Direct binding of auxin by TIR1 increases its affinity for the Aux/IAA proteins.

Auxin Transport

IAA is synthesized at shoot and root apical meristems. An elaborate transport system in all higher plants directs the flow of IAA to its target sites in a cell-to-cell fashion (Feraru and Friml, 2008). This transport system consists of two sets of transporters. Influx carriers, such as AUX1/LAX complexes, move extracellular IAA into the cells, while efflux carriers, such as PGP/PIN complexes, move IAA out to surrounding spaces (Mravec et al., 2008). The localization of these carriers is directional and can be changed upon various stimuli, allowing the hormone to alter the course of plant development in response to environmental and stress signals.

Since extracellular IAA exists as the protonated charge-neutral form in the acidic apoplast, it can diffuse into cells passively. Hence, the auxin efflux process is generally considered the more important regulatory step in auxin flux. Much evidence suggests that the directionality of auxin transport is governed by the subcellular localization of efflux carrier proteins. Two sets of efflux carriers have been identified. The first is a group of ABC transporters belonging to the multi-drug resistance (MDR)-like protein family. These MDRs may directly transport IAA or aid in the polar localization of the second set of transporters, the PIN efflux facilitators.

The past few years have seen much progress in our understanding of PIN genes (Blakeslee et al., 2005; Weijers and Friml, 2009). In Arabidopsis, eight PINs have been identified, with different tissue-specific expression patterns and subcellular localizations (Feraru and Friml, 2008). Among them, PIN1 localizes in xylem parenchyma cells and is thought to transport auxin from shoot to root tips, governing acropetal transport (Blakeslee et al., 2005). PIN2 localizes to root epidermal tissues and apical cortex cells, and may function to redistribute auxin in the gravitropism response. PIN3 localizes laterally in pericycle cells, and may also be involved in lateral re-distribution of auxin. Other PIN genes have their own tissue specificity and functions. Upon reaching the root tip, auxin is believed to move laterally by a PIN3-mediated process before entering the basipetal stream that flows toward the base of the root through the epidermal cells. A small amount of auxin is also produced by the root apical meristem, and enters the basipetal transport stream as well.

One set of transcription factors that might influence auxin flow through affecting PIN localizations consists of members of the HDZIPIII family (Byrne, 2006) whose role in lateral root development is well established from genetic studies in Arabidopsis. Gain-of-function HDZIPIII alleles led to increased lateral roots, while loss-of-function alleles led to a reduced number of lateral roots. The latter phenotype is not a result of decreased auxin sensitivity since lateral root induction in response to exogenous auxin was unaltered in these mutants. Further studies identified mis-localized components of the auxin transport machinery in HDZIPIII mutants (Izhaki and Bowman, 2007). Ectopic localization of PIN1 :: GFP fusions were observed in the embryos of these mutants (Hawker and Bowman, 2004). Additionally, expression of HDZIPIII proteins mirrors the predicted flow of auxin in plants. This and other evidence indicate that HDZIPIII proteins might influence auxin flow directly or indirectly—a flow which is a chief determinant of plant development and organ differentiation.

Flavonoids Regulate Auxin Transport

How HDZIPIII regulates auxin flux is still unclear. However, there is extensive evidence for the role of flavonoids in regulating auxin transport (Santelia et al., 2008). Most evidence stems from combined *in vitro* and genetic studies in Arabidopsis. Flavonoids act similar to synthetic chemical inhibitors (such as NPA) in affecting polar auxin transport, with the flavonols quercetin and kaempferol being the most significant regulators (Brown et al., 2001). In the flavonoid-null mutants of Arabidopsis, root auxin transport is increased compared to the wild type, and this phenotype can be corrected by an external application of flavonols (Peer and Murphy, 2007). Quercetin accumulates at the root tip when auxin concentrations increase. Similar aggregations were also found in roots of mutants of *Atpgp4* (Geisler and Murphy, 2006). Flavonols have also been used to inhibit mammalian P-glycoproteins (PGP) based on their regulatory effects on the catalytic sites.

Genetic studies not only identified two proteins that may bind to flavonoids and regulate polar auxin transport (AtAPM and AtMDR), but they also discovered that the most likely route of interaction between flavonoids and auxin transporters is through PIN homologs (Muday and Murphy, 2002). In *tt4*, the CHS-null mutant (Figure 15.2), PIN1 was mis-localized in many parts of the root, while in mutants, over-accumulating kaempferol and quercetin (*tt3*, Figure 15.2), PIN4 was mis-localized in the root tips and root transition zone. Application of nanomolar concentrations of flavonols to *pin2* mutants was sufficient to restore partial root gravitropism responses. Further evidence suggested that flavonoids promoted asymmetric PIN shifts during gravity stimulation, thus redirecting basipetal auxin streams necessary for root bending (Bandyopadhyay et al., 2007).

Auxin Role in Nodulation

Physiological and pharmacological evidence supports an auxin role in nodulation, at least for indeterminate nodulating plants. There are two major types of nodules formed in legume roots: indeterminate and determinate (Hirsch, 1992). Indeterminate nodules are characterized by a persistent nodule meristem. Initial cell divisions and meristem formation are similar between lateral roots and indeterminate nodules. Indeterminate nodules are elongated due to the addition of new cells to the distal end of the nodule from the meristem. Examples of plants that form indeterminate nodules include temperate legumes such as pea, *Medicago truncatula,* and clover.

Determinate nodules are spherical and lack a nodule meristem. Examples of plants producing determinate nodules include tropical legumes such as soybean, common bean, and *Lotus japonicus*. There is no sustained cell division during determinate nodule development, and cell expansion rather than cell division results in nodule growth. Additionally, determinate nodules arise from outer cortical cell layers, whereas indeterminate nodules arise from inner cortical cell layers. It should be noted that so far as is known, the signal transduction pathways responding to rhizobial Nod factors are highly similar between these two types of nodulating plants. Both types of legumes have similar Nod factor receptors, and a similar set of kinases that can initiate a Ca^{2+}-dependent signal cascade leading to nodulation-specific gene expressions. Also, both types of legumes react to a shoot-derived inhibitory signal upon rhizobial infection and suppress excessive nodule formation. Mutations in this signal transduction pathway lead to either significantly reduced nodule numbers or, in the case of a loss of inhibition, to a supernodulation phenotype with much higher than normal number of nodules. Many components of the overall nodulation signaling cascade are functional in arbuscular mycorrhizal fungus colonization as well, suggesting that rhizobial symbiosis evolved from the more ancient AM symbiosis signaling.

Auxin Has Major Effects on Plants with Indeterminate Nodules

Despite general similarity in Nod Factor recognition and signaling, auxin appears to have different effects on plants with determinate versus indeterminate nodules. In 1992, Hirsch showed that auxin transport inhibitors could induce nodule-like structures in indeterminate legume roots (Hirsch, 1992). Indeed, evidence for the inhibition of auxin transport prior to nodule primordia cell division was obtained later (Boot et al., 1999). In two indeterminate nodulating plants, white clover and vetch, Nod factors and plant flavonoids were reported to induce changes in auxin-inducible gene expression similar to the effect of synthetic auxin transport inhibitors (Pacios-Bras et al., 2003). Examination of auxin accumulation and transport during nodulation has primarily exploited auxin-inducible reporter gene constructs. They indicate different auxin homeostasis during nodulation between the two types of legumes nodules. In indeterminate nodule-forming plants, such as white clover and pea, an absence of auxin-inducible reporter gene expression was observed below the root site of rhizobial inoculation (Mathesius et al., 1998). This suggested a transient block in auxin transport at the rhizobial inoculation site. In contrast, no such block was observed in the determinate plant *L. japonicus*. Microscopic examinations for flavonoid distribution (using a fluorescent dye) in root cross sections were compared to auxin reporter gene expression patterns, and both indeterminate nodulating plants showed consistent correlations between auxin and flavonoid levels at the site of nodule primordia initiation (Mathesius et al., 1998). In contrast, no clear inhibition of auxin transport was observed in the determinate nodulating legumes soybean and *Lotus japonicus* after *Rhizobium* inoculation (Pacios-Bras et al., 2003). More recent genetic evidence is consistent with flavonoid-mediated inhibition of auxin transport being critical to nodulation in the indeterminate nodulating legume *M. truncatula* (see next section).

Using ^{3}H-IAA as a tracer, Mathesius' group measured auxin transport during nodulation in *M. truncatula*. Their results established that indeed a block in auxin transport occurred at the site of rhizobial inoculation (Mathesius et al., 1998). RNAi silencing of flavonoid biosynthesis in

M. truncatula led to increased auxin transport in the transformed roots of composite plants, providing genetic evidence that flavonoids act as auxin transport inhibitors in this indeterminate species (Wasson et al., 2006). Flavonoid-deficient roots lacked the ability to block auxin transport at the site of rhizobial inoculation whereas control roots showed a clear block in auxin transport. That the flavonoid-deficient roots did not form nodule primordia strongly suggested that it was due to the inability to block auxin transport in response to rhizobial inoculation. Further support for a role of auxin in nodule initiation also came from nodulation inhibition by the anti-auxin p-chlorophenoxyisobutyric acid and by RNAi silencing of several PIN genes (Huo et al., 2006).

Auxin plays further multiple roles during subsequent nodule development: It accumulates in the inoculation zone before cortical cell divisions and in the dividing cortical cells, as revealed by the expression of an auxin-responsive promoter (GH3) fused to GUS or GFP. AUX1 (MtLAX) gene expression was localized in nodule primordia in cells close to the vascular bundle of the primary root, suggesting that auxin might be channeled into a growing nodule primordium by AUX1 (Billy et al., 2001; Schnabel and Frugoli, 2004). While many homologs of Arabidopsis *PIN* genes have been identified in *M. truncatula*, only the *PIN2* homolog was induced by rhizobial inoculation. MtPIN2 is localized to the base of young nodules. A detailed expression analysis of *MtPIN2* promoter-driven GUS showed this gene was expressed at the base of lateral roots and nodule primordia close to primary root vasculatures (Huo et al., 2006). RNAi targeted suppression of *MtPIN1* to *PIN4* all significantly reduced nodule numbers, suggesting that auxin transport may play important roles at various stage of nodule development. However, there is as yet no direct evidence that flavonoids are involved in later stages of nodule development. The primary activity of flavonoids appears to be in the initiation of the nodule primordia.

Flavonoid Function in Nodulation

Both the Mathesius group and ours used a combination of metabolic engineering and genetic approaches to address the role of flavonoids in nodulation, in soybean (determinate nodules), and *M. truncatula* (indeterminate nodules) (Subramanian et al., 2006; Wasson et al., 2006; Zhang et al., 2008; Grunewald et al., 2009; Wasson et al., 2009). We silenced expression of several key flavonoid and isoflavone biosynthesis enzymes in transgenic roots of composite plants produced by *Agrobacterium rhizogenes*. "Hairy root composite plants," an excellent model system to study nodulations (Chabaud et al., 2006), are generated by *A. rhizogenes* transformation of cut explants. These explants can be cultured in nonsterile conditions. After emergence of transgenic hairy roots from the cut surface in a few weeks, these composite plants, with nontransgenic shoot tissues and transgenic root tissues, can be transplanted into soil or sand and subjected to *Rhizobium* inoculation. Even though the transgenic roots have altered levels of endogenous auxin and have lost the gravitropic response, nodulation signaling and nodule development do not appear to be disturbed. The nodule formation process, nodulation-induced gene expressions, and nodule numbers are almost identical between wild-type and transgenic roots (Subramanian et al., 2004). Additionally, each individual hairy root is an independent transformation event, allowing us to measure multiple transformed lines from one seedling, increasing the reliability of the phenotypic scoring.

Flavonoid Function in Determinate Nodulating Plants

When both soybean IFS genes were silenced by RNAi, there was an almost 90% reduction in flavonoids in transgenic roots and root exudates (Subramanian et al., 2006). IFS silencing also impaired nodulation by *B. japonicum* and reduced the nodule numbers. External daidzein supplementation of *B. japonicum* cultures, to induce Nod factor production prior to inoculation, did not restore nodulation in the isoflavone-knockdown roots. Therefore, we showed that, in addition to functioning as Nod gene inducers outside the root, isoflavones have essential activities inside the root as well.

To determine the internal function of specific isoflavonoids, we altered the ratios of various isoflavone compounds in soybean roots by CHR silencing (Subramanian et al., 2006). As a result, daidzein and its conjugates, which normally constitute more than 90% of root isoflavones, were reduced, and the isoflavone pathway flux was redirected to genistein, which now constituted 90% of the isoflavones. These alterations had no effect on nodulation. Thus, both internal genistein and daidzein could allow nodule development to proceed, a result consistent with both genistein and daidzein inducing Nod Factor production *in vitro*.

Tissue-specific expression of key isoflavone enzymes was examined using promoter-GUS fusions and more sensitive *in situ* hybridization analysis, revealing that isoflavone biosynthetic enzymes were specifically activated in root hairs and xylem parenchyma cells (Subramanian et al., 2006). Most other tissues under the epidermal layers maintained the same level of isoflavone gene expression during the nodulation process. Since the majority of soybean nodules are formed in the cortex tissues across from xylem poles, we believe that isoflavones are internal Nod gene inducers after initial legume-rhizobia signaling. We hypothesize that xylem-specific synthesis of isoflavone may provide chemo-attractants or a signal gradient directing the developing infection thread to the correct location inside the cortical tissue.

Crucial evidence in support of the above hypothesis is that isoflavone-hypersensitive *Bradyrhizobium* mutants could nodulate isoflavone-RNAi knockdown roots (Subramanian et al., 2006). The bacterial mutants produce Nod factors in the presence of very low concentrations of isoflavones. That these mutants were capable of nodulating isoflavone-knockdown roots suggested that Nod factor production after infection thread formation is essential for the nodulation process to continue beyond initial signaling stages.

Surprisingly, auxin transport was significantly altered in the isoflavone knockdown lines. As indicated by the activity of the DR5-GUS auxin reporter construct, isoflavone-knockdown lines had significantly higher auxin levels compared to vector-transformed controls, suggesting isoflavone played an important role in regulating auxin levels in soybean root (Subramanian et al., 2006). Addition of the synthetic auxin transport inhibitor, NPA, to the vector transformed controls, resulted in an elevation of auxin levels almost to that of the IFS RNAi plants. Apparently, isoflavones in wild-type plants reduce the flux of auxin entering into root elongation regions and root tips. However, this activity is not essential for nodulation. Consistent with previous reports that nodulation did not block normal auxin flow in determinate nodulating plants, our results suggested that isoflavones indeed regulate auxin transport in soybean roots, but that this activity plays a minor role during nodulation when compared to the activity of isoflavone as internal Nod gene inducers (Figure 15.3).

Flavonoid Function in Indeterminate Nodulating Plants

To determine the function of flavonoids and isoflavonoids in indeterminate nodulating plants, we first silenced the IFS genes of *M. truncatula*, just as we did with soybean (Zhang et al., 2008). Silencing of IFS once again led to significant reduction of total (iso)flavonoid compounds: the isoflavones formonectin and biochanin A comprise more than 90% of flavonoid compounds in *Medicago* roots. Surprisingly, this drastic elimination of flavonoids from *Medicago* roots had no effect on nodulation (Zhang et al., 2008), and the nodule numbers and nitrogen

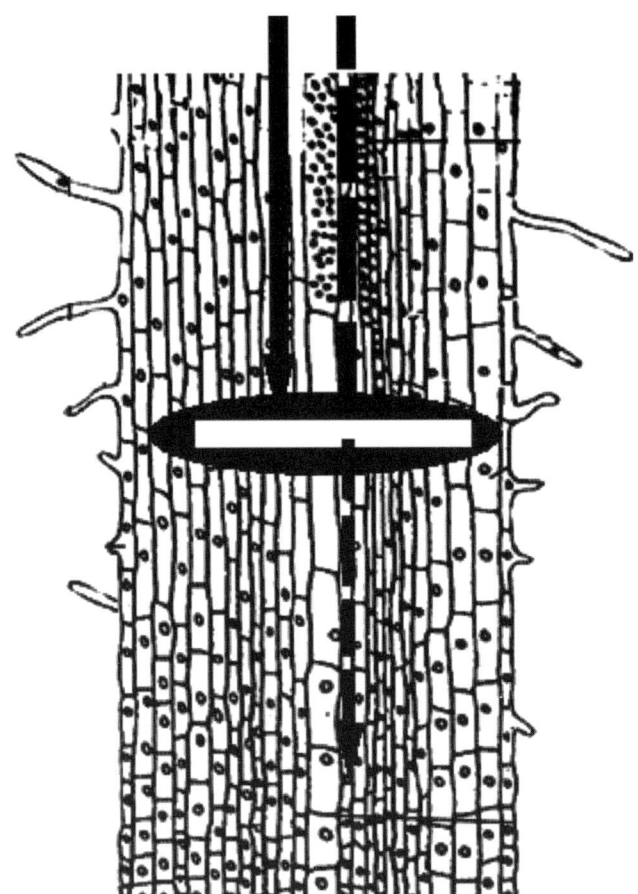

Figure 15.3. Nodulation induced auxin flux changes. The solid arrow on the left indicates that when rhizobia induce localized flavonoid accumulation in indeterminate nodulating plants (such as *Medicago*), auxin transport is blocked by flavonoids, particularly flavonol, leading to primordia development. The dashed arrow on the right indicates that when rhizobia induce localized isoflavonoid accumulation in determinate nodulating plants (such as soybean), polar auxin transport is slightly changed, but not enough to alter primordia development. In both cases, some flavonoids will serve as specific endogenous Nod factor biosynthesis inducers.

fixation activities remained the same. Compounds essential to soybean nodulation proved to be inconsequential in *M. truncatula*. However, when we silenced FNS genes in *Medicago*, nodule numbers were reduced by half (Zhang et al., 2007; Zhang et al., 2008). Dihydroxyflavone, produced by FNS (Figure 15.2), is a minor component of the Medicago root flavonoid profile. Interestingly, silencing of CHR, which removed 5-deoxyflavonoids and 5-deoxyisoflavonoids, showed a similar reduction in nodule numbers (Zhang et al., 2008). CHR silencing had a much more profound impact on the total flavonoid profile than did FNS silencing. The CHR-silenced lines had reductions in both isoflavone formonectin and dihydroxyflavones. Since formonectin was not important for *Medicago* nodulation, as shown in the above IFS RNAi experiment, CHR silencing confirmed that dihydroxyflavone was essential for the nodulation process.

Our silencing of the CHS gene in *Medicago* root confirmed earlier findings by the Mathesius group—that CHS silencing was detrimental to nodulation. The nodule numbers were reduced to almost zero, significantly lower than the FNS- or CHR-silenced lines (Zhang et al., 2008). CHS-silenced lines provided an important "baseline" resource to assay specific flavonoid compounds important for nodulation, simply by feeding the specific metabolites back to the CHS-silenced roots. Addition of dihydroxyflavones to the CHS RNAi root restored the nodule number to half that of normal whereas, upon feeding the same compound to the FNS RNAi root, nodule numbers were fully restored. Upon flavonol kaempferol addition to the CHS RNAi root, nodule numbers again increased to half normal, but flavonol failed to restore FNS RNAi nodule numbers. This discrepancy suggested that flavones and flavonols played distinct roles during nodulation in *Medicago,* and their functions did not overlap. When flavonols were added to the nodulating bacterium, *S. meliloti* and flavones were added to the CHS root, the nodulation was restored to the wild-type levels (Zhang et al., 2008).

We further demonstrated that auxin transport was altered in all three silenced lines, suggesting that both flavones and flavonols regulate auxin transport (Zhang et al., 2008). However, only flavonol regulation of auxin is essential for nodulation. This conclusion is based on the inoculation of CHS RNAi and FNS RNAi lines with a mutant *Rhizobium*, constitutive for Nod factor production: Only FNS RNAi lines were restored to normal nodule numbers. In the CHS RNAi lines that lack all root flavonoids, the constitutive Nod factor-producing rhizobial mutant depended on an external supply of kaempferol to restore normal nodulation (Zhang et al., 2008).

This set of experiments suggested that flavones function as internal Nod factor inducers in *Medicago*, and flavonols or related compounds function as auxin transport regulators during the nodulation (Figure 15.3). The physiological activities of both types of compounds are essential for this process. These requirements are very different from determinate nodulating plants, in which the Nod factor-inducing function is the most important issue.

Perspectives

We have come a long way in our understanding of the functions of phenylpropanoid compounds over the past few decades. We know they are essential for biotic and abiotic stress responses. We know they are essential for legume-*Rhizobium* interactions as well. However, the best-studied natural product pathway still contains many black boxes regarding the role of flavonoid compounds in plants. For example, we still cannot explain why leaves turn yellow during autumn. Similarly, during plant-microbe interactions, it is still not clear which components of auxin sensing and transport are regulated by flavonoid compounds and how they are regulated. Our grasp of auxin signaling in nodule development is particularly limited, primarily due to the lack of auxin signaling or transport mutants in legume species. Clear genetic evidence is emerging primarily from the use of reverse-genetics—the ablation of specific genes and examination of their nodulation and phenotypes. The role of flavonoid-mediated auxin transport inhibition and the role of the auxin efflux carrier PIN proteins in nodulation were identified this way. Although it has been postulated that nodule development might utilize preexisting mechanisms for lateral root initiation, it is not clearly understood if and what role components of the auxin signaling pathway play in nodule development or how their roles differ between the developmental programs of lateral roots and nodules. All of these questions require continued efforts recruiting the combined tools of genetics, molecular biology, cellular biology, biochemistry, and of course, the various –omics.

We believe that the near future will bring in new discoveries and deeper understanding of flavonoid biosynthesis and functions. And, in light of nodulation and its importance to plant strategies for acquiring nitrogen, these discoveries will tell a lot about how plants recruit stress responses to satisfy a basic nutritional need.

References

Akashi, T., Aoki, T., Ayabe, S.-I. 2005. Molecular and biochemical characterization of 2-hydroxyisoflavanone dehydratase. Involvement of carboxylesterase-like proteins in leguminous isoflavone biosynthesis. Plant Physiol 137: 882–891.

Akashi, T., Aoki, T., Ayabe, S. 1998. Identification of a cytochrome P450 cDNA encoding (2S)-flavanone 2-hydroxylase of licorice (*Glycyrrhiza echinata* L.; Fabaceae) which represents licodione synthase and flavone synthase II. FEBS Lett 431: 287–290.

Akashi, T., Aoki, T., Ayabe, S. 1999. Cloning and functional expression of a cytochrome P450 cDNA encoding 2-hydroxyisoflavanone synthase involved in biosynthesis of the isoflavonoid skeleton in licorice. Plant Physiol 121: 821–828.

Akashi, T., Sasaki, K., Aoki, T., Ayabe, S., Yazaki, K. 2009. Molecular cloning and characterization of a cDNA for pterocarpan 4-dimethylallyltransferase catalyzing the key prenylation step in the biosynthesis of glyceollin, a soybean phytoalexin. Plant Physiol 149: 683–693.

Alper, H., Stephanopoulos, G. 2009. Engineering for biofuels: exploiting innate microbial capacity or importing biosynthetic potential? Nat Rev Microbiol 7: 715–723.

Appert, C., Logemann, E., Hahlbrock, K., Schmid, J., Amrhein, N. 1994. Structural and catalytic properties of the four phenylalanine ammonia-lyase isoenzymes from parsley (Petroselinum crispum Nym.). Eur J Biochem 225: 491–499.

Ballance, G.M., Dixon, R.A. 1995. Medicago sativa cDNAs encoding chalcone reductase. Plant Physiol 107: 1027–1028.

Bandyopadhyay, A., Blakeslee, J.J., Lee, O.R., Mravec, J., Sauer, M., Titapiwatanakun, B., Makam, S.N., Bouchard, R., Geisler, M., Martinoia, E., Friml, J., Peer, W.A., Murphy, A.S. 2007. Interactions of PIN and PGP auxin transport mechanisms. Biochem Soc Trans 35: 137–141.

Banfalvi, Z., Nieuwkoop, A., Schell, M., Besl, L., Stacey, G. 1988. Regulation of nod gene expression in *Bradyrhizobium japonicum*. Mol Gen Genet 214: 420–424.

Barbour, W.M., Hattermann, D.R., Stacey, G. 1991. Chemotaxis of *Bradyrhizobium japonicum* to soybean exudates. Appl Environ Microbiol 57: 2635–2639.

Beld, M., Martin, C., Huits, H., Stuitje, A.R., Gerats, A.G. 1989. Flavonoid synthesis in Petunia hybrida: partial characterization of dihydroflavonol-4-reductase genes. Plant Mol Biol 13: 491–502.

Bell-Lelong, D.A., Cusumano, J.C., Meyer, K., Chapple, C. 1997. Cinnamate-4-hydroxylase expression in Arabidopsis. Regulation in response to development and the environment. Plant Physiol 113: 729–738.

Bellato, C.M., Balatti, P.A., Pueppke, S.G., Krishnan, H.B. 1996. Proteins from cells of Rhizobium fredii bind to DNA sequences preceding nolX, a flavonoid-inducible nod gene that is not associated with a nod box. Mol Plant Microbe Interact 9: 457–463.

Billy, F.D., Grosjean, C., May, S., Bennett, M., Cullimore, J.V. 2001. Expression studies on AUX1-like genes in *Medicago truncatula* suggest that auxin is required at two steps in early nodule development. Mol Plant Microbe Interact 14: 267–277.

Blakeslee, J.J., Peer, W.A., Murphy, A.S. 2005. Auxin transport. Curr Opin Plant Biol 8: 494–500.
Boerjan, W., Cervera, M.T., Delarue, M., Beeckman, T., Dewitte, W., Bellini, C., Caboche, M., Van Onckelen, H., Van Montagu, M., Inze, D. 1995. Superroot, a recessive mutation in Arabidopsis, confers auxin overproduction. Plant Cell 7: 1405–1419.
Boot, K.J.M., Van Brussel, A.N., Tak, T., Spaink, H.P., Kijne, J.W. 1999. Lipochitin oligosaccharides from *Rhizobium leguminosarum* bv. *viciae* reduce auxin transport capacity in *Vicia sativa* subsp. *nigra* roots. Mol Plant Microbe Interact 12: 839–844.
Brazier-Hicks, M., Evans, K.M., Gershater, M.C., Puschmann, H., Steel, P.G., Edwards, R. 2009. The C-glycosylation of flavonoids in cereals. J Biol Chem 284: 17926–17934.
Brown, D.E., Rashotte, A.M., Murphy, A.S., Normanly, J., Tague, B.W., Peer, W.A., Taiz, L., Muday, G.K. 2001. Flavonoids act as negative regulators of auxin transport in vivo in arabidopsis. Plant Physiol 126: 524–535.
Byrne, M.E. 2006. Shoot meristem function and leaf polarity: the role of class III HD-ZIP genes. PLoS Genet 2: e89.
Carroll, B.J., McNeil, D.L., Gresshoff, P.M. 1985. Isolation and properties of soybean [Glycinemax (l) Merr] mutants that nodulate in the presence of high nitrate concentrations. Proc Natl Acad Sci USA 82: 4162–4166.
Casimiro, I., Beeckman, T., Graham, N., Bhalerao, R., Zhang, H., Casero, P., Sandberg, G., Bennett, M.J. 2003. Dissecting Arabidopsis lateral root development. Trends Plant Sci 8: 165–171.
Casimiro, I., Marchant, A., Bhalerao, R.P., Beeckman, T., Dhooge, S., Swarup, R., Graham, N., Inze, D., Sandberg, G., Casero, P.J., Bennett, M. 2001. Auxin transport promotes Arabidopsis lateral root initiation. Plant Cell 13: 843–852.
Celenza, J.L., Jr., Grisafi, P.L., Fink, G.R. 1995. A pathway for lateral root formation in *Arabidopsis thaliana*. Genes Dev 9: 2131–2142.
Chabaud, M., Boisson-Dernier, A., Zhang, J., Taylor, C.G., Yu, O., Barker, D.G. 2006. Agrobacterium rhizogenes-mediated root transformation. In: Mathesius, U., Ed., The *Medicago Truncatula* Handbook. http://www.noble.org/MedicagoHandbook/
Charrier, B., Coronado, C., Kondorosi, A., Ratet, P. 1995. Molecular characterization and expression of alfalfa (Medicago sativa L.) flavanone-3-hydroxylase and dihydroflavonol-4-reductase encoding genes. Plant Mol Biol 29: 773–786.
Cone, K.C., Burr, F.A., Burr, B. 1986. Molecular analysis of the maize anthocyanin regulatory locus C1. Proc Natl Acad Sci USA 83: 9631–9635.
Dixon, R.A. 2005. Engineering of plant natural product pathways. Curr Opin in Plant Biol 8: 329–336.
Dixon, R.A., Lamb, C.J., Masoud, S., Sewalt, V.J., Paiva, N.L. 1996. Metabolic engineering: prospects for crop improvement through the genetic manipulation of phenylpropanoid biosynthesis and defense responses—A review. Gene 179: 61–71.
Downie, J.A., Walker, S.A. 1999. Plant responses to nodulation factors. Curr Opin Plant Biol 2: 483–489.
Dubos, C., Le Gourrierec, J., Baudry, A., Huep, G., Lanet, E., Debeaujon, I., Routaboul, J.M., Alboresi, A., Weisshaar, B., Lepiniec, L. 2008. MYBL2 is a new regulator of flavonoid biosynthesis in *Arabidopsis thaliana*. Plant J 55: 940–953.
Faktor, O., Kooter, J.M., Dixon, R.A., Lamb, C.J. 1996. Functional dissection of a bean chalcone synthase gene promoter in transgenic tobacco plants reveals sequence motifs essential for floral expression. Plant Mol Biol 32: 849–859.
Feraru, E., Friml, J. 2008. PIN polar targeting. Plant Physiol 147: 1553–1559.

Ferguson, B.J., Mathesius, U. 2003. Signaling interactions during nodule development. J Plant Growth Regul 22(1): 47–72.

Ferrer, J.L., Austin, M.B., Stewart, C., Jr., Noel, J.P. 2008. Structure and function of enzymes involved in the biosynthesis of phenylpropanoids. Plant Physiol Biochem 46: 356–370.

Fisher, R.F., Long, S.R. 1992. Rhizobium—plant signal exchange. Nature 357: 655–660.

Franke, R., Humphreys, J.M., Hemm, M.R., Denault, J.W., Ruegger, M.O., Cusumano, J.C., Chapple, C. 2002. The Arabidopsis REF8 gene encodes the 3-hydroxylase of phenylpropanoid metabolism. Plant J 30: 33–45.

Fukaki, H., Tasaka, M. 2009. Hormone interactions during lateral root formation. Plant Mol Biol 69: 437–449.

Geisler, M., Murphy, A.S. 2006. The ABC of auxin transport: the role of p-glycoproteins in plant development. FEBS Lett 580: 1094–1102.

Graham, T.L., Graham, M.Y., Yu, O. 2008. Genomics of secondary metabolism in soybean. In: Stacey, G., Ed., Genetics and Genomics of Soybean. Springer, New York, pp. 211–242.

Grunewald, W., van Noorden, G., Van Isterdael, G., Beeckman, T., Gheysen, G., Mathesius, U. 2009. Manipulation of Auxin Transport in Plant Roots during Rhizobium Symbiosis and Nematode Parasitism. Plant Cell 21: 2553–2562.

Hawker, N.P., Bowman, J.L. 2004. Roles for Class III HD-Zip and KANADI genes in Arabidopsis root development. Plant Physiol 135: 2261–2270.

Hernandez, J.M., Feller, A., Morohashi, K., Frame, K., Grotewold, E. 2007. The basic helix loop helix domain of maize R links transcriptional regulation and histone modifications by recruitment of an EMSY-related factor. Proc Natl Acad Sci USA 104: 17222–17227.

Hirsch, A.M. 1992. Developmental biology of legume nodulation. New Phytol 122: 211–237.

Hirsch, A.M., LaRue, T.A. 1997. Is the legume nodule a modified root or stem or an Organ sui generis? Crit Rev Plant Sci 16: 361–392.

Hobbie, L., Estelle, M. 1995. The axr4 auxin-resistant mutants of Arabidopsis thaliana define a gene important for root gravitropism and lateral root initiation. Plant J 7: 211–220.

Hobbie, L., McGovern, M., Hurwitz, L.R., Pierro, A., Liu, N.Y., Bandyopadhyay, A., Estelle, M. 2000. The axr6 mutants of Arabidopsis thaliana define a gene involved in auxin response and early development. Development 127: 23–32.

Holton, T.A., Brugliera, F., Tanaka, Y. 1993. Cloning and expression of flavonol synthase from Petunia hybrida. Plant J 4: 1003–1010.

Huo, X., Schnabel, E., Hughes, K., Frugoli, J. 2006. RNAi phenotypes and the localization of a protein::GUS fusion imply a role for *Medicago truncatula* PIN genes in nodulation. J Plant Growth Regul 25(2): 56–165.

Izhaki, A., Bowman, J.L. 2007. KANADI and class III HD-Zip gene families regulate embryo patterning and modulate auxin flow during embryogenesis in Arabidopsis. Plant Cell 19: 495–508.

Jung, W., Yu, O., Lau, S.M., O'Keefe, D.P., Odell, J., Fader, G., McGonigle, B. 2000. Identification and expression of isoflavone synthase, the key enzyme for biosynthesis of isoflavones in legumes. Nat Biotechnol 18: 208–212.

Kimura, Y., Aoki, T., Ayabe, S. 2001. Chalcone isomerase isozymes with different substrate specificities towards 6′-hydroxy- and 6′-deoxychalcones in cultured cells of *Glycyrrhiza echinata*, a leguminous plant producing 5-deoxyflavonoids. Plant Cell Physiol 42: 1169–1173.

Kobayashi, H., Naciri-Graven, Y., Broughton, W.J., Perret, X. 2004. Flavonoids induce temporal shifts in gene-expression of nod-box controlled loci in Rhizobium sp. NGR234. Mol Microbiol 51: 335–347.

Korkina, L.G. 2007. Phenylpropanoids as naturally occurring antioxidants: from plant defense to human health. Cell Mol Biol (Noisy-le-grand) 53: 15–25.

Lee, D., Douglas, C.J. 1996. Two Divergent Members of a Tobacco 4-Coumarate:Coenzyme A Ligase (4CL) Gene Family (cDNA Structure, Gene Inheritance and Expression, and Properties of Recombinant Proteins). Plant Physiol 112: 193–205.

Loh, J., Garcia, M., Stacey, G. 1997. NodV and NodW, a second flavonoid recognition system regulating nod gene expression in Bradyrhizobium japonicum. J Bacteriol 179: 3013–3020.

Magori, S., Kawaguchi, M. 2009. Long-distance control of nodulation: molecules and models. Mol Cells 27: 129–134.

Marchant, A., Bhalerao, R., Casimiro, I., Eklof, J., Casero, P.J., Bennett, M., Sandberg, G. 2002. AUX1 promotes lateral root formation by facilitating indole-3-acetic acid distribution between sink and source tissues in the Arabidopsis seedling. Plant Cell 14: 589–597.

Markmann, K., Parniske, M. 2009. Evolution of root endosymbiosis with bacteria: How novel are nodules? Trends Plant Sci 14: 77–86.

Martin, C., Paz-Ares, J. 1997. MYB transcription factors in plants. Trends Genet 13: 67–73.

Martinez-Romero, E. 2009. Coevolution in Rhizobium-legume symbiosis? DNA Cell Biol 28: 361–370.

Mathesius, U. 2009. Comparative proteomic studies of root-microbe interactions. J Proteom 72: 353–366.

Mathesius, U., Schlaman, H.R.M., Spaink, H.P., Sautter, C., Rolfe, B.G., Djordjevic, M.A. 1998. Auxin transport inhibition precedes root nodule formation in white clover roots and is regulated by flavonoids and derivatives of chitin oligosaccharides. Plant J 14: 23–34.

McIver, J., Djordjevic, M.A., Weinman, J.J., Bender, G.L., Rolfe, B.G. 1989. Extension of host range of Rhizobium leguminosarum bv. trifolii caused by point mutations in nodD that result in alterations in regulatory function and recognition of inducer molecules. Mol Plant Microbe Interact 2: 97–106.

Memelink, J. 2005. The use of genetics to dissect plant secondary pathways. Curr Opin Plant Biol 8: 230–235.

Morohashi, K., Grotewold, E. 2009. A systems approach reveals regulatory circuitry for Arabidopsis trichome initiation by the GL3 and GL1 selectors. PLoS Genet 5: e1000396.

Mravec, J., Kubes, M., Bielach, A., Gaykova, V., Petrasek, J., Skupa, P., Chand, S., Benkova, E., Zazimalova, E., Friml, J. 2008. Interaction of PIN and PGP transport mechanisms in auxin distribution-dependent development. Development 135: 3345–3354.

Muday, G.K., Murphy, A.S. 2002. An emerging model of auxin transport regulation. Plant Cell 14: 293–299.

Nishimura, R., Ohmori, M., Kawaguchi, M. 2002. The novel symbiotic phenotype of enhanced-nodulating mutant of Lotus japonicus: astray mutant is an early nodulating mutant with wider nodulation zone. Plant Cell Physiol 43: 853–859.

Pacios-Bras, C., Schlaman, H.R.M., Boot, K., Admiraal, P., Langerak, J.M., Stougaard, J., Spaink, H.P. 2003. Auxin distribution in Lotus japonicus during root nodule development. Plant Mol Biol 52: 1169–1180.

Peer, W.A., Murphy, A.S. 2007. Flavonoids and auxin transport: modulators or regulators? Trends Plant Sci 12: 556.

Peters, N.K., Frost, J.W., Long, S.R. 1986. A plant flavone, luteolin, induces expression of Rhizobium meliloti nodulation genes. Science 233: 977–980.

Rabinowicz, P.D., Grotewold, E. 2000. A novel reverse-genetic approach (SIMF) identifies Mutator insertions in new Myb genes. Planta 211: 887–893.

Ralston, L., Subramanian, S., Matsuno, M., Yu, O. 2005. Partial reconstruction of flavonoid and isoflavonoid biosynthesis in yeast using soybean type I and type II chalcone isomerases. Plant Physiol 137: 1375–1388.

Robert HS, Friml J 2009. Auxin and other signals on the move in plants. Nat Chem Biol 5: 325–332.

Ryder, T.B., Hedrick, S.A., Bell, J.N., Liang, X.W., Clouse, S.D., Lamb, C.J. 1987. Organization and differential activation of a gene family encoding the plant defense enzyme chalcone synthase in Phaseolus vulgaris. Mol Gen Genet 210: 219–233.

Santelia, D., Henrichs, S., Vincenzetti, V., Sauer, M., Bigler, L., Klein, M., Bailly, A., Lee, Y., Friml, J., Geisler, M., Martinoia, E. 2008. Flavonoids redirect PIN-mediated polar auxin fluxes during root gravitropic responses. J Biol Chem 283: 31218–31226.

Schnabel EL, Frugoli J 2004. The PIN and LAX families of auxin transport genes in *Medicago truncatula*. Mol Genet Genomics 272: 420–432.

Sprent, J.I. 2007. Evolving ideas of legume evolution and diversity: a taxonomic perspective on the occurrence of nodulation. New Phytol 174: 11–25.

Sprent, J.I. 2008. 60Ma of legume nodulation. What's new? What's changing? J Exp Bot 59: 1081–1084.

Steele, C.L., Gijzen, M., Qutob, D., Dixon, R.A. 1999. Molecular characterization of the enzyme catalyzing the aryl migration reaction of isoflavonoid biosynthesis in soybean. Arch Biochem Biophys 367: 146–150.

Subramanian, S., Stacey, G., Yu, O. 2006. Endogenous isoflavones are essential for the establishment of symbiosis between soybean and *Bradyrhizobium japonicum*. Plant J 48: 261–273.

Subramanian, S., Xu, L., Lu, G., Odell, J., Yu, O. 2004. The promoters of the isoflavone synthase genes respond differentially to nodulation and defense signals in transgenic soybean roots. Plant Mol Biol 54: 623–639.

Sullivan, M. 2009. A novel red clover hydroxycinnamoyl transferase has enzymatic activities consistent with a role in phaselic acid biosynthesis. Plant Physiol 150: 1866–1879.

Tamagnone, L., Merida, A., Parr, A., Mackay, S., Culianez-Macia, F.A., Roberts, K., Martin, C. 1998. The AmMYB308 and AmMYB330 transcription factors from antirrhinum regulate phenylpropanoid and lignin biosynthesis in transgenic tobacco. Plant Cell 10: 135–154.

van Tunen, A.J., Koes, R.E., Spelt, C.E., van der Krol, A.R., Stuitje, A.R., Mol, J.N. 1988. Cloning of the two chalcone flavanone isomerase genes from Petunia hybrida: coordinate, light-regulated and differential expression of flavonoid genes. Embo J 7: 1257–1263.

Vanneste, S., Friml, J. 2009. Auxin: a trigger for change in plant development. Cell 136: 1005–1016.

Ververidis, F., Trantas, E., Douglas, C., Vollmer, G., Kretzschmar, G., Panopoulos, N. 2007. Biotechnology of flavonoids and other phenylpropanoid-derived natural products. Part II: Reconstruction of multienzyme pathways in plants and microbes. Biotechnol J 2: 1235–1249.

Vierstra, R.D. 2009. The ubiquitin-26S proteasome system at the nexus of plant biology. Nat Rev Mol Cell Biol 10: 385–397.

Wagner, A., Ralph, J., Akiyama, T., Flint, H., Phillips, L., Torr, K., Nanayakkara, B., Te Kiri, L. 2007. Exploring lignification in conifers by silencing hydroxycinnamoyl-CoA:shikimate hydroxycinnamoyltransferase in Pinus radiata. Proc Natl Acad Sci USA 104: 11856–11861.

Wasson, A.P., Pellerone, F.I., Mathesius, U. 2006. Silencing the flavonoid pathway in *Medicago truncatula* inhibits root nodule formation and prevents auxin transport regulation by Rhizobia. Plant Cell 18: 1617–1629.

Wasson, A.P., Ramsay, K., Jones, M.G., Mathesius, U. 2009. Differing requirements for flavonoids during the formation of lateral roots, nodules and root knot nematode galls in *Medicago truncatula*. New Phytol 183: 167–179.

Weijers, D., Friml, J. 2009. SnapShot: Auxin signaling and transport. Cell 136: 1172–1172el.

Wen, F., Nair, N.U., Zhao, H. 2009. Protein engineering in designing tailored enzymes and microorganisms for biofuels production. Curr Opin Biotechnol 20: 412–419.

Weng, J.K., Li, X., Bonawitz, N.D., Chapple, C. 2008. Emerging strategies of lignin engineering and degradation for cellulosic biofuel production. Curr Opin Biotechnol 19: 166–172.

Winkel-Shirley, B. 2001. Flavonoid biosynthesis. A colorful model for genetics, biochemistry, cell biology, and biotechnology. Plant Physiol 126: 485–493.

Yu, O., Jez, J.M. 2008. Nature's assembly line: biosynthesis of simple phenylpropanoids and polyketides. Plant J 54: 750–762.

Zhang, J., Subramanian, S., Stacey, G., Yu, O. 2008. Flavones and flavonols play distinct critical roles during nodulation of *Medicago truncatula* by *Sinorhizobium meliloti*. Plant J 57: 171–183.

Zhang, J., Subramanian, S., Zhang, Y., Yu, O. 2007. Flavone synthases from *Medicago truncatula* are flavanone-2-hydroxylases and are important for nodulation. Plant Physiol 144: 741–751.

Zhao, J., Dixon, R.A. 2009. MATE transporters facilitate vacuolar uptake of epicatechin 3'-O-glucoside for proanthocyanidin biosynthesis in *Medicago truncatula* and *Arabidopsis*. Plant Cell.

Zhao, M., Morohashi, K., Hatlestad, G., Grotewold, E., Lloyd, A. 2008. The TTG1-bHLH-MYB complex controls trichome cell fate and patterning through direct targeting of regulatory loci. Development 135: 1991–1999.

Zhao, Y., Christensen, S.K., Fankhauser, C., Cashman, J.R., Cohen, J.D., Weigel, D., Chory, J. 2001. A role for flavin monooxygenase-like enzymes in auxin biosynthesis. Science 291: 306–309.

Chapter 16
Plant Hormones and Initiation of Legume Nodulation and Arbuscular Mycorrhization

Arijit Mukherjee and Jean-Michel Ané

Abbreviations

ABA	Abscisic acid
AM	Arbuscular mycorrhizae
AMF	Arbuscular mycorrhizal fungi
bit1	Branching infection threads 1
CCaMK	Calcium/calmodulin-dependent protein kinase
CRE1	Cytokinin receptor
DMI1/2/3	Doesn't make infections
DNF1/4/5/7	Does not fix nitrogen
EFD	Ethylene response factor required for nodule differentiation
EIN	Ethylene insensitive
ENOD	Early nodulation gene
ERN1/2/3	ERF required for nodulation
GA	Gibberellins
HCL	Hair curling
HMGR1	3-hydroxyl 3-methylglutaryl coenzyme A reductase 1
IGN1	Ineffective greenish nodules
IPD3	Interacting protein of DMI3
IT	Infection thread
itd1/3/4	Infection thread deficient
JA	Jasmonic acid
LATD	Lateral root organ-defective
LCO	Lipochitooligosaccharide
LHK1	Lotus histidine kinase
lin	Lumpy infections

Ecological Aspects of Nitrogen Metabolism in Plants, First Edition. Edited by Joe C. Polacco and Christopher D. Todd.
© 2011 by John Wiley & Sons, Inc. Published 2011 by John Wiley & Sons, Inc.

LNP	Lectin nucleotide phosphohydrolase
lot1	Low nodulation and trichome distortion
LPC	Lysophosphatidylcholine
LRR-RLK	Leucine rich repeat – receptor like kinase
LTP	Lipid transfer protein
LYK	LysM domain-containing receptor-like kinases
mcbep	Mycorrhizal colonization blocked at epidermis
MeJA	Methyl jasmonate
NF	Nodulation factors
NFBS1/2/3	Nod factor binding sites
NFP	Nod factor perception
NFR1/5	Nod factor receptor
NIN	Nodule inception
nip	Numerous infections and polyphenolics
Nod	Nodulation
nope1	No perception 1
NORK	Nodulation receptor-like kinase
NSP1/2	Nodulation signalling pathway
NUP85/133	Nucleoporin
PAM	Peri-arbuscular membrane
pdl	Poodle
PiT	Pre-infection thread
pmi1/2	Pre-mycorrhizal infection
PPA	Pre-penetration apparatus
PT4	Phosphate transporter 4
rmc	Reduced mycorrhizal colonization
RNS	Root nodule symbiosis
rpg	Rhizobium-directed polar growth
SA	Salicylic acid
SKL	Sickle
SYMRK	Symbiosis receptor kinase
taci1	Taciturn 1

Introduction

The availability of food increasingly depends on the availability of water and plant nutrients such as nitrogen, phosphate, and potassium. In most natural and agronomic ecosystems, the availability of nitrogen is a major limitation for plant growth. Unfortunately, in intensive agricultural settings, nitrogen is often provided in the form of nitrate fertilizers that leach into groundwater, form volatile nitrogen oxides, and pose serious health and environmental issues. The most abundant form of nitrogen, atmospheric dinitrogen gas (N_2), contains a triple covalent bond that makes N_2 unavailable to most organisms (Smil, 1997; Graham and Vance, 2003). The conversion of N_2 into nitrate requires a significant amount of energy provided, in industrial settings, by the use of natural gas. One ton of natural gas is necessary to produce one ton of fertilizer; therefore, the cost of nitrogen fertilizers is directly correlated with the increasing cost of fossil fuels. Thus, ecological, health, and economical factors make imminent the development of alternatives to nitrogen fertilizers for sustainable agriculture.

Plants often overcome their nutritional constraints by their ability to form symbiotic associations with soil microbes such as rhizobia and mycorrhizal fungi. Leguminous plants are able to develop a root nodule symbiosis (RNS) with nitrogen-fixing rhizobia that provide plants with nitrogen in return for carbon. Similarly, filamentous bacteria from the genus *Frankia* are able to develop RNS with actinorhizal plants (Gherbi et al., 2008). Nitrogen fixation is the conversion of N_2 into forms assimilable by the plant. It occurs in RNS through the activity of a bacterial nitrogenase expressed in specialized organs, the root nodules (Hirsch, 1992; Mylona, Pawlowski, and Bisseling, 1995). Associations between plant roots and fungi, called mycorrhizae, also help plants take up water and nutrients such as nitrogen and phosphate (Hawkins, Johansen, and George, 2000; Hodge, Campbell, and Fitter, 2001; Govindarajulu et al., 2005; Guether et al., 2009). In addition, mycorrhizal associations protect host plants against biotic and abiotic stresses (Subramanian and Charest, 1995; Trotta et al., 1996; El-Tohamy et al., 1999; Matsubara, Ohba, and Fukui, 2001). This ancient symbiosis probably allowed the colonization of land by plants more than 460 million years ago (Simon et al., 1993). Arbuscular mycorrhizae (AM) associations are present throughout the plant kingdom including angiosperms, gymnosperms, pteridophytes, and some bryophytes (Smith and Gianinazzi-Pearson, 1988; Harrison, 2005; Wang and Qiu, 2006). This association results in the formation of tree-shaped subcellular structures, known as arbuscules, which are the main sites of nutrient exchange between plant roots and AM fungi.

The establishment of RNS and AM depends on a coordinated signal exchange between host plants and their microbial partners. Plant strigolactones play a major role in AM initiation (Gomez-Roldan et al., 2007). Plant hormones are also involved in the establishment and regulation of these symbiotic processes in response to developmental and environmental factors. Striking similarities between RNS and AM emerged with respect to mutual recognition, infection process, genetic and hormonal regulation, similarities suggesting that RNS evolved from the ancestral AM symbiosis (Bonfante and Genre, 2008). This review focuses on signals exchanged between plants and their microbial symbionts, the perception of microbial signals by host plants, and the regulation of these processes by plant hormones.

Host Plant Signals

Root exudates often mediate the interactions between plants and their environment (Bais et al., 2006). This is particularly true for beneficial plant-microbe interactions, like RNS and AM, where the release of signals into the rhizosphere initiates such associations.

Flavonoids and Isoflavonoids

Flavonoids and isoflavonoids are a wide range of molecules that regulate interactions between plants and symbiotic or pathogenic microbes. More than 4,000 flavonoids are produced by vascular plants (Perret, Staehelin, and Broughton, 2000). Several flavonoids and isoflavonoids are found exclusively in legumes and play multiple roles in RNS.

Rhizobia can recognize specific flavonoids and isoflavonoids produced by host plants. For instance, *Bradyrhizobium japonicum* responds to daidzein and genistein, isoflavonoids secreted by soybean whereas *Sinorhizobium meliloti* responds to luteolin, secreted by alfalfa. Flavonoids and isoflavonoids allow rhizobia to recognize potential host plants (Peters, Frost, and Long, 1986). They act as inducers of *nod* gene expression and stimulate rhizobial chemotaxis which concentrates compatible rhizobia on the root surface (Rolfe, 1988; Stougaard, 2000)

Plant Hormones and Initiation of Legume Nodulation and Arbuscular Mycorrhization 357

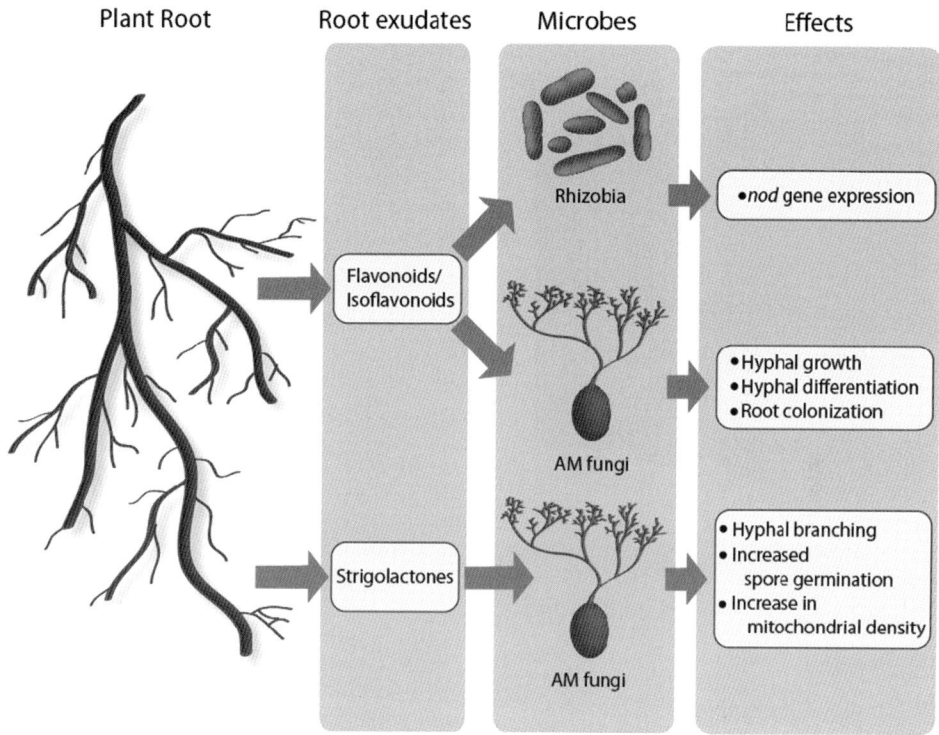

Figure 16.1. Effect of root exudates on rhizobia and AMF. In RNS, flavonoids and isoflavonoids secreted by host plant roots are perceived by interacting compatible rhizobia. Following their perception, *nod* gene expression is induced in rhizobia, and this leads to production of Nod factors. Similarly, in AM, flavonoids and isoflavonoids from root exudates result in hyphal growth, hyphal differentiation in AMF, and increased root colonization. Strigolactones have been identified as "branching factor" signals, secreted by host plant roots, which cause hyphal branching and increases both spore germination and mitochondrial density in AMF. **RNS**, root nodule symbiosis; **AM**, arbuscular mycorrhizae; **AMF**, arbuscular mycorrhizal fungi.

(Figure 16.1). Genetic analyses identified the bacterial NodD protein, a LysR-type regulator, as the major target of flavonoid action. Flavonoids interact with NodD and induce a conformational change that allows this protein to bind *nod* box elements in the *nod* promoters (Perret, Staehelin, and Broughton, 2000; Long, 2001). Expression of *nod* genes leads to the synthesis and excretion of Nod factors (NF), which are a major determinant of host specificity and required for nodule development (Perret, Staehelin, and Broughton 2000; Dénarié, Debellé, and Promé, 1996; Kobayashi et al., 2004). NFs induce the accumulation of flavonoids resulting in an increased secretion of flavonoids, which, in turn, stimulates NF production (Recourt et al., 1992; Dakora, Joseph, and Phillips, 1993; Schmidt, Broughton, and Werner, 1994).

Besides initiating this symbiotic dialogue, flavonoids play a crucial role as auxin transport regulators during nodule development. Cortical cell divisions are necessary for the formation of nodule primordia. This step is often preceded by an inhibition of auxin transport, which is mediated by flavonoids (Mathesius et al., 1998b). Endogenous flavonoids are required after penetration of rhizobia when exogenous flavonoids are no longer available (Wasson, Pellerone, and Mathesius, 2006; Subramanian, Stacey, and Yu, 2006). Nodules can be grouped into two categories, indeterminate and determinate, based on their respective abilities to maintain a functional meristem or not. The type of nodule is dependent on the host plant and is correlated with the location of cortical cell divisions: outer cortex for determinate nodules and inner cortex for indeterminate nodules (Hirsch, 1992). Silencing of the flavonoid pathway in roots of the indeterminate model legume, *Medicago truncatula*, affects auxin transport at the site of rhizobial inoculation. Flavonoid-deficient plants are unable to form nodules upon rhizobial inoculation indicating that a flavonoid-regulated inhibition of auxin transport is required for indeterminate nodule formation (Wasson, Pellerone, and Mathesius, 2006). In a similar study on soybean, a determinate nodulating legume, isoflavonoid-depleted roots are deficient in both auxin transport inhibition and nodulation (Subramanian, Stacey, and Yu, 2006). However, auxin transport inhibition by isoflavones is not required for determinate nodule formation.

Flavonoids and isoflavonoids are involved in communication between plants and arbuscular mycorrhizal fungi (AMF), which also recognize the presence of compatible host plants through root exudates (Chabot et al., 1992; Morandi, 1996; Scervino et al., 2005a, 2005b, 2005c). These signals affect hyphal growth, hyphal differentiation, and root colonization (Siqueira, Safir, and Nair, 1991; Tsai and Phillips, 1991). For instance, formononetin and biochanin A stimulate colonization and growth of clover roots, and quercetin-3-O-galactoside from *Medicago sativa* roots promotes spore germination of *Glomus etunicatum* and *Glomus marocarpum in vitro* (Siqueira, Safir, and Nair, 1991; Tsai and Phillips, 1991) (Figure 16.1). The stimulatory effect on hyphal growth is dependent on the flavonoid structure (Chabot et al., 1992; Bécard, Douds, and Pfeffer, 1992). In addition, flavonoids display a species- and genus-specific stimulatory effect on the pre-symbiotic growth of AMF (Scervino et al., 2005b). Flavonoids also play regulatory roles once host plants are colonized by AMF. Flavonoid levels vary with the phosphorus (P) status of the plant (Guenoune et al., 2001; Larose et al., 2002). For instance, in alfalfa plants with a low P status, flavonoids such as coumestrol, that stimulate hyphal growth and root colonization, accumulated at higher levels than the inhibitory phytoalexin-like medicarpin (Guenoune et al., 2001; Larose et al., 2002). The P status of the plant determines the level of flavonoid accumulation, which in turn regulates the accumulation of phytoalexins during AM symbiosis (Larose et al., 2002; Harrison and Dixon 1993; Catford et al., 2006).

Strigolactones

Hyphal branching in the presence of host plants is one of the most crucial stages in the life cycle of AMF. Hyphal branching enables AM fungi to establish contact with host roots and subsequently form a successful symbiosis (Buée et al., 2000). Since this process does not occur in the vicinity of roots from nonhosts such as white lupin and rapeseed, hyphal branching is considered to be tightly linked to host recognition (Giovannetti et al., 1993b). The "branching factor" is a plant signal that induces hyphal morphogenesis prior to root colonization (Buée et al., 2000). This signal has been detected in a wide range of host plants including maize, pea, tomato, and carrots. Root exudates of maize mutants affected in a chalcone synthase gene, which encodes a key enzyme of flavonoid biosynthesis, display hyphal branching similar to the wild-type plants that ruled out flavonoids as branching factor candidates (Buée et al. 2000). Plant sesquiterpenes have been identified as inducers of AMF hyphal branching in the root exudates of the model legume *Lotus japonicus* using spectroscopic analysis and chemical synthesis. This branching factor was identified as a strigolactone, 5-deoxy-strigol (Akiyama, Matsuzaka, and Hayashi, 2005).

Strigolactones are a group of sesquiterpene lactones that are exuded by plant roots and had been initially identified as seed-germination stimulants for parasitic weeds such as *Striga* and *Orobanche* (Bouwmeester et al., 2003). The strigolactones are derived from the carotenoid pathway, and root colonization by *G. rosea* was reduced significantly in maize mutants (y9) affected in this biosynthetic pathway (Matusova et al., 2005). Treatment of maize plants with fluridone, an inhibitor of the carotenoid pathway, also affected root colonization by AMF indicating that the production of strigolactones has a direct effect on the establishment of AM (Gomez-Roldan et al., 2007). Pea and rice mutants affected in strigolactone synthesis also show similar phenotypes (Umehara et al., 2008).

Strigolactones have been detected in root exudates of not only mycorrhizal plants but also in non-mycorrhizal ones such as white lupin and *Arabidopsis thaliana* (Besserer et al., 2006; Goldwasser et al., 2008). However, the amount of strigolactones exuded by the roots of white lupin is considerably lower than in mycotrophic plants (Yoneyama et al., 2008). Orobanchyl acetate and 5-deoxystrigol are the major strigolactones produced, and most plants exude a mixture of at least two strigolactones (Yoneyama et al., 2008). Orobanchol and orobanchyl acetate are distributed among eudicots while 5-deoxy strigol is widely distributed in both eudicots and monocots (Yoneyama et al., 2008).

Natural strigolactones (5-deoxy-strigol, sorgolactone, and strigol) as well as GR24, a synthetic analogue, induce extensive hyphal branching in *Gigaspora margarita* at $0.2\,\mu M$ after only 24 hours of treatment (Akiyama and Hayashi, 2006). These compounds are exuded by roots at very low concentrations and are highly unstable (Akiyama and Hayashi, 2006). Strigolactones from sorghum induce rapid changes in density, shape, and movement of mitochondria of *G. rosea* (Besserer et al., 2006). The strong and rapid response of AMF to low strigolactone concentrations suggests the existence of a sensitive perception mechanism for strigolactones in AMF. Spore germination was also enhanced in *G. intraradices* and *G. clarroideum* (Besserer et al., 2006) (Figure 16.1.). In addition, strigolactones induce gene expression of *G. margarita* Cu/Zn superoxide dismutase and chemotropic growth of *G. mosseae* hyphae (Lanfranco, Novero, and Bonfante, 2005; Sbrana and Giovannetti, 2005).

The P-status of plants regulates the branching effect of strigolactones in AMF. Hyphal branching is reduced in the presence of root exudates from high-P plants when compared to low-P plants and strigolactone production is reduced in high-P plants (Buée et al., 2000; Nagahashi, Douds, and Abney, 1996; Yoneyama et al., 2007a). Similarly, nitrogen (N) deficiency in sorghum promotes the production and exudation of strigolactones (Yoneyama et al., 2007b). Therefore, strigolactone exudation depends on the plant nutrient status, and exudation signals may vary among plant species (Yoneyama et al., 2007b).

Strigolactones also play nonsymbiotic roles. A major role of strigolactones is the regulation of aerial plant architecture (Umehara et al., 2008; Gomez-Roldan et al., 2008). Strigolactones act downstream of auxin to regulate bud outgrowth in pea and in Arabidopsis (Brewer et al., 2009). In addition, an iron-containing protein required for the biosynthesis of strigolactones regulates rice tiller bud outgrowth (Hao et al., 2009). It remains to be determined if strigolactones first evolved as endogenous plant hormones or as rhizosphere signals for AMF.

Other Plant and Environmental Signals

Several nonflavonoid molecules are inducers of rhizobial *nod* gene expression. Betaine, stachydrine, and trigonelline secreted from *Medicago sativa* seeds co-induce *nod* genes in the presence of flavonoids in *S. meliloti* (Phillips, Joseph, and Maxwell, 1992). Demethylation of stachydrine by *S. meliloti* increases its gene-inducing activity (Goldmann et al., 1994). Xanthones can also induce *nod* gene expression in *B. japonicum* (Yuen et al., 1995). Jasmonates, plant hormones induced in response to wounding or defense against pathogens, can stimulate *nod* gene expression in *R. leguminosarum* and *B. japonicum* (Rosas et al., 1998). Genistein along with JA or MeJA increased NF production in *B. japonicum* (Mabood et al., 2006).

Aldonic acids, erythronic and tetronic acids, secreted from lupin seeds induce NF production in *Rhizobium lupini, Mesorhizobium loti* and in *S. meliloti* (Gagnon and Ibrahim, 1998). A mixture of two different flavonoids or a flavonoid plus a nonflavonoid increased *nod* gene expression more than a single flavonoid, suggesting a synergy among *nod* gene activators for maximal induction (Mabood et al., 2006; Gagnon and Ibrahim 1998; Begum et al., 2001). Other inducers of *nod* gene expression include phenolics, vanillin, and isovanillin, secreted from nonlegumes such as wheat (Le Strange et al., 1990). Besides flavonoids, rhizobia come under the influence of chemo attractants such as simple sugars, amino acids, dicarboxylic acids, and hydroxyaromatic acids, which are found in the legume rhizosphere (Aguilar et al., 1988; Kape, Parniske, and Werner, 1991; Dharmatilake and Bauer 1992; Cooper, 2007). Biotin secreted from *Medicago sativa* roots increases *S. meliloti* growth rate and root colonization (Streit, Joseph, and Phillips, 1996; Heinz, Phillips, and Streit, 1999).

Following the initial recognition of plant signals in the rhizosphere, plant lectins probably play some role in rhizobial attachment to root hairs (Dazzo et al., 1984b; Smit, Kijne, and Lugtenberg, 1986). Lectins may confer some level of host specificity in the rhizobia-legume symbiosis by binding rhizobia specifically to the root hair surface of their host plants (Hamblin and Kent, 1973; Bohlool and Schmidt, 1974; Dazzo and Hubbell, 1975; Kato, Maruyama, and Nakamura, 1980; Stacey, Paau, and Brill, 1980; Dazzo and Hollingsworth, 1984a; Hirsch, 1999). However, many other studies do not support this lectin-mediated binding of homologous rhizobia to legume root hairs (Badenoch-Jones, Flanders, and Rolfe, 1985; Mills and Bauer, 1985; Vesper, Malik, and Bauer, 1987). Lectins may also play a role in successful infection-thread formation (Lodiero et al., 1995; Brelles-Mariño and Boiardi 1996; Kijne, Bauchrowitz, and Diaz, 1997).

In AM symbiosis, besides flavonoids and strigolactones, carbon dioxide seems an important factor for extensive hyphal growth of AMF (Chabot et al., 1992; Bécard, Douds, and Pfeffer, 1992). Hyphal growth rate and duration of growth are increased in *G. margarita* in the presence of 10 µM quercetin (a flavonoid) and 2% CO_2 (Bécard, Douds, and Pfeffer, 1992). Diffusible signals from bacteria in the rhizosphere also enhance *in vitro* germination of spores and hyphal growth in AMF without direct contact between the two organisms (Fortin et al., 2002). Various studies show that diverse bacteria (e.g., *Streptomyces orientalis*, *Pseudomonas sp.*) present in the rhizosphere enhance spore germination and hyphal growth prior to contact with the host (Mayo, Davis, and Motta, 1986; Mugnier and Mosse, 1987). Rhizobial Nod factors (NF) stimulate mycorrhizal colonization of nodulating and non-nodulating soybeans (Xie et al., 1995). To date, many bacterial strains belonging to different genera have been identified as so-called "mycorrhiza helper bacteria" (MHB) (Frey-Klett, Garbaye, and Tarkka, 2007).

The above-mentioned rhizosphere signals are instrumental in initiating the symbiotic dialogue that will lead to secretion of diffusible signals from the microbes. The perception and the effects of these microbial signals on host plants are discussed in the following section.

Epidermal Responses to Symbiotic Microbes

Microbial Signals and Their Plant Receptors

Nod Factors and Plant Responses to Nod Factors

During the initiation of legume RNS, expression of rhizobial *nod* genes is induced in response to flavonoids, isoflavonoids, and other signals secreted by host plants. Nod proteins allow the production of NF (Dénarié, Debellé, and Rosenberg, 1992). NF are lipochitooligosaccharidic

(LCO) signals that have an oligomeric backbone of β-1,4 N-acetylglucosamine residues, N-acylated at the nonreducing terminal residue (Dénarié, Debellé, and Promé, 1996). NF from different rhizobial species vary in the substitutions on the sugar residues and in the nature of the acyl chain. Variations in structure and amount of NF released by rhizobia are the major determinants of host specificity (Perret, Staehelin, and Broughton, 2000). However, an example of legume RNS in the absence of *nod* genes and NF production has been reported in photosynthetic Bradyrhizobia (Giraud et al., 2007).

Despite their bacterial origin, NF show striking similarities with plant hormones in their mode of action (Relić et al., 1993). NFs, like plant hormones, are recognized very specifically by host plants and are potent at very low concentrations (10^{-9}–10^{-12} M). Turnover of the signals is also required to ensure specificity. Chitinases modulate NF levels and suppress defense responses induced by high NF concentrations (Staehelin et al., 1995; Savouré et al., 1997). Ethylene, which has multiple roles in plant development and defense reactions, induces the expression of plant chitinases and thereby limits the extent of nodule initiation (Staehelin et al., 1994a; Staehelin et al., 1994b; Mellor and Collinge, 1995).

In most cases, NF are absolutely necessary for successful infection and nodule development. Isolated NFs are able to induce, in legume hosts, a variety of responses at the epidermal, cortical, and pericycle levels similar to those induced by rhizobia themselves. The root epidermis is the first barrier that rhizobia encounter during RNS. Bacterial invasion can occur either in root hair curls or through cracks in the epidermis. *Sesbania rostrata* is a unique model to study these two modes of penetration since it can be colonized via both entry modes: Well-aerated roots form nodules near root tips following infection via curled root hairs whereas waterlogged roots usually form nodules at the base of lateral roots, and infection occurs through root cracks and intercellular invasion (D'Haeze et al., 2003; Goormachtig, Capoen, and Holsters, 2004). Elegant studies in *S. rostrata* have shown that NF signalling is required for both penetration modes and that ethylene produced in waterlogged conditions is a major determinant for crack entry infection (D'Haeze et al., 2003; Goormachtig, Capoen, and Holsters, 2004).

In the root epidermis, NFs cause a variety of morphological changes including root hair swelling (has), root hair branching (hab), and root hair curling (hac), which are associated with cytoskeleton rearrangements (Timmers, Auriac, and Truchet, 1999; Esseling, Lhuissier, and Emons, 2003). In Arabidopsis, ethylene promotes root hair initiation and elongation (Tanimoto, Roberts, and Dolan, 1995; Schiefelbein, 2000). However, ethylene is not required for NF-induced root hair deformations (Heidstra et al., 1997). In epidermal cells, NF also induce several ionic responses including a rapid influx of calcium at the root hair tip, an efflux of potassium and chloride ions, a membrane depolarization, and "calcium spiking," which are oscillations of nuclear and perinuclear calcium concentration and occur 15–20 minutes after application of NF (Ehrhardt, Wais, and Long, 1996; Wais et al., 2000; Walker, Viprey, and Downie, 2000; Sieberer et al., 2009). Only a subset of epidermal cells is able to develop infection threads (IT) and induce expression of early nodulin genes such as *ENOD12, ENOD11* in response to NF or rhizobia (Scheres et al., 1990; Pichon et al., 1992; Journet et al., 1994, 2001). This "competence zone" broadly corresponds to the region of the root supporting actively growing root hairs. However, epidermal cells all along the root seem to induce calcium spiking in response to NF, thus suggesting that specific developmental factors also regulate the ability of epidermal cells to induce gene expression both in response to NF and in accommodation of rhizobial infection (Miwa et al., 2006). Ethylene decreases the sensitivity of epidermal cells to NF, which results in a reduced number of root hairs that are able to induce calcium spiking and expression of *ENOD11* (Oldroyd, Engstrom, and Long, 2001). One mode

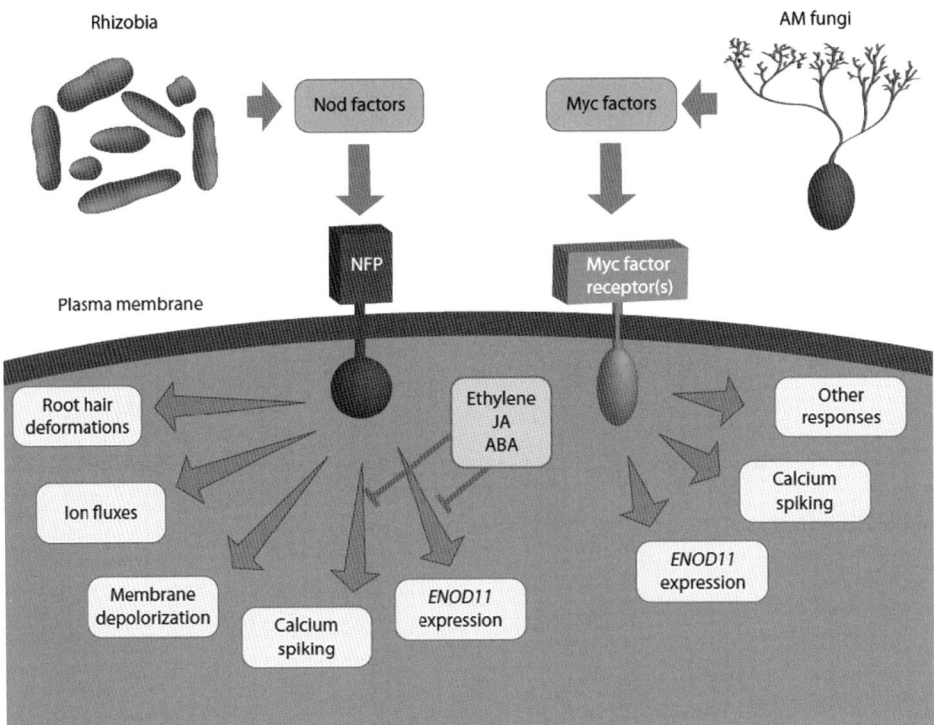

Figure 16.2. Early plant responses to Nod and Myc factor(s). Initial NF perception occurs at the epidermis by the proposed signalling receptor, NFP in *M. truncatula*. In the root epidermis, NFs cause a variety of morphological changes including root hair deformations. In epidermal cells, NFs also induce several ionic responses, a membrane depolarization, and calcium-spiking oscillations of perinuclear calcium concentration that occur 15–20 minutes after application of NFs. Only a subset of epidermal cells is able to induce expression of symbiotic genes such as *ENOD11* in response to NFs. Plant hormones, ethylene, and JA act in combination to downregulate calcium spiking and *ENOD11* expression. ABA also negatively regulates calcium spiking and *ENOD11* expression. Similarly in AM, diffusible signalling molecules released from AMF, Myc factor(s), also induce *ENOD11* expression, however, mainly in the root cortex. Calcium spiking in soybean cell cultures is also an early response to Myc factor(s) perception. The chemical structure of Myc factor(s) and the corresponding host plant molecular receptor(s) are still unknown. **NF**, Nod factor; **NFP**, Nod factor perception; ***ENOD***, early nodulation genes; **JA**, jasmonic acid; **ABA**, abscisic acid; **AM**, arbuscular mycorrhizae; **AMF**, arbuscular mycorrhizal fungi.

of ethylene action is therefore upstream or at the point of calcium spiking in the NF signal transduction pathway. Ethylene also determines the NF concentration required for calcium spiking (Oldroyd, Engstrom, and Long, 2001). Besides ethylene, JA affects the frequency of calcium spiking as well as the pattern of cells that are able to induce calcium spiking (Sun et al., 2006). JA and ethylene act in concert to regulate NF signal transduction (Sun et al., 2006). Abscisic acid (ABA) probably acts in a similar way for the regulation of NF signalling. It affects the occurrence and nature of NF-induced calcium spiking as well as *ENOD11* expression (Ding et al., 2008). Interestingly, it seems that ABA regulates NF signal transduction in an ethylene-independent manner (Figure 16.2).

The initial NF perception is likely to occur at the epidermal level by way of specific receptors (Goedhart et al., 2000). Genetic screens for mutants deficient in many NF-induced responses led to the successful identification and cloning of NF receptors: NFR1 and NFR5 in *L. japonicus,* and NFP in *M. truncatula* (Madsen et al., 2003; Radutoiu et al., 2003; Limpens et al., 2003). Both *L. japonicus* and *M. truncatula* genes encode receptor-like kinases with LysM domains. These LysM receptor-like kinases control host specificity in RNS (Radutoiu et al., 2007). Other potential candidates for NF receptors have been investigated through their ability to bind to NF. The lectin nucleotide phosphohydrolase (LNP) from *Dolichos biflorus* or lectin apyrase is one such candidate because of its preferential binding to NF (Etzler et al., 1999). Treatment of roots with an antiserum directed toward the LNP inhibits root hair deformations and nodule development upon inoculation with rhizobia (Etzler et al., 1999; Day et al., 2000). Similarly, silencing of the soybean ectoapyrase, *GS52*, led to a reduction in the number of mature nodules, indicating its significance in the nodulation process (Govindarajulu et al., 2009). LNPs are characterized by their increased apyrase activity (ATP and ADP hydrolysis) upon NF binding, localization to root hair tips, and induction upon rhizobial inoculation (Kalsi and Etzler, 2000; Kim, Sivaguru, and Stacey, 2006). Three independent NF binding sites (NFBS1, NFBS2, and NFBS3) have also been identified in *Medicago truncatula*, but their biochemical identity and biological role are still unknown (Bono et al., 1995; Gressent et al., 1999; Hogg et al., 2006).

Myc Factors and Corresponding Plant Responses

The identification of NFs and their corresponding receptors in legumes has been important to understanding RNS. Similarities between RNS and AM have prompted the scientific community to investigate the existence of a similar molecular dialog between plant roots and AMF. Diffusible signals, termed Myc factors, are produced by AMF and induce *ENOD11* expression in *M. truncatula* roots (Kosuta et al., 2003). Direct contact between AMF and *Medicago truncatula* roots was prevented using a fungus-impenetrable membrane. This response was elicited by various AMF, including *Gigaspora rosea, Gigaspora margarita, Gigaspora gigantea,* and *G. intraradices*. However, unlike NF that induces *ENOD11* expression mainly in the root epidermis, these diffusible signals induce *ENOD11* expression mainly in the root cortex (Kosuta et al., 2003). Diffusible signals from *G. margarita* elicit a transient cytosolic calcium elevation in soybean cell cultures within a few minutes, which is reminiscent of the early NF-induced calcium influx (Navazio et al., 2007) (Figure 16.2). AMFs induce calcium oscillations in *M. truncatula* root hair cells before direct contact with fungi. These oscillations may be an effect of Myc factors produced by AMF. However, the calcium signature of these oscillations is different from that of the NF-induced oscillations (Kosuta et al., 2008).

Myc factors are thermostable and amphiphilic, and have a mass lower than 3 kDa (Navazio et al., 2007). Arabidopsis, a nonhost of AM symbiosis, does not seem to perceive this fungal signal. Myc factors appear to be released by germinating AMF spores even in the absence of the host plant (Navazio et al., 2007). Similarly, low levels of NF are also produced by rhizobia in the absence of legume hosts, and strigolactones are present in root exudates in the absence of AMF (Akiyama, Matsuzaka, and Hayashi, 2005; Besserer et al., 2006). NFs as well as Myc factors stimulate lateral root formation (Olah et al., 2005). The identification of Myc factors and their corresponding molecular receptor(s) in host plants will be significant in our understanding of AM symbiosis.

Downstream Epidermal Signalling

A Shared Symbiotic (Sym) Signalling Pathway

Forward genetic approaches in *M. truncatula* and *L. japonicus* led to the identification of components downstream of NF receptors. Many corresponding mutants are not only deficient for RNS but also for AM indicating the existence of a conserved symbiotic (Sym) pathway required for the establishment of both symbioses (Szczyglowski et al., 1998; Catoira et al., 2000; Mitra, Shaw, and Long, 2004; Kistner et al., 2005). The first gene identified in this Sym pathway encodes a receptor-like kinase with leucine-rich repeats (LRR-RLK) named DMI2/NORK/SYMRK (Endre et al., 2002; Stracke et al., 2002). This protein is localized to the plasma membrane and especially in the IT membrane (Limpens et al., 2005). DMI2 was found to be required for actinorhizal RNS suggesting that the Sym pathway may be required for the establishment of RNS with nonlegumes (Gherbi et al., 2008). A 3-hydroxyl 3-methylglutaryl coenzyme A reductase (MtHMGR1) interacts with DMI2 in *M. truncatula,* and its activity is required for nodule development (Kevei et al., 2007). HMGRs are key enzymes of the mevalonate (MVA) biosynthetic pathway, which suggests a role for isoprenoids in NF signalling. A novel DNA-binding protein, SIP1, also interacts with SYMRK in *L. japonicus* and binds to the promoter of the early nodulin, *NIN* (Zhu et al., 2008).

Several nuclear proteins have been identified in the Sym pathway: two cation channels localized to the nuclear envelope—CASTOR and DMI1/POLLUX (Ané et al., 2004; Imaizumi-Anraku et al., 2005; Riely et al., 2007; Peiter et al., 2007); two nucleoporins—NUP85 and NUP133 (Kanamori et al., 2006; Saito et al., 2007); and one calcium/calmodulin-dependent protein kinase—DMI3/CCaMK (Lévy et al., 2004; Mitra et al., 2004; Tirichine et al., 2006a) (Figure 16.3). Most mutants in these genes are impaired for both nodulation and AM symbiosis. However, the R38 allele of *dmi2* is affected in RNS but not in AM development. The R38 allele encodes a kinase-deficient DMI2 receptor-like kinase and may represent a weak allele (Endre et al., 2002; Kevei et al., 2007). Another nuclear protein with coiled-coil domains, named IPD3/CYCLOPS, was found to interact with DMI3 (Messinese et al., 2007; Yano et al., 2008). In *L. japonicus*, CYCLOPS is required for both RNS and AM. However, while most mutants of the Sym pathway are blocked at the stage of hyphopodia formation, *cyclops* mutants exhibit inter- and intra-cellular hyphae with an abnormal morphology and development of a few arbuscules. *Cyclops* mutants are strongly impaired in IT development but less in nodule development (Yano et al., 2008). CYCLOPS is the last known protein of the common Sym cascade before nodulation- or mycorrhization-specific components.

Mutants in the Sym pathway are affected in most responses to NF and are unable to initiate IT development in the presence of rhizobia in plant roots. None of them is able to induce nodulin gene expression, such as *ENOD11* or *ENOD40* in *M. truncatula* and *NIN* in *L. japonicus*, in response to NF. Calcium spiking is blocked in *dmi1* and *dmi2* mutants in *M. truncatula* and in their *L. japonicus* homologs (*pollux, castor,* and *symrk*). Calcium spiking is also blocked in *L. japonicus nup85* and *nup133* mutants. However, *dmi3* and *ipd3/cyclops* mutants exhibit NF-induced calcium spiking indicating that DMI3 and IPD3 act downstream of this event (Wais et al., 2000; Miwa et al., 2006; Catoira et al., 2000; Stracke et al., 2002). NF-induced lateral root development is also defective in *M. truncatula dmi* mutants (Olah et al., 2005).

AMF diffusible Myc factors activate many responses similar to those induced by NF such as calcium spiking and lateral root development in a *DMI1* and *DMI2*-dependent manner (Kosuta et al., 2008; Olah et al., 2005). Besides legumes, components of the Sym pathway have been found to be conserved in rice as well (Yano et al., 2008; Chen, Ané, and Zhu, 2008; Banba et al., 2008; Gutjahr et al., 2008; Chen et al., 2009). Activation of many genes by rhizobia or AMF is dependent on the Sym pathway. However accumulating evidences points toward

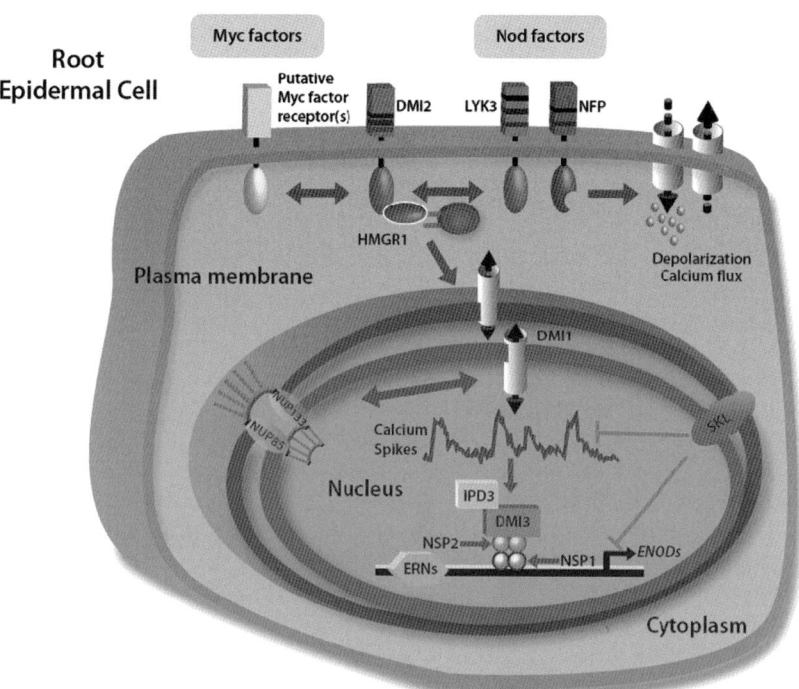

Figure 16.3. Molecular model of the Sym pathway in the root epidermis. Forward genetic approaches in *M. truncatula* have led to the identification of components downstream of NF receptors (NFP and LYK3). Many of these mutants are not only deficient for RNS but also for AM indicating the existence of a conserved symbiotic (Sym) pathway required for the establishment of both symbioses. Protein interactions have identified HMGR1 and IPD3 as proteins interacting with DMI2 and DMI3, respectively, in *M. truncatula*. Subcellular localization of the proteins involved in the Sym pathway indicates that NFs are perceived at the plasma membrane. This perception is linked to calcium spiking in the nucleus. Perception and transduction of calcium spiking involves calcium/calmodulin-dependent protein kinase (DMI3) and transcription factors like NSP1, NSP2, and ERNs. It has been proposed that NSP1, NSP2, and ERNs act in synergy to regulate expression of *ENOD*s. Effects of ethylene on calcium spiking and *ENOD* expression suggest a possible role of the plant hormone in the Sym pathway at or upstream of calcium spiking. Interestingly, the Arabidopsis EIN2 is localized to the nuclear envelope like DMI1, NUP133 and NUP85. **NF**, Nod factor; **NFP**, Nod factor perception; **LYK**, LysM domain-containing receptor-like kinases; **RNS**, root nodule symbiosis; **AM**, arbuscular mycorrhizae; **HMGR1**, 3-hydroxyl 3-methylglutaryl coenzyme A reductase 1; **IPD3**, interacting protein of DMI3; **DMI**, doesn't make infections; **NSP**, nodulation signalling pathway; **ERN**, ERF required for nodulation; **ENOD**, early nodulation genes; **SKL**, sickle; **EIN2**, ethylene insensitive 2; **NUP**, nucleoporin.

parallel pathways mediating AM. For example, Myc factor-induced *ENOD11* expression was found to be independent of the *DMI* genes in *M. truncatula* (Kosuta et al., 2003) even though *ENOD11* expression was not detected in epidermal cells of *dmi2* mutants in direct contact with AMF (Chabaud et al., 2002). Transcriptome analyses in *L. japonicus* revealed the existence of signal transduction pathways that are independent of the Sym pathway (Kistner et al., 2005). Similarly signalling pathways operate in parallel or independently of the Sym pathway in rice for the establishment of AM (Gutjahr et al., 2008).

The regulation of these pathways by plant hormones is also a feature common to RNS and AM. Mutants in *SKL*, a *Medicago truncatula* ortholog of Arabidopsis *EIN2*, are hyperinfected by rhizobia as well as by AMF (Penmetsa and Cook, 1997; Penmetsa et al., 2008). EIN2 is a well-characterized component of the ethylene signalling pathway (Alonso et al., 1999), and these mutants are insensitive to ethylene. Effects of ethylene on calcium spiking suggest that a possible site of ethylene action is at the calcium spiking level (Oldroyd, Engstrom, and Long, 2001; Penmetsa et al., 2008). Therefore, SKL might play a role at this point in the Sym pathway. Interestingly, EIN2 was localized to the nuclear envelope like DMI1/CASTOR/POLLUX, NUP133, and NUP85 (Alonso et al., 1999) (Figure 16.3).

Regulation of nodulin gene expression by the Sym pathway Two putative GRAS domain transcription factors, NSP1 and NSP2, function in the NF signalling pathway downstream of calcium spiking and DMI3 (Kaló et al., 2005; Smit et al., 2005; Heckmann et al., 2006; Murakami et al., 2006). Members of this family, such as gibberellin-insensitive (GAI), repressor of gal-3 (RGA) and SCARECROW (SCR), are present throughout the plant kingdom and play multiple roles in signal transduction, meristem maintenance, and plant development (Bolle, 2004). Mutants in *NSP1* and *NSP2* are affected in expression of early nodulins such as *ENOD11* and *NIN* (Oldroyd and Long, 2003). These transcription factors seem specific to the NF signalling pathway, and the corresponding mutants are not affected in AM. NSP1 and NSP2 interact to form a complex, and NSP1 binds directly to the promoter regions of *ENOD11* and *NIN* (Hirsch et al., 2009). This DNA-protein interaction is enhanced upon activation of the NF signalling pathway.

Three AP2/ERF-like transcription factors, ERN1, ERN2, and ERN3, play an important role in NF signalling downstream of DMI3 (Middleton et al., 2007; Andriankaja et al., 2007). *M. truncatula ERN1* was also identified via positional cloning of *bit1* (*branching infection threads 1*), a nodulation deficient mutant (Middleton et al., 2007). The ERN proteins bind to the *ENOD11* promoter slightly upstream of the NSP1 binding site (Andriankaja et al., 2007). ERN3 is a repressor of ERN1/ERN2-mediated transcriptional activation (Andriankaja et al., 2007). Another negative regulator of *ENOD11* expression is a transcription factor encoded by *NIN* since *nin* mutants in *M. truncatula* display an extended spatial distribution of *ENOD11* expression (Schauser et al., 1999; Marsh et al., 2007). *Nin* mutants in *L. japonicus* and *M. truncatula* display normal calcium spiking but are deficient for nodule organogenesis (Miwa et al., 2006; Marsh et al., 2007; Borisov et al., 2003; Schauser, Wieloch, and Stougaard, 2005). NSP1 and SIP1 bind to the *NIN* promoter, and NSP1 is required for *NIN* induction (Zhu et al., 2008; Hirsch et al., 2009). Therefore NSP1, NSP2, ERNs, SIP1, and NIN probably act in concert to regulate expression of early nodulin genes.

Infection Thread Formation in Legume Nodulation and Arbuscular Mycorrhizae

RNS and AM share striking similarities in their intercellular and intracellular infection process through the formation of tube-like structures named infection threads (IT) and intraradical hyphae, respectively. These structures are formed by the invagination of the plant cell membrane with remains of cell wall material deposited within them. Not surprisingly, the matrix surrounding the IT shares biochemical and immunological similarities with intraradical hyphae (Perotto, Brewin, and Bonfante, 1994).

Legume Nodulation

In RNS, the internal space of IT contains a matrix that is synthesized by both the plant and rhizobia (Kijne, 1992). Rhizobia travel from cell to cell in the host plant via IT. Their path is precisely guided in cortical cells through cytoplasmic structures known as pre-infection threads (PiT) (Timmers, Auriac, and Truchet, 1999; van Brussel et al., 1992). Rhizobial exopolysaccharides (EPS) and lipo-polysaccharides (LPS) are often required, in addition to NF, for IT formation (Jones et al., 2007). A high degree of NF structural specificity is required for initiating the infection process when compared to "signalling" responses such as calcium spiking, nodulin gene expression, or cortical cell divisions (Ardourel et al., 1994; Walker and Downie, 2000). The *SYM2* gene in *P. sativum* is involved in an NF perception mechanism that controls this "entry" process through the epidermis but not in signalling responses such as *ENOD12* induction or nodule primordia formation (Guerts et al., 1997). The *M. truncatula HCL/LYK3* gene is probably an ortholog of pea *SYM2* (Smit et al., 2007). Interestingly, it encodes a LysM receptor-like kinase homologous to *L. japonicus* NFR1 (Limpens et al., 2003; Smit et al., 2007). These observations led to a model with two types of NF receptors in *Medicago truncatula* and pea: (1) a signalling receptor controlling epidermal nodulin expression and cortical cell divisions and (2) an entry receptor (HCL/LYK3 in *M. truncatula* and SYM2 in pea) controlling IT formation. There is no evidence of such a dual receptor system in *Lotus japonicus* or soybean suggesting that it may be specific to legumes of the Galegoid clade.

Legume mutants in the Sym pathway are unable to initiate IT in the presence of rhizobia indicating that IT formation is directly or indirectly dependent on NF signalling. In addition, mutants with an altered *NIN* gene are characterized by their inability to form IT and initiate primordia formation (Schauser et al., 1999; Marsh et al., 2007; Borisov et al., 2003). Thus, even though NIN represses the spatial pattern of *ENOD11* expression as mentioned previously, it also acts as a positive regulator of IT formation and nodule primordia initiation. Genetic studies have identified additional mutants that are affected at later stages of IT development. For instance, *bit1* and *pdl* mutants, which are both affected in the *ERN1* gene, are able to initiate some IT, but their growth is severely affected (Middleton et al., 2007). Other *M. truncatula* mutants affected in IT formation include *lin, nip/latd, rpg,* and *rit1* (Kuppusamy et al., 2004; Veereshlingam et al., 2004; Starker et al., 2006). Similarly, mutants affected in IT growth have been identified in *L. japonicus* including *lot, crinkle, alb1, itd1/3/4* (Starker et al., 2006; Schauser et al., 1998; Kawagachi et al., 2002; Ooki et al., 2005; Sandal et al., 2006). *EFD*, an ERF transcription factor, negatively regulates *S. meliloti* infections in *M. truncatula* (Vernié et al., 2008).

Arbuscular Mycorrhization

A process similar to IT formation takes place for fungal infection in AM. Following contact with the root, AMF form swollen hyphal infection structures on the root epidermis called hyphopodia through which the fungi enter the root (Harrison, 2005). *G. margarita* form hyphopodia on isolated epidermal cell wall fragments but not on fragments of vascular or cortical cells or on the surface of nonhost roots (Giovannetti et al., 1993a, 1993b; Giovannetti and Citernesi, 1993; Nagahashi and Douds 1997). Hyphopodium formation is accompanied by the assembly of a pre-penetration apparatus (PPA) in host plants. The PPA is a subcellular structure that shares strong similarities with the PiT in legume nodulation and determines the path of hyphae growth through the root cells. It is composed of endoplasmic reticulum and cytoskeletal material and develops 4–5 hours after formation of the hyphopodium (Genre et al., 2005, 2008). The PPA was used as a morphological marker for transcriptome analysis in *M. truncatula* colonized by *G. margarita* (Siciliano et al., 2007). An expansin-like gene is expressed in the epidermis during PPA development and is therefore a convenient early host marker for successful mycorrhization (Siciliano et al., 2007).

Genetic analysis revealed several plant mutants in nonlegumes that are stage-defective in AM symbiosis. Pre-mycorrhizal infection mutants (*pmi1* and *pmi2*) in tomato are characterized by a decreased hyphopodia formation and resistance to root colonization when compared to wild-type plants (David-Schwartz et al., 2001, 2003). In contrast, hyphopodia are completely absent in the *nope1* mutant of maize (Paszkowski, Jakovleva, and Boller, 2006). The maize *taci1* mutant has a reduced number of hyphopodia; *taci1* may be a weak allele of *nope1* (Paszkowski, Jakovleva, and Boller, 2006). Expression of a lipid transfer protein (LTP) gene increases when *G. mosseae* penetrates the epidermis of *O. sativa* roots. However, its expression is decreased upon intercellular colonization of the root cortex (Blilou, Ocampo, and Garcia-Garrido, 2000). In legumes, the *L. japonicus mcbep* mutant is characterized by multiple and unusually swollen hyphopodia (Senoo et al., 2000).

Several legume mutants that form normal hyphopodia but are blocked prior to cortex invasion have been isolated in *M. truncatula* and *L. japonicus* (Catoira et al., 2000; Wegel et al., 1998). Genes of the common Sym pathway in *M. truncatula* and in *L. japonicus* are required for fungal penetration (Endre et al., 2002; Stracke et al., 2002; Ané et al., 2004; Imaizumi-Anraku et al., 2005; Kanamori et al., 2006; Saito et al., 2007; Lévy et al., 2004; Mitra et al., 2004; Tirichine et al., 2006a; Messinese et al., 2007; Yano et al., 2008). In rice, in addition to legumes, components of the Sym pathway play a similar role in fungal penetration after hyphopodia formation (Yano et al., 2008; Chen, Ané, and Zhu, 2008; Banba et al., 2008; Gutjahr et al., 2008; Chen et al., 2009). Similarly the *rmc* (reduced mycorrhizal colonization) tomato mutant supports hyphopodia formation but is affected in intraradical invasion (Barker et al., 1998). In summary, genetic studies allowed tremendous progress in the identification of key components that are required for hyphopodia formation and subsequent colonization of plant roots by AMF.

Hormone Regulation of Infection Thread Development

Not all infection events lead to the successful release of rhizobia in symbiosomes or in the formation of cortical arbuscules. In fact, most IT abort either in root hair cells or in the cortex (Penmetsa and Cook, 1997). This process often involves an HR-like reaction in alfalfa (Vasse, de Billy, and Truchet, 1993). In the *M. truncatula* ethylene-insensitive *skl* mutant, the number of aborted IT is lower than in wild-type plants and the number of infection events is higher, which indicates the role of ethylene as a negative regulator of IT development (Oldroyd, Engstrom, and Long, 2001; Penmetsa and Cook, 1997). Similarly, application of ACC, an ethylene precursor, decreases the number of infection events whereas application of AVG, an ethylene synthesis inhibitor, increases IT development (Oldroyd, Engstrom, and Long, 2001). Transgenic *L. japonicus* lines with reduced ethylene sensitivity showed a higher number of IT and nodule primordia than wild-type plants (Nukui, Ezura, and Minamisawa, 2004). Ethylene also affects cytoskeleton, PiT and IT formation in pea (Guinel and Geil, 2002). Thus, ethylene acts as a negative regulator of bacterial root hair entry in the root epidermis of many legumes. High ethylene levels also inhibit root hair entry in *Sesbania rostrata* but still allow infection pocket formation, an important step for crack entry at the base of lateral roots (Goormachtig, Capoen, and Holsters, 2004). Interestingly ethylene-insensitive soybean mutants and soybean plants treated with an inhibitor of ethylene perception displayed normal nodulation levels indicating that inhibition of nodulation by ethylene may not be a general rule across legumes (Schmidt et al., 1999).

Ethylene negatively regulates colonization of *Medicago sativa* by *G. mosseae* (Azcon-Aguilar, Rodriguez-Navarro, and Barea, 1981). Exogenous ethylene applications inhibit colonization of *Pisum sativum* and *Allium porrum* by *G aggregatum* (Geil, Peterson, and Guinel, 2001; Geil and Guinel, 2002). Although the number of hyphopodia formed was not affected, abnormal, swollen, and highly branched hyphopodia were formed, and fungal entry into the root was reduced (Geil, Peterson, and Guinel, 2001).

Besides ethylene, ABA acts as a negative regulator of rhizobial infection in *M. truncatula* (Ding et al., 2008). Methyl jasmonate (MeJA) and gibberellins (GA) also inhibit IT formation and *NIN* expression in *L. japonicus* (Nakagawa and Kawaguchi, 2006; Maekawa et al., 2009). GA also plays an important role in IT formation in nodulation of *S. rostrata* (Lievens et al., 2005). Auxins stimulate AM root colonization by increasing lateral root formation during the early stages of AM (Ludwig-Müller, 2000). Exogenous salicylic acid (SA) applications reduce root colonization in *Oryza sativa* at the onset of AM, but have no effect on hyphopodia formation (Blilou, Ocampo, and Garcia-Garrido, 2000). Interestingly, a reduction in endogenous SA levels in *L. japonicus* and *M. truncatula* leads to an increase in IT formation (Stacey et al., 2006). Thus, many plant hormones regulate IT formation and intraradical hyphae in RNS and AM.

Cortical Accommodation of Symbiotic Microbes and Nodule Development

Following penetration through the host plant epidermis during RNS and AM, root cortical cells need to accommodate symbiotic microbes in their cytoplasm as bacteroids in RNS or arbuscules/vesicles in AM. In addition, in the case of RNS, cortical cells start dividing and are reprogrammed to form a new primordium that will ultimately give rise to a functional nodule.

Regulation of Cortical Cell Divisions and Root Nodule Development

NFs induce several cellular responses in the root cortex. In particular, cortical cell divisions mark the initiation of nodule organogenesis. These cell divisions always occur across from protoxylem poles (Timmers, Auriac, and Truchet, 1999). Dividing cortical cells give rise to a nodule primordium. Early nodulins like *ENOD40*, *ENOD2*, *ENOD12*, and *ENOD20* are specific markers for NF-induced responses in cortical cells (Journet et al., 1994; Crespi et al., 1994; Charon et al., 1999). Expression of *MtHAP2-1*, a transcriptional regulator, is controlled by microRNA169 and is specific to meristem development in *M. truncatula* (Combier et al., 2006).

Genetics of Root Nodule Organogenesis

Bacterial IT formation and nodule organogenesis in the cortex are clearly uncoupled in several nodulation mutants and are therefore independent processes (Tirichine et al., 2006a; Gleason et al., 2006; Tirichine et al., 2006b; Murray et al., 2007; Tirichine et al., 2007). Mutants have been isolated in alfalfa, *L. japonicus*, and white clover that form spontaneous nodules even in the absence of rhizobia (Tirichine et al., 2006b; Truchet et al., 1989; Blauenfeldt et al., 1994). Spontaneous nodules are characterized by the absence of rhizobia, bacteroids, or IT but bear morphological features similar to those of functional nodules (Tirichine et al., 2006b; Blauenfeldt et al., 1994). Purified NFs are also able to induce cortical cell divisions and even lead to the formation of empty nodules in alfalfa (Truchet et al., 1989). Gain of function mutations of *DMI3* also lead to spontaneous nodulation and induce *ENOD11* expression in the epidermis as well as in the cortical cells associated with nodule primordia (Tirichine et al., 2006a; Gleason et al., 2006). DMI3 can activate nodule organogenesis, but root hair deformations or PiT structures are not observed in the absence of rhizobia. Therefore, although activation of DMI3 can coordinate early epidermal and cortical responses to NF, additional components are necessary for the formation of functional nodules (Tirichine et al., 2006a; Gleason et al., 2006).

Phytohormones are important determinants in nodule organogenesis. Gradients of both auxin and cytokinins are required for cortical cell divisions and nodule initiation (Libbenga et al., 1973). The auxin/cytokinin ratio is lowered in supernodulation mutant *nts386* in soybean suggesting a role for this gradient in nodule formation (Caba, Recalde, and Ligero, 1998). Inhibitors of polar auxin transport also induce pseudo-nodules and expression of early nodulins such as *ENOD2, ENOD12,* and *ENOD40* associated with nodule primordia (Hirsch et al., 1989; Hirsch and Fang, 1994; Fang and Hirsch, 1998). In white clover, expression of the auxin-responsive promoter of *GH3* is reduced just below the site of inoculation, between 12 and 24 hours after spot inoculation with rhizobia or NF (Mathesius et al., 1998a). After 24–48 hours, an increase in *GH3* expression was observed at the site of nodule initiation suggesting that auxin transport inhibition in the early stages of nodule development lowers the auxin/cytokinin ratio and triggers cell division. This decrease seems transient, since *GH3* expression resumes after 72 hours, presumably indicating an increase in auxin flow, which suppresses further cell divisions (Mathesius et al., 1998a). These results were confirmed later in *M. truncatula* (Huo et al., 2006). Flavonoids are often located in dividing and meristematic tissues of nodules (Mathesius et al., 1998b). Flavonoids can regulate auxin transport and therefore indirectly regulate cortical cell divisions by controlling auxin transport or accumulation (Brown et al., 2001). Flavonoid roles in nodulation are covered in detail in Chapter 15.

Legume nodules most frequently arise across from protoxylem poles. In the *M. truncatula* ethylene-insensitive mutant, *skl*, positioning of nodules is random indicating that ethylene regulates the localization of nodule primordia (Penmetsa and Cook, 1997). The ACC oxidase, an enzyme-producing ethylene, is expressed in cortical cells between the protoxylem poles. This observation suggests that ethylene prevents cortical cell divisions from occurring between protoxylem poles and allows nodule positioning across from the protoxylem poles (Heidstra et al., 1997). S*kl* mutants also display a reduced sensitivity toward cytokinins for root growth, which may indicate that cytokinin signalling is dependent on the ethylene pathway (Penmetsa et al., 2008). It also remains to be seen if ethylene pathways in the epidermis and in the cortex are co-regulated or are autonomous.

Cytokinins can induce nodule initiation, transcription of early nodulin genes, and cortical cell divisions. Application of cytokinins to a rhizobia- and NF-resistant mutant plant restored its ability to form nodules (Hirsch et al., 1997). A β-glucuronidase (*GUS*) reporter gene driven by the promoter of the Arabidopsis cytokinin response regulator *ARR5* is induced in nodule primordia (Lohar et al., 2004). Synthesis of the cytokinin zeatin in a nodulation-deficient mutant of *S. meliloti* was sufficient to induce nodule-like structures in alfalfa (Cooper and Long, 1994). A gain of function mutation in *LHK1*, a cytokinin receptor gene of *L. japonicus*, is able to elicit spontaneous nodules in the absence of rhizobia (Murray et al., 2007; Tirichine et al., 2007). Reciprocally, *lhk1* mutants can initiate bacterial infection but are unable to develop nodule primordia and loss of function mutations in *LHK1* can suppress the supernodulation phenotype of *har1* mutants. RNAi-mediated knockdown of *CRE1, LHK1* ortholog in *M. truncatula*, results in a dramatic reduction in nodule formation (Gonzalez-Rizzo, Crespi, and Frugier 2006). Altogether these data indicate that cytokinins are necessary and often sufficient to induce nodule organogenesis. ABA probably controls this process by regulating cytokinin levels in nodule primordia (Ding et al., 2008).

Perception of NF takes place at the epidermal level but there is as yet no evidence for NF perception at the cortical level. Therefore, following perception by receptors, NFs trigger a signal transduction cascade in epidermal cells, which includes calcium spiking and nodulin gene expression. Calcium spiking is perceived and transduced by DMI3, a key component of the Sym pathway. Activation of DMI3 is sufficient to trigger nodule organogenesis in the cortex in the absence of bacteria, similar to the gain of function mutants of LHK1. Two major hypotheses have emerged to define how cytokinins and NF signalling pathways interact during nodulation (Frugier et al., 2008; Oldroyd and Downie, 2008). According to one hypothesis, a linear pathway exists from NF perception through DMI3 and LHK1 to nodule organogenesis (Frugier et al., 2008). The alternative hypothesis argues that signalling pathways involving DMI3 and LHK1 are separate, but they converge at NSP2 to mediate nodulation (Oldroyd and Downie, 2008)(Figure 16.4).

Nodulation and Lateral Root Development: Similarities and Differences

Nodules and lateral roots both develop from periclinal divisions beneath the epidermis. However, lateral roots emerge from divisions of the pericycle whereas nodules originate from inner or outer cortical cells (Libbenga and Bogers, 1974). Moreover, nodules lack a root cap and exhibit a peripheral arrangement of vascular tissues as opposed to a central arrangement in lateral roots (Hirsch, 1992; Mukherjee and Frugoli, 2007). The *M. truncatula LATD/NIP* gene is required for both nodule and lateral root development (Veereshlingam et al., 2004; Bright et al., 2005). The *latd* mutant initiates nodule formation but does not complete it resulting in the formation of immature, non-nitrogen-fixing nodules. Similarly, lateral roots initiate but remain as short stumps. The lateral root phenotype but not the nodule phenotype can be rescued by ABA application indicating that ABA signalling and *LATD* activity control lateral root formation in concert (Veereshlingam et al., 2004; Liang, Mitchell, and Harris, 2007). The role of auxin in nodulation also overlaps with lateral root development. For example, auxin is necessary for lateral root induction and accumulates in lateral root primordia in a process similar to root nodule development (Bhalerao et al., 2002; Himanen et al., 2002). Correct auxin transport regulation is necessary for both lateral root initiation and nodulation (Bhalerao et al., 2002; Boot et al., 1999; Casimiro et al., 2001) Although, flavonoids may be regulators of root organogenesis via their effects on auxin transport and localization, flavonoid-deficient roots are able to form normal lateral roots but not nodules (Taylor and Grotewold, 2005; Peer and Murphy, 2007; Wasson et al., 2009). Nitric oxide also plays a critical role in lateral root growth (Pii et al., 2007). Nitric oxide (NO) interacts with the auxin-signalling pathway that controls indeterminate nodule formation. Cytokinin signalling is also an important regulator of both nodule organogenesis and lateral root development (Gonzalez-Rizzo, Crespi, and Frugier, 2006; Nishimura et al., 2004; Laplaze et al., 2007). However, response patterns are distinct between the two as relative increase in cortical cytokinin levels induce nodulation but inhibit lateral root formation (Gonzalez-Rizzo, Crespi, and Frugier, 2006; Laplaze et al., 2007). In Arabidopsis, three histidine kinases (AtLHK1, AtLHK2, AtLHK3), including a putative ortholog of MtCRE1, are involved in lateral root development (Nishimura et al., 2004). Thus, hormonal balances are key regulators of both nodule organogenesis and lateral root formation, but their modes of action differ in the two processes.

Accommodation of Symbiotic Microbes in the Cytoplasm of Host Cells, and Exchange Structures

Specialized Structures for Nutrient Exchanges: Symbiosomes and Arbuscules

Rhizobia progress through the root cortex within IT and are released into the cytoplasm of dividing cells in the nodule primordium. Once in these cells, rhizobia are released from

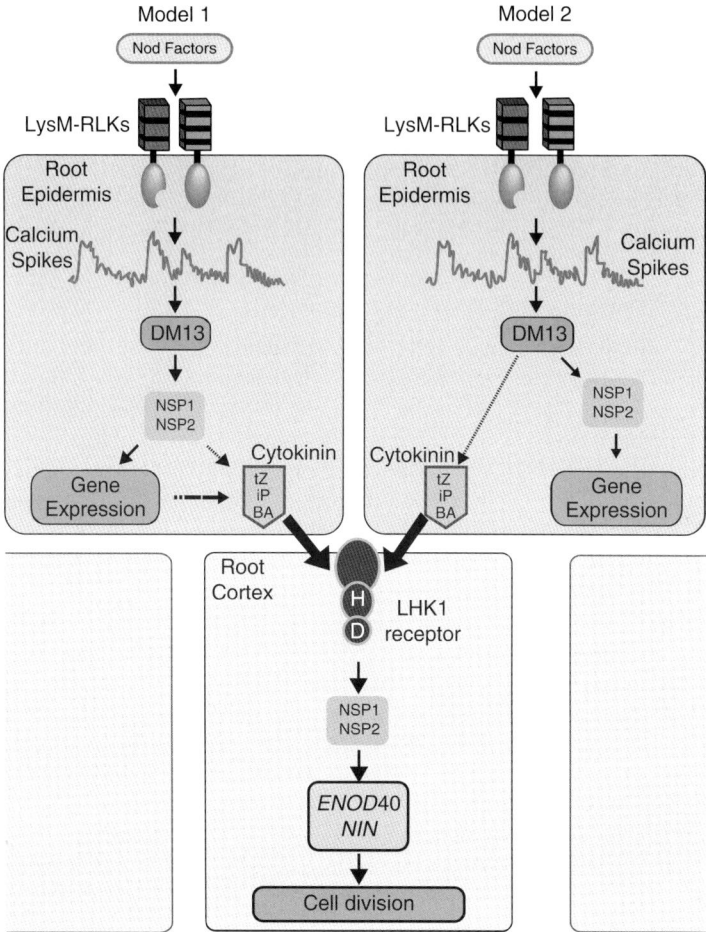

Figure 16.4. Coordination of epidermal and cortical responses during nodule development. Two lines of thought have emerged to define how cytokinins and NF signalling pathways interact during nodulation. In **Model 1**, Oldroyd and Downie (2008) propose that the signalling pathway involving DMI3 is separate from that involving LHK1, since NF induction of DMI3 leads to localized changes in cortical cytokinin levels that activate LHK1. However, both DMI3 and LHK1 require NSP1 and NSP2 to induce spontaneous nodulation. Thus, even though two independent pathways exist, they converge through the mobile GRAS domain proteins that have roles in both epidermal and cortical responses. Conversely in **Model 2**, Frugier et al. (2008) propose that a linear pathway exists from NF perception through DMI3 and LHK1 to nodule organogenesis. Following activation of LHK1, transcription regulators like NSP1, NSP2 as well as NIN and ENOD40 are required for cortical cell divisions for either pathway. Further experimentation needs to be done to have a clearer understanding of how epidermal and cortical responses are integrated during nodulation. **DMI**, doesn't make infections; **LHK**, lotus histidine kinase; **NSP**, nodulation signalling pathway; **NIN**, nodule inception; **ENOD**, early nodulation gene; **H**, histidine-kinase domain of LHK1 receptor; **D**, receiver domain of LHK1 receptor; **tZ**, *trans*-zeatin; **iP**, isopentenyladenine; **BA**, benzyladenine.

wall-less branches of IT, termed infection droplets, via an endocytosis-like process. Released bacteria start to differentiate into nitrogen-fixing bacteroids, which are separated from the plant cytoplasm by a plant-derived membrane known as the peribacteroid membrane (Hirsch, 1992; Schultze and Kondorosi, 1998). This membrane is initiated from the plasmalemma of infection droplets and is completed with ER and Golgi vesicles (Roth and Stacey, 1989). The peribacteroid membrane acts as a structural and functional interface between symbionts (Udvardi and Day, 1997). The organelle-like structure comprised of the peribacteroid membrane, bacteroids, and the space in between is known as the symbiosome. Thus, the peribacteroid membrane is often termed the symbiosome membrane. Several variations exist on this theme. For instance, in primitive legumes, like *Andira spp.*, rhizobia are accommodated in intracellular IT, which, in contrast to symbiosomes, are bound by a cell wall (Limpens et al., 2005). In determinate legumes, a symbiosome contains many bacteroids whereas, in indeterminate legumes, single bacteroids are present in a symbiosome (Prell and Poole, 2006).

The symbiosome membrane regulates the transport of metabolites between symbionts and is essential for stable symbiosis. Its degradation is usually correlated with senescence of the symbiosis (Udvardi and Day, 1997). The major metabolites exchanged across this interface are carbon, which increases nitrogenase activity in the bacteroid, and reduced nitrogen, which is assimilated in the plant cytoplasm. Several transporters found on the symbiotic interface are probably required for nutrient exchanges and include a dicarboxylate transporter, ammonium channels, and transporters for inorganic anions, iron, calcium, and auxin (Prell and Poole, 2006; Blumwald et al., 1985; Rosendahl, Dilworth, and Glenn, 1992; Moreau, Meyer, and Puppo, 1995; LeVier, Day, and Guerinot, 1996; Andreev et al., 1998; Panter et al., 2000; Day et al., 2001; Lodwig et al., 2003). The *L. japonicus* sulfate transporter Sst1 is present in symbiosome membranes and absolutely required for nitrogen fixation (Wienkoop and Saalbach, 2003; Krusell et al., 2005). Other proteomics experiments have revealed more than 200 legume proteins that are localized to the symbiosome membrane including an H^+-ATPase, ENOD16, ENOD8, and a multifunctional aquaporin (Wienkoop and Saalbach, 2003; Saalbach, Erik, and Wienkoop, 2002; Catalano, Lane, and Sherrier, 2004). GmN70 and LjN70 are inorganic anion transporters of the symbiosome membrane with enhanced preference for nitrate (Vincill, Szczyglowski, and Roberts, 2005). *M. truncatula* NOD25, the product of a nodulin, is exclusively translocated to the symbiosomes of infected cells, and the N-terminal signal peptide of this protein is sufficient for protein targeting to symbiosomes (Hohnjec et al., 2009).

Similarly, AM infection leads to the formation of arbuscules. These highly branched hyphal structures develop specifically in inner cortical cells (Harrison, 2005). Root meristems, differentiating tissues, endodermis, or vascular tissues are never infected by AM fungi. The structure of arbuscules is dependent on the host plant and fungal genotype. Arbuscule accommodation requires changes in host cell architecture including the formation of a periarbuscular membrane (PAM) that is produced by plasma membrane proliferation and cell wall deposition all around the fungus (Genre and Bonfante, 1998; Genre and Bonfante, 2005; Balestrini and Bonfante, 2005; Reinhardt, 2007). Similar to the peribacteroid membrane in RNS, the PAM separates fungal structures from the host cytoplasm (Harrison, 2005). PAM, fungal plasma membrane, and interstitial space between these two membranes are a specialized interface for nutrient and signal exchange between plants and their fungal endosymbiont. The *M. truncatula* phosphate transporter PT4 localizes specifically to the PAM confirming the significance of this structure for nutrient exchange (Harrison, Dewbre, and Liu 2002). Knocking-out expression of *MtPT4* leads to a premature degradation of arbuscules indicating that ability to transport phosphate, and probably other nutrients, are required for arbuscule maintenance

(Javot et al., 2007). Similarly, knocking down an AM-inducible phosphate transporter in *L. japonicus* suppresses the symbiotic association (Maeda et al., 2006). Several studies have shown that N is transferred to the plant as ammonium, although the transporters involved in N-acquisition are still unknown (Govindarajulu et al., 2005; Jin et al., 2005). A proton-ATPase and an ammonium transporter are upregulated in colonized cortical cells (Krajinski et al., 2008; Gomez et al., 2009). This ammonium transporter is a strong candidate for mediating ammonium uptake in AM (Gomez et al., 2009). Identification of additional components will help in our understanding of how symbiosomes and arbuscules mediate nutrient exchanges in RNS and AM, respectively.

Genetics of Symbiosome and Arbuscule Development

Symbiosome Development   Symbiosome formation is essential for nitrogen fixation. It represents a major step in the evolution of legume nodulation. For instance, symbiosomes are not found in nodules of primitive legume species (e.g., *Andira* spp.) and the non-legume *Parasponia* (Faria, Sutherland, and Sprent, 1986; Trinick, 1979). However, in more advanced legumes like *Medicago truncatula* and *Lotus japonicus*, symbiosome formation allows rhizobia to fix nitrogen efficiently (Limpens et al., 2005). Mutants in NF receptors or in the common Sym pathway do not develop symbiosomes probably because of the lack of IT. However, knock-down experiments in some of these genes revealed specific roles of this pathway in symbiosome formation. Knocking-down *DMI2* expression leads to numerous IT in the nodule central tissue but no symbiosomes (Limpens et al., 2005). A minimal level of *DMI2* expression is therefore required to release bacteria from IT into symbiosomes. The subcellular localization of DMI2 to IT membranes also suggests a specific function of DMI2 in rhizobial internalization. Similarly, SrSYMRK, *S. rostrata* ortholog of DMI2, is involved in the control of IT development and symbiosome formation (Capoen et al., 2005).

M. truncatula nip mutants are characterized by numerous wide IT in the nodule central tissue without the formation of any symbiosome (Veereshlingam et al., 2004). Since, DMI2 also acts early in the NF signalling pathway, symbiosome formation is either regulated by DMI2 through NIP or DMI2 and NIP co-regulate symbiosome formation through independent pathways.

In *Mtsym1 (TE7)* mutants, rhizobia are endocytosed into host cells but fail to elongate into bacteroids (Bénaben et al., 1995). Other key genes required for bacteroid development include *M. truncatula ENOD40-1* and *ENOD40-2* (Wan et al., 2007). Silencing of the *GS52* soybean ecto-apyrase also results in aberrant nodules with impaired bacteroid development, similar to the phenotype observed after *MtENOD40-1* and *MtENOD40-2* silencing (Govindarajulu et al., 2009). Transcription factors such as EFD (ethylene response factor for nodule differentiation) and *MtHAP2-1* are also required for bacteroid differentiation (Vernié et al., 2008; Combier et al., 2006). Additional genes involved in bacteroid differentiation in *M. truncatula* include *MtIRE*, *MtDNF1*, −4, −5, and −7 and *MtZPT2* (Mitra, Shaw, and Long, 2004; Starker et al., 2006; Frugier et al., 2000; Pislariu and Dickstein, 2007). In *L. japonicus*, the *ign1* (*ineffective greenish nodules 1*) mutants develop irregularly enlarged symbiosomes with multiple bacteroids at early stages of nodule development (8–9 days post inoculation). However, *ign1* mutants are arrested in nitrogen fixation due to subsequent disruption of symbiosomes and disintegration of infected cells. *IGN1* is therefore required for differentiation and persistence of symbiosomes (Kumagai et al., 2007).

Arbuscule Development   Many plant mutants affected in arbuscule development have been identified. Proteins of the Sym pathway in *M. truncatula, L. japonicus,* and rice are required for arbuscule formation, a conclusion based on the absence of intraradical hyphae in these mutants (Kistner et al., 2005; Novero et al., 2002; Demchenko et al., 2004). In addition, *sym36* of pea (Gianinazzi-Pearson et al., 1991) and R69 of *P. vulgaris* (Shirtliffe and Vessey, 1996) are affected in arbuscule development.

Gene expression patterns associated with arbuscule development provide evidence for the existence of cell autonomous and cell non-autonomous signals (Harrison, 2005). The cell autonomous signal induces the expression of genes (e.g., *MtPT4*) specific to cells containing arbuscules, whereas the non-autonomous signal induces the expression of genes (e.g., *MtSCP1*) in cells with arbuscules as well as in adjacent cells (Wulf et al., 2003; Liu et al., 2003). A carbon gradient may be involved since arbuscules often develop in cortical cells that are the closest to the vascular tissue, source of carbon transport to the root (Blee and Anderson, 1998). In arbuscule-containing cells, expression of genes associated with the urea cycle, amino acid biosynthesis and cellular autophagy is induced. Transcriptome analyses of colonized cortical cells indicate an upregulation of signalling components such as a LysM receptor-like kinase and transcription factors of the GRAS family (Gomez et al., 2009).

Hormonal Regulation of Symbiosome and Arbuscule Development

Plant hormones play many regulatory roles in bacteroid development. GA biosynthetic enzymes have been identified in *B. japonicum* under conditions similar to those inside the symbiosome (Tully et al., 1998). GA levels are regulated both before and after bacteroid differentiation either by rhizobia or by the host plant. Nitric oxide (NO) is a plant hormone and, at the same time, an inhibitor of nitrogenase activity (Trinchant and Rigaud, 1982; Kanayama and Yamamoto, 1990; Mathieu et al., 1998). *Bradyrhizobium* bacteroids overcome the inhibitory action of NO by inducing nitric oxide reductase (Meakin et al., 2006). In addition, NO and auxin synthesized by rhizobia positively affect nodule formation in indeterminate legumes but not in determinate legumes (Pii et al., 2007).

Cytokinins and NFs induce expression of *ENOD40* genes (Fang and Hirsch, 1998). *MtENOD40-1* and *MtENOD40-2* are both required for bacteroid development, which suggests a possible role for cytokinins in bacteroid development via *ENOD* genes in *M. truncatula* (Wan et al., 2007). EFD is a transcriptional regulator of both symbiosome formation and bacteroid differentiation. It participates in an ethylene-independent feedback inhibition of nodulation and regulates the expression of *MtRR4,* a primary cytokinin response regulator (Vernié et al., 2008; Gonzalez-Rizzo, Crespi, and Frugier 2006; Lohar et al., 2006). This also suggests a key role for cytokinins in symbiosome formation.

Similarly, several hormones play regulatory roles in arbuscule development. For example, JA biosynthetic enzymes and a protein encoded by a JA-responsive gene are specifically expressed in arbuscule-containing cells (Hause et al., 2002; Isayenkov et al., 2005). Enhanced JA levels in mycorrhizal roots are correlated with an increased number of collapsed arbuscules (Vierheilig, 2004). The use of *sitiens* tomato mutants with reduced ABA concentrations showed that ABA contributes to the susceptibility of tomato to infection by AMF and plays an important role in arbuscule development (Herrera-Medina et al., 2007). ABA also acts by negatively modulating the ethylene pathway (Herrera-Medina et al., 2007). Several cytokinin-induced early nodulin genes (*ENOD2/ENOD40*) are also upregulated in mycorrhizal roots (van Rhijn et al., 1997). High auxin levels in *G. intraradices*-inoculated roots correlate with a high number of arbuscules (Jentschel et al., 2007). Lysophosphatidylcholine (LPC), product of phospholipid metabolism, of unknown origin (plant or fungal) also seems to be an important signalling molecule involved in arbuscule development and expression of the *StPT3* and *StPT4* phosphate transporters in potato (Drissner et al., 2007).

Cortical cells of host roots accommodate symbiotic microbes in RNS and AM. These cells also undergo special reprogramming to develop functional nodules in RNS. Specialized structures, symbiosomes, and arbuscules are formed in RNS and AM and are the sites for nutrient exchanges. Genetic studies have identified several host components that are required for the formation of these structures. In addition, phytohormones closely regulate nodule organogenesis as well as formation of specialized exchange structures.

Perspectives

Sustainable food and fuel production are two major challenges currently facing our world. It is imperative to minimize ecological, health, and economic drawbacks associated with our dependence on chemical fertilizers. Using beneficial plant-microbe interactions has emerged as one rational and promising solution to these crises. Over the last decade, the scientific community has made tremendous strides to unravel signalling pathways involved in the establishment and maintenance of legume nodulation. Accumulating evidences point toward the existence a complicated network involving plant and microbial signals, genetic control by host plants, and a precise regulation of these mechanisms by phytohormones. Some of the unresolved questions in AM are the identification of Myc factors, the nature of their receptors, and mechanisms that operate independently of the common Sym pathway. Understanding how plant hormones regulate symbiotic development will be essential toward improving host-symbiont interactions and perhaps engineering novel host-microbe associations. Such a synergistic, multidisciplinary approach will have tremendous potential for crop improvement and the sustainability of our agricultural systems.

Acknowledgements

The authors gratefully acknowledge the DOE Great Lakes Bioenergy Research Center (DOE BER Office of Science DE-FC02-07ER64494) at the University of Wisconsin, Madison for their financial support. They would also thank Dr. Désirée den Os (University of Wisconsin, Madison) for critical reading of the manuscript and helpful suggestions.

References

Aguilar, J.M.M., Ashby, A.M., Richards, A.J.M., Loake, G.J., Watson, M.D., Shaw, C.H. 1988. "Chemotaxis of *Rhizobium leguminosarum* biovar *phaeoli* towards flavonoid inducers of the symbiotic nodulation genes," *Journal of General Microbiology,* vol. 134, pp. 2741–2746.

Akiyama, K., Hayashi, H. 2006. "Strigolactones: Chemical signals in fungal symbionts and parasitic weeds in plant roots", *Annals of Botany,* vol. 97, pp. 925–931.

Akiyama, K., Matsuzaka, K., Hayashi, H. 2005. "Plant sesquiterpenes induce hyphal branching in arbuscular mycorrhizal fungi," *Nature,* vol. 435, no. 7043, pp. 750–751.

Alonso, J.M., Hirayama, T., Roman, G., Nourizadeh, S., Ecker, J.R. 1999. "*EIN2*, a bifunctional transducer of ethylene and stress responses in Arabidopsis," *Science,* vol. 284, pp. 2148–2152.

Andreev, I.M., Dubrovo, O.N., Krylova, K.Y., Izmailov, S.F. 1998. "Calcium uptake by symbiosomes and the peribacteroid membrane vesicles isolated from yellow lupin root nodules," *Journal of Plant Physiology,* vol. 153, pp. 610–614.

Andriankaja, A., Boisson-Dernier, A., Frances, L., Sauviac, L., Jauneau, A., Barker, D.G., de Carvalho-Niebel, F. 2007. "AP2-ERF transcription factors mediate Nod factor dependent *Mt ENOD11* activation in root hairs via a novel cis-regulatory motif," *Plant Cell,* vol. 19, no. 9, pp. 2866–2885.

Ané, J.M., Kiss, G.B., Riely, B.K., Penmetsa, R.V., Oldroyd, G.E., Ayax, C., Lévy, J., Debellé, F., Baek, J.M., Kaló, P., Rosenberg, C., Roe, B.A., Long, S.R., Dénarié, J., Cook, D.R. 2004. "*Medicago truncatula DMI1* required for bacterial and fungal symbioses in legumes," *Science,* vol. 303, no. 5662, pp. 1364–1367.

Ardourel, M., Demont, N., Debellé, F., Maillet, F., de Billy, F., Promé, J.C., Dénarié, J., Truchet, G. 1994. "*Rhizobium meliloti* lipooligosaccharide nodulation factors: different structural requirements for bacterial entry into target root hair cells and induction of plant symbiotic developmental responses," *Plant Cell,* vol. 6, no. 10, pp. 1357–1374.

Azcon-Aguilar, C., Rodriguez-Navarro, D., Barea, J. 1981. "Effects of ethrel on the formation and responses to VA mycorrhiza in *Medicago* and *Triticum*," *Plant Soil,* vol. 60, pp. 461–468.

Badenoch-Jones, J., Flanders, D.J., Rolfe, B.J. 1985. "Association of Rhizobium strains with roots of *Trifolium repens,*" *Applied and Environmental Microbiology,* vol. 49, pp. 1511–1520.

Bais, H.P., Weir, T.L., Perry, L.G., Gilroy, S., Vivanco, J.M. 2006. "The role of root exudates in rhizosphere interactions with plants and other organisms." *Annual Review of Plant Biology,* vol. 57, pp. 233–266.

Balestrini, R. and Bonfante, P. 2005. "The interface compartment in arbuscular mycorrhizae: a special type of plant cell wall?" *Plant Biosystems,* vol. 139, pp. 8–15.

Banba, M., Gutjahr, C., Miyao, A., Hirochika, H., Paszkowski, U., Kouchi, H., Imaizumi-Anraku, H. 2008. "Divergence of evolutionary ways among common *sym* genes: CASTOR and CCaMK show functional conservation between two symbiosis systems and constitute the root of a common signaling pathway," *Plant and Cell Physiology,* vol. 49, no. 11, pp. 1659–1671.

Barker, S.J., Stummer, B., Gao, L., Dispain, I., O'Connor, P.J., Smith, S.E. 1998. "A mutant in *Lycopersicon esculentum* Mill. with highly reduced VA mycorrhizal colonization: isolation and preliminary characterization," *Plant Journal,* vol. 15, pp. 791–797.

Bécard, G., Douds, D.D., Pfeffer, P.E. 1992. "Extensive in vitro hyphal growth of vesicular-arbuscular mycorrhizal fungi in the presence of CO_2 and flavonols," *Applied and Environmental Microbiology,* vol. 68, pp. 1260–1264.

Begum, A.A., Leibovitch, S., Migner, P., Zhang, F. 2001. "Specific flavonoids induced nod gene expression and pre-activated nod genes of *Rhizobium leguminosarum* increased pea (*Pisum sativum L.*) and lentil (*Lens culinaris L.*) nodulation in controlled growth chamber environments," *Journal of Experimental Botany,* vol. 52, pp. 1537–1543.

Bénaben, V., Duc, G., Lefebvre, V., Huguet, T. 1995. "TE7, an inefficient symbiotic mutant of *Medicago truncatula* Gaertn. cv Jemalong," *Plant Physiology,* vol. 107, pp. 53–62.

Besserer, A., Puech-Páges, V., Kiefer, P., Gomez-Roldan, V., Jauneau, A., Roy, S., Portais, J.C., Roux, C., Bécard, G., Séjalon-Delmas, N. 2006. "Strigolactones stimulate arbuscular mycorrhizal fungi by activating mitochondria," *PLoS Biology,* vol. 4, pp. 1239–1247.

Bhalerao, R.P., Eklöf, J., Ljung, K., Marchant, A., Bennett, M., Sandberg, G. 2002. "Shoot derived auxin is essential for early lateral root emergence in Arabidopsis seedlings," *Plant Journal,* vol. 29, pp. 325–332.

Blauenfeldt, J., Joshi, P.A., Gresshoff, P., Caetano-Anollés, G. 1994. "Nodulation of whiteclover (*Trifolium-repens*) in the absence of rhizobium," *Protoplasma,* vol. 179, pp. 106–110.

Blee, K.A. and Anderson, A.J. 1998, "Regulation of arbuscule formation by carbon in the plant," *Plant Journal,* vol. 16, pp. 523–530.

Blilou, I., Ocampo, J.A., Garcia-Garrido, J.M. 2000. "Induction of Ltp (lipid transfer protein) and Pal (phenylalanine ammonia-lyase) gene expression in rice roots colonized by the arbuscular mycorrhizal fungus *Glomus mosseae*," *Journal of Experimental Botany,* vol. 51, no. 353, pp. 1969–1977.

Blumwald, E., Fortin, M.G., Rea, P.A., Verma, D.P.S., Poole, R.J. 1985. "Presence of host plasma membrane type H^+-ATPase in the membrane envelope enclosing the bacteroids in soybean root nodules," *Plant Physiology,* vol. 78, pp. 665–672.

Bohlool, B.B., Schmidt, E.L. 1974. "Lectins- possible basis for specificity in *Rhizobium*-legume root nodule symbiosis," *Science,* vol. 185, pp. 269–271.

Bolle, C. 2004. "The role of GRAS proteins in plant signal transduction and development," *Planta,* vol. 218, pp. 683–692.

Bonfante, P. and Genre, A. 2008. "Plants and arbuscular mycorrhizal fungi: an evolutionary-developmental perspective," *Trends in Plant Science,* vol. 13, pp. 492–498.

Bono, J.J., Riond, J., Nicolaou, K.C., Bockovich, N.J., Estévez, V.A., Cullimore, J.V., Ranjeva, R. 1995. "Characterization of a binding site for chemically synthesized lipo-oligosaccharidic NodRm factors in particulate fractions prepared from roots," *Plant Journal,* vol. 7, pp. 252–260.

Boot, K.J.M., van Brussel, A.A.N., Tak, T., Spaink, H.P., Kijne, J.W. 1999. "Lipochitinoligosaccharides from *Rhizobium leguminosarum* bv. *viciae* reduce auxin transport capacity in *Vicia sativa* subsp. *nigra* roots," *Molecular Plant-Microbe Interactions,* vol. 12, pp. 839–844.

Borisov, A.Y., Madsen, L.H., Tsyganov, V.E., Umehara, Y., Voroshilova, V.A., Batagov, A.O., Sandal, N., Mortensen, A., Schauser, L., Ellis, N., Tikhonovich, I.A., Stougaard, J. 2003. "The *Sym35* gene required for root nodule development in pea is an ortholog of *NIN* from *Lotus japonicus*," *Plant Physiology,* vol. 45, pp. 427–435.

Bouwmeester, H.J., Matusova, R., Zhongkui, S., Beale, M.H. 2003. "Secondary metabolite signalling in host-parasitic plant interactions," *Current Opinion in Plant Biology,* vol. 6, pp. 358–364.

Brelles-Mariño, G., Boiardi, J.L. 1996, "Nitrogen limitation of chemostat-grown *Rhizobium etli* elicits higher infection-thread formation in *Phaeolus vulgaris*," *Microbiology,* vol. 142, pp. 1067–1070.

Brewer, P.B., Dun, E.A., Ferguson, B.J., Rameau, C., Beveridge, C.A. 2009. "Strigolactone acts downstream of auxin to regulate bud outgrowth in pea and Arabidopsis," *Plant Physiology,* vol. 150, pp. 482–493.

Bright, L.J., Liang, Y., Mitchell, D., Harris, J. 2005. "The *LATD* gene of *Medicago truncatula* is required for both nodule and root development," *Molecular Plant-Microbe Interactions,* vol. 18, pp. 521–532.

Brown, D.E., Rashotte, A.M., Murphy, A.S., Normanly, J., Tague, B.W., Peer, W.A., Taiz, L., Muday, G.K. 2001. "Flavonoids act as negative regulators of auxin transport *in vivo* in Arabidopsis," *Plant Physiology,* vol. 126, pp. 524–535.

Buée, M., Rossignol, M., Jauneau, A., Ranjeva, R., Bécard, G. 2000, "The pre-symbiotic growth of arbuscular mycorrhizal fungi is induced by a branching factor partially purified from plant root exudates," *Molecular Plant-Microbe Interactions,* vol. 13, pp. 693–698.

Caba, J.M., Recalde, L., Ligero, F. 1998, "Nitrate-induced ethylene biosynthesis and the control of nodulation in alfalfa," *Plant Cell and Environment,* vol. 21, pp. 87–93.

Capoen, W., Goormachtig, S., De Rycke, R., Schroeyers, K., Holsters, M. 2005, "SrSymRK, a plant receptor essential for symbiosome formation," *PNAS,* vol. 102, no. 29, pp. 10369–10374.

Casimiro, I., Marchant, A., Bhalerao, R., Beeckman, T., Dhooge, S., Swarup, R., Graham, N., Inzé, D., Sandberg, G., Casero, P.J., Bennett, M. 2001. "Auxin transport promotes Arabidopsis lateral root initiation," *Plant Cell,* vol. 13, pp. 843–852.

Catalano, C.M., Lane, W.S., Sherrier, D.J. 2004. "Biochemical characterization of symbiosome membrane proteins from *Medicago truncatula* root nodules," *Electrophoresis,* vol. 25, pp. 519–531.

Catford, J.G., Staehelin, C., Larose, G., Piché, Y., Vierheilig, H. 2006. "Systemically suppressed isoflavonoids and their stimulating effects on nodulation and mycorrhization in alfalfa split-root systems," *Plant Soil,* vol. 285, pp. 257–266.

Catoira, R., Galera, C., de Billy, F., Penmetsa, R.V., Journet, E.P., Maillet, F., Rosenberg, C., Cook, D., Gough, C., Dénarié, J. 2000. "Four genes of *Medicago truncatula* controlling components of a Nod factor transduction pathway," *Plant Cell,* vol. 12, no. 9, pp. 1647–1666.

Chabaud, M., Venard, C., Defaux-Petras, A., Bécard, G., Barker, D.G. 2002. "Targeted inoculation of *Medicago truncatula in vitro* root cultures reveals *MtENOD11* expression during early stages of infection by arbuscular mycorrhizal fungi," *New Phytologist,* vol. 156, pp. 265–273.

Chabot, S., Bel-Rhlid, R., Chênevert, R., Piché, Y. 1992. "Hyphal growth promotion *in vitro* of the VA mycorrhizal fungus, *Gigaspora margarita* by the activity of structurally specific flavonoids compounds under CO_2 enriched conditions," *New Phytologist,* vol. 122, pp. 461–467.

Charon, C., Sousa, C., Crespi, M., Kondorosi, A. 1999. "Alteration of *enod40* expression modifies *Medicago truncatula* root nodule development induced by *Sinorhizobium meliloti*," *Plant Cell,* vol. 11, pp. 1953–1965.

Chen, C., Ané, J.M., Zhu, H. 2008. "OsIPD3, an ortholog of the *Medicago truncatula* DMI3 interacting protein IPD3, is required for mycorrhizal symbiosis in rice," *New Phytologist,* vol. 180, pp. 311–315.

Chen, C., Fan, C., Gao, M., Zhu, H. 2009. "Antiquity and function of CASTOR and POLLUX, the twin ion channel-encoding genes key to the evolution of root symbioses in plants," *Plant Physiology,* vol. 149, pp. 306–317.

Combier, J.P., Frugier, F., de Billy, F., Boualem, A., El-Yahyaoui, F., Moreau, S., Vernié, T., Ott, T., Gamas, P., Crespi, M., Niebel, A. 2006. "*MtHAP2-1* is a key transcriptional regulator of symbiotic nodule development regulated by microRNA169 in *Medicago truncatula*," *Genes and Development,* vol. 20, no. 22, pp. 3084–3088.

Cooper, J.B., Long, S.R. 1994. "Morphogenetic rescue of *Rhizobium meliloti* nodulation mutants by trans-zeatin secretion," *Plant Cell,* vol. 6, pp. 215–225.

Cooper, J.E. 2007. "Early interactions between legumes and rhizobia: disclosing complexity in a molecular dialogue," *Journal of applied microbiology,* vol. 103, no. 5, pp. 1355–1365.

Crespi, M., Jurkevitch, E., Poiret, M., d'Aubenton-Carafa, Y., Petrovics, G., Kondorosi, E., Kondorosi, A. 1994. "*enod40*, a gene expressed during nodule organogenesis, codes for a non-translatable RNA involved in plant growth," *EMBO Journal,* vol. 13, pp. 5099–5112.

Dakora, F.D., Joseph, C.M., Phillips, D.A. 1993. "Alfalfa (*Medicago sativa* L.) root exudates contains flavonoids in the presence of *Rhizobium meliloti*," *Plant Physiology,* vol. 101, pp. 819–824.

David-Schwartz, D., Badani, H., Smadar, W., Levy, A.A., Galili, G., Kapulnik, Y. 2001. "Identification of a novel genetically controlled step in mycorrhizal colonization: plant resistance to infection by fungal spores but not extra-radical hyphae," *Plant Journal,* vol. 27, pp. 561–569.

David-Schwartz, R., Gadkar, V., Wininger, S., Bendov, R., Galili, G., Levy, A.A., Kapulnik, Y. 2003. "Isolation of a premycorrhizal infection (*pmi2*) mutant of tomato, resistant to arbuscular mycorrhizal fungal colonization," *Molecular Plant-Microbe Interactions,* vol. 16, no. 5, pp. 382–388.

Day, D.A., Poole, P.S., Tyerman, S.D., Rosendahl, L. 2001. "Ammonia and amino acid transport across symbiotic membranes in nitrogen-fixing legume nodules," *Cellular and Molecular Life Sciences,* vol. 58, pp. 61–71.

Day, R.B., McAlvin, C.B., Loh, J.T., Denny, R.L., Wood, T.C., Young, N.D., Stacey, G. 2000. "Differential expression of two soybean apyrases, one of which is an early nodulin.," *Molecular Plant-Microbe Interactions,* vol. 13, pp. 1053–1070.

Dazzo, F.B. and Hollingsworth, R.E. 1984a. "Trifoliin-A and carbohydrate receptors as mediators of cellular recognition in the *Rhizobium trifolii*- clover symbiosis," *Biology of the Cell,* vol. 51, pp. 267–274.

Dazzo, F.B., Hubbell, D.H. 1975. "Cross-reactive antigens and lectin as determinants of symbiotic specificity in *Rhizobium*-clover association," *Applied Microbiology,* vol. 30, pp. 1017–1033.

Dazzo, F.B., Truchet, G.L., Sherwood, J.E., Hrabak, E.M., Abe, M., Pankratz, S.H. 1984b. "Specific phases of root hair attachment in the *Rhizobium trifolii*-clover symbiosis," *Applied and Environmental Microbiology,* vol. 48, pp. 1140–1150.

Demchenko, K., Winzer, T., Stougaard, J., Parniske, M., Pawlowski, K. 2004. "Distinct roles of *Lotus japonicus SYMRK* and *SYM15* in root colonization and arbuscule formation," *New Phytologist,* vol. 163, pp. 381–392.

Dénarié, J., Debellé, F., Promé, J.C. 1996. "Rhizobium lipo-chitooligosaccharide nodulation factors: signaling molecules mediating recognition and morphogenesis," *Annual Review of Biochemistry,* vol. 65, pp. 503–535.

Dénarié, J., Debellé, F., Rosenberg, C. 1992. "Signaling and host range variation in nodulation," *Annual Review of Microbiology,* vol. 46, pp. 497–531.

D'Haeze, W., De Rycke, R., Mathis, R., Goormachtig, S., Pagnotta, S., Verplancke, C., Capoen, W., Holsters, M. 2003. "Reactive oxygen species and ethylene play a positive role in lateral root base nodulation of a semiaquatic legume," *PNAS,* vol. 100, pp. 11789–11794.

Dharmatilake, A.J., Bauer, W.D. 1992. "Chemotaxis of *Rhizobium meliloti* towards nodulation gene-inducing compounds from alfalfa roots," *Applied and Environmental Microbiology,* vol. 58, pp. 1153–1158.

Ding, Y., Kaló, P., Yendrek, C., Sun, J., Liang, Y., Marsh, J.F., Harris, J.M., Oldroyd, G.E.D. 2008. "Abscisic Acid coordinates Nod Factor and cytokinin signaling during the regulation of nodulation in *Medicago truncatula*," *Plant Cell,* vol. 20, no. 10, pp. 2681–2695.

Drissner, D., Kunze, G., Callewaert, N., Gehrig, P., Tamasloukht, M., Boller, T., Felix, G., Amrhein, N., Bucher, M. 2007. "Lyso-Phosphatidylcholine is a signal in the arbuscular mycorrhizal symbiosis," *Science,* vol. 318, pp. 265–268.

Ehrhardt, D.W., Wais, R., Long, S.R. 1996. "Calcium spiking in plant root hairs responding to Rhizobium nodulation signals," *Cell,* vol. 85, no. 5, pp. 673–681.

El-Tohamy, W., Schnitzler, W.H., El-Behairy, U., El-Beltagy, M.S. 1999. "Effect of VA mycorrhiza on improving drought and chilling tolerance of bean plants (Phaseolus vulgaris L.)," *Journal of Applied Botany,* vol. 73, pp. 178–183.

Endre, G., Kereszt, A., Kevei, Z., Mihacea, S., Kaló, P., Kiss, G.B. 2002. "A receptor kinase gene regulating symbiotic nodule development," *Nature,* vol. 417, no. 6892, pp. 962–966.

Esseling, J.J., Lhuissier, F.G., Emons, A.M. 2003. "Nod factors-induced root hair curling: continuous polar growth towards the point of Nod factor application," *Plant Physiology,* vol. 132, pp. 1982–1988.

Etzler, M.E., Kalsi, G., Ewing, N.N., Roberts, N.J., Day, R.B., Murphy, J.B. 1999. "A Nod Factor binding lectin with apyrase activity from legume roots," *PNAS,* vol. 96, no. 10, pp. 5856–5861.

Fang, Y., Hirsch, A.M. 1998. "Studying early nodulin gene *ENOD40* expression and induction by nodulation factor and cytokinin in alfalfa," *Plant Physiology,* vol. 116, pp. 53–68.

Faria, S., Sutherland, J.M., Sprent, J.I. 1986. "A new type of infected cell in root nodules of *Andira* spp. (Leguminosae).," *Plant Science,* vol. 45, pp. 143–147.

Fortin, J.A., Bécard, G., Declerck, S., Dalpé, Y., St.-Arnaud, M., Coghlan, A.P., Piché, Y. 2002. "Arbuscular mycorrhiza on root-organ cultures," *Canadian Journal of Botany,* vol. 80, pp. 1–20.

Frey-Klett, P., Garbaye, J., Tarkka, M. 2007. "The mycorrhiza helper bacteria revisited," *New Phytologist,* vol. 176, pp. 22–36.

Frugier, F., Kosuta, S., Murray, J.D., Crespi, M., Szczyglowski, K. 2008. "Cytokinin: secret agent of symbiosis," *Trends in Plant Science,* vol. 13, no. 3, pp. 115–120.

Frugier, F., Poirier, S., Satiat-Jeunemaître, B., Kondorosi, A., Crespi, M. 2000. "A Krüppel-like zinc finger protein is involved in nitrogen-fixing root nodule organogenesis," *Genes and Development,* vol. 14, pp. 475–482.

Gagnon, H., Ibrahim, R.K. 1998. "Aldonic acids: a novel family of nod gene inducers of *Mesorhizobium loti, Rhizobium lupini* and *Sinorhizobium meliloti,*" *Molecular Plant-Microbe Interactions,* vol. 11, pp. 988–998.

Geil, R., Guinel, F. 2002. "Effects of elevated substrate-ethylene on colonization of leek (*Allium porrum*) by the arbuscular mycorrhizal fungi *Glomus aggregatum,*" *Canadian Journal of Botany,* vol. 80, pp. 114–119.

Geil, R., Peterson, R., Guinel, F. 2001. "Morphological alterations of pea (*Pisum sativum* cv. *Sparkle*) arbuscular mycorrhizas as a result of exogenous ethylene treatment," *Mycorrhiza,* vol. 11, pp. 137–143.

Genre, A., Bonfante, P. 2005. "Building a mycorrhizal cell: how to reach compatibility between plants and arbuscular mycorrhizal fungi," *Journal of Plant Interactions,* vol. 1, pp. 3–13.

Genre, A., Bonfante, P. 1998. "Actin versus tubulin configuration in arbuscule-containing cells from mycorrhizal tobacco roots," *New Phytologist,* vol. 140, pp. 745–752.

Genre, A., Chabaud, M., Facio, A., Barker, D.G., Bonfante, P. 2008. "Prepenetration assembly precedes and predicts the colonization patterns of arbuscular mycorrhizal fungi within root cortex of both *Medicago truncatula* and *Daucus carota,*" *Plant Cell,* vol. 20, pp. 1407–1420.

Genre, A., Chabaud, M., Timmers, T., Bonfante, P., Baker, G.D. 2005. "Arbuscular mycorrhizal fungi elicit a novel intracellular apparatus in *Medicago truncatula* root epidermal cells before infection," *Plant Cell,* vol. 17, pp. 3489–3499.

Gherbi, H., Markmann, K., Svistoonoff, S., Estevan, J., Autran, D., Giczey, G., Auguy, F., Péret, B., Laplaze, L., Franche, C., Parniske, M., Bogusz, D. 2008. "SymRK defines a common genetic basis for plant root endosymbioses with arbuscular mycorrhiza fungi, rhizobia, and Frankiabacteria," *PNAS,* vol. 105, no. 12, pp. 4928–4932.

Gianinazzi-Pearson, V., Gianinazzi, S., Guillemin, G.P., Trouvelot, A., Duc, G. 1991. "Genetic and cellular analysis of resistance to vesicular-arbuscular (VA) endomycorrhizal fungi in pea mutants," *Advances in molecular genetics and plant–microbe interactions.* Kluwer, Dordrecht, pp. 336–342.

Giovannetti, M., Avio, L., Sbrana, S., Citernesi, A.S. 1993a. "Factors affecting appresorium development in the vesicular arbuscular mycorrhizal fungi *Glomus mosseae*," *New Phytologist,* vol. 123, pp. 115–122.

Giovannetti, M., Sbrana, C., Avio, L., Citernesi, A.S., Logi, C. 1993b. "Differential hyphal morphogenesis in arbuscular mycorrhizal fungi during pre-infection stages," *New Phytologist,* vol. 125, pp. 587–594.

Giovannetti, M., Citernesi, A.S. 1993. "Time course of appresorium formation on host plants by arbuscular mycorrhizal fungi," *Mycological Research,* vol. 97, pp. 1140–1142.

Giraud, E., Moulin, L., Vallenet, D., Barbe, V., Cytryn, E., Avarre, J.C., Jaubert, M., Simon, D., Cartieaux, F., Prin, Y., Bena, G., Hannibal, L., Fardoux, J., Kojadinovic, M., Vuillet, L., Lajus, A., Cruveiller, S., Rouy, Z., Mangenot, S., Segurens, B., Dossat, C., Frank, W.L., Chang, W.S., Saunders, E., Bruce, D., Richardson, P., Normand, P., Dreyfus, B., Pignol, D., Stacey, G., Emerich, D., Verméglio, A., Médigue, C., Sadowsky, M. 2007. "Legume symbioses: absence of *Nod* genes in photosynthetic bradyrhizobia," *Science,* vol. 316, pp. 1307–1312.

Gleason, C., Chaudhuri, S., Yang, T., Muñoz, A., Poovaiah, B.W., Oldroyd, G.E. 2006. "Nodulation independent of rhizobia induced by a calcium-activated kinase lacking autoinhibition," *Nature,* vol. 441, no. 7097, pp. 1149–1152.

Goedhart, J., Hink, M.A., Visser, A.J., Bisseling, T., Gadella, T.W. 2000. "*In vivo* fluorescence correlation microscopy (fcm) reveals accumulation and immobilization of Nod factors in root hair cell walls," *Plant Journal,* vol. 21, pp. 109–119.

Goldmann, L., Lecouer, L., Message, B., De La Rue, M., Schoonejans, E., Tepfer, D. 1994. "Symbiotic plasmid genes essential to the catabolism of proline betaine, or stachydrine, are also required for efficient nodulation by *Rhizobium meliloti*," *FEMS Microbiology Letters,* vol. 115, pp. 305–311.

Goldwasser, Y., Yoneyama, K., Xie, X., Yoneyama, K. 2008. "Production of Strigolactones by *Arabidopsis thaliana* responsible for *Orobanche aegyptiaca* seed germination," *Plant Growth Regulation,* vol. 55, pp. 21–28.

Gomez, S.K., Javot, H., Deewatthanawong, P., Torres-Jerez, I., Tang, Y., Blancaflor, E.B., Udvardi, M.K., Harrison, M.J. 2009. "*Medicago truncatula* and *Glomus intraradices* gene expression in cortical cells harboring arbuscules in the arbuscular mycorrhizal symbiosis," *BMC Plant Biology,* vol. 9, pp. 10.

Gomez-Roldan, V., Fermas, S., Brewer, P.B., Puech-Pagès, V., Dun, E.A., Pillot, J.P., Letisse, F., Matusova, R., Danoun, S., Portais, J.C., Bouwmeester, H., Bécard, G., Beveridge, C.A., Rameau, C., Rochange, S.F. 2008. "Strigolactone inhibition of shoot branching," *Nature,* vol. 455, pp. 189–194.

Gomez-Roldan, V., Roux, C., Girard, D., Bécard, G., Puech, V. 2007. "Strigolactones: Promising plant signals," *Plant Signaling and Behavior,* vol. 2, pp. 163–164.

Gonzalez-Rizzo, S., Crespi, M., Frugier, F. 2006. "The *Medicago truncatula CRE1* cytokinin receptor regulates lateral root development and early symbiotic interaction with *Sinorhizobium meliloti*," *Plant Cell,* vol. 18, pp. 2680–2693.

Goormachtig, S., Capoen, W., Holsters, M. 2004. "Rhizobium infection: lessons from the versatile nodulation behaviour of water-tolerant legumes," *Trends in Plant Science,* vol. 9, no. 11, pp. 518–522.

Govindarajulu, M., Kim, S.Y., Libault, M., Berg, R.H., Tanaka, K., Stacey, G., Taylor, C.G. 2009. "GS52 ecto-apyrase plays a critical role during soybean nodulation," *Plant Physiology*, vol. 149, pp. 994–1004.

Govindarajulu, M., Pfeffer, P.E., Jin, H.R., Abubaker, J., Douds, D.D., Allen, J.W., Bucking, H., Lammers, P.J., Shachar-Hill, Y. 2005. "Nitrogen transfer in the arbuscular mycorrhizal symbiosis," *Nature*, vol. 435, pp. 819–823.

Graham, P.H., Vance, C.P. 2003, "Legumes: Importance and constraints to greater use," *Plant Physiology*, vol. 131, pp. 872–877.

Gressent, F., Drouillard, S., Mantegazza, N., Samain, E., Geremia, R.A., Canut, H., Niebel, A., Driguez, H., Ranjeva, R., Cullimore, J., Bono, J.J. 1999. "Ligand specificity of a high-affinity binding site for lipo-chitooligosaccharidic Nod factors in *Medicago* cell suspension cultures," *PNAS*, vol. 96, pp. 4704–4709.

Guenoune, D., Galili, S., Phillips, D.A., Volpin, H., Chet, I., Okon, Y., Kapulnik, Y. 2001. "The defense response elicited by the pathogen *Rhizoctonia solani* is suppressed by colonization of VM-fungus *Glomus intraradices*," *Plant Science*, vol. 160, pp. 925–932.

Guerts, R., Heidstra, R., Hadri, A.E., Downie, J.A., Franssen, H., van Kammen, A., Bisseling, T. 1997. "*Sym2* of *Pisum sativum* is involved in a Nod factor perception mechanism that controls the infection process in the epidermis," *Plant Physiology*, vol. 115, pp. 351–359.

Guether, M., Neuhäuser, B., Balestrini, R., Dynowski, M., Ludewig, U., Bonfante, P. 2009. "A mycorrhizal-specific ammonium transporter from *Lotus japonicus* acquires nitrogen released by arbuscular mycorrhizal fungi," *Plant Physiology*, vol. 150, pp. 73–83.

Guinel, F.C., Geil, R.D. 2002. "A model for the development of the rhizobial and arbuscular mycorrhizal symbioses in legumes and its use ito understand the roles of ethylene in the establishment in these two symbioses," *Canadian Journal of Botany*, vol. 80, pp. 695–720.

Gutjahr, C., Banba, M., Croset, V., An, K., Miyao, A., An, G., Hirochika, H., Imaizumi-Anraku, H., Paszkowski, U. 2008. "Arbuscular Mycorrhiza-specific signaling in rice transcends the common symbiosis signaling pathway," *Plant Cell*, vol. 20, no. 11, pp. 2989–3005.

Hamblin, J., Kent, S.P. 1973. "Possible role of phytohemagglutinin in *Phaseolus* L," *Nature New Biology*, vol. 245, pp. 28–29.

Hao, L., Renxiao, W., Qian, Q., Meixian, Y., Xiangbing, M., Zhiming, F., Cunyu, Y., Biao, J., Zhen, S., Jiayang, L., Yonghong, W. 2009. "DWARF27, an iron-containing protein required for the biosynthesis of strigolactones, regulates rice tiller bud outgrowth," *Plant Cell*, vol. 21, pp. 1512–1525.

Harrison, M.J., Dewbre, G.R., Liu, J. 2002. "A phosphate transporter from *Medicago truncatula* involved in the acquisition of phosphate released by arbuscular mycorrhizal fungi," *Plant Cell*, vol. 14, pp. 2413–2429.

Harrison, M.J. 2005. "Signaling in the arbuscular mycorrhizal symbiosis," *Annual Review of Microbiology*, vol. 59, pp. 19–42.

Harrison, M.J., Dixon, R.A. 1993. "Isoflavonoid accumulation and expression of defense gene transcripts during the establishment of vesicular-arbuscular mycorrhiza in roots of *Medicago truncatula*," *Molecular Plant-Microbe Interactions*, vol. 6, pp. 643–654.

Hause, B., Maier, W., Miersch, O., Kramell, R., Strack, D. 2002. "Induction of jasmonate biosynthesis in arbuscular mycorrhizal barley roots," *Plant Physiology*, vol. 130, pp. 1213–1220.

Hawkins, E.J., Johansen, A., George, E. 2000. "Uptake and transport of organic and inorganic nitrogen by arbuscular mycorrhizal fungi," *Plant and Soil*, vol. 226, pp. 275–285.

Heckmann, A.B., Lombardo, F., Miwa, H., Perry, J.A., Bunnewell, S., Parniske, M., Wang, T.L., Downie, J.A. 2006. "*Lotus japonicus* requires two GRAS domain regulators, one of which is functionally conserved in a non-legume," *Plant Physiology*, vol. 142, pp. 1739–1750.

Heidstra, R., Yang, Y.C., Yalcin, Y., Peck, S., Emons, A.M., van Kammen, T., Bisseling, T. 1997. "Ethylene provides positional information on cortical cell division but is not involved in Nod factor-induced root hair tip growth in *Rhizobium* legume interaction," *Development*, vol. 124, pp. 1781–1787.

Heinz, E.B., Phillips, D.A., Streit, W.R. 1999. "BioS, a biotin-induced, stationary phase, and possible LysR-type regulator in *Sinorhizobium meliloti*," *Molecular Plant-Microbe Interactions*, vol. 12, pp. 803–812.

Herrera-Medina, M.J., Steinkellner, S., Vierheilig, H., Bote, J.A.O., Garrido, J.M.G. 2007. "Abscisic acid determines arbuscule development and functionality in the tomato arbuscular mycorrhiza," *New Phytologist*, vol. 175, pp. 554–564.

Himanen, K., Boucheron, E., Vanneste, S., de Almeida, E.J., Inzé, D., Beeckman, T. 2002. "Auxin-mediated cell cycle activation during early lateral root initiation," *Plant Cell*, vol. 14, pp. 2339–2351.

Hirsch, A.M. 1999. "Role of lectins (and rhizobial exopolysaccharides) in legume nodulation," *Current Opinion in Plant Biology*, vol. 2, pp. 320–326.

Hirsch, A.M. 1992. "Developmental biology of legume nodulation," *New Phytologist*, vol. 122, pp. 211–237.

Hirsch, A.M., Bhuvaneswari, T.V., Torrey, J.G., Bisseling, T. 1989. "Early nodulin genes are induced in alfalfa root outgrowths elicited by auxin transport inhibitors," *PNAS*, vol. 86, pp. 1246–1248.

Hirsch, A.M., Fang, S.Y., Asad, S., Kapulnik, Y. 1997. "The role of phytohormones in plant microbe symbioses," *Plant Soil*, vol. 194, pp. 171–184.

Hirsch, A.M., Fang, Y. 1994. "Plant hormones and nodulation: what's the connection?" *Plant Molecular Biology*, vol. 26, pp. 5–9.

Hirsch, S., Kim, J., Muñoz, A., Heckmann, A.B., Downie, J.A., Oldroyd, G.E.D. 2009. "GRAS proteins form a DNA binding complex to induce gene expression during nodulation signaling in *Medicago truncatula*," *Plant Cell*, vol. 21, pp. 545–557.

Hodge, A., Campbell, C.D., Fitter, A.H. 2001. "An arbuscular mycorrhizal fungus accelerates decomposition and acquires nitrogen directly from organic material.," *Nature*, vol. 413, pp. 297–299.

Hogg, B.V., Cullimore, J.V., Ranjeva, R., Bono, J.J. 2006. "The *DMI1* and *DMI2* early symbiotic genes of *Medicago truncatula* are required for a high-affinity nodulation factor-binding site associated to a particulate fraction of roots," *Plant Physiology*, vol. 140, no. 1, pp. 365–373.

Hohnjec, N., Lenz, F., Fehlberg, V., Vieweg, M.F., Baier, M.C., Hause, B., Küster, H. 2009. "The signal peptide of the *Medicago truncatula* modular nodulin *MtNOD25* operates as an address label for the specific targeting of proteins to nitrogen-fixing symbiosomes," *Molecular Plant-Microbe Interactions*, vol. 22, pp. 63–72.

Huo, X., Schnabel, E., Hughes, K., Frugoli, J. 2006. "RNAi phenotypes and the localization of a protein::GUS fusion imply a role for *Medicago truncatula PIN* genes in nodulation," *Journal of Plant Growth Regulation*, vol. 25, pp. 155–165.

Imaizumi-Anraku, H., Takeda, N., Charpentier, M., Perry, J., Miwa, H., Umehara, Y., Kouchi, H., Murakami, Y., Mulder, L., Vickers, K., Pike, J., Downie, J.A., Wang, T., Sato, S., Asamizu, E., Tabata, S., Yoshikawa, M., Murooka, Y., Wu, G.J., Kawaguchi, M., Kawasaki, S., Parniske, M., Hayashi, M. 2005. "Plastid proteins crucial for symbiotic fungal and bacterial entry into plant roots," *Nature*, vol. 433, no. 7025, pp. 527–531.

Isayenkov, S., Mrosk, C., Stenzel, I., Strack, D. & Hause, B. 2005. "Suppression of allene oxide cyclase in hairy roots of *Medicago truncatula* reduces jasmonate levels and the degree of mycorrhization with *Glomus intraradices*," *Plant Physiology,* vol. 139, pp. 1401–1410.

Javot, H., Penmetsa, R.V., Terzaghi, N., Cook, D.R., Harrison, M.J. 2007. "A *Medicago truncatula* phosphate transporter indispensable for the arbuscular mycorrhizal symbiosis," *PNAS,* vol. 104, pp. 1720–1725.

Jentschel, K., Thiel, D., Rehn, F., Ludwig-Müller, J. 2007. "Arbuscular mycorrhiza enhances auxin levels and alters auxin biosynthesis in *Tropaeolum majus* during early stages of colonization," *Physiologia Plantarum,* vol. 129, pp. 320–333.

Jin, H., Pfeffer, P.E., Douds, D.D., Piotrowski, E., Lammers, P.J., Shachar-Hill, Y. 2005. "The uptake, metabolism, transport and transfer of nitrogen in an arbuscular mycorrhizal symbiosis," *New Phytologist,* vol. 168, pp. 687–696.

Jones, K.M., Kobayashi, H., Davies, B.W., Taga, M.E., Walker, G.C. 2007. "How rhizobial symbionts invade plants: the *Sinorhizobium-Medicago* model," *Nature Reviews in Microbiology,* vol. 5, no. 8, pp. 619–633.

Journet, E.P., El-Gachtouli, N., Vernoud, V., de Billy, F., Pichon, M., Dedieu, A., Arnould, C., Morandi, D., Barker, D.G., Gianinazzi-Pearson, V. 2001. "*Medicago truncatula ENOD11*: A novel RPRP-encoding early nodulin gene expressed during mycorrhization in arbuscule-containing cells," *Molecular Plant-Microbe Interactions,* vol. 14, no. 6, pp. 737–748.

Journet, E.P., Pichon, M., Dedieu, A., de Billy, F., Truchet, G., Barker, D.G. 1994. "*Rhizobium meliloti* Nod factors elicit cell-specific transcription of the *ENOD12* gene in transgenic alfalfa," *Plant Journal,* vol. 6, pp. 241–249.

Kaló, P., Gleason, C., Edwards, A., Marsh, J., Mitra, R.M., Hirsch, S., Jakab, J., Sims, S., Long, S.R., Rogers, J., Kiss, G.B., Downie, J.A., Oldroyd, G.E. 2005. "Nodulation signaling in legumes requires NSP2, a member of the GRAS family of transcriptional regulators," *Science,* vol. 308, no. 5729, pp. 1786–1789.

Kalsi, G., Etzler, M. 2000. "Localization of a nod factor-binding protein in legume roots and factors influencing its distribution and expression," *Plant Physiology,* vol. 124, pp. 1039–1048.

Kanamori, N., Madsen, L.H., Radutoiu, S., Frantescu, M., Quistgaard, E.M.H., Miwa, H., Downie, J.A., James, E.K., Felle, H.H., Haaning, L.L., Jensen, T.H., Sato, S., Nakamura, Y., Tabata, S., Sandal, N., Stougaard, J. 2006. "A nucleoporin is required for induction of Ca^{2+} spiking in legume nodule development and essential for rhizobial and fungal symbiosis," *PNAS,* vol. 103, no. 2, pp. 359–364.

Kanayama, Y., Yamamoto, Y. 1990. "Inhibition of nitrogen fixation in soybean plants supplied with nitrate II. Accumulation and properties of nitrosylleghemoglobin in nodules," *Plant Physiology,* vol. 31, pp. 207–214.

Kape, R., Parniske, M., Werner, D. 1991. "Chemotaxis and *nod* gene activity of *Bradyrhizobium japonicum* in response to hydroxycinnamic acids and isoflavonoids," *Applied and Environmental Microbiology,* vol. 57, pp. 316–319.

Kato, G., Maruyama, Y., Nakamura, M. 1980. "Role of bacterial polysaccharides in the adsorption process of the *Rhizobium*- pea symbiosis," *Agricultural and Biological Chemistry,* vol. 44, pp. 2843–2855.

Kawagachi, M., Imaizumi-Anraku, H., Koiwa, H., Niwa, S., Ikuta, A. 2002. "Root, root hair, and symbiotic mutants of the model legume *Lotus japonicus*," *Molecular Plant-Microbe Interactions,* vol. 15, pp. 17–26.

Kevei, Z., Lougnon, G., Mergaert, P., Horvath, G.V., Kereszt, A., Jayaraman, D., Zaman, N., Marcel, F., Regulski, K., Kiss, G.B., Kondorosi, A., Endre, G., Kondorosi, E., Ané, J.M.

2007. "3-hydroxy-3-methylglutaryl coenzyme a reductase 1 interacts with NORK and is crucial for nodulation in *Medicago truncatula*," *Plant Cell,* vol. 19, no. 12, pp. 3974–3989.

Kijne, J. 1992. "The Rhizobium infection process," *In Biological Nitrogen Fixation.* Stacey, G., Burris, R.H., and Evans, H.J. (eds). Chapman and Hall, London, pp. 349–398.

Kijne, J.W., Bauchrowitz, M.A., Diaz, C.L. 1997. "Root lectins and rhizobia," *Plant Physiology,* vol. 115, pp. 869–873.

Kim, S.Y., Sivaguru, M., Stacey, G. 2006. "Extracellular ATP in plants: Visualization, localization, and analysis of physiological significance in growth and signaling," *Plant Physiology,* vol. 142, pp. 984–992.

Kistner, C., Winzer, T., Pitzschke, A., Mulder, L., Sato, S., Kaneko, T., Tabata, S., Sandal, N., Stougaard, J., Webb, K.J., Szczyglowski, K., Parniske, M. 2005. "Seven *Lotus japonicus* genes required for the transcriptional reprogramming of the root during fungal and bacterial symbiosis," *Plant Cell,* vol. 17, pp. 2217–2219.

Kobayashi, H., Naciri-Graven, Y., Broughton, W.G., Perret, X. 2004. "Flavonoids induce temporal shifts in gene-expression of nod-box controlled loci in *Rhizobium* sp. NGR234," *Molecular Microbiology,* vol. 51, pp. 335–347.

Kosuta, S., Chabaud, M., Lougnon, G., Gough, C., Dénarié, J., Barker, D.G., Bécard, G. 2003. "A diffusible factor from arbuscular mycorrhizal fungi induces symbiosis-specific *MtENOD11* expression in roots of *Medicago truncatula*," *Plant Physiology,* vol. 131, no. 3, pp. 952–962.

Kosuta, S., Hazledine, S., Sun, J., Miwa, H., Morris, R.J., Downie, J.A., Oldroyd, G.E.D. 2008. "Differential and chaotic calcium signatures in the symbiosis signaling pathway of legumes," *PNAS,* vol. 105, no. 28, pp. 9823–9828.

Krajinski, F., Hause, B., Gianinazzi-Pearson, V., Franken, P. 2008. "*Mtha1*, a plasma membrane H^+-ATPase gene from *Medicago truncatula*, shows arbuscule-specific induced expression in mycorrhizal tissue," *Plant Biology,* vol. 4, pp. 754–761.

Krusell, L., Krause, K., Ott, T., Desbrosses, G., Krämer, U., Sato, S., Nakamura, Y., Tabata, S., James, E.K., Sandal, N., Stougaard, J., Kawaguchi, M., Miyamoto, A., Suganuma, N., Udvardi, M.K. 2005. "The sulfate transporter SST1 is crucial for symbiotic nitrogen fixation in *Lotus japonicus* root nodules," *Plant Cell,* vol. 17, pp. 1625–1636.

Kumagai, H., Hakoyama, T., Umehara, Y., Sato, S., Kaneko, T., Tabata, S., Kouchi, H. 2007. "A novel ankyrin-repeat membrane protein, IGN1, is required for persistence of nitrogen-fixing symbiosis in root nodules of *Lotus japonicus*," *Plant Physiology,* vol. 143, pp. 1293–1305.

Kuppusamy, K.T., Endre, G., Prabhu, R., Penmetsa, R.V., Veereshlingam, H., Cook, D.R., Dickstein, R., VandenBosch, K.A. 2004. "*LIN*, a *Medicago truncatula* gene required for nodule differentiation and persistence of rhizobial infections," *Plant Physiology,* vol. 136, pp. 3682–3691.

Lanfranco, L., Novero, M., Bonfante, P. 2005. "The mycorrhizal fungus *Gigaspora margarita* possesses a CuZn superoxide dismutase that is up-regulated during symbiosis with legume hosts," *Plant Physiology,* vol. 137, pp. 1319–1330.

Laplaze, L., Benkova, E., Casimiro, I., Maes, L., Vanneste, S., Swarup, R., Weijers, D., Calvo, V., Parizot, B., Herrera-Rodriguez, M.B., Offringa, R., Graham, N., Doumas, P., Friml, J., Bogusz, D., Beeckman, T., Bennett, M. 2007. "Cytokinins act directly on lateral root founder cells to inhibit root initiation," *Plant Cell,* vol. 19, pp. 3889–3900.

Larose, G., Chênevert, R., Moutoglis, P., Gagnéb, S., Pichéc, Y., Vierheilig, H. 2002. "Flavonoid levels in roots of *Medicago sativa* are modulated by the developmental stage of the symbiosis and the root colonizing arbuscular mycorrhizal fungus," *Journal of Plant Physiology,* vol. 159, pp. 1329–1339.

Le Strange, K.K., Bender, G.L., Djordjevic, M.A., Rolfe, B.G., Redmond, J.W. 1990. "The *Rhizobium* strain NGR234 *nodD1* gene product responds to activation by the simple phenolic compounds vanillin and isovanillin present in wheat seedling extracts," *Molecular Plant-Microbe Interactions,* vol. 3, pp. 214–220.

LeVier, K., Day, D.A., Guerinot, M.L. 1996. "Iron uptake by symbiosomes from soybean root nodules," *Plant Physiology,* vol. 111, pp. 893–900.

Lévy, J., Bres, C., Geurts, R., Chalhoub, B., Kulikova, O., Duc, G., Journet, E.P., Ané, J.M., Lauber, E., Bisseling, T., Dénarié, J., Rosenberg, C., Debellé, F. 2004. "A putative Ca^{2+} and calmodulin-dependent protein kinase required for bacterial and fungal symbioses," *Science,* vol. 303, no. 5662, pp. 1361–1364.

Liang, Y., Mitchell, D.M., Harris, J.M. 2007. "Abscisic acid rescues the root meristem defects of the *Medicago truncatula latd* mutant," *Developmental Biology,* vol. 304, pp. 297–307.

Libbenga, K.R., Bogers, R.J. 1974. "Root nodule morphogenesis," *The Biology of Nitrogen Fixation.* North-Holland Publishing Co., Amsterdam, pp. 430–472.

Libbenga, K.R., Van Iren, F., Bogers, R.J., Schraag-Lammers, M.F. 1973. "The role of hormones and gradients in the initiation of cortex proliferation and nodule formation in *Pisum sativa L,*" *Planta,* vol. 114, pp. 29–39.

Lievens, S., Goormachtig, S., Den Herder, J., Capoen, W., Mathis, R., Hedden, P., Holsters, M. 2005. "Gibberellins are involved in nodulation of *Sesbania rostrata,*" *Plant Physiology,* vol. 139, no. 3, pp. 1366–1379.

Limpens, E., Franken, C., Smit, P., Willemse, J., Bisseling, T., Geurts, R. 2003. "LysM domain receptor kinases regulating rhizobial Nod factor-induced infection," *Science,* vol. 302, no. 5645, pp. 630–633.

Limpens, E., Mirabella, R., Fedorova, E., Franken, C., Franssen, H., Bisseling, T., Geurts, R. 2005. "Formation of organelle-like N_2-fixing symbiosomes in legume root nodules is controlled by DMI2," *PNAS,* vol. 102, no. 29, pp. 10375–10380.

Liu, J., Blaylock, L., Endre, G., Cho, J., Town, C.D., VandenBosch, K.A., Harrison, M.J. 2003. "Transcript profiling coupled with spatial expression analyses reveals genes involved in distinct developmental stages of the arbuscular mycorrhizal symbiosis," *Plant Cell,* vol. 15, pp. 2106–2123.

Lodiero, A.R., Lagares, A., Martinez, E.N., Favelukes, G. 1995. "Early interactions of *Rhizobium legumniosarum* bv. *phaseoli* and bean roots—Specificity in the process of adsorption and its requirement of Ca^{2+} and Mg^{2+} ions," *Applied and Environmental Microbiology,* vol. 61, pp. 1571–1579.

Lodwig, E.M., Hosie, A.H.F., Bourdès, A., Findlay, K., Allaway, D., Karunakaran, R., Downie, J.A., Poole, P.S. 2003. "Amino-acid cycling drives nitrogen fixation in the legume-*Rhizobium* symbiosis," *Nature,* vol. 422, pp. 722–726.

Lohar, D.P., Schaff, J.E., Laskey, J.G., Kieber, J.J., Bilyeu, K.D., Bird, D.M. 2004. "Cytokinins play opposite roles in lateral root formation, and nematode and *Rhizobial* symbioses," *Plant Journal,* vol. 38, pp. 203–214.

Lohar, D.P., Sharopova, N., Endre, G., Peñuela, S., Samac, D., Town, C., Silverstein, K.A., VandenBosch, K.A. 2006. "Transcript analysis of early nodulation events in *Medicago truncatula,*" *Plant Physiology,* vol. 140, pp. 221–234.

Long, S.R. 2001. "Genes and signals in the rhizobium-legume symbiosis," *Plant Physiology,* vol. 125, pp. 69–72.

Ludwig-Müller, J. 2000. "Hormonal balance in plants during colonization by mycorrhizal fungi," *Arbuscular Mycorrhizas: Physiology and Function.* Kluwer Academic Publishers, Amsterdam, pp. 263–283.

Mabood, F., Souleimanov, A., Khan, W., Smith, D.L. 2006. "Jasmonates induce Nod factor production by *Bradyrhizobium japonicum*," *Plant Physiology and Biochemistry,* vol. 44, pp. 759–765.

Madsen, E.B., Madsen, L.H., Radutoiu, S., Olbryt, M., Rakwalska, M., Szczyglowski, K., Sato, S., Kaneko, T., Tabata, S., Sandal, N., Stougaard, J. 2003. "A receptor kinase gene of the LysM type is involved in legume perception of rhizobial signals," *Nature,* vol. 425, no. 6958, pp. 637–640.

Maeda, D., Ashida, K., Iguchi, K., Chechetka, S.A., Hijikata, A., Okusako, Y., Deguchi, Y., Izui, K., Hata, S. 2006. "Knockdown of an arbuscular mycorrhiza-inducible phosphate transporter gene of *Lotus japonicus* suppresses mutualistic symbiosis," *Plant and Cell Physiology,* vol. 47, pp. 807–817.

Maekawa, T., Maekawa-Yoshikawa, M., Takeda, N., Imaizumi-Anraku, H., Murooka, Y., Hayashi, M. 2009. "Gibberellin controls the nodulation signaling pathway in *Lotus japonicus*," *Plant Journal,* vol. 58, pp. 183–194.

Marsh, J.F., Rakocevic, A., Mitra, R.M., Brocard, L., Sun, J., Eschstruth, A., Long, S.R., Schultze, M., Ratet, P., Oldroyd, G.E. 2007. "*Medicago truncatula NIN* is essential for rhizobial-independent nodule organogenesis induced by autoactive calcium/calmodulin-dependent protein kinase," *Plant Physiology,* vol. 144, no. 1, pp. 324–335.

Mathesius, U., Bayliss, C., Weinman, J.J., Schlaman, H.R.M., Spaink, H.P., Rolfe, B.G., McCully, M.E., Djordjevic, M.A. 1998a. "Flavanoids synthesized in cortical cells during nodule initiation are early developmental markers in white clover," *Molecular Plant-Microbe Interactions,* vol. 11, pp. 1223–1232.

Mathesius, U., Schlaman, H.R., Spaink, H., Sautter, C., Rolfe, B., Djordjevic, M.A. 1998b. "Auxin transport inhibition precedes root nodule formation in white clover roots and is regulated by flavanoids and derivatives of chitin oligosaccharides," *Plant Journal,* vol. 14, pp. 23–34.

Mathieu, C., Moreau, S., Frendo, P., Puppo, A., Davies, M.J. 1998. "Direct detection of radicals in intact soybean nodules: presence of nitric oxide-legehemoglobin complexes," *Free Radic Biol Med,* vol. 24, pp. 1242–1249.

Matsubara, Y., Ohba, N., Fukui, H. 2001. "Effect of arbuscular mycorrhizal fungus infection on the incidence of fusarium root rot in asparagus seedlings," *Journal of the Japanese Society for Horticultural Science,* vol. 70, pp. 202–206.

Matusova, R., Rani, K., Verstappen, F.W.A., Franssen, M.C.R., Beale, M.H., Bouwmeester, H.J. 2005. "The strigolactone germination stimulants of the plant-parasitic *Striga* and *Orobanche spp.* are derived from the carotenoid pathway," *Plant Physiology,* vol. 139, pp. 920–934.

Mayo, K., Davis, R., Motta, J. 1986. "Stimulation of germination of spores of *Glomus versiforme* by spore-associated bacteria," *Mycologia,* vol. 78, pp. 426–431.

Meakin, G.E., Jepson, B.J.N., Richardson, D.J., Bedmar, E.J., Delgado, M.J. 2006. "The role of *Bradyrhizobium japonicum* nitric oxide reductase in nitric oxide detoxification in soybean root nodules," *Biochemical Society Transactions,* vol. 34, pp. 195–196.

Mellor, R.B., Collinge, D.B. 1995. "A simple-model based on known plant defense reactions is sufficient to explain most aspects of nodulation," *Journal of Experimental Botany,* vol. 46, no. 282, pp. 1–18.

Messinese, E., Mun, J.H., Yeun, L.H., Jayaraman, D., Rouge, P., Barre, A., Lougnon, G., Schornack, S., Bono, J.J., Cook, D.R., Ané, J.M. 2007. "A novel nuclear protein interacts with the symbiotic DMI3 calcium- and calmodulin-dependent protein kinase of *Medicago truncatula*," *Molecular Plant-Microbe Interactions,* vol. 20, no. 8, pp. 912–921.

Middleton, P.H., Jakab, J., Penmetsa, R.V., Starker, C.G., Doll, J., Kaló, P., Prabhu, R., Marsh, J.F., Mitra, R.M., Kereszt, A., Dudas, B., VandenBosch, K. 2007. "An ERF transcription factor in *Medicago truncatula* that is essential for Nod factor signal transduction," *Plant Cell,* vol. 19, pp. 1221–1234.

Mills, K.K., Bauer, W.D. 1985. "*Rhizobium* attachment to clover roots," *Journal of Cell Science,* vol. 2, pp. 333–345.

Mitra, R.M., Gleason, C.A., Edwards, A., Hadfield, J., Downie, J.A., Oldroyd, G.E., Long, S.R. 2004. "A Ca^{2+}/calmodulin-dependent protein kinase required for symbiotic nodule development: Gene identification by transcript-based cloning," *PNAS,* vol. 101, no. 13, pp. 4701–4705.

Mitra, R.M., Shaw, S.L., Long, S.R. 2004. "Six nonnodulating plant mutants defective for Nod factor-induced transcriptional changes associated with the legume-rhizobia symbiosis," *PNAS,* vol. 101, no. 27, pp. 10217–10222.

Miwa, H., Sun, J., Oldroyd, G.E.D., Downie, J.A. 2006. "Analysis of calcium spiking using a cameleon calcium sensor reveals that nodulation gene expression is regulated by calcium spike number and the developmental status of the cell," *Plant Journal,* vol. 48, no. 6, pp. 883–894.

Morandi, D. 1996. "Occurrence of phytoalexins and phenolic compounds on endomycorrhizal interactions, and their potential role in biological control," *Plant Soil,* vol. 185, pp. 241–251.

Moreau, S., Meyer, J.M., Puppo, A. 1995. "Uptake of iron by symbiosomes and bacteroids from soybean nodules," *FEBS Letters,* vol. 361, pp. 225–228.

Mugnier, J. and Mosse, B. 1987. "Vesicular-arbuscular infections in Ri T-DNA transformed root grown axenically," *Phytopathology,* vol. 77, pp. 1045–1050.

Mukherjee, A., Frugoli, J. 2007. "Advances in legume/rhizobial symbiosis research" in *Recent Developments in Medicinal Plant Research*, Capasso, A., Ed., Research Signpost Press, Trivendrum, India pp. 373–392.

Murakami, Y., Miwa, H., Imaizumi-Anraku, H., Kouchi, H., Downie, J.A., Kawaguchi, M., Kawasaki, S. 2006. "Positional cloning identifies *Lotus japonicus* NSP2, a putative transcription factor of the GRAS family, required for *NIN* and *ENOD40* gene expression in nodule initiation," *DNA Research,* vol. 13, pp. 255–265.

Murray, J.D., Karas, B.J., Sato, S., Tabata, S., Amyot, L., Szczyglowski, K. 2007. "A cytokinin perception mutant colonized by rhizobium in the absence of nodule organogenesis," *Science,* vol. 315, no. 5808, pp. 101–104.

Mylona, P., Pawlowski, K., Bisseling, T. 1995. "Symbiotic nitrogen fixation," *Plant Cell,* vol. 7, pp. 869–885.

Nagahashi, G., Douds, D.D., Abney, G.D. 1996. "Phosphorus amendment inhibits hyphal branching of the VAM fungus *Gigaspora margarita* directly and indirectly through its effect on root exudation," *Mycorrhiza,* vol. 6, pp. 403–408.

Nagahashi, G., Douds, D.D.J. 1997. "Appresorium formation by AM fungi on isolated cell walls of carrot roots," *New Phytologist,* vol. 136, pp. 299–304.

Nakagawa, T., Kawaguchi, M. 2006. "Shoot-applied MeJA suppresses root nodulation in *Lotus japonicus*," *Plant and Cell Physiology,* vol. 47, no. 1, pp. 176–180.

Navazio, L., Moscatiello, R., Genre, A., Novero, M., Baldan, B., Bonfante, P., Mariani, P. 2007. "A diffusible signal from arbuscular mycorrhizal fungi elicits a transient cytosolic calcium elevation in host plant cells," *Plant Physiology,* vol. 144, no. 2, pp. 673–681.

Nishimura, C., Ohashi, Y., Sato, S., Kato, T., Tabata, S., Ueguchi, C. 2004. "Histidine kinase homologs that act as cytokinin receptors possess overlapping functions in the regulation of shoot and root growth in *Arabidopsis*," *Plant Cell,* vol. 16, pp. 1365–1377.

Novero, M., Faccio, A., Genre, A., Stougaard, J., Webb, K.J., Mulder, L., Parniske, M., Bonfante, P. 2002. "Dual requirement of the *LjSym4* gene for mycorrhizal development in epidermal and cortical cells of *Lotus japonicus* roots," *New Phytologist,* vol. 154, pp. 741–749.

Nukui, N., Ezura, H., Minamisawa, K. 2004. "Transgenic *Lotus japonicus* with an ethylene receptor gene *Cm-ERS1/H70A* enhances formation of infection threads and nodule primordia," *Plant and Cell Physiology,* vol. 45, pp. 427–435.

Olah, B., Briere, C., Bécard, G., Dénarié, J., Gough, C. 2005. "Nod factors and a diffusible factor from arbuscular mycorrhizal fungi stimulate lateral root formation in *Medicago truncatula* via the *DMI1/DMI2* signaling pathway," *Plant Journal,* vol. 44, no. 2, pp. 195–207.

Oldroyd, G.E., Downie, J.A. 2008. "Coordinating nodule morphogenesis with rhizobial infection in legumes," *Annual Review of Plant Biology,* vol. 59, pp. 519–546.

Oldroyd, G.E.D., Engstrom, E.M., Long, S.R. 2001. "Ethylene inhibits the Nod factor signal transduction pathway of *Medicago truncatula*," *Plant Cell,* vol. 13, pp. 1835–1849.

Oldroyd, G.E.D., Long, S.R. 2003. "Identification and characterization of *nodulation-signaling pathway 2*, a gene of *Medicago truncatula* involved in Nod factor signaling," *Plant Physiology,* vol. 131, pp. 1027–1032.

Ooki, Y., Banba, M., Yano, K., Maruya, J., Sato, S., Tabata, S., Saeki, K., Hayashi, M., Kawaguchi, M., Izui, K., Hata, S. 2005. "Characterization of the *Lotus japonicus* symbiotic mutant *lot1* that shows a reduced nodule number and distorted trichomes," *Plant Physiology,* vol. 137, pp. 1261–1271.

Panter, S., Thomson, R., de Bruxelles, G., Laver, D., Trevaskis, B., Udvardi, M. 2000. "Identification with proteomics of novel proteins associated with the peribacteroid membrane of soybean root nodules," *Molecular Plant-Microbe Interactions,* vol. 13, pp. 325–333.

Paszkowski, U., Jakovleva, L., Boller, T. 2006. "Maize mutants affected at distinct stages of the arbuscular mycorrhizal symbiosis," *Plant Journal,* vol. 47, pp. 165–173.

Peer, W.A., Murphy, A.S. 2007. "Flavonoids and auxin transport: modulators or regulators?," *Trends in Plant Science,* vol. 12, pp. 556–563.

Peiter, E., Sun, J., Heckmann, A.B., Venkateshwaran, M., Riely, B.K., Otegui, M.S., Edwards, A., Freshour, G., Hahn, M.G., Cook, D.R., Sanders, D., Oldroyd, G.E., Downie, J.A., Ané, J.M. 2007. "The *Medicago truncatula* DMI1 protein modulates cytosolic calcium signaling," *Plant Physiology,* vol. 145, no. 1, pp. 192–203.

Penmetsa, R.V., Cook, D. 1997. "A legume ethylene-insensitive mutant hyperinfected by its rhizobial symbiont," *Science,* vol. 275, pp. 527–530.

Penmetsa, R.V., Uribe, P., Anderson, J., Lichtenzveig, J., Gish, J.C., Nam, Y.W., Engstrom, E., Xu, K., Sckisel, G., Pereira, M., Baek, J.M., Lopez-Meyer, M., Long, S.R., Harrison, M.J., Singh, K.B., Kiss, G.B., Cook, D.R. 2008. "The *Medicago truncatula* ortholog of Arabidopsis *EIN2*, *sickle*, is a negative regulator of symbiotic and pathogenic microbial associations," *Plant Journal,* vol. 55, pp. 580–595.

Perotto, S., Brewin, N.J., Bonfante, P. 1994. "Colonization of pea roots by the mycorrhizal fungus *Glomus versiforme* and by *Rhizobium* bacteria—immunological comparison using monoclonal antibodies as probes for plant cell surface components," *Molecular Plant-Microbe Interactions,* vol. 7, pp. 91–98.

Perret, X., Staehelin, C., Broughton, W.J. 2000. "Molecular basis of symbiotic promiscuity," *Microbiology and Molecular Biology Reviews,* vol. 64, no. 1, pp. 180–201.

Peters, N.K., Frost, J.W., Long, S.R. 1986. "A plant flavone, luteolin, induces expression of *Rhizobium meliloti* nodulation genes," *Science,* vol. 233, no. 4767, pp. 977–980.

Phillips, D.A., Joseph, C.M., Maxwell, C.A. 1992. "Trigonelline and stachydrine released from alfalfa seeds activate NodD2 protein in *Rhizobium meliloti*," *Plant Physiology,* vol. 99, pp. 1526–1531.

Pichon, M., Journet, E.P., Dedieu, A., de Billy, F., Truchet, G., Barker, D.G. 1992. "*Rhizobium meliloti* elicits transient expression of the early nodulin gene *ENOD12* in the differentiating root epidermis of transgenic alfalfa," *Plant Cell,* vol. 4, pp. 1199–1211.

Pii, Y., Crimi, M., Cremonese, G., Spena, A., Pandolfini, T. 2007. "Auxin and nitric oxide control indeterminate nodule formation," *BMC Plant Biology,* vol. 7, pp. 21.

Pislariu, C., Dickstein, R. 2007. "An IRE-like AGC kinase gene, *MtIRE*, has unique expression in the invasion zone of developing root nodules in *Medicago truncatula*," *Plant Physiology,* vol. 144, pp. 682–694.

Prell, J., Poole, P. 2006. "Metabolic changes in rhizobia," *Trends in Microbiology,* vol. 141, pp. 161–168.

Radutoiu, S., Madsen, L.H., Madsen, E.B., Felle, H.H., Umehara, Y., Gronlund, M., Sato, S., Nakamura, Y., Tabata, S., Sandal, N., Stougaard, J. 2003. "Plant recognition of symbiotic bacteria requires two LysM receptor-like kinases," *Nature,* vol. 425, no. 6958, pp. 585–592.

Radutoiu, S., Madsen, L.H., Madsen, E.B., Jurkiewicz, A., Fukai, E., Quistgaard, E.M., Albrektsen, A.S., James, E.K., Thirup, S., Stougaard, J. 2007. "LysM domains mediate lipochitin-oligosaccharide recognition and Nfr genes extend the symbiotic host range," *The EMBO Journal,* vol. 26, no. 17, pp. 3923–3935.

Recourt, K., Van Tunen, A.J., Mur, L.A., Van Brussel, A.A.N., Lugtenberg, B., Kijne, J.W. 1992. "Activation of flavonoid biosynthesis in roots of *Vicia sativa* subsp. *nigra* plants by inoculation with *Rhizobium leguminosarum* biovar *viciae*," *Plant Molecular Biology,* vol. 19, pp. 411–420.

Reinhardt, D. 2007. "Programming good relations—development of the arbuscular mycorrhizal symbiosis," *Current Opinion in Plant Biology,* vol. 10, no. 1, pp. 98–105.

Relić, B., Talmont, F., Korsinska, J., Golinowski, W., Promé, J.C., Broughton, W.J. 1993. "Biological activity of *Rhizobium* sp. NGR234 Nod factors on *Macroptilium atropurpureum*," *Molecular Plant-Microbe Interactions,* vol. 6, pp. 764–774.

Riely, B.K., Lougnon, G., Ané, J.M., Cook, D.R. 2007. "The symbiotic ion channel homolog *DMI1* is localized in the nuclear membrane of *Medicago truncatula* roots," *Plant Journal,* vol. 49, no. 2, pp. 208–216.

Rolfe, B.J. 1988. "Flavones and isoflavones as inducing substances of legume nodulation," *Biofactors,* vol. 1, pp. 3–10.

Rosas, S., Soria, R., Correa, N., Abdala, G. 1998, "Jasmonic acid stimulates the expression of nod genes in *Rhizobium*," *Plant Molecular Biology,* vol. 38, pp. 1161–1168.

Rosendahl, L., Dilworth, M.J., Glenn, A.R. 1992. "Exchange of metabolites across the peribacteroid membrane in pea root nodules," *Journal of Plant Physiology,* vol. 139, pp. 635–638.

Roth, L.E., Stacey, G. 1989. "Cytoplasmic membrane systems involved in bacterium release into soybean nodule cells as studied with two *Bradyrhizobium japonicum* mutant strains," *European Journal of Cell Biology,* vol. 49, pp. 13–23.

Saalbach, G., Erik, P., Wienkoop, S. 2002. "Characterisation by proteomics of peribacteroid space and peribacteroid membrane preparations from pea (*Pisum sativum*) symbiosomes," *Proteomics,* vol. 2, pp. 325–337.

Saito, K., Yoshikawa, M., Yano, K., Miwa, H., Uchida, H., Asamizu, E., Sato, S., Tabata, S., Imaizumi-Anraku, H., Umehara, Y., Kouchi, H., Murooka, Y., Szczyglowski, K., Downie, J.A., Parniske, M., Hayashi, M., Kawaguchi, M. 2007. "NUCLEOPORIN85 is required for calcium spiking, fungal and bacterial symbioses, and seed production in *Lotus japonicus*," *Plant Cell,* vol. 19, no. 2, pp. 610–624.

Sandal, N., Petersen, T.R., Murray, J., Umehara, Y., Karas, B., Yano, K., Kumagai, H., Yoshikawa, M., Saito, K., Hayashi, M., Murakami, Y., Wang, X.W., Hakoyama, T., Imaizumi-Anraku, H., Sato, S., Kato, T., Chen, W.L. 2006. "Genetics of symbiosis in *Lotus japonicus*: Recombinant inbred lines, comparative genetic maps, and map position of 35 symbiotic loci," *Molecular Plant-Microbe Interactions,* vol. 19, no. 1, pp. 80–91.

Savouré, A., Sallaud, C., El-Turk, J., Zuanazzi, J., Ratet, P., Schultze, M., Kondorosi, A., Esnault, R., Kondorosi, E. 1997. "Distinct response of *Medicago* suspension cultures and roots to Nod factors and chitin oligomers in the elicitation of defense-related responses," *Plant Journal,* vol. 11, pp. 277–287.

Sbrana, C., Giovannetti, M. 2005. "Chemotropism in the arbuscular mycorrhizal fungus *Glomus mosseae*," *Mycorrhiza,* vol. 15, pp. 539–545.

Scervino, J.M., Ponce, M.A., Erra-Bassels, R., Vierheilig, H., Ocampo, J.A., Godeas, A. 2005a. "Arbuscular mycorrhizal colonization of tomato by *Gigaspora* and *Glomus* species in presence of roots flavonoids," *Journal of Plant Physiology,* vol. 162, pp. 625–633.

Scervino, J.M., Ponce, M.A., Erra-Bassels, R., Vierheilig, H., Ocampo, J.A., Godeas, A. 2005b. "Flavonoids exhibit fungal species and genus specific effects on the presymbiotic growth of *Gigaspora* and *Glomus*," *Mycological Research,* vol. 109, pp. 789–794.

Scervino, J.M., Ponce, M.A., Erra-Bassels, R., Vierheilig, H., Ocampo, J.A., Godeas, A. 2005c. "Flavonoids exclusively present in mycorrhizal roots of white clover exhibit different effects on arbuscular mycorrhizal fungi than flavonoids exclusively present in non-mycorrhizal roots of white clover," *Journal of Plant Interactions,* vol. 15, pp. 22–30.

Schauser, L., Handberg, K., Sandal, N., Stiller, J., Thykjaer, T., Pajuelo, E., Nielsen, A., Stougaard, J. 1998. "Symbiotic mutants deficient in nodule establishment identified after T-DNA transformation of *Lotus japonicus*," *Molecular Genetics and Genomics,* vol. 259, pp. 414–423.

Schauser, L., Roussi, A., Stiller, J., Stougaard, J. 1999. "A plant regulator controlling development of symbiotic root nodules," *Nature,* vol. 402, pp. 191–195.

Schauser, L., Wieloch, W., Stougaard, J. 2005. "Evolution of NIN-like proteins in *Arabidopsis*, rice and *Lotus japonicus*," *Journal of Molecular Evolution,* vol. 60, pp. 229–237.

Scheres, B., Van De Wiel, C., Zalensky, A., Horvath, B., Spaink, H., Van Eck, H., Zwartkruis, F., Wolters, A.M., Gloudemans, T., Van Kammen, A., Bisseling, T. 1990. "The *ENOD12* gene product is involved in the infection process during the pea-*Rhizobium* interaction," *Cell,* vol. 60, pp. 281–294.

Schiefelbein, J.W. 2000. "Constructing a plant cell: the genetic control of root hair development," *Plant Physiology,* vol. 124, pp. 1525–1531.

Schmidt, J.S., Harper, J.E., Hoffman, T.K., Bent, A.F. 1999. "Regulation of soybean nodulation independent of ethylene signaling," *Plant Physiology,* vol. 119, pp. 951–959.

Schmidt, P.E., Broughton, W.J., Werner, D. 1994, "Nod factors of *Bradyrhizobium japonicum* and *Rhizobium* sp. NGR234 induce flavonoid accumulation in soybean root exudates," *Molecular Plant-Microbe Interactions,* vol. 7, pp. 384–390.

Schultze, M., Kondorosi, A. 1998, "Regulation of symbiotic root nodule development," *Annual Review of Genetics,* vol. 32, pp. 33–57.

Senoo, K., Solaiman, M.Z., Kawagachi, M., Imaizumi-Anraku, H., Akao, S., Tanaka, A., Obata, H. 2000, "Isolation of two different phenotypes of mycorrhizal mutants in the model legume plant *Lotus japonicus* after EMS-treatment," *Plant Cell Physiology,* vol. 41, pp. 726–732.

Shirtliffe, S.J., Vessey, J.K. 1996. "A nodulation (Nod$^+$/Fix$^-$) mutant of *Phaseolus vulgaris* L. has nodule-like structures lacking peripheral vascular bundles (Pvb–) and is resistant to mycorrhizal infection," *Plant Science,* vol. 118, pp. 209–220.

Siciliano, V., Genre, A., Balestrini, R., Cappellazzo, G., deWitt, P.J.G.M., Bonfante, P. 2007, "Transcriptome analysis of arbuscular mycorrhizal roots during development of the pre-penetration apparatus," *Plant Physiology,* vol. 144, pp. 1455–1466.

Sieberer, B.J., Chabaud, M., Timmers, A.C., Monin, A., Fournier, J., Barker, D.G. 2009, "A nuclear-targeted cameleon demonstrates intranuclear Ca^{2+} spiking in *Medicago truncatula* root hairs in response to rhizobial nodulation factors," *Plant Physiology*, 151: 1197–1206.

Simon, L., Bousquet, J., L'evesque, R.C., Lalonde, M. 1993. "Origin and diversification of endomycorrhizal fungi and coincidence with vascular land plants," *Nature,* vol. 363, pp. 67–69.

Siqueira, J.O., Safir, G.R., Nair, M.G. 1991, "Stimulation of vesicular-arbuscular mycorrhiza formation and growth of white clover by flavonoid compounds," *New Phytologist,* vol. 118, pp. 87–93.

Smil, V. 1997, "Global population and the nitrogen cycle," *Scientific American,* vol. 277, pp. 76–81.

Smit, G., Kijne, J.W., Lugtenberg, B.J.J. 1986. "Correlation between extracellular fibrils and attachment of *Rhizobium leguminosarum* to pea root hair tips," *Journal of Bacteriology,* vol. 168, pp. 821–827.

Smit, P., Limpens, E., Geurts, R., Fedorova, E., Dolgikh, E., Gough, C., Bisseling, T. 2007. "*Medicago* LYK3, an entry receptor in rhizobial nodulation factor signaling," *Plant Physiology,* vol. 145, no. 1, pp. 183–191.

Smit, P., Raedts, J., Portyanko, V., Debellé, F., Gough, C., Bisseling, T., Geurts, R. 2005. "NSP1 of the GRAS protein family is essential for rhizobial Nod factor-induced transcription," *Science,* vol. 308, no. 5729, pp. 1789–1791.

Smith, S.E., Gianinazzi-Pearson, V. 1988, "Physiological interactions between symbionts invesicular-arbuscular mycorrhizal plants," *Annual Review of Plant Physiology,* vol. 39, pp. 221–244.

Stacey, G., McAlvin, C.B., Kim, S.Y., Olivares, J., Soto, M.J. 2006. "Effects of endogenous Salicylic Acid on nodulation in the model Legumes *Lotus japonicus* and *Medicago truncatula*," *Plant Physiology,* vol. 141, pp. 1473–1481.

Stacey, G., Paau, A.S., Brill, W.J. 1980. "Host recognition in the *Rhizobium*- soybean symbiosis," *Plant Physiology,* vol. 66, pp. 609–614.

Staehelin, C., Schultze, M., Kondorosi, E., Kondorosi, A. 1995. "Lipo-chitooligosaccharide nodulation signals from *Rhizobium meliloti* induce their rapid degradation by the host plant alfalfa," *Plant Physiology,* vol. 108, pp. 1607–1614.

Staehelin, C., Granado, J., Müller, J., Wiemken, A., Mellor, R.B., Felix, G., Regenass, M., Broughton, W.J., Boller, T. 1994a. "Perception of *Rhizobium* nodulation factors by tomato cells and inactivation by root chitinases," *PNAS,* vol. 91, pp. 2196–2200.

Staehelin, C., Schultze, M., Kondorosi, E., Mellor, R.B., Boller, T., Kondorosi, A. 1994b, "Structural modifications in *Rhizobium meliloti* Nod factors influence their stability against hydrolysis by root chitinases," *Plant Journal,* vol. 5, pp. 319–330.

Starker, C.G., Parra-Cohnenares, A.L., Smith, L., Mitra, R.M., Long, S.R. 2006. "Nitrogen-fixing mutants of *Medicago truncatula* fail to support plant and bacterial symbiotic gene expression," *Plant Physiology,* vol. 140, pp. 671–680.

Stougaard, J. 2000. "Regulators and regulation of legume root nodule development," *Plant Physiology,* vol. 124, pp. 531–540.

Stracke, S., Kistner, C., Yoshida, S., Mulder, L., Sato, S., Kaneko, T., Tabata, S., Sandal, N., Stougaard, J., Szczyglowski, K., Parniske, M. 2002. "A plant receptor-like kinase required for both bacterial and fungal symbiosis," *Nature,* vol. 417, no. 6892, pp. 959–962.

Streit, W.R., Joseph, C.M., Phillips, D.A. 1996. "Biotin and other water-soluble vitamins are key growth factors for alfalfa root colonization by *Rhizobium meliloti* 1021," *Molecular Plant-Microbe Interactions,* vol. 9, pp. 330–338.

Subramanian, K.S., Charest, C. 1995, "Influence of arbuscular mycorrhizae on the metabolism of maize under drought stress," *Mycorrhiza,* vol. 5, pp. 273–278.

Subramanian, S., Stacey, G., Yu, O. 2006, "Endogenous isoflavones are essential for the establishment of symbiosis between soybean and *Bradyrhizobium japonicum*," *Plant Journal,* vol. 48, pp. 261–273.

Sun, J., Cardoza, V., Mitchell, D.M., Bright, L., Oldroyd, G.E.D., Harris, J.M. 2006. "Crosstalk between jasmonic acid, ethylene and Nod factor signaling allows integration of diverse inputs for regulation of nodulation," *Plant Journal,* vol. 46, pp. 961–970.

Szczyglowski, K., Shaw, R.S., Woperis, J., Copeland, S., Hamburger, D., Kasiborski, B., Dazzo, F.B., de Bruijn, F.J. 1998. "Nodule organogenesis and symbiotic mutants of the model legume *Lotus japonicus*," *Molecular Plant-Microbe Interactions,* vol. 11, pp. 684–697.

Tanimoto, M., Roberts, K., Dolan, L. 1995, "Ethylene is a positive regulator of root hair development in *Arabidopsis thaliana*," *Plant Journal,* vol. 8, pp. 943–948.

Taylor, L.P., Grotewold, E. 2005. "Flavonoids as developmental regulators," *Current Opinion in Plant Biology,* vol. 8, pp. 317–323.

Timmers, A.C.J., Auriac, M.C., Truchet, G. 1999. "Refined analysis of early symbiotic steps of the *Rhizobium-Medicago* interaction in relationship with microtubular cytoskeleton rearrangements," *Development,* vol. 126, pp. 3617–3628.

Tirichine, L., Imaizumi-Anraku, H., Yoshida, S., Murakami, Y., Madsen, L.H., Miwa, H., Nakagawa, T., Sandal, N., Albrektsen, A.S., Kawaguchi, M., Downie, J.A., Sato, S., Tabata, S., Kouchi, H., Parniske, M., Kawasaki, S., Stougaard, J. 2006a. "Deregulation of Ca^{2+}/calmodulin-dependent kinase leads to the spontaneous nodule development," *Nature,* vol. 441, pp. 1153–1156.

Tirichine, L., James, E.K., Sandal, N., Stougaard, J. 2006b. "Spontaneous root-nodule formation in the model legume *Lotus japonicus*: a novel class of mutants nodulates in the absence of rhizobia," *Molecular Plant-Microbe Interactions,* vol. 19, pp. 373–382.

Tirichine, L., Sandal, N., Madsen, L.H., Radutoiu, S., Albrektsen, A.S., Sato, S., Asamizu, E., Tabata, S., Stougaard, J. 2007. "A gain-of-function mutation in a cytokinin receptor triggers spontaneous root nodule organogenesis," *Science,* vol. 315, no. 5808, pp. 104–107.

Trinchant, J.C., Rigaud, J. 1982. "Nitrite and nitric oxide as inhbitors of nitrogenase from soybean bacteroids," *Applied and Environmental Microbiology,* vol. 44, pp. 1385–1388.

Trinick, M.J. 1979. "Structure of nitrogen-fixing nodules formed by *Rhizobium* on roots of *Parasponia andersonii* Planch," *Canadian Journal of Microbiology,* vol. 25, pp. 565–578.

Trotta, A., Varese, G.C., Gnavi, E., Fusconi, A., Sampò, S., Berta, G. 1996, "Interactions between the soilborne root pathogen *Phytophthora nicotianae* var. parasitica and the arbuscular mycorrhizal fungus *Glomus mosseae* in tomato plants," *Plant Soil,* vol. 185, pp. 199–209.

Truchet, G., Barker, D.G., Camut, S., de Billy, F., Vasse, J., Huguet, T. 1989. "Alfalfa nodulation in the absence of rhizobia," *Molecular and General Genetics,* vol. 219, pp. 65–68.

Tsai, S.M., Phillips, D.A. 1991. "Flavonoids released naturally from alfalfa promote development of symbiotic *Glomus* spores *in vitro*," *Applied and Environmental Microbiology,* vol. 57, pp. 1485–1488.

Tully, R.E., van Berkum, P., Lovins, K.W., Keister, D.L. 1998. "Identification and sequencing of a cytochrome P450 gene cluster from *Bradyrhizobium japonicum*," *Biochimica Et Biophysica Acta,* vol. 1398, pp. 243–255.

Udvardi, M.K., Day, D.A. 1997. "Metabolite transport across symbiotic membranes of legume nodules," *Annual Review of Plant Physiology,* vol. 48, pp. 493–523.

Umehara, M., Hanada, A., Yoshida, S., Akiyama, K., Arite, T., Takeda-Kamiya, N., Magome, H., Kamiya, Y., Shirasu, K., Yoneyama, K., Kyozuka, J., Yamaguchi, S. 2008. "Inhibition of shoot branching by new terpenoid plant hormones," *Nature,* vol. 455, pp. 195–200.

van Brussel, A.A.N., Bakhuizen, R., van Spronsen, P.C., Spaink, H.P., Tak, T., Lugtenberg, B.J.J., Kijne, J.W. 1992. "Induction of pre-infection thread structures in the leguminous host plant by mitogenic lipo-oligosaccharidess of *Rhizobium*," *Science,* vol. 257, pp. 70–72.

van Rhijn, P., Fang, Y., Galili, S., Shaul, O., Atzmon, N., Wininger, S., Eshed, Y., Lum, M., Li, Y., To, V., Fujishige, N., Kapulnik, Y., Hirsch, A.M. 1997. "Expression of early nodulin genes in alfalfa mycorrhizae indicates that signal transduction pathways used in forming arbuscular mycorrhizae and *Rhizobium*-induced nodules may be conserved," *PNAS,* vol. 94, no. 10, pp. 5467–5472.

Vasse, J., de Billy, F., Truchet, G. 1993. "Abortion of infection during the *Rhizobium meliloti*-alfalfa symbiotic interaction is accompanied by a hypersensitive reaction," *Plant Journal,* vol. 4, pp. 555–566.

Veereshlingam, H., Haynes, J.G., Penmetsa, R.V., Cook, D.R., Sherrier, D.J., Dickstein, R. 2004. "*nip,* a symbiotic *Medicago truncatula* mutant that forms root nodules with aberrant infection threads and plant defense-like response," *Plant Physiology,* vol. 136, pp. 3692–3702.

Vernié, T., Moreau, S., de Billy, F., Plet, J., Combier, J.P., Rogers, C., Oldroyd, G., Frugier, F., Niebel, A., Gamas, P. 2008. "EFD is an ERF transcription factor involved in the control of nodule number and differentiation in *Medicago truncatula*," *Plant Cell,* vol. 20, no. 10, pp. 2696–2713.

Vesper, S.J., Malik, N.S.A., Bauer, W.D. 1987. "Transposon mutants of *Bradyrhizobium japonicum* altered in attachment to host roots," *Applied and Environmental Microbiology,* vol. 53, pp. 1959–1961.

Vierheilig, H. 2004. "Regulatory mechanisms during the plant-arbuscular mycorrhizal fungus interaction," *Canadian Journal of Botany,* vol. 82, pp. 1166–1176.

Vincill, E.D., Szczyglowski, K., Roberts, D.M. 2005. "GmN70 and LjN70: anion transporters of the symbiosome membrane of nodules with a transport preference for nitrate," *Plant Physiology,* vol. 137, pp. 1435–1444.

Wais, R.J., Galera, C., Oldroyd, G., Catoira, R., Penmetsa, R.V., Cook, D., Gough, C., Dénarié, J., Long, S.R. 2000. "Genetic analysis of calcium spiking responses in nodulation mutants of *Medicago truncatula*," *PNAS,* vol. 97, no. 24, pp. 13407–13412.

Walker, S.A., Downie, J.A. 2000. "Entry of *Rhizobium leguminosarum* bv. *viciae* into root hairs requires minimal Nod factor specificity, but subsequent infection thread growth requires *nodO* or *nodE*," *Molecular Plant-Microbe Interactions,* vol. 13, pp. 754–762.

Walker, S.A., Viprey, V., Downie, J.A. 2000. "Dissection of nodulation signaling using pea mutants defective for calcium spiking induced by Nod factors and chitin oligomers," *PNAS,* vol. 97, no. 24, pp. 13413–13418.

Wan, X., Hontelez, J., Lillo, A., Guarnerio, C., van de Peut, D., Fedorova, E., Bisseling, T., Franssen, H. 2007. "*Medicago truncatula ENOD40-1* and *ENOD40-2* are both involved in nodule initiation and bacteroid development," *Journal of Experimental Botany,* vol. 58, pp. 2033–2041.

Wang, B., Qiu, Y.L. 2006. "Phylogenetic distribution and evolution of mycorrhizas in land plants," *Mycorrhiza,* vol. 16, pp. 299–363.

Wasson, A.P., Pellerone, F.I., Mathesius, U. 2006. "Silencing the flavonoid pathway in *Medicago truncatula* inhibits root nodule formation and prevents auxin transport regulation by rhizobia," *Plant Cell,* vol. 18, pp. 1617–1629.

Wasson, A.P., Ramsay, K., Jones, M.G., Mathesius, U. 2009. "Differing requirements for flavonoids during the formation of lateral roots, nodules and root knot nematode galls in *Medicago truncatula*," *New Phytologist,* vol. 183, pp. 167–179.

Wegel, E., Schauser, L., Sandal, N., Stougaard, J., Parniske, M. 1998. "Mycorrhiza mutants of *Lotus japonicus* define genetically independent steps during symbiotic infection," *Molecular Plant-Microbe Interactions,* vol. 11, pp. 933–936.

Wienkoop, S., Saalbach, G. 2003, "Proteome analysis: novel proteins identified at the peribacteroid membrane from *Lotus japonicus* root nodules," *Plant Physiology,* vol. 131, pp. 1080–1090.

Wulf, A., Manthey, K., Doll, J., Perlick, A.M., Linke, B., Bekel, T., Meyer, F., Franken, P., Küster, H., Krajinski, F. 2003. "Transcriptional changes in response to arbuscular mycorrhiza development in the model plant *Medicago truncatula*," *Molecular Plant-Microbe Interactions,* vol. 16, pp. 306–314.

Xie, Z.P., Staehelin, C., Vierheilig, H., Wiemken, A., Jabbouri, S., Broughton, W.J., Vogeli-Lange, R., Boller, T. 1995. "Rhizobial Nodulation factors stimulate mycorrhizal colonization of nodulating and nonnodulating soybeans," *Plant Physiology,* vol. 108, pp. 1519–1525.

Yano, K., Yoshida, S., Muller, J., Singh, S., Banba, M., Vickers, K., Markmann, K., White, C., Schuller, B., Sato, S., Asamizu, E., Tabata, S., Murooka, Y., Perry, J., Wang, T., Kawagachi, M., Imaizumi-Anraku, H., Hayashi, M., Parniske, M. 2008. "CYCLOPS, a mediator of symbiotic intracellular accommodation," *PNAS,* vol. 105, pp. 20540–20545.

Yoneyama, K., Yoneyama, K., Takeuchi, Y., Sekimoto, H. 2007a. "Phosphorus deficiency in red clover promotes exudation of orobanchol, the signal for mycorrhizal symbionts and germination stimulant for root parasite," *Planta,* vol. 225, pp. 1031–1038.

Yoneyama, K., Xie, X., Kusumoto, D., Sekimoto, H., Sugimoto, Y., Takeuchi, Y., Yoneyama, K. 2007b. "Nitrogen deficiency as well as phosphorus deficiency in sorghum promotes the production and exudation of 5-deoxystrigol, the host recognition signal for arbuscular mycorrhizal fungi and root parasites.," *Planta,* vol. 227, pp. 125–132.

Yoneyama, K., Xie, X., Sekimoto, H., Takeuchi, Y., Ogasawara, S., Akiyama, K., Hayashi, H., Yoneyama, K. 2008 "Strigolactones, host recognition signals for root parasitic plants and arbuscular mycorrhizal fungi, from Fabaceae plants," *New Phytologist,* vol. 179, pp. 484–494.

Yuen, J.P.Y., Cassini, S.T., De Oliveira, T.T., Nagem, T.J., Stacey, G. 1995. "Xanthone induction of nod gene expression in *Bradyrhizobium japonicum*," *Symbiosis,* vol. 19, pp. 131–140.

Zhu, H., Chen, T., Zhu, M., Fang, Q., Kang, H., Hong, Z., Zhang, Z. 2008. "A novel ARID DNA-binding protein interacts with SymRK and is expressed during early nodule development in *Lotus japonicus*," *Plant Physiology,* vol. 148, no. 1, pp. 337–347.

Chapter 17
Nitric Oxide as a Signal Molecule in Intracellular and Extracellular Bacteria-plant Interactions

Andrés Arruebarrena Di Palma, Lorenzo Lamattina, and Cecilia M. Creus

Beneficial Plant-Microorganism Interactions

All plants live in intimate association with many microorganisms. These microorganisms colonize the surfaces, the intercellular spaces within tissues or even the inside of plant cells (Brencic and Winans, 2005). Some of the most complex interactions that terrestrial plants experience occur between the roots and their surrounding environment (Bais et al., 2006). It is precisely in the soil where these interactions are subjected to a plethora of conditions that determine the success or failure of root colonization, leading to effects on plant developmental processes.

Soil has been divided into three main zones according proximity to the root: (1) rhizoplane or the root surface, (2) rhizosphere or the soil under root "influence," and (3) bulk soil (Manthey et al., 1994). Bacteria that inhabit soils can affect plant growth and development. The interactions they establish with roots vary from beneficial, deleterious, or neutral effects on plants (Hirsch et al., 2003).

The distribution of microorganisms in the rhizosphere can be classified into four main categories related to root proximity and intimacy: (1) bacteria living in the soil near roots, using metabolites "leaked" from roots as carbon (C) and nitrogen (N) sources, (2) bacteria colonizing the rhizoplane, (3) bacteria residing in root tissue, inhabiting spaces between cortical cells, and (4) bacteria living inside cells in specialized root structures or nodules (Gray and Smith, 2005). The latter are generally represented by two groups: the legume-rhizobia and the woody plant-*Frankia* associations (Hallmann et al., 1997; Gray and Smith, 2005). A more simple and convenient classification based on the preferred colonizing site is proposed by Gray and Smith (2005) dividing plant growth-promoting rhizobacteria (PGPR) into extracellular PGPR (ePGPR), and intracellular PGPR (iPGPR), the latter existing inside root cells generally in specialized nodular structures.

Ecological Aspects of Nitrogen Metabolism in Plants, First Edition. Edited by Joe C. Polacco and Christopher D. Todd.
© 2011 by John Wiley & Sons, Inc. Published 2011 by John Wiley & Sons, Inc.

Extracellular PGPR are free-living bacteria (Kloepper et al., 1989). Some of them invade the tissues of living plants and cause unapparent and asymptomatic infections (Sturz et al., 2000). Extracellular PGPR represent a wide variety of soil bacteria that may induce plant growth by direct or indirect modes of action (Beauchamp, 1993; Glick, 1995). Four different mechanisms may be included within the direct mode: (1) the production of phytohormones (Barbieri and Galli, 1993; Dobbelaere et al., 1999) and stimulatory bacterial volatiles (Ryu et al., 2003); (2) the lowering of plant ethylene levels (Glick et al., 1998); (3) the improvement of plant nutrient status by either making available macro- and micronutrients from insoluble sources (Delvasto et al., 2006; Rodríguez et al., 2006) or by nonsymbiotic nitrogen fixation (Boddey et al., 2008); and (4) the stimulation of disease-resistance mechanisms such as induced systemic resistance (ISR) (van Loon, 2007). Indirect effects arise when ePGPR act as bio-control agents leading to reduced diseases (Compant et al., 2005), when they stimulate other beneficial symbioses (Burdman et al., 2000; Tokala et al., 2002) or when they protect the plant by degrading xenobiotics in inhibitory contaminated soils (Jacobsen, 1997).

As was mentioned above, the ePGPR colonize the rhizosphere and rhizoplane. However, some of them can also enter the root interior and establish endophytic populations. Many of them are able to traverse the endodermal barrier, crossing from the root cortex to the vascular system, and subsequently thrive as endophytes in stem, leaves, tubers, and other organs (Compant et al., 2005). The extent of endophytic colonization of host plant organs and tissues reflects the ability of bacteria to adapt selectively to these specific ecological niches and to establish an intimate association with the host plant.

As the primary target for rhizobacteria, the root is usually the first organ to show stimulatory effects. However, according to the specific PGPR-plant interaction and management practices, the effects could range from no promotion, only root promotion, or both root and shoot growth promotion, including improved agronomic yield.

Azospirillum is perhaps the most researched PGPR (Bashan et al., 2004 addressed later in this chapter), and one of the first observations regarding its plant growth promotion was on root morphology (Okon, 1985). Soon after *Azospirillum* inoculation, the root displays significant increases in the number and the length of root hairs (RH), the rate of appearance and the number of lateral roots (LR), the diameter and length of lateral and adventitious roots (AR), and the total root surface area (Jain and Patriquin, 1984; Kapulnik et al., 1985; Fallik et al., 1994; Dobbelaere et al., 1999; Creus et al., 2005). In addition, an increase in cell division in the root tips of inoculated wheat was reported (Levanony and Bashan, 1989).

An increase in the branching degree of roots, leading to a change in root architecture, and its associated enhanced capacity to explore soil for water and nutrients, would contribute to a better-hydrated status of plants exposed to water deficit. It was reported that *Azospirillum*-inoculated wheat seedlings subjected to osmotic stress developed a significantly higher coleoptile and better water status than non-inoculated seedlings (Alvarez et al., 1996; Creus et al., 1998; Pereyra et al., 2006). Plants exposed to salt stress also suffer water deficit and, indeed, *Azospirillum*-inoculated wheat seedlings were able to survive when exposed to up to 320 mM NaCl for 3 days (Creus et al., 1997).

Roots in soil encounter physical, chemical, and biological conditions that influence their growth. In the complex scenario of the rhizosphere, plant roots offer a niche for the proliferation of soil bacteria that use root lysates and exudates as nutrient sources. Bacteria, in turn, also secrete metabolites into the rhizosphere, several of which can act as signalling compounds perceived by neighboring cells within the same bacterial micro-colony, by other bacterial cells present in the rhizosphere, or by root cells of the host plant (Gray and Smith, 2005; Bais et al., 2006; van Loon, 2007). Some of these signals are molecules common to bacteria, roots, or nodules, auxins being a notable example (Brown, 1972; Mathesius et al., 1998). Other signals, such as N-acylhomoserine lactones, quorum-sensing molecules produced by Gram-negative bacteria, are perceived by the roots of some species, which in turn produce exudates that mimic the bacterial signals (Bauer and Mathesius, 2004). On the other hand, other signals show a high degree of host specificity, and the exchange between specific partners leads to the establishment of symbiotic relationships. Flavonoids, signal molecules exuded by the roots, activate the expression of bacterial nodulation genes, resulting in the production of rhizobial lipochitooligosaccharide signals (Nod factors). Under appropriate conditions, Nod factors induce root cortex cells to divide and to develop into nodule primordia (Schultze and Kondorosi, 1998).

It was stated above that microbes represent the largest bio-diverse community in soils (Torsvik and Øvreås 2002; Bartelt-Ryser et al., 2005). However, they are unevenly distributed, mainly congregated around nutrient sources and organic matter. The processes occurring in the rhizosphere involving community level interactions under field natural conditions are as yet poorly understood (Watt et al., 2006).

The mechanisms by which PGPR elicit plant growth promotion, from the viewpoint of signal transduction pathways within plants, is a matter of debate. Ryu and others (2005) proposed that elicitation of growth promotion by PGPR strains involves brassinosteroid, IAA, salicylic acid, gibberellin, and ethylene signalling. They suggested that growth promotion by PGPR in *Arabidopsis* is associated with several different signal transduction pathways. We here propose that NO is a signal molecule produced by bacteria and that it participates in the signalling cascades in different types of plant root responses.

Soil Nitrogen Forms Affect Biological Processes in the Rhizosphere

Nitrogen Forms That Influence Root Development

Roots growing in soil run into physical, chemical, and biological environments that influence the rhizosphere and affect plant growth. Roots may encounter open spaces including biopores that contain live roots and dead root residues from previous plants, different pH, oxygen (O_2) levels, water contents, nutrient concentrations, and a wide range of soil organisms among other factors (Watt et al., 2006). The main sources of N in the soil are ammonium (NH_4^+), nitrite (NO_2^-), and nitrate (NO_3^-) (Osmont et al., 2007), although dissolved organic N is also present in numerous soils, even at higher concentrations than NH_4^+ (Christou et al., 2005), depending on soil pH (Kemmitt et al., 2006).

Most plants obtain the vast majority of their nitrogen through root absorption of the inorganic ions NH_4^+ and NO_3^- from the soil solution. The spatial and temporal availability of soil NH_4^+ and NO_3^- is highly heterogeneous, within centimeters from the roots and over the course of a day. Due to this heterogeneous environment with N-poor and N-rich soil patches, plants must optimize mechanisms that enable roots to locate and acquire N resources (Bloom et al., 2003).

NH_4^+ and NO_3^- have different effects on root development. Tomato seedlings growing in hydroponics in the presence of NH_4^+ showed higher mass, length, and root branching than those growing with NO_3^- (Bloom et al., 2003). As NH_4^+ uptake makes the rhizoplane acidic and NO_3^- uptake changes it to slightly alkaline, the authors proposed that pH changes and redox potentials generated by absorption of different ions in the rhizosphere might directly affect growth via cell expansion in acid conditions. However, the predominant and preferred

N-form in agricultural soils is NO_3^- (Christou et al., 2005) as high concentrations of NH_4^+ are toxic (Mehrer and Mohr, 1989).

Numerous studies have focused on the effects of NO_3^- on *Arabidopsis thaliana* root development as a model system. When *A. thaliana* seedlings were grown in a medium containing a high, uniformly distributed NO_3^- concentration (>10 mM), the production of LR was strongly inhibited, even though neither the number of LR primordia nor the rate of primary root growth was affected (Zhang and Forde, 1998). When 10 mM NO_3^- was applied in patches there was no enhancement in LR number, but the preformed LR had a higher elongation rate (Zhang et al., 1999). This response was NO_3^--specific, no differences were observed when the N-form was NH_4^+. Using the *A. thaliana* double mutant *nia1-nia2* mutants, with greatly reduced activity for nitrate reductase, the same authors showed that the accumulation of NO_3^- itself within the plant was capable of generating an inhibitory effect systemically (Zhang et al., 1999). Remans and others (2006) have shown that another *A. thaliana* mutant defective in the NO_3^- transporter NRT1.1 also showed less lateral root elongation rates resulting in changed root architecture when grown in high (10 mM) or low (0.05 mM) NO_3^-.

Taken together, these results indicate that NO_3^- can regulate root morphology in two different ways. The localized stimulatory effect of NO_3^- allows for autonomous responses by individual lateral roots, whereas the systemic inhibitory effect ensures that these responses are modulated according to the plant nutritional needs, optimizing resource allocation within the plant as a whole (Zhang et al., 1999; Walch-Liu et al., 2006; Walch-Liu and Forde, 2008).

Bacterial Nitrogen Transformations That Influence Root Development

As previously mentioned, soil complexity and dynamic changes determine the way plants will grow and develop. In this context, N-form transformations produced by PGPR will strongly influence root responses. For example, in oxygenated soils, the toxic effects of high NH_4^+ concentrations can be ameliorated by nitrifying bacteria, which transform it to NO_2^- or NO_3^-. Additionally, one could envisage that soil bacteria could interfere with NO_3^- effects on lateral roots via changes in NO_3^- concentration in the rhizosphere or changes in plant N status. Mantelin and Touraine (2004) and Mantelin and others (2006) analyzed the role of NO_3^- in combination with PGPR effects on root morphology. A study performed in *Arabidopsis* inoculated with *Phyllobacterium* sp. suggested that high N status and bacterization affected LR development by different mechanisms. The rhizobacteria promoted LR growth independently of the NO_3^- concentration. Conversely, increased NO_3^- concentration repressed LR growth whether the plants were inoculated or not (Mantelin et al., 2006).

A third possibility is that PGPR could change the N forms that the plant senses, thus affecting the root response. Mantelin and others (2006) showed that *Phyllobacterium* sp. produce NO_2^- and NH_4^+ in NO_3^--containing medium and proposed that this N-transformation could be involved in the *Arabidopsis* root response. They also stated that the strain used was able to reduce NO_3^- and NO_2^-. Based on these results and those from our labs, we propose that bacteria can produce NO from NO_3^- in the medium and that NO can then influence root growth. In fact, NO effects on root morphology are now well established (Durner and Klessig, 1999; Lamattina et al., 2003; del Rio et al., 2004; Besson-Bard et al., 2008).

Diverse signals regulate LR formation, including environmental and endogenous factors (Malamy, 2005). Among environmental signals, nutrients are one of the major regulators of LR development, and, among internal signals, auxin plays a major role. In addition, adventitious rooting was shown to have a link to auxin via NO (Pagnussat et al., 2002, 2003). NO acts at earlier stages of LR formation through the activation of cell division in the pericycle (Correa-Aragunde et al., 2004). Complex interactions between different phytohormones, mainly auxins, determine the AR formation process. Several observations support a link between auxin- and NO-dependent signalling pathways during AR formation (Pagnussat et al., 2002, 2003, 2004; Lanteri et al., 2006). The first evidence showed that auxin induces AR formation through an increase of NO at the base of cucumber hypocotyl explants (Pagnussat et al., 2002).

Specialized epidermal cells form RH that increase the soil area exploitable by the plant (Peterson and Farquhar, 1996). NO has also been shown to be involved in both the initiation and the elongation processes of RH development (Lombardo et al., 2006).

Nitric Oxide Production in Bacteria

NO is a simple molecule with a unique and fascinating chemistry. It is water-soluble but also a lipophilic free radical reactive gas. It is produced by prokaryotes and eukaryotes. It profoundly influences cell physiology, with roles in several different processes including defense signalling, stress responses, and physiological and developmental programs (Lamattina et al., 2003, Grün et al., 2006).

In mammals, NO synthases (NOS) are the primary sources of NO. They are complex, highly regulated enzymes that oxidize L-arginine to NO and L-citrulline. Animal NOS is a dimeric protein formed by an N-terminal oxygenase domain (NOSoxy) that binds protoporphyrin IX (heme), 6R-tetrahydrobiopterin (H4B), and the substrate L-arginine, and by a C-terminal reductase domain (NOSred) that binds FMN, FAD, and NADPH (Stuehr, 1997).

Plant NO synthesis, however, appears more complex and includes both NO_2^- and arginine-dependent mechanisms. One of the primary sources of NO synthesis is accomplished by cytoplasmic nitrate reductase, which not only catalyzes NO_3^- reduction to NO_2^- but also NO_2^- to NO (Yamasaki et al., 1999; Lamattina et al., 2003). Evidence for the existence of a plant NOS is controversial, though many reports demonstrate NO production in an L-arginine-dependent manner (Chandok et al., 2003; Guo et al., 2003; Hu et al., 2005; Corpas et al., 2006; Jasid et al., 2006; Zemojtel, 2006). Based on the available evidence, the source of arginine-dependent NO production in plants had been presumed to be via a mammalian-type NOS. However, no such homologue has been found in plant genomes to date. In 2003, two types of plant NOS (AtNOS1 and iNOS) were reported; neither shared sequence similarity with mammalian-type NOS but were thought to have unique mechanisms for arginine-dependent NO synthesis (Chandok et al., 2003; Guo et al., 2003). One year later, the two papers describing iNOS were retracted (Travis, 2004). At present, the NOS activity of AtNOS1 seems to be in doubt (Crawford, 2006; Zemojtel, 2006). Although the molecular basis of nos1 mutant is unknown, Crawford (2006) proposed to rename this gene as AtNOA1 for nitric oxide associated one so that it is not confused with animal NOS genes. Recently Moreau and others (2008) presented experimental evidence showing that AtNOA1 is unable to bind arginine and oxidize it to NO and demonstrated it is a cGTPase. This gap in the knowledge of an arginine-dependent NO generation pathway in plants is still to be resolved.

Mechanisms of bacterial NO production are more diverse than those of animals and plants. In contrast to eukaryotes, the formation of NO in prokaryotes has mainly been attributed to catabolic processes, NO being an intermediate of both denitrification (Zumft, 1997) and nitrification pathways (Kuenen and Robertson, 1994; Jetten, 2001; Wrage et al., 2001) (Figure 17.1).

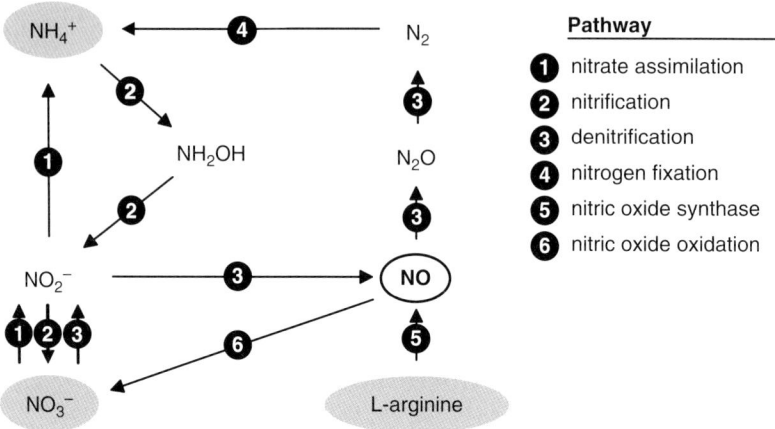

Figure 17.1. Metabolic pathways involved in bacterial N-transformations leading to NO production. Gray ovals indicate key N forms used by bacteria to synthesize nitric oxide (NO). Nitrite (NO_2^-) is produced as an intermediary by catabolic processes such as denitrification or nitrification in both autotrophic and heterotrophic bacteria. NO_2^- indirectly (2) or directly (3) is transformed to NO. Similarly, nitrate (NO_3^-) assimilation (1) produces NO_2^- which is reduced to NO. In bacteria that harbor the nitrogenase enzymatic complex NH_4^+ can be produced by biological nitrogen fixation (4). Some Gram-positive bacteria harbor an NO synthase-like enzyme (5). NO oxidation detoxifies excess NO, which can be carried out by flavohemoglobins (6).

All of these pathways represent a general picture on N metabolism in bacteria and are not necessarily present in a single microorganism.

Denitrification is the stepwise dissimilative reduction of NO_3^- to NO_2^-, NO, nitrous oxide (N_2O) and dinitrogen (N_2) by the corresponding N oxide reductases. In this process, NO_3^- is used instead of O_2 as a final electron acceptor in respiration. This pathway allows denitrifiers to generate energy and to grow under low-oxygen or anaerobic conditions (Zumft, 1993; Cutruzzolá, 1999). Although the denitrification pathway has been known for years as a strictly anaerobic process, it is now accepted that it can also take place under aerobic conditions (Meiberg et al., 1980; Zumft, 1997). Essential in this pathway is the periplasmic nitrate reductase (Nap), which is neither repressed nor inactivated by O_2 (Steenhoudt et al., 2001a).

NO is also produced during the nitrifying process (Anderson et al., 1993; Jetten, 2001; Wrage et al., 2001), an aerobic NH_4^+ oxidation pathway yielding hydroxylamine (NH_2OH), NO_2^- and finally, NO_3^- (Wrage et al., 2001). Nitrification is a two-step process that involves two different groups of bacteria: ammonia-oxidizing bacteria (AOB), which catalyze NH_4^+ oxidation to NO_2^-, and nitrite-oxidizing bacteria (NOB) that oxidize NO_2^- to NO_3^-. So far, no autotrophic bacterium is known to oxidize NH_4^+ directly to NO_3^- (Bothe et al., 2000). Particularly, under low O_2 tension, AOB can use the NO_2^- they generate as electron acceptor instead of O_2, generating NO as a sub-product (Jetten, 2001).

Heterotrophic nitrification is yet another pathway in which NH_4^+ is oxidized by heterotrophic microorganisms (Paul and Clark, 1996). Contrary to autotrophic nitrification, heterotrophic nitrification can be completely carried out by a single species of bacterium and provides no energy (Wrage et al., 2001). Most of the heterotrophic nitrifiers are also aerobic denitrifiers (Castignetti and Hollocher, 1984), and precisely these combined pathways render NO as a subproduct once NO_3^- and/or NO_2^- produced by heterotrophic nitrification is reduced to NO by the enzymes of the aerobic denitrification pathway (Kuenen and Robertson, 1994).

In addition to NO production as a product or a subproduct of the above-mentioned N-metabolism pathways, many Gram-positive bacteria also harbor a specific NOS-like enzyme. These bacteria include *Nocardia* sp., *Streptomyces turgidiscabies*, *Deinococcus radiodurans*, and some *Bacillus* species (Chen and Rosazza, 1994; Sudhamsu and Crane, 2009). Sequence analysis of the genes encoding bacterial NOS (bNOS) reveal great similarity with the N-terminal NOSoxy domain of mammalian NOS, whereas the NOSred domain is completely absent in bNOS. Nevertheless, bNOS are functional and effectively synthesize NO from L-arginine, also showing inhibition by the mammalian NOS inhibitors (Chen and Rosazza, 1994; Choi et al., 1997, Sari et al., 1998; Adak et al., 2002; Midha et al., 2005). Regarding the reductase domain, absent in NOS from Gram-positive bacteria, it is hypothesized that some proteins belonging to flavodoxin family could function as electron donors to bNOS (Zemojtel et al., 2003; Wang et al., 2007).

A recent review on bNOS (Sudhamsu and Crane, 2009) reports the existence of a novel NOS in the myxobacterium *Sorangium cellulosum* (Schneiker-Bekel et al., 2007). To our knowledge this is the first reported Gram-negative microorganism with an NOS-type sequence. In contrast to Gram-positive bNOS, the *S. cellulosum* NOS has a domain, which could have a redNOS function, in a manner similar to mammalian NOS (Sudhamsu and Crane, 2009).

The NO-dioxygenase activity is operative in bacteria and regulates NO levels. Under nitrosative stress the transformation of NO to NO_3^- can ameliorate toxic effects. Some of the enzymes that harbor this NO-dioxygenase capability are related to hemoglobins and flavohemoglobins (Gardner et al., 2000; Frey et al., 2002).

Nitric Oxide as a Signal Molecule In Bacteria and Plant-microbe Interactions

Nitric Oxide Functions in Bacteria Other Than PGPR

Many bacterial physiological processes involve NO participation both as an intermediate in metabolic pathways and as a regulatory signal molecule. Denitrification involves NO as an obligate intermediate and as a signal molecule that positively affects NO_2^- reductase (Nir) and NO reductase (Nor) expression in *Rhodobacter sphaeroides* (Kwiatkowski and Shapleigh, 1996), *Paracoccus denitrificans* (Van Spanning et al., 1999), and *Pseudomonas stuzeri* (Vollack and Zumft, 2001), among others. The role of NO as signal molecule in the nitrifying bacteria is less known.

Though scarce in number, there are reports on a function of NO in biofilm formation. Schmidt and others (2004) showed that treating *Nitrosomonas europaea* cultures with gaseous NO-induced changes in growth characteristics in culture, turning cells into nonmotile forms that produced biofilms on the reactor walls. Nevertheless, in *Pseudomonas aeruginosa* growing in aerobic conditions, it was shown that a rise in the NO content in the preformed biofilm induced its dispersion and stimulated swarming motility (Barraud et al., 2006). This process occurred when the biofilm-dominating conditions became anaerobic, inducing respiratory Nir. In addition, *P. aeruginosa* Δ*nirS* mutants, which produce less NO, show an increased dispersion of the biofilm, while Δ*NorCB* mutants, which accumulate NO, show a high degree of biofilm formation (Barraud et al., 2006). These results point to a distinct regulatory mechanism for biofilm formation or dispersion in AOB and denitrifiers. Since plant roots are common sites for biofilm formation (Danhorn and Fuqua, 2007) the importance of NO as a regulator of the process, and the mechanisms involved, are worthy areas of research.

The role of NO in the stimulation of oxidative and nitrosative stress defenses in bacteria has been widely studied (Nakano, 2002; Mukhopadhyay et al., 2004; Gusarov and Nudler, 2005; Spiro, 2007; Ren et al., 2008). In *Bacillus subtilis,* NO was reported to exert rapid protection from oxidative stress generated by hydrogen peroxide (Gusarov and Nudler, 2005). That the response occurred in as little as 5 seconds, led the authors to state that the protective effect could be related to direct activation of catalase by NO action on its heme cluster and not to the induction of catalase structural gene (*KatA*) expression. However, they also proposed an alternative mechanism whereby NO transiently interrupts enzymatic reduction of free Cys, thereby suppressing the Fenton reaction (Gusarov and Nudler, 2005).

Pathogenic bacteria use their own NO as a key defense against the immune oxidative burst, thereby establishing bNOS as an essential virulence factor. The survival of the soil Gram-positive *Bacillus anthracis*, a human pathogen, is critically dependent on its own bNOS activity (Shatalin et al., 2008). The mechanism underlying bNOS-dependent resistance to macrophage killing relies on NO-mediated activation of bacterial catalase and suppression of the Fenton reaction.

In addition to the rapid responses in the two previous examples, there is evidence that NO acts as a signal molecule regulating the expression of genes involved in oxidative and nitrosative responses in *B. subtilis* (Nakano, 2002) and in *Escherichia coli* (Mukhopadhyay et al., 2004). The ability of NO to bind iron (Fe)-containing proteins could explain its role in mediating and/or regulating oxidative stress damage prevention, when ferric (Fe^{3+}) ions produced via the Fenton reaction are present.

Also related to oxidative stress is the global ferric uptake regulator (*fur*), whose expression and activity are modified by treatment with NO-donors in the Gram-negative bacterium *E. coli* (Mukhopadhyay et al., 2004). The same occurs with the *fur* homologue, *yqkL* (Bsat et al., 1998), in the Gram-positive *B. subtilis* (Moore et al., 2004). This genetic regulator also controls numerous genes related to Fe homeostasis, such as those involved in siderophore synthesis and siderophore-Fe^{3+} complex transporters (Andrews et al., 2003). *Fur* is also implicated in numerous other roles (Escolar et al., 1999). By both *in vivo* and *in vitro* experiments, it was shown that NO modulates *fur* activity through its binding to Fe^{3+} in the Fe-*fur* complex (D'Autréaux et al., 2002). However, there is as yet no conclusive evidence supporting regulation of bacterial Fe uptake by NO modulation of *fur*.

Finally, for a detailed discussion of how some bacterial transcription factors are affected by NO, as well as how Fe-proteins could function as NO-sensors, the excellent review of Spiro (2007) should be consulted.

Nitric Oxide in Intracellular PGPR (iPGPR)

In leguminous plants, infection with *Rhizobium* and bacteria of other genera leads to the formation of new organs, the root nodules. Inside this structure, the bacteroids fix atmospheric N_2 producing NH_4^+ and provide it as amides or ureides to the host plant. Functional nodules result from the combination of developmental and infectious processes that ensure organ formation and involve exchange of specific signal molecules between the two partners (Schultze and Kondorosi, 1998). This complex process is extensively covered in Chapters 15 and 16.

Nodule development requires local accumulation of auxin (Spaepen et al., 2007). It is known that rhizobia synthesize auxin (see Chapter 11). Most *Rhizobium* species have been shown to produce indole-3 acetic acid (IAA) via pathways different from plants (Theunis et al., 2004), and many studies have indicated that changes in auxin balance in the host plant are a prerequisite for nodule organogenesis (Mathesius et al., 1998 and Chapter 15). It was shown that NO content was enhanced in nodules of indeterminate legumes (*Medicago* species) initiated by IAA-overproducing rhizobia (Pii et al., 2007). The development of LR was also augmented, a process that is known to be tightly regulated by both IAA and NO (Correa-Aragunde et al., 2004). In fact, the specific NO scavenger 2-(4-carboxyphenyl)-4,4,5,5,-tetramethylimidazoline-1-oxyl-3 oxide (cPTIO) markedly reduced nodulation induced by wild type and IAA-overproducing strains (Pii et al., 2007). Conversely, no effect was observed in plants bearing determinate nodules (Pii et al., 2007). Considering the evidence of both the inhibition of polar auxin transport during the first steps of nodulation (Mulder et al., 2005) and the requirement of NO in auxin-induced root developmental processes (Pagnussat et al., 2002, 2003; Correa-Aragunde et al., 2004; Lombardo et al., 2006), it was hypothesized that NO could have a signalling role in the establishment of legume-rhizobia interactions (Baudouin et al., 2006; Pii et al., 2007).

Biological N_2 fixation is a tightly regulated process in the microaerobic nodule interior. N_2 reduction is catalyzed by nitrogenase, a complex of two proteins, a MoFe protein and a Fe protein, with marked O_2 sensitivity. The finding that nodules accumulate significant levels of NO raises the question of its possible physiological role during N_2 fixation. Several proteins that are synthesized by the plant or the bacterial partner have strong affinity for NO, especially leghemoglobin (Lb), the most abundant protein in N_2-fixing nodules. The major form of Lb *in vivo* is ferrous-Lb, which binds NO to form NO-Lb complexes (Harutyunyan et al., 1996). Based on this evidence and given that soybean nodules have a plentiful supply of arginine, the substrate of NOS activity, Streeter (1987) and Hérouart and others (2002) speculated that NO might have a regulatory role in nodule development. In fact, the pioneering paper of Cueto and others (1996) provided preliminary functional evidence for the existence of a NOS-like enzyme in the roots and the nodules of *Lupinus albus*. Later Baudouin and others (2006) found that functional nodules formed by *Sinorhizobium meliloti* in *Medicago truncatula* synthesized NO by a mechanism that is related neither to denitrification nor to N_2 fixation. Both reports showed that mammalian NOS inhibitors were effective in inhibiting NO synthesis, suggesting that an NOS-like activity could be responsible for NO production.

The formation of NO-Lb complexes has to date been viewed as a means of Lb inactivation, leading to the impairment of the optimal N_2 fixation capacity (Mathieu et al., 1998). However, as new information accumulates regarding NO as a signalling molecule, this negative aspect of NO-Lb complex formation could be losing devotees.

In addition to Lb, the *fixL* gene product, a bacterial protein responsible for O_2 sensing, can also readily bind NO instead of O_2 (Gilles-Gonzalez and Gonzalez, 2005). FixL-like proteins regulate microaerobic adaptation by rhizobia and many other bacteria. Mesa and others (2003) suggested that NO is a signal that activates bacterial responses to low O_2 tension in soybean nodules, formed by *Bradyrhizobium japonicum*, by regulating the bacterial NnrR transcription factor, which controls denitrification gene expression.

N_2 fixation is particularly dependent on Fe-containing proteins. When free-living rhizobia eventually become bacteroids in the symbiotic interaction, there is an associated 3-fold increase in accumulated cytochrome. Nitrogenase is massively synthesized, constituting more than 10% of total bacteroid protein. Nitrogenase may contain more than 30 atoms of Fe (Verma and Long, 1983; Sangwan and O'Brian, 1992). Nitrogenase, together with heme-containing Lb, may account for up to 40% of total soluble protein of the infected cell, making the nodule one of the most Fe-loaded plant organs (Kuzma et al., 1993). The required Fe is "provisioned" by the ferritin complex, able to store thousands of Fe atoms (Briat and Lobreaux, 1997). Ferritin levels increase early in nodule development (Ragland and Theil, 1993). Interestingly, it was shown that NO induces accumulation of ferritin protein and mRNA in *Arabidopsis thaliana* (Murgia et al., 2002).

Cohen and others (2006) proposed that, in the rhizosphere, *Streptomyces*-derived NO, in concert with that endogenously produced in roots, might have a role in increasing iron bioavailability in plants. Their "NO-iron bridge" hypothesis is based on the evidence that many *Streptomyces* possess a NOS-like activity, and that supplementation of Fe-deficient plants with NO can reverse chlorosis. Although NO did not increase the total Fe content of plants, Graziano and Lamattina (2005) pointed out that NO may increase Fe availability by reducing Fe^{3+} to ferrous (Fe^{2+}) and forming iron nitrosyl complexes, enabling the passage of iron from the apoplast to the cell.

In an interaction involving a nodulating bacterium and a root surface streptomycete, Tokala and others (2002) showed that the surface of emerging nodules in *Pisum sativum* was colonized by *Streptomyces lydicus* WYEC108, which then sporulated within root surface cell layers. This streptomycete promoted nodulation, enhanced nodule growth and bacteroid differentiation, and aided the bacteroids in Fe assimilation.

Finally, since Fe^{3+} catalyses the formation of hydroxyl radical via a Fenton-type reaction, a finely tuned regulatory system must be operative to control Fe^{3+} reactivity. In this system, NO participation cannot be ruled out. Elucidation of possible modes of iron management in the regulation of symbiotic interactions seems to be essential for understanding biological N_2 fixation and legume plant development.

Nitric Oxide in Extracellular PGPR (ePGPR)

We hypothesize that NO, a recognized ubiquitous signalling molecule, plays a key role in regulating plant interactions with intimately associated microorganisms. Many of the functions mentioned in previous sections for NO could be also operating in bacteria that interact from "outside" the plant cell. However, the few reports that show a specific role for NO in plant-ePGPR interaction usually address bio-control and pathogenesis mechanisms.

The relationship between NO production by ePGPR and their protective ability as bio-control agents was examined by Wang and others (2001). They reported that a *P. fluorescens* NO-overproducing mutant was more effective than the wild type for biological control of *Ralstonia solanacearum* in infected tomato. Cohen and others (2005) and Cohen and Mazzola (2006) suggested that NO production by soil bacteria via nitrification could have a potential role in the induction of plant systemic resistance. They investigated the influence of bacterially emitted NO on plant disease resistance. Split root experiments demonstrated that when roots were grown in soils amended with low-glucosinolate *Brassica napus* seed meal there was a resultant systemic protection of the other roots of the same plant challenged with *Rhizoctonia solani* (Cohen et al., 2005). This specific amendment changed the structure of the microbial community increasing nitrifiers, with the concomitant increase in NO production in the soil. Nitrapyrin (inhibitor of nitrification)-treated or pasteurized amendments did not generate increased NO content and did not result in crop protection.

In pathogenic interactions, bacterial NO also plays a role. Plant pathogenic *Streptomyces* species are the causal agents of potato scab disease. In infected tissues, some cell wall components induce bNOS activity. The production of NO by *S. turgidiscablies* via an NOS-like enzyme enables the nitration of the phytotoxic dipeptide thaxtomin A, leading to an efficient infection (Kers et al., 2004; Wach et al., 2005).

Bacterial NO production could be involved in protecting bacteria from the plant oxidative burst. Cohen and Yamasaki (2003) showed that the APG1 strain of *Rhodococcus sp.*, an *Azolla pinnata* leaf endophyte that harbors NOS-like activity, has a high degree of tolerance to oxidative stress generated by H_2O_2. They correlated NO production with H_2O_2 resistance. NO synthesis is apparently involved in the bacterial acclimation to the harsh oxidative conditions in leaves. In bacteria, NO activates the global regulators of genes involved in oxidative stress tolerance, genes encoding catalase, superoxide dismutase, etc. (Spiro, 2007).

Azospirillum brasilense as a Model ePGPR for Studying NO Function in Beneficial Plant-microbe Interactions

Bacteria belonging to the genus *Azospirillum* are known to promote plant development and improve yield under appropriate conditions (Okon, 1994; Bashan, 1999; Dobbelaere et al., 2001). The genus *Azospirillum* is included in the alpha subclass of Proteobacteria belonging to the IV rRNA superfamily (Xia et al., 1994).

As the most intensively studied associative bacterium, *Azospirillum* has become a cornerstone of rhizosphere research in addition to its agricultural applications. *Azospirillum* is an excellent model for studies on plant-associated bacteria in general (Bashan and Levanony, 1990). Although *Azospirillum* was first isolated from cultivated cereals, it can be found in association with roots of any plant species in which it has been sought, establishing it as a ubiquitous root colonizer (Bashan and Holguin, 1997). Early studies with *Azospirillum* spp. aimed to assess whether their effect on plant growth derives from N_2-fixation ability. It was concluded that *Azospirillum* mainly acts by influencing the morphology, architecture, and physiology of the root system, rather than as an N_2-fixing bacterium (Dobbelaere et al., 2001). Increased root development, in turn, leads to an increased root surface that could improve plant nutrition and water uptake and be a key factor for plant growth promotion by ePGPR (Okon, 1985). Developmental changes promoted in roots must be triggered prior to the changes in uptake of nutrients, which would be increased over time concomitant with increased root surface. Thus, nutritional improvement by ePGPR is an indirect consequence of their effect on root development (Mantelin and Touraine, 2004). Nevertheless, direct effects on root transport systems cannot be ruled out. Bertrand and others (2000) showed that an *Achromobacter* sp. enhanced NO_3^- uptake rate per unit of root area in *Brassica napus* roots, and Becker and others (2002) reported that inoculation with *A. brasilense* increased the expression of the high affinity NH_4^+ transporter LEAMT1; 2 of tomato.

Phytohormone production by *Azospirillum* has been proposed to be the means by which it alters the host root growth pattern (Dobbelaere et al., 1999). Phytohormones, mainly auxins, cytokinins, and gibberellins, are the most commonly invoked means of plant growth promotion exerted by PGPR. Among them, auxins are thought to play the major role. To evaluate the involvement of bacterial IAA in the promotion of root development, mutants altered in IAA production were employed. Some experiments showed a reduced ability to promote root system development associated with different kinds of auxin mutants (Barbieri and Galli, 1993; Kundu et al., 1997; Dobbelaere et al., 1999).

Azospirillum can synthesize IAA by at least three different pathways (Dobbelaere et al., 1999). The main enzyme for IAA synthesis is indole-3-pyruvate decarboxylase, encoded by *ipdC*, which has an auxin responsive element in its promoter region (Lambrecht et al., 2000). Expression of *A. brasilense ipdC* is upregulated by IAA. Since it is possible that auxins may be a root exudate component (Rovira, 1965), Lambrecht and others (2000) proposed that plant-derived IAA in the rhizosphere could enhance expression of *A. brasilense ipdC*, increasing bacterial synthesis of IAA, which, in turn, would stimulate plant root proliferation. However, they also stated that as root development is inhibited above a certain concentration of exogenously supplied IAA, this IAA amplification loop, must be "contained" by other compensating factors (Dobbelaere et al., 1999). There are no reports showing to what extent IAA is produced in the rhizosphere by *Azospirillum* (Lambrecht et al., 2000; Steenhoudt and Vanderleyden, 2000). Moreover, the role of chemical signals in mediating rhizospheric interactions is just beginning to be understood (Bais et al., 2006). Among these signals N species such as NO_3^- and NO are implicated in root growth and development (Lamattina et al., 2003; Mantelin and Touraine, 2004; Molina-Favero et al., 2007).

The analogies between the observed *Azospirillum* stimulation of plant root development and the capability of NO to act as a plant growth regulator (Santner and Estelle, 2009) that promotes AR formation, LR development, and RH formation, led us to explore whether *Azospirillum* promotes root growth and modifies root architecture via NO. *A. brasilense* Sp245 is able to produce NO in aerobic cultures using NH_4^+ or NO_3^- as N source (Creus et al., 2005; Molina-Favero et al., 2008). Three different NO production pathways could be operating under these conditions (Molina-Favero et al., 2007). First, under NO_3^- and aerated conditions, aerobic denitrification is the main source of NO as an isogenic Nap knockout mutant (Faj164, Steenhoudt et al., 2001a) produced NO at 5% the wild-type level (Molina-Favero et al., 2008).

Another pathway operating when NH_4^+ is the N source is heterotrophic nitrification. Although less NO is produced compared to aerobic denitrification, the addition of intermediary compounds (NH_2OH and NO_2^-) enhanced NO synthesis in a dosis-dependent manner (Arruebarrena Di Palma, 2008; Molina-Favero et al., 2008). Finally, the existence of an NOS-like activity in *A. brasilense* is consistent with increased NO upon addition of L-arginine to pure cultures (Creus et al., 2005). Except for the report of Sudhamsu and Crane (2009) on *S. cellulosum* NOS, no evidences had been presented for the existence of NOS-like activity in Gram-negative bacteria.

The involvement of NO in the *Azospirillum*-plant interaction has been demonstrated in only two reports (Creus et al., 2005; Molina-Favero et al., 2008). When NO was removed from the interaction between tomato root and *A. brasilense* Sp245, both lateral and adventitious root formations were inhibited and attained the level of non-inoculated seedlings (Molina-Favero et al., 2008). Inoculation of tomato with the wt Sp245 strain modified root architecture while the mutant Faj164 had no effect. Additionally Faj009, another Sp245 isogenic strain with reduced ability to synthesize IAA (*ipdC*$^-$) (Costacurta et al., 1994), produced an intermediate phenotype in roots (Molina-Favero et al., 2008). Thus, while bacterial IAA has a prominent role in root development, our results clearly point to a strong involvement of NO in *Azospirillum*-induced root branching.

Considering the versatile N metabolic capability of *Azospirillum* and its ability to produce the key plant growth regulators auxins and NO, a possible scenario of the way that it promotes plant growth is depicted in Figure 17.2.

Successful *Azospirillum* colonization of the root is crucial to achieve plant growth enhancement (Bashan et al., 2004). In addition to providing a C-rich environment, plant roots initiate cross talk with soil microbes by producing signals that are recognized by the microbes, which in turn initiate colonization (Bais et al., 2006). Bacteria physically interact with surfaces to form complex multicellular assemblies including biofilms and smaller aggregates. Biofilm development and the resulting intimate interaction with plants often require cell-cell

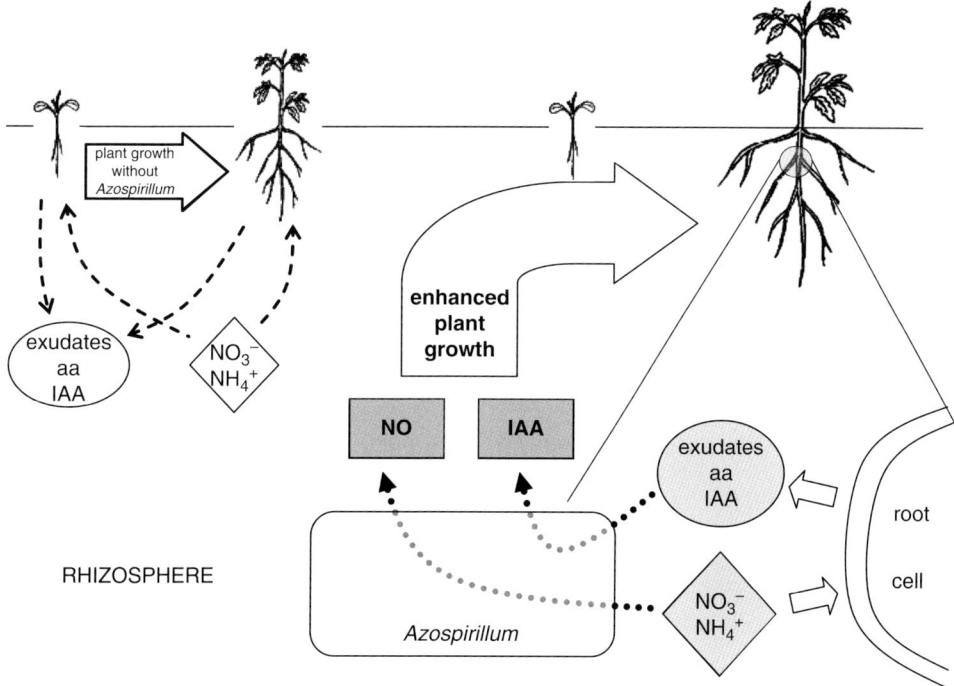

Figure 17.2. *Azospirillum* influences on plant growth.
During seed germination and the initial growth stages of the seedling, the developing root releases exudates providing rhizospheric bacteria with C-sources such as amino acids (aa) and hormonal compounds such as auxins (IAA). The model PGPR *Azospirillum* is able to perceive IAA in the rhizosphere upregulating its own indole-3-pyruvate decarboxylase gene, enhancing bacterial IAA synthesis. This produces a positive loop between root and bacteria as bacterial IAA can in turn stimulate root development (Lambrecht et al., 2000). In addition, *Azospirillum* is able to produce nitric oxide (NO) by changing nitrogen forms present in the rhizosphere through denitrification, heterotrophic nitrification, and/or by a NOS-like activity (Molina-Favero et al., 2007). It is not yet known if *Azospirillum* can trigger a mechanism that induces NO synthesis by root cells. Both signal molecules, NO and IAA, in concert, and/or separately, produce changes in root development changing its architecture and leading to enhanced mineral and water uptake, which in turn increase plant growth.

communication among the colonizing bacteria (Danhorn and Fuqua, 2007). Pothier and others (2007) showed that wheat seed extracts induced numerous genes in *A. brasilense* Sp245 including dissimilatory nitrite reductase (NirK), which is the major source of NO in bacteria (Cutruzzolá, 1999). They also showed that NirK was upregulated when *A. brasilense* Sp245 colonized wheat roots (Pothier et al., 2008). Steenhoudt and others (2001b) showed that root colonization ability was drastically impaired in an *A. brasilense* Sp245 mutant unable to reduce NO_3^- because both its periplasmic (Nap) and cytoplasmic (Nas; assimilatory) NO_3^- reductases were inactive. Similar results were obtained in *P. fluorescens* YT101 deficient in respiratory nitrate reductase (Nar) (Ghiglione et al., 2000). In light of these results it would be interesting to study the role of NO in colonization and biofilm formation on roots by *Azospirillum*.

Plants are often exposed to perturbations in their immediate environment. In this regard, *Azospirillum* exerts beneficial effects in stressful as well as "normal" conditions (Okon and Labandera-Gonzélez, 1994; Creus et al., 1998; Casanovas et al., 2002, 2003; Bacilio et al., 2004; Creus et al., 2004). Abiotic factors can also directly influence PGPR activity, their effect on plant growth, and the dynamics of root microbial communities. Previous studies demonstrated that the positive effects of *Azospirillum* were mainly derived from morpho-physiological changes of the inoculated plant roots resulting in an enhanced capacity for water and mineral uptake (Okon and Kapulnik, 1986). A general response of plants to *Azospirillum* inoculation is an improved water status, mainly expressed under water and/or salinity stress conditions (Creus et al., 1997; Casanovas et al., 2002, 2003; Creus et al., 2004; Barassi et al., 2006; Pereyra et al., 2006).

It is known that NO participates in plant responses to several abiotic stresses. Oxidative damage can be ameliorated by treatment with an NO-donor in wheat and cucumber under salinity (Ruan et al., 2002; Shi et al., 2007). Garcia-Mata and Lamattina (2001) demonstrated stomatal closure in water-stressed detached wheat leaves pretreated with an NO-donor. A specific NO scavenger reverted NO action by restoring the transpiration rate and stomatal aperture. Kolbert and others (2008) found a significant increase in endogenous NO production in roots of polyethylene-glycol-stressed pea seedlings.

Further strong support for NO being a key molecule in abiotic stress responses was the finding of a mutant in *A. thaliana* AtNOS1 (renamed AtNOA1) (Crawford, 2006), with a reduced NO level, and which was shown to be more sensitive to saline stress (Zhao et al., 2007).

The remarkable analogies between NO functions in plant responses to stress conditions and the ameliorative effects that *Azospirillum* exerts on stressed plants are consistent with the hypothesis that bacterial NO is involved in plant protection by PGPR.

Conclusion and Perspectives

An overarching goal of root and rhizosphere research is to understand how roots function in the field, either to understand how natural ecosystems work or to develop better crops and agricultural or horticulture practices (Welbaum et al., 2004; Malamy, 2005). Microorganisms in the rhizosphere react to the many metabolites released by plant roots. They also interact with plant roots by means of their own secreted metabolites. In this scenario, signals derived from changes in the soil environment trigger selective root and shoot responses, as well as dynamic bacterial changes (Bais et al., 2006). Moreover, a sustainability analysis of crop production in terms of modern technological agriculture requires a detailed knowledge of the interrelationships between the microorganisms added to the system and those present in the soil. Plant-microbe interactions in the rhizosphere are also responsible for important processes such as carbon sequestration, ecosystem functioning, and nutrient cycling (Singh et al., 2004). NO is a central component in the N cycle, and several biological and chemical pathways are involved in regulating NO steady state levels in soils, including denitrification, N mineralization, N_2 fixation, and nitrification (Jetten, 2001; Wrage et al., 2001). NO has many of the properties of a typical signalling molecule. NO also has roles in symbiotic as well as ePGPR-plant interaction, although there is not yet a good mechanistic understanding.

Roots have to sense, decode, and coordinately interpret numerous stimuli. A possible ecological role for soil NO has been suggested as an exogenous information carrier that provides the seeds with integral information about external environmental factors affecting germination and plant growth (Giba et al., 2007). Because there is no sterile soil in nature, bacteria in the rhizosphere can be changing the different N-forms that regulate plant development. In this way, ePGPRs, compared to the rest of the soil flora, act in intimate association with roots and could have a major impact on root responses. NO emission by soils is attributed mainly to this bacterial activity. Soil bacteria could be the intermediary sensors of environmental conditions for plants to interpret, in real time, rhizospheric heterogeneity.

References

Adak, S., Bilwes, K.S., Panda, K., et al. 2002. Cloning, expression, and characterization of a nitric oxide synthase protein from *Deinococcus radiodurans*. *Proceedings of the National Academy of Sciences,* 99, 107–112.

Alvarez, M.I., Sueldo, R.J., Barassi, C.A. 1996. Effect of *Azospirillum* on coleoptile growth in wheat seedlings under water stress. *Cereal Research Communications,* 24, 101–107.

Anderson, I.C., Poth, M., Homstead, J., et al. 1993. A comparison of NO and N_2O production by the autotrophic nitrifier *Nitrosomonas europaea* and heterotrophic nitrifier *Alcaligenes faecalis*. *Applied and Environmental Microbiology,* 59, 3525–3533.

Andrews, S.C., Robinson, A.K., Rodriguez-Quiñones, F. 2003. Bacterial iron homeostasis. *FEMS Microbiological Reviews,* 27, 215–237.

Arruebarrena Di Palma, A. 2008. Producción de oxido nítrico por nitrificación heterotrófica en *Azospirillum*. Graduate Thesis. Mar del Plata University, Buenos Aires, Argentina.

Bacilio, M., Rodríguez, H., Moreno, M., et al. 2004. Mitigation of salt stress in wheat seedlings by a gfp-tagged *Azospirillum lipoferum*. *Biology and Fertility of Soils,* 40, 188–193.

Bais, P.H., Weir, L.T., Perry, L.G., et al. 2006. The role of root exudates in rhizosphere interactions with plants and other organisms. *Annual Review in Plant Biology,* 57, 233–266.

Barassi, C.A., Ayrault, G., Creus, C.M., et al. 2006. Seed inoculation with *Azospirillum* mitigates NaCl effects on lettuce. *Scientia Horticulturae,* 109, 8–14.

Barbieri, P. and Galli, E. 1993. Effect on wheat root development of inoculation with an *Azospirillum brasilense* mutant with altered indole-3-acetic acid production. *Research in Microbiology,* 44, 69–75.

Barraud, N., Hassett, D.J., Hwang, S.-H., et al. 2006. Involvement of nitric oxide in biofilms dispersal of *Pseudomonas aeruginosa*. *Journal of Bacteriology,* 188, 7344–7353.

Bartelt-Ryser, J., Joshi, J., Schmid, B., et al. 2005. Soil feedbacks of plant diversity on soil microbial communities and subsequent plant growth. *Perspectives in Plant Ecology, Evolution and Systematics,* 7, 27–49.

Bashan, Y., Levanony, H. 1990. Current status of *Azospirillum* inoculation technology: *Azospirillum* as a challenge for agriculture. *Canadian Journal of Microbiology,* 36, 591–608.

Bashan, Y., Holguin, G. 1997. *Azospirillum*-plant relationships: Environmental and physiological advances (1990–1996). *Canadian Journal of Microbiology,* 43, 103–121.

Bashan, Y. 1999. Interactions of *Azospirillum* spp. in soils: a review. *Biology and Fertility of Soils,* 29, 246–256.

Bashan, Y., Holguin, G., de-Bashan, L.E. 2004. *Azospirillum*-plant relationships: physiological, molecular, agricultural, and environmental advances (1997–2003). *Canadian Journal of Microbiology,* 50, 521–577.

Baudouin, E., Pieuchot, L., Engler, G., et al. 2006. Nitric oxide is formed in *Medicago truncatula-Sinorhizobium meliloti* functional nodules. *Molecular Plant-Microbe Interactions,* 19, 970–975.

Bauer, W.D., Mathesius, U. 2004. Plant responses to bacterial quorum-sensing signals. *Current Opinion in Plant Biology,* 7, 429–433.

Beauchamp, C.J. 1993. Mode of action of plant growth-promoting rhizobacteria and their potential use as biological control agents. *Phytoprotection,* 71, 19–27.

Becker, D., Stanke, R., Fendrik, I., et al. 2002. Expression of the NH_4^+-transporter gene *LEAMT1;2* is induced in tomato roots upon association with N_2-fixing bacteria. *Planta,* 215, 424–429.

Bertrand, H., Plassard, C., Pinochet, X., et al. 2000. Stimulation of the ionic transport system in *Brassica napus* by a plant growth-promoting rhizobacterium (*Achromobacter* sp). *Canadian Journal of Microbiology,* 46, 229–236.

Besson-Bard, A., Pugin, A., Wendehenne, D. 2008. New insights into nitric oxide signalling in plants. *Annual Review of Plant Biology,* 59, 21–39.

Bloom, A.J., Meyerhoff, P.A., Taylor, A.R., et al. 2003. Root development and absorption of ammonium and nitrate from the rhizosphere. *Journal of Plant Growth Regulation,* 21, 416–431.

Boddey, R.M., Jantalia, C.P., Zotarelli, L., et al. 2008. Techniques for the quantification of plant-associated biological nitrogen fixation. In: *Biological Nitrogen Fixation: Towards Poverty Alleviation Through Sustainable Agriculture* (Samson, D.D., Chimphango, B.M., Valentine, A.J., Elmerich, C., Newton, W.E., Eds.), pp. 37–41. Springer, Berkeley.

Bothe, H., Jost, G., Schloter, M., et al. 2000. Molecular analysis of ammonia oxidation and denitrification in natural environments. *FEMS Microbiology Reviews,* 24, 673–690.

Brown, M.E. 1972. Plant growth substances produced by microorganisms of soil and rhizosphere. *Journal of Applied Bacteriology,* 43, 443–451.

Brencic, A., Winans, S.C. 2005. Detection of and response to signals involved in host-microbe interactions by plant-associated bacteria. *Microbiology and Molecular Biology Reviews,* 69, 155–194.

Briat, J.F., Lobreaux, S. 1997. Iron transport and storage in plants. *Trends in Plant Science,* 2, 187–193.

Bsat, N., Herbig, A., Casillas-Martinez, L., et al. 1998. *Bacillus subtilis* contains multiple Fur homologues: identification of the iron uptake (Fur) and peroxide regulon (PerR) repressors. *Molecular Microbiology,* 29, 189–198.

Burdman, S., Jurkevitch, E., Okon, Y. 2000. Recent advances in the use of plant growth promoting rhizobacteria (PGPR) in agriculture. In: *Microbial Interactions in Agriculture and Forestry,* Vol. 2, (Subba Rao, N.S. and Dommergues, Y.R., Eds.), pp. 227–248. Science Publishers, Enfield, UK.

Casanovas, E.M., Barassi, C.A., Sueldo, R.J. 2002. *Azospirillum* inoculation mitigates water stress effects in maize seedlings. *Cereal Research Communications,* 30, 343–350.

Casanovas, E.M., Barassi, C.A., Andrade, F.H., et al. 2003. *Azospirillum* inoculated maize plant responses to irrigation restraints imposed during flowering. *Cereal Research Communications,* 31, 395–402.

Castignetti, D., Hollocher, T.C. 1984. Heterotrophic nitrification among denitrifiers. *Applied and Environmental Microbiology,* 47, 620–623.

Chandok, M.R., Ytterberg, A.J., van Wijk, K.J., et al. 2003. The pathogen-inducible Nitric oxide synthase (iNOS) in plants is a variant of the P protein of the glycine decarboxylase complex. *Cell,* 113, 469–482.

Chen, Y., Rosazza, P.N. 1994. A bacterial nitric oxide synthase from a *Nocardia* species. *Biochemical and Biophysical Research Communications,* 203, 1251–1258.

Choi, W-S., Chang, M-S., Han, J-W., et al. 1997. Identification of nitric oxide synthase in *Staphylococcus aureus. Biochemical and Biophysical Research Communications,* 237, 554–558.

Christou, M., Avramides, E.J., Robrets, J.P., et al. 2005. Dissolved organic nitrogen in contrasting agricultural ecosystems. *Soil Biology and Biochemistry,* 37, 1560–1563.

Cohen, M.F. and Yamasaki, H. 2003. Involvement of nitric oxide synyhase in sucrose-enhanced hydrogen peroxide tolerance of *Rhodoccocus* sp. strain APG1, a plant-colonizing bacterium. *Nitric Oxide,* 9, 1–9.

Cohen, M.F., Yamasaki, H., Mazzola, M. 2005. *Brassica napus* seed meal soil amendment modifies microbial community structure, nitric oxide production and incidence of *Rhizoctonia* root rot. *Soil Biology and Biochemistry,* 37, 1215–1227.

Cohen, M.F., Mazzola, M. 2006. Resident bacteria, nitric oxide emission and particle size modulate the effect of *Brassica napus* seed meal on disease incited by *Rhizoctonia solani* and *Pythium* spp. *Plant and Soil,* 286, 75–86.

Cohen, M.F., Yamasaki, H., Mazzola, M. 2006. Nitric oxide research in agriculture: Bridging the plant and bacterial realms. In: *Abiotic Stress Tolerance in Plants: Toward the Improvement of Global Environment and Food* (Rai, A.K. and Takabe, T., Eds.), pp. 71–90. Springer Verlag, Berlin, Germany.

Compant, S., Duffy, B., Nowak, J., et al. 2005. Use of plant growth-promoting bacteria for bio-control of plant diseases: principles, mechanisms of action, and future prospects. *Applied and Environmental Microbiology,* 71, 4951–4959.

Corpas, F.J., Barroso, J.B., Carreras, A., et al. 2006. Constitutive arginine-dependent nitric oxide synthase activity in different organs of pea seedlings during plant development. *Planta,* 224, 246–254.

Correa–Aragunde, N., Graziano, M., Lamattina, L. 2004. Nitric oxide plays a central role in determining lateral root development in tomato. *Planta,* 218, 900–905.

Costacurta, A., Keijers, V., Vanderleyden, J. 1994. Molecular cloning and sequence analysis of an *Azospirillum brasilense* indole-3-acetic pyruvate decarboxylase gene. *Molecular Genetics and Genomics,* 243, 463–472.

Crawford, N.M. 2006. Plant nitric oxide synthase: back to square one. *Trends in Plant Science,* 11, 526–527.

Creus, C.M., Sueldo, R.J., Barassi, C.A. 1997. Shoot growth and water status in *Azospirillum*-inoculated wheat seedlings grown under osmotic and salt stresses. *Plant Physiology and Biochemistry,* 35, 939–944.

Creus, C.M., Sueldo, R.J., Barassi, C.A. 1998. Water relations in *Azospirillum* inoculated wheat seedlings under osmotic stress. *Canadian Journal of Botany,* 76, 238–244.

Creus, C.M., Sueldo, R.J., Barassi, C.A. 2004. Water relations and yield in *Azospirillum*-inoculated wheat exposed to drought in the field. *Canadian Journal of Botany,* 82, 273–281.

Creus, C., Graciano, M., Casanovas, E., et al. 2005. Nitric oxide is involved in the *Azospirillum brasilense*-induced lateral root formation in tomato. *Planta,* 221, 297–303.

Cueto, M., Hernández-Perera, O., Martín, R., et al. 1996. Presence of nitric oxide synthase activity in roots and nodules of *Lupinus albus. FEBS Letters,* 398, 159–164.

Cutruzzolá, F. 1999. Bacterial nitric oxide synthesis. *Biochimica et Biophysica Acta,* 1411, 231–249.

D'Autréaux, B., Touati, D., Bersch, B., et al. 2002. Direct inhibition by nitric oxide of the transcriptional ferric uptake regulation protein via nitrosylation of the iron. *Proceedings of the National Academy of Sciences,* 99, 16619–16624.

Danhorn, T., Fuqua, C. 2007. Biofilm Formation by Plant-Associated Bacteria. *Annual Reviews in Microbiology,* 61, 401–422.

del Rio, L.A., Corpas, F.J., Barroso, J.B. 2004. Nitric oxide and nitric oxide synthase activity in plants. *Phytochemistry,* 65, 783–792.

Delvasto, P., Valverde, A., Ballestera, A., et al. 2006. Characterization of brushite as a recrystallization product formed during bacterial solubilization of hydroxyapatite in batch cultures. *Soil Biology and Biochemistry,* 3, 2645–2654.

Dobbelaere, S., Croonenborghs, A., Thys, A., et al. 1999. Phytostimulatory effect of *Azospirillum brasilense* wild type and mutant strains altered in IAA production on wheat. *Plant and Soil,* 212, 155–164.

Dobbelaere, S., Croonenborghs, A., Thys, A., et al. 2001. Responses of agronomically important crops to inoculation with *Azospirillum. Australian Journal of Plant Physiology,* 28, 871–879.

Durner, J., Klessig, D.F. 1999. Nitric oxide as a signal in plants. *Current Opinion in Plant Biology,* 2, 369–374.

Escolar, L., Pérez-Martín, J., de Lorenzo, V. 1999. Opening the iron box: Transcriptional Metalloregulation by the Fur Protein. *Journal of Bacteriology,* 181, 6223–6229.

Fallik, E., Sarig, S., Okon, Y. 1994. Morphology and physiology of plant roots associated with *Azospirillum.* In: *Azospirillum Plant Associations* (Okon, Y., Ed.), pp 75–85. CRC Press, Boca Raton.

Frey, A.D., Farrés, J., Bollinger, C.J.T., et al. 2002. Bacterial hemoglobins and flavohemoglobins for alleviation of nitrosative stress in *Escherichia coli. Journal of Microbiology,* 68, 4835–4840.

Garcia-Mata, C., Lamattina, L. 2001. Nitric oxide induces stomatal closure and enhances the adaptive plant responses against drought stress. *Plant Physiology,* 126, 1196–1204.

Gardner, P.R., Gardner, A.M., Martin, L.A., et al. 2000. Nitric-oxide dioxygenase activity and function of flavohemoglobins. *Journal of Biological Chemistry,* 275, 31581–31587.

Ghiglione, J.-F., Gourbiere, F., Potier, P., et al. 2000. Role of respiratory nitrate reductase in ability of *Pseudomonas fluorescens* YT101 to colonize the rhizosphere of maize. *Applied and Environmental Microbiology,* 66, 4012–4016.

Giba, Z., Grubisic, D., Konjevic, R. 2007. Seeking the role of NO in breaking seed dormancy. In: *Nitric Oxide in Plant Growth, Development and Stress Physiology,* Plant Cell Monographs (Lamattina, L. and Polacco, J., Eds.), pp. 91–112. Springer-Verlag GmbH & Co, Berlin and Heidelberg, Germany.

Gilles-Gonzalez, M.A., Gonzalez, G. 2005. Heme-based sensors: defining characteristics, recent developments, and regulatory hypotheses. *Journal of Inorganic Biochemistry,* 99, 1–22.

Glick, B.R. 1995. The enhancement of plant growth by free-living bacteria. *Canadian Journal of Microbiology,* 41, 109–117.

Glick, B.R., Penrose, D.M., Li, J. 1998. A model for the lowering of plant ethylene concentrations by plant growth promoting bacteria. *Journal of Theoretical Biology,* 190, 63–68.

Gray, E.J., Smith, D.L. 2005. Intracellular and extracellular PGPR: commonalities and distinctions in the plant-bacterium signalling processes. *Soil Biology and Biochemistry,* 37, 395–410.

Graziano, M., Lamattina, L. 2005. Nitric oxide and iron in plants: an emerging and converging story. *Trends Plant Science,* 10, 4–8.

Grün, S., Lindermayr, C., Sell, S., et al. 2006. Nitric oxide and gene regulation in plants. *Journal of Experimental Botany,* 57, 507–516.

Guo, F.-Q., Okamoto, M., Crawford, M.J. 2003. Identification of a plant nitric oxide synthase gene involved in hormonal signalling. *Science,* 302, 100–103.

Gusarov, I., Nudler, E. 2005. NO-mediated cytoprotection: instant adaptation to oxidative stress in bacteria. *Proceedings of the National Academy of Sciences,* 102, 13855–13860.

Hallmann, J., Quadt-Hallmann, A., Mahaffee, W.F., et al. 1997. Bacterial endophytes in agricultural crops. *Canadian Journal of Microbiology,* 43, 895–914.

Harutyunyan, E.H., Safonova, T.N., Kuranova, I.P., et al. 1996. The binding of carbon monoxide and nitric oxide to leghaemoglobin in comparison with other haemoglobins. *Journal of Molecular Biology,* 264, 152–161.

Hérouart, D., Badouin, E., Frendo, P., et al. 2002. Reactive oxygen species, nitric oxide and glutathione: a key role in the establishment of the legume-*Rhizobium* symbiosis? *Plant Physiology and Biochemistry,* 40, 619–624.

Hirsch, A.M., Bauer, W.D., Bird, D.M., et al. 2003. Molecular signals and receptors: controlling rhizosphere interactions between plants and other organisms. *Ecology,* 84, 858–868.

Hu, X., Neill, S.J., Tang, Z., et al. 2005. Nitric oxide mediates gravitropic bending in soybean roots. *Plant Physiology,* 137, 663–670.

Jacobsen, C.S. 1997. Plant protection and rhizosphere colonization of barley by seed inoculated herbicide degrading *Burkholderia* (*Pseudomonas*) *cepacia* DBO1 (pRO101) in 2,4-D contaminated soil. *Plant and Soil,* 189, 139–144.

Jain, D.K., Patriquin, D.G. 1984. Root hair deformation, bacterial attachment, and plant growth in wheat-*Azospirillum* associations. *Applied and Environmental Microbiology,* 48, 1208–1213.

Jasid, S., Simontacchi, M., Bartola, C.G., et al. 2006. Chloroplasts as a nitric oxide cellular source. Effect of reactive nitrogen species on chloroplastic lipids and proteins. *Plant Physiology,* 142, 1246–1255.

Jetten, M.S.M. 2001. New pathways in ammonia conversion in soils and aquatic systems. *Plant and Soil,* 230, 9–19.

Kapulnik, Y,, Okon, Y., Henis, Y. 1985. Changes in root morphology of wheat caused by *Azospirillum* inoculation. *Canadian Journal of Microbiology,* 31, 881–887.

Kemmitt, S.J, Wright, D., Goulding, K.W.T, et al. 2006. pH regulation of carbon and nitrogen dynamics in two agricultural soils. *Soil Biology & Biochemistry,* 38, 898–911.

Kers, J.A., Wach, M.J., Krasnoff, S.B., et al. 2004. Nitration of a peptide phytotoxin by bacterial nitric oxide synthase. *Nature,* 429, 79–82.

Kloepper, J.W., Lifshitz, R., Zablotowicz, R.M. 1989. Free-living bacterial inocula for enhancing crop productivity. *Trends Biotechnology,* 7, 39–44.

Kolbert, Z., Bartha, B., Erdei, L. 2008. Osmotic stress- and indole-3-butyric acid-induced NO generation are partially distinct processes in root growth and development in *Pisum sativum. Physiologia Plantarum,* 133, 406–416.

Kuenen, J.G., Robertson, L.A. 1994. Combined nitrification-denitrification processes. *FEMS Microbiology Reviews,* 15, 109–117.

Kundu, B.S., Sangwan, P., Sharma, P.K., et al. 1997. Response of pearl-millet to phytohormones produced by *Azospirillum brasilense. Indian Journal of Plant Physiology,* 2, 101–104.

Kuzma, M.M., Hunt, S., Layzell, D.B. 1993. Role of oxygen in the limitation and inhibition of nitrogenase activity and respiration rate in individual soybean nodules. *Plant Physiology,* 101, 161–169.

Kwiatkowski, A.V., Shapleigh, J.P. 1996. Requirement of Nitric Oxide for Induction of Genes Whose Products Are Involved in Nitric Oxide Metabolism in *Rhodobacter sphaeroides* 2.4.3. *Journal of Biological Chemistry,* 271, 24382–24388.

Lamattina, L., Garcia-Mata, C., Graciano, M., et al. 2003. Nitric oxide: the versatility of an extensive signal molecule. *Annual Review in Plant Biology,* 54, 109–136.

Lambrecht, M., Okon, Y., van de Broek, A., et al. 2000. Indole-3-acetic acid: a reciprocal signalling molecule in bacteria-plant interactions. *Trends in Microbiology,* 8, 298–300.

Lanteri, M.L., Pagnussat, G.C., Lamattina, L. 2006. Calcium and calcium dependent protein kinases are involved in nitric oxide- and auxin-induced adventitious root formation in cucumber. *Journal of Experimental Botany,* 57, 1341–1351.

Levanony, H., Bashan, Y. 1989. Enhancement of cell division in wheat root tips and growth of root elongation zone induced by *Azospirillum brasilense* Cd. *Canadian Journal of Botany,* 67, 2213–2216.

Lombardo, M.C., Graziano, M., Polacco, J.C., et al. 2006. Nitric oxide functions as a positive regulator of root hair development. *Plant Signalling and Behaviour,* 1, 28–33.

Malamy, J. 2005. Intrinsic and environmental factors regulating root system growth. *Plant, Cell and Environment,* 28, 67–77.

Mantelin, S., Touraine, B. 2004. Plant growth-promoting bacteria and nitrate availability: Impacts on root development and nitrate uptake. *Journal of Experimental Botany,* 55, 27–34.

Mantelin, S., Desbrosses, G., Larcher, M., et al. 2006. Nitrate-dependent control of root architecture and N nutrition are altered by a plant growth-promoting *Phyllobacterium* sp. *Planta,* 223, 591–603.

Manthey, J.A., McCoy, D.L., Crowley, D.E. 1994. Stimulation of rhizosphere iron reduction and uptake in response to iron deficiency in citrus rootstocks. *Plant Physiology and Biochemistry,* 32, 211–215.

Mathesius, U., Schlaman, H.R.M., Spaink, H.P., et al. 1998. Auxin transport inhibition precedes root nodule formation in white clover roots and is regulated by flavonoids and derivatives of chitin oligosaccharides. *The Plant Journal,* 14, 23–34.

Mathieu, C., Moreau, S., Frendo, P., et al. 1998. Direct detection of radicals in intact soybean nodules: Presence of nitric oxide-leghemoglobin complexes. *Free Radical Biology and Medicine,* 24, 1242–1249.

Mehrer, I., Mohr, H. 1989. Ammonium toxicity: description of the syndrome in *Sinapis alba* and the search of it causation. *Physiologia Plantarum,* 77, 545–554.

Meiberg, J.B., Bruinenberg, M.P.M., Harder, W. 1980. Effect of dissolved oxygen tension on the metabolism of methylated amines in *Hyphomicrobium* X in the absence and presence of nitrate: evidence for aerobic denitrification. *Journal of General Microbiology,* 120, 453–463.

Mesa, S., Bedmar, E.J., Chanfon, A., et al. 2003. *Bradyrhizobium japonicum* NnrR, a Denitrification Regulator, Expands the *FixLJ-FixK2* Regulatory Cascade. *Journal of Bacteriology,* 185, 3978–3982.

Midha, S., Mishra, R., Aziz, M.A., et al. 2005. Cloning, expression, and characterization of a recombinant nitric oxide synthase-like protein from *Bacillus anthacis*. *Biochemical and Biophysical Research Communications,* 336, 346–356.

Molina-Favero, C., Creus, C., Barassi, C.A., et al. 2007. Nitric oxide and plant growth promoting rhizobacteria: Common features influencing root growth and development. In: *Advances in Botanical Research,* Vol. 46 (Kader, J.C. and Delseny, M., Eds.), pp. 1–33. Academic Press, London.

Molina-Favero, C., Creus, C.M., Simontacchi, M., et al. 2008. Aerobic nitric oxide production by *Azospirillum brasilense* Sp245 and its influence on root architecture in tomato. *Molecular Plant-Microbe Interaction,* 21, 1001–1009.

Moore, C.M., Nakano, M.M., Wang, T., et al. 2004. Response of *Bacillus subtilis* to nitric oxide and the nitrosating agent sodium nitroprusside. *Journal of Bacteriology,* 186, 4655–4664.

Moreau, M., Lee, G.I., Wang, Y. et al. 2008. AtNOS/AtNOA1 is a functional *Arabidopsis thaliana* cGTPase and not a nitric-oxide synthase. *Journal of Biological Chemystry*, 283, 32957–32967.
Mukhopadhyay, P., Zheng, M., Bedzyk, L.A., et al. 2004. Prominent roles of the NorR and Fur regulators in the *Escherichia coli* transcriptional response to reactive nitrogen species. *Proceedings of the National Academy of Sciences,* 101, 745–750.
Mulder, L., Hogg, B., Bersoult, A., et al. 2005. Integration of signaling pathways in the establishment of the legume-rhizobia symbiosis. *Physiologia Plantarum,* 123, 207–218.
Murgia, I., Delledonne, M., Soave, C. 2002. Nitric oxide mediates iron-induced ferritin accumulation in *Arabidopsis*. *The Plant Journal,* 30, 521–528.
Nakano, M.M. 2002. Induction of ResDE-Dependent Gene Expression in *Bacillus subtilis* in Response to Nitric Oxide and Nitrosative Stress. *Journal of Bacteriology,* 184, 1783–1787.
Okon, Y. 1985. *Azospirillum* as a potential inoculant for agriculture. *Trends in Biotechnology,* 3, 223–228.
Okon, Y., Kapulnik, Y. 1986. Development and function of *Azospirillum*-inoculated roots. *Plant and Soil,* 90, 3–16.
Okon, Y. 1994. *Azospirillum*/plant associations. CRC Press, Boca Raton, FL.
Okon, Y., Labandera-González, C.A. 1994. Agronomic applications of *Azospirillum*: An evaluation of 20 years worldwide field inoculation. *Soil Biology and Biochemistry,* 26, 1591–1601.
Osmont, K.S., Sibout, R., Hardtke, C.S. 2007. Hidden Branches: Developments in Root System Architecture. *Annual Review of Plant Biology,* 58, 93–113.
Pagnussat, G.C., Simontacchi, M., Puntarulo, S., et al. 2002. Nitric oxide is required for root organogenesis. *Plant Physiology,* 129, 954–956.
Pagnussat, G.C., Lanteri, M.L., Lamattina, L. 2003. Nitric oxide and cyclic GMP are messengers in the IAA-induced adventitious rooting process. *Plant Physiology,* 132, 1241–1248.
Pagnussat, G.C., Lanteri, M.L., Lamattina, L. 2004. Nitric oxide mediates the indole acetic acid induction activation of a mitogen-activated protein kinase cascade involved in adventitious root development. *Plant Physiology,* 135, 279–286.
Paul, E.A., Clark, F.E. 1996. Ammonification and nitrification. In: *Soil Microbiology and Biochemistry* (Paul, E.A. and Clark, F.E., Eds.) 2nd ed. pp. 191–197. Academic Press, New York.
Pereyra, M.A., Zalazar, C.A., Barassi, C.A. 2006. Root phospholipids in *Azospirillum*-inoculated wheat seedlings exposed to water stress. *Plant Physiology and Biochemistry,* 44, 873–879.
Peterson, P.J.L., Farquhar, M.L. 1996. Roots hairs: Specialized tubular cells extending root surfaces. *The Botanical Review,* 62, 1–40.
Pii, Y., Crimi, M., Cremonese, G., et al. 2007. Auxin and nitric oxide control indeterminate nodule formation. *BMC Plant Biology,* 7, 21–31.
Pothier, J.F., Wisniewski-Dyé, F., Weiss-Gayet, M., et al. 2007. Promoter-trap identification of wheat seed extract-induced genes in the plant-growth-promoting rhizobacterium *Azospirillum brasilense* Sp245. *Microbiology,* 153, 3608–3622.
Pothier, J.F., Prigent-Combaret, C., Haurat, J., et al. 2008. Duplication of plasmid-borne nitrite reductase gene *nirK* in the wheat-associated plant growth-promoting rhizobacterium *Azospirillum brasilense* Sp245. *Molecular Plant-Microbe Interactions,* 21, 831–42.
Ragland, M., Theil, E.C. 1993. Ferritin (mRNA, protein) and iron concentrations during soybean nodule development. *Plant Mololecular Biolology,* 21, 555–560.

Remans, T., Nacry, P., Pervent, M., et al. 2006. The *Arabidopsis* NRT1.1 transporter participates in the signalling pathway triggering root colonization of nitrate-rich patches. *Proceedings of the National Academy of Sciences,* 103, 19206–19221.

Ren, B., Zhang, N., Yang, J., et al. 2008. Nitric oxide-induced bacteriostasis and modification of iron-sulfur proteins in *Escherichia coli. Molecular Microbiology,* 70, 953–964.

Rodriguez, H., Fraga, T., Bashan, Y. 2006. Genetics of phosphate solubilization and potential applications for improving plant growth-promoting bacteria. *Plant and Soil,* 287, 15–21.

Rovira, A.D. 1965. Plant root exudates and their influence upon soil microorganisms. In: *Ecology of Soil-Borne Plant Pathogens – Prelude to Biological Control* (Baker, K.F. and Snyder, W.C., Eds.), pp. 170–186. John Murray, London.

Ruan, H., Shen, W., Ye, M., et al. 2002. Protective effects of nitric oxide on salt stress-induced oxidative damage to wheat (*Triticum aestivum* L.) leaves. *Chinese Science Bulletin,* 47, 677–681.

Ryu, C.-M., Farag, M.A., Hu, C.-H., et al. 2003. Bacterial volatiles promote growth in *Arabidopsis. Proceedings of the National Academy of Sciences,* 100, 4927–4932.

Ryu, C-M., Hu, C-H., Locy, R.D., et al. 2005. Study of mechanisms for plant growth promotion elicited by rhizobacteria in *Arabidopsis thaliana. Plant and Soil,* 268, 285–292.

Sangwan, I., O'Brian, M.R. 1992. Characterization of delta-amino levulinic acid formation in soybean root nodules. *Plant Physiology,* 98, 1074–1079.

Santner, A. and Estelle M. 2009. Recent advances and emerging trends in plant hormone signalling. *Nature,* 459, 1071–1078.

Sari, M.-A., Moali, C., Boucher, J.-L., et al. 1998. Detection of a nitric oxide synthase possibly involved in the regulation of the *Rhodococcus* sp R312 nitrite hydratase. *Biochemical and Biophysical Research Communications,* 250, 364–368.

Schmidt, I., Steenbakkers, P.J.M., op den Camp, H.J.M., et al. 2004. Physiologic and proteomic evidence for a role of nitric oxide in biofilm formation by *Nitrosomonas europaea* and other ammonia oxizers. *Journal of Bacteriology,* 186, 2781–2788.

Schneiker-Bekel, S., Perlova, O., Kaiser, O., et al. 2007. Complete genome sequence of the mixobacterium *Sorangium cellulosum. Nature Biotechnology,* 25, 1281–1289.

Schultze, M., Kondorosi, A. 1998. Regulation of symbiotic root nodule development. *Annual Review of Genetics,* 32, 33–57.

Shatalin, K., Gusarov, I., Avetissova, E., et al. 2008. *Bacillus anthracis*-derived nitric oxide is essential for pathogen virulence and survival in macrophages. *Proceedings of the National Academy of Sciences,* 105, 1009–1013.

Shi, Q., Ding, F., Wang, X., et al. 2007. Exogenous nitric oxide protect cucumber roots against oxidative stress induced by salt stress. *Plant Physiology and Biochemistry,* 45, 542–550.

Singh, B.K., Millard, P., Whiteley, A.S., et al. 2004. Unravelling rhizosphere-microbial interactions: opportunities and limitations. *Trends Microbiology,* 12, 386–393.

Spaepen, S., Vanderleyden, J., Remans, R. 2007. Indole-3-acetic acid in microbial and microorganism-plant signalling. *FEMS Microbiology Reviews,* 31, 1–24.

Spiro, S. 2007. Regulators of bacterial responses to nitric oxide. *FEMS Microbiology Reviews,* 31, 193–211.

Steenhoudt, O., Vanderleyden, J. 2000. *Azospirillum* a free-living nitrogen fixing bacterium closely associated with grasses: genetic, biochemical and ecological aspects. *FEMS Microbiology Reviews,* 24, 487–506.

Steenhoudt, O., Keijers, V., Okon, Y., et al. 2001a. Identification and characterization of periplasmic nitrate reductase in *Azospirillum brasilense* Sp245. *Archive in Microbiology,* 175, 344–352.

Steenhoudt, O., Ping, Z., Vande Broek, A., et al. 2001b. A spontaneous chlorate-resistant mutant of *Azospirillum brasilense* Sp245 display defects in nitrate reduction and plant colonization. *Biology and Fertility of Soils*, 33, 317–322.

Streeter, J.G. 1987. Carbohydrate, organic acid, and amino acid composition of bacteroids and cytosol from soybean nodules. *Plant Physiology*, 85, 768–773.

Stuehr, D.J. 1997. Structure-function aspects in the nitric oxide synthases. *Annual Review of Pharmacology and Toxicology*, 37, 339–359.

Sturz, A.V., Christie, B.R., Nowak, J. 2000. Bacterial endophytes: Potential role in developing sustainable systems of crop production. *Critical Review Plant Science*, 19, 1–30.

Sudhamsu, J., Crane, B.R. 2009. Bacterial nitric oxide synthases: what are they good for? *Trends in Microbiology*, 17, 212–218.

Theunis, M., Kobayashi, H., Broughton, W.J., et al. 2004. Flavonoids, NodD1, NodD2, and Nod-box NB15 modulate expression of the y4wEFG locus that is required for indole-3-acetic acid synthesis in *Rhizobium* sp. strain NGR234. *Molecular Plant-Microbe Interaction*, 17, 1153–1161.

Tokala, R.K., Strap, J.L., Jung, C.M., et al. 2002. Novel plant-microbe rhizosphere interaction involving *Streptomyces lydicus* WYEC108 and the pea plant (*Pisum sativum*). *Applied and Environmental Microbiology*, 68, 2161–2171.

Torsvik, V., Øvreås, L. 2002. Microbial diversity and function in soil: from genes to ecosystems. *Current Opinion in Microbiology*, 5, 240–245.

Travis, J. 2004. NO-making enzyme no more: Cell, PNAS papers retracted. *Science*, 306, 960.

van Loon, L.C. 2007. Plant responses to plant growth-promoting rhizobacteria. *European Journal of Plant Pathology*, 119, 243–254.

Van Spanning, R.J.M., Houben, E., Reijnders, W.N.M., et al. 1999. Nitric oxide is a signal for NNR-mediated transcription activation in *Paracoccus denitrificans*. *Journal of Bacteriology*, 181, 4129–4132.

Verma, D.P.S., Long, S. 1983. The molecular biology of *Rhizobium*-legume symbiosis. In: *International Review of Cytology*, Suppl. 14. (Jeon, K., Ed.), pp. 211–245. Academic Press, New York.

Vollack, K.-I., Zumft, W.G. 2001. Nitric oxide signalling and transcriptional control of denitrification genes in *Pseudomonas stuzeri*. *Journal of Bacteriology*, 183, 2516–2526.

Wach, M.J., Kers, J.A., Krasnoff, S.B., et al. 2005. Nitric oxide synthase inhibitors and nitric oxide donors modulate the biosynthesis of thaxtomin A, a nitrated phytotoxin produced by *Streptomyces* spp. *Nitric Oxide*, 12, 46–53.

Walch-Liu, P., Forde, B.G. 2008. Nitrate signalling mediated by the NRT1.1 nitrate transporter antagonises L-glutamate-induced changes in root architecture. *The Plant Journal*, 54, 820–828.

Walch-Liu, P., Ivanov, I.I., Filleur, S., et al. 2006. Nitrogen regulation of root branching. *Annals of Botany*, 97, 875–881.

Wang, Y., Yang, Q., Tosa, Y., et al. 2001. Nitric oxide-overproducing transformants of *Pseudomonas fluorescens* with enhanced bio-control of tomato bacterial wilt. *Journal of General Plant Pathology*, 71, 33–38.

Wang, Z.-Q., Lawson, R.J., Buddha, M.R., et al. 2007. Bacterial flavodoxins support nitric oxide production by *Bacillus subtilis* nitric-oxide synthase. *Journal of Biological Chemistry*, 282, 2196–2202.

Watt, M., Silk, W.K., Passioura, J.B. 2006. Rates of root and organism growth, soil conditions, and temporal and spatial development of the rhizosphere. *Annals of Botany*, 97, 839–855.

Welbaum, G.E., Sturz, A.V., Dong, Z., et al. 2004. Managing soil micro-organisms to improve productivity of agroecosystems. *Critical Reviews in Plant Sciences,* 23, 175–193.

Wrage, N., Velthof, G.L., van Beusichem, M.L., et al. 2001. Role of nitrifier denitrification in the production of nitrous oxide. *Soil Biology and Biochemistry,* 33, 1723–1732.

Xia, Y., Embley, T.M., O'Donnell, A.G. 1994. Phylogenetic analysis of *Azospirillum* by direct sequencing of PCR-amplified 16S rDNA. *Systematic and Applied Microbiology,* 17, 197–201.

Yamasaki, H., Sakihama, Y., Takahashi, S. 1999. An alternative pathway for nitric oxide production in plants: new features of an old enzyme. *Trends in Plant Science,* 4, 128–129.

Zemojtel, T., Frohlich, A., Palmieri, M.C., et al. 2006. Plant nitric oxide synthase: a never-ending story? *Trends in Plant Science,* 11, 524–525.

Zemojtel, T., Wade, R.C., Dandekar, T. 2003. In search of prototype of nitric oxide synthase. *FEBS Letters,* 554, 1–5.

Zhang, H.M., Forde, B.G. 1998. An *Arabidopsis* MADS box gene that controls nutrient-induced changes in root architecture. *Science,* 279, 407–409.

Zhang, H.M., Jennings, A., Barlow, P.W., et al. 1999. Dual pathways for regulation of root branching by nitrate. *Proceedings of the National Academy of Sciences,* 96, 6529–6534.

Zhao, M.-G., Tian, Q.-Y., Zhang, W.-H. 2007. Nitric oxide synthase-dependent nitric oxide production is associated with salt tolerance in *Arabidopsis. Plant Physiology,* 144, 206–217.

Zumft, W.G. 1993. The biological role of nitric oxide in bacteria. *Archives of Microbiology,* 160, 253–264.

Zumft, W.G. 1997. Cell biology and molecular basis of denitrification. *Microbiological and Molecular Biology Reviews,* 61, 533–616.

Index

Page references followed by f denote figures; those followed by t denote tables.

Abscisic acid
   as negative regulator of rhizobial infection, 369
   plant responses to nod factors, 362
   regulation of symbiosome and arbuscule development, 375
   role in arbuscular mycorrhizal infection in tomato, 362
Acacia
   A. constricta, ant associations with, 310, 312
   arbuscular mycorrhizal symbiosis, 201, 202f
   use in revegetation of degraded areas, 201
Acaulospora morrowiae, 201
ACC, 239–40
ACC deaminase, 238–40
Acceleration hypothesis, 267–8
Acetylene reduction activity, 242, 243f
Acidic phosphatases (APases), in the rhizosphere, 103
Actinorhizal plants, 121–2
Actinorhizal symbioses, 117–30, 143
   gas exchange and oxygen protection in nodules, 125–6
   macrosymbionts, 121–9
   microsymbionts, 117–21
      Frankia carbon and nitrogen metabolism, 120–121
      Frankia host specificity, 119–20, 119f
      nodule induction and structure, 122–4, 123f
   nodule metabolism, 125–9, 128f
   overview, 117
Adventitious roots

ant-plant symbiosis, 314–15, 315f
   auxin and nitric oxide linked to formation of, 400–401
Agapeta zoegana, 265
Aglaophyton major, 20
Agrobacterium, 169, 170
   A. rhizogenes, 169, 172
   A. tumefaciens, 171
   cytokinin production by, 222
   opine production by, 224
Agropyron, resource allocation after herbivory, 260, 261
AHL (*N*-acyl-homoserine lactone) system, 235
Alaska
   alder *(Alnus)* in, 149–59
   map of, 150f
Alder *(Alnus)*
   actinorhizal symbioses, 118, 122, 124–6
   in Alaska, 149–59
      alder in Alaskan ecosystems, 151–2
      distribution, habits, and habitats of Alaskan alder, 149–51
      ecology of alder-*Frankia* interaction, 152–9
   arbuscular-mycorrhizal (AM) interactions, 158
   distribution, 140
   ectomycorrhizal (EM) interactions, 158
   general biology of, 140
   herbivory, 158–9
   in interplanting systems, 129
   nitrogen-fixation, 140–142
      regulation of, 142
   taxonomy, 140, 141f

Ecological Aspects of Nitrogen Metabolism in Plants, First Edition. Edited by Joe C. Polacco and Christopher D. Todd.
© 2011 by John Wiley & Sons, Inc. Published 2011 by John Wiley & Sons, Inc.

Alder-*Frankia* symbiosis, 138–59
  alder-mycorrhizae interactions, effects of, 158
  ecology of alder-*Frankia* interaction, 152–9
    ecosystem effects, 157–9
    field associations, 152–3, 154*t*, 155*f*–6*f*
    host physiology, 152–3, 157*t*
  Frankia phylogenetics and host specificity, 143, 144*t*, 145
  host fitness, effects on, 147–8
  host intraspecific variation in *Frankia* compatibility, 147
  as mutualism, 145
Alkaloids, 224
Allocasuarina, actinorhizal symbioses and, 118–19, 127
Allorhizobium, 169, 170
Alnobetula, 140
α-Proteobacteria, 198, 199
Amanita muscaria, 71, 75–8
AMF. *See* Arbuscular mycorrhizal fungi
Amino acid transporter, 75
Amino acid utilization, ectomycorrhizal fungi, 75
1-aminocyclopropane-1-carboxylate (ACC) deaminase, 238–40
Ammonia
  leaves as source or sink for atmospheric, 219
  in nitrogen cycle, 7
  oxidation to nitric acid, 8
  produced by Haber-Bosch process, 5, 8, 167
  use for biosynthesis, 7
  as waste product of phylloplane bacteria, 224
Ammonification, in nitrogen cycle, 7
Ammonium
  in actinorhizal symbioses, 127, 129
  in ectomycorrhizal (EM) interactions, 76–8, 78*f*
  effects on root development, 399–400
Ammonium transporters, 22
  in arbuscular mycorrhizal symbiosis, 55–7, 58*f*, 59
  in ectomycorrhizal fungi, 77, 78*f*
Andira spp., 373
Angoumois grain moth *(Sitotroga cerealella)*, 285
Ant gardens, 314
Anthocyanin
  functions of, 334
  produced by phenylpropanoid pathway, 334, 336, 336*f*
  structure of, 334*f*
Anthoxanthum odoratum, resource allocation after herbivory, 260

Ant-symbiosis, 308–23
  ecological factors influencing importance and quality of provided nutrients, 318–19, 319*f*
  evidence for uptake of ant-provided nutrients, 309–12
    in desert systems, 312
    isotope tracer studies, 310, 312
    soil improvements by ant activities, 309–10
    table of, 311*t*
  as evolutionary adaptations for nutrient concentration and provisioning, 319–22
  mechanisms of nutrient concentration and assimilation, 313–18
    nitrogen uptake in leaves, 315–17
    nitrogen uptake in roots, 313–15, 315*f*
    nitrogen uptake in stems, 317–18
  myrmecophytes, 309, 310, 312–13
  phylogenetics of, 321*f*
  specializations for nutrient uptake in roots, 314–15
    adventitious roots, 314–15, 315*f*
    ant gardens, 314
    colony waste and ground-nesting organisms, 314
  specializations for nutrient uptake through leaves, 316–17
    leaf lamina, 316
    petiole, 316
    trichomes, 317
APases (acidic phosphatases), in the rhizosphere, 103
Apoplast
  as environment for nitrogen-fixing bacteria, 221
  nitrogen compounds in, 219
  proteins in, 98, 100
Appressorium, 21
Arabidopsis thaliana
  infection with *Pseudomonas*, 100, 102*f*
  NO_3^- ion effects on root development, 400
  TT12 transporter, 337
Arabinogalactan proteins, in the rhizosphere, 104
Arachis hypogaea, nodulation of, 179
Arbuscular mycorrhizal fungi (AMF)
  diversity of, 201
  hyphal branching in, 358–9
Arbuscular mycorrhizal (AM) symbiosis, 52–61
  alder *(Alnus)*, 158

bacterial roles in, 61
benefits of, 200–201
diversity of, 356
genomic analysis of nitrogen nutrition, 60–61
nitrogen as signaling molecule, 59–60
nitrogen movement, 53–7
  nitrogen acquisition and assimilation, 53–6
  nitrogen transfer from fungus to host plant, 56–9
nutrient exchange processes, 22–3, 54f
occurrence of, 200
phosphorus acquisition, 52
phosphorus transporters, 60–61
plant hormones and, 354–76
  flavonoids and isoflavonoids, 358
  strigolactones, 358–9
specificity in symbioses with higher plants, 200–201
Arbuscular mycorrhization, 20–29
  evolution of, 20
  genetic network controlling, 23–9, 24t–6t, 27f, 28f
  mycorrhizal infection process, 20–22
  nutrient exchange between fungus and plant, 22–3
  process, 367–8
Arbuscule
  described, 21–2, 373–4
  genetics of development, 375
  hormonal regulation of development, 374–5
Arginase, 55
Arum maculatum, 22
Arum-type arbuscules, 22
Assimilatory nitrate reduction, 7
Atmosphere, nitrogen gas concentration in, 5
Auxin
  auxin/cytokinin ratio, 370
  effect of flavonoids on transport, 340–342, 357–8, 370
  effect on strigolactone synthesis, 21
  endophyte production of, 238–9
  production by plant-associated microbes, 222
  regulation of arbuscule development, 375
  role in nodulation, 342–4
  signaling, 341
Azoarcus, 168
  A. BH72, 231, 232
  type IV pili, 234
Azorhizobium, 169–71
Azospirillum, 168
  A. brasilense
    colonization by, 234–7

plant growth promotion by, 238–9, 407–10, 409f
  A. irakense, 236
  industrial uses of, 244
  plant growth promotion by, 398
Azotobacter, 168
Azteca ants association with *Cecropia* trees, 313, 319

Bacillus amyloliquefaciens, volatile organic compound production by, 240
Bacillus anthracis, 404
Bacillus subtilis, 240, 404
Bacillus thuringiensis, 293
Bacteria
  endophytes of grasses and cereals, 231–44
  nitric oxide (NO)
    functions in bacteria, 403–4
    production in bacteria, 401–3, 402f
  nitrogen transformations that influence root development, 400–401
  roles in arbuscular mycorrhizal (AM) symbiosis, 61
Bacteroids
  description of, 30–31
  metabolic differentiation in, 182–3
Barley bifunctional inhibitor, 284–5
Betaine, 359
β-Proteobacteria, 198–9
Bifunctional inhibitors, 284–5
Biochanin A, 358
Biocontrol, 240
Biodiversity, of nitrogen strategies, 10
Biofilm, nitric oxide (NO) function in formation of, 403–4
Biological nitrogen fixation (BNF)
  endophytes, 237, 238
  in nitrogen cycle, 6–7
  selection of *Rhizobium* strains for high efficiency, 199–200
Bosch, Carl, 5
Bouteloua gracililis, 264
Bowman-Birk inhibitors, 285
Brachiaria decumbens, 202
Bradyrhizobium, 168–70, 198
  B. betae, 170
  B. japonicum, 179–83
    determinate nodule formation with soybean, 179–83
  nodABC operon of, 33
Branching factor, 358
Brassica napus, apoplant nitrogen content, 219

Brazil
  seriously degraded soils
    case studies, 203–9
    extent of problem, 195–6
    tree seedling preparation for recovery of degraded areas, 202–3
  surveys of the nodulation of legume species in, 200
Broccoli, foliar fertilization of, 219
Burkholderia, 171–2, 199

Caesalpinioideae, 197
Calcium calmodulin kinase *(CCAMK)*, 23, 26–9, 364
Callosobruchus maculatus, 293
Campsiandra comosa, indeterminate nodules from, 198
Carbamate kinase, in the rhizosphere, 97*t*
Carbon allocation following insect herbivory, 258–65
  after damage to aboveground tissues, 260–263
  after damage to belowground tissues, 263–5
  constrained or flexible allocation responses?, 259
Carbon dioxide
  fixation in actinorhizal symbioses, 127
  role in hyphal growth, 360
Carbon:nitrogen (C:N) ratio, 84*f*
Carboxypeptidase inhibitor, potato, 286–7
Carrot *(Daucus carota)*, 22
Casaurina, 197
Cascade effect of nitrogen, 9–11
CASTOR (cation channel), 26–9, 364
Casuarina
  actinorhizal symbioses, 118–20, 118*f*, 122–3, 123*f*, 125, 127
  in interplanting systems, 129
Catalase, in the rhizosphere, 97*t*
CCaMK, 26–9, 364
Ceanothus, actinorhizal symbioses and, 118, 143
Cecropia trees, ant association with, 313, 317, 319, 320
Centaurea, herbivory effects on, 263, 265
Cereals, nitrogen-fixing endophytes of, 231–44
C1 factor, maize, 338
Chalcone isomerase, 335, 336*f*
Chalcone reductase, 336, 336*f*
Chalcone synthase, 335, 336*f*
Cheater hypothesis, 39–41
Chitinases
  of ectomycorrhizal fungi, 73–4
  endophyte production of, 241
  nodulation and, 35
  secreted in the rhizosphere, 96, 96*t*, 103, 106
Cicadas, 264
Cinnamate 4-hydroxylase, 335, 336*f*
Clethropsis, 140
Co-adaptation of partners in rhizobial symbiosis, 41–2
Colorado potato beetle *(Leptinotarsa decemlineata)*, inhibition by plant peptidase inhibitors, 286, 289, 292, 296
Competence zone, 361
Comptonia, actinorhizal symbioses and, 125
Cordia alliadora, 318, 319*f*
Coriaria, actinorhizal symbioses and, 118, 122, 126
Cortical cell divisions, regulation of, 369–71
4-Coumaroyl-CoA ligase, 335, 336*f*
Coumestrols, 335
Cupriavidus, 172
Cuticle, diffusion of water across plant, 218–19
Cyanobacteria, 168
Cycling of elements, 6. *See also* Nitrogen cycle
Cyclopia, 172
CYCLOPS, 26–9, 364
Cynosurus cristatus, resource allocation after herbivory, 260
Cyphocleomus achates, 265
Cystatins, 286
Cytisus scoparius, 172–3
Cytokinin
  auxin/cytokinin ratio, 370
  effect on infection thread development, 176
  induced nutrient mobilization, 222
  production by plant-associated microbes, 222, 223
  receptor, 34
  regulation of symbiosome and arbuscule development, 375
  signaling role in nodulation by nitrogen-fixing bacteria, 222

Daidzein, 336, 336*f*, 356
Dalbergia louvelli, 172
Datisca, actinorhizal symbioses and, 118, 122, 123*f*, 126, 127
Defense protein secretion in the rhizosphere, 105

Denitrification, 402
Determinate nodule, 197, 197f
 auxin transport and, 358
 flavonoid function in determinate nodulating
  plants, 344–5
 formation between *Bradyrhizobium japonicum*
  and soybean, 179–83
 metabolic differentiation in bacteroids,
  182–3
 metabolic differentiation of soybean root
  nodule cells, 180–181
 metabolic differentiation of symbiosome
  membrane and symbiosome space,
  181–2
 soybean nodule formation and morphology,
  179–80
 spherical, 197, 197f
Devosia, 172, 179
Diazotrophs, 167–8
 endophytes of grasses and cereals, 231–44
Dihydroflavonol reductase, 336, 336f
Dihydroflavonols, 336
Dischidia major, 314–15, 315f
Dissolved organic nitrogen (DON), 71–2
DM12, 364, 374
DM13, 364, 369–71, 372f
Dolichos biflorus, lectin nucleotide
 phosphohydrolase of, 363
Dryas, actinorhizal symbioses and,
 118, 122
Duroia, 316

Ectomycorrhizal (EM) interactions, 69–88
 alder *(Alnus)*, 158
 arbuscular mycorrhizal fungi compared,
  69–70
 mycelium photograph, 70f
 nitrogen provision to host tree, 69–88
  inorganic nitrogen mobilization, 76–80
   ammonium, 76–8, 78f
   nitrate utilization, 78–9
   urea utilization, 80
  nitrogen transfer at the mycorrhizal
   interface, 80–86
    evidence from ^{15}N data, 83–4, 83t, 84f
    in field studies, 80–81, 82f
    in microcosms, 81–2
    molecular data supporting exchanges,
    84–6
  organic nitrogen mobilization, 73–6
   amino acid utilization, 74–5
   hydrolysis by ECM fungi, 73–4

   peptide utilization, 75–6
  origin of, 69–70
 EF-Tu, in the rhizosphere, 97t
 Ehrlich, Paul *(The Population Bomb)*, 195
 Elaeagnus, in interplanting systems, 129
 Endomycorrhizae. *See also* Arbuscular
  mycorrhization
  evolution of, 20
  symbiosis, specificity in, 38
 Endophytes
  auxin production by, 238–9
  common features of, 231–2
  defined, 231
  ethylene modulation by, 239–40
  extracellular plant growth-promoting
   rhizobacteria (ePGPR), 398
  genes related to colonization and plant
   response, 241–2
  genomes of, 231
  identification of, steps required in, 232–3
  mechanisms of plant growth promotion,
   237–41, 237t
   direct, 238–40
   indirect, 240–241
  mode of colonization, 234–7
  nitrogen-fixing of grasses and cereals, 231–44
  occurrence, distribution, and plant
   localization, 232–3, 233f
 ENOD11, 361–2, 362f, 364–7, 369
 ENOD12, 361
 ENOD40, 364, 372f, 374
 Ensifer, 169, 170
 EPGPR. *See* Extracellular plant
  growth-promoting rhizobacteria
 Epidermal responses to symbiotic microbes,
  360–369
 Epiphytic tank bromeliads, 220
 Eriogonum inflatum, 317
 ERN proteins/genes, 33, 366, 367
 Erosion gulleys, recuperation of, 203–5, 204f
 Erwinia hericola, IAA synthesis in, 222
 Esterases, secreted in the rhizosphere, 96
 Ethylene
  effect on root hair invasion, 176
  modulation by endophytic bacteria, 239–40
  as negative regulator of rhizobial infection,
   368–9
  plant responses to nod factors, 361–2
  root hair initiation in *Arabidopsis*, 361
 European corn borer *(Ostrinia nubilalis)*,
  inhibition by plant peptidase
  inhibitors, 288

Eutrophication, 10–11
Extensin, 104
Extracellular plant growth-promoting
      rhizobacteria (ePGPR)
   described, 397–8
   nitric oxide in, 406–7

Fabaceae. See also Legumes
   evolution of symbiosis in, 35–6
   occurrence of *Rhizobium* symbiosis in, 29–37
   phylogenetics of, 29, 30f
Fabids, 121
FAO (Food and Agriculture Organization), 9, 195
Fatty acid epoxide hydroxylase, 181
Ferric uptake regulator *(fur)*, 404
Fertilizer
   loss of synthetic from croplands, 167
   population growth matched to use of, 9
   produced by Haber-Bosch process, 5, 8
   subsidizing, 9
Festuca rubra, resource allocation after
      herbivory, 260
FISH (fluorescent *in situ* hybridization) analysis, 237
Fixation genes in *Rhizobium*, 31–2
Fix gene, 31, 405
Flagellin, in the rhizosphere, 97t
Flavan 3-ol
   functions of, 334
   structure of, 334f
Flavanol, structure of, 334f
Flavanones
   in phenylpropanoid pathway, 335, 336
   structure of, 334f
Flavone, structure of, 334f
Flavone-C-glucosyltransferase, 337
Flavone synthase, 336
Flavonoids
   as chemoattractants for rhizobia, 173
   effect on auxin transport, 340–342,
      357–8, 370
   as endogenous plant sunscreens, 334
   function in nodulation, 344–7, 346f
   functions in legume-rhizobia interactions,
      333–48
   as *nod* gene inducers, 338–40, 356–7,
      357f, 360
   phosphorus status of plant and, 358
   phytoalexins, 334, 335
   recognition by NodD proteins, 32
   root exudation of, 357f
   stress responses of, 334

Flavonol synthase, 336, 336f
FlgK, in the rhizosphere, 97t
Fluorescent *in situ* hybridization (FISH)
      analysis, 237
Foliar fertilization, nitrogen resources from,
      219–20
Food and Agriculture Organization (FAO),
      9, 195
Forest soils, nitrogen availability in, 71–3, 72f
Formica purpilosa association with *Acacia
      constricta*, 310, 312
Formononetin, 358
Fossil fuels
   energy efficiency of nitrogen in, 9
   release of oxidized nitrogen from, 5, 8
Founder effect, 39
Frankia
   actinorhizal symbioses, 117–18, 118f,
      123–9, 123f
   alder-*Frankia* symbiosis, 138–59
      ecology of alder-*Frankia* interaction in
         Alaska, 152–9
      ecosystem effects, 157–9
      field associations, 148–9, 152–3, 154t,
         155f–6f
      host physiology, 152–3, 157t
   carbon and nitrogen metabolism, 120–121
   free-living, 117, 120–121, 125, 142
   general biology of, 142–3
   genomes, 120, 129
   glycogen granules, 126
   host specificity, 119–20, 119f, 143–5, 144t
   ineffective strains, 146–7
   nitrogenase of, 117–18, 125
   phylogenetics, 143–4
   spore⁺ and spore⁻ strains, 118, 148
   variation in host interactions, 145–8
      effects on host fitness, 147–8
      host intraspecific variation in *Frankia*
         compatibility, 147
      variation in symbiotic behavior, 146–7
Fumarases, secreted in the rhizosphere, 96
Fur (ferric uptake regulator), 404

GDH/GS (glutamate dehydrogenase and
      glutamine synthetase) pathway,
      53, 78, 127
Genetic drift, 39
Genetics
   of root nodule organogenesis, 369–71
   of symbiosome and arbuscle development,
      374–5

Genistein, 336, 336f, 356, 359
Genomic analysis of nitrogen nutrition in arbuscular mycorrhizal (AM) symbiosis, 60–61
GH3 gene, 370
Gibberellins
  inhibition of infection thread formation by, 369
  regulation of symbiosome and arbuscule development, 375
Gigaspora clarroideum, strigolactones and, 359
Gigaspora intraradices, strigolactones and, 359
Gigaspora margarita, 56, 201, 202
  hyphal branching in, 359
  hyphopodia of, 367
  strigolactones and, 359
Gigaspora mosseae
  lipid transfer protein gene, 368
  strigolactones and, 359
Gigaspora rosea, strigolactones and, 359
GintAMT1 gene, 53
Glomalin, 96
Glomeromycota, 19, 20, 200
Glomus clarum, 201, 202
Glomus hoi, 56
Glomus intraradices, 23, 56, 60
Glomus mosseae, 23, 56
Glucanases, secreted in the rhizosphere, 96, 97t, 106
Gluconacetobacter diazotrophicus, 219, 231, 235
  acetylene reduction activity, 243, 243f
  carbon sources utilized by, 233, 234f
  genes related to colonization, 241
  occurrence in sugarcane varieties, 233, 233t
  transmission of, 233
Glucosinolates, 224
Glutamate dehydrogenase and glutamine synthetase (GDH/GS) pathway, 53, 78, 127
Glutamine synthetase, 180
Glutamine synthetase and glutamine oxoglutarate amino-transferase (GS/GOGAT), 53, 78, 127
Glycine max. See Soybean
Glycogen granules, in *Frankia,* 126
GmosAAP1 gene, 56
Grasses, nitrogen-fixing endophytes of, 231–44
Grazing optimization, 260
GS/GOGAT (glutamine synthetase and glutamine oxoglutarate amino-transferase), 53, 78, 127

Gunnera, cyanobacterial colonization of, 308
Gymnostoma, actinorhizal symbioses and, 125

Haber, Fritz, 5
Haber-Bosch process, 5, 8, 167
Hakea actites, 74
Hartig net, 70f, 71
HDZIPIII proteins, 342
Hebeloma spp., 71, 74–7, 79, 81, 82f, 85
Helicoverpa zea, inhibition by plant peptidase inhibitors, 288, 290, 294
Hemoglobin, in actinorhizal symbioses, 125–6
Herbaspirillum, 241–2
  H. rubrisubalbicans, 235–7
  H. seropedicae, 231, 236, 237
Herbivory
  of alder *(Alnus),* 158–9
  insect herbivory effects on nitrogen economy of plants, 257–70
    ecological consequences of N loss from plants following herbivory, 265–8, 268f
    ecological effects, 258
    evolutionary effects, 258
    nitrogen deposition and the effects on N economy, 268–9
    patterns of N and C allocation following herbivory, 258–65
  patterns of N and C allocation following, 258–65
    after damage to aboveground tissues, 260–263
    after damage to belowground tissues, 263–5
    constrained or flexible allocation responses?, 259
Hippophae rhamnoides, actinorhizal symbioses and, 122
Hopanoids, 118
Hormonal regulation
  of infection thread formation, 368–9
  of nodule organogenesis, 370
  of symbiosome and arbuscule development, 375–6
Hydathodes, 218
Hydnophytum spp., ant associations with, 310, 317, 320
Hydrolysis by ectomycorrhizal fungi, 73–4
Hydroxycinnamoyl transferase, 335, 336f
3-hydroxyl 3-methylglutaryl coenzyme reductase, 364, 365f
Hyphal branching, in arbuscular mycorrhizal fungi, 358–9

Hyphopodia, 367–8
Hypopodium, 21

IAA. *See* Auxin; Indole-3 acetic acid
Indeterminate nodule, 197, 198*f*
   auxin transport and, 358
   flavonoid function in indeterminate nodulating plants, 345–7
   infection thread growth, 178
Indole-3-acetamide, 239
Indole-3 acetic acid (IAA), 341
   endophyte production of, 238–9
   production by *Azospirillum,* 407–8
   production by plant-associated microbes, 222
   production by *Rhizobium,* 404–5
Indole-3-pyruvate, 239
Induced resistance, 101, 102*f*
Infection thread, 30, 33
   in actinorhizal nodulation, 124
   development process, 176–8, 177*f*
   signaling in formation of, 366–9
      arbuscular mycorrhization, 367–8
      hormone regulation of, 368–9
      legume nodulation, 366–7
   structure of, 176
Insects
   ant-plant symbiosis, 308–23
      ecological factors influencing importance and quality of provided nutrients, 318–19, 319*f*
      evidence for uptake of ant-provided nutrients, 309–12
      as evolutionary adaptations for nutrient concentration and provisioning, 319–22
      mechanisms of nutrient concentration and assimilation, 313–18
      myrmecophytes, 309, 310, 312–13
      nitrogen uptake in leaves, 315–17
      nitrogen uptake in roots, 313–15, 315*f*
      nitrogen uptake in stems, 317–18
      phylogenetics of, 321*f*
      specializations for nutrient uptake in roots, 314–15
      specializations for nutrient uptake through leaves, 316–17
   herbivory effects on nitrogen economy of plants, 257–70
      ecological consequences of N loss from plants following herbivory, 265–8, 268*f*
      ecological effects, 258
      evolutionary effects, 258
      nitrogen deposition and the effects on N economy, 268–9
      patterns of N and C allocation following herbivory, 258–65
   plant peptidase inhibitors, 280–298
      adaptations of insects to, 296–7, 298*f*
      as defensive compounds against phytophagous insects, 288
      effects on gut physiology of insects, 290–295
      families of, 281–8
      induction in response to insect attack, 288–90, 291*f*
Intracellular plant growth-promoting rhizobacteria (iPGPR)
   described, 397
   nitric oxide in, 404–6
IPD3, 364, 365*f*
Iron, nitrogenase and, 405–6
Iron mining waste, revegetation of, 208, 209*f*
Isoflavanone dehydratase, 336
Isoflavones
   in legumes, 335
   structure of, 334*f*
Isoflavone synthase, 336, 336*f*
Isoflavonoids
   as inducers of *nod* gene expression, 356, 357*f*
   role in defense against pathogens, 335
   root exudation of, 357*f*
Isoliquiritigenin, 336, 336*f*

Jasmonates, 359
Jasmonic acid
   plant responses to nod factors, 362
   regulation of symbiosome and arbuscule development, 375
   signaling, 101, 102*f*

Klebsiella pneumoniae, 342, 231, 235, 238
Kunitz inhibitors, 283

Lacccaria bicolor, 53, 86
   amino acid transport, 75
   ammonium transporter, 77–8, 78*f*
   genome sequencing, 71
   nitrate transporter, 79
   peptide utilization, 75–6
Laskowski mechanism of protease inhibition, 283, 284, 296

Lateral root development
 auxin and, 340–341, 400–401
 nodulation, similarities and differences with, 371
Leaf
 nitrogen uptake in ant-symbiosis, 315–17
 phylloplane microbes, 217–24
Lectin nucleotide phosphohydrolase, 363
Lectins
 in the rhizosphere, 97*t*, 360
 roles in rhizobia-legume symbiosis, 360
Leghemoglobin, 31, 168, 180, 405
Legumes
 Arum-type arbuscules, 22
 flavonoid functions in legume-rhizobia interactions, 333–48
 nodulation
  nodule types and classification, 197, 197*f*, 198*f*
  plant hormones and, 354–76
   flavonoids and isoflavonoids, 356–8
  non-rhizobia able to induce nodules, 171–3
  phylogenetics of, 29, 30*f*
  in recovery of seriously degraded soils, 195–210
  rhizobial diversity and, 168–70
  rhizobial symbiosis, 29–37
   host specificity and promiscuity, 198–9
Leucine aminopeptidase, 74
Leucoanthocyanindins, 336, 336*f*
LHK1, 370–371, 372*f*
Lignins, produced by phenylpropanoid pathway, 333–4
Limiting factor for growth, nitrogen as, 10
Lipase, secreted in the rhizosphere, 96*t*
Lipid transfer proteins, 96*t*, 368
Lipoxygenase, 181
Lon protease, 103
Lotus japonicus
 ammonium transporter of, 22, 56
 genetic studies, 23, 24*t*, 27–9
Lupinus, nodulation of, 172, 173
Luteolin, 356
Lysophosphatidylcholine, 375–6

Macaranga, 322
Machaerium lunatum, β-Proteobacteria and, 198
Magicicada, 264
Maieta, 320
Maize
 resource allocation after herbivory, 262
 rhizobial endophytes of, 171

Major intrinsic proteins (MIPs), in mycorrhizal plants, 57, 59
Manure management, improvements in, 12
Medicago sativa, 96*t*–7*t*, 100–101, 102*f*
Medicago truncatula
 ammonium transporter genes, 57
 auxin effect on nodulation, 343–4
 flavonoid function in nodulation, 345–7
 genetic studies, 23, 25*t*, 27–9
 MtPT4 orthophosphate transporter, 22
 nitric oxide (NO) accumulation in roots of, 60
Mertensia ciliata, 317
Mesorhizobium, 170
Methionine, overproducing PPFM mutants, 223
Methyl jasmonate, inhibition of infection thread formation by, 369
Methylobacterium
 cytokinin production by, 222
 M. nodulans, 171
 as phylloplane symbionts, 222–3
Mimosa, nodulation of, 171, 172
Mimosoideae, 197
MIPs (major intrinsic proteins), in mycorrhizal plants, 57, 59
MtPT4, 22
Mutualism, alder-*Frankia* symbiosis as, 145
Myc factors, 363
Mycorrhiza helper bacteria, 360
Mycorrhizal fungi. *See also specific fungi*
 arbuscular (AMF)
  diversity of, 201
  hyphal branching in, 358–9
 arbuscular mycorrhizal (AM) symbiosis, 20–29, 52–61
 description of, 13
 ectomycorrhizal (EM) interactions, 69–88
 in recovery of seriously degraded soils, 195–210
Myrica, actinorhizal symbioses and, 122, 125–6
Myrmecodia spp., ant associations with, 310
Myrmecophytes, 309, 310, 312–13
Myrmelachista flavocotea association with *Ocotea atirrensis*, 312
Myzus persicae, inhibition by plant peptidase inhibitors, 284, 289

N-acyl-homoserine lactone (AHL) system, 235
Nanjing declaration on nitrogen management (2004), 11
Naringenin, 336, 336*f*
Naringenin chalcone, 335, 336*f*

Nematodes, densities in coniferous forest soils, 263
Neptunia natans, 172, 179
Net primary productivity (NPP), estimates of global terrestrial, 6
Nif genes, 31, 244
    of *Methylobacterium*, 171
    of *Neptunia*, 172
    of *Ochrobacterium*, 172
NIN gene, 366, 367, 372f
Nitrate reductase, 79
Nitrate transporter genes, in AM symbiosis, 57
Nitrate utilization, ectomycorrhizal fungi, 78–9
Nitric acid, formation in Oswald process, 8
Nitric oxide (NO)
    in arbuscular mycorrhizal symbiosis, 60
    Azospirillum brasilense influence on plant growth, 407–10, 409f
    in extracellular PGPR (ePGPR), 406–7
    functions in bacteria, 403–4
        biofilm formation, 403–4
        stimulation of oxidative and nitrosative stress defenses, 403–4
    in intracellular PGPR (iPGPR), 404–6
    mammals, synthesis in, 401
    nitrogenase inhibition by, 375
    NO-lb (leghemoglobin) complexes, 405
    production in bacteria, 401–3, 402f
    scavenger, 405
    synthesis in plants, 401
Nitrification
    autotrophic, 402
    heterotrophic, 402–3, 408
Nitrite reductase, 79
Nitrogen, earth reserves of, 167
Nitrogen allocation following insect herbivory, 258–65
    after damage to aboveground tissues, 260–263
    after damage to belowground tissues, 263–5
    constrained or flexible allocation responses?, 259
Nitrogenase
    of *Frankia*, 117–18, 125
    inhibition by nitric oxide, 375
    metabolic requirements of, 167, 182
    of *Methylobacterium*, 171
    molybdenum-containing, 182
    oxygen sensitivity of, 167–8, 405
    phylogeny of genes for, 35
    protection from oxygen, 31, 125
    of rhizobia, 31
    structure, 405
Nitrogen cycle, 5–13
    herbivore effects on cycling in ecosystems, 266–8
    human-induced changes, 8–9
    illustration of, 7f
    preindustrial, 6–7
Nitrogen fixation
    in actinorhizal symbioses, 117–18, 121, 125–7
    in alder-*Frankia* symbiosis, 140–142
        regulation of, 142
    biological (BNF)
        endophytes, 237, 238
        in nitrogen cycle, 6–7
        selection of Rhizobium strains for high efficiency, 199–200
    by endophytes of grasses and cereals, 231–44
    industrial, 8, 167
    by leaf-associated bacteria, 221, 221t
    O_2 dilemma of, 117–18
    in rhizobial symbiosis, 30–33
        molecular basis of, 31–3
Nitrogenous wastes, plant storage of, 224
Nitrogen use efficiency (NUE)
    increase, need for, 12
    present level of, 12
    technological developments concerning, 11
Nitrosomonas europaea, 403
NO. *See* Nitric oxide
NodC, 32
NodD, 32, 35, 38
Nod factors, 105–6
    genetics of, 32
    plant responses to nod, 360–363, 362f
    receptors, 37, 174
    regulation of symbiosome and arbuscule development, 375
    rhizobial specificity determined by, 37
    signaling network, 28f, 33–4, 34t, 174
    signaling role, 360–363, 362f
    strain-specific, 40
    variation in, 173
Nod genes, 31–2
    of *Burkholderia*, 171–2
    of *Cupriavidus*, 172
    of *Methylobacterium*, 171
    of *Neptunia*, 172
    of *Ochrobacterium*, 172
NO dioxygenase, 403
Nod operon

flavonoids as inducers, 338–40, 356–7, 357f, 360
induction of, 339–40
nodABC operon, 32–3
Nodulation
  actinorhizal symbioses
    Frankia carbon and nitrogen metabolism, 120–121
    Frankia host specificity, 119–20, 119f
    gas exchange and oxygen protection in nodules, 125–6
    macrosymbionts, 121–9
    microsymbionts, 117–21
    nodule induction and structure, 122–4, 123f
    nodule metabolism, 125–9, 128f
  in alder *(Alnus)*, 142
  auxin role in, 342–4
  coordination of epidermal and cortical responses, 372f
  flavonoid function in, 344–7, 346f
    determinate nodulating plants, 344–5
    indeterminate nodulating plants, 345–7
  genes in *Rhizobium*, 31–2
  genetics of root nodule organogenesis, 369–71
  "hairy root composite plants" for study of, 344
  infection thread formation, 366–7
  lateral root development, similarities and differences with, 371
  process, 338–9
  in rhizobial symbiosis, 30
  signaling network in rhizobial-plant symbiosis, 33–4
Nodules
  evolution of, 35–6
  nitrogen fixation within rhizobial symbioses, 30–31
  Parasponia, 36
  types and classification of, 197, 197f, 198f
Nodulins, 180
Noe gene, 31
Nol gene, 31
NO reductase (Nor), 375, 403
NO synthase (NOS), 401, 403
NPP (net primary productivity), estimates of global terrestrial, 6
NSP proteins, 366, 372f
Nuclear porin genes, 27–8
NUE. *See* Nitrogen use efficiency (NUE)

Ochrobacterium, 172
Ocotea atirrensis, ant associations with, 312
Oligogalacturonides, 289

Opines, 224
Orthophosphate transporters, 22
Oryza sativa. See Rice
Oswald, Wilhelm, 8
Oswald process, 8
Oxidative stress defenses in bacteria, role of NO in stimulation of, 404
Oxidized nitrogen (NOx), released from burning fossil fuels, 5, 8
Oxygen protection, in actinorhizal symbioses, 125–6

Pachycentria, 313
Paenibacillus polymyxa, 234
Pantoea
  cytokinin production by, 222
  P. agglomerans YS19, 222
Papilionoideae, 197
Paracoccus denitrificans, 403
Parasponia, 35, 36, 124, 143, 171
Paris quadrifolia, 22
Paris-type arbuscules, 22
Pathogenesis-related (PR) proteins, in the rhizosphere, 98, 99f, 106
Paxillus involutus, 76, 80–82
Peperomia, 314
Peptidase inhibitors, 280–298
  clans of, 281, 283t
  classification, 281
  as defensive compounds against phytophagous insects, 288
    adaptations of insects to plant peptidase inhibitors, 296–7, 298f
    effects on gut physiology of insects, 290–295
    induction in response to insect attack, 288–90, 291f
  families of, 281–8
    family I3 (Kunitz-P inhibitors), 283
    family I4 (serpins), 284
    family I6 (cereal bifunctional trypsin/α-amylase inhibitors), 284–5
    family I12 (Bowman-Birk inhibitors), 285
    family I13 (potato inhibitor I), 285–6
    family I20 (potato inhibitor II), 286
    family I25, subfamily I25B (phytocystatins), 286
    family I37 (potato carboxypeptidase inhibitor), 286–7
  MEROPS families, 287
  overview, 287–8
  table of, 282t

Peptidase inhibitors (*cont'd*)
  mode of action, 280–281
  occurrence, 281
  transgenic plants over-expressing, 295–6
Peptide utilization, ectomycorrhizal fungi, 75
Peroxidases, secreted in the rhizosphere, 96, 96*t*, 100, 104
Petiole, ant domatia of, 316
PGPR. *See* Plant growth-promoting rhizobacteria
Phaselic acid synthase, 337
Phaseolus vulgaris
  genetic studies, 25*t*
  nodulation of, 172
Pheidole bicornis ants, 316
Phenylalanine, in phenylpropanoid pathway, 335, 336*f*
Phenylalanine ammonia lyase, 335, 336*f*
Phenyl-benzopyran ring, 334
Phenylpropanoid pathway
  description of, 335–7
  lignins produced by, 333–4
  newly identified structural genes, 337
  schematic of, 336*f*
  structures of representative compounds, 334*f*
  transcriptional regulation of, 338–47
    auxin role in nodulation, 342–4
    flavonoid function in nodulation, 344–7, 346*f*
    flavonoids as auxin transport inhibitors, 340–342
    flavonoids as nod operon inducers, 338–40
Phleum pratense, 262
Phosphatases, secreted in the rhizosphere, 96, 96*t*, 103
Phospholipase, promotion of root development by, 59
Phosphorus status of plant
  flavonoids, 358
  hyphal branching, 359
Phosphorus transporters, 60–61
Phyllobacterium sp., 170, 172, 400
Phylloplane microbes
  Methylobacterium spp. as phylloplane symbiont, 222–3
  nitrogen fixation by, 221, 221*t*
  nitrogen sources, 218–20
    atmosphere, 219
    foliar fertilization, 219–20
    plant host, 218–19
    nutrient-imposed limitations, 220
  overview, 217–18
  pathogens *versus* nonpathogens, 224
  plant growth regulators, production of, 222
  positive influences on plant nitrogen economy, 220–221
  as waste managers, 224
Physcomitrella patens, 21
Phytoalexins, 334, 335
Phytocystatins, 286
Phytohormones, 238
Phytosulfokine, 104
PIN genes, 341–2
Pink-pigmented facultatively methylotrophic (PPFM) bacteria, 222–3
Pioneer species, 122
Piper spp., ant associations with, 310, 312, 316
Pisum sativum, genetic studies in, 25*t*–6*t*, 29
Pitcher plants, 220
Plantago lanceolata, wireworm damage to, 265
Plant growth-promoting rhizobacteria (PGPR), 397–8
  extracellular (ePGPR)
    described, 397–8
    nitric oxide in, 406–7
  intracellular (iPGPR)
    described, 397
    nitric oxide in, 404–6
  production of phytohormones by, 238
Plant peptidase inhibitors. *See* Peptidase inhibitors
Plants
  actinorhizal, 121–2
  ant-symbiosis, 308–23
    ecological factors influencing importance and quality of provided nutrients, 318–19, 319*f*
    evidence for uptake of ant-provided nutrients, 309–12
    as evolutionary adaptations for nutrient concentration and provisioning, 319–22
    mechanisms of nutrient concentration and assimilation, 313–18
    myrmecophytes, 309, 310, 312–13
    nitrogen uptake in leaves, 315–17
    nitrogen uptake in roots, 313–15, 315*f*
    nitrogen uptake in stems, 317–18
    phylogenetics of, 321*f*
  arbuscular mycorrhizal (AM) symbiosis, 20–29, 52–61
  ectomycorrhizal (EM) interactions, 69–88
  insect herbivory effects on nitrogen economy, 257–70

ecological consequences of N loss from
  plants following herbivory, 265–8,
  268f
ecological effects, 258
evolutionary effects, 258
nitrogen deposition and the effects on N
  economy, 268–9
patterns of N and C allocation following
  herbivory, 258–65
patterns of N and C allocation following
  herbivory, 258–65
after damage to aboveground tissues,
  260–263
after damage to belowground tissues, 263–5
constrained or flexible allocation
  responses?, 259
Poa pratensis, resource allocation after
  herbivory, 262
Podocarpus macrophyllus, 171
POLLUX (cation channel), 26–9, 364
Polygalacturonase, in the rhizosphere, 97t, 105
Polyhydroxybutyrate granules, 31
Polyphosphate granules, 22
Polysaccharides, rhizobial production of, 175
The Population Bomb (Ehrlich), 195
Potato carboxypeptidase inhibitor, 286–7
Potato inhibitor I, 285–6
Potato inhibitor II, 286
PPFM (pink-pigmented facultatively
  methylotrophic) bacteria, 222–3
Pre-infection threads, 367
Pre-penetration apparatus, 367
Proanthocyanindins
  functions of, 334
  in phenylpropanoid pathway, 336, 336f
Proteases
  of ectomycorrhizal fungi, 73–4
  secreted in the rhizosphere, 96, 96t
Proteinase K, in the rhizosphere, 97t
Protoxylem poles, 370
PR (pathogenesis-related) proteins, in the
  rhizosphere, 98, 99f, 106
Pseudomonas
  IAA production by, 239
  P. aeruginosa, 403
  P. chlororaphis, 240
  P. stutzeri, 231, 403
  P. syringae, 96t–7t, 101
  volatile organic compound production by,
  240
Pterocarpan 4-dimethylallyltransferase, 337
Pterocarpans, 335

Quercetin-3-O-galactoside, 358
Quorum sensing, 235

Ralstonia solanacearum, 406
Reactive nitrogen
  defined, 5
  fates of, 9–10, 10f
  human-induced changes in production of,
  8–9, 8f
  negative impacts of, 10–11
  in nitrogen cycle, 5, 7–12
  strategies to limit, 11–12
  too much or too little of a good thing, 9–10
  yearly natural production levels, 7
Red-backed shrike, 10
Red clover *(Trifolium pretense)*, phaselic acid
  synthase in, 337
Remijia, 316
Rhizobia
  diversity of, 198–9
    legume rhizobial sources, 168–70
    nonlegume rhizobial sources, 170–171
  free-living, 29, 31
  nomenclature, 199
  taxonomy, 168–70
Rhizobiaceae, 168–9
Rhizobiales, 169
Rhizobial infection process
  crack entry, 178–9
  root hair entry, 173–8
    bacterial release, 178
    exchange of molecular signals between
      plant and microorganism, 173–4
    infection thread development, 176–8, 177f
    root colonization, 174–5
    root hair invasion, 176
Rhizobial symbiosis
  bacterial uptake and infection, 30
  determinate nodule formation in soybean,
    179–83
  evolutionary explanations for specificity
    cheater hypothesis, 39–41
    functional co-adaptation of partners, 41–2
    vicariance, 39
  evolution of
    in host plants, 35–6
    in rhizobia, 35
  flavonoid functions in legume-rhizobia
    interactions, 333–48
  infection process
    crack entry, 178–9
    root hair entry, 173–8

Rhizobial symbiosis *(cont'd)*
  infection thread formation
    hormonal regulation, 368–9
    process, 366–7
  nitrogen fixation within the nodule, 30–31
  occurrence in *Fabaceae*, 29–37
  signaling network in plants, 33–4, 34*t*
  specificity, 36–7
    evolutionary explanations for, 38–42
    plant determinants of, 37
Rhizobium. *See also* Rhizobial symbiosis
  genome, 31
  host specificity, 32, 198–9
  indole-3 acetic acid (IAA), 404–5
  plant defense, 103
  promiscuity of legumes with, 198–9
  R. alamii, 170
  R. cellulosilyticum, 170
  R. daejeonense, 170
  R. etli, 171, 234
  R. leguminosarum, 168, 170, 174–5, 175*f*
  R. oryzae, 170
  R. radiobacter, 171
  R. rhizogenes, 171, 172
  R. selenitireducens, 170
  R. tropici, 169, 171
  in recovery of seriously degraded soils, 195–210
  selection of strains for high BNF efficiency, 199–200
  symbiosis
    bacterial uptake and infection, 30
    nitrogen fixation within the nodule, 30–31
    occurrence in *Fabaceae*, 29–37
Rhizodeposition
  described, 95, 261–2
  herbivory and, 262
Rhizopogon roseolus, 75, 78–9
Rhizosphere
  bacteria-plant interactions in, 397–411
  categories of distribution of microorganisms, 397
  defined, 95
Rhizosphere proteins, 95–107
  model of plausible mechanisms of protein exudation, 99*f*
  orchestration of communication between plants and soil microbes, 100–101, 102*f*
  pathogenesis-related (PR) proteins, 98, 99*f*
  signaling and regulation, 101–6

cross talk between roots and pathogens, 105
  symbiosis, 105–6
  table of extracellular proteins of plant-microbe interaction, 96*t*–7*t*
Rhodobacter sphaeroides, 403
Rhododendron, foliar applications of urea to, 219
Ribosome-inactivating proteins, secreted in the rhizosphere, 96
Rice *(Oryza sativa)*
  flavone-C-glucosyltransferase (C-GT), 337
  genetic studies, 23, 25*t*
  rhizobial associations with, 170–171
Rmc (reduced mycorrhizal colonization) tomato mutant, 368
Root hairs
  infection by rhizobia, 30
  in nodulation process, 33
  rhizobial entry by, 173–8
Root herbivores, 263–5
Roots
  adventitious
    ant-plant symbiosis, 314–15, 315*f*
    auxin and nitric oxide linked to formation of, 400–401
  colonization, rhizobial, 174–5
  development
    auxin role in lateral organ development, 340–341
    bacterial nitrogen transformations that influence, 400–401
    nitrogen forms that influence, 399–400
    phospholipase promotion of, 59
  epidermal responses to symbiotic microbes, 360–369
  lateral root development
    auxin role in, 340–341, 400–401
    nodulation, similarities and differences with, 371
  nitrogen uptake in ant-symbiosis, 313–15, 315*f*
R2R3-Myb-like transcription factors, 338

Salicylic acid, 28, 101, 102*f*, 369
SAR (systemic acquired resistance), 101, 102*f*
Scutellospora spp., 201
Secondary metabolites, as antibiotic compounds, 240
Serpins (serine peptidase inhibitors), 284
Sesbania rostrata
  nod factor signaling in, 361
  nodulation of, 179

Sewage treatment, improvements in, 12
Shepherdia canadensis, actinorhizal symbioses and, 122
Siderophores, 240
Signaling. *See also specific signaling molecules*
　accommodation of symbiotic microbes in cytoplasm of host cells, 371–6
　downstream epidermal signaling, 364–6, 365f
　epidermal responses to symbiotic microbes, 360–369
　genetics of symbiosome and arbuscle development, 374–5
　hormonal regulation of symbiosome and arbuscle development, 375–6
　host plant signals, 356–60, 357f
　　flavonoids and isoflavonoids, 356–8
　　strigolactones, 358–9
　infection thread formation, 366–9
　　arbuscular mycorrhization, 367–8
　　hormone regulation of, 368–9
　　legume nodulation, 366–7
　microbial signals and their plant receptors, 360–363
　　myc factors and corresponding plant responses, 363
　　nod factors and plant responses to nod factors, 360–363, 362f
　nitrogen as signaling molecule in arbuscular mycorrhizal (AM) symbiosis, 59–60
　regulation of cortical cell divisions and root nodule development, 369–71
　　genetics of root nodule organogenesis, 369–71
　　nodulation and lateral root development, 371
　Rhizobium symbiotic signaling networks in plants, 33–4
　specialized structures for nutrient exchange, 371–4
　symbiotic (Sym) pathway, 364–6, 365f
Sinorhizobium
　S. meliloti, 32, 96t–7t, 100–101, 102f, 103
　taxonomy, 169, 170
SKL gene, 366, 370
Soil
　nitrogen availability in forest soils, 71–3, 72f
　rate of flux through nitrogen pools, 73
　zones according to proximity to root, 397
Soil degradation, recovery of
　case studies

　　recuperation of erosion gulleys at a rural site, 203–5, 204f
　　rehabilitation of a decapitated hillside in the Atlantic Forest Region, 205–7, 207t
　　revegetation of iron mining waste, 208, 209f
　extent of problem, 195–6
　use of Mycorrhizae and *Rhizobium* in, 195–210
　　case studies, 203–9
　　selection of *Rhizobium* strains for high BNF efficiency, 199–200
　　specificity of mycorrhizal symbiosis, 200–201
　　tree seedling preparation, 202–3
Sorangium cellulosum, 403
Southern corn rootworm *(Diabrotica undecimpunctata howardi)*, inhibition by plant peptidase inhibitors, 295
Soybean *(Glycine max)*
　determinate nodule formation, 179–83
　　metabolic differentiation in bacteroids, 182–3
　　metabolic differentiation of soybean root nodule cells, 180–181
　　metabolic differentiation of symbiosome membrane and symbiosome space, 181–2
　　soybean nodule formation and morphology, 179–80
　peptidase inhibitors, 285, 288
　pterocarpan 4-dimethylallyltransferase (G4DT), 337
　spherical determinate nodules of, 197, 197f
Spodoptera exigua, inhibition by plant peptidase inhibitors, 288, 290, 293, 294, 297
Stachydrine, 359
Stem, nitrogen uptake in ant-symbiosis, 317–18
Streptomyces
　chitinase production by, 241
　nitric oxide (NO) production by, 406
Strigolactones
　description of, 358–9
　root exudation of, 21, 357f
Sucrose synthase, 180, 181
Sugarcane
　Gluconacetobacter diazotrophicus and, 233, 233t, 234f, 241–2
　Herbaspirillum rubrisubalbicans and, 236–7, 241–2
　Herbaspirillum seropedicae and, 236–7, 241–2

Sugar cane stalk borer *(Diatracea saccharalis)*, inhibition by plant peptidase inhibitors, 285, 294
Suillus bovinus, 75
Superoxide dismutase, secreted in the rhizosphere, 96*t*
Symbiosis. See also specific types of symbiosis
 actinorhizal symbioses, 117–30
 alder-*Frankia,* 138–59
 ant-plant, 308–23
 arbuscular mycorrhization, 20–29, 52–61
 Methylbacterium spp. as phylloplane symbionts, 222–3
 rhizobial, 20–29, 29–37
 root-secreted proteins involved in, 105–6
 specificity in endomycorrhizal, 38
Symbiosis genes, 31–3
Symbiosomes
 described, 373
 genetics of development, 374
 hormonal regulation of development, 374–5
 metabolic differentiation of symbiosome membrane and symbiosome space, 181–2
Symbiotic (Sym) pathway, 364–6, 365*f*
SYMBIOTIC RECEPTOR KINASE *(SYMRK),* 23, 26, 27*f,* 28–9
Systemic acquired resistance (SAR), 101, 102*f*
Systemic resistance, induction by rhizobacteria, 240–241
Systemin, 104

Taxonomy, rhizobial, 168–70
Thaumatin-like proteins, secreted in the rhizosphere, 96, 97*t*
Thaxtomin A, 406
Tillandsia spp., ant associations with, 310, 317
TolC, in the rhizosphere, 97*t*
Tomato *(Solanum lycopersicum)*
 abscisic acid role in arbuscular mycorrhizal infection in, 362
 arbuscular mycorrhization, 22
 genetic studies, 23
 mycorrhizal colonization mutants of, 368

Tomentella sp., 74
Transcriptional regulation of phenylpropanoid pathway, 338–47
 auxin role in nodulation, 342–4
 flavonoid function in nodulation, 344–7, 346*f*
 flavonoids as auxin transport inhibitors, 340–342
 flavonoids as nod operon inducers, 338–40
Transpiration, apoplast nitrogen and, 219
Transport inhibitor response 1, 341
Tribolium confusum, inhibition by plant peptidase inhibitors, 288
Trichomes
 ant domatia of, 317
 secretions from, 218
Trifolium pretense, phaselic acid synthase in, 337
Trigonelline, 359
Tryptophan, IAA biosynthesis from, 239
TT12 transporter, 337
Tuber borchii, 71, 77, 79

Ubiquitin, in the rhizosphere, 97*t*
United Nations Food and Agriculture Organization (FAO), 9, 195
Urease, in the rhizosphere, 97*t*
Urea utilization, ectomycorrhizal fungi, 80
Uricase, 180

Vicariance, 39
Vitamin B12, overproducing PPFM mutants, 223
Volatile organic compounds (VOCs), endophyte production of, 240

Western corn rootworm *(Diabrotica virgifera),* inhibition by plant peptidase inhibitors, 292
Wireworm, 265

Xanthones, 359
Xerocomus pruinatus, 74

Zea maize. See Maize